Emelyan M. Emelyanov

The Barrier Zones in the Ocean

Emelyan M. Emelyanov

The Barrier Zones in the Ocean

Translated into English by L.D. Akulov, E.M. Emelyanov and I.O. Murdmaa

With 245 Figures and 107 Tables

Springer

Dr. Emelyan M. Emelyanov

Research Professor in Marine Geology

Prospekt Mira 1
Kaliningrad 236000
Russia

E-mail: abio@atlas.baltnet.ru, ioran@atlas.baltnet.ru

Translated into English by – Leonid D. Akulov, Emelyan M. Emelyanov, Ivar O. Murdmaa (**specific geological** terminology)

The book was originally published in Russian under the title **"The barrier zones in the ocean. Sedimentation, ore formation, geoecology". Yantarny Skaz, Kaliningrad, 1998, 416 p. (with some new chapters, Tables and Figures)**

Library of Congress Control Number: 2005922606

ISBN-10 3-540-25391-2 Springer Berlin Heidelberg New York
ISBN-13 978-3-540-25391-4 Springer Berlin Heidelberg New York

Springer is a part of Springer Science+Business Media
Springeronline.com

Printed in The Netherlands

Cover design: E. Kirchner, Heidelberg
Typesetting: SPI Publisher Services, Pondicherry, India
Production: Almas Schimmel
Printing: Krips bv, Meppel
Binding: Stürtz AG, Würzburg

Printed on acid-free paper 30/3141/as 5 4 3 2 1 0

Dedication

This book is dedicated to Immanuel Kant
Professor of Albertina University
on the occasion of Königsberg's 750th
and Kaliningrad's 60th Anniversary.

Preface

Processes of sedimentation and ore formation in seas and oceans are commonly considered from the standpoint of the facial or climatic zonations. Regularities of climatic lithogenesis, which have been developed by N.M.Strakhov for land and continental seas, were later adopted by P.L.Bezrukov and A.P.Lisitzin in application to oceanic aquatories. Processes of sedimentation in all oceans were considered from this point of view. Later, a proper understanding of processes of sedimentogenesis at smaller oceanic areas – in lithologic-geochemical provinces – became increasingly important. Traditionally, the description of processes of sedimentogenesis in Russian-language literature, was usually terminated by the formation of bottom sediment types and their composition. In this case, the ore formation was as if beyond the scope of these processes. The same is true for geoecological problems. In particular, ore-forming processes and geoecology are commonly studied by scientists as different disciplines.

All this led the author, which have studied processes of sedimentogenesis and geochemistry of single elements for about 45 years, to conclusion that processes of sediment transformation, change of forms of migration of chemical elements and formation of sediments with anomalously high concentrations of chemical elements in seawater and at the bottom of seas and ocean are especially active at the boundaries between different water environments and at the boundaries between areals of different sediment zones or their lithologic-geochemical zones or provinces (Emelyanov, 1982). In the course of time, the author pay particular attention to processes occurring just at the boundaries and within boundary zones. Following A.I.Perelman (1979), the author called these boundaries as geochemical barriers and geochemical barrier zones. Special symposiums and conferences with the aim to evaluate the role of processes occurring at these barriers and boundary zones were held in the USSR (Russia) (Emelyanov, 1980, 1981, 1982, 1984, 1986, 1998, 2003; Kasimov and Vorobyov (eds.), 2002).

Studies and generalization of experience enabled the author to synthesize knowledge in all these scientific fields. The book based on this synthetic approach was published in 1998 in Russian language. It had a wide response in Russia and in other countries. The author suggested the term limology (from the Latin word "limes", meaning boundary) to identify the field of science related to boundary processes.

Following the theory of geochemical barriers and GBZs, the processes of sedimentogenesis were considered for the Atlantic Ocean (Emelyanov, 1982), Baltic (Emelyanov, 1986, 1995), Black (Mitropolskiy et al., 1982) and Mediterranean seas

(Emelyanov and Shimkus, 1986). Attempts have been undertaken to classify the stages and models of evolution of sedimentary manganese (Emelyanov, 1986, 1995) and ferruginous pseudo-oolitic ores (Emelyanov, 1982).

When writing this book, the author was based on data collected in the course of his own 45-year long studies, and also scientific publications in Russian language. Readers in other countries are hardly aware of most of them, if at all. Also, the author used principal works of overseas authors on the subject (see References).

The given (English language) book is considerably improved and complemented version of Russian edition (approx. 15% more than its Russian version): text, tables and especially Figures were subjected to considerable renewing. The reason for this is that new generalizing publications on the Arctic basin have been issued (Lisitzin, Vinogradov, Romankevich, 201; Kassens et al., 1999; Shulz and Zabel (eds.), 2000; Aibulatov, 2000) as well as many other publications, open the way for this edition. Unfortunately, problems with using scientific English language journals from other countries, including Internet libraries, prevent us from wider using of literature published after 2000 year.

I gratefully acknowledge all scientists whose publications were used in this book.

This book consists of 4 parts and describes various theoretical and application aspects of the concept of geochemical barriers and barrier zones in oceans, which are interpreted as portions of water or sedimentary column, where the intensity of migration of chemical elements abruptly changes in one or another direction within a short space interval.

The first chapter of the book describes general characteristics of barrier and barrier zones in oceans, and the hierarchy of them is established. Here, the author provides a detailed substantiation of the concept, according to which GBZs means a discontinuity of continuity for medium properties, both in vertical and horizontal directions. This general point is illustrated by various examples of sediment fluxes in nature: from the time when sediments enter the ocean and until they are buried to be finally transformed in bottom sediments. According to the author, there are more than 40 geochemical barriers and barrier zones in the ocean.

The second chapter is called as "Barriers as transformation areas of sedimentary material". Here, the author describes specific processes of delivery, transformation and deposition of sedimentary material onto bottom, where the Atlantic Ocean, Baltic, Black and Mediterranean seas taken as an example.

This chapter describes the early stage of processes of sedimentogenesis, with River-Sea GBZs taken as an example. For this purpose, data obtained from rivers including Amazon, Congo, Columbia, Ob, Lena, Yenisei, Nile, Neva and Neman, where the formation of thick sedimentary bodies occurs, are used. Hereinafter, the effects of mechanical barriers occurring in zones such as "Shore-Sea", "Front of coastal upwelling" and "Ice-Water boundary", are considered with provision for their importance in terms of geology.

Fate of sedimentary materials crossing Ocean-Atmosphere barrier is traced until they are buried after the early stage of diagenesis. The sequence of environmental conditions where sediments are exposed to various effects is as following: a layer of photosynthesis, thermoicline, halocline, oxygen minimum layer, CCD, water-bottom GBZs and upper active layer of sediments. Description of processes occurring in each of cited vertical GBZs is terminated by an estimation of its importance in terms of geology.

The third chapter is called as "Lithogeochemical barriers as leading factors in ore formation". This chapter describes the present-day processes of ore formation in coastal zone (placers) and also ferruginous pseudo-oolitic ores (in river-sea GBZ) and phosphorites (in zone of coastal upwelling). Also, role of pycnocline in formation of ferromanganese crusts in platform seas, role of oxygen minimum layer in formation of cobalt-bearing ore crusts, role of CCD in formation of Fe-Mn nodules, role of redox (Eh) barrier in water in formation of carbonate-manganese sediments are considered. Model of genesis of carbonate-manganese ores from Oligocene deposits in the southern Russian platform and in the Western Europe, which was developed on the basis of concept suggested by the author for studying sediments in deeps of the Baltic sea, is suggested.

The forth chapter is called "Some geological aspects of variations in marine ecological systems". According to the author, there are boundaries of first and second order as well as internal boundaries in marine ecosystems. Boundaries such as water-atmosphere, shore-sea, river-sea and water-bottom belong to the first type, whereas troposphere, land watersheds and underground waters flowing into sea belong to the second type. Some facts illustrating pollution of coastal areas with heavy metals, products of decomposition of mineral fertilizes, detergents, oil hydrocarbons, etc. are given. One may easily see that accumulation of polluting materials is most intensive just in barrier zones, where they are partly exposed to fossilization, contributing thus to self-cleaning of waters areas.

Chapter IV considers specific aspects of contamination of marine and ocean ecosystems by oil products, biogenic materials (nutrients), heavy metals, radionuclides and other polluting materials. Evidences that all these polluting materials are confined to geochemical barriers and barrier zones are given.

The book will be purposeful for scientists, postgraduate students, geoecologists. This book was used by the author to read lectures in the university in application to the disciplines: 1) marine geochemistry and 2) oceanic environmental management.

In general, this book is an effective instrument for further developing the theory of lithogenesis and ore-genesis and can be used for practical needs in wide spectra of human activities in seas and oceans.

References

Aibulatov, N.A., *Ekologicheskoe ekho kholodnoy voyny v moryakh rossiyskoy Arktiki* (An Ecological Response of the Cold War's Impact in the Russian Arctic Seas), Moscow: GEOS, 2000.

Emelyanov, E.M., The Role of the Geochemical Barrier Zones in the Sedimentology on the Basis of Atlantic Ocean, *Tezisy dokladov 4–oy vsesouznoy shkoly morskoy geologii* (Abstract of the 4th All–Union Conference on Marine Geology), Moscow, 1980, vol. 2, pp. 30–31.

Emelyanov, E.M., Function on the Geochemical Barrier Zones in the Sedimentogenesis, in *Geograficheskiye aspekty izucheniya gidrologii i gidrokhimii Azovskogo morya* (Geographic Aspects of Studying of the Hydrology and Hydrochemistry of the Azov Sea), Leningrad: Geograph. Society of the USSR, 1981, pp. 137–151.

Emelyanov, E.M., Le Role des Zones de Barriers Geochemiques dans la Sedimentation (Exemple du Basin Nord–Atlantic), (The Role of Geochemical Barrier Zones in Sedimentation), *Bull. de l'Institut de Geologie du Bassin d'Aquitaine* (Bordeaux), 1982, nos. 31–32, pp. 361–364.

Emelyanov, E.M., The Most Important Geochemical Barrier Zones in the Ocean (on the Basis of Atlantic Ocean), *Izv. Akad. Nauk SSSR* (Proc. AS USSR), geogr. Ser., 1984, no. 3, pp. 39–53.

Emelyanov, E.M., The Role of the Geochemical Barrier Zones during Formation of the Mineral Resources in the Seas and Oceans, in Geology of the Oceans and Seas, *Tezisy 6–oy vsesouznoy shkoly morskoy geologii* (Abstracts of 6th All–Union Marine Geol. School), Moscow, 1984, vol. 1, pp. 218–219.

Emelyanov, E.M., Geochemical Barriers and Barrier Zones and their Function in the Sedimentogenesis, in *Geokhimiya osadochnogo protsessa v Baltiyskom more* (Geochemistry of the Sedimentary Process in the Baltic Sea), Emelyanov, E.M. and Lukashin, V.N., Eds., Moscow: Nauka, 1986, pp. 3–25.

Emelyanov, E.M., Baltic Sea: Geology, Geochemistry, Paleoceanograraphy, Pollution, P.P. Shirshov Institute of Oceanology RAS, Atlantic Branch, Kaliningrad: Yantarny Skaz, 1995.

Emelyanov, E.M., *Bariernye zony v okeane:osadko–i rudoobrazovaniye, geoecologiya* (The Barrier Zones in the Ocean: Sedimentation and Ore Formation, Geoecology), Kaliningrad: Yantarny skaz, 1998.

Emelyanov E.M. Barrier zones in the ocean, In: Kasimov N.S., and Vorobyov A.E. (eds.). Geochemical barriers in the hypergenic zone, MSU, Moscow, 2002, pp. 77-107.

Emelyanov, E.M. and Shimkus, K.M., *Geochemistry and Sedimentology of the Mediterranean Sea*, Dordrecht: D. Reidel Publ. Co., 1986.

Marine Geochemistry, Schulz, H.D. and Zabel, M., Eds., Springer, 2000.

Mitropolskiy, A.Yu., Bezborodov, A.A., Ovsianyi, E.I., *Geokhimiya Chornogo morya* (Geochemistry of the Black Sea), Kiev: Naukova Dumka, 1982.

Kasimov N.S., and Vorobyov A.E. (eds.). *Geochemical barriers in the hypergenic zone,* MSU, Moscow, 2002, 394 p.

Kassens, H., Bauch, H.A., Dmitrenko, I., Eicken, H., Hubberten, H.–W., Melles, M., Thiede, I. and Timokhov, L., *Land–Ocean Systems in the Siberian Arctic: Dynamics and History*, Berlin: Springer–Verlag, 1999, pp. 189–195.

Lisitzin, A.P., Vinogradov, M.E., Romankevich, E.A., Eds., *Opyt sistemnykh okeanologicheskikh issledovaniy v Arktike* (Experience of System Oceanologic Studies in the Arctic), Moscow: Scientific World, 2001, pp. 385–393.

Perelman, A.I., *Geokhimiya* (Geochemistry), Moscow: Visshaya Shkola, 1979.

Contents

Introduction

At the beginning of 20th century the ocean was considered to be a huge stationary body of water, where sedimentation processes followed a particle-by-particle model and sedimentary strata were continuous along great distances (Twenhofel, 1932). However, even the first monographs devoted to single seas (Arkhangelsky and Strakhov, 1938) and the World Ocean as a whole (Klenova, 1948; Shepard, 1963) gave grounds for regarding the sedimentation process as a complex and versatile phenomenon, which depends greatly on shapes of basins, tendencies in their evolution and features of life in these basins. Factors such as the discovery of mid-oceanic ridges (MORs), the theory of plate tectonics, recognition of living forms at the greatest depths and other facts led scientists to the conclusion that processes such as mobilization, transportation, differentiation, sedimentation and burial of sedimentary material during the Phanerozoic were the result of events involving magmatism, shifting of continents, lowering or uplifting of certain locales of lithospheric plates, climatic changes, evolution of biota and eustatic fluctuations of the level of the World Ocean. It has become clear that sedimentation processes in the World Ocean are closely related to physicochemical processes of the earth as a planet.

The laws of climatic types of lithogenesis (Strakhov, 1961–1963), which were initially developed for continents and continental seas, were subsequently transformed for oceans (Bezrukov, 1959, 1962, 1964). At the same time, close relations between sedimentation processes and circumcontinental and vertical types of zonation were revealed. A.P. Lisitzin (1974, $1977_{1,2}$, 1978, 1979, 1981, 1983) distinguished and described types of sedimentogenesis in the World Ocean that corresponded to climatic zones on continents: glacial, temperate humid, arid and equatorial humid. It was found that the properties of primary sedimentary matter acquired in various water and ice catchment areas of the mainland in different climatic zones remain practically unchanged during its transportation and precipitation. In the ocean, each climatic zone is responsible for the development of specific types of sediments typical for that zone only. In light of the concept of ocean zonation, some global regularities of terrigenous, biogenic, vulcanogenic (Lisitzin, 1978, 1991) and glacial sedimentation (Lisitzin, 1993) have been described. Sediment types have been described for the Atlantic (Emelyanov et al., 1975; Emelyanov, 1982_1) and Pacific (Murdmaa, 1987) oceans with respect to ocean zonation.

Biogeochemical research (Romankevich, 1977, 1994; Lisitzin, Vinogradov, 1982; Lisitzin, 1983) gave a powerful incentive to the development of sedimentology. It is early works of V.I. Vernadsky that provided the foundation for biogeochemical studies. In his fundamental works *Biosphere* and *Essays on Geochemistry* (quotation after Vernadsky, 1954-1960), he stated for the first time that living organisms are the main factor determining the migration of chemical elements on the earth.

Living matter controls a great many geochemical functions in the ocean.

Now it is clear that the factor of life must be considered for a proper understanding of processes occurring in the ocean. Ignorance of living organisms and organic matter present in the ocean will lead to the result that the picture of how sedimentary material develops in the ocean will be totally distorted. Organisms, their remnants and organic matter not only transport sedimentary material to the ocean floor, but also play a crucial role in the process of sediment formation (Vinogradov, 1967, p.89). Concentrations of chemical elements, especially heavy metals, in organisms is several orders of magnitude higher than those in seawater.

Passing of metals through food chains is one of the most important means of exporting these metals to bottom sediments. A considerable proportion of sediments settles onto the bottom by means of so-called "pellet transport," that is, when sedimentary material passes through the digestive tracts of organisms and reaches the seabed without considerable biogeochemical changes (Lisitzin, 1983).

Over the last decades, attempts are becoming more and more frequent to classify processes occurring in the aquatic shell of the earth and to consider the World Ocean as an integral part of the nature. However, the science of sedimentogenesis, which is based on aspects involving climatic, tectonic, and circumcontinental zonation of types of sediments, makes it possible to bring to light only common, generally global, regularities concerning the formation of the types and properties of sediments and the development of mineral resources. Many interesting processes and their details remain unclear. In order to fill the gaps in these regularities, hypotheses have been suggested that geochemical zonation of sedimentogenesis in the ocean is of great importance (Strakhov, 1976; Emelyanov, 1982_1) and that some processes are not uniformly active everywhere in the ocean, but only in some relatively narrow zones, strips or layers, where the intensity of fluxes, transformation and redistribution of sedimentary material is the highest. As far back as 1926, Vernadsky (1954-1960) distinguished two types of film—planktonic and benthic—and also drew lines between two centers of life—coastal (marine) and Sargasso. The atmosphere–water and water–bottom barriers were distinguished, and processes within them were briefly described by Horne (1972). Later, some other surfaces and interfaces, including water– atmosphere, coastal–sea, hydrological boundaries,etc., were identified (Aizatulin et al., 1976; Lebedev, 1986; Monin and Romankevich, 1979; Chester, 1990; Romankevich, 1994). Over the last 10–20 years, research into processes in boundary zones intensified because of the growing need to quantify the processes on such borders as water–bottom (Bischoff and Piper, 1979; Fanning and Manheim, 1982; Chester, 1990), water–atmosphere (Duce et al., 1972; Cheng, 1975; Mclntyre, 1974), river–sea (Lisitzin, 1974, 1978, 1988; Gordeev, 1983; Artemiev, 1993; Monin and Gordeev, 1988; Lisitzin et al., 1983), hydrofronts (Horne, McIntyre, Wood, G. Bogorov, K.N. Fedorov, T.A. Aizatulin, V.L. Lebedev, Chester), the oxygen minimum layer, calcium carbonate compensation depth (I. Berger, A. P. Lisitzin, I.0. Murdmaa), and others.

Is it possible to say that bottom sediments reflect processes occurring within the ocean boundary zones? If yes, in what major aspects can this occur? Proceeding from this, what consequences regarding lithology, ore formation and geoecology may be expected? Does this information about sedimentogenesis in boundary zones give another incentive to formulate new hypotheses and construct new theoretical models in the field of lithology, geochemistry and geoecology? Does information on boundary processes contribute to the prediction of mineral and biological resources in the ocean? Is it possible to solve the reverse problem in the field of palaeoceanology—that is, proceeding from geochemical data collected for various bottom sediments, to reconstruct the location, intensity and time spans of particular boundary zones in seas and oceans in the geological past? These and many other questions still remain unanswered, or else they have been answered insufficiently in the literature. The natural barriers in the ocean, which are "interfaces" that provide for contact between phases (substances and media) with different physical and chemical characteristics, have become the subject of research only recently. In addition to such long-known boundaries as sea–land, air–sea and sea–bottom, various interfaces have been found within the ocean body itself, including the so-called polar front, subarctic front, subtropical convergence, etc. These interfaces, named after their regular localities in the ocean, are characterized by wide changes in their physical and chemical characteristics. Three main types of these boundaries can be distinguished: hydrological front, which divides warm and cold waters; equatorial front, which divides waters having different signs of the Coriolis parameters; and layer of discontinuity (or "jump"), which is a border between waters of different density and turbulent conditions. This layer, which can be considered as an interface for model purposes, is a lower border of the upper thermally quasi-homogeneous layer, which varies from a few tens to hundred of meters in thickness.

The apparent physical, physicochemical land biological effects on the boundaries are confined to media and substances in the ocean involved in contacts and interactions. At the same time, the boundary structures or active surfaces of each pair of different substances (structures), each of which is on own level of organization, directly participate in a variety of complex chemical cycles. Among these features, of special importance for proper understanding of ocean structures and the mechanism of transformation are the following types of boundaries: macroscopic—atmosphere–water, water–bottom, water–ice, hydrofronts, etc.; microscopic—water–suspended matter, water–sedimentary particles, water–microorganisms, etc.. Although the scale of contacting pairs is markedly different, there is the interesting regularity that active processes are focused in thin interlayers (laminae), where substances and energy are exposed to rapid transformations, processes of biological exchange become much more intensive, and the activity of microorganisms is the highest. Although each pair participating in interphase contacts repeats all processes which were realized at the earlier levels of the established hierarchy, beginning from the atomic level, each subsequent level is responsible for contacts and the scale of interactions typical for the given level only. This appears to be related to events involving the change in the mechanisms of transport and transformation of substances in the course of contacts and interactions occurring at different hierarchical levels. At the microscopic level—the motion of electrons on the peripheral structures of atoms and the motion of ions, molecular diffusion, coagulation, adsorption, etc.; at the macroscopic level—turbulent diffusion, movement of water masses and suspended matter, and environmental metabolism of biogeocenoses caused by the activity of organisms.

Over the last 10–15 years, the science of geochemical barriers has become a booming field. It is landscape geochemistry that laid the foundations of this discipline (Perelman, 1966, 1979, 1989). The concept of geochemical barriers came as the result of the development and generalization of scientific ideas with respect to concentration of chemical elements in the earth's crust. The lassification of geochemical barriers established by Perelman is based on ideas about the types of chemical-element migration. Also, mechanical, physicohemical, biogeochemical and anthropogenic barriers were identified within the scope of this classification. The following classes of barriers were distinguished: anaerobic, sulphate-reducing, sulphurous, carbonaceous, alkaline, acidic, evaporative, and adsorptive. Complex barriers (e.g., precipitation of Fe–Mn gels on the oxidation-reduction barrier and the sorption of chemical elements by these metals on the complex oxygen-sorbing barrier) can be distinguished by the superposition of different geochemical processes (Fig. 1). In addition, two-sided barriers can be recognised, i.e., barriers that formed as a result of movement of waters of different types towards this barrier from different sides (two-sided acidic and alkaline barriers); and also various lateral barriers, which are the result of subhorizontal movements of water, and radial barriers that formed as a result of vertical (downward or upward) migration of solutions.

Progress in theory of geochemical barriers in application to the ocean led the author (Emelyanov, $1979_{1,2}$, 1980, 1981, 1982, 1984) to the conclusion that the essence of oceanic geochemical barriers and boundary zones is that they are sections (or layers) of water or sedimentary columns of seas and oceans where the rates of migration of certain chemical elements rapidly decrease within a very short spacial interval (the same is true for their concentrations), whereas values for other elements tend to increase. As is widely accepted (Lebedev, 1986, p.116), the limit of a boundary layer is considered to be the place where the effect of a boundary on a surrounding reduces to a value of 10–20%.

The situation is very common when several (instead of one) barriers exist side by side at a narrow section or in a thin sea (ocean) layer. To identify these sections (layers), the author introduced the concept of "geochemical barrier zones" (GBZs) A GBZ in an ocean (sea) is a natural boundary (layer, strip) where the depositional conditions (hydrodynamic, physicochemical, etc.) are different on opposite sides of this boundary, and this leads to rapid variations in the forms and strength of migration of a certain group (association) of chemical elements and, consequently, to variations in their concentrations. The difference in absolute concentrations (in mg per litre or μg per litre) of macrocomponents and microelements in water strata, and also in their relative contents (in %) in suspended matter and sediments, between the opposite sides of a GBZ commonly ranges from a factor of 2 to 10 and is periodically as large as 1000–100000! Thus, rapid variations in absolute concentrations and relative contents of elements, and also the change in the form of their migration, are the major criteria for identifying GBZs and boundary zones. GBZs can be considered discontinuities in environmental conditions (both in the horizontal and vertical directions).

The main quantitative characteristic of GBZs is commonly a gradient G = (m1-m2/l), where m1 is the value of the given element (or of pH, Eh, etc.) "in front" of a barrier, m2 is the value of this element "behind" a barrier, and l is the width of the barrier (Fig. 1). In contrast to land-based GBZs, where the flux of a substance moving towards the barrier is commonly unidirectional, in seas and oceans, especially within the water strata, the flux of a substance towards the barrier is frequently bidirectional: a group of elements (or forms of them) approach the barrier from one side, whereas another group of elements (or modified forms of them) approach the barrier from the other side. As evidenced by analysis of boundary phenomena having various scales (levels of established hierarchy) and forms, "anomalies of opposite sign are typical for such phenomena: these anomalies are caused by the effects of boundary surfaces and are a source of energy for contact zones"(Lebedev, 1986, pp.157–158).

This work is mainly concerned with various aspects of a new scientific discipline—the science of lithogeochemical limology of the ocean. Limology (from the Latin *limes*—boundary) is the name given to the science that is mainly concerned with a complex system of boundaries and interfaces of various origin which are present in the natural environment and also with the study of physical and geochemical processes within these boundaries.

Processes of input and transformation of sediments have been described by the author as a succession of various cycles, with reference to both horizontal (source—supply to the centre of the ocean—transformation) and vertical (atmosphere–water— the entire ocean strata—ocean–bottom surface—burial and transformation in the uppermost active layer of bottom sediments) distribution of sedimentary material.

In addition to aspects such as classification of barriers and barrier zones and description of the processes of sedimentogenesis, the aim of this work is also to understand the role they play in 1) ore formation and 2) geoecology (pollution and self-cleaning of sea and ocean environmental systems). How far does a single particle (or sedimentary material on the whole) penetrate the ocean? What happens to it on its way? What barriers must be overcome? What does it look like? To answer these questions, the book begins by describing the delivery and transformation of sedimentary material with respect to river–sea barrier zone, where land-based sources of particles that enter this barrier may be 1000–5000 km from the mouth of a river.

A considerable portion of sedimentary material escapes the river–sea and coastal–sea interfaces and settles onto the upper microlayer of the ocean as a result of atmospheric fallout (this microlayer also gives rise to a certain portion of sedimentary material). Particles (or sedimentary material) should come through the upper microlayer, thermocline, photic layer, halocline, oxygen minimum layer, etc., before finally settling on the bottom. Thus, the author performs an important job in tracking the vertical and horizontal movement of particles from their source regions until they are completely buried. Traveling through the various marine bio- and geochemical cycles is a good way for the reader to trace the route of particles, from the very beginning until the end, which is full of unexpected obstacles, "pleasures" and "grief." In the course of these "travels," the author focuses on describing the behavior and evolution of particles within particular, often very narrow boundaries and interfaces, where particles (sedimentary material) undergo maximum changes, instead of oceanic water strata as a whole—a problem that has been considered in detail by many authors.

The aim of this book is to classify boundaries, paying particular attention to their dynamic functioning and mechanisms that cause them. Among a great many of natural boundaries/barriers, which are found at all levels of the global hierarchy of matter—from cells and smallest colloidal particles to the global sphere's boundaries and largest ecosystems—up until now we have succeeded in revealing only those barriers where the composition of substances and forms of their cross-boundary migration exhibit dramatic changes upon crossing these boundaries. Undoubtedly, research efforts aimed at determining common features for all natural barriers in the ocean and new information concerning the position of these boundaries in the established classification scheme, as well as new methods for studying boundary effects, will make it possible to predict probable geologo-geochemical and ecological consequences in the ocean, mineral and biological resources, polluted and self-purifying zones, recreation areas, etc.

Following the theory of geochemical barriers and GBZs, the processes of sedimentogenesis were considered for the Atlantic Ocean (Emelyanov, 1982_1), Baltic (Emelyanov, 1982_1, 1986, 1995), Black (Mitropolskiy et al., 1982) and Mediterranean seas (Emelyanov and Shimkus, 1986). Attempts have been undertaken to classify the stages and models of evolution of sedimentary manganese (Emelyanov, 1986, 1995; Emelyanov and Lisitzin, 1986) and ferruginous pseudo-oolitic ores (Emelyanov, 1974, 1982).

In preparing this text the author drawn information from many published sources and from many informal discussions with my colleagues with respect to processes of sedimentation and ore formation in the World Ocean, and this information provided the basis for writing this manuscript. The author would like to acknowledge the contribution of the entire staff of the Department of Geology of the Atlantic (Atlantic Branch of the P.P. Shirshov Institute of Oceanology, RAS, Kaliningrad—headed by the author), who participated in the collection of materials.

The author also acknowledges the following people for enlightening discussions in various aspects: K.M. Shimkus, A. Aibulatov, G.N. Baturin, U.A. Bogdanov, B.A. Bubnov, I.I. Volkov, V. Gordeev, B.N. Lukashin, A.U. Mitropolskiy, I.O. Murdmaa, A.G. Rozanov, E.S. Trimonis, E.A. Romankevich, G.S. Kharin, U.P. Khrustalev, and many other scientists.

Special thanks to A.P. Lisitzin, who was the author's consultant for many years. The manuscript was improved by the helpful comments and suggestions of K.M. Shimkus, to whom the author is grateful for his careful reviews and critical remarks. Some chapters were read by A.I. Ainemer, G.N. Baturin, V.V. Gordeev, I.O Murdmaa, V.A. Kravtsov, E.S.Trimonis, V. M. Slobodianik, G.S. Kharin and B. Winterhalter. Their general assistance is gratefully acknowledged.

The English translation was made by L.D. Akulov, E.M. Emelyanov and I.O. Murdmaa (specific geological terminology). The translation was improved by Aaron Carpenter, editor of the English version of this book.

The Figures were designed by author's assistants T.G. Konovalova and Y.E. Olhova, the text and Tables were retyped by G.V. Tishinskaya. Their assistance is gratefully acknowledged also.

Russian (original) book was broaden: there were added some new chapters, Tables and Figures (10-15% of total book).

The monograph will be of interest to marine geologists, oceanographers, geochemists and geoecologists, and also to many students, mostly advanced in their undergraduate training or beginning graduate work, who are specializing in these disciplines. It is designed so that it can be read on its own, like any other textbook: it was used by the author as part of a course for Kaliningrad University students in the field of marine geochemistry, and oceanic environmental management and geoecology, as well as for training Baltic Research-Teaching Centre staff.

This work was funded by the Russian Academy of Sciences, Moscow, and the Russian Federal programs "Integration" and "World Ocean."

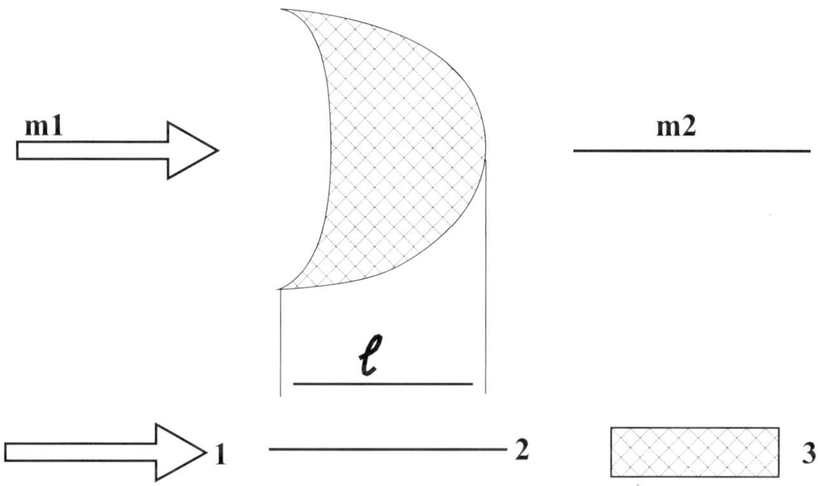

Geochemical Barriers

Part I

Brief Characteristics and Classification of Barriers and Barriers Zones in the Ocean

"Nature is striking perfection, which is the summation of limits.... If we know the limits of nature, we will know how its mechanism works. The main thing is to know the limits."

A. Barriko

Natural objects, the scales of which range from colloid forms to oceanic environmental systems, are very variable, and this is an inherent feature of these objects. Processes of transformation of material occurring in the ocean are commonly transitive, and the distribution of physical, chemical and biological processes is neither random nor uniform throughout the whole length of the water strata: instead, they are focused in relatively thin zones displaying active transformation of material and energy, which are called boundary regions of the ocean (Fig. I.1). These active zones are responsible for abrupt, jumplike changes of natural processes, i.e., when the processes on one side of such a boundary are substantially different from those on its other side (Emelyanov, 1982).

This is why these active surfaces may be regarded as natural barriers. In contrast to barriers, the remaining areas (those occurring between boundaries) of the ocean are thought to be relatively homogeneous, nongradient, chemically inert, and biologically inactive areas. Such a picture of the ocean's structure, which is a subject of current, considerable interest, is based on the existence of outer and inner active boundaries related to fronts, dispersions, and environmental and geological (sedimentary, hydrothermal, volcanic, etc.) systems. Boundary phenomena in the ocean are very different in size and nature, ranging from physical and physicochemical processes to biological ones, and from planetary fronts to the smallest suspended particles. However, this does not preclude the availability of certain essentially important properties com-

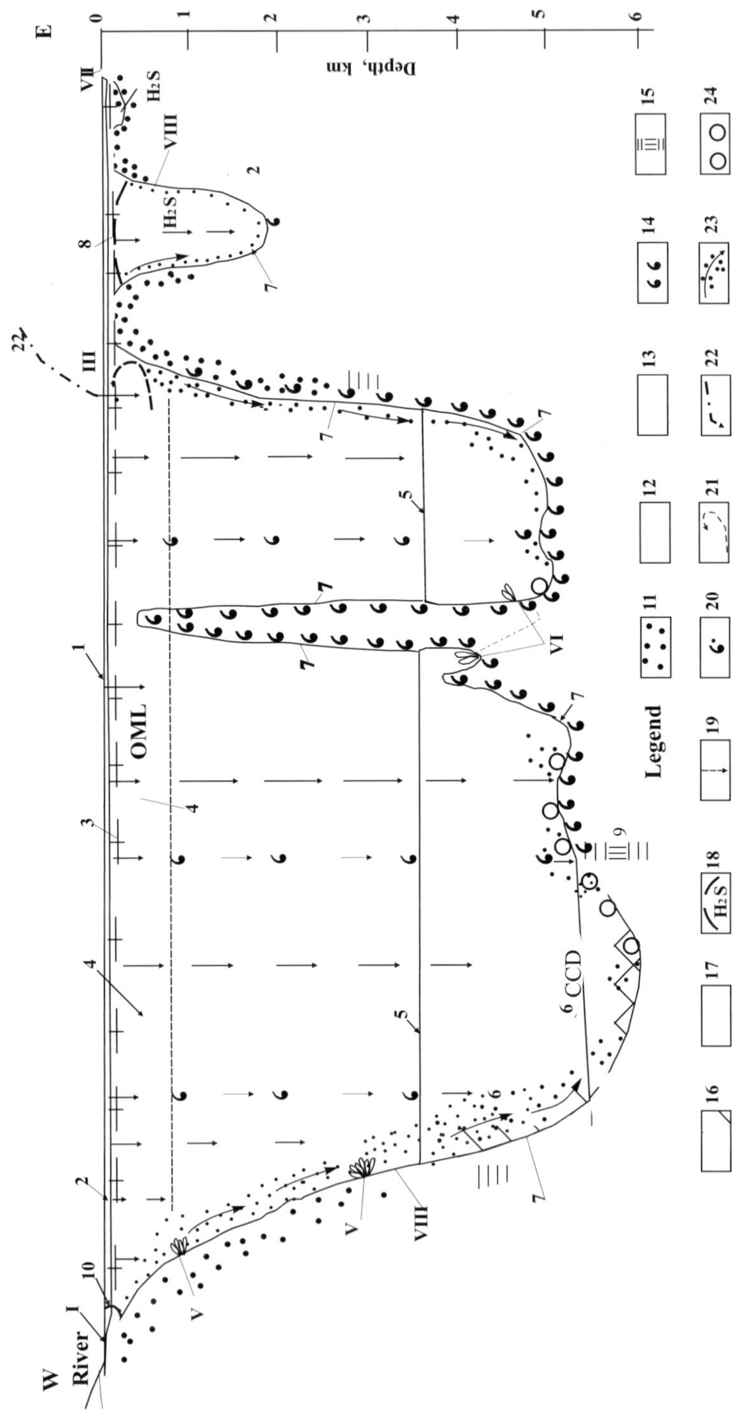

Caption see page 3

mon to all these outer and inner boundaries. A major feature common to such boundaries/barriers is that they are zones of increased transformation of substance and energy, controlling the functioning of the ocean as an integral system.

The variety and complexity of all types of natural barriers in the ocean that exist at all levels of the established hierarchy, from a cell or the smallest colloidal particle to the largest environmental systems, hinders presenting a list of these natural phenomena. Moreover, only a small portion of such boundary effects had been found up to now. However, research efforts are required to reveal the essential features common to all natural barriers in the ocean, as well as new methods for studying boundary effects, in particular, in order to develop the theory of lithogenesis in the ocean and its related ore formation.

The geochemical barriers (GBs), geochemical barrier zones (GBZs), and lithogeochemical barrier zones (LGBZs) recognized by Emelyanov (1986_2) (Table I.1) can be divided into the following subdivisions: (1) universal (global), regional and local (i.e., with reference to spatial scaling of these phenomena); (2) continuous; discrete and periodical; diagenetic, and epigenetic, i.e., showing their steplike nature and the stage of lithogenesis, etc.

Earlier, GBs and GBZs were subdivided by the author into horizontal and vertical GBZs (Emelyanov, 1979, 1982). GBs that can be recognized by an observer moving horizontally were called horizontal GBs, and those that can be recognized by an observer moving vertically, from surface to bottom—or vice versa, from bottom to surface—were called vertical GBs. Thus, vertical GBs are most frequently represented by a plane (or flat layer) extending horizontally (or quasi-horizontally) to the ocean surface or seafloor.

With reference to types of migration of chemical elements and the prevailing mode of sedimentogenesis, GBZs as a whole and GBs in particular have been divided by the author (Emelyanov, 1982, 1984) into the following subdivisions: mechanical (hydrodynamic), physicochemical, and biogeochemical (see Table I.1).

Fig. I.1. The most important barrier zones in the ocean and the processes of the redistribution and accumulation of sedimentary material and chemical elements on the bottom. According to E.M. Emelyanov, 1982_1, with additions.
Horizontal barrier zones: I—hydrofront or river–sea barrier zone; II—near-shore upwelling front; III—front of divergences (the zone is not shown in the figure); IV—ice–water zone (not shown in the figure); V—underwater springs of groundwaters; VI—hydrotherm–seawater barrier zone; VII—shore–sea barrier zone (1st, 2nd and 3rd mechanical zones); VIII—border where clastic sediments change into clay sediments (3rd mechanical zone).
Vertical barrier zones: 1—ocean–atmosphere (or upper microlayer); 2—photic layer; 3—discontinuity layer (halocline, pycnocline); 4—oxygen minimum layer; 5—start of lysocline; 6—Carbonate Compensation Depth (CCD) (for $CaCO_3$); 7—water–bottom barrier zone; 8–redox barrier in the water strata (O_2—H_2S layer); 9—redox (Eh) barrier in muds (layer where reduced sediments change into oxidized ones); 10—salinity barrier (river–sea zone).
Legend shown by special symbols 11–23: 11—clastic sediments (sand, aleurites); 12—terrigenous mud; 13—pelagic red clay; 14—biogenic calcareous sediments; 15—interlayering of the oxidized (red clay) and reduced (terrigenous) muds in the pelagic area of the ocean; 16—the layer between the lysocline and the CCD; 17—the layer below the CCD; 18—the layer of O_2 disappearance and H_2S appearance in the water strata; 19—rain of organic detritus, (soft particles of organisms); 20—rain of biogenic colcareous detritus; 21—upwelling; 22—flux of aeolian material; 23—near-bottom turbidity layer (nepheloid layer); 24—FMNs.

Table I.1 Cassification of the Geochemical Barriers and Geochemical Barrier Zones in Seas and Oceans

A. Geochemical Barriers That Exist Depending on Their Position in Space

I. Mechanical (hydromechanic)
- I.1. Coastal zone or littoral (shore–sea", 1st mechanical barrier)
- I.2. Zone of sharp decrease in repetition of asymmetry phenomena of wave near-bottom currents (2nd mechanical or hydrodynamic barrier)
- I.3. Zone of strong near bottom currents of the main basin (the boundary of replacement of clastic sediments by clayey sediments) (3rd hydrodynamic barrier)

II. Physicochemical and biogeochemical
- II.1. Alkaline–acidic barriers (strongly acidic, weakly acidic, neutral, weakly alkaline and alkaline)
- II.2. Oxidation–reduction barriers
  - a. Redox Eh barrier in water and the O_2–H_2S layer in water
  - b. Redox Eh barriers in sediments (usually +200 to +400 mV)

III. Salinity barriers
- III.1. River–sea
- III.2. Seawater–brines
- III.3. Springs of ground water–sea water
- III.4. Halocline

IV. Temperature and dynamic barriers
- IV.1. Thermocline (T), freezing point, Mendeleyev temperature (40°C), evaporation barrier, etc.
- IV.2. Dynamic (P) barriers—the changes of the state of matter with changing pressure

V. Light barriers

B. Geochemical Barriers That Exist Independently of Their Position in Space

VI. Authonomic barriers
- VI.1. Water–living matter
- VI.2. Water–suspended matter
- VI.3. Electrochemical barriers

C. Geochemical Barrier Zones

VII. Barrier zones
- VII.1. Ice–sea water
- VII.2. River–sea
- VII.3. Hydrofronts and divergences
- VII.4. Centres of submarine discharge
- VII.5. Hydrotherm–sea water
- VII.6. Sea–atmosphere
- VII.7. Photic layer
- VII.8. Discontinuity (jump) layer (thermocline–halocline–picnocline)
- VII.9. Oxygen minimum layer
- VII.10. Critical levels of carbonates
  - a. Lysocline (for aragonite, calcite, etc.)
  - b. Calcium Carbonate Compensation Depth
- VII.11. Water–bottom
- VII.12. Upper active sediment layer

C. Others Barriers and Barrier Zones

GBs, irrespective of their spatial positioning, are found everywhere organisms or suspended matter are present.

The author suggested that areas and layers of water where GBs cannot be easily distinguished against the background of environmental conditions but whose influence is seen in bottom sediments may be called lithogeochemical barrier zones (LGBZs) (Lisitzin and Emelyanov, 1984). LGBZs, just like any other boundary between two hydrochemically and geochemically different environments, also experience variations in the intensity of migration of substances and in its forms, but these changes are not as distinct as those for GBZs. In contrast to some types of GBZ, the influence of LGBZs is frequently observed in bottom sediments as distinct boundaries between sediments of different types or as boundaries between various lithological sedimentary provinces (Emelyanov, 1982, 1986_2; Emelyanov and Shimkus, 1986).

Mechanical (Hydrodynamic) Barriers

Mechanical (hydrodynamic) barriers are found at the boundary between two media, each of which has its own hydrodynamic conditions. The essence of a barrier is as follows: under conditions of sharply decreased hydrodynamic activity of seawaters, certain heavy and large particles, which are responsible for the transportation of a certain group of chemical elements, cannot penetrate from one medium into another to be deposited there.

The contemporary notions on mechanical (hydrodynamic) barriers were developed by A.I. Ainemer and G.I. Belyaev (1982) and Shilo (1985). It was assumed that a hydrodynamic barrier in a water medium is a situation typical for a mass field, which appears under conditions when its main parameters undergo sharp changes by a certain value. The result of this is a redistribution of dynamic regimes on both sides of the barrier and, consequently, compositional and grain-size redistribution of sediments. On the whole, a hydrodynamic barrier is a function of four variables: atmospheric processes, relief and bottom roughness, shape of coastlines, and internal interaction of field masses. Depending on the type of origin, hydrodynamic barriers have three main subdivisions: 1) geomorphological (resulting from lithosphere–hydrosphere interactions), 2) hydrological (caused by contacts between different water masses and various parameters), and 3) atmospheric or synoptic (the cause of which is atmosphere–hydrosphere interactions).

The coastal zone is commonly divided into the main elements as follows (Fig. I.2, A, B, C). The zone of destruction of nonregular waves is part of a profile where nonregular waves are destroyed. Nonregular waves collapse at one point to form a line of alongshore degradation. The surf zone is the area between the shoreline (oversplashing of waves) and the outermost limit of the breakers. The outer area of the coastal zone is that part located seaward of the surf zone. The oversplash zone is that near the water's edge, which is dominated by return (translational) movements of water after the final collapse of waves. The upper limit of this zone is the line of maximum oversplash, and the lower line is the place of collision between an occurring wave and water outflow (Kosyan and Pykhov, 1991, p. 13).

Some other principles may be used to recognize the coastal zone. In such a way, S. Neshyba (1991) recognized the inner and outer zones of the coastal zone (Fig. I.3), and his recognition of the boundaries was based on physical and biological factors. According to Neshyba (1991, p. 289), "The outer zone is controlled by such factors as solar energy, water turbidity, or photosynthetic processes." Neshyba believes that the benthic community, which is at the smallest distance from coast, becomes an integral part of the coastal community, and vice versa.

The river–seawater mixing zone is one of a variety of mechanical (hydrodynamic) barriers.

N.A. Aibulatov (1990, p. 257) identified three types of hydrodynamic barriers in the coastal zone (Fig. I.2 C): (1) the coastal (surf) or littoral zone; (2) the zone of reduced repetitiveness of a phenomenon, such as asymmetry of wave velocities in the near-bottom water layer; (3) the zone within the influence of strong currents of the main stream of the basin (or the boundary where clastic sediments are replaced by clayey sediments).

1. The coastal (surf) zone and zone where the asymmetry of wave velocities in the near-bottom layer is characterized by a decreased repetitiveness. Zones within the influence of the first and second barriers are areas where the interaction between seawater and land-based rocks is the most intensive. Wave processes occurring in this zone are very intense, and there are also strong tidal and alongshore currents.

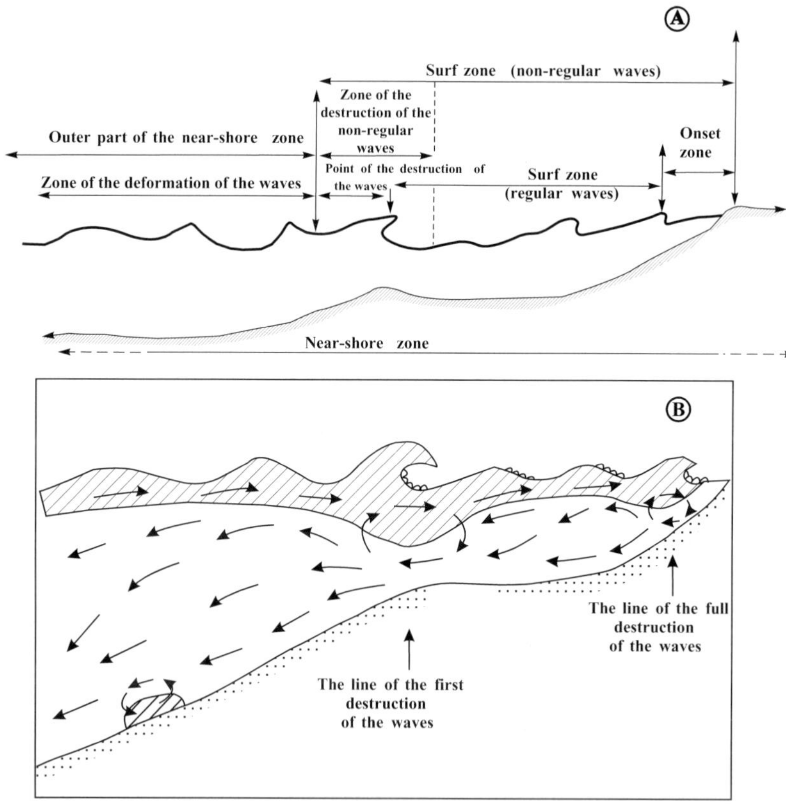

Fig. I.2. Shore–Sea GBZ
A—Division of the near-shore sedimention area into characteristic parts (after Kosyan and Pyhov), 1991, p. 12.
B—General scheme of the cross water circulation in the shore zone. Scheme is compiled on the basis of natural observations. After Leontyev, 1989. Dashed parts—the transfer to the shore side.

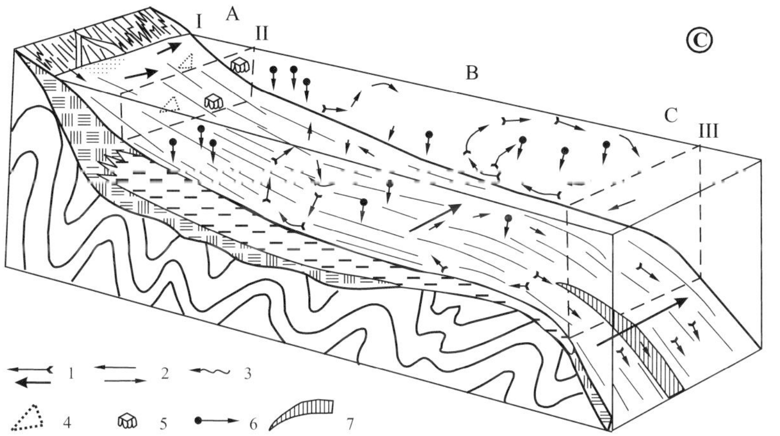

Fig. I.2. *Continued*

C—Block scheme of the transfer of sedimentary material in water on the shelf in a non-high-tide sea. After Aibulatov, 1990.
1—direction of the transfer (the thickness of arrows corresponds the intensity of the transfer).
2—transfer to the seaward side by rip currents
3—evacuation to the seaward side;
4—transfer in the form of accumulative forms;
5—transfer by the ice;
6—gravitational deposition;
7—canyon.
I, II, III—hydrodynamic barriers on the shelf during diametrical transfer of sediments;
A, B, C—upper, middle and lower dynamic part of the shelf.

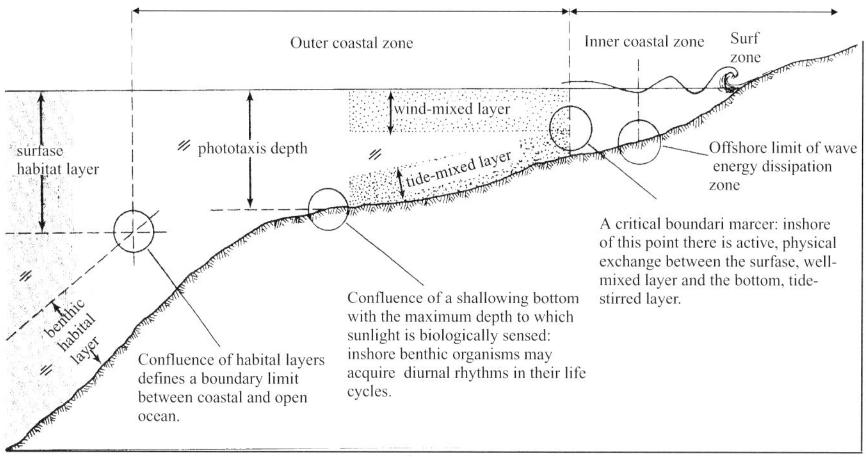

Fig. I.3. Several possible definitions of the boundaries of a coastal zone. Neshyba, 1991.

The author believes that the limits of the coastal–sea zone coincide with the limit of maximum oversplash (top) and with the lower limit of the so-called "wave reworking zone" (i.e., the zone of active resuspension and redeposition of sediments due to the effects of waves). The lower limit also coincides with a zone where the repetitiveness of a phenomenon (asymmetry of near-bottom velocities) is much smaller than that observed upward the slope (Aibulatov, 1990, p. 13). V.V. Longinov (1973, p. 16) suggested that the lower boundary may be represented by the isobath, the depth of which equals one-third of the length of the largest annual waves, or about ten heights of these waves. This boundary is the zone of initial transformation of wave motions and the appearance of asymmetry of wave velocities in the near-bottom layer. This geochemical barrier is dominated by processes of mechanical destruction of land rocks due to the effects of mechanical energy. Also, this boundary is responsible for chemical leaching of rocks and for various biogeochemical processes that have resulted from the activity of marine (fouling) plants, bottom fauna, and flora. Active processes of mechanical distribution of sediments take place according to the grain size of particles, resulting in the accumulation of clastic and sand material on the seafloor and removal of clayey and pelitic sediments beyond the limits of the littoral zone.

The main activity in the coastal–sea zone is mechanical workings that involve the destruction of the coast and the seabed by hydrodynamic processes and sediment redistribution. About 60% of the solid material delivered to the littoral part of the sea by these effects are then transported seaward to be deposited beyond the limits of the littoral zone.

Deposition of particles from water flow depends on such factors as current velocities and depth (below the sea surface). In waves, small parcels of water experience circular (rotational) motion. Solid particles are also involved in this motion. The trajectories of these particles are quasi-circles the radii of which rapidly decrease with depth. So, at a depth corresponding to a factor of 0.2 of the length of the wave, the speed of the wave is a factor of 0.3 of the velocity of the surface particles; at a depth corresponding to a factor of 0.4 of the length of a wave, this value is 0.1; and when the depth is greater than a factor of 0.5 of the length of the wave, the motion of waves is almost absent (Neshyba, 1991, p.263).

At the coastal–sea barrier, the flux of sedimentary material flows in two opposite directions: it moves seaward or landward, or, if it moves along the shore, in both directions together with the isobath. The direction of this flux depends on many factors, including: (1) the profile and depth of the submarine slope, (2) petrographic composition of rocks, (3) climatic factors (heat, humidity, wind waves, ice conditions), (4) direction and strength of sea currents, (5) tidal currents, (6) occurrence of life and its forms within the coastal–sea barrier zone, and many other factors. Depending on the direction of the flux of a substance, the coastal–sea zone may be of the abrasion or accumulative type. However, the general (summarized) flux of a substance in the World Ocean is seaward (Zenkovich, 1962; Nevessky, 1967; Aibulatov, 1987, 1990).

Specialists involved in studying coastal currents and alongshore transport of sedimentary material traditionally subdivide the study area into small sections of coast (these may be from hundreds of meters to kilometers across), or cells, where the movement of material occurs (Neshyba, 1991). These cells are limited either by rocky prominences of land or canyons (Fig. I.4). Within the limits of one cell, sedimentary material is delivered to the sea, where it is distributed and finally deposited. Most of the material that enters the sea ultimately ends up in deep, hydrodynamically passive recesses. Transport of sand sediments is especially intense in cells containing canyons (cells off the Caucasian coast in the Black Sea, a part of the ocean off the mouth of the Congo river, etc.). Exceptions are shelf areas where river mouths are not associated with canyons (for example, rivers of the Baltic and North seas) (Neshyba, 1991, p. 285). In these areas, much of the alluvial material is deposited along the shoreline due to the effects of coastal currents and only fine (pelitic) material eventually reaches the deeps.

The coarsest sediments are deposited in surf zone, which is characterized by the highest hydrodynamic activity. At greater depths, i.e., between hydrodynamic (mechanical) barriers N.1 and N.2, there is well sorted sand and silt, and sometimes gravel.

The coastal–sea GB is responsible for the enrichment of sediments in terrigenous SiO_2 and, under more favorable conditions, in Fe, Ti, Zr, Sn, Au, Pt, and other elements to amounts that make them attractive for industry. The enrichment process occurs predominantly at depths ranging from 20 to 50 m, i.e., within

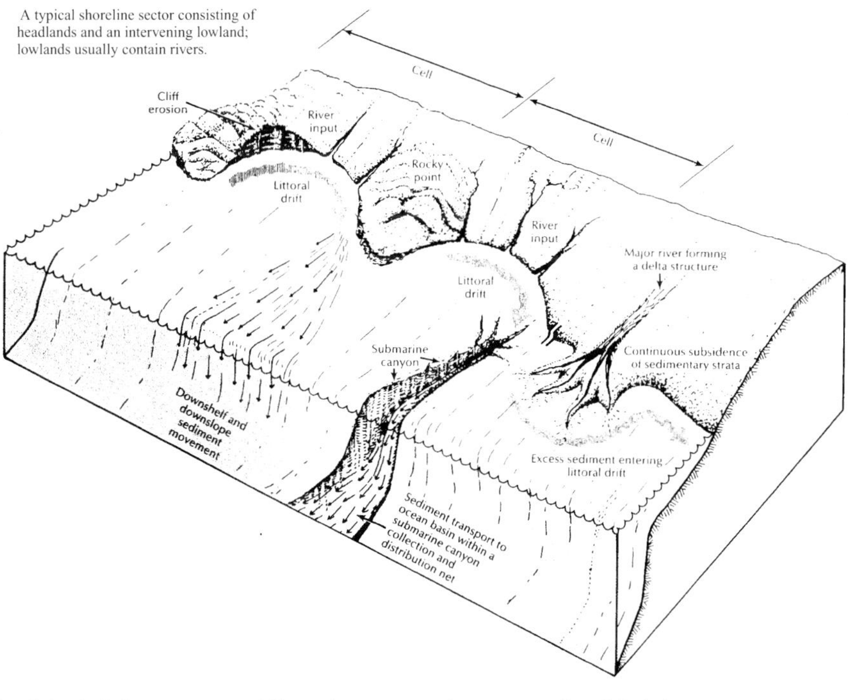

Fig. I.4. Cellular structure of littoral transport along a coastline. Neshyba, 1991.

the band where processes of natural mineralogical sorting are the most intense. This band may be indentified as a subtype of the coastal–sea GBZ and may be called a mineralogenic GBZ.

The constituents of this coastal–sea zone also include such GBs as water–bottom, air–water, water–suspended matter, and water–living substance, but their role in sedimentogenesis (the process of transformation and migration of material) is much smaller, by one to several orders of magnitude, than that of mechanical activity.

The coastal–sea GBZ plays an essential role in supplying sediments to water basins (seas, oceans) and in accumulating them in specific depositional environments (Zenkovich, 1962; Longinov, 1973; Ainemer and Belyaev, 1980). However, the total mass of abrasion material entering the World Ocean is very small compared with the sediment load supplied by rivers and comprises only 150 million to 300 million tons per year (Lisitsin, 1974). The amount of sediments that reach pelagic parts of the World Ocean from land (i.e., from the coastal–sea zone) makes up only 5–10% of the total solid material of terrigenous origin delivered to the ocean[*)]. So we can say that the coastal–sea barrier zone is of minor importance in terms of pelagic sedimentogenesis in the ocean (especially in the Pacific Ocean).

2. The zone within the influence of strong currents. This zone commonly coincides with that part of the bottom where clastic material is replaced by clayey material (the third mechanical barrier, see Fig. I.2 C). This mechanical barrier is found at various depths below the coastal–sea barrier, extending from the second barrier to the shelf edge (i.e., as deep as the third barrier and periodically even deeper). In the water layer that is bathymetrically higher than this geochemical barrier, active hydrodynamic processes take place that prevent the main portion of pelitic material (grain size <0.01 mm) from settling: only coarser particles with a grain size greater than 0.01 mm are deposited there from sea currents. At greater depths, where the current velocity is greatly diminished, settling of pelitic particles commences. Thus, this barrier is observed on the bottom as the "debris (sand–silt) sediments–pelitic (clayey) mud" interface and represents the last stage of sedimentogenesis, which is of the greatest importance for seas and oceans: coarse sediment (sand-sized and clay-sized particles) is progressively sorted from fine sediment containing the smallest nonclayey substance. It is worth mentioning that the World Ocean receives as much as 25.1 billion tons of solid material annually from land (Lisitzin, 1991, p.124). Over the course of a year, this material must be sorted (differentiated) into sand-sized and pelitic-sized fractions (40% of the total material), which are deposited mainly in shelf areas and in the upper part of the continental slope, and pelitic fractions (about 60% of the total), which are deposited at great depths. This fractionation of sediments is accompanied also by differentiation of chemical elements. The space between the third mechanical barrier and the shore is the area where deposition of clastic (debris) sediments enriched in SiO_{2terr}, Sn, Zr, Ti, Au, Pt, and Ag occurs, and seaward of

[*)] For the Atlantic ocean, this value is much more than 10% (Emelyanov, 1982)

this barrier (towards the center of the water basin), clayey mud enriched in Al, K, Na, Fe, Mn, P, Ti, W, and other microelements is deposited.

Also, the role of the debris forms of elements abruptly decreases in this direction, whereas the proportion of sorbed and hydrogenous forms of these elements increases. Thus, the hydrodynamic activity is responsible for realization in the water strata of one of the most spectacular geochemical processes—mechanical differentiation (sorting according to grain size) of sedimentary material. This process is found everywhere in the ocean, but its role in the distribution of chemical elements is of crucial importance only in those areas where particles with grain sizes larger than 0.01 mm cannot be supported in the water by currents. This natural boundary caused by the hydrodynamic regime ecompasses all water basins of the Earth along their peripheries and therefore can be reliably defined as a universal GB. This barrier is frequently found at depths of 10–50 m or may even rest against the shore (equatorial humid zone of the ocean, small seas, lakes), but sometimes it lies more deeply, down to the shelf edge or, only under extraordinary environmental conditions (ice and moderate humid climatic zones of the ocean), at depths of several kilometers (Emelyanov et al., 1975, Emelyanov, 1982).

Physicochemical Barriers

Acidic–alkaline barriers. Many chemical elements are found in the ocean both in dissolved and particulate (suspended) forms, depending on the surrounding conditions. So, cationogenic elements are found in dissolved form in acidic waters, whereas in neutral and reducing conditions they barely dissolve, if at all. Elements such as Ca, Sr, Ba, Ra, Cu, Zn, Cd, Fe^{2+}, Mn, Ni and some others actively migrate in acid and weakly oxidizing conditions, whereas in alkaline waters these elements are nearly immobile. The mobility of anionogenic elements (Cr^{6+}, Se^{6+}, Mo^{6+}, V^{5+}, As^{5+}) is markedly better in alkaline waters than that under different natural conditions (Perelman, 1979, p. 126). Precipitation of elements from water depends not only on the pH value but also on temperature. Tables and plots have been established for many chemical elements which describe the conditions under which hydroxide precipitation from weak solutions commences (Garrels and Christe 1968). For example, when the water temperature is 25°C, hydroxides of Fe (III), Al, Cr, $Cu(OH)_2$, $Fe(OH)_2$, $Co(OH)_3$, $Mn(OH)_2$, and $Mg(OH)_2$ precipitate quite readily from solutions with pH values of 2.48, 4.1, 5.3, 5.4, 5.5, 6.8, 9.0, and 10.5, respectively (Fig. I.5). When metals precipitate at acidic–alkaline barriers, they change into a colloidal form (into solution) and this leads to the increased migration ability of many chemical elements. For the waters of the World Ocean, a neutral (pH = 6.5–7.5) or weakly reducing (pH = 7.5–8.5) environment predominates. This environment is unfavorable for the active migration of many metals. To make this migration more active, an environment with pH = 6.5 or >8.5 is required. Strongly or weakly oxidizing environments are typical for hydrothermal springs on the bottom of seas and oceans. Abrupt changes in the pH values are found also at brine–seawater (see Part II.12) and O_2–H_2S interfaces (see Part II.11). A pH gradient also appears in the upper water layer, as a result of acid rainfall to the ocean.

In general, four classes of diluted solutions can be identified by acidic and alkaline regimes, each of which is associated with precipitation of a particular hydroxide (Table I.2). It is interesting that the most abundant hydroxides, Fe and Mn, established according to their geochemical properties, are far from each other in the classification scheme: precipitation of $Fe(OH)_2$ occurs at pH = 5.5, whereas $Mn(OH)_2$ precipitates at pH = 9.0. Owing to this, not only are the mechanisms that cause increased concentrations of Fe and Mn different, but also the mechanisms that cause the development and localization of the ores of these elements. In general, as seen from Table I.2, elements such as Fe, Cu, Al, Zn, and a variety of other metals easily migrate in class I solutions, metals in the form of bicarbonates and complex compounds with organic acids easily migrate in class II solutions, and metals that

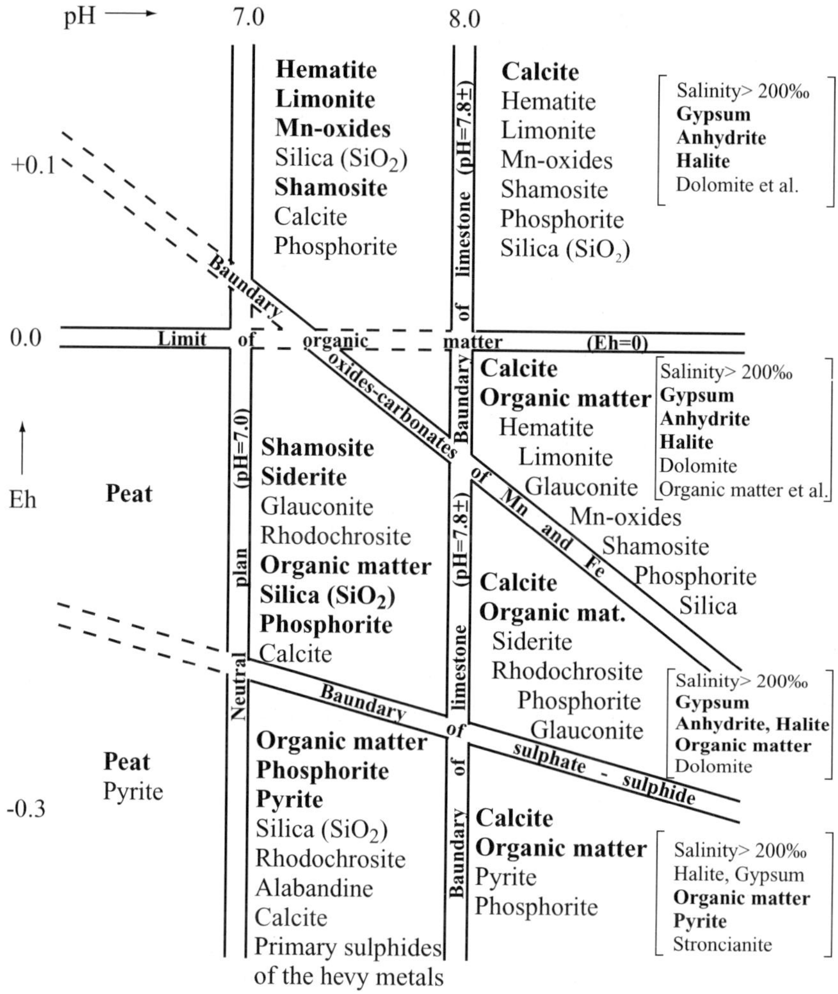

Fig. I.5. End associations in chemical sediments and characteristic environments for them in the dependence of Eh and pH. After V.S. Krumbein and R.M. Garrels (from Shcherbina, 1972, p. 248) with additions by the author (stroncianite). Shamosite – sedimentary silicates of Fe. Associations in straight brackets are typical for a solution of high salinity (>200‰)

precipitate as insoluble hydroxides and carbonates, as well as anionogenic elements (Si, Ge, As, V, Mo, Se, etc.), easily migrate in class III solutions. A strongly alkaline environment (class IV) is favorable for migration of Na, Li, F, Mo, U, Y, Be, B, silica, humic, Al, Mo, and others, but Ca, Mg, Ba, Sr, and Fe are hindered (Perelman, 1989, p.152).

Reduction–oxidation (redox) barriers. There are two main groups of barriers of this type in the ocean: (1) in water and (2) in sediments. In water, the redox barrier

Table I.2 pH at the beginning of precipitation of hydroxides from diluted solutions and PR of hydroxides at T° 25°C. After Perelman, 1989, p. 50.

Hydroxides	pH	SP
I. Highly acidic (pH < 3)		
NbO_2OH	0.4	-
$Co(OH)_3$	-	2.5×10^{-4}
$Sb(OH)_3$	0.9	4×10^{-42}
$Ti(OH)_4$	1.4-1.6	1×10^{-30}
$Sn(OH)_4$	2	1×10^{-57}
$Zr(OH)_4$	2	8×10^{-52}
$Fe(OH)_3$	2.48	4×10^{-38}
II. Weakly acid (pH = 3–6.5)		
$Sn(OH)_2$	3.0	1×10^{-27}
$Th(OH)_4$	3.5	1×10^{-50}
$Ga(OH)_3$	3.5	5×10^{-37}
III. Neutral and weakly alkaline (pH = 6.5–8.5)		
$In(OH)_3$	3.7	1×10^{-33}
$Al(OH)_3$	4.1	1.9×0^{-42}
$UO_2(OH)_2$	4.2	-
$Bi(OH)_3$	4.5	1×10^{-30}
$Se(OH)_3$	4.9	1×10^{-27}
$Zn(OH)_2$	5.2	4.5×10^{-17}
$Cr(OH)_3$	5.3	7×10^{-31}
$Cu(OH)_2$	5.4	1.6×10^{-19}
$Fe(OH)_2$	5.5	4.8×10^{-16}
$Be(OH)_2$	5.7	1×10^{-20}
$Pb(OH)_2$	6.0	7×10^{-16}
$Co(OH)_2$	6.8	1.3×10^{-15}
$Ni(OH)_2$	6.7	8.7×10^{-19}
$Cd(OH)_2$	6.7	2.3×10^{-14}
$Y(OH)_3$	6.8	1×10^{-24}

(Continued)

Table I.2 pH at the beginning of precipitation of hydroxides from diluted solutions and PR of hydroxides at T° 25°C. After Perelman, 1989, p. 50.—*cont'd*

Hydroxides	pH	SP
$Hg(OH)_2$	7.0	3×10^{-26}
$Nd(OH)_2$	7	-
$Ge(OH)_3$	7.4	-
$La(OH)_3$	8	1×10^{-20}
IV. Strong alkaline (pH < 8.5)		
$Mn(OH)_2$	9.0	4.10×10^{-14}
Ag(OH)	9.0	2×10^{-8}
Mg(OH)2	10.5	5×10^{-12}

*) I–IV: Classes according to acidity–alkalinity. SP—solubility product.

is found under conditions when the water strata divides into aerated and stagnated waters, as well as when there is a mixing of thermal waters with marine waters, polluted (waste) waters with river (lake) waters, and in many other cases.

Here we examine only two variants of the redox barrier: (1) at the boundary between oxygenated and stagnant waters in the water strata (O_2–H_2S interface in water) and (2) between oxidized and reduced sediments on the bottom (Eh barrier).

The redox Eh barrier or the O_2–H_2S layer in the water strata appears under conditions of sharp layering of the water strata into two layers: an upper hydrodynamically active layer of relatively lower salinity and a lower layer characterized by low dynamic activity (and frequently by higher salinity). The vertical mixing is largely absent at the boundary between these two layers. In the lower layer, which is dominated by stagnancy conditions, values of O_2 reach zero, whereas H_2S appears (Fig. I.6). The organic matter of the bottom sediments is the source of hydrogen sulfide. In seas, oceans, and lakes, extinct bodies of phytoplankton, or organic detritus, sink from the upper (photic) layer towards the seabed. More than half of this "rainfall" decomposes as it descends through the water column. Only a small portion of the organic matter remains intact to reach the sea bottom, where the process of its decomposition continues on the seabed or within the bottom sediments. The lesser the depth of the sea or ocean, the greater the amount of organic matter that reaches the seabed and accumulates in bottom sediments under the same sedimentary environments. Decomposition of organic matter is strongly affected by bacteria, which reduce sulfates (SO_4^{2-}), which have accumulated in muds as a result of complex biochemical interactions, to sulfides. The model shown in Fig 1.7 is an example of

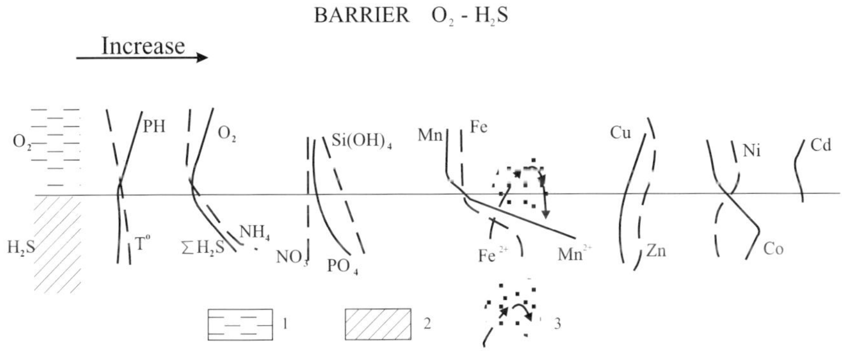

Fig. I.6. Redox barrier at the O_2–H_2S layer in the water strata and tendency to increase (or decrease) the main physical or chemical parameters and microelements. After Emelyanov, 1986_2. 1—water with O_2; 2—water with H_2S; 3—the direction of the diffusion of microelements and precipitation of the suspended gel (eternal diffusional flux).

microbiological processes. According to the authors of this model, "Upon comparison of calculated equilibrial redox conditions and the real concentrations in active water media, it has been supposed that microorganisms use reversible reactions taking place at each stage, which leads to a metastable state corresponding to the quasi-equilibrium of the model under consideration" (Vershinin and Rosanov, 2002, p. 16). The scheme in Fig. I.7 gives the succession of redox processes occurring in the O_2–H_2S zone, as well as various probable combinations of these processes. The characteristics of these processes are given in other publications (Stumm and Bacini, 1978; Vershinin and Rozanov, 2002). The larger the amount of organic matter in the sediments, the higher the chances for free H_2S to develop there. Hydrogen sulfide is initially accumulated in interstitial waters of the sediments, which makes up 60–80% of the total sediment volume (a layer of 0–10 cm). When the sediments are enriched in hydrogen sulfide, this gas begin to diffuse to the near-bottom waters. If the near-bottom environment is exposed to the effects of currents and the near-bottom water experiences continuous renewal, H_2S is oxidized and then disappears completely. But if the effects of mixing in the near-bottom water are weak, then all the oxygen present in this water is operatively used to oxidize organic matter, and the result is that hydrogen sulfide, which is constantly supplied from sediment, appears in near-bottom waters, extending to increasingly higher levels above the seabed. Accumulation of hydrogen sulfide is especially active in near-bottom waters, where the water strata is characterized by sharp salinity stratification. The mixing process of deep waters is hindered by the density transient layer (pycnocline). Water stratification in tropical waters is also caused by the difference in temperature. Here there is a distinct temperature jump (thermocline).

The values of Eh in water at the boundary of the redox barrier abruptly change from positive (top) to sharply negative (below) values. In the water strata, the O_2–H_2S boundary is found in the Black Sea (at depths of 100–200 m), in the Carjaco Deep of the Caribbean Sea (at a depth of 375 m), in the deeps of the Baltic

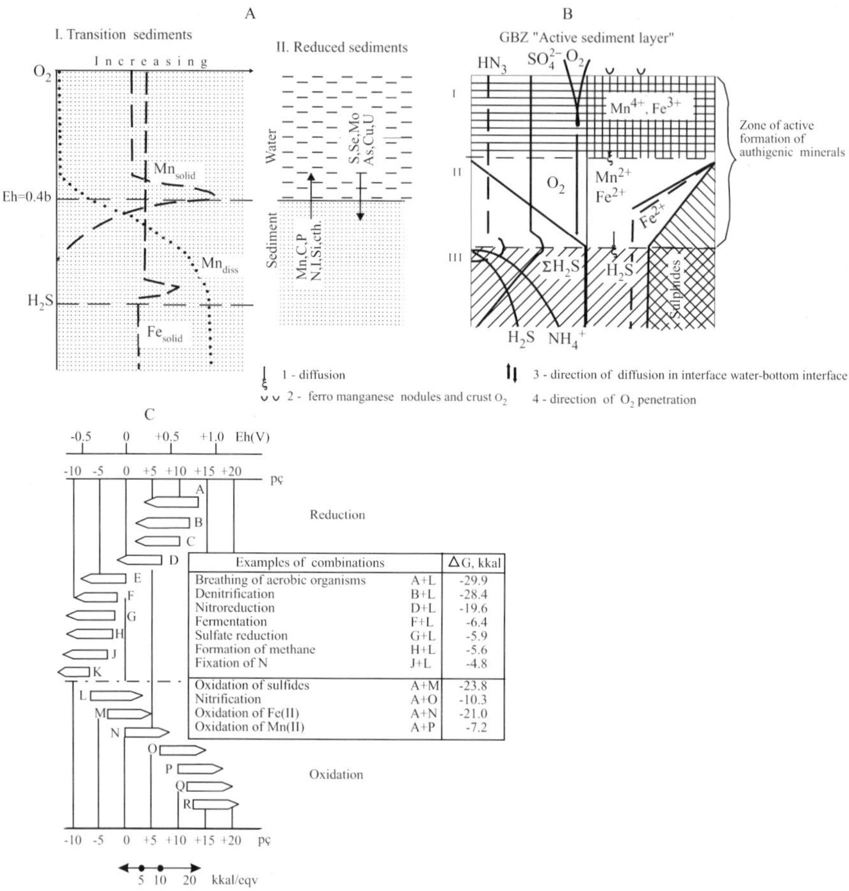

Examples of combinations		ΔG, kkal
Breathing of aerobic organisms	A+L	-29.9
Denitrification	B+L	-28.4
Nitroreduction	D+L	-19.6
Fermentation	F+L	-6.4
Sulfate reduction	G+L	-5.9
Formation of methane	H+L	-5.6
Fixation of N	J+L	-4.8
Oxidation of sulfides	A+M	-23.8
Nitrification	A+O	-10.3
Oxidation of Fe(II)	A+N	-21.0
Oxidation of Mn(II)	A+P	-7.2

Fig. I.7. Redox (Eh) barrier in the sediments (A) (after Rozanov, 1984) and diagenetic zone (B) in the Active Layer of Sediments GBZ (after Volkov, 1984) with some changes by the author.

C. Scheme of microbiological redox processes in water. After Stumm and Baccini, 1978. Numbers of oxidation–reduction potentials Eh and p̄, calculated from the meaning of the Gibbs free energy (ÄG°) for pH 7.

A—reduction of O_2; B—denitrification; C—oxides of Mn (IV) → Mn(II); D—reduction of nitrates; E—oxides of Fe(III) → Fe(II); F—reduction of organic matter; G—reduction of sulfates; H—formation of hydrogen; L—oxidation of organic matter, M—S^{2-} → SO_4^{2-}; N—oxidation of Fe(II); O—NH_4^+ → NO_3^-; P—oxidation of Mn(II); Q—N_2 → NO_3; R—formation of O_2.

Sea (at depths of 80–200 m), in Norwegian fjords (at various depths), and also in the deeps of the Red Sea, which are filled with hot brines (at depths of around 2000 m). At the O_2–H_2S boundary, there is diffusive exchange by reacting forms of Fe, Mn, and, to a lesser degree, Ni, Co, Cu, Zn, etc. (Brewer and Spencer, 1974). Above the O_2–H_2S boundary, large amounts of Mn and Fe gels are accumulated and this leads to the development of a deep turbidite layer. Dissolved forms of transitive elements (Fe, Mn, and, to a lesser degree, Co) accumulate in the O_2–H_2S zone. Elements such as Cu, Zn, Mo, Se, and As are actively removed (eliminated) from water along with sulfides and organic aggregates, to be deposited in increased amounts in muds. The Fe content in muds in the hydrosulfuric zone is commonly equal to the clarke value, and the Mn content is usually less than the clarke of this element. Mn is represented mainly by clastic forms, and reactionable forms of Fe are represented by sulfides. Sulfides are enriched in Cu, Zn, Mo, Se, and As. The sediments of a stagnated zone accumulate C_{org} (Strakhov, 1976; Emelyanov, 1982). When the hydrosulfuric contamination is not continuous but only sporadic (the deeps of the Baltic Sea, some fjords) and when the physicochemical O_2–H_2S barrier occasionally touches the bottom, the development of manganese–carbonate mud occurs (Emelyanov, 1979, 1981). Under conditions of the influx of O_2-enriched water into deeps, Mn^{2+} transforms to Mn^{4+} and precipitates as very fine particles of gels (which leads to the development of the near-bottom turbidity layer in water) to be eventually deposited on the bottom. A layer of gel-like Mn up to 1–1.5 cm in thickness forms on the bottom. Mn^{2+} remains in the sediments, where it is partially found as carbonates. It is assumed that Mn carbonates, and maybe vivianite (and barite?), must be accumulated both in muds of the continental slope of the Black Sea and in Carjaco Deep (in areas where the O_2–H_2S barrier zone comes into contact with the bottom).

The Eh barrier is measured in volts (V), and in the sediments of the World Ocean (at least in the uppermost 10 m of the sediment column), this value most commonly ranges from +0.7 to −0.35 V. If the environment is characterized by an Eh potential that falls within a range from −0.35 to 0.00 V, then this environment is strongly reducing (H_2S is present); if the Eh values range from 0.0 to +0.2 V, the environment is reducing (sulfides are present); if Eh ranges from +0.2 to +0.4V, then the environment is weakly reducing (there are reactions that reduce Mn^{4+}, and sulfides are present); if the Eh values are +0.4V or higher, then this environment is oxidizing (Mn^{4+} and Fe^{3+} are present) (Rozanov, 1988, 1995)(Fig. I.7).

At the water–bottom boundary, and periodically also in the sediment sequence, the thickness of the Eh barrier varies from a few millimeters to several centimeters. In many other cases, the thickness of the Eh layer is so large it is almost indistinct. The thickness of the Eh-barrier layer varies from meters in sediments to several tens of meters in the water strata (the O_2–H_2S interface).

The Eh barrier is found in sediments occurring at various depths in seas and oceans. The occurrence of the Eh barrier in sediments of the World Ocean and the thickness of this barrier depend on a number factors, such as bioproductivity of the water strata, the depth of the water basin, the hydrodynamics, and the content of organic matter in bottom sediments.

The Eh barrier commonly extends horizontally, and fluxes of matter towards this barrier are vertical (typically, water is carried upwards). However, in some cases this barrier may extend vertically (in such circumstances, fluxes of material towards this barrier are horizontal, as it is, for example, near steep outcrops of hard rocks). In zones where there are several Eh barriers in the sediment sequence, the fluxes of material derived from a reduced layer move not only upwards but also downwards.

Organic matter is the most important factor contributing to the development of oxidation–reduction processes in seas and oceans. The C_{org} content in the upper sediment layer in the World Ocean ranges from 0.05 to 16% (Fig.I.8), and within strata of Quaternary sediments, these values are from 0.1 to 22%. The maximum C_{org} contents were found in seas and lagoons and in coastal upwelling areas, where the primary production is maximum. Up to 13.08% C_{org} was found in Holocene sediments of the Baltic Sea (Emelyanov, 1981); up to 22% in the Black Sea; up to 13.56% in the Mediterranean Sea (in sapropelic beds of Pleistocene sediments) (Emelyanov and Shimkus, 1986, p. 302); and up to 16.06% in sediments of Walfish Bay (Emelyanov, 1973). In the open ocean, increased C_{org} contents in sediments were found in equatorial and also in northern and southern humidic zones. Minimal C_{org} contents (0.1%) were found in sediments of arid climatic zones, especially in red clays and foraminiferal oozes in the areas surrounding the North Atlantic Ridge and in pelagic clays (0.05–0.20%).

A direct relationship between the location of the Eh barrier and contents of C_{org} in sediments is very often evident: the lesser the C_{org} values, the greater the depth of the Eh barrier below the sediment surface. Because sediments with increased, or even maximum, C_{org} contents are predominantly abundant in continental periphery areas, and sediments with minimal C_{org} contents are found in deeps (see Fig. I.8), the thickness of the oxidized layer generally shows the same distribution: oxidized film is essentially absent in the continental zone of the World Ocean, where C_{org} contents in sediments are high, and it is found at the greatest depths (below the sediment surface) in pelagic sediments (first and foremost, in arid climatic zones), where the C_{org} contents and rates of sediment accumulations are minimal.

The main agent in oxidizing organic matter is O_2. Oxygen is responsible for the oxidation of 90% of the total organic matter delivered to the bottom. Under anaerobic conditions, oxidizing agents are also nitrates (NO_3^-), Mn^{4+} and Fe^{3+} compounds, sulfates, and CO_2 (Rozanov, 1988). About one-half of the total organic matter left after oxidation by O_2 undergoes oxidation under such conditions.

The position of the Eh barrier relative to the water–bottom boundary is very different for various areas of the World Ocean. This barrier commonly coincides with the water–bottom boundary on continental shelves. In pelagic zones of the oceans (below the CCD level), the Eh barrier sinks into sediment strata, where it is found at depths of 1–5 meters or greater below the sediment surface (Fig. I.9).

The Eh barrier is one of the most important factors controlling mineral formation. Processes occurring at this barrier are mostly related to a change in the valence state of elements and their authigenous compounds. The formation of these compounds depends on the electrochemical potentials of Eh reactions occurring at this barrier, which in turn depend on chemical properties of elements. Reactions at the Eh barrier act to bring the system back into equilibrium. In solutions, the state of equilibrium is established instantaneously (Ryzhenko, 1982). In bottom sediments, reactions predominantly run through interstitial water, which in turn tends to be in a state of equilibrium with the sediments mixing with it. Therefore, a state of equilibrium does not come about immediately: several days, months, years, thousands, or even millions of years may be required to bring the system into equilibrium. In

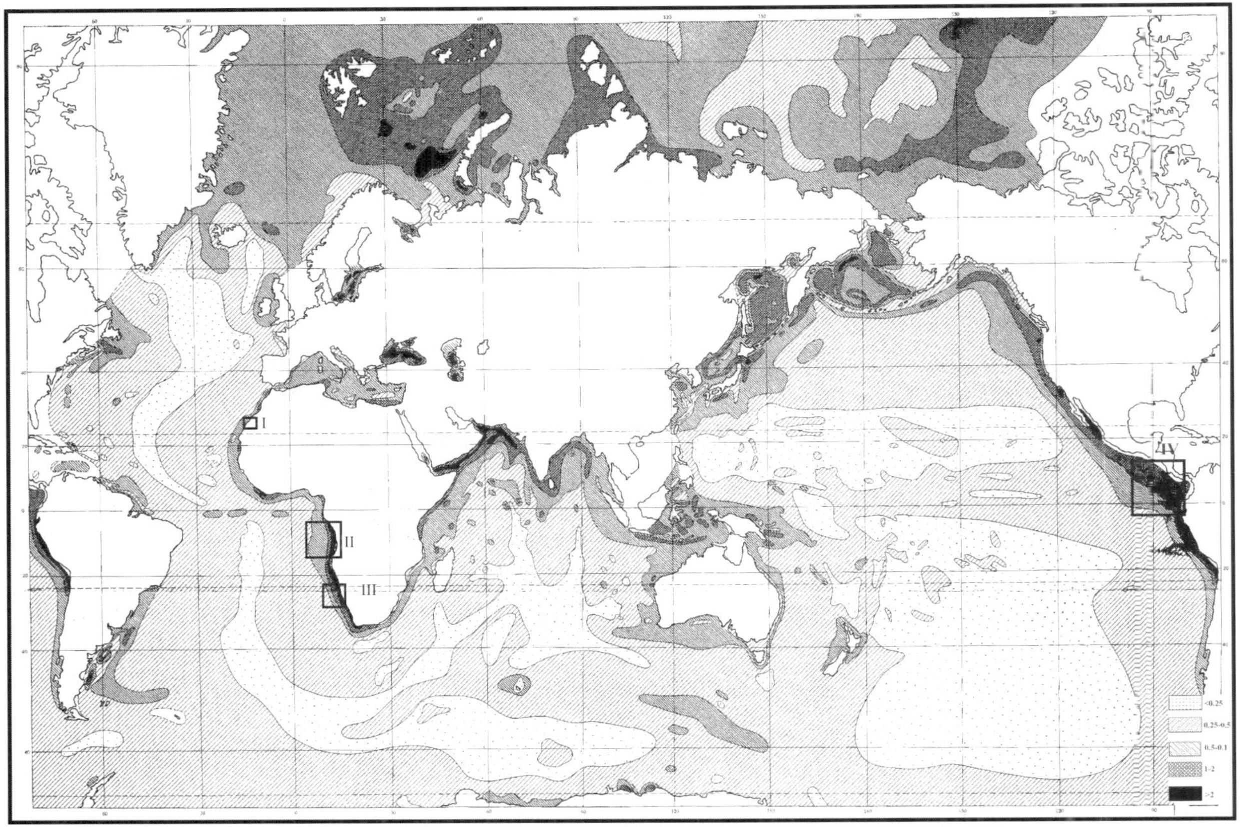

Fig. I.8. Distribution of total organic carbon (C_{org}, in %) in sediments of the World Ocean (0–5cm layer). After Bezrukov et al., 1977. Distribution of C_{org}. In individual small areas see Figs. II.1.7; II.1.10; II.4.3; II.4.5; II.4.8; II.6.5.

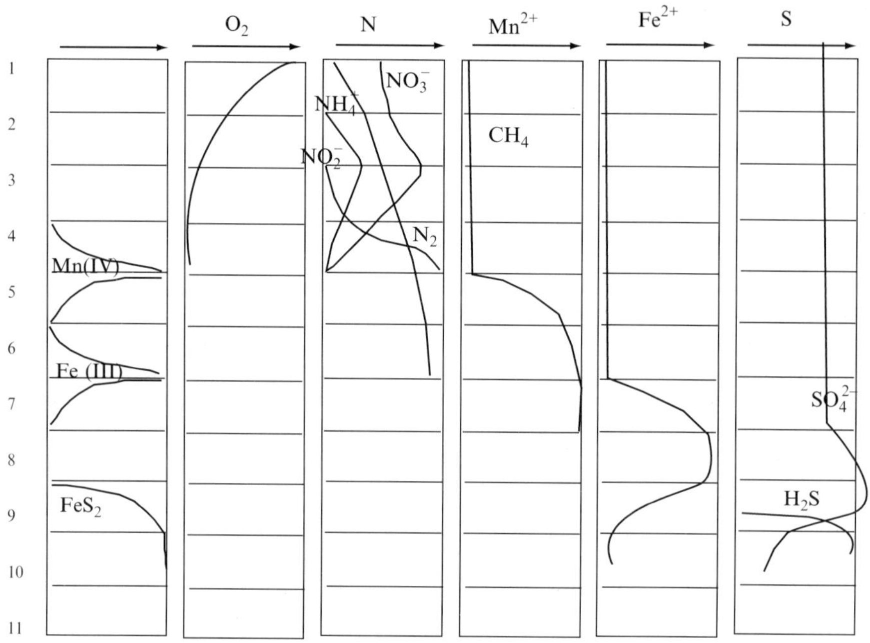

Fig. I.9. Changes of composition of interstitial water on different redox horizons of ocean bottom sediments, the layer of the accumulation of solid phases of Fe and Mn, accumulation of CH_4, H_2S, SO_4^{2-}, and forms of N and the oxygen (Table II.15.1). After Rozanov, 1988.

addition, a nonequilibrium system is brought to equilibrium through several intermediate (metastable) stages.

Above the Eh redox barrier, sediments commonly have a brown coloration, whereas below this barrier, mostly in zone of negative Eh values, sediments are of various hues, from gray to dark gray. Interstitial waters of these sediments often contain H_2S, HS^-, or S^{2-}. Under such conditions, chalkophile (Cu, Zn, Pb, Ag) and siderophile (Fe, Ni, Co) elements can form nonsoluble sulfides.

The Eh barrier is thus responsible for chemogenic–diagenetic differentiation of material in sediments. Mn^{4+} and Fe^{3+} move from the reducing to oxidizing zone to be precipitated then as a solid phase. Owing to a wide range of Eh values, elements of various groups participate in this migration. In the specific facial conditions present at this barrier, sediments are strongly enriched in manganese and iron, as well as in Ni, Co, Cu, Mo and other microelements. Very frequently, all this leads to the development of polymetallic concretions.

Salinity Barriers

The following types of salinity barrier can be recognized in the ocean: the river–sea barrier; (2) seawater–brines barrier, if there are brines in deep-sea deeps; (3) salt barrier conditioned in an ice environment, that is, when melted fresh waters enter the sea; (4) hydrotherm–seawater barrier zone; and (5) spots of submarine discharges. The initiation of a salt barrier in an ice environment is caused by melting of snow and ice in Antarctica, Greenland and in certain other ice-covered land and ocean areas. In this case, meltwater flows into the ocean to form the upper freshwater layer, which has a thickness that ranges from several centimeters to several meters. This water is occasionally found below an ice cover, separating thus the bottom surface of ice from typical seawater. Also, an upper fresh water layer, which is of considerable thickness, develops in the Arctic Ocean near the mouths of large rivers. In this case, river waters break through the river–sea barrier to extend well away from a river mouth.

Three zones are commonly recognized in the river–sea system: I—freshwater zone; II—zone of water with low salt content; III—altwater zone (Fig. I.10). In the freshwater zone, there are high concentrations of water suspension, high current velocities, etc. The main variations in environmental conditions in the river–sea system are listed in Tables II.1.1, II.1.2.b, and II.1.3.

At the river–sea GBZ, the salinity most commonly varies from 1‰ to 35‰, and at the seawater–brines boundary, from 30‰ to 260‰. In some deeps, the temperature of brines is very variable: from 20–25 to 40–60°C.

The most important features related to river–sea GBZs are salinity fronts. Fronts of estuaries are frequently visually recognized by streaks of foam, piles of garbage, and the color and transparency of the water. A streak of foam is a sign of frontal convergence. Distribution of freshened waters over the sea surface is bounded by this front. A strip of garbage is found at a shorter distance from the mouth, and even closer to the mouth is a color boundary (Fig. I.11). In estuaries, there are extraordinarily sharp lateral and transverse gradients of velocities and density. These gradients are related to frontal systems and play an important role in the dynamics of estuarine circulation (Fedorov, 1983, p. 208).

There are several types of estuarine fronts. In estuaries which have the shape of a channel, a "salinity wedge" is commonly present. In this case, near-bottom salt waters move upstream in rivers, creating a near-bottom salinity wedge, which terminates in the near-bottom front (Fedorov, 1983, p. 201).

Salinity fronts are present not only near the mouths of rives, but also at great distances from them, in the open ocean. These fronts are especially distinct in the vicini-

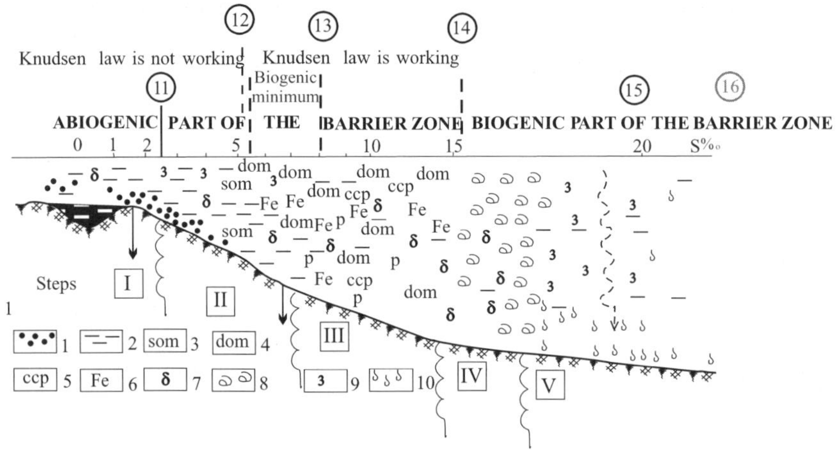

Fig. I.10. Principal scheme of sedimentation in the river–sea (in the marginal filter). scheme is compiled on the basis of Arctic rivers by Lisitzin, 2001, p. 68, with some changes by the author.

I, II, III, IV and V—stages of sedimentation and geochemical processes (abiogenic, gravitational, deposition not according to the Knudsen law, river waters and deposition according to the Knudsen law, colloidal-chemical or sorbtional, biological). Steps change gradually with increasing salinity and deposition of suspended particular matter (SPM) during the mixing of river and sea waters.
1–2—gravitational stage of GBZs: 1—deposition of the sandy-silty part of river SPM under input of river waters; 2—deposition of the main part of fine silty and pelitic material of river SPM (coagulation); 3—deposition of suspended organic matter (SOM); 4–6—sorptional stage of GBZs: 4—flocculation of dissolved organic matter (DOM); 5—coagulation of colloidal part; 6—coagulation of Fe hydroxides; 7–10—biological part of GBZ: 7—maximum of bacterioplankton; 8—phytoplankton–transformation of dissolved forms of elements and (pollutants) into biogenic SPM (at stages I and II after transparency becoming better); 9—zooplankton-forced filtration of the remains of river SPM, estuary snow and phytoplankton, formation of coarse pellets from them; 10—flux of pellets of zooplankton from water strata to the bottom.
On the horizontal axis—increasing salinity (S‰) in the surface layer of waters at the river–sea GBZ. Roman numerals below the section—consecutive stages of sedimentation in GBZ: I—gravitational stage with deposition of sandy and coarse aleurite fraction of SPM, formation of Fe-oxide covers on the grains—granular sorbent (S‰ <2); II—deposition of the main part of clay minerals, sorbtion on clay minerals (S‰ = 2–5–7); III—colloidal part of GBZ: flocculation of DOM and colloidal organic matter (COM), Fe-sorbents, remains of clay minerals and SOM. Formation of floccules of estuary snow covered by bacteria (δ) (S‰= 5—7 to 15); IV—phytoplankton part of GBZ. Transparency of water and high content of nutrients provides for increased phytoplankton production after deposition of main part of SPM. Movement of dissolved elements of river water into SPM: phytoplankton organisms, phytoplankton biopump (S‰=15—20); V—area of the maximum development of zooplankton filtrators (copepods and etc.). Bulk filtration of SPM and phytoplankton, powerful biopump, the movement of fine SPM into coarse pellets. Boundaries of GBZs can be displaced according to local conditions and seasons. Critical points of GBZs (numbers in circles on the upper side). 11—the area of murky river waters with transparency less than 1 m by Secchi disk: deficiency of light, lower first production of phytoplankton; 12—start of critical salinity interval (5–8‰), minimum biodiversity; the beginning of working of the Knudsen law, which determines the constant relation of the main salts in seawater; 13—the end of biological break , predominance of marine flora and fauna; 14—bulk coagulation (clay minerals, DOM, COM, Fe oxihydrates), river waters become more transparent with the help of sorbents (S‰ = 8–15); 15—sedimentation of colloidal-sorbtional part of GBZ is finishing: as waters become more transparent, phytoplankton production increases—the start of the main biogenic part of GBZ, phytoplankton part previously at the beginning, and then zooplankton part; extraction of residual part of SPM by zooplankton filtrators, transformation into coarse pellets, and then deposition of pellets on the bottom quickly (100–500 m/day) (S‰ = 15–20); 16—outer boundary of GBZ (approximately S‰ = 20).

ties of the Ganges, Brahmaputra, Iravadi, Amazon, Orinoco, Congo and other rivers. The drops in salinity across these fronts, which are hundreds of kilometers away from their river mouths, are variable and can reach 1‰ and even more. The effect of water freshening is especially clearly seen near the mouth of the Orinoco River: the salinity varies from 33.5‰ in the area of the island of Trinidad to 34.5‰ in the Caribbean Sea, at 15°N, 70°W (Fedorov, 1983, p. 71). Near the mouth of the Amazon River, the effect of water freshening can be traced as far as 1000–1200 km from the mouth (Gibbs, 1970).

Each river–sea GBZ has its own distinguishing features, which make them different from barrier zones in other areas of the World Ocean. The main feature characterizing each zone is the salinity value, which falls within a range from 2‰ to 15‰ and which varies depending on such factors as river strength, seawater salinity, the hydrodynamics of coastal waters, and weather conditions. The configuration of the salinity gradient zone near the mouth of a river is often very unusual owing to the above-mentioned factors. Nevertheless, there is a feature common to all near-mouth areas in seas and oceans: the shape of this boundary (salt barrier), both horizontally and vertically, runs along the 15–20‰ isohaline. The vertical extension of the most active zone of the salt barrier varies from hundreds of meters to several kilometers, and its vertical extension (by depth) is several tens of meters.

River–sea GBZs are characterized by abrupt variations not only in salinity, but also in such features as pH, Alk, the amount of nutrients, and the hydrodynamic situation. In addition, current velocities and the concentration of suspension are sharply diminished there (see Fig. I.5). The phenomenon of a "biological contour" (or "filter") is present (and very distinct) at the river–sea boundary. Also, the river–sea GBZ is a place where other GBs, such as land–sea, air–water, water–bottom, water–living substance, and acidic–alkaline (the value of pH may increase from 7 to 8), intersect, each of which grows stronger or weaker due to the influence of other barriers. Owing to such circumstances, the river–sea GBZ plays one of the most important roles.

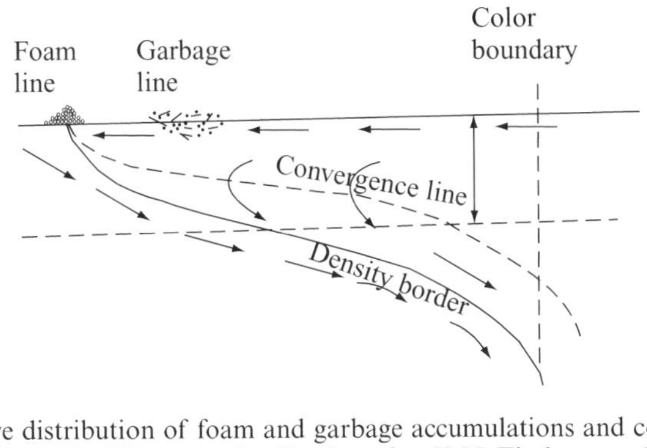

Fig. I.11. Relative distribution of foam and garbage accumulations and color boundary not far from estuary front. By Klemas, 1980. Citation after K.N. Fiodorov, 1980.

The occurrence of so-called "seals" is a very characteristic feature of the river–sea barrier zone. The concentration of suspended matter in a muddy seal is much higher than that in the river itself. In an organomineral seal, substances such as organic acids and dissolved forms of Fe and other components are removed from suspension by flocculation; as well, many microelements are coprecipitated (removed from solution) along with them (Sholkovitz, 1976). In such a biological seal, development of phyto- and zooplankton is the most active.

In the open ocean, the halocline (a layer of an increased vertical gradient of salinity) is the salinity barrier of greatest importance. The halocline in the ocean is usually identified as the surface of the 35‰ isohaline. In the ocean, it is found at depths of 300–700 m. The drop in salinity across the halocline in the ocean commonly comprises 1‰; in the South Atlantic this value is 2.5‰; and, very rarely, it is as large as 6‰ (Lebedev, 1986).

In the ocean, a halocline is usually not as distinct as a thermocline and it is several hundreds of meters deeper than the thermocline. In low latitudes (in arid climatic zones), the halocline may come to the surface of the ocean due to effects involving strong evaporation of ocean water and weak accumulation of atmospheric water: salinity of the upper water layer here is 2‰ higher than that near the equator (Horne, 1972, p. 128). Near the equator, just as in medium and high latitudes, the salinity of the upper water layer is decreased, as a result of atmospheric fallout and river runoff.

A salt barrier with a well-pronounced halocline and thermocline is also found below scouring lenses of freshened waters in the open ocean. Such lenses are found, for example, in the Guyana Deep of the Atlantic Ocean. They are formed as a result of strong influx of Amazonian waters (Monin and Gordeyev, 1988): freshened waters break through the river–sea barrier and the strong Guyana current on the shelf edge to form large lenses (up to hundreds of kilometers in diameter) of up to 10–15 m in thickness, which begin to travel over wide areas of the Atlantic Ocean until they are mixed with ocean waters.

At the seawater–brines boundary, active exchanges by chemical elements between brines and normal seawater occur as a result of diffusion. Large amounts of organic detritus and anthropgenic waste products accumulate on the surface of brines ("liquid bottom"), which is exposed to active biochemical (bacterial) processes. Above this liquid bottom layer, i.e., within an oxidizing environment, colloidal clots of Mn- and Fe hydroxides accumulate, whereas below the layer, dissolved forms of Fe, Mn and Co accumulate. On the bottom, above the seawater–brines boundary, accumulation of Fe- and Mn hydroxides, calcite, and aragonite occurs, and below this layer, Fe sulphides, montmorillonite, siderite, rhodochrosite, strontianite and other autigenic minerals accumulate in cold brines (for example, gypsum). In the Red Sea, in deeps filled up with brines, sediments become iron–montmorillonite, Fe–Mn sulfidic and manganese (mangano-sideritic) forms (Miller et al.,1966; Bishoff, 1969).

Temperature, Density and Dynamic Barriers

Temperature and density barriers, which are in close relation to each other, form the thermodynamic barriers (Perelman, 1979, p. 80). These barriers arise upon changes in temperature (T) and pressure (P). In seas and oceans, these parameters vary within a wide range: temperature changes from –2°C (in near-bottom polar waters) to +350 to +400°C (in hydrotherms), and pressure varies from 1 atm on the ocean surface to 1100 atm at maximum depths. The thermodynamic barriers are of great importance, because many minerals develop or dissolve due to the effects of these barriers. For example, solubility of some minerals decreases as temperature decreases, and many minerals can be formed only under conditions of strong pressure. In order to study conditions under which minerals can be formed, researchers often exploit the "phase rule," which follows Gibbs' law. This law says that "in an equilibrium heterogenous system, there is a direct relationship between the number of degrees of freedom for thermodynamic parameters (temperature, pressure, concentration) and the number of phase P and number of components L_0(K): C = K – P + n, where n is the number of outside factors contributing to equilibration of this system (most frequently, temperature and pressure)" (Perelman, 1979, p. 81).

The following critical temperatures are characteristic of World Ocean waters (including interstitial water in sediments): 0°C (or from 1°C to 0°C for saline waters), 40°C and 100°C. At the first critical point, water changes from the liquid to the solid state, and vice versa. The temperature of 40°C is "the typical Mendeleev temperature" (Perelman, 1979, p. 107). Upon reaching this temperature, water changes some of its physical properties (for example, electric conductivity), which is more than likely linked to a jumplike strengthening in the "dissolution" of the delicate water body. The temperature of 100°C is the point where water changes from the liquid to the gaseous state.

The thermocline (T) is a vertical temperature gradient in some layer of a body of water that is appreciably greater than the gradients above and below it. In the ocean, the thermocline is a water layer lying between the surface waters and intermediate and deep waters (*Oceanographic Encyclopedia*, 1974, p. 526). The main thermoclines in the oceans are permanent, seasonal or diurnal.

The most characteristic temperature range in the World Ocean is from –1°C to +30°C. This is why water in the ocean is found either in a liquid or solid state. However, the water temperature in hydrothermal systems increases to +400°C, and in the brines of the Red Sea, to +67°C. Within this temperature range, not only is there a point at which water changes to the gaseous state, but also a point at which the environmental conditions are such that the vital activities of particular groups of bacteria are impossible—this factor is of great importance for the migration and concentration of various groups of chemical elements.

A permanent thermocline is not actually related to the seasons of the year (Fig. I.12). The thermocline is shallowest at all points near the position of the equator; in medium latitudes, it is found at depths of up to 700 m; and it rises close to the surface in the vicinities of latitudes 50°–60°, both north and south.

The parameters of the seasonal thermocline do depend on the seasons of the year. The thermocline is very distinct in areas where interseasonal variations are great. In addition, this thermocline is essentially absent during the autumn and winter. The seasonal thermocline appears in spring and then becomes distinct in summer.

It is medium latitudes where the seasonal thermocline is most distinct. This thermocline is found at shallower depths than those of the permanent thermocline.

The diurnal thermocline appears during daylight hours and disappears at night. It is found at a very shallow depth below the sea surface.

The structure of the diurnal thermocline is complex and often involves subdivision into finer layers (fine structure) (Fedorov, 1983): abrupt drops in current speeds are observed in some single microlayers. Some layers of the thermocline, each of which has unique properties, can extend in various directions, so that suspended matter accumulated here may be carried away from this barrier.

In polar areas of the ocean, the temperature barrier at the ice–water boundary is very distinct. When the water temperature is negative, the physical state of seawater may take two different forms: the solid state, under conditions of increasingly negative temperatures, and the liquid state, typical for the positive part of the temperature spectrum. The result of such a difference in physical states of seawater is a variety of hydrometeorological, hydrochemical, biological and mechanical processes.

In the thermocline layer, there are processes caused by action and interaction of Rossby waves.

Density of seawater (ρ) is one of the main physical parameters of seawater. Density is a function of such parameters as temperature T°, salinity S and pressure P: $\rho = (T, S, g)$ (Burkov et al., 1978, p. 35). The density of seawater approximately varies from 1 for fresh surface water to 1.076 for deep ocean waters. When the pressure increases, the density of seawater decreases at a rate of about 45.10^{-4} g/cm^3 per 1000 dbar (or 1000 m). In contrast, when it is the salinity that increases, the density of seawater increases at a rate of 8.10^{-4} g/cm^3 per 1‰ (*Oceanographic Encyclopedia*, 1974, p. 380). The change in density due to the effects of temperature is a complex function. For example, if the salinity of water is 35‰, then the density of seawater decreases from 5.10^{-5} g/cm^3 per 1°C at a temperature of 0°C to 34.10^{-5} g/cm^3 per 1°C at 30°C.

The layer of seawater with the largest density gradient is called a pycnocline. A pycnocline is a layer separating light surface waters from heavier deep waters.

In the ocean, the depth of the halocline often coincides with that of the pycnocline, and periodically there is a situation when all three principal features—thermocline, halocline and pycnocline—are observed at the same depth. In this chapter, our task is to examine the role of the halocline and pycnocline in geological

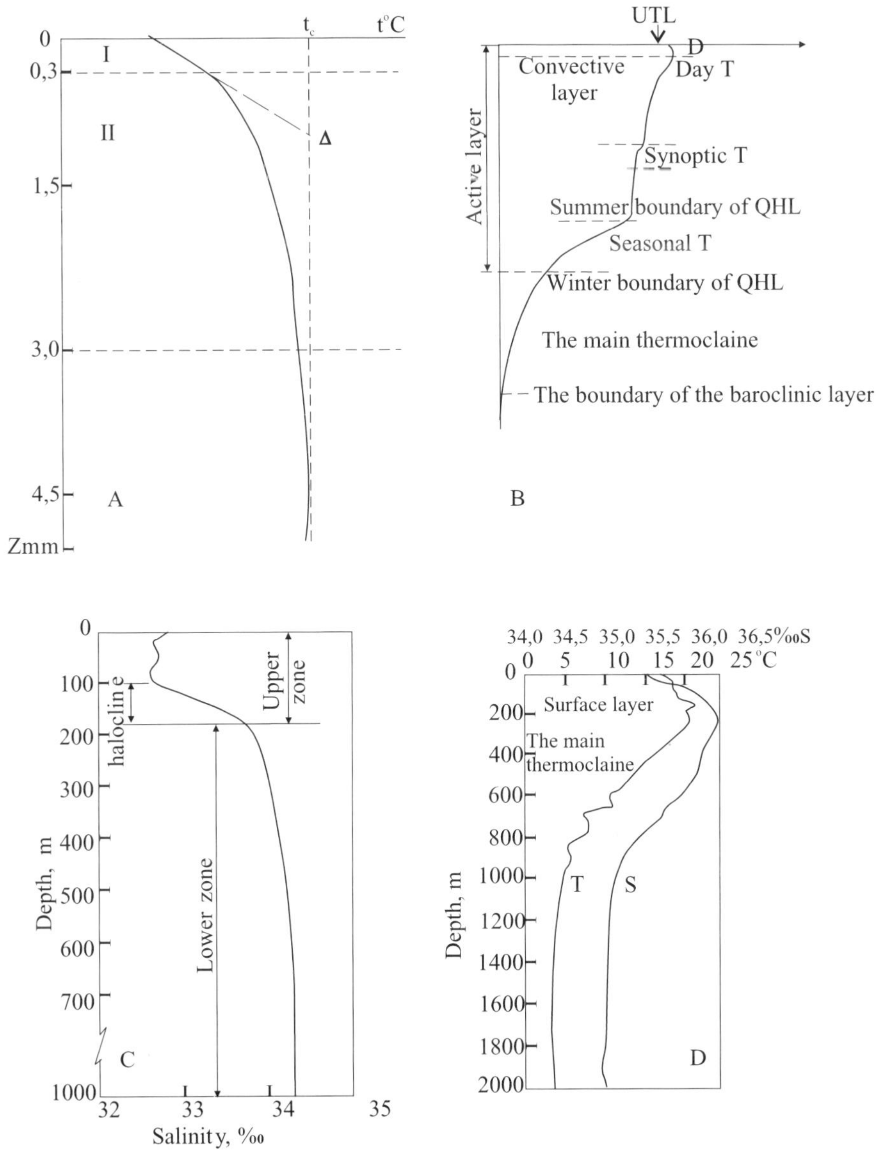

Fig. I.12. Typical profiles of ocean water.
A. Thermal structure of the boundary water layer (upper microlayer). After Hunarzhua et al., 1977.
B. System of the upper water layers of the ocean. After Lebedev, 1986. UTL—surface thermal layer; T—thermocline; QHL—quasihomogeneous layer.
C. Salinity profile for the subarctic zone of the Pacific Ocean, February, 1957. After Horne, 1970.
D. Temperature (T) and salinity (S) profiles. After Horne, 1972.

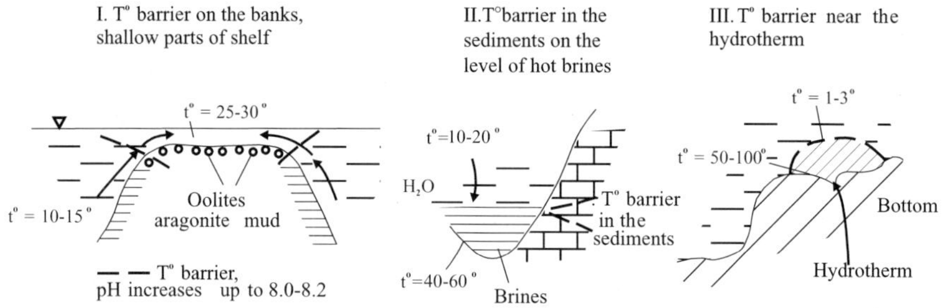

Fig. I.13. Principal examples of the temperature (T°) barrier in the sea and ocean.
I. Shallow bank in the arid climatic zone of the ocean (Bahamas).
II. Deeps with hot brines (Red Sea).
III. Thermal barrier at hot hydrothermal spring (T° to +350—+400°C) and cold sea water (T° –1°—+3°C).

development, mostly in application to the case when both of these features are at the same depth. For convenience, this layer will be periodically referenced to as a transition (or jump) layer.

In hydrophysical literature, the pressure $\rho = (T, S, \rho)$ is often replaced by density ((δ_t (T, S, pa), i.e., the density reduced to an atmospheric pressure of 1013.25 millibars (at T, S = const) (Burkov et al.,1978, p. 36). In general, the density δ_t of the ocean water varies from 2.3 in the surface layer to 27.87 in the near-bottom layer. The largest gradient of density δ_t is found at depths of 50–250 m.

Several examples of thermodynamic barriers acting in the near-bottom waters of seas and oceans are illustrated in Fig. I.13. When cold carbonate-rich water rises to shallow banks where the temperature of seawater increases to 25–30°C, aragonite precipitates from solution. In hydrothermal systems, i.e., in places displaying an

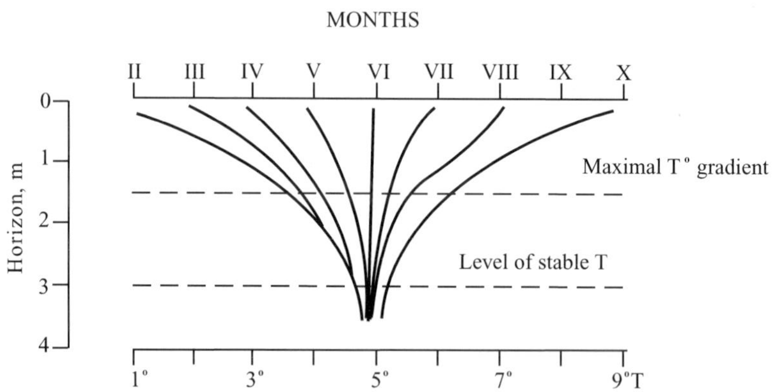

Fig. I.14. Changes in temperature of bottom sediments in shallow gulfs and seas during different seasons. Arkona Basin in the Baltic Sea. After Emelyanov, 1995_1.

abrupt decrease in temperature and CO_2, calcite and other carbonates precipitate. In general, as follows from the LeShatelier law, the changes in the parameters in such systems are equilibrated by outside forces acting in the opposite direction. This means, for example, that in a hydrothermal system, precipitation commences from those minerals the development of which leads to the release of heat, and this equilibrates the decrease in temperature that occurs. This principle is also true for mineral formation at the seawater–hot brines contact line, etc.

In shelf seas, at the turn of the seasons, a temperature barrier arises in the bottom sediments. Thus, for example, in the Arkona Deep (Baltic Sea) at depths of 40 m, the temperature of the bottom sediments at the 1-m level below the seafloor was 1°C in February, whereas in October it was as large as 9°C (Fig. I.14).

In oceans where at depths of more than 1 km the variation in temperature with depth is small and pressure increases by 1 atm per approximately 10 m of depth, it is the baric (dynamic) barrier that plays the most significant role in mineral ore formation (see Part II.6).

Geochemical Barriers That Exist Independently of Their Position in Space (Autonomic Barriers)

Water–living matter barrier. The effects caused by a water–living matter barrier are found everywhere in the World Ocean. Seawater, its composition, chemical and physical state of dissolved organic, nonorganic and organomineral matter—all of these are affected by living organisms. V.I. Vernadsky (1965) stated that in the ocean, five functions of living matter exist. The three functions that represent the major contribution to ocean environments, including processes of sedimentogenesis, are gaseous, concentrating, and oxidation–reduction. The gaseous function is of greatest importance for developing the O_2–CO_2 system in the ocean. The concentrating function is expressed in the selective extraction by organisms of certain groups of chemical elements (when organisms accumulate more of some elements than others). Concentrating functions fall into two categories: (1) the capture of such elements as H, O, C, N, and P, which are present in the body of any organism; (2) the capture by certain organisms of those elements which are absent in other organisms.

The largest effect that living organisms have on seawater takes place in certain regions called biological contours of natural waters (Zaitsev, 1979, p.21). These are the barrier zones water–atmosphere, water–coastal, water–bottom, river–sea, etc.

Water–suspended matter. Each suspended solid particle participates in a particular way in interactions with seawater. Dimensions of particles suspended in seawater vary in grain size from 0.1–0.4 μm to 1 mm, but most frequently are up to 10 μm. Owing to the small surface of a single particle, its sorption energy is negligibly small. However, if we consider that there are billions of particles suspended in the water column of the World Ocean, the total surface of these particles is enormous. The total surface of particles suspended in a water strata the vertical length of which is 1 km and cross-sectional area of which is $1 m^2$, is 10 thousand to 40,000 m^2 (Zenkevich, 1960). The total surface of suspended particles present in the Atlantic Ocean is 8642×10^9 m^2, and this is 86,418 times larger than the surfaced of the Earth (Emelyanov, 1982). This means that the sorption energy of the summarized surface of suspended particles present in seawater is very large.

The water–suspended matter boundary is characterized by the following processes: (1) active exchange by Ca^{2+}, Mo^{2+}, and Na^{2+}; (2) the sorption of microelements (such as Zn, Cu, Pb, Cd, Mg, Mo, Ag, Co, Ni, V, Sr, Y), as well as Fe, Mn and other elements, from seawater by solid particles. Some elements are adsorbed due to ionic exchange, whereas the sorption of other elements is an irreversible process. In many cases, processes of desorption of elements occur when elements pass from particles to solution.

Due to adsorption, many chemical elements are extracted from seawater and their concentrations in sediments increase to values which are of interest for commercial mining. The role of the water–suspended matter barrier is of great importance, especially in those areas of the World Ocean where new minerals form. These GBs are as follows: acidic–alkaline, oxidation–reduction, Eh barrier in sediments (especially on the surface of the seabed), the O_2–H_2S (or redox) boundary in seawater, and the following GBZs—river–sea, springs of submarine discharges, hydrothermal springs on the seabed, and water–bottom.

Electrochemical barriers. The appearance of barriers of this type stems from the effects of natural electric fields. Due the effect of natural electric current, charged particles move in the water medium towards electric poles (electrochemical barriers), where they accumulate. According to I.S. Goldberg (St. Petersburg, personal communication), ore concentrations of elements may develop at electrochemical barriers.

Electrochemical processes have been studied in detail, especially in application to the case of the interaction between solid phase and seawater (Horne, 1972, p. 275). This interaction plays an extraordinarily important role in the chemical-element exchange at the water–bottom barrier. The situation near the solution–solid matter surface is very complicated. A structure adjacent to solid body is strongly disturbed. To obtain a qualitative picture of this, one should imagine that there are "two layers made up of ions with opposite charges, the rank of which depends on the charge of the solid surface....If the solid surface is charged, then ions arising in solution form an electrical double-layer system....Physical and chemical properties of this interphase water are essentially different from the properties of "normal" water in the volume phase" (Horne,1972, pp. 276, 321). Action applied by electrochemical forces to the water–bottom boundary depends on the character of the seabed, i.e., on the type of sediments or solid rocks exposed on the seafloor. Because a major part of the ocean bottom is covered with sediments, writes Horne (1972, p. 283), the chemistry of the seawater–bottom interface is mostly associated with the chemistry of marine sediments.

Geochemical Barrier Zones

Geochemical barrier zones (GBZs) are those areas in seas and oceans where there are not one but several GBs, and each of which grows stronger or weaker due to the influence of other barriers. In some cases, the participation of these barriers in various processes is successive. Now we will consider certain GBZs that are of greatest importance in terms of sedimentology.

Ocean–atmosphere boundary. This boundary separates two different phases, air and liquid. The surface film of water (or surface microlayer, SML) has surprising physical, chemical and biochemical properties (Horne, 1972). One of such properties is surface tension, which is much higher than that for other liquids. The thickness of the SML is a few tenths of a millimeter (Table I.3). The ocean–atmosphere interface is characterized by sharp changes in its physical and chemical parameters, including the thermal structure (Fig.I.12), disordering of water structure (Fig. I.15), and other phenomena. "In the SML, the orientation of water molecules is such that negatively charged oxygen atoms move outward relative to the aqueous phase" (Horne, 1972, p. 246). Owing to these and some other properties, there is a tendency for the surface area of water to reach the minimum for the given volume and under the given external conditions.

In the SML "surfactant assemblies can function as templates for the deposition of silicates to form mesoporous silicas" (Yang et al., 1996, p. 589). Such films can be grown without a solid substrate by surfactant templating at the air–water interface (Fig. I.15). The films are continuous and have a root-mean-square surface roughness of about 3Å. They are resilient enough to withstand significant bending and are flexible enough to be transferred onto substrates of different shapes (Yang et al., 1996). It has been suggested that both a surfactant overstructure at the air–water interface and micellar aggregates in solution interact collectively with soluble, polymerizable, silicate building blocks.

At the ocean–atmosphere boundary, there is an active exchange of elements between air and water. The upper film is responsible for accumulation of great amounts of pleiston (Cheng, 1975) and bacterioplankton (Sorokin, 1977), which play an important role in the chemical exchange between the atmosphere and hydrosphere. The amount of neuston, which also includes saprophytic bacteria, is several orders of magnitude greater than those in any other horizons of the ocean (Zaitsev, 1979). Neuston is food for organisms belonging to the next level of the food chain—protozoa and invertebrates. The latter are food for larvae, young fish and larger invertebrates, whereas larvae and young fish are food for fish and pleistocenic physalia.

Table I.3. Approximate sizes (in mm) of the thin surface structure of water. After Lebedev, 1987

Macrolayer	Vertical size
Layer of surface active matter	$10^{-9} - 10^{-4}$
Layer of surface tension and associated anomalies of matter composition (10%)	$\sim 10^{-8}$
Structure-order zone	$\sim 10^{-2}$
Delta-layer of thermal (heat) output	$\sim 10^{-5}$
Layer of molecular conduction (laminar regime)	$\sim 10^{-4}$
Diffusive microlayer	$\sim 10^{-4}$
Conductive microlayer	$\sim 10^{-3}$
Viscous microlayer	
Equivalent thickness	$\sim 3 \times 10^{-3}$
Extra-turbulent layer	$\sim 10^{-2}$
Cold film	$\sim 10^{-2}$
Double thermal skin-layer	$\sim 10^{-2} - 10^{-1}$
Hyponeistal	5×10^{-2}

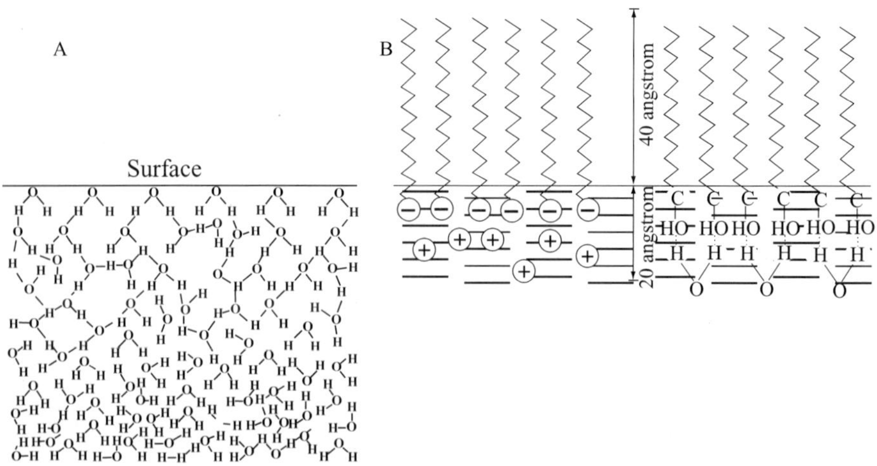

Fig. I.15. Changes in the water structure beneath the sea surface (A) and disposition of the molecules of surface-active matter in the upper water microlayer. Greater part of dry oil or fat molecules hover above the surface of the water (B). After Horne, 1972.

So, the upper film (SML) of water is the actual "nursery" of the ocean (Zaitsev, 1979).

It is characteristic that neuston comprises specimens that inhabit the lower (aqueous) part of the film—hyponeuston—and specimens that live in the upper (air) part of film—epineuston (Zaitsev, 1979). The upper film is a place where many chemical elements are involved in the food chain. At this barrier, aeolian material, which initially entered the ocean in "dry" form, experiences desorption (to a lesser degree, this material adsorbs microelements from water).

As a result of intensive biogeochemical processes, water suspension in the upper film (10–100 μm) is enriched in Fe, Cu, Ni, Pb, and other elements to values from 6 to 50 times (periodically, as much as up to 1000 times) greater than those present in deep-sea suspensions (McIntyre, 1974).

Photic zone. Photosynthesis is the process in which carbohydrates (saccharine, starch) are manufactured from abiotic material (which makes up the tissue of plants) and water in the presence of chlorophyll, which uses light energy and releases oxygen.

The activity of photosynthetic processes is the highest in the upper 0- to 50-m layer, and it becomes smaller in the zone extending from about 100 to 175 or more meters (Fig. I.16, I.16c). The lower boundary of the photic layer, as is widely

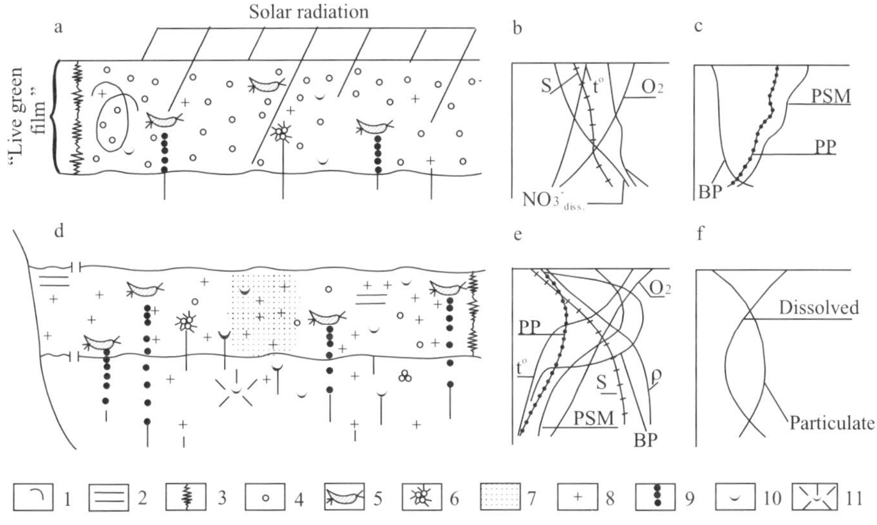

Fig. I.16. Principal scheme of the biogeochemical processes in the GBZs photic layer (a,b,c) and layer of discontinuity (halocline, pycnocline) (d,e,f). After Emelyanov, 1982_1.
a—scheme of biochemical processes;
b—distribution of t°, S (salinity), density (ρ), P_{diss} and NO_3 in the layer; c—distribution of phytoplankton (PP), suspended matter (PSM) and bacterioplankton (BP) in the photic layer; d—scheme of biogeochemical processes in the layer of discontinuity; e, f—distribution of t°, S, ρ, PSM, PP and BP (e) and the dissolved forms of the Cu, Mn, Ni, Co, Mo, Fe and others in the layer of discontinuity (f).
1—intensive water mixing; 2—horizontal mixing; 3—microturbulence; 4—phytoplankton; 5—zooplankton; 6—shells zooplankton; 7—gathering of the plankton; 8—organic detritus; 9—fecal pellets; 10—Fe—and Mn hydroxides; 11—sorbtion of elements.

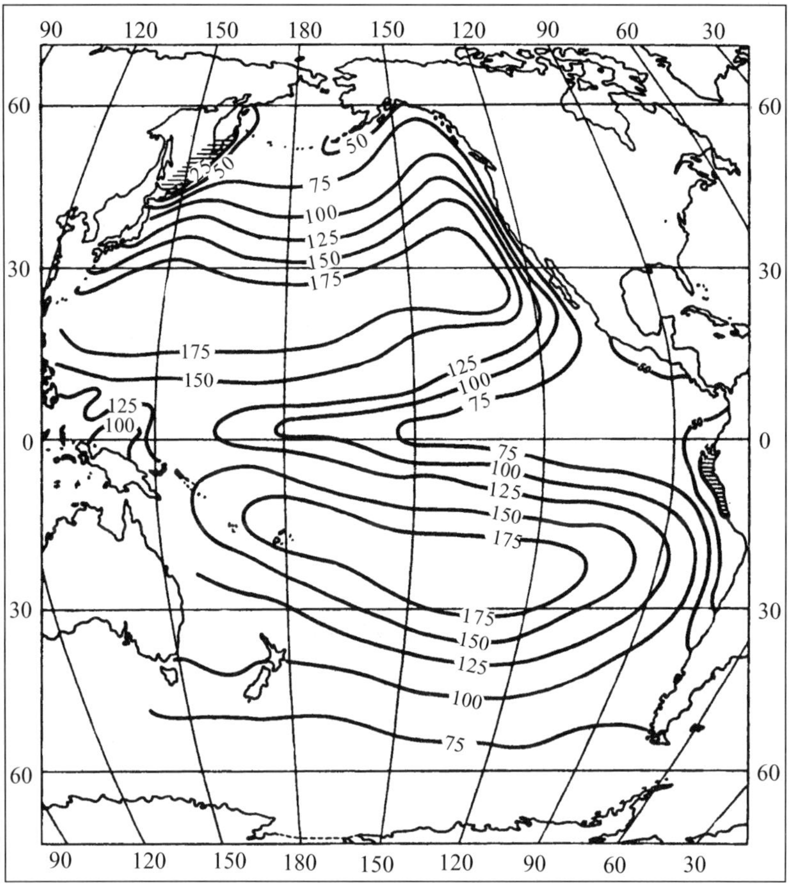

Fig. I.17. Position of the lower boundary of the photic layer (light barrier), m, in the Pacific Ocean. boundary was calculated at 0.1% surface illumination. After Sapozhnikov, 1981.

accepted, is found at the depth at which 0.1% of the light intensity from the sea surface is retained (Fig. I.17). Thus, the deficiency in light is an inhibitory barrier for organisms with photosynthesizing functions.

In addition to the light barrier, some other vertical boundaries can be identified within the photic layer, for example, the "compensation point," "critical depth" and "compensation depth" (Neshyba, 1991, p. 201). The compensation point, as defined by Neshyba (Fig. I.18), is the depth at which the light intensity is sufficient to bring about a balance between the oxygen produced by plant cells via photosynthesis and that consumed by them via metabolic processes. The compensation depth depends on how quickly the amount of light sufficient for photosynthesis decreases with depth. The depth at which the compensation point occurs is greater for clear waters than for greenish waters in coastal areas. Plants may exist even at greater depths, but

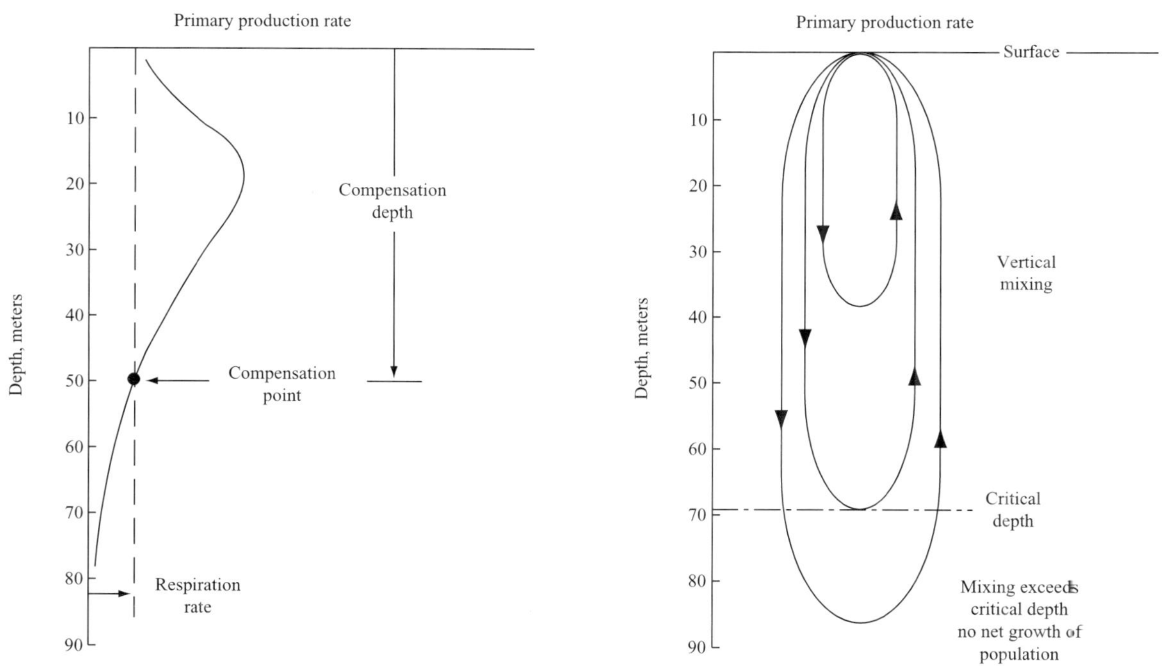

(a) Compensation point is the condition for which the rate of oxygen generation within the plant cell during photosynthesis just balances its rate of oxygen consumption in body metabolic processes. The depth at which this occurs depends on how rapidly the intensity of available light decreases with depth; thus clear water will yield a deeper compensation point than will the greenish waters of coastal areas. Plants can live at depths below these levels, but they must lose energy to the work of maintaining adequate oxygen intake.

(b) Critical depth pertains to the entire population of phytoplankters and the depth down to which the outside forces of winds and waves cause the surface layer to mix. Plankters near the surface and high light intensity can synthesize more new organic chemical energy than they consume in body metabolsim. Plant cells mixed deep into the water column suffer from the lack of adequate light; they tend to produce less new chemical energy than they consume. When vertical mixing is confined to depths shallower than "critical," the entire population produces a net surplus. But mixing to greater depth results in a net loss, with overall consumption outstripping overall production.

Fig. I.18. Definitions of compensation point, compensation depth, and critical depth. Neshyba, 1991.

in this case they are forced to expend their energy to maintain the corresponding consumption of oxygen.

The critical depth is an aspect of photosynthetic processes that refers to the entire phytoplankton population. This depth is that down to which the surface layer mixes due to the effects of external forces—wind and waves. In the euphotic zone, which is abundantly supplied with light, the amount of energy accumulated by planktonic organisms may be more than their energy losses for metabolism. When plant cells reach greater depths owing to the effects of mixing, they experience deficiency in light energy, the result of which is that the amount of chemical energy they produce is less than their energy consumption. When the vertical mixing is restricted to depths of not more than the critical depth, the total energy produced by the entire plant population is the "net surplus." Nevertheless, when the mixing zone extends to greater depths, the energy produced by organisms is less than their energy consumption; i.e., the result of their activity is the "net loss" (Neshyba, 1991, p. 201).

It is the photic layer (including the SML), where the exchange by material and energy between the hydrosphere and atmosphere is the strongest. Also, this layer is dominated by complex and intensive biogeochemical processes, including the transformation of one form of elements into another, origination of living organisms (including organisms with skeletons and shells, which occasionally form thick sedimentary sequences on the seabed) and their dying. The photic layer is impregnated with "living matter," which predetermines, in a direct or indirect way, all chemical properties of this layer (Vernadsky, 1965, p. 220). Vernadsky's law says that "the migration of chemical elements over the earth's surface and in the biosphere as a whole is accomplished either by direct participation of living matter (biogenic migration) or seeps in a medium the geochemical properties of which (oxygen, carbon dioxide, hydrogen sulfide, etc.) have been predominantly conditioned by living matter (both by living matter that was a contributory factor in the geological past and by the present-day inhabitants of this bioenvironmental system" (Perelman, 1989).

Solar radiation is responsible for manifestations of life activities both in the biosphere and, particularly, in the "upper living film" of the ocean. Living organisms in the photic layer convert solar energy into mechanical or biochemical energy. On a global scale, these processes in the ocean are either stimulated or hindered by solar energy. According to A.P. Chizhevsky (1976, p. 245), "In years of increased solar activity, the amount of energy delivered to the earth strongly increases." Factors such as variable solar energy flux and specific facial conditions influence the parameters of the biosphere (including phytoplankton), the changes of which, in turn, are recorded in sediments as various microlayers.

The photic layer of the World Ocean is responsible for about 43% of the total primary production of the earth's biosphere. Representing the primary production of the World Ocean by C yields a value of about 20 billion tons per year (Romankevich, 1977), or 110 billion tons of dry plankton. In the World Ocean, the annual production of organic matter makes up about 100 g of C per 1 m^2.

In the photic layer, there are some specific features: intensive horizontal and vertical mixing of waters, normal contents of oxygen (3–8 ml/l), and decreased contents of phosphates and nitrites (Fig. I.16). Processes that are the most intensive in the GBZ photic layer are also characteristic of self-contained GBs, such as water–living matter and water–suspended matter. The intensity of the photosynthetic processes is different for various places throughout the World Ocean. The intensity is highest in the biological contours (river–sea, coastal–sea, etc.), which are boundaries falling into category I according to the classification established for the marine environmental system (see Part IV).

In addition to active production of organic matter, the following processes occur within the photic layer: extraction of such elements as C, Si, P, N, Fe, Cu, Ni, Zn, Mn, Al, and Ti from seawater; migration of these elements by means of suspended biogenic material; use of the "living green film" by zooplankton; passing of elements into soft or hard parts of the zooplanktonic shell and formation of fecal pellets; rapid settling of chemical elements such as Cu, Mg, Si, Ca, C_{org}, Fe, and others onto the bottom with fecal pellets (pelletic transport).

The specific reaction common to all living matter, including, in particular, the photic layer, is the transition of chemical elements from soluble to less soluble forms (Korzh, 2003). Researchers have found that the behavior of the elements Cl, Na, B, Vr, Sr, Si, Rb, Cs, Mo, U, W, Tl, Re and Au, despite considerable difference in their chemical and physical properties, is governed by the law of transformation and migration of material as the ocean–atmosphere GB. A characteristic feature of these elements is "balance between fluxes of these elements in the ocean–land–atmosphere system" (ibid.). Korzh also concluded that the redistribution of mean elemental compositions of the elements in solutions and in particulate sediments, when they pass through GB systems, including lithosphere–living matter–hydrosphere; river–living matter–sea; ocean–living matter–atmosphere, depends non-linearly upon source concentrations of these elements (Korzh, 2003, p. 519). The pattern of biochemical behavior of elements, which are found in nature in microconcentrations, is generally different from that of their chemical forms occurring in lithosphere and hydrosphere in macroconcentrations (Korzh, 2003, p. 519). According to another conclusion, biochemical processes are responsible for "general relative growth in concentrations of elements in the solid phase, which increases with increasing abundance of these elements in the environment" (Korzh, 2003, p. 519). When elements are removed from solution, they participate in biogenic migration.

Solar energy accumulating in the photic zone via photosynthetic processes leaves the photic zone and moves downwards as the energy of organic material. The photic layer is a global factory which works to transform solar energy. This energy is accumulated both in living and nonliving matter and is used consequently not only for transformation of this material in the photic layer or throughout the whole length of the water strata of the World Ocean, but also for transformations in the sequence of bottom sediments, at all stages of their development, up to catagenesis.

Hydrofronts and divergences. A front in the ocean is a transition zone between two water masses which is characterized by steepened lateral gradients of the main thermodynamic parameters, in contrast to their usually gentle lateral gradients between stationary climatic or other extremities in the ocean (Fig. I.19). The frontal boundary is the interface within the frontal zone which coincides with the maximum gradients of one or more characteristics (temperature, salinity, density, velocity) (Fedorov, 1983, p. 22).

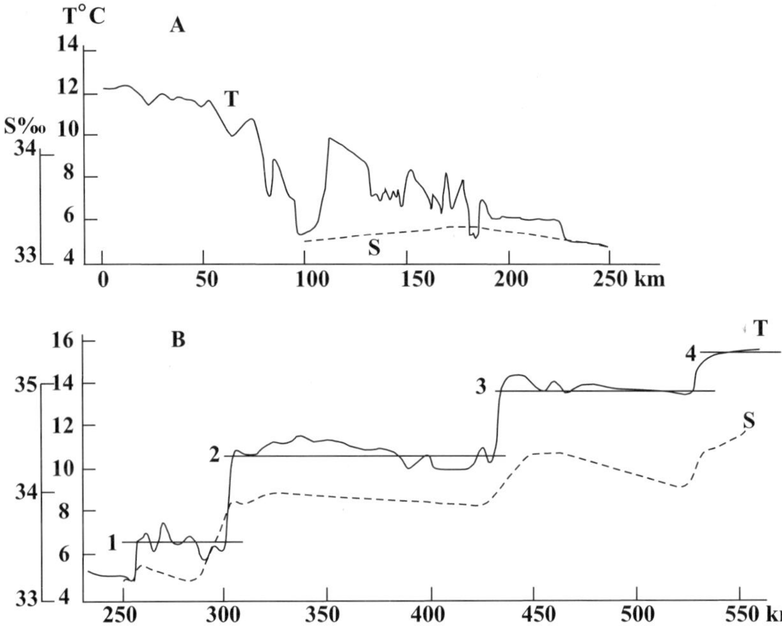

Fig. I.19. Temperature of seawater in the area where Kurosio-Kurilian current crosses the convergence zone. After Fedorov, 1983.
1,2,3 and 4—frontal steps, S—salinity.

The term front is short for frontal interface. K.N. Fedorov (1983) distinguished two classes of frontal zones in the World Ocean: climatic and synoptic. According to this author, frontal interfaces (fronts) can be placed into the following main subdivisions:

I. Geographic and semi-geographic:
   1. Climatic: (a) planetary; (b) circulation-intrusive, circulation-topographic; (c) discharge; (d) estuarine;
   2. Synoptic: (a) advective-vortical, (b) storm, (c) upwelling, coastal and equatorial.

II. Ageostrophic:
   1. Climatic: (a) runoff, (b) estuarine, (c) shelf-seasonal, (d) tidal, (e) circulation-topographic.
   2. Synoptic: a) shallow-water–under storm conditions, b) circulation-topographic.

As well, there is a simpler classification of fronts suggested by V.L. Lebedev (1986, p.186), according to which fronts fall into the following categories:

1. Planetary fronts of (a) moderate latitudes; (b) the equatorial zone; (c) frontal whirls;
2. Fronts of countercurrents;
3. Fronts of upwelling;

4. Tidal fronts in shallow-water areas;
5. Fronts of river mouths and straits: (a) estuarine fronts, (b) strait fronts.

"Planetary fronts may be considered unique thermal machines in which the energy of zonal temperature variations, via certain transformations, transforms into the energy of jet currents" (Lebedev, 1986, p. 159).

In frontal zones, there are processes such as vertical flows of water, active mixing of waters, and active hydrochemical processes; organisms are also concentrated there. Hydrochemical and climatic conditions on both sides of the front are the most contrasting. The prevailing conditions in these zones may by called "band marine landscapes"* (Lebedev, 1986, p. 159). Let us recall that zonal circulation of waters, and consequently many fronts, can be traced to depths of 800–1000 m (Stepanov, 1974) or even greater.

Planetary fronts in moderate latitudes are subdivided into polar (PFs) and subpolar fronts (SPFs) (Figs. I.20, I.21). In the equatorial ocean, the following types of fronts have been defined: tropical (TF), subequatorial (SEF) and equatorial (EF) fronts. In some of these fronts, at a depth of 50 m, the drop in temperature reaches 10° (Lebedev, 1986, p. 169), giving rise to strong frontal currents.

From space (at an altitude of 230 km above the surface of the earth), fronts appear as long, thin lines several hundreds of kilometers in length (Yoder et al., 1994). This is especially true for the front between the cold (24°C) south equatorial current (SEC) in the Pacific Ocean, moving westward from the coast of Central America, and the warm (28°C) north equatorial countercurrent (NECC), which flows eastward. There are cross flows of water in the jets, and this leads to the convergence of both currents. Over this convergence zone, there is a belt of atmospheric instability, which extends (some 1900 km) from east to west. On the southern edge of the SEC, water rises from the deep water layer toward the surface and continues to move westward at a speed of 0.8–0.9 m/s. Also, water in the SEC moves northward across the jet toward the convergence line, where it reaches a greater depth (about 40 m), to be overlain by the NECC. This convergence line, which is about 7 km wide, is clearly seen in satellite photos from 230 km. This zone of convergence is very rich in diatomic phytoplankton, in which *Rhisosolenia* is the predominant genus.

In moderate latitudes, fronts of countercurrents appear over the course of time between "inflows of cold waters near the western coasts of the ocean and flows of warm jet currents against the western coast" (Lebedev, 1986, p. 170). In the Pacific Ocean, the front of a countercurrent is the most distinctive in area stretching along the northwestern margin of the Gulf Stream. In the equatorial zone, fronts of countercurrents are found at the boundaries where tradewinds meet.

In areas of upwelling, there are alongshore temperature fronts (LTFUs). Under the action of a temperature front, which develops, for example, in the vicinities of

* Landscape (geographic) in the ocean—a province or geographical zone in the ocean with relatiely uniform bottom topography, which is qualitatively different from the topographical pattern of other areas owing to its characteristic geographical structure (i.e., the particular topographic components which make up this landscape).

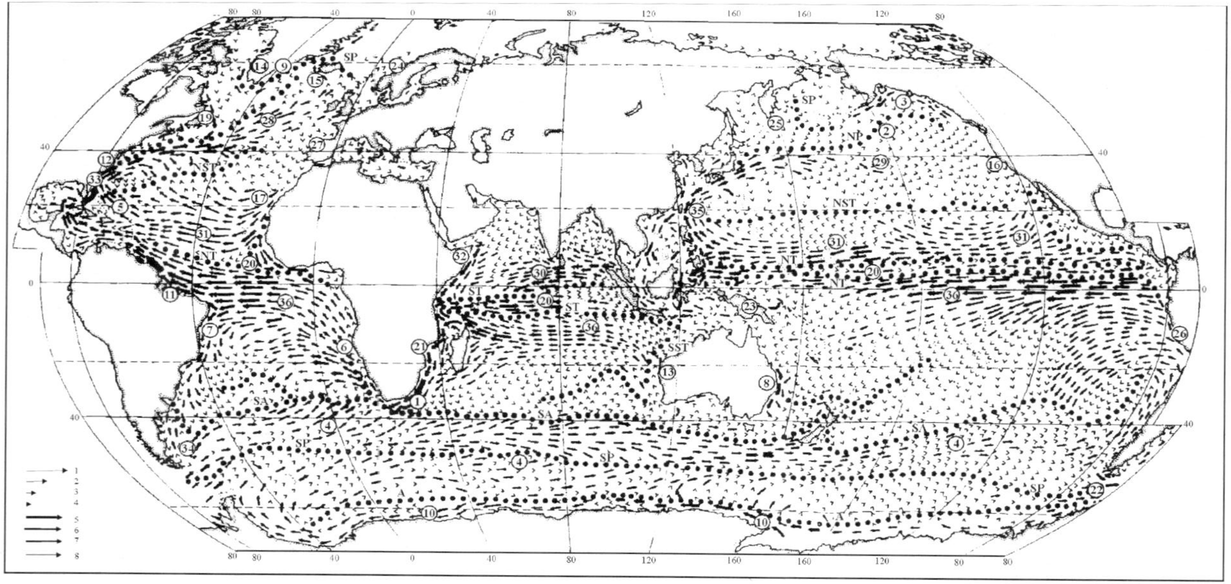

Fig. I.20. Scheme of mean long-term surface currents. After Schott, 1943

Fronts: NP—northern polar; SA—subantarctic (subtropical convergence); SP—southern polar (Antarctic convergence);

Divergences: SP—subpolar; NT—northern tropical; ST—southern tropical; A—Antarctic.

Convergences: NST—northern subtropical; NT—northern tropical; ST—southern tropical; SST—southern subtropical.

Currents: A—Agulhas; 2—Aleutian; 3—Alaskan; 4—Antarctic circular; 5—Antillian; 6—Bonguela; 7—Brazilian; 8—Eastern Australian; 9—Eastern Greenland; 10—Eastern winds; 11—Guyana; 12—Gulf Stream; 13—Western Australian; 14—Western Greenland; 15—Irminger; 16—Californian; 17—Canary; 18—Curosio; 19—Labrador; 20—Equatorial countercurrent; 21—Mozambique; 22—Cape Horn; 23—New Guinea; 24—Norwegian; 25—Oisio; 26—Peruvian; 27—Portugal; 28—North Atlantic; 29—North Pacific; 30—North Formosa; 31—Northern Equatorial; 32—Somali; 33—Florida, 34—Falkland, 35—Formosa, 36—southern Equatorial.

Stability of currents (shown by length of arrow):
1—stable; 2—more than average, 3—less than average, 4—unstable.

Speed of currents (shown by width of arrow), miles/day: 5—more than 36; 6—from 24 to 36; 7—from 12 to 24; 8—from 0 to 12.

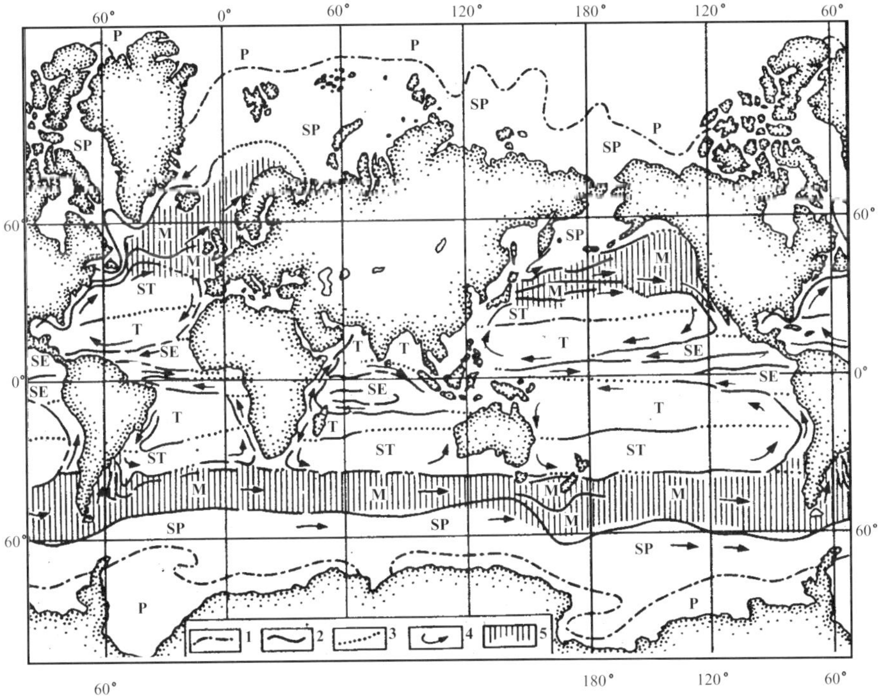

Fig. I.21. Distribution of fronts and ocean zonality.
1—border of permanent ice; 2—temperature fronts on the 50- to 100-m horizons; 3—the extrapolation of fronts; 4—currents; 5—mixing zone of cold-water and warm-water phytoplankton. Geographical zones and belts: P—polar and SP—subpolar zones (both form the cold belt); M—moderate belt; ST, T—subtropical and tropical zones (both form the warm belt); E, SE—equatorial and subequatorial zones (both form the equatorial belt).

the Somali Peninsula in the Indian Ocean, the current changes its direction and its velocity increases three- to fourfold (Lebedev, 1986, p. 170). There is a very distinct LTFU off southwestern Africa, where the strong Benguela current flows.

Physiochemical and biogeochemical barriers in deep waters. Barriers of this type are GBZs such as the transition (jump) layer, oxygen minimum layer, and calcium carbonate compensation depth. Each of these GBZs includes several other barriers: temperature, dynamic, oxidation–reduction, density, and others.

Transition (jump) layer. Barriers such as temperature (thermocline), salinity (halocline), and density (pycnocline) are especially distinct in the transition layer GBZ. Very frequently, these barriers are found at different depths: the depth of the thermocline is commonly several tens of meters shallower than that of the halocline and pycnocline. At the same time, in many other cases, these barriers function at essentially the same depth: about 30–80 m in closed seas and about 30–200 m in oceans.

The general distribution of the main physical, chemical and biogeochemical characteristics is shown in Fig. I.16, and the major processes and their role in terms of geological development are described in Part II.

Oxygen minimum layer. In the photic layer, the excess of replenishment over biological oxygen demand (BOD) reaches a maximum. As a result, the water of this layer is rich in oxygen. However, at greater depths, replenishment and consumption of oxygen become equilibrated. The depth at which these parameters are kept balanced is called the depth of oxygen compensation (DOC). An oxygen minimum layer (OML) forms and exists at a depth where the ratios of replenishment to consumption are minimum, in comparison to over- and underlying water layers, so that the water can become very oxygen-deficient (Ivanenkov and Chernyakova, 1979, p.157).

As for the water strata of the World Ocean, the OML is clearly identified in the concentration-depth profiles as an abrupt decrease in dissolved oxygen: from 6–8 ml/l in the upper mixed (photic) water layer to 0.5–0.2 ml/l O_2, or even to anoxic conditions, in the OML. Minimum oxygen contents (< 0.5–0.0 ml/l) are found in areas of upwelling, i.e., where the primary production is extremely high and the water strata is well stratified (Ivanenkov and Chernyakova, 1979). As is clearly seen on the map of the oxygen minimum, waters with low concentrations of oxygen, originating from the upwelling areas of continental shelves, spread out to form "tongues" of oxygen-deficient water (Fig. I.22). These tongues can easily be traced both in the Atlantic and Pacific oceans, where they spread westward to great distances on both sides of the equator.

The two main factors contributing to the development of the OML in the ocean are (1) the rate at which oxygen is supplied to the water by horizontal (or vertical) advection caused by turbulent mixing of waters and (2) BOD over the course of decomposition of organic matter sinking from the photic layer (Ivanenkov and Chernyakova, 1979).

Many of the above-mentioned barriers are active within the OML, including water–living matter; water–suspended matter; and certain others.

Carbonate compensation depth (CCD). The presence of vertical zonation in the distribution of $CaCO_3$ in the water column of the World Ocean has been observed and reported by many researchers (Murray and Renard, 1891; Murray and Chumley, 1924; Pia, 1933; Correns, 1937; Strakhov, 1960–1962; Lisitzin, 1971). The essence of such a phenomenon is that calcareous (carbonate) material begins to dissolve when it is still in the water column, at a depth of about 3000 m. This depth is called the lysocline (Berger, 1970) (Fig. I.23). Below the lysocline, the dissolution of calcareous remains of planktonic and benthic organisms is especially active on the seabed. At depths of about 4700–6000 m, the rate of accumulation of carbonates lags behind the dissolution of carbonates (and disintegration of shells into crystallites). In the domestic literature, this depth is called the critical depth of carbonate accumulation (Bezrukov, 1962). The CCD (Revell, 1944) is the depth at which the supply of Ca lags behind its dissolution. The CCD is the lower boundary of a sediment component, the dissolution of which is positively controlled by the extent to which the water is undersaturated (Berger, 1981, p. 1442).

The CCD is commonly found at depths of 4.5–6 km (see Fig. II.13.1). The depth of the CCD is shallower at high latitudes. Parameters characteristic of the given CCD are as follows: low temperatures (from +1 to −2°C), decreased salinity (34.5–35.0‰), increased density of waters (>1.02780 g/cm^3), decreased oxygen content (<0.40–0.50 mg-at O_2/l, increased SiO_3–Si content, strong undersaturation with

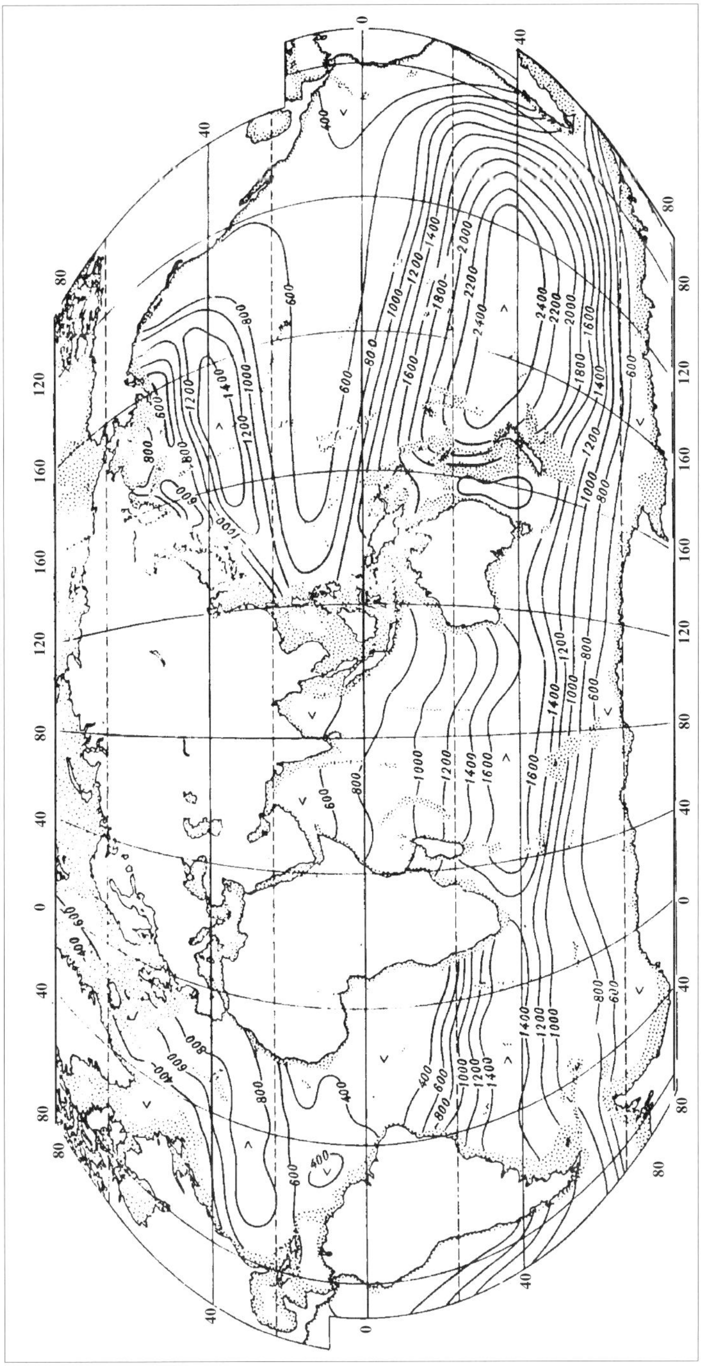

Fig. I.22. Distribution of oxygen in the oxygen minimum layer (m-at/l). After Ivanenkov, Tchernyaeva, 1979.

According to some authors (Bychkov et al., 1987, p. 52), there is also the of saturation depth (SD).

Recognition of the FL, CL and CCD levels is facilitated by variations in planktonic foraminifera associations in sediments (Lisitzin et al., 1977; Belyaeva and Burmistrova, 1987). So, the distinctive features of these levels are the number of species, their composition, compositional ratios of planktonic foraminifera vs. benthic foraminifera, and structural complexes for each group of foraminifera.

The positions of all these CLCs change regularly according to the climatic zonality of the ocean, and consequently, it depends on such factors as biological productivity of waters and their chemistry and hydrological pattern. Factors also contributing to the positioning of the CCD are the rate of vertical mixing of waters, the environmental conditions for deposition and burial of calcareous shell material, the exchange of water between the oceans, climatic changes, evolving variations in the properties of shells, etc. However, although the positions of the CCDs are different for each ocean (and also for different deeps), all of them are subject to the same regularity: these CCDs are found at the maximum depth near the equator, and they are everywhere shallowest at low (polar) latitudes. At high latitudes, some of these CLCs (for example, the PCD, ACD, and CL) reach the surface of the ocean. The CLC is a transition from facies of carbonate (>50% $CaCO_3$), calcareous (50–30% $CaCO_3$) and low-calcareous (30–10% $CaCO_3$) sediments, which are bathymetrically above the CCD, to facies of noncarbonaceous (10–0% $CaCO_3$) pelagic muds, which are below this level.

The water–bottom barrier. The water–bottom barrier zone is found at various depths and in different tectonic and facial environments. It is evident that this barrier zone encompasses the near-bottom water layer, which has a thickness ranging from 1 to 100 cm (depending on facial conditions), and the uppermost centimeter of the sediment column (or the millimetric film which covers solid rocks). The water–bottom GBZ can exist in a variety of physicochemical conditions: the seabed may either experience erosion by strong bottom currents or provide the main substrate for particle-by-particle deposition of sediments; the near-bottom waters can be saturated with oxygen, contain free H_2S, etc.

Among a variety of other boundary zones, the surface area of water–bottom barrier zone is the largest on the earth. It is a complicated submarine topography owing to which the surface area of this barrier zone is greater than that of the atmosphere–water boundary zone. Only when wave activity is very high can the surface area of the atmosphere–water boundary zone exceed that for water–bottom boundary zone. Whereas the total area of the World Ocean is 361.26×10^6 km^2, the area of the shelf (in $10^6 \times$ m^2) is 26.50; hummocky plains, 140; mid-ocean ridges, 55; and oceanic highs and seamounts, 30. The fraction of plain areas makes up 12–15% of the total surface area of the bottom of the World Ocean (Lisitzin, 1988). The area of the Atlantic Ocean is 91.66×10^6 km^2; however, with allowance for the surface area of the bottom topography, the total surface area of the ocean bottom will be as large as 144.59×10^6 km^2, or almost 1.5 times larger. From this it follows that the total surface area of the flattened bottom of the World Ocean may be estimated at about 540×10^6 km^2, instead of 361.26×10^6 km^2.

According to Vernadsky, the water–bottom boundary is another (second) living film. Various animals, ranging from microorganisms (microbenthos) to large mollusks, echinodermata, fish, etc., accumulate within this boundary. Organisms such as filter-feeders, mudeaters and burrowing animals play an important role in geochemical processes. These organisms disturb the sediments as they forage for food, with the result that the water–bottom GBZ extends down (into sediments) to a depth of about 10–20 cm or more below the sediment surface. The water bottom GBZ is dominated by processes including the disintegration of the main mass of fecal pellets, decomposition of the remains of bios that settled on the bottom or were buried there, and also the regeneration of biogenic elements. This boundary is the end point for the transfer of matter and energy from the photic layer down to the bottom sediments through various food chains.

The water–bottom GBZ is an intersection of other GBs (or their activities are evident in a particular place), but the most important among these barriers are hydrodynamic, reduction–oxidation (or redox) (Eh), and acidic–alkaline (pH). These and many other GBs tend to form strong gradients of particular properties of water, suspended matter, and sediment, leading to a variety of biochemical interactions (Figs. I.24a, I.24b, I.24c).

Water–bottom GBZs are characterized not only by abrupt variations in the values of pH, Eh, O_2, and Alk (which generally become smaller in the near-bottom layer), but also in concentrations of dissolved forms of metals in interstitial waters. Every bed surface within these GBZs is dominated by processes including the destruction of biological remains; decomposition of fecal pellets; intensive exchange between water and sediment due to diffusion; regeneration of sedimentary material in the digestive tracts of benthic organisms (filter-feeders, mudeaters) under the influence of microbial activity; development of microbenthos; generation of biogenic elements; the release of Si, C, P, N, Fe, Mn and other elements from sediment into surrounding waters; a decrease in C_{org} and SiO_{2m} contents in some places and a decrease in $CaCO_3$ and $MgCO_3$ in sediments (generally below the lysocline); and an increase in the contents of Al, Fe, K, Mn, Ti, Ni, Co, V, Cu, Zn and other elements in sediments in comparison to their values in deep-sea suspensions. Here, there is active biochemical and chemical differentiation of elements, as well as evidence of weak mechanical differentiation (sorting according to grain size) in the near-bottom turbidity layer. For some elements (K, Mo), their input due to the vertical diffusion at the water–bottom boundary is comparable to the total amount of these elements that the ocean receives from river runoff, whereas for some other elements (Na, Ca, CO_4^{2-}, HCO_3), the input makes up about 50% of river runoff (Bischoff and Ku, 1971). The rate of dissolution of biogenic (generally diatomic) opal on the bottom and the rate of diffusion of Si from interstitial waters of sediment into near-bottom waters are so great that they lead to the development of anomalies (with a very large silica content), which are very distinct in deep waters of the World Ocean (Edmond et al., 1979).

At the water–bottom boundary, under favorable facial conditions, either plane or round polymetallic concretions develop.

The upper active sediment layer. The regeneration of sedimentary material does not terminate at the water–bottom boundary, but can continue in the upper layer of the

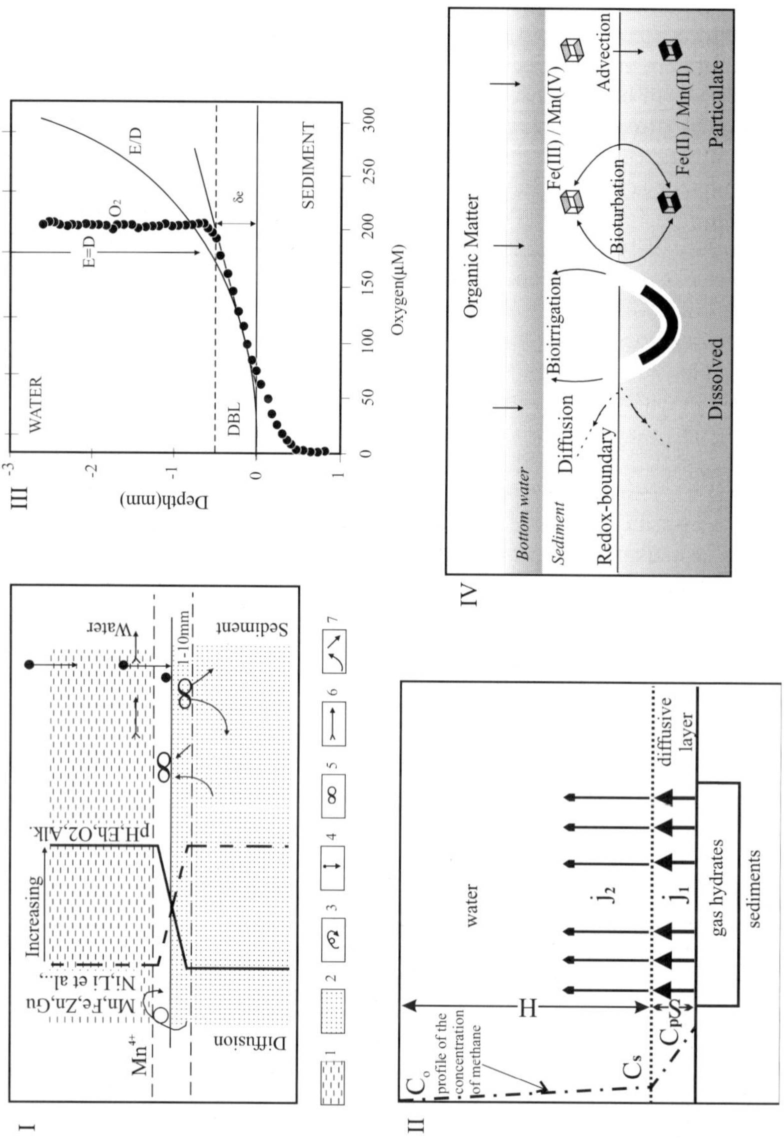

Caption see p.55

sediment sequence, i.e., after burial of sedimentary material and submersion to a certain depth below the seabed. This behavior is clearly seen in the quantitative and qualitative regeneration of biogenic elements, first of all, organic matter, which originates in the photic layer and sinks to the seabed to be buried in the bottom sediments (Table 1.4, Fig. I.24d; see Fig. I.16b).

Because the rates of decomposition of organic components during sedimentogenesis are different, the relative amounts and "reacting ability" of this organic matter in pelagic clays will be different from what they are in terrigenous muds on the ocean periphery. All this results in a variety of types and rates of diagenetic processes occurring on the seabed. Also, the depth and, consequently, the thickness of the upper active sediment layer (UASL) GBZ depend on a number of factors, such as (1) the rates of supply of sedimentary material to the bottom and the rates of its dissolution on the surface of seabed; (2) the rates of sedimentation–bottom erosion–redeposition of sediments; (3) the petrographic composition of sediments; (4) the position of the Eh redox GBZ within the sediments, i.e., the type of diagenesis.

Two main types of diagenesis can be distinguished: anaerobic and aerobic (see Fig. I.6). The former is typically found in the peripheral areas of the ocean, i.e., in areas with high concentrations of OM in the bottom sediments. This type of diagenesis has been studied in detail, for example, the Black Sea (Strakhov, 1976; Volkov, 1984, 1979) and also peripheral areas of the Pacific Ocean (*Geochemistry of Diagenesis...*, 1980). The thickness of the UASL GBZ is maximum (up to 5–6 m) under conditions of anaerobic diagenesis, minimum (20–50 cm) under conditions of aerobic diagenesis, and intermediate under anarobic–aerobic diagenesis. A major source of biogenic components (Si, NH_4, P) in porous waters is OM. Ammonium nitrate (NH_4) accumulates in the porous waters as a result of decomposition of organic matter (of albumen type) containing nitrogen (Pushkina, 1980, p. 76). In seawater, the content of NH_4 is about 0.01 mg/l, in porous water it is 0.1–20 mg/l.

Fig. I.24. Main processes in the water–bottom GBZ. I. Main parameters.
1—water; 2—bottom; 3—diffusion of Mn^{2+} and other elements; 4—deposition of fecal pellets from the photic layer; 5—FMNs; 6—near-bottom currents; 7—the main directions of the flux of the chemical elements during the formation of the nodules and their back current.
II. Oxygen microgradient (n) at the sediment–water interface compared to the ratio, E/D (logarithmic scale), between the vertical eddy diffusion coefficient, E, and the molecular diffusion coefficient, D. Oxygen concentration was constant in overflowing seawater.
It decreased linearly within the diffusive boundary layer (DBL) and penetrated only 0.7 mm into the sediment. DBL had a thickness of 0.45 mm. Its effective thickness δ_e is defined by the intersection between the linear DBL gradient and the constant bulk water concentration.
Diffusive boundary layer occurs where E becomes smaller than D, i.e., where E/D = I (arrow).
Data from Aarhus Bay, Denmark, at 15 m water depth during fall 1990. After Jorgensen, 2000.
III. Model of the exchange through the water–sediments interface. After Egorov, 2001 (from Lisitzin et al., 2001).
j_1—flux of molecular diffusion; j_2—flux of convective diffusion; S—thickness of the diffusional layer; H—characteristic scale of decreasing concentration; Cp—dissolution of methane; Cs—concentration on the upper border of the layer; Co—concentration of methane.
IV. Modes of transport in the sediment: molecular diffusion, bioirrigation, bioturbation and advection. After Hease, 2000, p. 243.

Table I.4 Average group composition of organic matter in plankton, suspended matter and sediments. After Romankevich (1977) and Volkov (1979)

Group of organic matter	Composition (% from OM)				
	Phyto-plankton	Zoo-plankton	Suspension	Terrigenous Sediment from oceanic margins	Pelagic red clays
Albumin-like deposits	30–40	55	44	0.1–0.6	Traces
Amino acids	40–53	72	50	4–5	1.5
Carbohydrates (carbon-like compounds)	20	10-30	18	7	15
—alcohol soluble saccharides*)	4.6	9	8.7	8.4	7
—water-soluble polysaccharides	46.2	8.1	11.8	26.3	3.8
—water-insoluble polysaccharides	46.5	76.7	76.2	65.3	89.2
Lipides	13–25	27	119.5–43	6.3	1
Humic matter					
—humic and fulvic acids	0	0	6	30	6
—humins	0	0	trace	36	74

*) In % of the total content of carbohydrate

Alkaline reserve (Alk) is an indicator of the type of water and the rate with which mineralization of OM occurs.

Maximum amounts of biogenic components (NH_4, P, Si) and Alk are found in porous waters in coastal areas of oceans and in hemipelagic muds, whereas their minimums are found in pelagic clays. This behavior is explained by the distribution of primary production (Bogorov, 1971; Koblentz-Mishke et al., 1977), as well as by the available biomass of benthos. The distribution of the biomass of benthos in the ocean is as follows: the shelf accounts for 82.0% of the total mass of benthos; the continental slope, 16.6%; and pelagic areas, 0.8%. The distribution of C_{org} is consistent with the pattern described for benthos: high contents of C_{org} are confined to sediments in the middle and outer parts of shelf areas and in the upper part of the continental slope, relatively low contents are restricted to the lower part of the continental slope and at its base, and very low contents are found in pelagic areas of oceans, especially in the deeps of arid climatic zones (see Fig. I.8). Destruction of biogenic matter depends on such factors as the rate of sedimentation, type of sediments, quantity and quality of biogenic matter available in the sediments, existence of oxygen in sediments and in the near-bottom water layer, temperature of sediments and water, and certain other parameters. This is evidence that the UASL is dominated by the following processes: intensive diffusive exchange by elements between interstitial water and seawater (across the water–bottom interface), active regeneration of sediments by bottom organisms (mudeaters), sulfate-reducing reactions (the initial stage of diagenesis), initiation of the development of chemogenic–diagenetic minerals (which, depending on Eh, pH and Alk, are hydroxides, sulfates, sulfides, carbonates), etc.

The processes of destruction of organic matter and redistribution of chemical elements continue until the time when physics and chemistry bring the system back into equilibrium.

Ice–water. This barrier zone is found predominantly in polar areas. The main distinguishing feature of this barrier zone is that there are three states of water: solid, liquid, and gaseous. This GB is an intersection of several temperature barriers (freezing point, ice melting, evaporation barrier) and certain other physical, chemical and biological factors. In the course of production of ice, water experiences cryogenic[*)] metamorphisation: non-frozen water become mineralized, and sulfates of Ca and Na precipitate from saline waters.

Because mineralization of seawater is low when it is found in the solid state (in the form of ice), freezing of saline waters in polar areas may be a contributing factor to the development of undercooled brines. There is evidence that brines with temperatures as low as −10°C may be found in shelf areas of the Polar Ocean (Perelman, 1989, p.239).

In seawater, the temperature gradient from the ice edge towards the open sea is commonly 0.2–2°C per mile. The gradients of other factors, such as cloudiness, degree of lightness, direction of atmospheric circulation over the sea surface, are also fairly strong.

[*)] Criogenic means that processes are going on in the environment with the negative temperature

Nowadays, the total amount of ice produced in the World Ocean annually is estimated at 3.33×10^9 km^2, and the total surface of ice-covered areas is 26 million km^2 (Lisitzin, 1994, p. 410). Also, a good deal of ice is supplied to the sea from land. Approximately 76% of this discharge (or about 3340 km^3 per year) reaches coastal areas of the World Ocean to form glaciers, and then it migrates over the world's seas and oceans as icebergs.

The "shoreline" of ice (or the ice–water boundary), if considering not only solid ice but also floating ice and icebergs, is enormous. It may be expected that the total extension of ice shoreline is comparable in length to or even more than that of, the World Ocean. This is why, in application to such aspects as biology, physics, chemistry and geology, processes occurring on the ice–water boundary are as important as they are at the shore–sea boundary.

The geographical positioning of the ice–water boundary in the World Ocean is not stable and varies with time (Kassens et al., 1999; Lisitzin et al., 2001): in winter, this boundary migrates towards the equator, whereas in summer it moves towards the poles. Poleward of this boundary is solid ice. The dynamic activity of waters (waves, drift currents) is low. Ice-covered environments are under cover of everlasting night, where photosynthetic processes and chemical weathering of rocks are suppressed. Phytoplankton is essentially absent there. In ice-covered waters, the role of bios and chemical processes in the extraction of chemical elements is reduced almost to zero.

Elements migrate in the form of clastic material under the influence of glaciers, ice, icebergs and sea currents. Phytoplankton actively develops near the ice edge and within ice itself. Below ice, fouling plants develop intensively.

In circumstances of winter cooling, heat is delivered to the atmosphere from the upper water layer, which mixes under the freezing point (see diagram in Fig. I.25). Further cooling gives rise to ice (diagram b). Jets of saltier water (i.e., brines which develop as a result of ice formation) reach greater depths, resulting in the increased density of the mixed layer. Under conditions of prolonged ice formation, brine (a water of greater density) may break through the stability layer to reach deeper layers. Then this water moves away (see diagram b). According to Neshyba (1991, p. 34), the reason for such behavior is the development of Antarctic bottom water, which then moves north. Sinking of cold, heavy water in the Weddel Sea is shown in diagram form in Fig. I.26.

Participation of water masses in a variety of interactions with shelf ice causes them to change their characteristics, including temperature, salinity, and concentrations of ^{4}He and δ^{18}o—this is also a contributing factor in the development of ice seawater (ISW), which is transported by bottom currents beyond the limits of shelf, where it then joins the Weddel Sea bottom water (WSBW), which has a temperature of about −1.5°C (Weppering et al., 1996). A similar process of development of cold, near-bottom waters occurs also in the Northern Hemisphere, for example, in Greenland, where the North Atlantic deep water (NADW) develops (diagram d in Fig. I.25).

The result of melting of sea ice is warmer and freshened surface waters, which spreads over the surface (diagram e).

"The distinguishing feature of the thin laminar water layer on the ice–water boundary is its complicated density stratification. Stratification of fresh water on the ice–water boundary is explained as being due to the existence of a layer of maximum density, which develops at a certain distance from this boundary. The reason for development of this layer is water cooling (and its densification) due to contact with ice. As a result, there are two water flows moving in opposite directions along a vertical wall of ice. Cold water, which is lighter in weight, moves upwards, along the ice surface, whereas warmer and heavier water moves downwards (diagram a in Fig. I.25) (Lebedev, 1986).

The structure of the density stratification of seawater that results from ice melting is also very complicated, similar to freshwater stratification. Cooling of seawater to such a degree accounts for the fact that the molecular thermal conductivity of

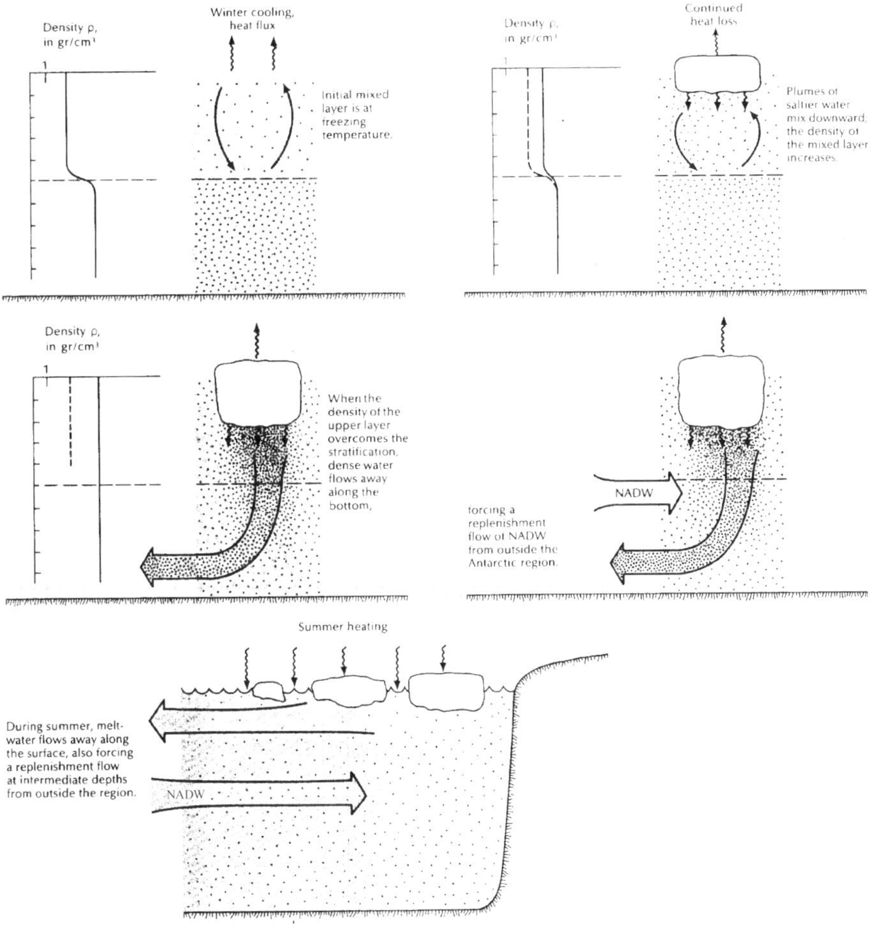

Fig. I.25. How vertical circulation, forced by sea ice forming and melting around Antarctica, replenishes surface water with upwelled water from intermediate depths. source is North Atlantic Deep Water (NADW). After Neshyba, 1991.

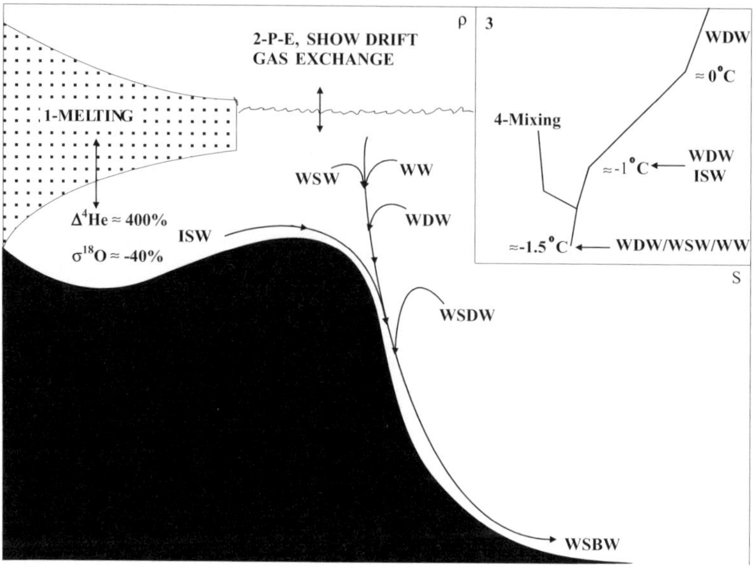

Fig. I.26. scheme showing the formation and sinking of cold water in the Weddel Sea and its transfer as near-bottom water to the equator. After Weppering et al., 1996.
On the scheme: black—bottom; dots—glacier hanging over the shelf.
Waters: WW—winter water of the ocean; WSW—the water of the west part of the shelf; WDW—Weddel deep water; ISW–ice shelf water; WSDW—Weddel deep water; WSBW—Weddel near-bottom water.
1—dissolution of the glacier; 2—transfer of snow and exchange by gases; 3—position of different waters and distribution of water temperature.
Density (ρ) and salinity (S) are shown in the scheme.

seawater (1.4×10^{-7} m^2/s) is two orders of magnitude greater than the coefficient of molecular diffusion for salt (1.3×10^{-9} m^2/s). This is why molecular transfer of heat in water is 10 times as fast as the molecular exchange of salt" (Lebedev, 1986 p. 130).

Springs of submarine discharge. Near the springs of ground waters, several types of barriers are clearly seen, which can be subdivided into the following: salinity, physicochemical, temperature, and certain other barriers (where variable parameters are the values of pH, Eh, alkalinity, etc.). Within the ground water–seawater mixing zone, there are strong gradients of some physical and chemical parameters and also of chemical element concentrations.

The springs of submarine discharge GBZs are related to processes involving the delivery of metals (Mn, Ni, V, Ti, Cu, Mo, Cr, Zr, Ba, Co, Cd, Ag, Ga, Fe, and others) into near-bottom waters, the distribution of these elements in the near-bottom waters and bottom sediments over certain areas (Fig. I.27, I.28), cementation of sediments by carbonates, anomalous diagenesis, and, episodically, accumulation of metalliferous brines and ore deposits.

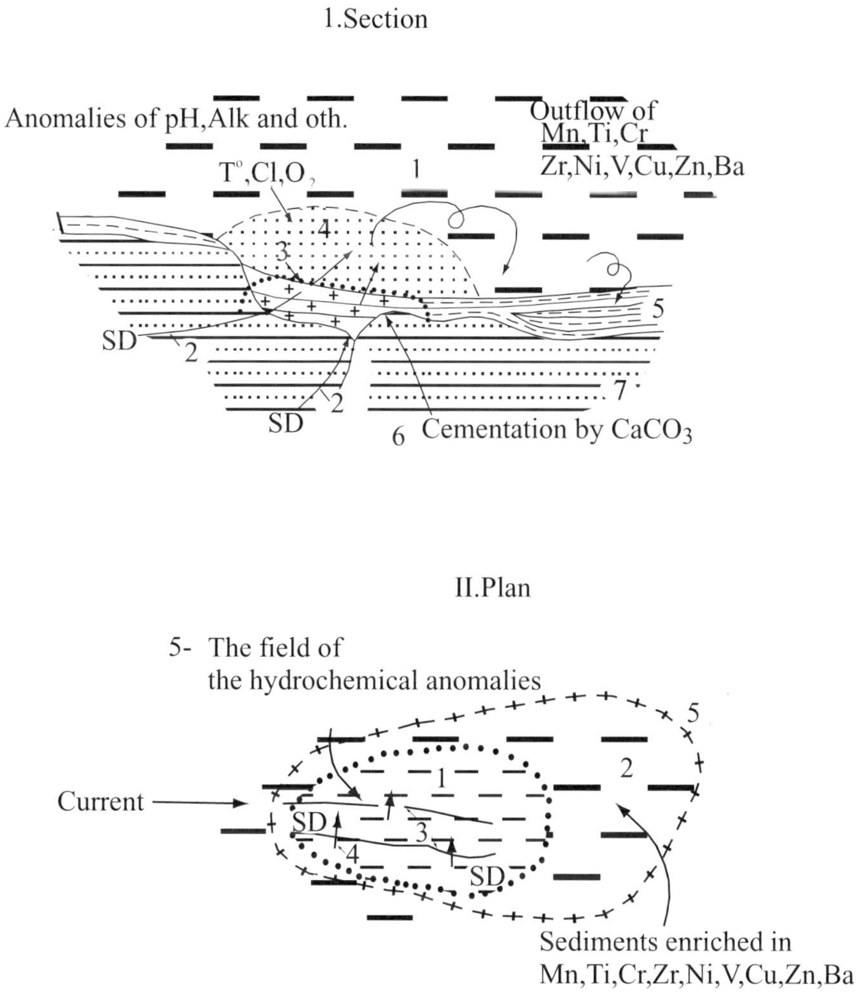

Fig. I.27. Geochemical barriers in the groundwater springs–seawater GBZ.

I. Section. 1—sea water; 2—direction of submarine discharge (SD); 3—mixed (ground and sea) water; 4—outflow of chemical elements from the zone of SD; 5—recent (young) bottom; 6—calcareous cementation of young sediments under the influence of SD; 7—hard rock.

II. Plan. 1—field of hydrochemical anomalies; 2—young bottom sediments; 3—fractures in hard rocks; 4—outflow of groundwater on bottom surface.

Hydrotherm–sea water GBZ. In this GBZ, thermal waters mix with near-bottom sea waters. The former are generally hot: their temperature often attains 100°C, and in oceanic hydrotherms, temperatures are as large as 300–400°C (Fig. I.29). The near-bottom waters of the ocean are cold: their temperature generally varies from −1 to +1°C. This is why the most distinguishable feature of the hydrotherm–seawater GBZ is the temperature barrier; also, there is an acidic

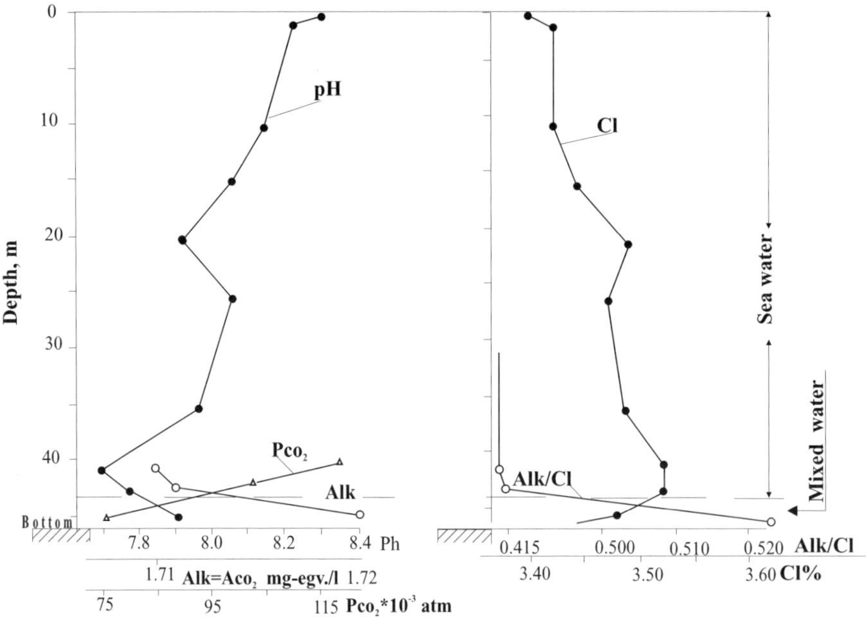

Fig. I.28. Changes in the chemical parameters and concentration of chemical elements in the groundwater springs–seawater GBZ (station Sh-1174, depth 46m, Baltic Sea).

barrier. Data on acidic–alkaline properties and chemical and gas compositions are commonly used to establish the classification scheme of thermal waters (Table I.5). In reference to alkaline–acidic conditions, hydrotherms can be subdivided into the following: (1) acidic, (2) weakly acidic, (3) neutral, and (4) alkaline. Thermal waters are commonly acidic: their pH is typically 3.3–5.9, whereas the pH of near-bottom seawater is about 7.8. Acidic hydrotherms are almost completely devoid of compounds of carbonate, bicarbonates and Mn: they are enriched in H^+, Al^{3+}, Fe^{2+}, and other elements (Baskov and Surikov, 1989). The majority hydrotherms are devoid of oxygen and contain H_2S.

Hydrothermal systems are thought to be barogradient (Letnikov et al.,1977; Perelman, 1989). Decreasing pressure is a driving force of fluid. Barogradients are generally strong in open structures, i.e,. those having canals running to the bottom surface or into big cavities. Thus, the reason for the ore ormation that develops in hydrothermal systems is the action of several (combined) barriers: temperature, baric (dynamic), and oxidation–reduction. Some other barriers can be found occurring in places near hydrothermal springs: saline, hydrodynamic, and biochemical.

There are two main types of hydrothermal springs: (1) high-temperature (T°C up to 350°C) and quick-growing "black smokers," which expel very fine pyrrhotite (in the form of very fine hexagonal plates), pyrite, chalkopyrite, sphalerite, Cu-contain-

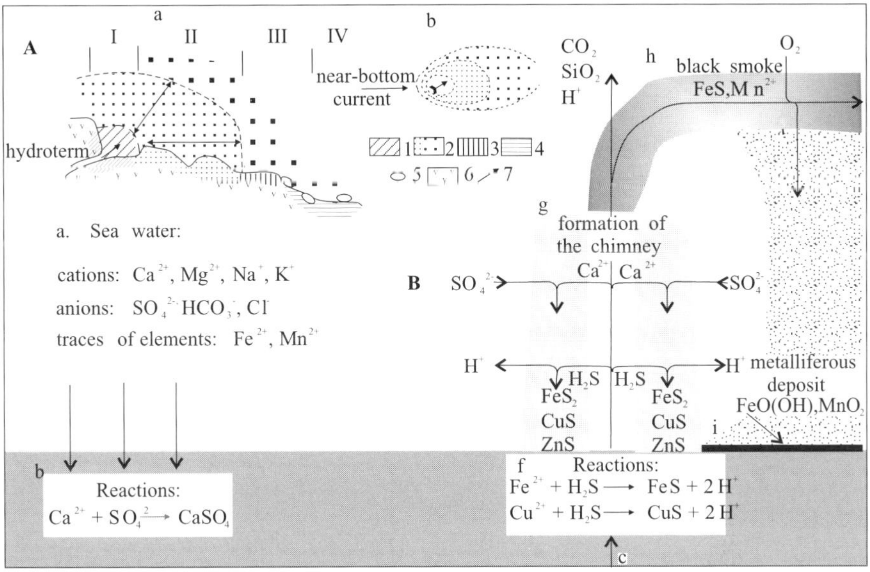

Fig. I.29. Principal scheme of ore formation and facies near the hydrotherm in the ocean bottom (A) and composition of smoke in the black smoker and chemical reactions (B). After Edmond and Damm, 1983.

A. a—profile, b—plan. After Emelyanov, 1982_1.

Bottom sediments: 1—enriched in sulfides; 2—enriched in Fe- and Mn oxides, ferromanganese crusts (the density of the points show the relative enrichment by Fe- and Mn oxides); 3—Mn crusts, metalliferous sediments; 4—pelagic red clay; 5—FMNs; 6—basalts; 7—exit of the hydrotherm.

I-IV—scales of zones: I—meters; II—hundreds meters-kilometers; III—tenths of kilometers; IV—hundreds of kilometers.

On scheme B the arrows show the directions of diffusion and the transportation of the chemical elements and components. Below—the main chemical reactions.

a. main cations of seawater;
b. reactions in sediments;
e. reactions in the hydrotherm;
f. reactions and deposits in the chimney;
g. formation of the chimney and formation of smoke;
h. black smoke;
i. reactions and deposition of metalliferous deposits.

ing pyrite and some other minerals (see Fig.I.29); and (2) low-temperature (T°C 32–330°C) "white smokers," which expel suspended particles of amorphous silica, rosette of barite, and globules of pyrite (Haymon and Kastner, 1981; Haymon, 1983; Hekinian et al., 1983).

When the activity of black smokers dies down, they evidently change to white smokers. In such circumstances, anhydrite dissolves and metastable sulfides become oxidized and change to oxyhydroxides. Probably, unstable sulfides react with silica to form iron-rich smectites.

Thermal waters are not initially pure: they are represented by solutions (which are well-mixed before reaching the surface of the bottom) consisting of an acidic and

Table I.5 Systematics of recent hydrotherms (in brackets—examples of types of waters). After Perelman, 1989.

Alkaline-acid conditions	Reduction-oxidation conditions		
	Oxidizing	Reduction without H_2S (thermal gley)	Reduction
Strongly acidic	I. Oxygen strongly acidic (acid therms of volcanism regions)	V. Gley strongly acidic (strongly acidic chloride brines of platforms and foredeeps) districts	IX. Hydrogen sulfide strongly acid (acid fumaroles of volcanic)
Subacidic	II. Oxygen subacidic	VI. Gley subacidic sulfide subacidic (hydrogen sulfide waters)	X. Hydrogen
Neutral	III. Oxygen neutral and subalkaline	VII. Gley neutral and subalkaline	XI. Neutral and subalkaline hydrogen sulfide and sulfidic
Strongly alkaline	IV. Oxygen strongly alkaline	VIII. Strongly alkaline hydrogen sulfide and sulfidic	XII. Strongly alkaline

high-temperature reducing solution enriched in metals and a low-temperature alkaline oxidizing solution (Edmond et al., 1979). In oceans, around the outlets of hydrothermal springs, there are conglomerations made up of authigenic-hydrothermal structures: sulfides, sulfates, hydroxides, bottom organisms, etc. The constructions most frequently occur in the form of pipes, equipped with one or several holes or "chimneys," which are used for high-velocity effusion of hot water with a large amount of suspended particles composed of fresh authigenic minerals ("fume").

The hydrotherm–seawater boundary is a place of intensive biological activity (biological contour). Around springs of hydrothermal water, there is intensive development of organisms such as bacteria, phytoplankton, and benthos (sponges, worms, mollusks, crabs, and shrimps). At the boundary of alkaline–acidic and oxidation barriers (Table I.3), Fe and Mn hydroxides precipitate from solution and the process is accompanied by sorption of microelements by these substances. As a result, bottom sediments are seen to have large concentrations (up to values of commercial importance) of zinc and copper (sulfidic Zn and Mn ores) and manganese (manganese crusts). Also, elements such as Fe, Co, Cu, Ni and many others are used for production of metalliferous sediments there (Zelenov, 1972; Rona et al., p. 975; Betzer et al., 1974; Bogdanov et al. 1979; Gurvich, 1998). A major factor contributing to the accumulation of metalliferous sediments, manganese crusts (Fig. I.29), and Cu sulfides are specific facial conditions, i.e., the occurrence of hydrotherm–seawater GBZs. Widths of barriers (and the barrier zone as a whole) and their position relative to a hydrothermal spring on the bottom surface depends on such factors as the yield of the given spring and its characteristics (pH, salinity, O_2 content). The horizontal dimensions of the field of hydrothermal sediments around the hydrothermal springs are approximately similar to that of the magmatic heating chamber (Crane, 1979) above which the hydrotherm occurs.

Chapter 7

Geochemical Landscapes

The participation of the above-described media in a variety of interactions, which are responsible for the appearance of boundary effects at the contacts between interacting media, have led us to think that these fronts (frontal interfaces) are very convenient means for dividing the ocean landscape into provinces*) (landscape zoning or geographical demarcation). Because these boundaries are distinguishable by specific natural conditions, the term "band" landscapes of the ocean may be applied (Aizatullin et al., 1984) to all oceanic landscapes of this type. These band landscapes are commonly found between various water masses, and the ocean is thus divided into relatively comparable parts, the peripheral areas of which are exposed to the effects of intense physical, chemical and biological processes, whereas their centers are occupied by a relatively inert (in terms of physics, chemistry and biology) water mass ("central" landscape). Band and central landscapes therefore must be considered as dynamical alternatives to each other. Central landscapes are areas of stationary, relatively uniform behavior of biogeochemical components, whereas band landscapes are boundaries where mechanisms and forms of migration of chemical elements might be disturbed, either gradually or suddenly. The effects of these boundaries are the discontinuity of a particular biogeochemical process and variations in the intensity of migration of elements, which leads either to the dilution or regeneration of a substance. In this respect, the band geochemical landscape is a GB of the ocean.

By analogy with continental geochemical landscapes, the geochemical landscape of the ocean is considered as being a part of the oceanic aquatoria, which consists of one or several morphologically similar areas where the migration of chemical elements in water, suspended matter, sediments and living matter occurs in a form not present in other parts of the ocean. According to E.P. Polynov (1956), the geochemical landscape is formed by paragenetic association of conjugated elementary landscapes, which are interconnected by migration of elements and also by the fact that they are confined to one type of mesorelief. Elementary geochemical landscapes of the ocean, which are a reflection of dynamically stable forms of migration of elements, can be subdivided according to the predominant pattern of behavior: removal, transportation, redeposition, or transformations of (postsedimentation) sedimentary material. The distinguishing feature of the elementary geochemical landscape is the uniformity of climatic, sedimentological, hydrodynamic and

*) See remark on p. 45

biocenous patterns. As a rule, such a landscape is restricted to a certain element of topography. The criteria for recognizing this landscape may be homogeneous sediment, as a result of interactions between the litho-, hydro- and biosphere.

Geochemical landscapes of the World Ocean can be subdivided into geochemical landscape belts (provinces) and geochemical landscape regions (surfaces). The former refer to global aspects of the World Ocean (shelf, continental slope, abyssal), whereas the latter refer to the largest morphostructures of the ocean. The third type of landscape is distinguished by regional particularities in the method of migration, forms of deposits, and increased concentrations of some biochemical components. Thus, geochemical landscapes (central landscapes) and geochemical barriers (band landscapes) are elements of spatial structure of the ocean and are the main units of landscape-regional zoning.

PART II

BARRIERS AS TRANSFORMATIONAL AREAS OF SEDIMENTARY MATERIAL

River-Sea

In this chapter we will examine the effects of barriers in application to the mouths of the Amazon, Congo and Colombia rivers, which flow into the ocean; the Nile, the sediment load of which is deposited in the deep part of a large sea; the Neva and Neman (Klaipeda Strait) rivers, which discharge into the shallow-water areas of the small (shelf) Baltic sea; and the "underwater" Stolpe River, the sediments of which are also deposited in the Baltic.

Amazon. The Amazon is the world's largest river according to its drainage area (6,150,000 km^2) and the average annual runoff (6000 billion m^3) (Table II.1.1, Fig. II.1.1). The denudation rates in the drainage area are as follows, in tons per year: dissolved material, 36.8 t/km^2 (total dissolved material runoff from the drainage area is 231.8 million t/yr); solid material, 79.0 t/km^2 (total of 498.5 million t/yr). The overall annual rate of denudation estimates of 116.0 t/km^2, with sedimentary material runoff from the entire drainage area, 734.4 million t/yr (Gibbs, 1967). The drainage area of South America delivers annually 990 million t/yr of terrigenous material into the Atlantic Ocean (Emelyanov and Trimonis, 1977). Dissolved material makes up 32% of the overall amount.

At the river mouth, three layers of water are distinguished: upper, middle (at the pycnocline) and lower. The saline waters of the ocean are not wedged into the mouth of the Amazon (Monin and Gordeyev, 1986). This distinguishes the river from others like the Mississippi and Congo (these rivers are characterized by circulation in the form of a saline wedge).

Two barriers are distinctly defined within the Amazon river–sea GBZ: salinity and hydrodynamical. The biofilter is also well expressed. The salinity barrier is located between the 2- to 14‰ isohalines This barrier demarcates the oceanic waters from those of the river. The 5‰ isohaline in the high water season is located at a distance of 150 km from the mouth; the 35‰ isohaline, at a distance of 230–270 km (Monin and Gordeyev, 1986). Absorbing energy from the Andes, water at the bed of the Amazon flows at an average velocity of 1.75 m $\times$ s^{-1} and maintains this approximate velocity (1.75–1.80 m $\times$ s^{-1}) until the salinity barrier. The two upper layers of river water (the upper and pycnocline layers) at the shelf edge come into contact with the Guyana current directed along the shore from southeast to northwest. The velocity of the current is partially reduced (hydrodynamic barrier), but a large portion of water is directed to the northwest with the Guyana current.

The Amazon annually discharges about 900–990 million tons of sedimentary materials (Emelyanov and Trimonis, 1977); according to Gibbs (1967), 498.5 million tons).

Table II.1.1 Some characteristics of the river loads of the Amazon, Congo and Niger

River	Water discharge, km^3/a^{-1}	Drainage area, 10^6 km^2	Amount of dissolved matter		Dissolved C_{org}			Particulate C_{org}			Dissolved C_{nonorg}, 10^6 $t \times a^{-1}$
			$mg \times l^{-1}$	10^6 $t \times a^{-1}$	$mg \times l^{-1}$	% from dissolved matter	Supply, 10^6 $t \times a^{-1}$	$mg \times l^{-1}$	% from dissolved matter	Supply, 10^6 $t \times a^{-1}$	
Amazon at Obidus	5780	4.69	50	290	3.3	6.6	19.1	2.25	1.44	14.0	31.7
Congo (Zaire) at Brazzaville	1300	3.5	28	36.6	7.8	27.7	10.15	2.15	5.8	2.8	1.81
Niger (Lokodga)	152	1.2	67	14.0	3.5	3.8	0.53	4.3	2.6	0.66	1.24

*) The Table is compiled from *Biogeochemistry of Major World Rivers*, 1991 (citation after Romankevich, 1994)

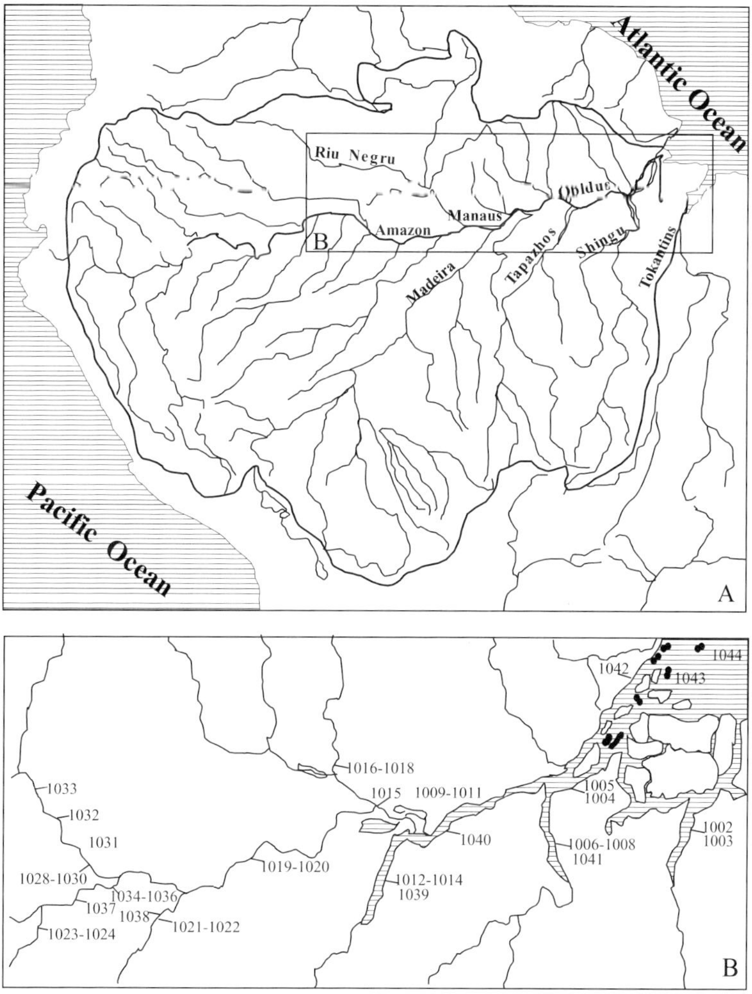

Fig. II.1.1. Amazon River basin (A) and position of the stations (B) of the R/V *Professor Shtokman* (March–April 1983, cruise 9).

Due to the high stream velocity, even at the mouth of this river (1.0–2.0 m × s^{-1}), high suspended matter concentrations are not only retained at the salinity barrier, but even increase here to 100–112 mg × l^{-1} (mud stopper, Fig. I.6). Distribution patterns of the suspended matter concentrations practically coincide with the isohalines. The northward Guyana current off the river mouth on the shelf has a velocity of about 35–75 cm × s^{-1} (Monin and Gordeyev, 1986); therefore it is capable of causing about 20–40% of Amazon water to drift northwest (including its suspended matter).

The salinity barrier is responsible for the intensive development of phytoplankton, and the level of bioproductivity there is maximum (Edmond et al., 1981; Sholkovitz and Price, 1980). In 1976, a rapid blooming of diatomic algae was found to occur between the 7– to 15‰ isohalines. However, the highest bioproductivity (biological stopper) is generally found immediately beyond the outer border of the salinity barrier. The area of maximum chlorophyll concentration is practically repeated by the salinity barrier in both its location and shape (Fig. II.1.2). Thus,

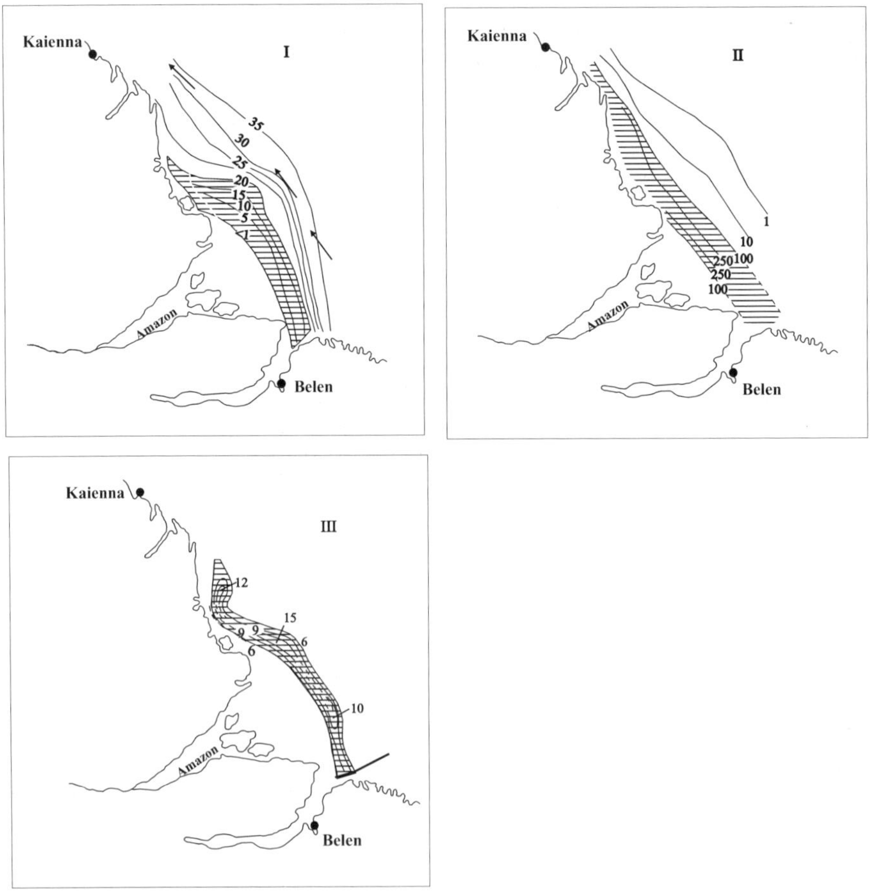

Fig. II.1.2. Salinity barrier, the PSM and chlorophyll distribution in the river–sea GBZ at the Amazon. After Monin and Gordeev, 1988; Artemyev, 1993.
I—Salinity (isohalines in ‰). Salinity barrier (1–15 ‰) is hatched. Arrows—strong sea current.
II—Concentration of PSM in the surface waters (0–1 m), mg/l. Barrier zone with high concentrations of PSM is hatched.
III—Distribution of chlorophyll in surface waters (0–1 m) during the season of high water (May 1983), mg/m^3. Area of high concentration is hatched. Numbers on the scheme represent the concentrations of chlorophyll, mg/m^3.

during the movement of river water towards the open ocean, four water types are distinguished: yellowish brown, yellow, green and blue (Fig. II.1.3, II.1.4). The first type consists mainly of Amazon water; the second is represented by water with the maximum concentration of suspended matter [here, the water contains 400–734 mg $\times$ l^{-1} of suspended matter (Trimonis et al., 1987)]; the third type of water is characterized by the highest biological productivity; and the fourth is represented by typical oceanic water.

Along with the salinity and hydrodynamic barriers, as well as the biological filter, more barriers are revealed, but their importance is less noticeable compared to the first three. For example, the acidic–alkaline barrier is insignificant: in river water, the pH level is 7.4, while in the ocean it is 8.2. A substantially lower pH value was measured only in the Rio Negro River (pH <5), but this river flows into the Amazon very far from its mouth.

What then are the lithogeochemical consequences of the river–sea GBZ at the Amazon?

Due to the effect of the Guyana current, the major amount of suspended matter from the Amazon is directed to the northwest (Fig. II.1.5) to be gradually deposited either in the coastal zone forming a muddy shore (Freidefond et al., 1988), or leaps gradually to the shelf edge and accumulates on the continental slope and at the foot of it in the Guyana basin (Emelyanov and Kharin, 1974; Lisitzin et al., 1975; Damuth and Kumar, 1975; Gibbs, 1976; Trimonis et al., 1987).

According to the data by Eisma et al. (1991, p. 182), about 52 ± 16% of the overall sediment transport by the Amazon is deposited on the shelf off its mouth (within a depth interval of no greater than the 100-m isobath), about 20–25% is transported by the alongshore current to the northwest towards the Orinoco River mouth to be deposited further offshore as mud banks (7–8.5% of all transported matter from the Amazon) or as nearshore mud sediments (12–14%). Mud banks were found to extend all the way along the shores of Kayena, Surinam, and Guyana, almost up to the Orinoco River mouth (Fig. II.1.5).

However, a certain portion of the material (about 5% of the total suspended matter from the Amazon) passes the hydrodynamic barrier represented by the Guyana current, reaches the shelf edge and spreads to the open ocean. This occurs in the bottom water, where the suspended matter concentration is much higher than in the upper layers (Emelyanov and Kharin, 1974). Especially high concentrations of suspended matter were observed in the bottom water above the mouth bar, where they range within 1500–4000 mg $\times$ l^{-1} (Monin et al., 1986, p. 28). Obviously, this is conditioned on two reasons: (1) the entry of muddy river water and (2) stirring up of the upper layer sediments by the strong turbulent motion at the point where river flows meet powerful tidal currents.

Comparing the salinity barrier of the Amazon with the vertical distribution scheme of suspended matter, Fig. II.1.3 shows that suspended matter overcomes hydrodynamic obstacles at the bottom layer and is directed towards the open ocean. This is related to rapid precipitation of the major amount of suspended matter from the upper water layers to the bottom due to reduction in the velocity of the current. Near-bottom turbid water proceeds beyond the limits of the shelf as a nepheloid layer, which is well expressed at the northern portion of

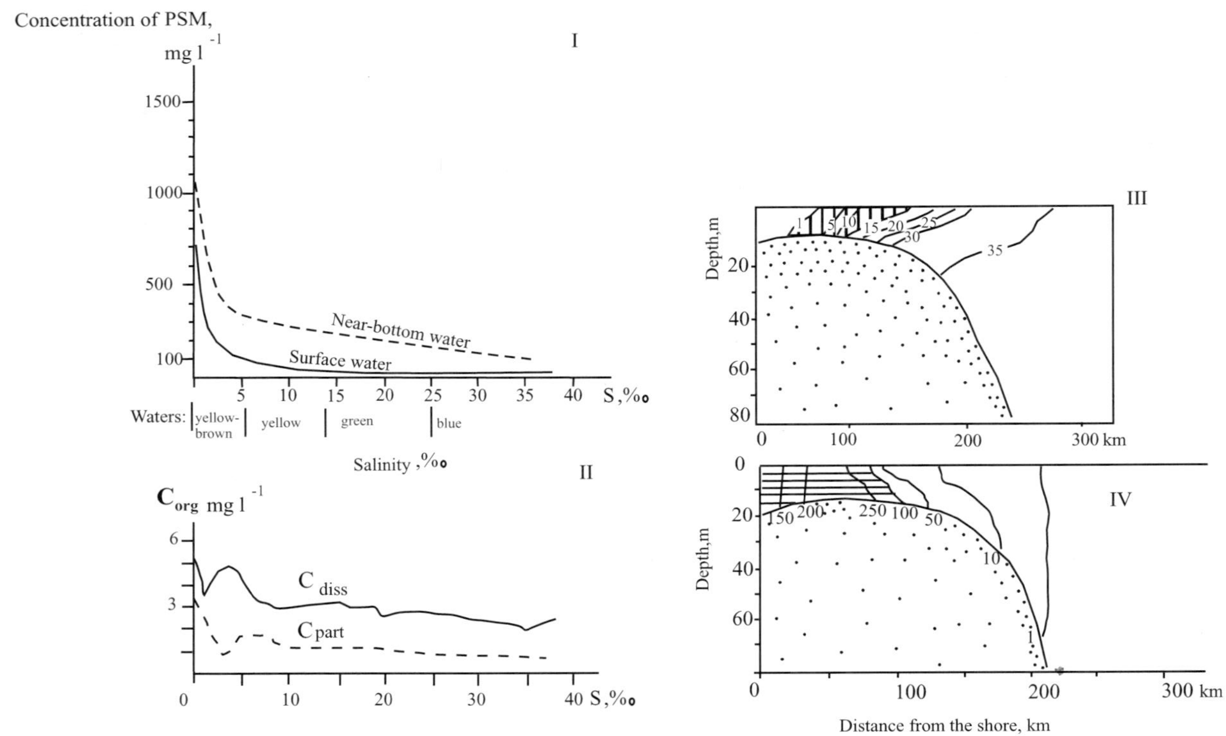

Fig. II.1.3. Distribution of PSM and the organic compounds in the water of the river–sea GBZ at the Amazon. After Monin et al., 1986; Artemyev, 1993. I—Distribution of the average concentrations of PSM on the dependence of salinity (March—April, 1983). Types of the waters are shown below. II—Distribution of organic C in surface waters (0–1 m) in the river–sea GBZ at the Amazon on the dependence of salinity. C_{diss}—dissolved; C_{part}—particulate. III—position of the salinity barrier on the vertical profile during the season of high water. Isohalines in ‰. Salinity barrier (1–15‰) is hatched. IV—Distribution of the PSM on the vertical profile in the river–sea GBZ at the Amazon, mg/l. Area with concentrations of more than 100 mg/l is hatched.

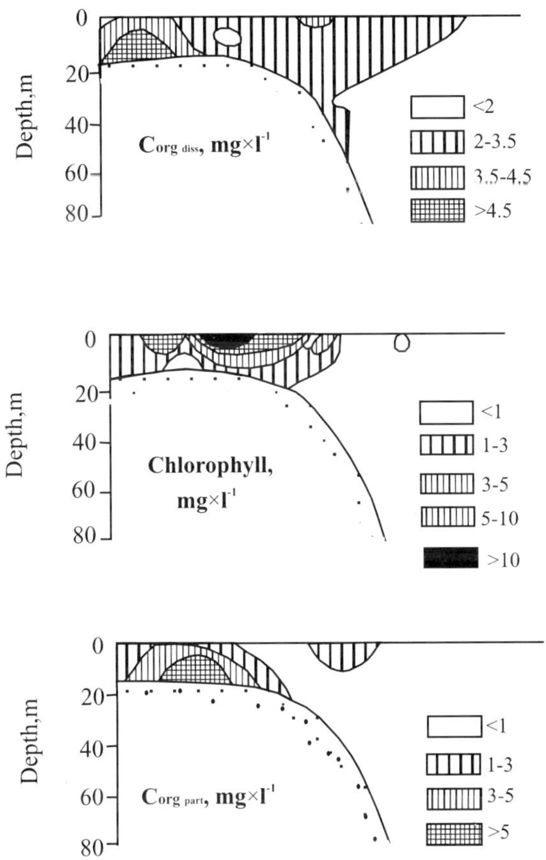

Fig. II.1.4. Distribution of dissolved (C_{diss}) and particulate (C_{part}) organic C and chlorophyll in the river–sea GBZ at the Amazon. After Monin, Gordeev et al., 1986.

the submarine fan and moves to the north of the Guyana Basin (Emelyanov and Kharin, 1974).

At the salinity barrier fresh river water is mixed up with saline seawater, which serves as an electrolyte. As a result of physicochemical processes, coagulation of dissolved components takes place, as well as ion-exchange reactions on the surface both of minerals carried from land and of new colloidal particles formed in the water mixing zone, first of all, on the surface of Fe- and Mn hydroxides. At the salinity barrier, dissolved organic forms of Fe, Al, Cu, Ni, Co, Cd, and Mn coagulate (flocculate) from solvents into suspended matter. According to the data of Sholkovitz (1976, 1977), the portion of the coagulated forms comprise (in percentage of gross output of dissolved forms): up to 95% Fe; 25–45% Mn; 19% Al; 41% Cu; 43% Ni; 11% Co; and 5% Cd.

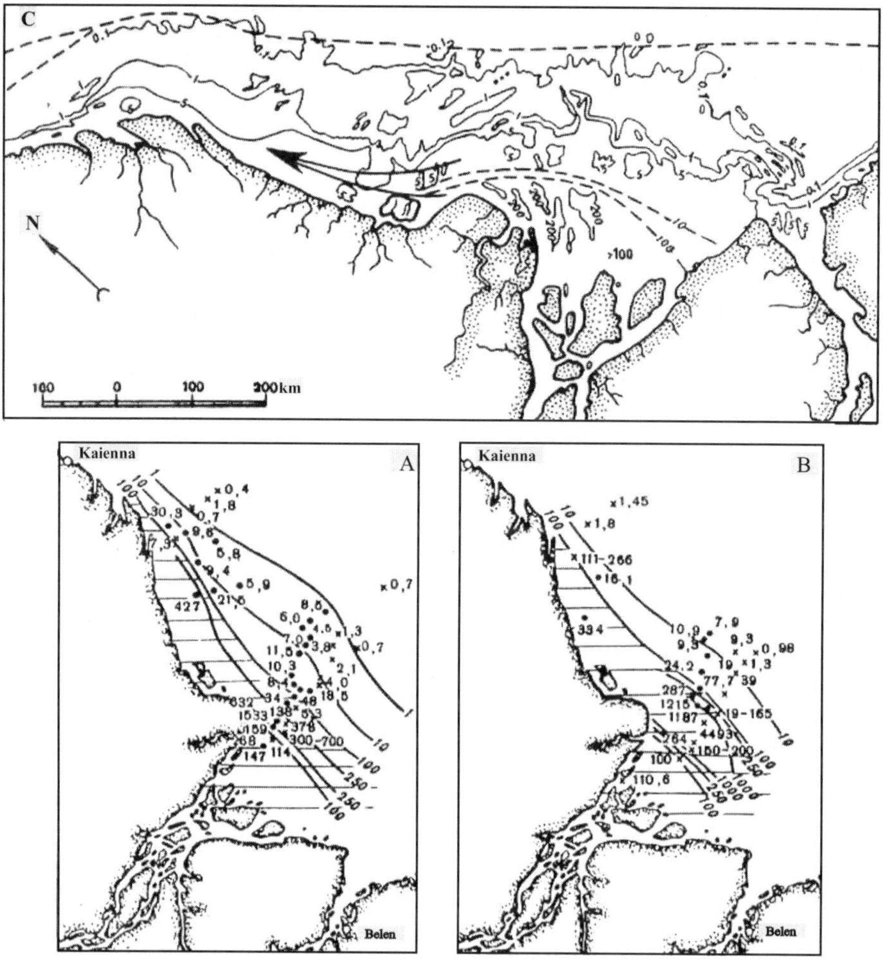

Fig. II.1.5. Distribution of PSM in the waters of the Amazon River.
Distribution of PSM (mg/l) in surface waters (0–1 m) (A) and in the near-bottom waters (B). After Monin and Gordeev, 1988.
Points—samples, cross—samples of the French expedition on the R/V *Calypso*. Area with concentrations >100 mg/l is hatched.
C—concentration of PSM in surface waters of Amazon River mouth according to Landsat data (Gibbs, 1982). Arrows show direction of sea current. Numbers of lines show concentrations of PSM, mg/l; the numbers at the dotted lines show the average concentrations, mg/l.

At the external border of the salinity barrier (the 14‰ isohaline), a noticeable enrichment of suspended matter in the following elements occurs: U, Yb, Ba, and Cr; to a lesser degree La, Ce, and Sm (Monin et al., 1986); as well as in some other elements (Sholkovitz, 1977).

On the other hand, approximately 70–95% of all dissolved Fe transported into the sea by rivers passes to solution as a result of its desorption at the mixing zone

(Figueres et al., 1978). Iron is intensely extracted from suspended matter to solution at the mouths of the Amazon, Congo and Khuankhe rivers. In contrast to iron, manganese is adsorbed at the mouths of some rivers, while at others, it is desorbed. This is determined by its high chemical mobility (valence change) depending on the biogeochemical condition both of rivers and of seawater. Usually Mn manifests itself as an nonconservative element, but in some conditions it behaves as a conservative element like boron. The oncentration of dissolved forms of conservative chemical elements (for example, B) rises as water salinity increases. Unconservative elements like Fe concentrate in waters with low salinity (0–5‰). In waters with S = 5–10‰, the dissolved Fe concentration remains low. The concentration of nonconservative elements (Mn, Cu) changes irregularly: the concentration of dissolved Mn strongly decreases at S = 2–5‰, whereas at S = 5–10‰, it strongly increases, approaching maximum values when S = 15–25‰. Thereupon, it again decreases in typical seawater (Chester, 1990, pp. 46–49]. The concentration of dissolved Cu increases when the salinity is 2–7 and 28–35‰. All this points to the different properties of elements. In order to understand the process of adsorption and desorption of elements or to correctly differentiate concentrations by different elements of water at the river–sea GBZ, it is necessary to take into consideration these properties, as well as the salinity, pH, concentration of suspended matter and other features. Thus, for example, in the behavior of Zn and Cd at the zone of water mixing, it is possible to differentiate three important features (Chester, 1990, p. 62): (1) an increase in the adsorption of both elements with increasing pH (higher than 7.0–8.5); (2) a decreases in the adsorption of Cd, and to a lesser degree of Zn, with increasing salinity; (3) an increase in the adsorption of Zn and Cd with increasing concentration of suspended matter. Beyond the salinity barrier, i.e., already in green water where phytoplankton reaches maximum development, bioassimilation of elements (C, P, N, Fe, Mn, Zn, Cu, and others) by organisms occurs, where they are converted from dissolved to particulate form.

Thus, river–sea GBZs appear to be the strongest filter located at the borders of continents. These marginal filters basically regulate the interaction between dissolved and particulate matter. Those elements which are desorbed in the mixing zone replenish their reserves in the sea (ocean), but those elements that are adsorbed or involved in the biological cycle may be precipitated onto the bottom with particulate material, thus being excluded from further biogeochemical recycling. Desorption of Ba, Mn, Ni and Cd at low salinity within the mixing zone is an important process in the behavior of these elements in the river–sea system (Chester, 1990, p. 79).

The Amazon has a large effect on sedimentation. Namely, off the Amazon River mouth in the Atlantic Ocean, one of the largest sedimentary fans in the world has evolved (Damuth and Kumar, 1975). The thickness of the terrigenous deposits of this fan is more than 5 km, and in some places up to 10 km. The rate of sediment accumulation on the river–sea GBZ is very high, greater than 60 mm/1000 yr. Recall that this is a hundred times greater than the sedimentation rate in the pelagic areas of the Atlantic Ocean.

Suspended matter of Amazon consists of quartz (15–40%), feldspar (2–3%) and clay minerals (75–90%). The latter are mainly represented by montmorillonite and kaolinite, and to a lesser degree, by illite and chlorite (Table II.1.2).

Table II.1.2 Mineral composition of particulate matter and bottom sediments of the Amazon and Congo

Tributary, sediments	Quartz	Feld-spars	Clay minerals	M	K	J	Ch	T	G
PSM of Amazon (Gibbs, 1967, 1973, 1976)									
Mountain	15–34	11–17	50–74	7–17	6–24	23–31	3–7	-	-
Mixed	15–36	2–3	76–79	0.5–51	1–5	12–43	0.1–12	-	-
Tropical	2–22	0–3	89–98	0–68	12–95	0–14	0–0.1	-	-
PSM of Congo (Eisma et al., 1978, p. 392)									
River	36	2–2	-	2	44	8	<5	0	2–3
Water over shelf	18	1–2	-	4	62	6	<5	<1	2–3
Water over the edge of shelf	20	2–3	-	3	60	8	<5	<1	1–2
Ocean	48	3–4	-	4	29	8	<5	2–3	29
Nile River mud (alluvium at Alexandria) (Rateev et al., 1966)									
Mud sample H-1	-	-	-	60–80	15–20	-	-	-	-

Minerals: M – montmorillonite; K – kaolinite; J – illite; Ch – chlorite; T – talc; G – glauconite

The bulk chemical composition of river suspended matter in the Amazon differs from the average composition of suspended matter in other rivers of the world in lower contents of SiO_2 and Fe_2O_3 and increaed contents of Al_2O_3, TiO_2, and P_2O_5 (Tables II.1.3, II.1.4, II.1.5, II.1.6).

Mineral composition of muds in the Amazon foredelta (Emelyanov, Kharin, 1974) is practically the same as the composition of suspended matter and alluvium of this river. Among the clastic minerals (>0.1 mm and 0.1–0.05 mm fractions), quartz, ore minerals, mica (especially yellow mica), zircon, green hornblende, epidote–zoisite, etc., predominate. Among clay minerals, illite (up to 40–60%), kaolinite, and montmorillonite (up to 20–40%) predominate. The mud contains trace amounts of chlorite and mixed-layer minerals (see stations 60 and 61 in Table II.1.7).

Brown shelf muds of the Amazon foredelta contain 0.55–0.72% C_{org}, 0.20–1.37% SiO_{2am}, 0.3–0.10% Mn, and 5–7% Fe; however, they are somewhat enriched in Al_2O_3 (up to 18.29–21.04%), K_2O (3.18–3.65%), Ti (0.53–0.69%), and P (0.06–0.10%), as well as in Ni and Zr (Emelyanov and Kharin; 1974; Emelyanov, 1982_1). The SiO_2/Al_2O_3 ratio in brown muds of the Amazon foredelta is the lowest for all muds of the Atlantic Ocean: 2.2–2.5. According to other data (Trimonis, 1995, p.156) the

Table II.1.3 Chemical composition of Amazon River particulate suspended matter (PSM) in comparison to the PSM of other rivers, and sedimentary deposits of continents, %. Citation after Monin and Gordeev, 1988, p. 42

Material	SiO_2	Al_2O_3	TiO_2	CaO	MgO	Fe_2O_3	MnO	Na_2O	K_2O	P_2O_5	Author
Amazon (PSM)	58.0	23.5	0.79	1.04	1.68	10.5	0.09	-	3.08	0.18	Sholkovitz et al., 1978
Amazon (PSM)	57.9	21.7	1.16	2.25	1.85	7.68	0.1	1.06	2.17	0.37	Martin and Meybeck, 1979
Congo (PSM)	50.7	30.0	0.82	0.80	1.20	13.6	0.15	-	1.1	0.48	Sholkovitz et al., 1978
PSM of World Ocean rivers	64.8 54.8 60.3	15.1 15.65 16.98	- 0.67 0.97	4.03 3.52 4.08	2.32 2.07 1.82	9.50 7.28 7.36	- 0.14 0.16	1.48 1.35 0.96	2.4 1.81 2.53	- 0.14 0.29	Garrels et al., 1973 Gordeev, 1983 Martin and Meybeck, 1973
PSM of the cold and tropical zones	56.65	21.5	1.22	1,06	1.6	8.81	0.14	0.69	2.2	0.36	Martin and Meybeck, 1973
PSM of the cold and moderate zones	62.87	13.58	0.82	6.0	2.0	6.40	0.17	1.16	2.73	0.25	Martin and Meybeck, 1978
Soil	70.8	13.45	0.76	1.93	1.04	5.43	0.11	0.85	1.64	0.18	Vinogradov, 1962
Sedimentary rocks of continents (clay and shales)	51.07	19.72	0.75	3.56	2.22	4.76	0.086	0.89	2.75	0.18	Vinogradov, 1962
Earth's crust	57.64	15.45	0.88	7.01	3.87	7.20	0.15	2.87	2.32	0.23	Ronov and Yaroshevsky, 1967

Table II.1.4 Content of chemical elements and components in alluvium (A) and in PSM ($CaCO_3$-P in %, Cu-Cr in 10^{-4}%). After Martin and Meyback, 1979; Emelyanov et al., 1978; Monin and Gordeev, 1988; Striuk, 1994.

	Amazon			Nile				Neva	Vistula	Neman
	PSM	A		A				A	A	A
	-	Limits	Average	H-1*)	9*)	10*)	Average	-	-	-
$CaCO_3$	-	0.0-4.3	1.74	2.05	5.00	5.74	5.81	0.4–6.5	8.6–10.0	12.0–29.0
SiO_{2am}	-	0.5–6.2	2.03	-	-	-	-	-	-	-
C_{org}	-	0.01–7.0	1.95	0.78	2.16	1.80	1.19	9.2–13.2	13.6–14.8	12.6–16.8
Fe	5.5	0.27–6.70	1.86	7.80	7.38	7.08	6.73	1.7–3.4	2.3–2.5	1.6–2.9
Mn	0.10	0.01–0.41	0.06	0.09	0.17	0.11	0.11	0.1–0.2	0.5	0.2
Ti	0.58	0.18–0.60[2*]	0.34[2*]	1.38	1.27	1.13	1.26	0.1–0.3	0.1	0.1–0.2
K	2.18	0.96–1.34[2*]	1.11[2*]	-	1.06	1.39	1.30	-	1.2	-
Na	0.79	0.52–1.10[2*]	0.85[2*]	-	1.08	2.53	1.22	-	0.3–0.6	-
Al	11.9	-	9.41[2*]	-	-	-	-	0.2–4.3	0.8–2.3	0.7–3.4
P	0.12	-	0.37[2*]	-	-	-	-	-	0,6	0,2
Cu	266	1–30	14	-	62	70	52	24–90	136–150	46–66
Ni	105	6–100	40	83	85	85	85	60–176	64– 76	40–60
Co	41	5–33	18	-	-	-	-	-	-	-
Zn	426	10–267	74	79	142	86	108	278–720	720–780	272–478
Cr	193	6–71	33	-	148	128	135	90–2080	200–320	62–150

*) Station numbers (Emelyanov and Shimkus, 1986)
2* See Tables II.1.3

Table II.1.5 Content of >0.1 mm fraction (in %) and chemical elements in the alluvium of the Amazon (Fe-Na, in %; Rb-Ni, in 10^{-4} %) (for the location of stations see Fig.II.1.1B)

Station	Depth, m	Horizon, cm	S type	Fr.>0.1	Fe	Mn	Ti	K	Na	Rb	Li	Cu	Zn	Cr	Ni
PSh-1009	11	0 – 4	Ca	18.4	3.00	0.05	0.60	1.34	1.10	82	102	19	70	40	60
PSh-1015	68	0 – 4	S	97.9	1.64	0.02	0.21	0.96	0.84	38	48	12	48	40	46
PSh-1019	17	0 – 3	S	95.0	1.34	0.02	0.18	1.12	0.94	54	54	12	30	20	56
PSh-1040	32	0 – 5	S	93.6	1.68	0.04	0.40	1.04	0.52	52	26	8	35	<6	–2
			Average	-	1.9	0.03	0.34	1.11	0.85	56.5	57	13	46	-	-

S type – sediment type: S – sand (fraction 1.0–0.1 mm prevail); Ca – coarse aleurite (silt) (0.1–0.05 and 0.05–0.01 mm fractions prevail).

Table II.1.6 Content of chemical components and elements in terrigenous sediments (in the alluvium) from the Amazon River estuary (core PSh-1044, depth 13 m.), ($CaCO_3$-Na in %, Rb-Ni in 10^{-4}%) (for the location of stations see Figs. II.1.1B, II.1.6)

Horizon, cm.	S type	$CaCO_3$	C_{org}	SiO_2	Al	Fe	Mn	Ti	N	Mg	P	K	Na	Rb	Li	Cu	Zn	Cr	Ni
0 – 5	S	0.92	0.69	60.00	9.52	4.42	0.09	0.54	0.28	0.97	0.39	1.66	0.94	112	64	42	112	80	50
5 – 10	Apm	0.83	0.71	62.00	10.05	4.70	0.07	0.53	0.20	1.00	0.39	1.78	0.82	142	78	36	100	96	46
50 – 55	Fam	0.50	0.32	58.00	7.93	3.32	0.05	0.49	0.19	0.77	0.37	1.50	1.00	86	48	14	88	60	26
70 – 75	Apm	0.67	0.60	-	-	4.40	0.06	0.44	0.20	1.09	0.35	1.74	1.20	146	72	28	112	80	24
95 – 100	Fam	0.33	0.32	63.00	8.99	3.48	0.04	0.45	0.16	0.84	0.35	1.66	0.98	168	68	16	88	70	26
120 – 125	Apm	1.33	0.58	53.00	9.52	4.80	0.09	0.47	0.22	1.07	0.37	1.66	0.92	142	68	170	120	90	40
180 – 185	Apm	0.67	0.30	68.00	9.52	3.48	0.07	0.53	0.16	0.93	0.35	1.66	1.04	124	60	20	94	64	32
205 – 210	Apm	0.83	0.96	62.00	9.63	3.70	0.07	0.47	0.18	1.00	0.39	1.74	0.94	150	72	34	112	74	40
235 – 240	Apm	1.08	0.74	58.00	9.84	6.14	0.08	0.32	0.20	1.42	0.37	2.14	1.06	168	112	40	146	124	40
290 – 295	Apm	1.00	0.85	55.00	9.95	4.44	0.08	0.44	0.20	1.27	0.31	1.90	0.98	176	88	30	122	74	26
323 – 328	Apm	0.92	0.61	56.00	9.57	4.52	0.07	0.43	0.18	1.61	0.39	1.84	1.36	132	62	28	114	84	26
345 – 350	Apm	0.50	0.53	58.00	9.36	3.60	0.07	0.60	0.19	0.91	0.44	1.84	0.90	120	86	28	114	60	42
375 – 380	Apm	0.83	0.41	56.00	8.62	4.70	0.07	0.45	0.18	1.00	0.37	1.80	1.06	172	74	28	112	74	50
425 – 430	Apm	0.41	0.40	56.00	9.79	4.70	0.07	0.45	0.16	1.06	0.39	1.70	0.96	134	88	30	114	80	40
Average		0.78	0.58	60.75	9.12	4.31	0.07	0.47	0.20	1.06	0.37	1.75	1.02	141	74	31	111	79	36

S type – sediment type: S – sand; Fam – fine-aleuritic (silty) mud; Apm – aleuro-pelitic mud. For the sediment types of the core PSh-1044 see Fig. II.1.6.

Table II.1.7 Grain size distribution (%) and content of chemical components and elements (%) in the mud of the shelf of the South American (between the Amazon River and Trinindad) and Guyana Basin (after Emelyanov et al., 1975)

Fraction, mm	60[1]	65	31	28	56	23	11	16	30	24	15	21	42	64	63	222	352–2	294	290	356–2
	27[2]	50	55	63	72	120	145	160	55	82	162	205	305	400	810	1144	1270	4033	4721	4800
Grain size distribution (fractions in mm)																				
>0.1	0.1	0.4	0.4	0.3	1.6	0.9	0.7	0.6	8.2	3.7	3.6	0.5	0.7	9.4	-	4.0	1.7	1.5	1.0	-
0.1–0.05	0.2	0.4	0.8	0.4	0.4	2.7	1.7	0.5	6.8	9.2	2.5	1.9	5.8	10.0	-	3.5	1.8	2.0	2.0	-
0.05–0.01	24.5	27.6	3.8	6.7	14.1	10.5	11.7	7.3	7.0	15.7	8.6	8.3	8.5	7.7	-	6.5	6.2	12.0	5.5	-
0.01–0.005	45.0	40.8	41.8	58.3	43.3	47.1	41.5	45.2	32.4	14.3	6.6	11.8	23.5	5.7	-	5.5	9.5	8.0	6.0	-
0.005–0.001	24.6	26.8	44.3	27.8	31.9	29.1	35.7	39.0	34.2	46.4	56.6	57.4	50.5	22.5	-	28.5	38.0	27.0	27.0	-
<0.001	5.6	4.0	8.9	6.5	8.7	9.7	8.7	8.0	11.4	10.7	22.1	20.1	11.0	44.8	-	52.0	42.7	49.5	60.0	-
0.1–0.01	24.7	28.0	4.6	7.1	14.5	13.2	13.4	7.8	13.8	24.9	11.1	10.2	14.3	17.7	-	10.0	8.0	14.0	6.0	-
0.01–0.001	69.6	67.6	86.1	86.1	79.2	76.2	77.2	83.2	66.6	60.7	63.2	69.2	74.0	28.2	-	34.0	47.5	35.0	33.0	-
<0.01	75.2	71.5	95.0	92.6	83.9	85.9	85.9	91.6	79.0	71.4	85.3	89.3	85.0	72.9	-	86.0	90.2	84.5	93.0	-
Md	0.007	0.007	0.005	0.006	0.006	0.006	0.005	0.005	0.004	0.004	0.004	0.002	0.004	0.001	-	0.0009	0.001	0.001	0.001	-
So	1.6	1.6	1.7	1.5	1.7	1.8	1.7	1.7	2.4	3.0	2.1	2.1	2.2	-	-	3.2	-	3.2	-	-
Content, %																				
$CaCO_3$	0.55	0.50	2.50	1.79	1.77	-	8.07	-	-	18.90	13.42	8.39	9.39	9.44	4.16	9.30	8.37	8.90	8.90	7.07
SiO_{2am}	0.20	1.02	0.88	0.72	1.60	1.37	0.87	0.69	-	0.82	0.81	0.79	1.19	1.77	2.60	-	1.10	-	-	0.97
Corg	0.72	0.63	0.92	0.82	0.55	-	1.28	-	-	0.93	-	1.29	1.21	1.32	1.16	-	0.82	-	-	0.62
Fe	5.84	5.49	5.58	5.88	6.33	4.84	4.95	5.21	-	4.75	4.74	4.84	4.10	5.08	4.61	0.82	4.80	1.03	6.24	5.42
Mn	0.10	0.10	0.05	0.07	0.07	0.04	0.04	0.04	-	0.05	0.03	0.04	0.03	0.02	0.03	0.03	0.08	0.08	0.03	0.08
Ti	0.68	0.58	0.52	0.40	0.53	0.43	0.38	0.47	-	0.41	0.44	0.42	0.46	0.43	0.46	0.20	0.42	0.04	0.05	0.43
P	0.08	0.10	0.07	0.08	0.08	0.07	0.10	0.07	-	0.09	0.07	0.06	0.04	0.06	0.05	0.05	0.05	0.07	0.08	0.04

[1] Sation number; [2] depth, m.

average chemical elements content in Quarternary (mainly of Holocene) muds of the Amazon submarine fan equals (in percent): SiO_2, 51.5; TiO_2, 0.82; Al_2O_3, 16.65; Fe_2O_3, 7.16; MnO, 0.35; CaO, 3.64; MgO, 2.64; Na_2O, 1.91; P_2O_5 0.23. Ratios are as follows: SiO_2/Al_2O_3 3.44; K_2O/Na_2O, 1.63; Al/Ti, 19.17; Fe/Ti, 10.50; Mn/Ti, 0.57; Fe/Al, 0.62. In contrast to the mud from the Equatorial African shelf, coprolites and hydrogoethite–shamosite (hydrobiotite) ooids are almost completely absent in the mud from the South American shelf. Markedly smaller contents of iron, amorphous silica and, especially, organic matter and phosphorus are present in these muds. Contents of manganese are higher in muds from the South American shelf than in those from the African shelf. These differences are conditioned mainly by the absence of upwelling at the shelf zone of South America, which leaves deep traces in the sediment composition of the African shelf.

Deep-sea muds (beyond the shelf edge) of the Amazon fan are finer than shelf muds in their grain size (Fig. II.1.6, Table II.1.8) and have practically the same chemical composition as hemipelagic mud from other regions of the Atlantic (Emelyanov, 1982_1).

Congo. In terms of drainage area and water discharge, this is second largest river in the world after the Amazon (Table II.1.1); in terms of the total of dissolved and suspended matter discharge, it is the eighteenth largest. The length of the river is 4700 km, from the East African Lake to the Atlantic Ocean. Just like the Amazon, the Congo drains vast areas of forest landscape. The tributaries Alim, Licule and Sanha drain ecosystems of tropical forest. The following specific features were discovered in them: a lower pH value, a higher concentration of dissolved and suspended organic matter, and a lower concentration of dissolved and suspended inorganic matter. These rivers have been called "black" due to the color of the water in them. The rate of mechanical denudation of the drainage area of the Congo River is lower, *8* t/km^2 per year; its rate of chemical denudation, *5* t/km^2 per year.

The salinity barrier is clearly distinguished. Ocean surface waters diluted by fresh water from the Congo River (with a salinity of about 21–23‰) can be traced to the edge of the shelf (depth, 200 m), whereas water of higher salinity (23–34‰) can be found at a distance of about 150–250 nautical miles from shore. These waters are directed from the 200-m isobath west-northwest (Eisma and van Bannekom, 1978). The thickness of the freshened lens is commonly 5–10 m. In comparison to ocean water, the lens is rich in oxygen, silicon, phosphorus and nitrates and differs in temperature (Romankevich, 1994).

The water column above the continental slope is clearly divided into five layers with different hydrophysical characteristics (Romankevich, 1994).

The surface waters are highly productive, and primary productivity is greater than 500 mgC × m^2/day (Koblents-Mishke et al., 1973). The concentration of suspended C_{org} in surface waters approaches 500 µg × l^{-1}, but quickly decreases with depth (Romankevich, 1994); the concentration of overall suspended matter lies in the range of 1–10 mg × l^{-1}. The turbid (enriched in suspended matter) waters of the Congo are spread over a distance of 300–500 km in the direction of the open ocean and 150–200 km in the meridional direction (Romankevich, 1994, p. 222).

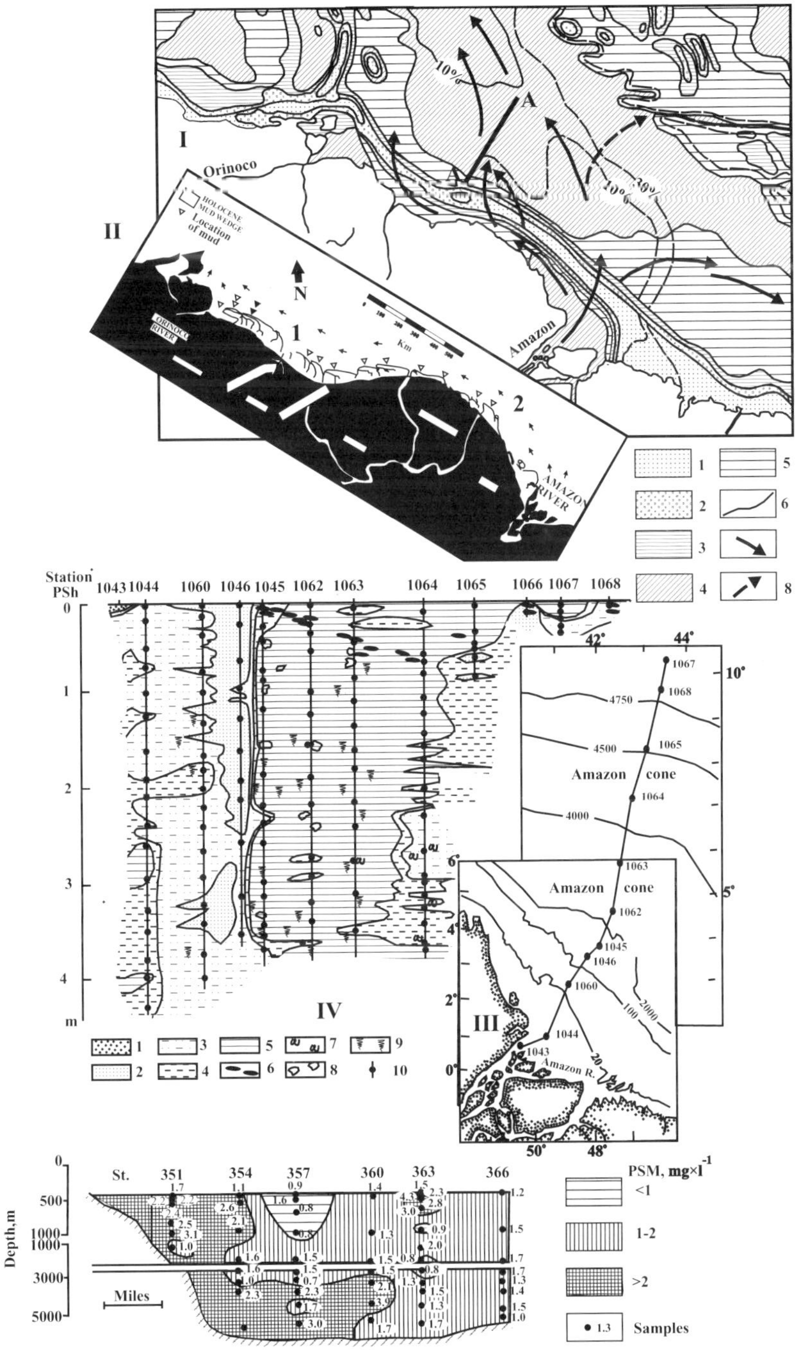

Fig. II.1.6. Sedimentation near the Amazon River mouth.

Continues

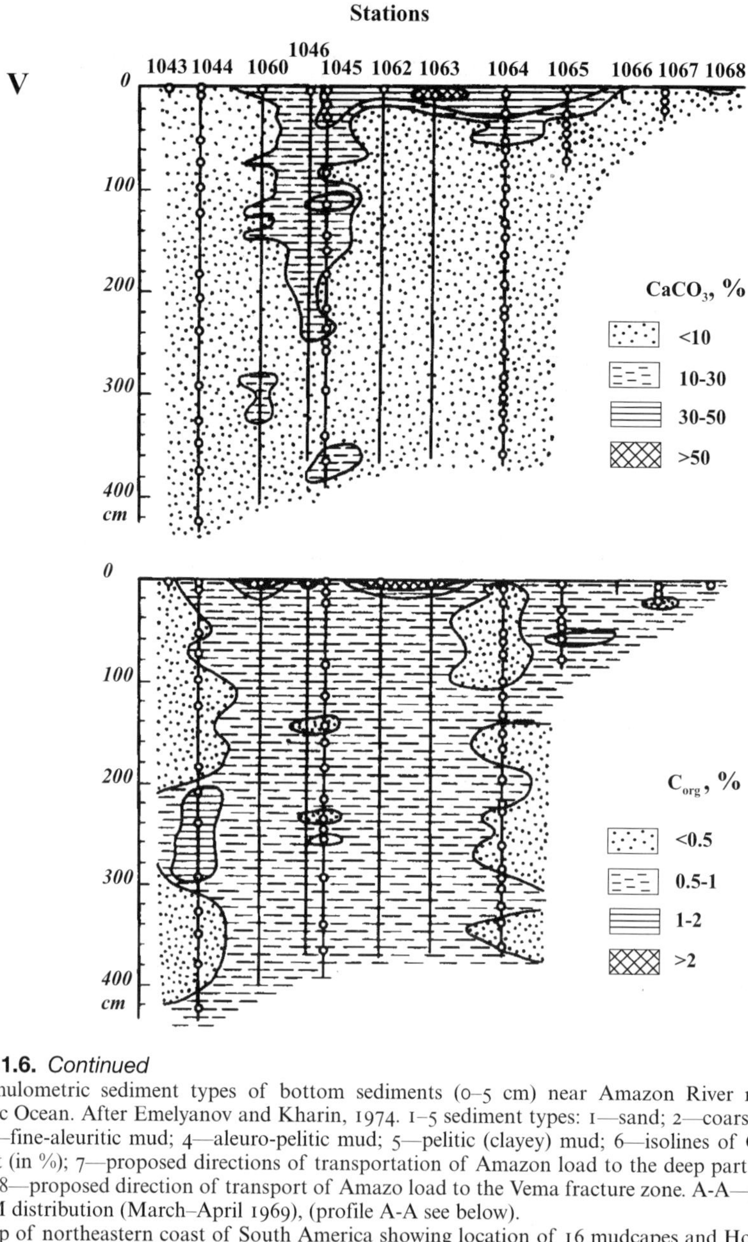

Fig. II.1.6. *Continued*

I. Granulometric sediment types of bottom sediments (0–5 cm) near Amazon River mouth, Atlantic Ocean. After Emelyanov and Kharin, 1974. 1–5 sediment types: 1—sand; 2—coarse-aleurite; 3—fine-aleuritic mud; 4—aleuro-pelitic mud; 5—pelitic (clayey) mud; 6—isolines of $CaCO_3$ content (in %); 7—proposed directions of transportation of Amazon load to the deep part of the ocean; 8—proposed direction of transport of Amazo load to the Vema fracture zone. A-A—profile of PSM distribution (March–April 1969), (profile A-A see below).

II. Map of northeastern coast of South America showing location of 16 mudcapes and Holocene deposits. Mudcapes do not occur downstream of large rivers (e.g., Essequibo, Corantijn, Marowijne). Note particularly the overlapping mudcapes in eastern Venezuela, which suggest that mudcape migration is a major mechanism of coastal-plain accretion. There is no evidence of mudcapes on the Holocene coastal plain south of the Cassiinterstitial mudcape and present rates of coastal erosion (0.5–1 km/1000 yr) are insufficient to have removed evidence of old mudcapes since the cessation of active progradation (500–1000 yr B.P.). Tidal bores (pororocas) that formed near and in the mouths of southern rivers (e.g., Araguari, Fleshal) are evidence that tidal currents produce energetic conditions that may prevent mudcape genesis. After Allison et al., 1995.

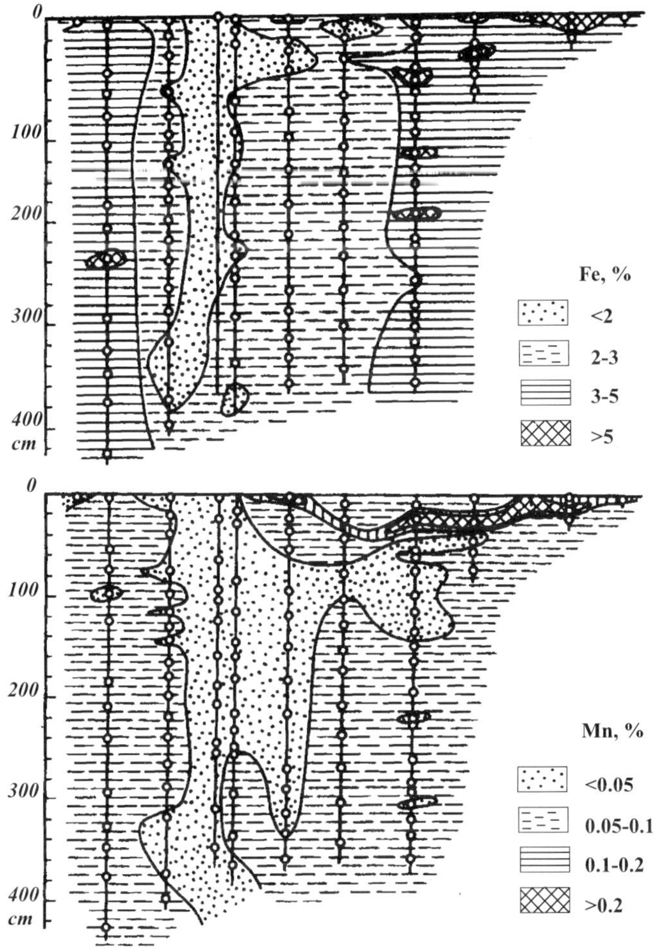

Fig. II.1.6. *Continued*
III. Location of geological stations of the profile IV (R/V *Professor Shtokman*, cruise 9, 1983). Isobaths in m.
IV. Lithological profile III from Amazon River mouth to the open ocean. After Emelyanov et al., 1989.
1–6—terrigenous sediment types: 1—fine sand; 2—coarse aleurite; 3—fine-aleuritic mud; 4—aleuro-pelitic mud; 5—pelitic (clayey) mud; 6—interlayers of dense brown clay and crusts (proposed hiatus before the Holocene transgression); 7—floral remains; 8–9—patchiness: 8—coarse, marble-like; 9—fine, spotty; 10—studied samples.
V. Distribution of $CaCO_3$, C_{org}, Fe and Mn in the sediment strata of Amazon fan (see profile IV, this figure). After Trimonis, 1995.

Table II.1.8 Average content of chemical elements in surficial sediments (0–5 cm) on the 6°S profile in the Angola Basin, Atlantic Ocean (C_{org}-Mn, in %; Ni-Sr, in 10^{-4}%) (after Lukashin et al., 1994)

Station, test area*1	C_{org}	Si	Al	Fe	Ti	Ba	P	Mn	Ni	Co	Cu	Zn	Cr	V	Pb	Sr
V-20-III	2.44	16.43	14.6	14.62	0.61	0.077	0.22	0.08	53	41	43	185	128	223	25	232
V-20-II	2.27	23.26	15.3	6.75	0.66	0.061	0.12	0.35	37	27	55	124	102	154	24	183
B-3174	2.26	21.63	14.7	6.39	0.53	0.044	0.17	0.07	40	28	70	133	98	132	27	210
V-20-I	0.79	24.77	10.6	7.40	0.59	0.82	0.11	0.65	81	37	133	155	103	170	30	204
Average*2)	0.56	20.04	7.05	6.2	0.45	0.363	0.15	1.02	182	105	387	248	91	154	89	1088
Average*3)	-	26.2	8.58	4.37	0.51	0.058	0.05	0.04	68	19	45	95	90	130	20	300

*1) For the test area location see Fig. II.1.7.

*2) For ocean sediments

*3) For clay and shales of the continental platform

The concentration of suspended matter in the waters decreases with increasing salinity: water with a salinity of 0–3‰ contains 15–45 mg × l^{-1}; with 10‰, 10–24 mg × l^{-1}; with 15–20‰, 2–8 mg × l^{-1}; and with 25–30‰, 0.5–10 mg/l (Eisma et al., 1978). The total volume of suspended particles in water with a salinity of 5.6‰ is about 2.4 mm^3/l, whereas for waters with salinities of 19.6‰, 27.5‰ and 33.7‰, these values are 0.81, 0.47 and 0.17 mm^3 × l^{-1}, respectively. The quantity of particles with sizes of 2.37–4.74 and 18.1–57.5 µm also falls sharply at the salinity barrier (10–20‰). The quantity of particles with sizes of 2.37–4.74 µm falls from 60,000 at S‰ = 10–15‰ to 5000 at 20‰, while the quantity of particles with sizes of 18.1–57.5 µm falls from 200,000 to 10–15,000 (Eisma et al., 1978).

One of the most characteristic features of the geology in the area off the Congo's mouth is a deep (600–800 m) submarine canyon well expressed in the bottom topography. The width of the canyon on the upper continental slope is about 15 km. Deeper, it becomes narrower, and at the foot of the continental slope its width is about 7 km. The canyon serves as a continuation of the Congo River by which terrigenous material from the river is carried to the pelagic area of the ocean, i.e., to the central Angola basin. Therefore, the river's terrigenous material intensively accumulates not only at the shelf edge and on the upper continental slope, but also in the deep-water (central) areas of the Angola basin down to the flank of the Mid-Atlantic Ridge (Fig. II.1.7).

Nearly one-half of the river's suspended matter is deposited at the Congo canyon head and possibly at the mangroves north and south of the river's mouth (Eisma et al.,1978). Mainly coarse-grained material (sand–aleurite–pelite) is deposited, whereas fine particles aggregated in the process of coagulation are transported forward. Because of this, suspended matter in the river mouth area is enriched in fine-grained particles and organic matter. Transported material is mainly carried out in the 0- to 1-m-thick bottom water layer (and evidently along the bottom of the canyon). However, at the border of the 20‰ isohaline (i.e., over shelf depths of 100–150 m), suspended matter settles to deeper water layers, begins to precipitate in the canyon, and is carried along it to the pelagic area of the Angola basin. Particles with a diameter less than 10 µm are mainly carried to the open ocean. According to the data of Eisma (1978, p.404), only a small amount (<5%) of the river's suspended matter reaches the deep-water area of the Angola basin.

At test area V-20-III, located on the shelf (Fig. II.1.7), the maximum sedimentation rates exceed 300 mm/1000 yr, and at one station, even 1600 mm/1000 yr (Kuptsov, 1994), about 3–16 times greater than at the the Amazon River mouth. The sedimentation rates here, especially at the canyon bed, fluctuate in the range of 0.5–6 mm per year and commonly equal 1–3 mm. High sedimentation rates are fixed on the continental slope and at its foot. So, at test area V-20-II, the rates are 250–462 mm/1000 yr; at V-20-I, 177–393 mm/1000 yr. The sedimentation rates sharply decrease just on the flank of the Mid-Atlantic Ridge, where mixed (10–50% $CaCO_3$) or typical biogenic (50–80% $CaCO_3$) oozes occur.

The fan is composed of terrigenous (<10% $CaCO_3$) or low-calcareous (10–30% $CaCO_3$) terrigenous grey muds. The terrigenous material discharge from the Congo River is the determining factor in the formation of sediments within the fan. However, the role of Congo River is not limited to this. The terrigenous material

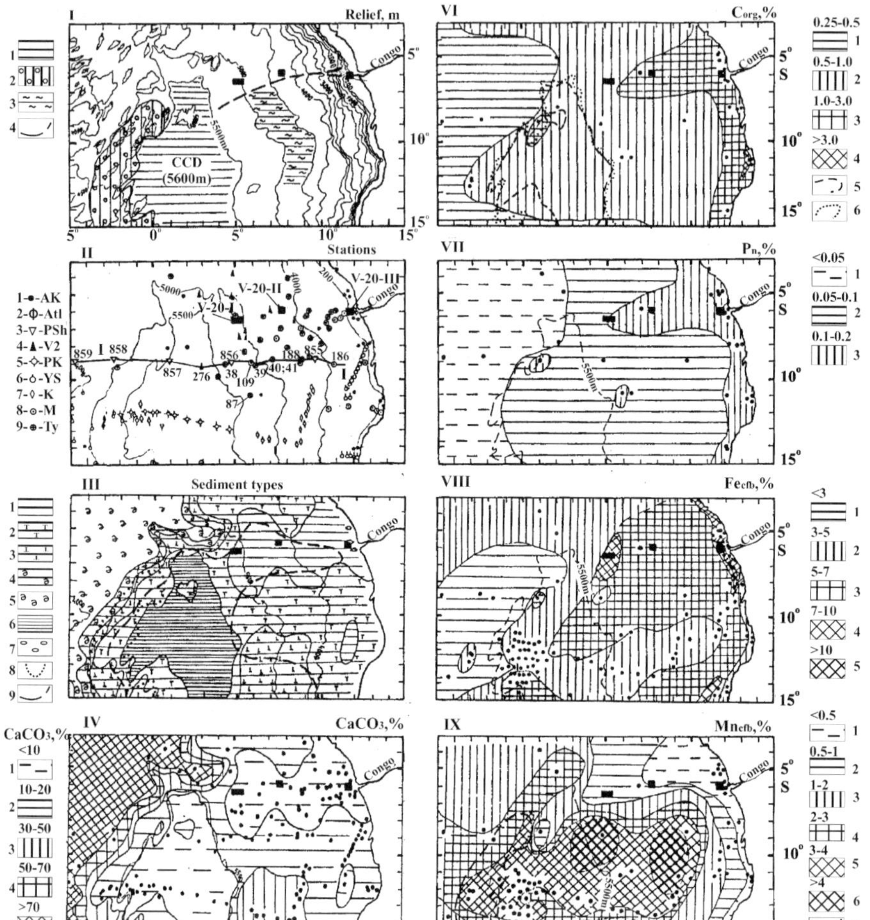

Fig. II.1.7. Sedimentation in river–sea GBZ at Congo River mouth and in the Angola Basin, Atlantic Ocean.

I. Bottom relief (isobaths in m). 1—pelagic area below the isobath 5500 m, where occur pelagic red clay; 2—the area on the flank of Mid-Atlantic Ridge, where FMNs are distributed in the mixed terrigeneous–foraminiferal mud (oozes), 10–50 % $CaCO_3$; 3—area where in the sediment strata are distributed interlayers enriched in Mn (up to 1–10 % Mn) and flat ferromanganese concretions (see Fig III.10.2); 4—Congo (Zaire) canyon and main direction of transport of Congo River's load to the Angola Basin. Black blocks—test areas investigated in detail (see scheme II).

derived from the Congo River discharge, although somewhat diluted by pelagic biogenic carbonate, is traced over the entire pelagic area of the Angola basin up to the flank of the Mid-Atlantic Ridge (to the 5100-m contour, or the 0^0 meridian). Its presence is identified by diluting effects of river-derived terrigenous material, by an abnormally high organic carbon content for pelagic sediments, and by distribution patterns of amorphous SiO_2, iron, manganese (Fig. II.1.7), copper, and nickel (Emelyanov, 1992).

Along the entire profile—from the estuary of the Congo River to test area V-20-I and further west, even to station PSh-857 (Table II.1.7)—high or increased (compared to ocean sediments) concentrations of C_{org}, Fe, Ti, and P occur in sediments; Mn, Cu, Zn, V and some other elements also show local increased concentrations. The highest contents of C_{org} and bulk nitrogen have been found in sediments washed by the OML (Table II.1.9). The C/N ratio decreases from 12.5 to 6.4 with increasing depth below the sediment surface.

Core PSh-855, which was collected at the foot of the continental slope (depth 4297 m), shows an occurrence of weakly laminated terrigenous mud. At the 0–6 cm layer they are colored in brown tones: they contain little C_{org} but are very rich in manganese (up to 2.2%) and nickel (up to 250 ppm). According to the data of foraminiferal analysis carried out by N.P. Lukashina, the mud is of Late Pleistocene age (oxygen isotope stage 3). Consequently, suspended matter from the Congo did not enter at area of station PSh-855 in the Holocene. Transitional, brownish grey mud occurs in the 6- to 30-cm interval. Below 30 cm and up to 485 cm, the mud is grey, smells of H_2S, and contains up to 2.5% C_{org}. In additiion to the surface layer, there are two more interlayers enriched in manganese, and two enriched in $CaCO_3$. Sediments throughout the core are enriched in copper, nickel, zinc, and lithium (Fig. II.1.8A). A considerable admixture of diatomic frustules has been noted in the core (Fig. II.1.8D).

In core PSh-856 (depth 5305 m), located somewhat south of the main thalweg of the Congo canyon and 300 m deeper than test area V-20-I, the composition of

---

II. Geological stations. 1–9—stations of different research vessels; letters—abreviations of the names of the R/V. Test areas V-20-1, V-20-II and V-20-III, investigated in detail during cruise 20 of R/V *Vytiaz* (Romankevich, 1994). I-I—lithological profile (see Fig. II.1.8.). Isobaths 200, 4000, 5000, 5500 m are shown.

III. Sediment types (0–5 cm): 1—terrigenous aleuro-pelitic and pelitic (clayey) mud (>50% of the <0.01 mm fraction, <10% $CaCO_3$); 2—the same, but $CaCO_3$=10–30%; 3—mixed biogenic–terrigenous mud (30–50 % $CaCO_3$); 4—nano-foraminiferal ooze (50–70 % $CaCO_3$); 5—high-carbonate foraminiferal ooze (70–90% $CaCO_3$); 6—red clay (<10% $CaCO_3$); 7—hydrogoethite–shamosite sand and aleurite, low-ferrugineous (6–10% Fe) and ferrugineous (10–40 % Fe); 8—border of the FMN field; 9—border of low-siliceous (diatomic) mud (10–20% SiO_{2am}). Thin isoline—isobath 4000 m.

IV. Distribution of $CaCO_3$ in surficial sediments (0–5 cm), %.

V. Distribution of SiO_{2am}×1.6 in surficial sediments (0–5 cm), %.

VI. Distribution of C_{org} in surficial sediments (0–5 cm), %. 1–4—C_{org} content, %; 5—border of red clay; 6—isobaths 5500 m.

VII. Distribution of phosphorus in (natural dry-P_n) surficial sediments (0–5 cm), %.

VIII. Distribution of iron in surficial sediments (0–5 % cm) (recalculated on cfb), %.

IX. Distribution of manganese in surficial sediments (0–5 cm) (recalculated on cfb), %. 7—border of FMN field.

X. Distribution of Ni in surficial sediments (0–5 cm) (recalculated on cfb), %.

Table II.1.9 Content of C_{org} and N (in %) and their ratio (C/N) in surficial (0–5 cm) sediments of the Congo River estuary and cone (after Romankevich, 1994)

Depth, m	C_{org}	N	C/N
Estuary			
<6	2.76	0.23	11.8
6	1.44	0.13	11.3
Cone (the profile to the west from the Congo River estuary)			
45	3.34	0.27	12.5
106	3.43	0.29	11.7
237	2.84	0.28	10.3
682	4.51	0.48	9.5
1533	3.22	0.35	9.3
3100	2.31	0.28	8.2
4074	1.25	0,6	7.8
4474	0.96	0.15	6.4

muds is different. Just the upper (Holocene) cap (0–20 cm) of sediments appears to be weakly oxidized (brown in colour) and low calcareous (10-30% CaCO3). Below, along the entire recovered 3-m-long core ((Fig. II.1.8B), sediments are represented by terrigenous (non-carbonate) pelitic and aleuritic–pelitic grey muds. They contain 0.5–1.5% C_{org}. At the bottom of the core there is a strong smell of H_2S. The mud is diffusely layered and only a single turbidite interbed occurs at the 150- to 160-cm interval. The muds contain less Fe, Mn, Ti, and Cr (as compared to the muds from test areas V-20-II and V-20-I). The muds in core PSh-856 contain significantly more diatoms than those in core PSh-855, and very few foraminiferal remains. The SiO_{2am} content in the muds of the entire core exceeds 10%.

Although core PSh 857 was retrieved at a water depth of 5680 m (Fig. II.1.7, diagram II), which is below the CCD (which here is located at 5600 m), and the sediment has brown coloration, it contains from 2.14 to 2.68% C_{org} and from 11.87 to 13.37% $CaCO_3$, values atypical of a red (pelagic) clay. This sample of mud is the deepest one taken in the Atlantic Ocean (Emelyanov, 1982), where the C_{org} content is more than 1%. Consequently, the suspended matter of the Congo River, traveling through the canyon as a high-energy flow, is not completely deposited on the continental slope, but much of it does reach the central part of the basin to be gradually deposited there. Undoubtedly, the organic matter is mainly allochthonous and has been brought here from the the Congo River mouth. The high C_{org} contents (more than 3%) in the muds of the Congo River valley and of test area V-20-I confirm this conclusion. Large organic matter fluxes have been also fixed in the canyon valley (Romankevich, 1994). Core AK-87, which was collected at a depth of 5320 m, 280 m above the CCD level, is a significant feature of the Angola basin. The recovered sediment is characterized

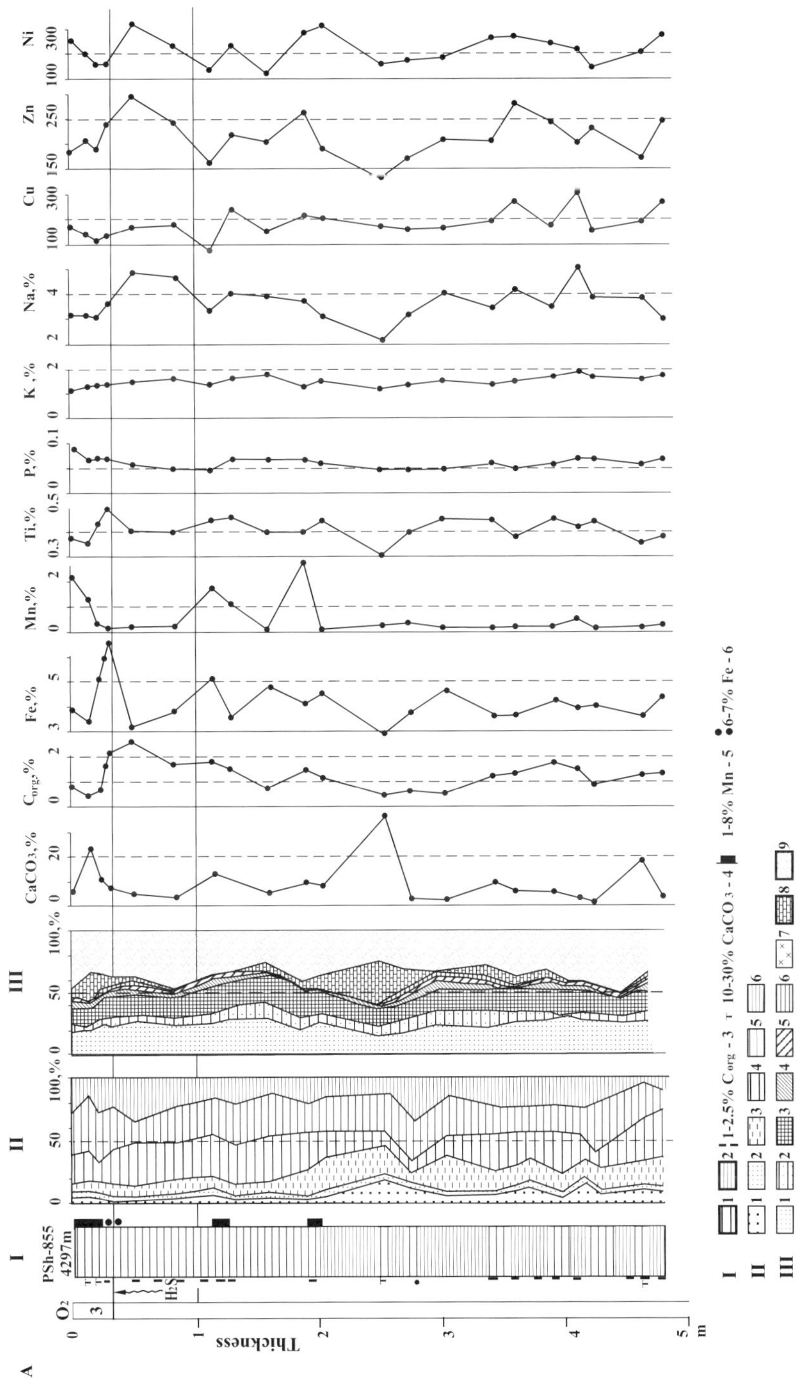

Caption see page 97

sediments of the Angola Basin are highest in the Central Atlantic: from 0.01 to 0.1 mg × cm^{-2} × ka^{-1} (Emelyanov, 1998).

The water of the Congo River contains, after filtration 255 μg × l^{-1} Fe on average (Figueres et al., 1978 p. 332; Table II.1.10), whereas an average value 23 μg × l^{-1} Fe has been estimated in the Amazon River mouth (Monin and Gordeyev, 1988, p.102, Tables II.1.1, II.1.4). After filtration of river water through filters with pores from 1.2 up to 0.025 μm, the concentration of "dissolved" Fe in the Congo River water sharply decreases with increasing salinity: 400–455 μg × l^{-1} have been measured at 0‰ salinity; 71–170 μg × l at 6‰; 50–100 μg × l^{-1} at 12‰; 30–70 μg × l^{-1} at 16‰; 20–40 μg × l^{-1} at 24‰; and 10–15 μg × l^{-1} at 32‰,.

At the mixed water zone, about 50–85% of all dissolved Fe coagulates on terrigenous (silicate) particles and is precipitated into sediments (Figueres et al., 1978, pp. 335–336). Recall that the suspended matter of the Congo River on average contains 13.6% Fe_2O_3, 25–30% more than in the suspended matter of the Amazon or Vistula rivers (see Table II.1.6). Suspended matter of the Congo River is also enriched in Al and P and contains 0.15% MnO_2 and low contents of Si, Ca, K, Mg, in comparison to the suspended matter of Amazon and oceanic sediments (Sholkovitz et al., 1978, p. 412). This is related to the soil composition of the drainage basin and the character of pronounced chemical weathering.

Distribution of ore components in the offshore area of the Angola basin is characterized by the following: (1) intensive delivery of Fe and Mn from the Congo River (Eisma and van Benekom, 1978); (2) a pronounced separation of these two elements: iron accumulates predominantly on the shelf (up to 42.2% of sediment) and at the upper part of the continental slope, whereas Mn (together with Cu and Ni) accumulates on the continental slope and in the abyssal zone of the Angola basin. These two elements jointly accumulate only in Fe–Mn nodules on the flank of the Mid-Atlantic Ridge. The average contents of elements in these nodules are 11.95% Fe,

Table II.1.10 The concentration of dissolved forms of microelements in river waters, mg × l^{-1}

River, location	Al	Fe	Mn	Zn	Cu	Ni	Co	Author
Amazon, estuary	90.0	23.0	1.4	-	1.60	-	-	Monin and Gordeev, 1988
Congo, estuary	-	255.0	-	-	0.38–0.47			Eisma et al., 1978
Neman (Klaipeda Strait)	-	64.0	3.7	12.7	4.2	-	-	Emelyanov and Lukashin, 1986
Average for rivers of the World ocean	50	40	8,2	30	1.5	0.5	0.2	Monin and Gordeev, 1988

20.73% Mn, 0.75% Ni, 0.38% Cu, and 0.20% Co (for more detail see Part III). The arrangement of the vast nodule field on the flank of the Mid-Atlantic Ridge (depth 5000–4500 m) indicates that very tiny particles of suspended matter (Mn- and Fe hydroxides), as well as ore elements in dissolved and colloidal form supplied by the Congo River, should reach the edge of this ore field and serve as starting material for the formation of nodules.

Intensive delivery of Mn from the Congo River took place earlier (in the Pleistocene). Three to four manganese-rich interbeds with a Mn content of up to 8% occur in the upper strata (3–5 m) of the Late Quaternary section on the continental slope (depth 4000–4500m). In core AK-87-2 (depth 5320 m), the manganese-rich interbeds contain 3–4% Mn. Their thickness is 3–20 cm. Four of such interbeds are distinguished in the 131-cm-long core (Fig. II.1.8C). The manganese interbeds contain up to 650 ppm Ni and up to 350 ppm Cu. Enrichment in Mn of pelagic clay (manganese interbeds) is a result of diagenetic Mn^{2+} migration from the interstitial waters of underlying reduced sediments.

Columbia River. The Columbia River mouth represents a spectacular example of river and ocean waters mixing (Conomos and Gross, 1968; Neshyba, 1991, p. 295) (Fig. II.1.9). In these publications the processes here are considered from physical (Fig. II.1.9A), chemical (Fig. II.1.9B) and biological (Fig. II.1.9C) aspects. It is very characteristic that the salinity barrier expressed as the halocline is clearly traced at the river mouth. Through it, saline waters are involved into the mixing process in the estuary. Mixing is facilitated in the sense that seawater at the zone of mixing rises from a deeper layer rich in nitrates. A low-salinity aureole spreads far away to sea. Consequently, the estuary zone is enriched in nutrients as a result of estuarine upwelling. Deep ocean water is a major reservoir and source of nitrates during summer seasons. The zone where the deep water is involved into mixing is extraordinarily productive—it exceeds the open sea 20 times in productivity .

Nile. The waters of the Nile River, before being regulated by the High Aswan Dam (1964), entered east and northeast, not mixing with seawater for a long time. Before the dam came into being, sailors could replenish their ships with fresh water directly from the sea. At present, the current of the Nile is not as strong. However, the salinity and hydrodynamic barriers are as clearly traceable as before. This is seen by the different hydrodynamic and visual characteristics, as well as by the distribution of suspended matter and other features (Emelyanov and Shimkus, 1972, 1986).

Areals in the Levant Sea, to where the largest input of sediments arrives from the Nile, are clearly outlined by the distribution of terrigenous ($<30\% CaCO_3$ and >65–70% of terrigenous material) muds; by increased contents in muds of the heavy, 0.1- to 0.05-mm fraction of clinopyroxenes and of montmorillonite; by the presence of certain chemical elements (C_{org}, Fe and Ti); and by other features (Fig. II.1.10) (Emelyanov et al., 1979; Emelyanov and Shimkus, 1986; Emelyanov, 1994). The sediment discharge from the Nile is the source of these components. The alluvium of this river, which drains basalts in the area of the Abyssinian plateau in the equatorial humid zone, is rich in those products formed during the process of deep chemical

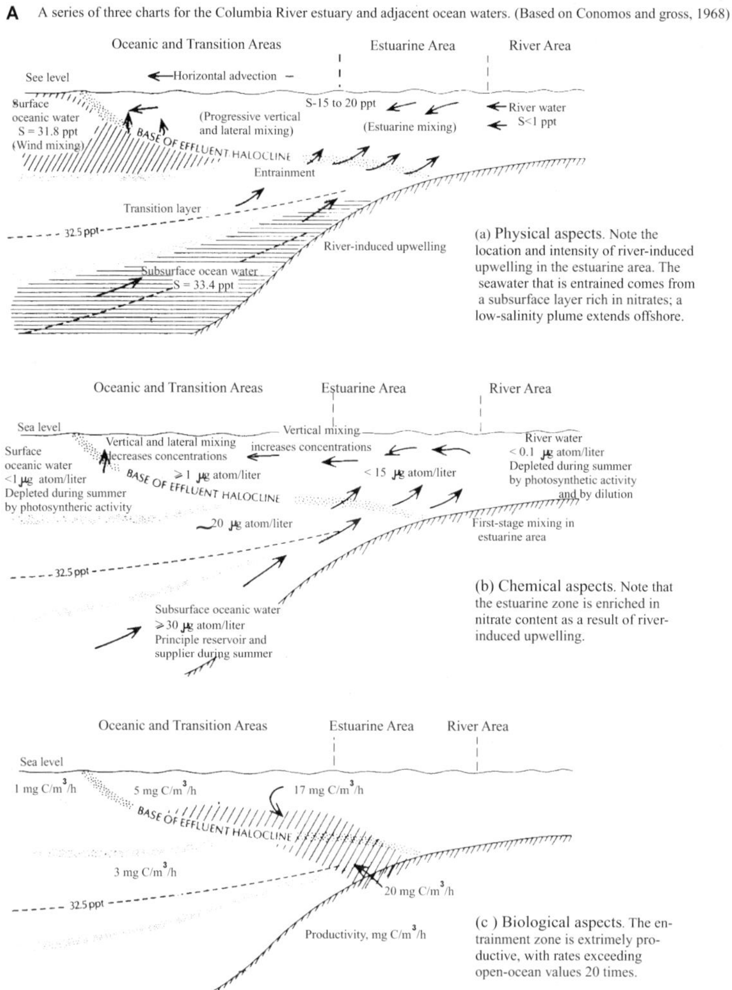

Fig. II.1.9. A series of three charts for the Columbia River estuary and adjacent ocean waters. (Based on Conomos and Gross, 1968). (A) and the river's plume (B) and radionuclide distribution (C) in the ocean near the river mouth.
A. A series of three charts. After Neshyba, 1991.

Continues

weathering. These are precisely the minerals (clinopyroxene, biotite, montmorillonite, and kaolinite) and chemical elements (Fe, Ti, Cr, and Mn) (Table II.1.2, II.1.4) enriching the muds of the pre-Nile province. Biogenic components (C_{org}, partly P) are represented, for the most part, by allochthonous matter that forms in the Nile delta.

Before 1902 (before the appearance of the first dams near Aswan) the Nile discharged about 10^{11} m^3 of water into the sea (Summerhayes et al., 1978), but after

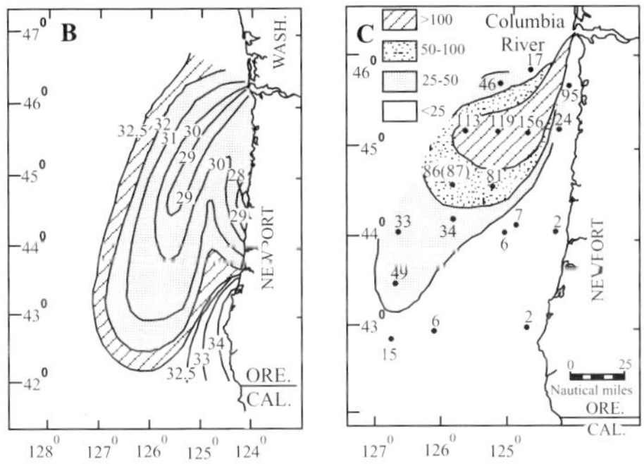

Fig. II.1.9. *continued* B. During summer, the south-flowing California current dominates nearshore circulation, so that the plume (salinity, ‰) is displaced southward as it simultaneously diffuses outward from shore. C. C of a radionuclides, chromium-51, during summer 1965. Close similarity of patterns in part B and C demonstrates that dispersion of dissolved chemicals or suspended particulates can be modeled quite accurately just by tracing salinity contours of surface water, a measure of dispersion of river fresh water. After Neshyba, 1991.

1964, when the High Aswan Dam was built, runoff decreased to 3.5×10^{10} m^3. Before the construction of the dam, it delivered 2×10^8 tons of sediment into the sea annually, 98–99% of which was transported in suspended matter and 1–2% as bottom load.

Terrigenous fine sands have accumulated near the shores within the outer delta of the Nile; terrigenous sandy-mud and mud are found mainly in the middle parts of the shelf and at both of its branches; terrigenous and shelly sands (mainly relict >50% $CaCO_3$) are present on the outer shelf, and they are replaced again by terrigenous muds (mud and clay) on the slope of the outer shelf edge (Fig. II.1.10A). The greatest quantity of suspended matter (up to 3.8 mg × l in the 5-m layer and up to 41.3 mg × l^{-1} near the bottom) is delivered by the left, Rosetta, branch of the Nile Rosetta. The high concentrations of suspended matter in the upper water layer (up to 1 mg × l^{-1}) can be traced as far as the shelf edge in the Rosetta canyon (northwest of the mouth). Another mass of turbid water is transported by a coastal marine current which moves east-northeast. The bulk of suspended matter is transported not by the surface current, but by a near-bottom current.

The Nile delivers large amounts of denudation products of basic rocks, including basalt. In the light fraction, 0.1–0.05 mm, quartz makes up 18–46%; feldspars, 9–33%; micas, 1–12%; weathered particles and fragments of basalts, 18.55% (Emelyanov et al., 1978). Predominant among heavy minerals and particles are clinopyroxenes (22–41%), common hornblende (9–14%), epidote–clinozoisite (5–10%) and ore minerals (ilmenite, chromite, leucoxene) (5–15%). The concentration of basaltic fragments in the heavy fraction is also very high: 12–43%. The fraction (<0.001 mm) is dominated by montmorillonite (the letter M on the map in Fig. II.1.10A means that the clay is montmorillonitic). Closest to the Nile estuary ore minerals and basaltic fragments deposited; farther away, quartz, feldspars, clinopyroxenes, common hornblende and biotite. Terrigenous mud accumulates in the outer foredelta, on the continental slope

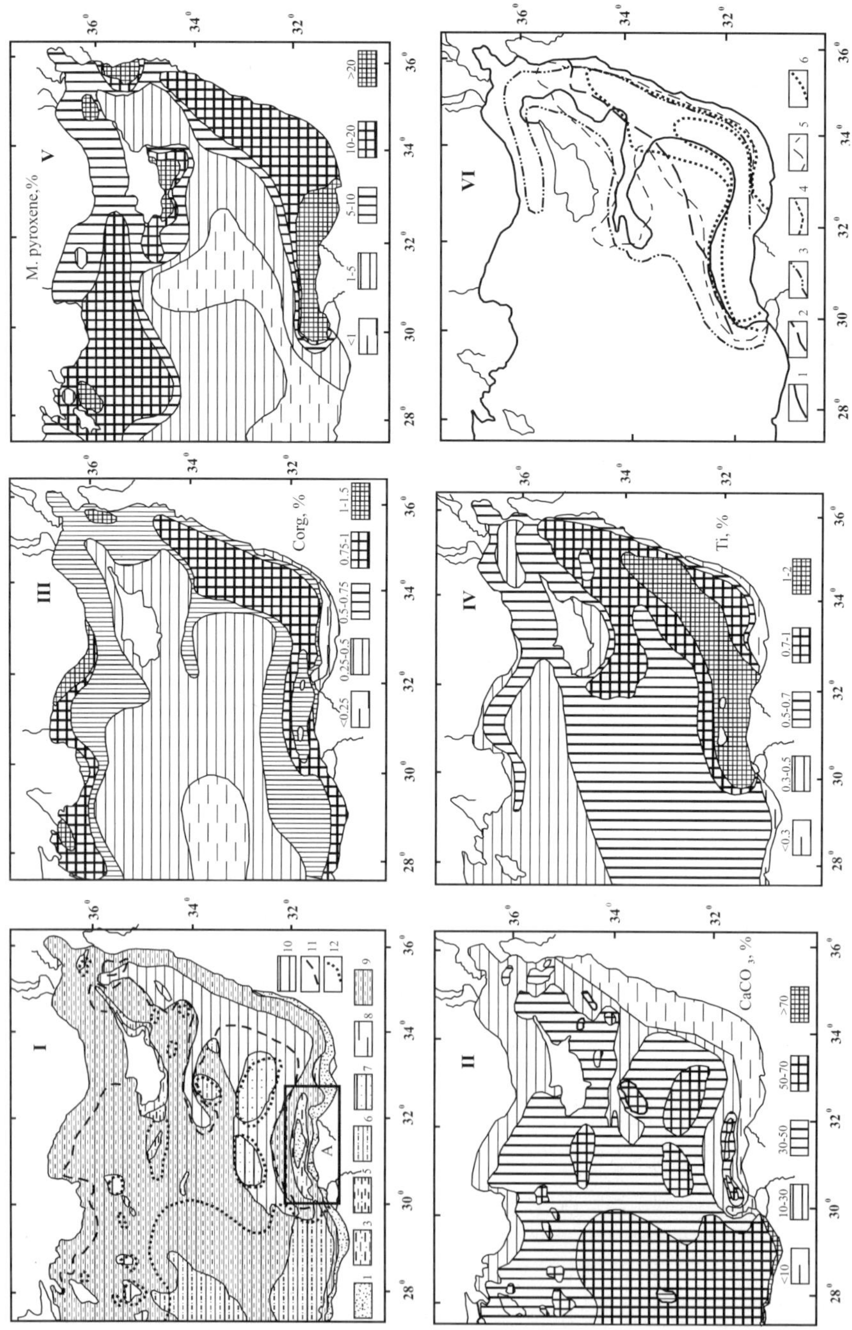

Caption see page 104

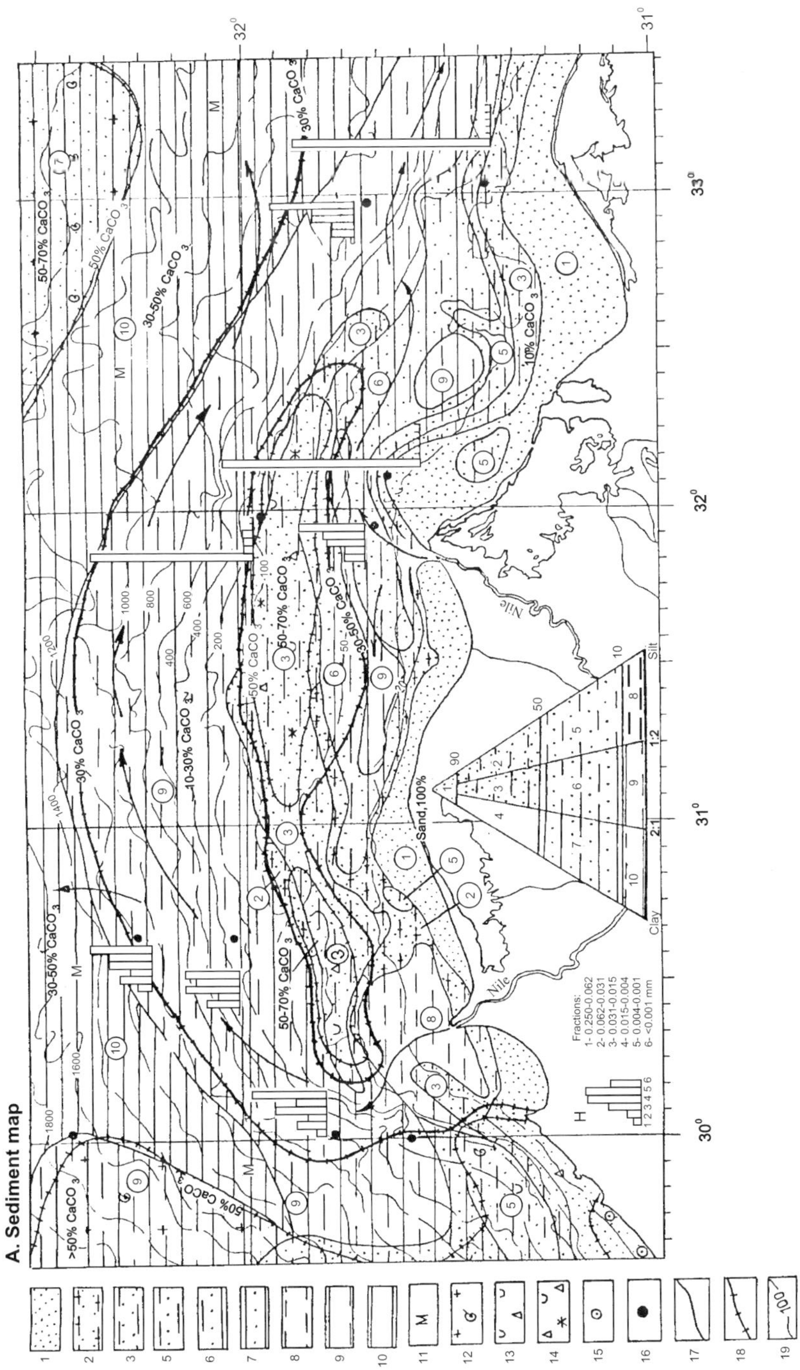

Caption see page 104

and in the eastern part of the Levant Sea until Iskenderun Bay and further west, all the way to Rhodos Island (Fig. II.1.10A).

Sedimentation rates are very high—up to >400 mm/1000 yr (Emelyanov, 1994_1). This is about three to four times higher than on the shelf off the Amazon River mouth and about the same as off the Congo River. Grey noncarbonate (<10% $CaCO_3$) mud from the river runoff load has accumulated in the region of the Nile foredelta and also close to the Israeli–Lebanese coast.

The clastic component of the mud is represented by quartz, plagioclase, clinopyroxene, green hornblende, epidote, ore minerals and other grains. In the pelitic fraction (<0.01 mm), in addition to clay minerals (mainly montmorillonite), iron hydroxide, in the form of small globules, and rutile needles are present in considerable quantities. A significant admixture of this material increases the Fe content in muds of the pre-Nile province up to 8.5%, and Ti, up to 1.57%. The titanium content in the slightly ferrugineous (6–10% Fe) mud of this province is at the maximum for sediments of the entire Mediterranean Sea. After drying, the mud becomes very compact and grey- to rye-colored. By virtue of its high content of minerals and chemical elements of the basalt group (Fe, Ti) (Fig. II.1.10), the pre-Nile province is one of the most striking in the Mediterranean Sea (Emelyanov, 1968_2, 1979_1; Emelyanov, 1994_1).

Ob, Lena, and Yenisei. The Ob, Lena, and Yenisei are the three great rivers of Siberia that flow into the Arctic Ocean. These very large rivers, with their heads in the southern parts of Siberia, exist almost at the center of Asian continent. Some parameters of these rivers are given in Table II.1.11.

Fig. II.1.10. Recent sediment types (0–5 cm) of the Nile River's foredelta, Eastern Mediterranean (A) and recent sedimentation in Levantine Sea (I–VI).
A. Recent sediments (for location, see inset A to map I, this figure (after Emelyanov et al., 1996)). Types of sediments according to grain-size (see Folk's triangle): 1—sand; 2—silty sand; 3—muddy sand; 5—sandy silt; 6—sandy mud; 7—sandy clay; 8—silt; 9—mud (silt and clay); 10—clay; 11—montmorillonitic clay; 12—coccolith–foraminiferal or foraminiferal–coccolithic ooze; 13—sediments containing a considerable admixture of detritus (2–4 mm); 14—sediments containing abundant bioclastic fragments (coquina, coral, algae); 15—oolithic, spherolitic sand; 16—stations, for which granulometric histograms (H) are given (fractions in mm, %); 17—boundary between granulometric sediment types; 18—isolines of contents of 30 and 50% $CaCO_3$; 19—isobaths, in m.
I–VI—Recent sedimentation in eastern part of Levantine Sea under the influence of the Nile River's load. After Emelyanov, 1994_1.
I. Types of bottom sediments in Levantine Sea (0–5 cm layer): 1—sand; 2—silty sand (not shown); 3—muddy sand; 4—clayey sand (not shown); 5—sandy silt; 6—sandy mud; 7—sandy clay; 8—silt; 9—mud (silt and clay); 10—clay; 11–12—isolines of $CaCO_3$ content (11–30%; 12–50%). A—sediment map (see inset A, this figure).
II,III,IV. Distribution of $CaCO_3$ (II), C_{org} (III), Ti (IV) in upper sediment layer (0–5 cm), in % of dry sediment.
V. Distribution of monoclinic pyroxene in heavy subfraction of 0.1–0.05 mm of upper bottom sediment layer (0–5 cm), in % of subfraction.
VI. Recent sedimentation in Nile lithogeochemical province (0–5 cm layer). Accumulation of sediments composed of: 1—<30% $CaCO_3$; 2—>10% monoclinic pyroxene in heavy subfraction of 0.1–0.05 mm; 3—>50% montmorillonite in fraction <0.001 mm; 4—>70% Ti (in cfb); 5—>7% Fe (in cfb); 6—>0.75% C_{org}.

Table II.1.11 Supply of sedimentary material by three Siberian Rivers into Arctic Ocean (yearly) (after Lisitzin, 1994, p. 67)

River	Drainage area, $\times 10^3 km^2$	Water discharge	Washout sedimentary material from drainage area, $t \times km^{-2}$	Average turbidity, $g \times m^{-3}$	Sediment supply, $10^6 t$
Ob	2975	400	5.31	39.50	15.8
Yenisei	2580	610	5.12	21.63	13.2
Lena	2490	511	6.18	301.4	15.4

In two monographic reviews (Kassens et al., 1999; Lisitzin et al., 2001), more exact information concerning the amount of sediments supplied to the ocean by these three rivers is presented, as well as a great deal of data on riverborne sediment transport routes in the Arctic Ocean, regeneration of this sedimentary material due to the effects of river–sea barrier, oceanic waters and ices.

In the mouth of Ob River, fresh water of lowered salinity (<1‰) extends about 120–150 miles into the ocean, whereas at the mouth of Enisei river, river water extends only some 30 miles away from land (Fig. II.1.11). Concentration of suspended matter in river waters, before they reach the salinity barrier (S = 5–15‰), is much more than 5 mg/l, and its values become as large as 20–28 mg/l during floods (Lisitzin, 2001). In past years, these values were even greater, exceeding 400 mg/l. Then after the salinity barrier (whose salinity is higher than 20‰), the concentration of suspended material in water does not exceed 0.5 mg/l. Important general points are illustrated in Fig.II.1.11: (1) Seaward of mixed river–ocean water, the concentration of suspended material rapidly decreases with distance from shore, approximately 100 times or more, and (2) in the ocean, suspension-rich material is carried to very great distances from the source area (river mouths) by near-bottom currents, instead of by surface layer circulation. The third feature to characterize the distribution of suspended matter are increased concentrations of suspended matter at the salinity barrier itself ("oozy lid"). Concentrations of suspended matter begin to decrease in the interval of the 15- to 20‰-isohalines. Observations have shown that it is in this zone where the flux of suspended material to the bottom is highest (up to 1321 mg/m^2/day) (in contrast to 9–65 mg/m^2/day in the Kara Sea), and, consequently, the largest thickness of the bottom sediment cover can be found here (Lisitzin, 2001). At present, in areas adjacent to the mouth of the Lena River, sediments accumulate at rates of up to 200 mm/1000 yr, and they occur here in the form of a lens about 400 km in diameter.

Three stages of transformation and deposition of sedimentary material in the river–sea GBZ can be indentified (after Lisitzin, 2001, in terms of the marginal filter): (1) mechanical sorting and deposition of sedimentary material; (2) regeneration of remains of fine (pelitic) suspended matter due to the effects of saline water, which acts as an electrolyte (development of colloids, flocculation and sorption of microelements by these small particles). An especially important role in the sorption

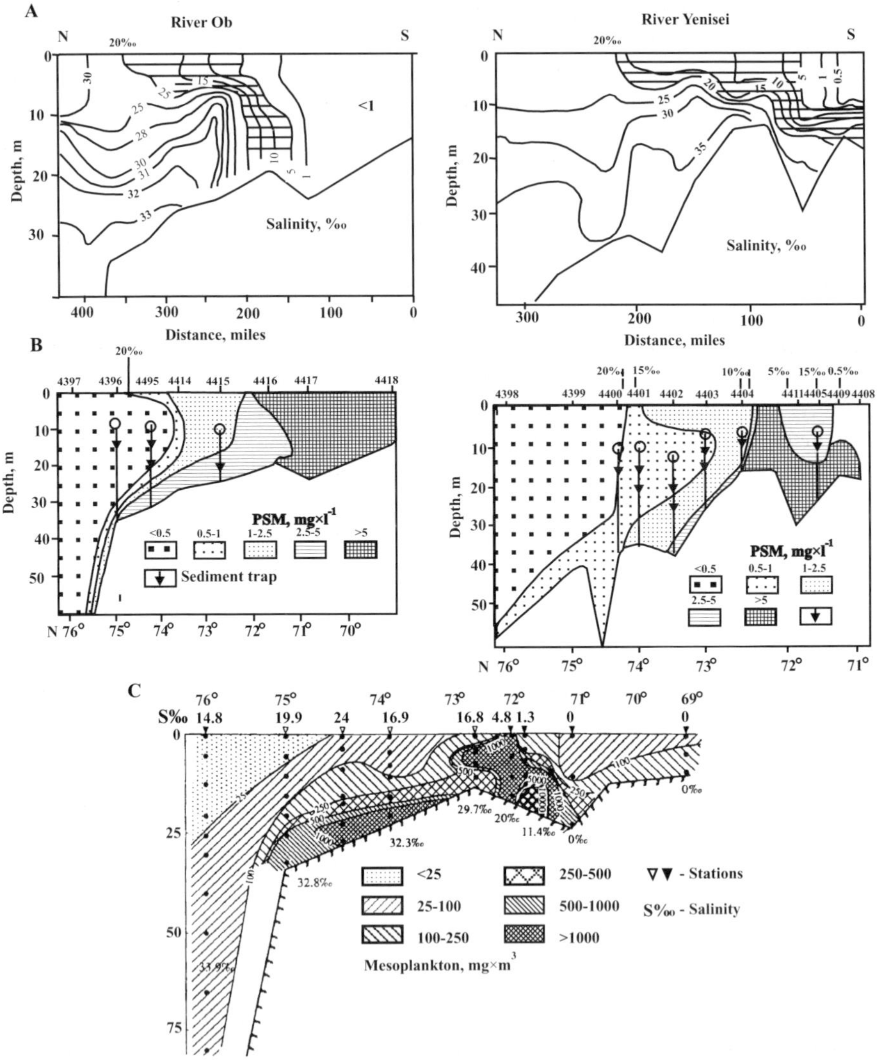

Fig. II.1.11. Salinity barrier in mouths of Ob and Yenisei (A) and concentration of particulate suspended matter (PSM) (in mg/l) (B) (after Lisitzin, 2001, p.60) and mesoplankton (mainly *Calanus finmarchicus*) biomass (mg/m^3) (C) in water of Ob's mouth (after Vinogradov and Shushkina, 2001, p. 287).

of microelements from seawater and flocculation is played by dissolved organic matter, concentrations of which are very high in river waters and low in ocean waters (Lisitzin, 2001, p. 60). Newly formed floccules, as well as aggregates consisting of these floccules and terrigenous and biogenic particles, are food for plankton and bacteria. The biomass of phytoplankton and zooplankton in a given part of the salinity barrier sharply increases (Vinogradov and Shushkina, 2001, p. 287). Take the mouth of the Ob River for example, where this phenomenon is especially evident (Fig. II.1.10C). In this region, the maximum of mesoplankton is found within the depth interval of the 4- to 17‰ isohalines on the sea surface and the 10- to 25‰ isohalines near the bottom. In addition, at the second stage, there is sorption of chemical elements from seawater by newly formed floccules and aggregates. Seawater becomes cleaner due to removal of pollutants delivered by river waters. The third stage begins when seawater becomes markedly cleaner and transparent, with the result that the thickness of the photic layer increases. These effects are caused by small concentrations of suspended material in the water column due to chemical conditioning of waters by the two former stages. The photic layer, in turn, is responsible for rapid growth of the algal biomass, which is dominated by phytoplankton. This is the biological part of the marginal filter. This filter can be subdivided into two parts. Carbon dioxide, biogenic elements (P, N, Si, etc.) and microelements are extracted from seawater by passing through this phytoplanktonic filter. Arctic phytoplankton is dominated by diatoms. They are food for zooplankton, which is represented mainly by *Copepoda*. *Copepoda* extract food from seawater by means of filtering water. In more general terms, they are an effective means of extracting suspended matter from seawater, making it clearer. According to some estimates, only 1–1.5 days are required for crustaceans to filter the total volume of the waters of these three great rivers. In addition to crustaceans, seawater is also filtered by bottom organisms, especially mollusks.

The third (biological) stage the point where marginal filter of the oceans ends.

Neman and Vistula. These rivers discharge into one and the same sedimentary basin, the Gdansk basin (Baltic Sea). Therefore, it is convenient to examine them together (Table II.1.12). The Neman discharges into the Curonian Gulf, which is a good settling basin for river suspended matter. Sedimentation rates are very high here, up to 1000 mm/1000 yr. The water of Curonian Gulf is fresh; therefore, the salinity barrier is located only at the exit of the Klaipeda Strait (Fig. II.1.13). The salinity barrier is significantly weaker than that at the Amazon or other rivers that discharge into the ocean, and so the salinity of the Baltic Sea in the strait is 8–10‰. Therefore, the barrier is somewhat extended: it is more or less clear between the 2- to 8‰ isohalines. Saline water in the form of a bottom wedge not only approaches Klaipeda's port, but also enters the strait; at the same time, surface river water in the form of a surface layer flows away from the shore for 10–15 km. The

Table II.1.12 Annual discharge and load of the Neman, Vistula and Neva rivers, Baltic Sea

River	Discharge, km^3	Average concentration of PSM, $mg \times l^{-1}$	Sediment supply, $10^3 t$
Neman	22.0	36.0	792.0
Vistula	29.0	35.0	1015.0
Neva	73.0	6.3	459.9

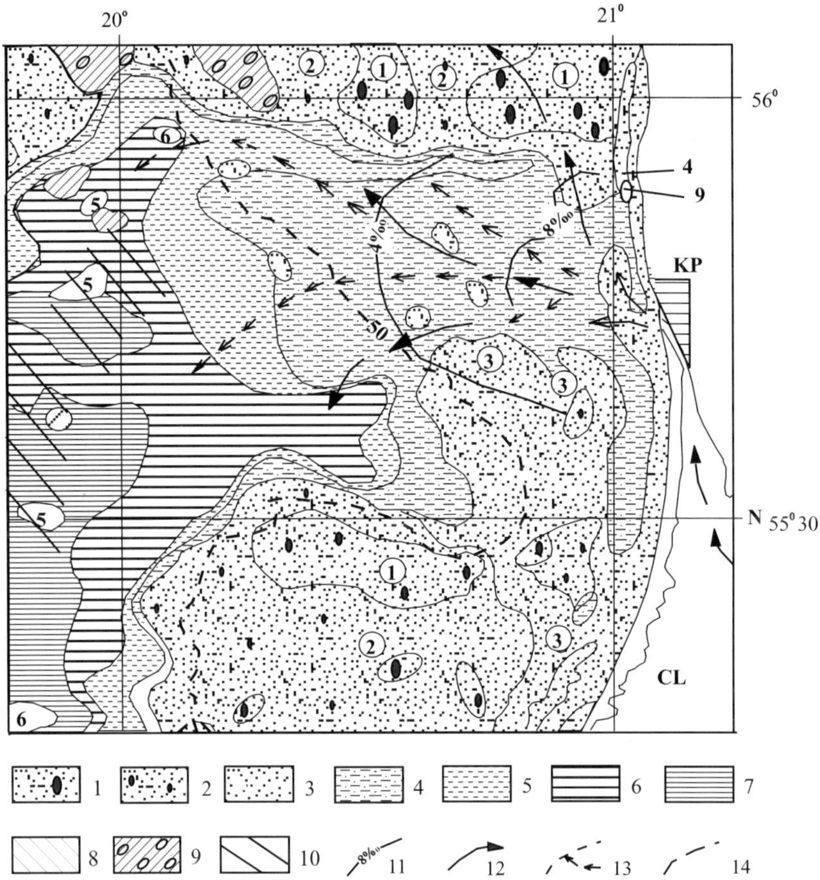

Fig. II.1.12. Sediments and sedimentation at Klaipeda Strait (at outflow of Neman River), Baltic Sea (see Fig II.2.1.)

Bottom sediments: 1–10—terrigenous bottom sediment: 1—sand, gravel, boulders; 2—sand, mainly coarse, with gravel; 3—sand; 4—coarse aleurite (silt); 5—fine-aleuritic (silty) mud; 6—aleuro-pelitic mud; 7—pelitic (clayey or illitic) mud; 8—homogenous clay, lacustrine (Hl_1); 9—loam and till with boulders and pebble; 10—sapropel like mud (3–5% C_{org}); 11—isohalines 4 and 8‰ 12—main paths of water; 13—Distribution (slow slump) of near-bottom turbidite waters with fine (pelitic) material; 14—50-m isobath; KP—Klaipeda port; CL—Curonian lagoon (trap for Neman River load); N—Neman River. Numbers on map—numbers of legend.

thickness of this mixed layer is only about 5–10 m (Lukashin et al., 1986). The maximum vertical salinity gradient is 5‰/m ; the lateral one (at the sea surface) is 4‰ per nautical mile. A sharp reduction in the concentration of suspended matter, and also the particulate form of chemical elements, practically coincides with the salinity barrier. The concentration is maximum between the 3 and 6‰ isohalines, up to 42.5 mg × l^{-1}. The bottom saline wedge is enriched in suspended matter. The max-

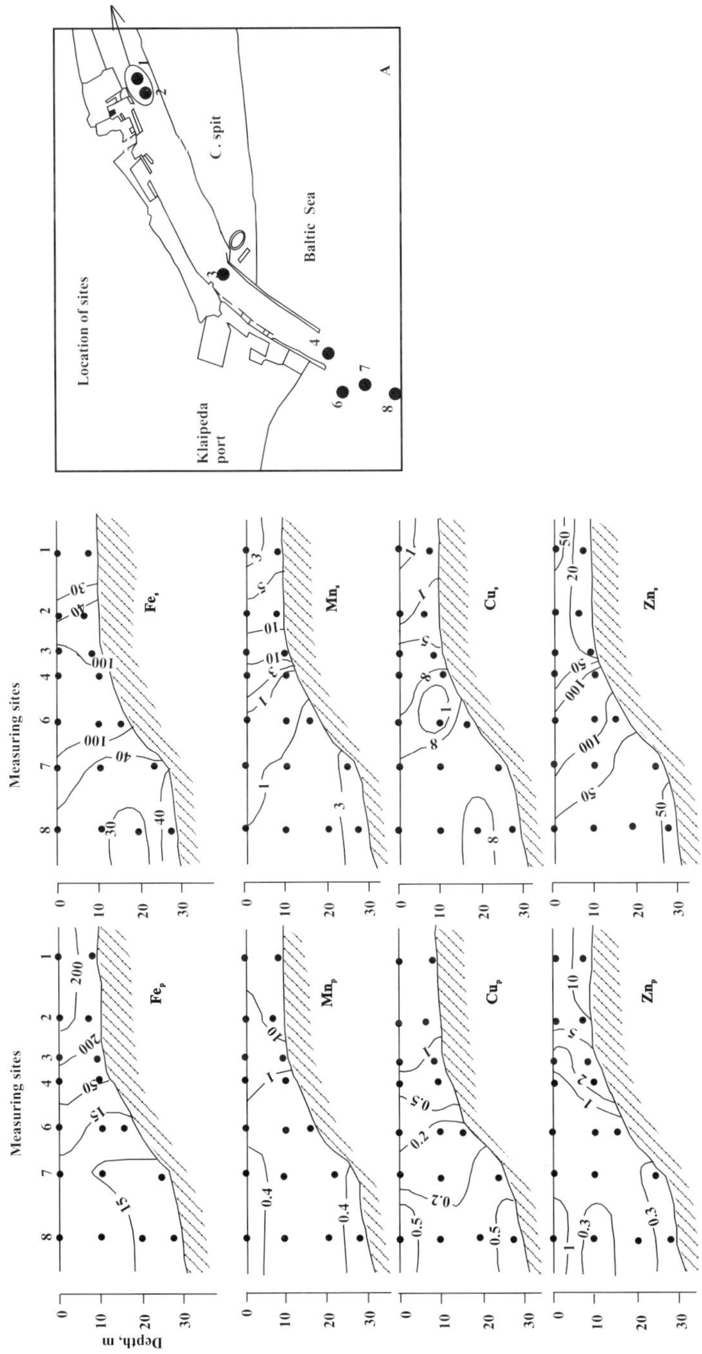

Fig. II.1.13. Concentrations (in µg/l) particulate (p) and dissolved (s) forms of Fe, Mn, Zn and Cu at the river–sea salinity barrier in Klaipeda port. C—Curonian. After Galkus and Jokšas, 1997. Location of stations are shown in inset A.

imum concentration of suspended matter is contingent upon three reasons: (1) movement of suspended matter from the Curonian Gulf; (2) the rapid blooming here of diatomaceous phytoplankton [the "biofilter" (Pustelnikov, 1976; Lukashin et al., 1986)]; and (3) coagulation of dissolved matter.

The transformation process of suspended and dissolved matter in the zone of the salinity barrier at the estuary of the Klaipeda Strait occurs stage by stage (Emelyanov and Pustelnikov, 1975; Emelyanov, 1982_1, 1986_1; Lukashin et al., 1986). At the beginning, in connection with the spring load of the river flow (hydrodynamic barrier), the concentration of suspended matter decreases (a large portion, in connection with a decrease in current velocity, is precipitated) and desorption of metals from river suspended matter occurs, along with an increase in the concentration of their dissolved forms as a result. Then, beyond the 5‰ isohaline, hydroxide coagulation of Fe, Al, Mn and other elements, and also organic matter, occurs, forming floccules of hydroxides and organic matter. Fresh hydroxides, as well as colloidal organic particles, possess an increasing ability to adsorb elements.

The main mass of pelitic and fine-aleuritic materials is transported into the nearby Gdansk and Gotland basins (Emelyanov, 1968_1). In the Neman sedimentation province of the Gdansk basin, mud accumulation starts at depths of 63–66 m. Four-to-five-meter-thick, gray, banded terrigenous muds are abundant in the outermost (from the Klaipeda Strait) and deepest (83–105 m) part of the Neman province. This bedding passes locally into lamination. The $CaCO_3$ content does not normally exceed 10%, and C_{org} does not exceed 1.5–5%. It should be noted that up to 3–6% is contained in laminated muds. The pre-Neman valley is filled with aleuropelitic and pelitic muds starting from water depths of 70 m or deeper (Fig. II.1.13). The thickness of these sediments is as much as 9.7 m (borehole D-2-1) and reaches 10.2 m if the aleuritic sediments which lay at the foot of the Hl_{2-3} sedimentary strata are taken into consideration (the age of these aleuritic sediments in the valley are not known exactly). The muds are terrigenous and contain up to 15% $CaCO_3$ (normally up to 5%), up to 6% C_{org}, up to 6% Fe, 0.06–0.08% P, and small amounts of Mn (up to 0.07%; in places, up to 0.11%).

The Vistula River is much bigger than the Neman. It delivers annually about 1,015,000 tons of solid matter (Table II.1.12). Up until 1895 it discharged into the settling basin of the Vistula lagoon. Now, it discharges directly into Gdansk Bay (Fig. II.1.14). The thickness of recent sediments in frontal zone of the delta varies from 0.5 to 3 m. Side-scan sonar images show a flat seafloor with a pattern of sandy ribbonlike features orientated in the direction of the bottom current. In front of the fan, the sediments consist of fine-grained sand with dispersed clay and they are intercalated with mud laminae; the sediments are intensely bioturbated mainly by *Nereis diversicolor* (Müller). The sand also contains a few mollusks such as *Macoma baltica* (Linnaeus) and *Mya arenaria* (Linnaeus), some of which are alive. The sediments also contain a small amount of very fine gravel (Zachowich et al., 2002).

The recent muds of the Gdansk basin (area of stations 2673, 2674, and 3675, Fig. II.1.14) were deposited at water depths starting from 60–65 m. The thickness of the Litorina deposits ranges from 1 to 3 m. In the lower part of the strata of the deposits, there is evidence of acoustic laminations in seismic records. The sediment

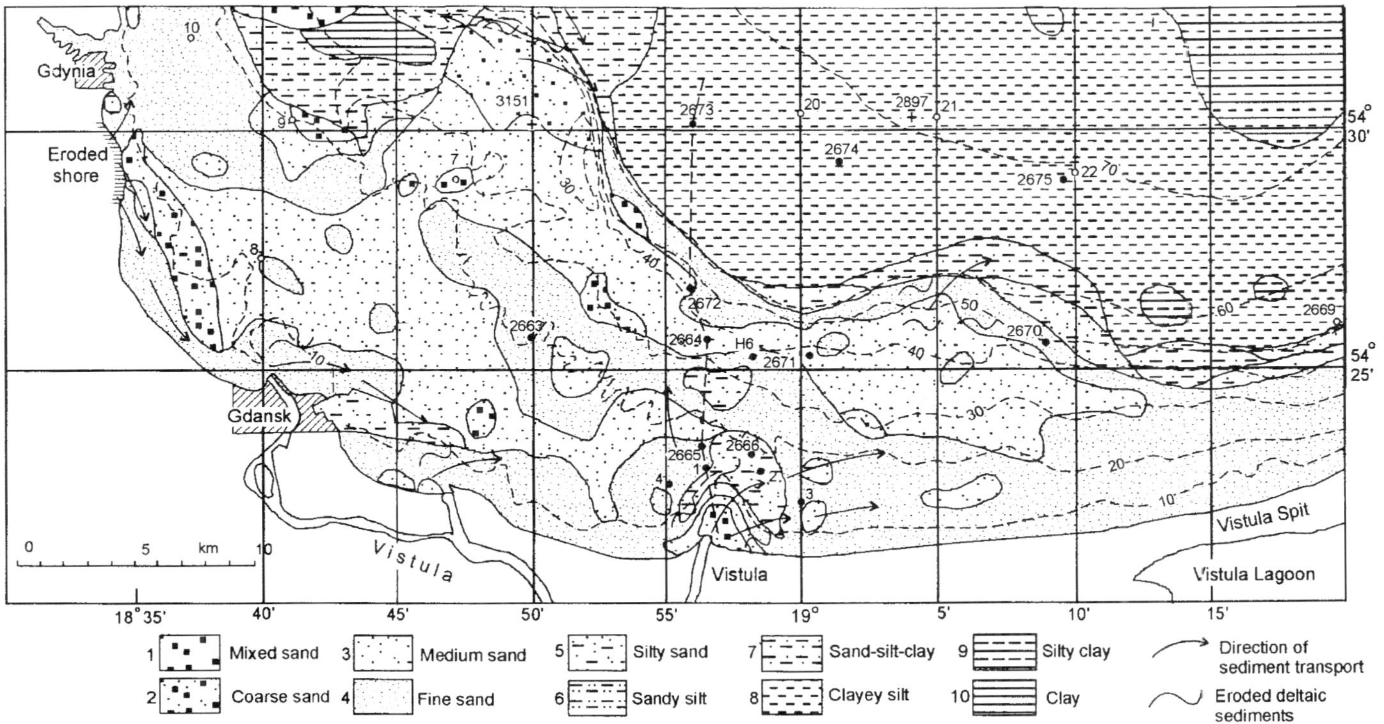

Fig. II.1.14. Recent sediment types (0–5 cm) in Gulf of Gdansk (according to Shepard's classification), location of studied samples of sediments, areas of erosion and sediment transport. After Zachowich et al., 2002. Isobaths in m.

is composed of olive-grey to dark olive-grey mud with a small sand content, displaying common bioturbation, which is probably caused by mollusks such as *Macoma baltica* (Linnaeus). In X-ray photographs, the deposits show several erosion surfaces caused by the effects of internal waves within the pycnocline.

Silts in cores 22 and 2675 show interesting sediment sequences of various origin. Particles from the upper layer (0–5 cm) are very similar to fluvial grains from the Vistula River. There is a clear relationship between grains from the recent sediments of the lower sample (20–30 cm) and the grains from shore Miocene deposits and from near-shore glacial cliff deposits. Such a difference between the upper and lower samples of recent sediments (between the 0- to 5- and 20- to 30-cm layers) likely reflects changes in sediment sources. Since 1895, when the artificial outlet of the Vistula River was opened, inputs of sand in this area substantially increased, and they built up a large fan over the last 105 years. Formerly, when the old outlet still functioned, the hydrological situation was much different (and more complicated) than it is now. Probably, sediment transport from the west predominated. The sand fraction is thought to have been transported by ice in winter from the western coast, where glacial and fluvioglacial deposits occur, to the deeper part of the bay.

A thick aleurite layer of Litorina age (7800–0 years) has accumulated at the base of the underwater part of the foredelta of the Vistula—up to 780 cm. The sedimentation rates are rising here, up to 1.0–1.4 mm × y^{-1}. A light subfraction of sand (1.0–0.1 mm) in marine Holocene aleurites (0–34 cm) from core AK-2669 (see Fig. II.1.15) contain 51–8% plant remains (spores, pollen grains, scraps of wood, and grass), 13.6–21.4% quartz, 1.5–3.0% feldspar with N=1.545, 1.0–3.0% green mica, and also rock fragments and aggregates (Fig. II.1.13). A coarse aleurite interlayer (334–122 cm) present in the core AK-2669 appeared to consist of quartz. The light subfraction of sand (1.0–0.1 mm) contains 45.6–84.6% of quartz. These sediments contain acidic feldspars (4.1–11.3%), mica (muscovite and biotite), glauconite (up to 6%), rock fragments and aggregates, plant remains, and biogenic siliceous particles. The heavy sandy subfraction of coarse aleurites in the 34 to 122-cm layer consist of biotite and muscovite (up to 60–80% of the total). The contents of weathered grains and hydrogoethite are much smaller. The concentration of iron sulfides in sediments is also considerable (1.3–43.2% pyrite). The light subfraction of coarse aleurite (0.10–0.05 mm) is suprisingly dominated by quartz (70.4–76.5%) and acidic feldspars. In lesser quantities present are glauconite (1–9%), mica, biogenic siliceous particles, plant remains (spores, pollen grains, scraps of vegetation) and other substances. The heavy coarse aleurite subfraction is dominated by ilmenite, epidote–clinozoisite, common hornblende, garnet, zircon, and rutile. Authigenic minerals (3–10% of the subfraction) are represented by pyrite, glauconite, and carbonate. Remains of fish teeth and bones have been found in this sediment.

A layer of terrigenous fine aleurite, 45–60% of which is represented by the 0.05–0.01 mm fraction, occurs in the 122- to 405-cm interval (station AK-2669). Relationships between the main minerals are essentially the same as in the 0- to 122-cm layer. Nevertheless, some noticeable exceptions exist. First, spores, pollen and other plant remains are very common throughout the fine aleurite interval. The C_{org} content in the mud is not less than 1% and is 1.92% in the 200- to 205-cm layer.

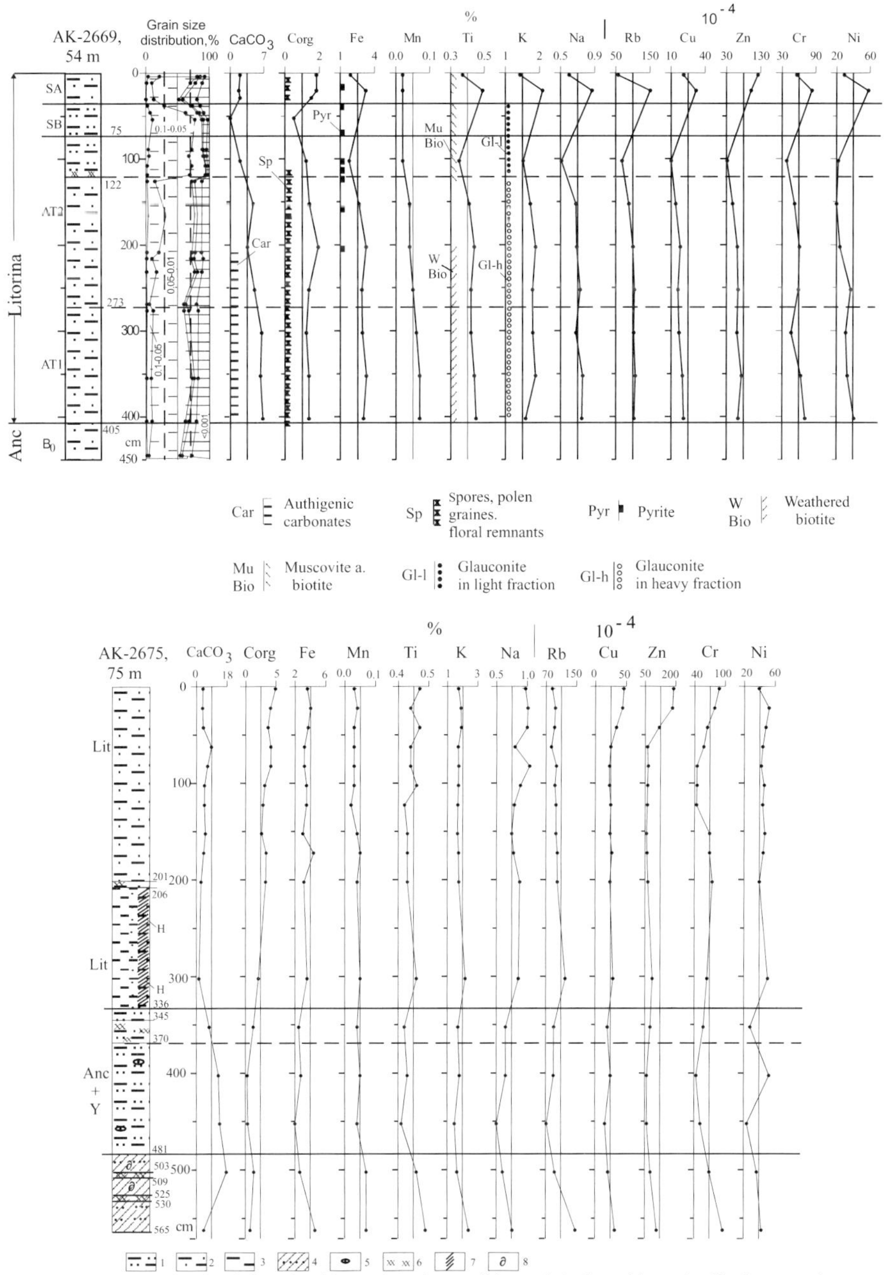

1-coarse aleurite; 2-fine aleuritic mud; 3-aleuropelitic mud; 4-clay with sandy-silty lenses and with the peat remnants (lagoonal deposits); 5-sandy-silty lenses; 6-remnants of peat; 7-mud with hydrotroilite (H); 8-shelly remnants.

Fig. II.1.15. Lithology, grain-size distribution and content of some components, elements and minerals in sediment cores AK-2669 and AK-2675 from Vistula foredelta region, Gdansk Basin (Baltic Sea). For core location see Fig. II.1.14.

Second, glauconite persists in the heavy subfraction, but not in the light subfraction, as in the 3 to 122-cm layer. Third, biotite is constantly present as a weathered mineral (but not as fresh as in younger sediments of the core). This weathered glauconite prevails in the core deeper than 200 cm. Authigenic clayey-calcareous aggregates are very abundant in the same layer (200–405 cm). This is circumstantial evidence that sedimentation conditions in the vicinity of the AK-2669 core changed at least four times. Material (heavy glauconite, weathered biotite) was supposedly intensively transported from the Sambian Peninsula side during development of the 200 to 405-cm layer (the age is from the Atlantic 1 [AT1] until the mid-Atlantic 2 [AT2] chronozone). The Vistula River transported a large amount of pollen and spores and is believed to be responsible, indirectly, for the development of clayey-calcareous aggregates. A 200- to 122-cm-thick layer (AT2) has formed under transitional conditions. The hydrodynamic situation changed dramatically at the end of AT2, when the current pattern changed. This led to accumulation of sediments with coarser grain-size (coarse aleurites) containing light glauconite, fresh muscovite and biotite. Conditions again changed at the boundary between the Subboreal (SB) and Sub-Atlantic chronozones (SA), and attained the present-day situation.

$CaCO_3$ content in the cores is commonly no greater than 10%. Carbonates are represented mainly by clastic dolomite, rare clastic calcite and diagenetic calcite crystals. C_{org} content in the core increases progressively upward. Fe content also rises up to 4.92%. The content of Mn is very insignificant, less than 0.05%. The upper layer (0–5 cm) contains 5.66% SiO_{2am}.

In the central part of the Gdansk basin, the accumulation of aleuro-pelitic and pelitic muds took place at the beginning of the Litorina transgression. Near-bottom water contained small amounts of oxygen; the C_{org} content increased to 2.13–3.36% and hydrotroilite appeared. A change in the environmental situation (termination of sedimentation) and an increase in hydrodynamic activity approximately in the middle of the Litorina stage (the 206- to 20-cm layer) led to replacement of muds by accumulation of coarser-grained sediments, namely, fine-aleuritic muds with peat remains. More recent (Atlantic–Sub-Atlantic stages), sedimentation conditions were similar to that of present day.

The mud is very soft (water content 60–75%), greenish grey and grey, sometimes with black hydrotroilite (mottled or striped). The muds contain a large amount of pollen grains, spores and other plant remains, rare mollusk shells, marine diatoms, and the dinoflagellate cysts *Operculodinium centrocarpum*. One of these samples contains as much as 6% C_{org} (Emelyanov, 2002).

It is very typical for muds to have strips of other (usually black–grey hydrotroilitic) or clear microlamination. These features were found in cores collected from water depths of more than 80 m. Some cores are made up of two or three intervals with microlamination. Microstratified muddy beds in cores collected in the area of gas craters (pockmarks) are very typical. Cores obtained in the activity zone of gas craters (Fig. II.1.16) contain grey, banded (strongly reduced) muds. These muds contain free H_2S (up to 2 ml × l^{-1} of sediment). On the whole, from 5 to 6% Fe is found in these muds, while reactiontable Fe^{2+} accounts for up to 1% in sediments, or up to 15–20% in total Fe content. Aleurites containing 2–3% C_{org} are notably enriched in phosphorus (0.10–0.50% P), vanadium and other trace elements.

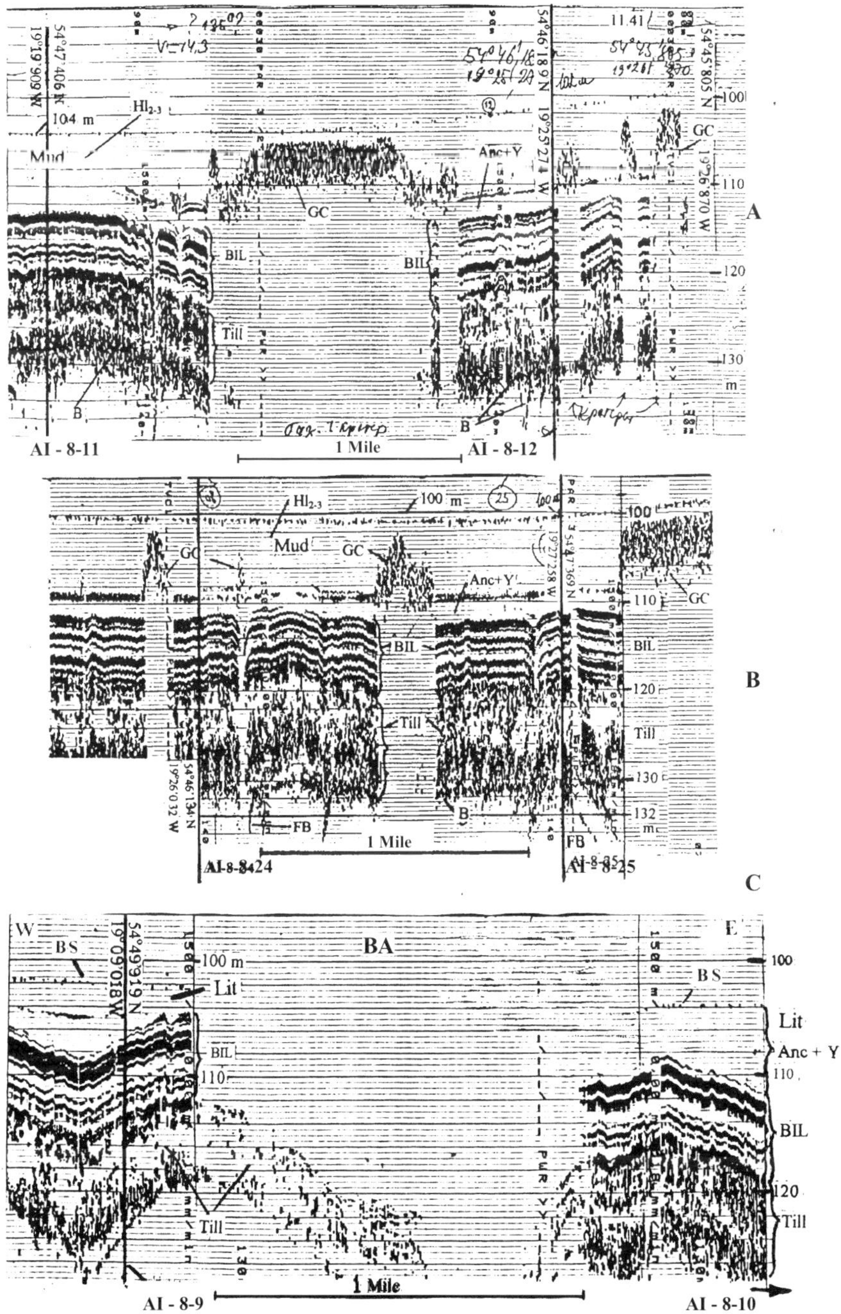

Caption see page 116

The presence of varved and microlaminated sediments is evidence that processes of bioturbation were absent during the period of accumulation of these layers and later on. However, during this time, benthic organisms were absent. Such a situation is possible only when H_2S is present at the bottom and in upper layers of mud. The stratigraphic position of microlaminated horizons is different for different cores, and this leads us to the inference that contamination of bottom water by H_2S was only local. In addition, gas craters may serve as sources of H_2S.

The thickness of marine Holocene mud in the Vistula sedimentation province (Fig. II.1.16) ranges from 0 to 10 m. The maximum thickness has been discovered in Gdansk Bay, i.e., in close proximity to the Vistula River mouth. There is the moraine sill, which intersects the Gdansk trench from south-southeast to north-northwest and divides it into two sedimentary basins: the Vistula basin in the south is the deepest (up to 115 m); the Neman basin in the north is shallower. The thickness of muds is 1.5–3 times greater in the Vistula basin than in the Neman basin. Consequently, the sedimentation rate here is higher (Fig. II.1.17) than at the mouths of the above rivers (Emelyanov, 2002).

Neva (Fig. II.1.18). Owing to the low salinity in the Gulf of Finland (3–7‰), the salinity barrier of the Neva River is weakly expressed. If we take into consideration that biogeochemical processes occurs most intensively between the 3- to 15‰ isohalines, then this barrier does not end at the Gulf of Finland; its border can be found only in the Arkona basin of Baltic Sea, where salinity increases up to 15–19‰. Therefore, the processes of changes in the form of elements in the Gulf of Finland at river mouths are weakly expressed. Nevertheless, at the beginning of the salinity barrier (between the 3 and 4‰ isohalines, i.e., at the Neva River estuary), the concentration of suspended matter falls sharply: from 9–8 mg × l^{-1} near Kotlin Island up to 3–2 mg × l^{-1} in the area of the 3.5‰ isohaline (Emelyanov, $1995_{2,3}$). Here, because of the decrease in the current velocity (hydrodynamic barrier), a significant part of the suspended matter (including suspended matter of the Neva River and Neva Bay) precipitated from the surface to the bottom layer, partly precipitated in Kotlin depression (deep), and partly with the bottom water, migrates further in a deeper and hydrodynamically less active area of the Gulf of Finland. Along the way, suspended matter enriched in elements (P, Fe, Mn, Zn, Cu, Pb, Cd) is diluted with typically marine suspended matter. The content of C_{org} in phytoplankton of the Gulf of Finland is 38.03%; Ca, 0.85%; Mg, 1.00%; Na, 3.3%; K, 1.04%; P, 0.26%. Phytoplankton comprise practically all the trace elements studied (Baturin et al.,

Fig. II.1.16. Structure of Quaternary deposits of central part of Gdansk Basin in points AI-8–11—AI-8–12 (A) and AI-8-24–25 (B), in area of pockmarks (depth 100–101 m), according to parasound profiles. Hl_{2-3}—marine Holocene (Litorina) mud (thickness 8.2 m); Anc+Y—Ancylus and Yoldia clay (2.5 m); BIL—varved clay of Baltic Ice Lake (8.0–8.5 m). Till—10.5 m; B—basement; FB—proposed fractures in basement; GC—gas craters (acoustical anomalies).
Structure of sedimentary cover in deepest part of Gdansk Basin according to profiler parasound record. Reason for absence of bottom surface (BS) at area A (proposed depth—110–115 m) is unknown.

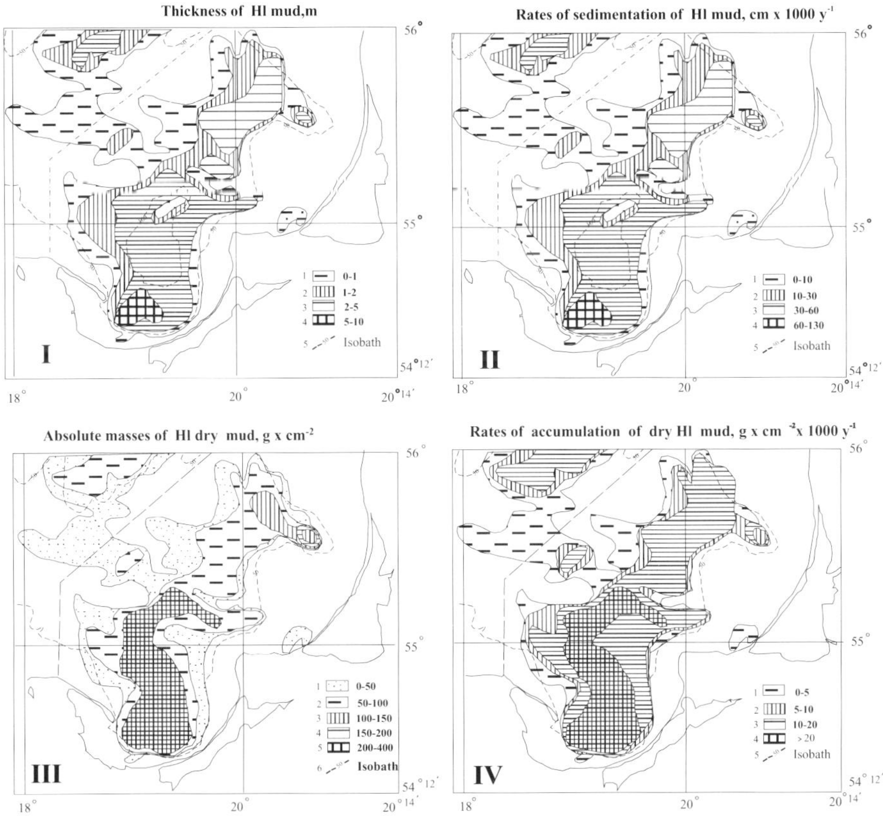

Fig. II.1.17. Thickness (in mm) of marine Holocene (Hl_{2-3}) mud (I), rates of sedimentation of mud (II) masses of dry marine Holocene mud (III), and rates of accumulation of dry marine Holocene mud (IV) in Gdansk Basin. Isobaths 50 and 100 in m.

1993), and their concentration in phytoplankton is 1000–10,000 times greater than in seawater. Thus, at the Neva River mouth (as at the mouths of many other rivers), phytoplankton is the strongest concentrator of not only biogenic elements (C, Na, Si, Ca, Mg, P) but also of metals. Contents of Al, Ti, Na, K, Co, Li, Rb, Cs, Hf, Sc, U, Th, La, Sm, Ce, Tb, Ib, Ln, and Ag in plankton of the Gulf of Finland are even higher than in marine suspended matter of the Baltic Sea.

In the polluted area of the Gulf of Finland, the annual primary production may be more than 150 $gC \times m^{-2}$; at the same time, in areas with "normal" concentrations of biogenic matter, it varies within 15–60 gC/m^2. Daily primary production in the eutrophic area approaches 1.0–1.2 $gC \times m^{-2}$ and is observed in June–August, while in the oligotrophic water it is much lower: 0.07–0.28 $gC \times m^{-2}$. In the eutrophic area, blue–green algae predominate; in the oligotrophic area, diatoms and Dinaphyceae predominate.

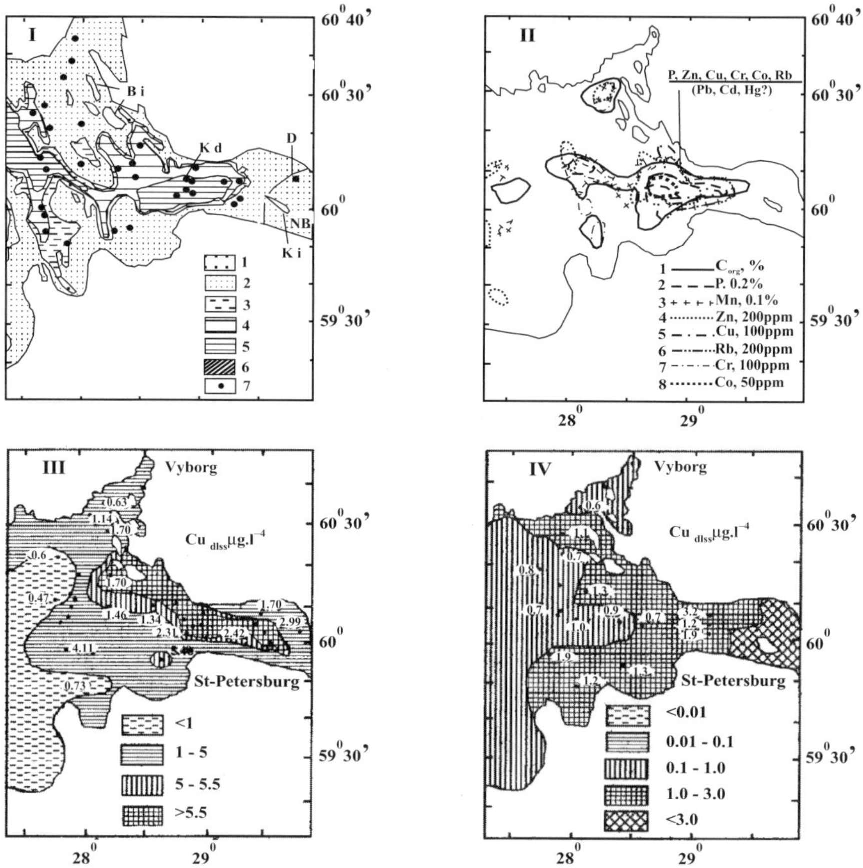

Fig. II.1.18. Bottom sediments and concentrations of chemical elements in sediments and in surface waters (0–1 m) of eastern part of Gulf of Finland near Neva mouth. After Emelyanov, $1995_{2,3}$, 1997.
I. Surficial (0–5 cm) terrigenous bottom sediments: 1—sand, gravel, stones; 2—sand, mainly fine; 3—coarse aleurite (silt) and fine silty mud; 5—pelitic mud; 6—moraine deposits; 7—geological stations. D—dam; NB—Neva Bay; Ki—Kotlin Island; Kd—Kotlin deep (40–45 m); BI—Berezovyje Islands.
II. Isolines of equal contents of chemical elements in surficial bottom sediments (0–5 cm), C_{org}–Mn—in %, Zn–Co in 10^{-4}%.
III, IV, V and VI. Concentrations of dissolved forms of Cu, Pb and Cd and particulate forms of Cu in surface waters of eastern part of Gulf of Finland, in µg/l .

Continues

The interval between the 2- to 3‰ isohalines is responsible for rapid development of both freshwater and brackish forms of zooplankton. The weighed biomass of zooplankton in this part of the "hydrofront" amounts to 1.4–2.3 g × m^{-3}, on average. Adjacent to the Neva River mouth, the amount of biomass was 0.4–1.0 g × m^3, and beyond the limits of the 3‰ isohaline it was 0.6–1.3 g × m^{-3}. Zooplankton binds

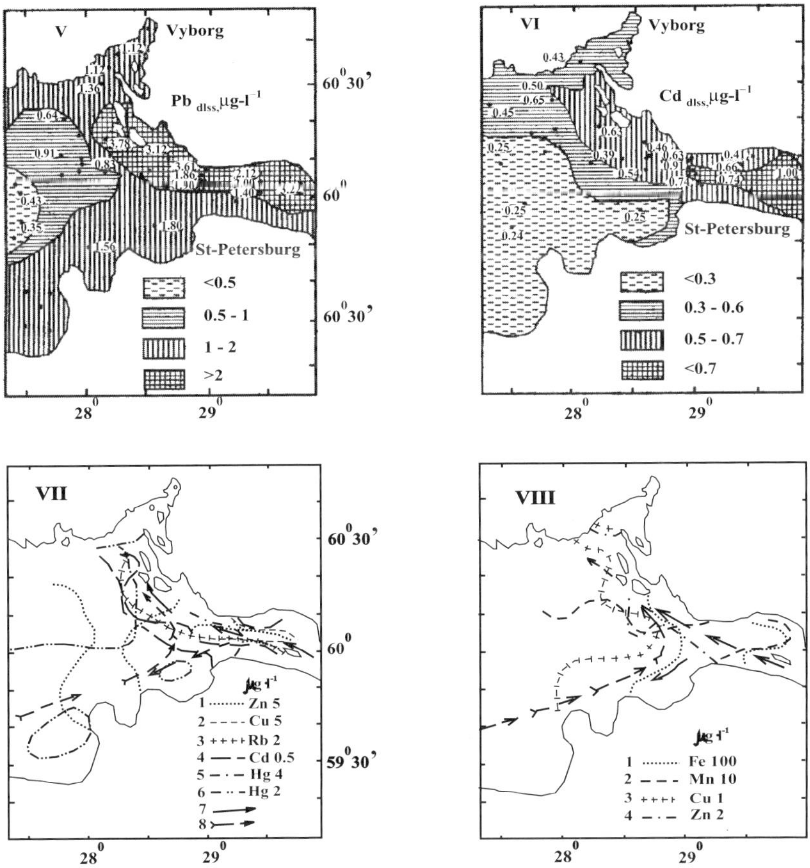

Fig. II.1.18. *Continued*

VII. Isolines of equal concentrations of dissolved forms of elements in surface waters, in µg/l; 7—flux of polluted waters; 8—general currents of seawater.

VIII. Isolines of equal concentrations of particulate forms of elements in surface waters, in µg/l.

food (mainly organic detritus and clay particles) in their stomachs to form fecal pellets, which precipitate to the bottom in large quantities. In this way, zooplankton accelerate precipitation of suspended matter to the bottom. Dissolution and mineralization of phytoplankton starts already in the photic layer, but occurs more intensively at the pycnocline (where the sinking of organic detritus is somewhat delayed), in the bottom water and on the bottom surface. As a result, more mobile elements, such as mercury, phosphatic and organic phosphorus, iron, manganese, and other heavy metals are accumulated in the bottom water (Emelyanov, 1993, $1995_{2,3}$). It is as though a layer with higher Hg concentration "hovers" over layers of maximum concentration of other elements, owing to its quickest release from phytoplankton remains (as a result of destruction and oxidation during contact with dissolved oxygen). After mercury, Fe, Mn, Zn, Cu, Pb, Cd and other elements are released into solution. As a result, water (mainly bottom or interstitial water) becomes enriched in these elements. Iron and manganese are released into solution as unstable or stable Fe-organic complexes. Ligands of these dissolved complexes are subsequently oxidized; iron and manganese hydrolize; and Fe^{2+} and Mn^{2+} ions partially precipitate into solid phase to form tiny particles of $Fe(OH)_3$ and MnO_2. Fresh iron and manganese hydroxides are good sorbents of certain trace elements. Therefore, the complicated transformation of these two transitional metals is accompanied by changes in trace elements migration forms: trace elements are sorbed by Fe- and Mn hydroxides, and during dissolution of these hydroxides, together with Fe^{2+} and Mn^{2+}, trace elements are released into solution.

In the Gulf of Finland, where the path of phytoplankton remains from the photic layer to the bottom is short due to small depths (30–80 m), destruction of organic detritus and release of metals into solution occur mainly on the bottom surface or within the upper sediment layer. In this case, only interstitial waters are enriched in metals. Because the oxidized film on the sediment surface is absent in the studied area, we suggest that intensive diffusion of dissolved Mn^{2+} ions and, to a lesser degree, of Fe^{2+} from interstitial water into bottom water takes place here. Insofar as the oxidation rate of Mn^{2+} ions is significantly lower than that of Fe^{2+} ions, the diffusion process leads to chemical separation of these two elements. In the weakly oxidizing conditions of the Gulf of Finland, iron remained mainly in the bottom sediments, whereas manganese migrated in significant quantities from the sediments into bottom water. In the conditions of the Gulf of Finland, dissolved (in the form of Mn^{2+}) and particulate manganese (in the form of very small gels of its hydroxides) were observed in bottom depressions, where a decrease in oxygen was observed (lower than 6–5 ml $\times$ l^{-1}). However, oxygen content in the bottom water is still sufficient for intensive oxidation of Mn^{2+} and the gel formation of its hydroxide. As a result, Mn content in suspended matter of the bottom water in places increases up to 10–15%. The iron content in the suspended matter is the same as that in the upper water column (2–5%).

The processes of flocculation, desorption and sorption are the most important at the river–sea salinity barrier. In the Gulf of Finland, flocculation and sorption on hard particles of the dissolved form of metals occurs most intensively between the 3- to 6‰ isohalines (Emelyanov, 1995). In spite of the weak expression of the salinity barrier, this tendency of reduction of the dissolved form of metals is well recognized, especially if the given concentration of metals in water with a salinity of 8–9‰ is plotted on a graph. The correlation coefficient r appears to be highest (negative) between salinity and both forms of Mn and the suspended form of Fe. Nevertheless, if we exclude samples from the Bornholm basin, r increases for Zn, Cu and Pb as well. The reason why concentration of metals decreases as salinity increases is not only due to the effects of flocculation and consumption of trace elements by phytoplankton, but also due to dilution of river waters by cleaner sea waters.

Floccules and metallorganic compounds, both carried by rivers and formed in the mixing zone, sooner or later are affected by the hydrolysis reaction to take the form of hydroxides. Micelles of these hydroxides obviously become enlarged with time (up to floccules 1–50 μm in size) and are precipitated to the bottom, starting with the area of Kotlin deep. However, the process of mechanical precipitation of suspended matter, biochemical processes, flocculation and sorption processes are incapable of extracting excess (relative to seawater) quantities of transitional and heavy metals immediately in the zone of the hydrological front. Metalloorganic compounds (in particular, iron–organic) in seawater are sufficiently standard. In dissolved and colloidal states, they may be preserved for a long time. Sea currents transport them a great distance (Fig. II.1.18). In the water itself, in the eastern part of the Gulf of Finland and Neva Bay, the concentrations of dissolved forms of heavy metals, in comparison to their average concentration in open areas of the Baltic Sea, are higher: for Cu, 10 times; Pb, 8–10 times; for Hg, 5–7 times; Cd, 3–4 times; and Zn, 2–3 times. The concentrations of their particulate forms are approximately just as high. Areals of polluted water by heavy metals run from Kotlin Island west-northwest to a distance of about 40–45 miles (up to 80 km). The presence of a counterclockwise current explains why water of the southern region of the eastern part of the Gulf of Finland is cleaner than that in the north. The waters of Vyborg Bay, especially in the northern part, are less polluted by metals than waters near Kotlin Island and the Berezovye Islands. This confirms the assumption that water polluted by dissolved metals enter from Neva Bay and, possibly, polluted at Kronstadt port (Kotlin Island) is picked up by currents; it is directed along the northern shores and approaches the northern Berezovye Islands (Emelyanov, $1995_{2,3}$).

A significant feature is that areals of increased concentrations of dissolved and particulate forms of elements in seawater coincide with areals of bottom sediments, displaying increased contents of the same elements (Fig. II.1.18). Ferromanganese nodules (FMNs) form here (see Part III).

The "underwater" Stolpe River, Baltic Sea. In a semi-isolated epicontinental sea like the Baltic, hydrodynamics plays the leading role in sediment accumulation. This is not only related to the shallow depth of the sea (for the most part, less than 200 m), but also to sharp division of the water column into two layers: the upper hydrodynamically, very active layer, and the deep, less active layer. The border that demarcates the two layers is clearly manifested at the Baltic halocline, which almost fully coincides with the pycnocline and is situated at a depth of about 60–70 m in the Bornholm deep, and about 70–80 m in the Central Baltic (in the Gotland deep) (Fig. II.1.19). The halocline is the border that demarcates the facies situation of sediment accumulation: above the halocline, basically clastic sediments (gravel, sand, aleurite) are deposited; below it, muds occur (Blashchishin and Emelyanov, 1977). The facies situation above the halocline is divided into two subzones: the coastal zone of sediment accumulation, bottom erosion, and redeposition, related to wave and alongshore current activity (a depth of 30–40 m); and the zone of nondeposition or transit of sedimentary material from land into the deeps (a depth of 30–40 m up to 70–80 m, i.e., up to the halocline level).

The sediment cover in the Baltic Sea deeps bathymetrically below the halocline is also discontinuous instead forming a "blanket" (Emelyanov, 1986_1, 1995_1). Sediments are in places absent altogether not only on slopes of the deeps, but also

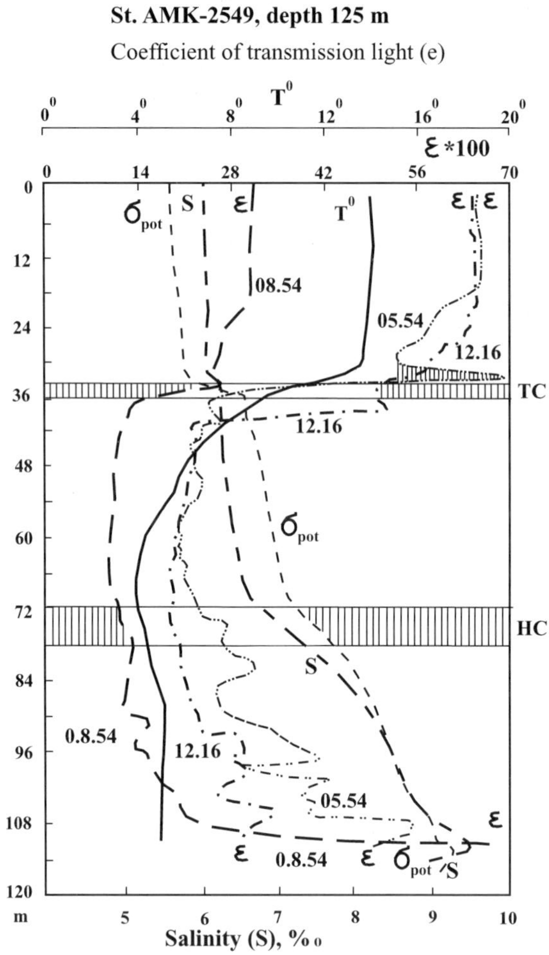

Fig. II.1.19. Vertical structure of the water column in region of the Stolpe underwater river fordelta, South Gotland Basin of Baltic Sea: Distribution of temperature (T°), salinity (S‰), density (σ_{pot}) and coefficient of transmission light (ε) for different times of day (05:54; 08:54 and 12:16) (station AMK–2549, depth 125, 55°53′80N, 18°42′60E.
TC—thermocline; HC—halocline.
Maximum of transmission light on curve ε is shown for 05:54 over the thermocline. It is proposed that this maximum is formed by phytoplankton.

in their central, deepest parts. This indirectly points to the presence of strong bottom currents in the deeps, in spite of periodic stagnation. The presence of such currents was also confirmed by immediate observations aboard the Mir submersible (Emelyanov et al., 1996, 1994): in the northern Baltic, on the slopes of deeps and banks, and around lumps of granite and gneiss, "rings" of nondeposition have been observed with "tails" of mud ridges (lumps) along the direction of the bottom currents.

Major contributing factors in maintaining the balance of the Baltic Sea are as follows: river runoff (430 $km^3 \times y^{-1}$), atmospheric fallout (230 $km^3 \times y^{-1}$), and evaporation (185 $km^3 \times y^{-1}$). Excess inflow over evaporation amounts to 475 $km^3 \times y^{-1}$. (Anonymous, 1986). This results in an abundant freshwater outflow from the Baltic Sea and produces strong deep currents of the saline (North Sea) water into the Baltic that, in turn, leads to the constant stratification of the water column and entry of oxygen into deep water due to space-convective mixing. The most effective mechanism of Baltic deep-water renovation is the periodic advective entry of a large volume of saline and oxygen-rich water from the North Sea, the so-called "main inflow of the Baltic". Inflows are conditioned by intensification of the atmospheric processes in the North Atlantic, North Sea and Baltic Sea, namely, the long-term, strongly zonal atmospheric circulation between the North Atlantic and Eastern Europe (Matthaus, 1995). As from 1880 to 1994 (excluding the period of the First and Second World wars) 111 inflows of saline Atlantic water occurred.

The saline North Sea water enters the Gotland deep from the Bornholm basin, through the latitudinally oriented Stolpe trench (Fig. II.1.19). The depth of this trench is 70–80 m, and the height of its banks (slopes) ranges from 20 to 30 m. These slopes are composed either of moraine or of overlying sands. Sand also covers the trench floor, and coarse aleurite occurs in its lows (depressions). This points to periodic bottom currents in the trench with a velocity of 100–150 cm/s.

There have been practically no direct velocity measurements of the bottom currents below the halocline level in the Gotland deep. Therefore, for in evaluating them, one author used a theoretical graph (Postma, 1967). According to this graph, a consolidated clay, here represented by moraine with water content less than 50%, is washed away at velocities not less than 80–100 cm/s. Unconsolidated clay, represented by Ancylus and Yoldia clays, with a water content of about 70%, is washed away when the velocity is 60–70 cm/s. However, loose clays with a water content of about 80% represented by Litorina muds, are washed away at velocities of approximately 50 cm/s. As well, transportation of particles with a size of about 1 μm occurs when the velocities are about 10–15 cm/s; for particles of 5 μm, the velocity is 15–20 cm/s; and for those of 10 μm, it is 25–30 cm/s.

Therefore, the southern "bank" (slope) of the Stolpe trench composed of moraine with erosion features is washed by currents with a velocity not less than 50–70 cm/s. Judging by the presence of two narrow valleys (with relative depths of 3–10 m) at the eastern end of the floor of the Stolpe trench, saline water flows out of the trench and into the southern Gotland deep in two streams: a strong northwestern and a weak southeastern one (Emelyanov and Gritsenko, 1999; Emelyanov, 2001). Because of the deep extension, as compared with the Stolpe trench, the current loses velocity (hydrodynamic barrier) and starts to unload sand, aleurite, and later, pelite (Figs. II.1.20, II.1.21; see also Fig. II.14.5) This is the unusual foredelta of the "underwater" Stolpe river. The pelite-size material transported via the right (southern) branch of the Stolpe trench is also discharged into this foredelta (settling basin). It reaches the branch from the Polish–Lithuanian side, where the main unloading of sedimentary material from currents occurs, including particles <0.001 mm.

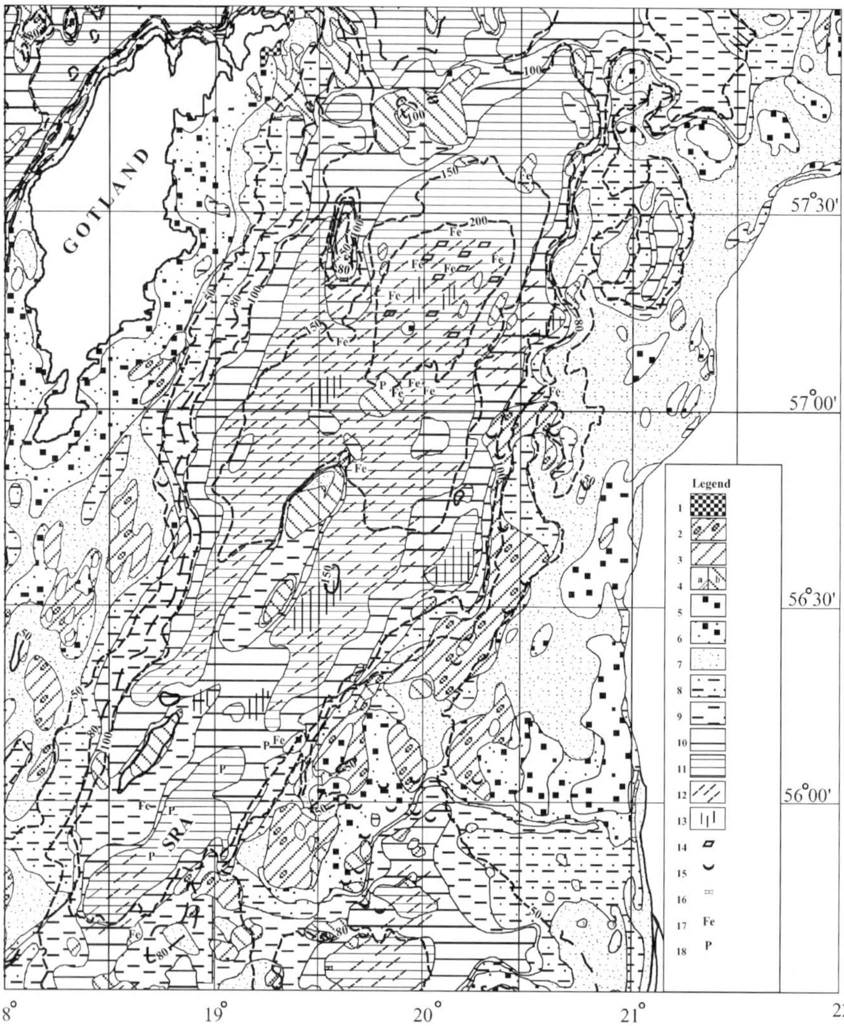

Fig. II.1.20. Bottom sediments of Gotland Basin (0–5 cm layer). Legend: 1—Bedrock (with a thin cover of sand and gravel); 2—Till (with a thin cover of sand, gravel and stones); 3—subaquatic loam and varved clay with a thin cover of recent mixed sediments (sand–gravel), BIL; 4—Clay (a) (Late Pleistocene or Early Holocene, Yoldia+Ancylus) with a thin cover of recent mud or sand (b); 5—Sand, gravel, stones; 6—Mixed, medium or coarse sand, locally with gravel; 7—Fine sand; 8—Coarse silt (aleurite) (>50 % of fraction 0.1–0.01 mm); 9—Fine silty (aleuritic) mud (>50 % of fraction 0.1–0.01 mm); 10—Mud (aleuro-pelitic) (50–70 % of fraction <0.01 mm); 11—Clayey (pelitic) mud (>70 % of fraction <0.01 mm); 12—Sapropel-like mud (3–5 % Corg); 13—Sapropelitic mud (5–10 % Corg); 14—$MnCO_3$ (rhodochrosite); 15-FMNs and crusts; 16—Sediments with hightened (up to 10–20%) of $CaCO_3$; 17—Sediments enriched in Fe (>6%); 18—Sediments enriched in P (>0.2% P). SRA—Stolpe River's foredelta.

At a depth of 75–80 m, at the Stolpe River mouth, there are no muds; they begin to deposit at depths of 80–90 m on the hard moraine substrate. The thickness of marine muds here is about 10–100 cm. However, at a distance of 10–20 miles from the mouth of the Stolpe trench, the thickness increases up to 400 cm, and in some small areas it attains 500 cm, or even thicker than 590 cm (Fig. II.1.20; cf. Fig. II.14.5). This is the greatest thickness of marine Holocene muds (Hl_{2-3}) in the entire Gotland deep. The rate of mud accumulation (0.8–0.9 mm/yr) here is nearly twice as high as the sedimentation rate in the deepest part of the Central Gotland deep.

The upper sediment layer (0–10 cm) is represented by semiliquid (water content 70–85%), reduced, dark greenish-grey mud. Below 10 cm, the water content somewhat decreases (to 60%) and mud becomes soft and plastic, but the colour remains the same—dark greenish-grey, in places, with black hydrotroilitic spots or interbeds. The redox potential Eh is usually negative, up to −200 mV. The mud is well saturated with gas, mainly methane (NH_4), and contains free H_2S. Craters of mud volcanoes or pockmarks, expressed as significant hollows in the bottom surface, occur at sites of gas seeps. Within the mud field of the Stolpe River foredelta, they number more than 300 (Blashchishin et al., 1990). The pockmarks are valley-like and sinuous in shape. The length of some pockmarks are up to 5–20 km; in width, 50–100 m; depth, 1–5 m (from the bottom surface). Gas flux from such pockmarks into the bottom water is very high.

Muds contain from 1.60% to 2.86% C_{org}; moreover, high C_{org} are traced throughout the sequence of marine muds (Tables II.1.13–II.1.15). Along with organic matter, muds are enriched with N and contain on average 0.12% P—1.5 times more than the marine mud from the Gdansk basin. Sediments of the southern Gotland deep are characterized by mean (clarke) concentrations for reduced muds of the Baltic of Fe, Mn, Ti, K, Na, Mg; of trace elements Ba, V, Sn, Cr, Li, Rb, Z; and somewhat increased average contents of Zn, Cu, Ni, and Co (Table II.1.15).

However, the presence of sapropel-like laminated muds in many cores of the southern Gotland deep, usually forming in periodically stagnating deeps, allows us to assume that hydrogen sulfide appeared periodically in the bottom water of local depressions within these deeps.

A distinguishing feature for muds of the underwater Stolpe River foredelta are pillowlike forms (clinoforms) composed of mud, which are caused by the effects of the near-bottom contour current (Fig. II.1.21). To the left of the bottom current stream, suspended matter from the flow occurs; under the current and to the right of it, erosion of moraine and nondeposition of sediments take place.

The hydrodynamic conditions in the muddy part of the Stolpe River foredelta are rather active. Numerous erosion channels occur on the mud field surface. Many mud volcanic craters are oriented southwest to northeast, i.e., by the direction of the bottom current (Blazhchishin et al., 1990). At the bottom of the foredelta there is an almost constant, well-expressed light dispersion layer related to increased suspended matter content near the bottom (Table II.1.16, Fig. II.1.19). At test area III AK-44 (depth 110–130 m) located in the eastern part of the foredelta of the Stolpe River, stagnation phenomena at the bottom are not observed. The bottom water commonly contains oxygen. Evidence of this is the brown (oxidized) film on the mud surface, despite the fact that under the film (likely several millimeters thick), black

Table II.1.16. Concentration of PSM ($\mu g \times l^{-1}$) oxygen, dissolved forms of nitrogen and phosphorus (µg-at/l) and dissolved forms of heavy metals ($\mu g \times l^{-1}$) in the southern part of the Gotland Deep (station AMK-2549, depth 125 m)

Horizon, m	PSM, $\mu g \times l^{-1}$	O_2	N_{min}	N_{org}	SiO_3	$P_{Ban.}$	$P_{org.}$	Cu	Zn	Cd	Pb	Hg, $ng \times l^{-1}$
1	0.49	6.83	2.03	19.13	7.6	0.69	0.52	2.40	8.20	0.43	2.33	1
5	0.53	6.78	2.45	19.09	7.8	0.75	0.52	2.00	5.70	0.32	1.15	5
10	-	6.86	2.11	20.47	7.5	0.79	0.76	-	-	-	-	-
15	0.60	6.72	2.13	20.73	7.5	0.84	0.65	2.10	8.33	0.23	2.21	6
20	0.85	6.62	1.83	20.77	7.7	0.65	-	-	-	-	-	-
25	-	6.65	1.96	20.37	7.8	0.68	-	-	-	-	-	-
30	0.31	6.63	2.07	20.46	7.7	0.63	2.96	8.20	0.13	1.80	-	-
34	0.47	6.57	2.61	22.27	9.3	-	-	-	-	-	-	-
50	0.10	6.69	3.66	17.24	12.0	0.97	0.63	1.90	6.05	0.70	2.30	-
67	0.19	6.08	-	-	-	1.52	-	4.60	12.5	1.56	2.30	2
75	0.21	6.5	3.16	20.58	14.1	1.55	0.85	2.86	10.8	0.30	1.30	7
84	0.31	-	-	-	-	-	-	1.41	5.03	0.24	0.53	1
100	0.43	2.64	9.05	19.93	27.7	2.31	0.98	1.80	19.2	0.25	1.70	-
106	18.63	2.73	-	-	-	2.43	6.81	3.68	10.0	0.50	3.50	3
117	1.60	3.34	9.90	21.99	27.3	3.12	1.79	2.70	30.0	1.43	2.40	-

For salinity, density and other parameters see Fig. II.1.15.

contributing to (1) mechanical differentiation of suspended matter according to grain size of particles, (2) chemical (coagulation) and biological conditioning of very fine particles in water suspended matter and solutions, and (3) purification of river waters from many of the pollutants entering the sea or ocean from rivers. According to some estimates, only 102 days are required for zooplankton to filter the total volume of all the rivers of the world! (Bogorov, 1974; Lisitzin et al., 2001). Beyond the limits of the river–sea GBZ, seawater becomes clear, in contrast to river water. This is especially true for areas neighboring the mouths of rivers flowing into the Arctic Ocean (Kassens et al., 1999; Lisitzin et al., 2001).

The river–sea GBZ is the main factor for sedimentary material in the world. Sedimentary material prepared in drainage areas is packed together, concentrated, and enters the narrowest region between the drainage area and sea—a river mouth; then, having rushed past all barriers until the end of its compressed path, it is redistributed, but already according to other (in comparison to drainage) laws, characterized only for border areas of seas and oceans. Among a large number of narrow passages, river–sea GBZs play the most important role (by virtue of the amount of transported sedimentary material); in all, there are about 12 giant rivers (including the Amazon and Congo) with suspended matter discharge of more than 60–100 million tons per year. Namely, at the estuaries of these giant rivers about 30% of the entire terrigenous matter at the first level (at the shelf) of the earth's avalanche sedimentation accumulates (Lisitzin, 1988, 1991, pp. 175–176).

At estuaries and also in the ocean (sea) close to the mouths of rivers a large variety of sedimentation schemes have been observed, but in all instances there is a characteristic increase in the rates of sedimentation, formation of huge, in terms of thickness and area, sedimentary fans and avalanche accumulation of terrigenous muds. Only in rivers discharging directly into a shallow sea (for example, at the mouths of the Neman and Neva rivers, Baltic Sea) sedimentary fans are either not formed or are of insignificant size. The schemes of distribution of bottom sediments near estuary and mouth areas are different. In the Baltic Sea, for example, the sediments are represented by sands and aleurites; at a river mouth like that of the La Plata, the following types of sediment successively accumulate: sands (at the upper part of the La Plata), aleurites, estuary muds, and sands of the open ocean shelf (Emelyanov et al., 1975). At the shore of the Amazon, terrigenous brown muds are distributed; at the shelf edge, grey terrigenous muds occur; the middle part of the shelf is a nondeposition area.

Terrigenous (river) sediment plays an important role in ocean sedimentology, especially in the Atlantic Ocean. A.P. Lisitzin (1974, 1978, 1986, p. 53) writes that only 7.8% of terrigenous material delivered to the river–sea barrier reaches pelagic areas of the ocean (depths greater than the 3000-m isobath). However, the author believes (Emelyanov, 1982_1) that the value of 7.8% may be characteristic only of the Pacific Ocean. For the Atlantic Ocean it is too low (Emelyanov and Trimonis, 1977; Trimonis, 1995): riverborne terrigenous sediments have accumulated here—and in pelagic areas, even up to the flanks of the Mid-Atlantic Ridge. Even beneath the CCD level (4700–6000 m) either terrigenous material predominates in the content of pelagic clay (North American and Angola basins), or the muds as a whole are terrigenous (Argentine Basin) (Figs. II.1.22, II.1.23).

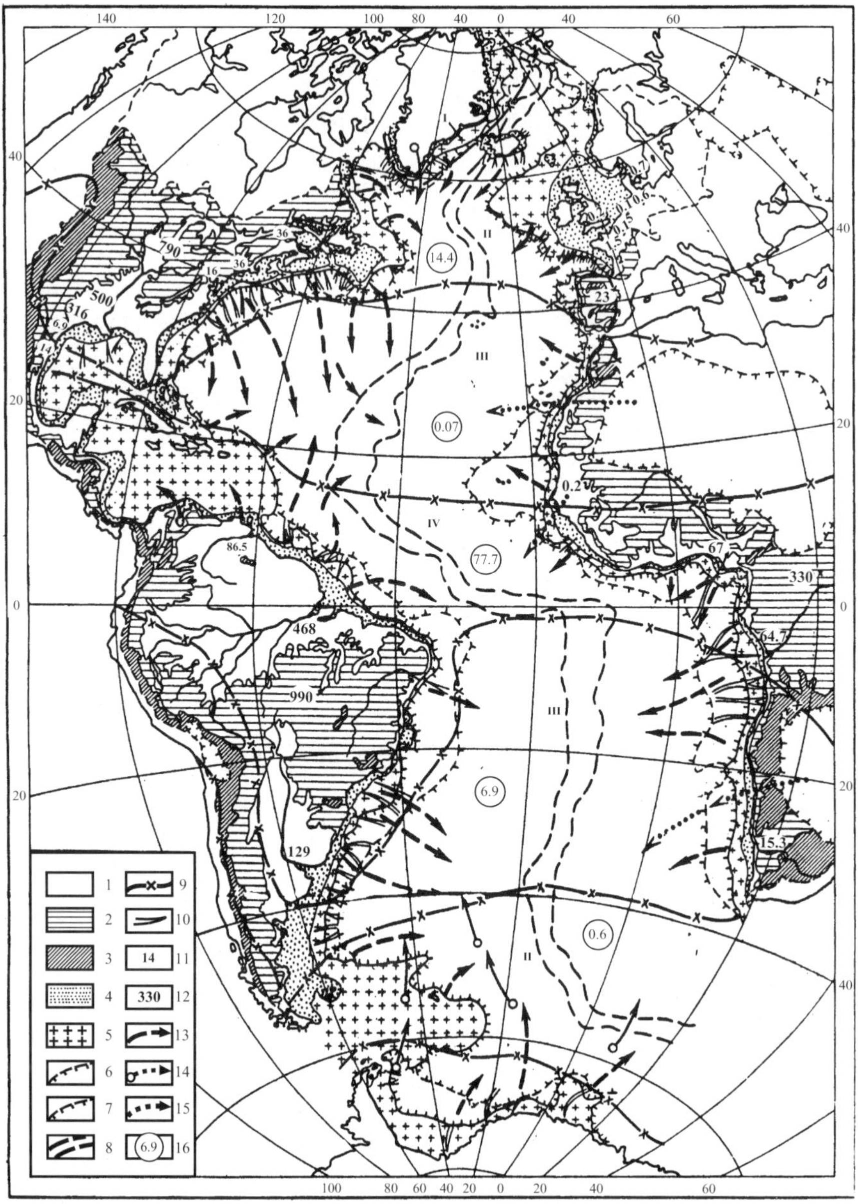

Fig. II.1.22. Intensity of mechanical denudation of drainage area of Atlantic Ocean and modulus of accumulation of river sedimentary material in different climatic zones. After Emelyanov and Trimonis, 1977.

1–3—heights on continents, m: 1—0–200, 2—200–1000, 3—>1000; 4—shelf; 5—zone of continental slope, including seas of transition zone; 6—foot of continental slope; 7—accumulative plains; 8—border of crest of Mid-Atlantic Ridge; 9—borders of climatic zones; 10—canyons; 11—river solid load, 10^6 t/yr; 12—mass of sedimentary material washed from each continent, 10^6 t/year; 13—paths of supply of main part of riverine material; 14—paths of supply of main part of glacial and iceberg material; 15—paths of supply of aaeolian material; 16—average modulus of accumulation of sedimentary material in each climatic zone, t/km²/yr. I–IV—climatic zones: I—ice (north and south); II—moderately humid (north and south);III—arid (north and south) and IV—equatorial humid.

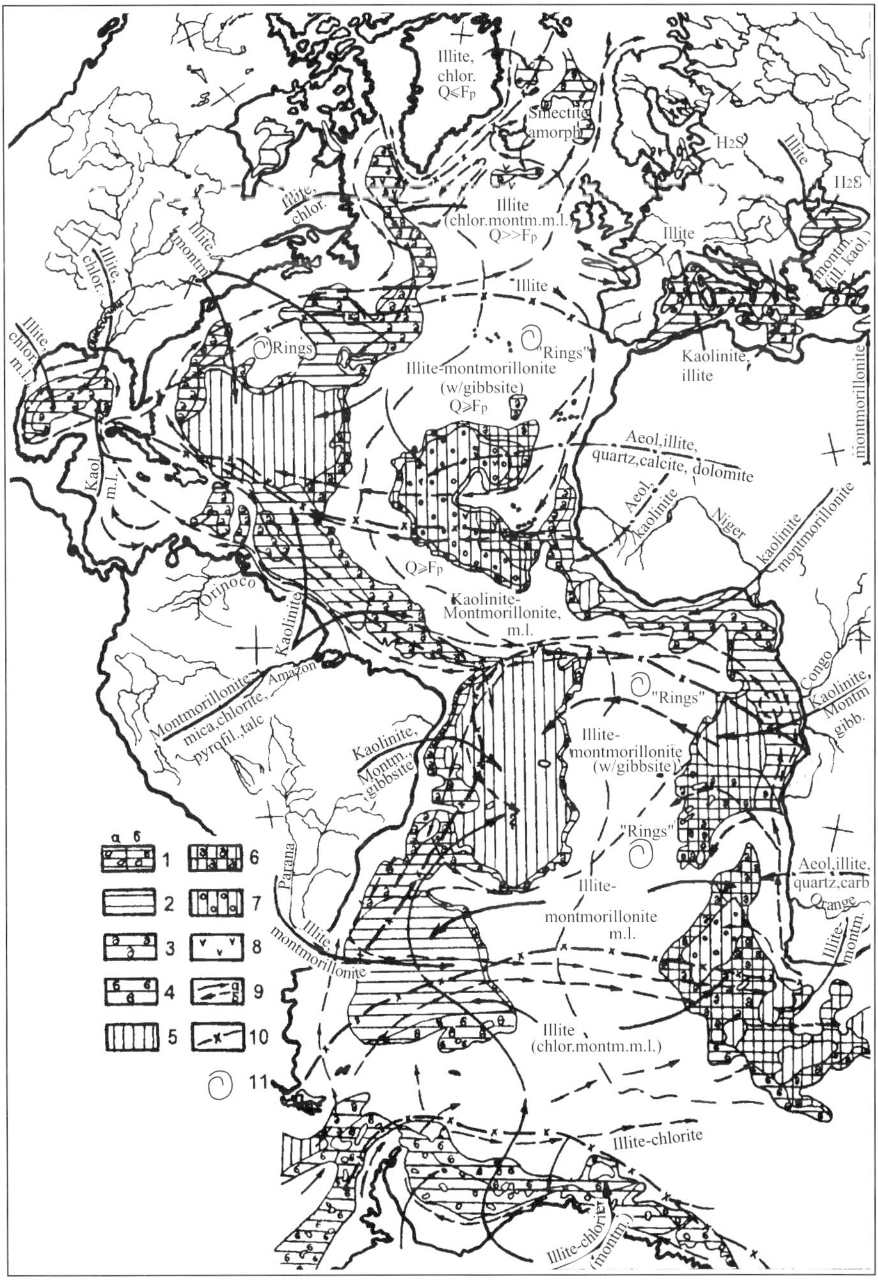

Caption see page 134

The main transport pathway for the riverborne load on its way across the limits of the river–sea barrier is not the surface water layer, but the near-bottom water layer, which is responsible for the transport of a major portion of suspended matter into shelf areas, and even beyond its limits (more than 30–50% in the Atlantic Ocean) (Emelyanov, 1968_1; Emelyanov and Kharin, 1974; Lisitzin et al., 1975; Emelyanov et al., 1987; Aibulatov, 1990). As a result, in large rivers, starting mainly in the Miocene (i.e., for an approximate period of 15 million years), sedimentary fans with thicknesses of about 10–15 km accumulated; and in the Ganges and Brachmaputra rivers, up to 19 km (Lisitzin, 1982, 1988, 1991; Konyukhov, 1983).

Therefore, estuaries and continental areas, including pelagic (situated at depths of 5–6 km) areas of the World Ocean close to large rivers are geomorphological and hydrodynamic traps of river terrigenous material. 1.729 billion tons of terrigenous material per year (Lisitzin, 1977) penetrate these traps; according our data, the amount is much larger. Only the dissolved part of riverborne material penetrates the pelagic areas with practically no loss (Lisitzin, 1986, p. 54). This is the basis for the appearance of biogenic sedimentary material in the open ocean; i.e., it provides for the appearance of a giant biogenic province on the ocean surface. In the ocean, in the photic zone, about 110 billion tons of biogenic matter is produced annually. However, according to calculations (Lisitzin, 1986, pp. 34–43) only 37% of the SiO_{2am} produced (or 0.172 billion tons) precipitates to the bottom of the World Ocean and is fixed in sediments; only 79.4% of biogenic $CaCO_3$ produced (or 1.079 billion tons); only 0.4% of the organic carbon product at the photic layer (i.e., 0.085 billion tons C_{org}, or 0.16 billion tons of organic matter). In all, in such a way, the pelagic sediments of the World Ocean retain 0.172 + 1.079 + 0.160 = 1.411 billion tons of biogenic material per year; of terrigenous material, 1.729 billion tons per year. The relationship of terrigenous and biogenic materials retained in pelagic sediments of the World Ocean changes radically: it is approximately 1 : 2.

In addition to terrigenous and biogenic matter, as well as anthropogenic matter, river arteries supply large quantities of ore elements. As a result, they create favorable conditions for the formation of iron pseudoolitic ores under the same conditions (usually those of the shelf close to river estuaries); in others—usually at the very edge of the distant, deepest-sea areas of the fans of transported materials—for the formation of iron–manganese nodules (see part III).

Fig. II.1.23. Flux of river's terrigenous material and its distribution in Atlantic Ocean. 1–6—types of mud: 1—mixed marine-glacial (iceberg) and clayey (illite–chlorite): (a) without considerable amount of amorphous silica (<10% SiO_{2am}), (b) with considerable amount of amorphous silica (10–30% SiO_{2am}); 2—clayey (<30% $CaCO_3$); 3—nano-foraminiferal-clayey (30–50% $CaCO_3$); 4—diatomic–clayey (10–30% SiO_{2am}); 5—pelagic red clay (oxidized) (<10% $CaCO_3$); 6—mixed pelagic foraminiferal–clayey (10–50% $CaCO_3$); 7—considerable admixture of aeolian material in red clay; 8—admixture of pyroclastic; 9—sources (a) and path (b) of terrigeneous–pelitic material; 10—boundaries of climatic zones of sedimentogenesis (see Fig. II.1.18.); 11—circulation of water (cyclonic or anticyclonic rings) (see Fig. II.3.5.). Minerals: Q—quartz; Fp—feldspar; m.l.—mixed layered; chlor.—chlorite; mont.—montmorillonite.

Strong near-bottom currents play an important role in the redistribution of sedimentary materials in the sea (or ocean)—in manner different from underwater rivers. At places where the velocity of the current of these "rivers" fades away, vast sedimentary fields (foredeltas) are formed. This type of foredelta forms clearly at the exit of the near-bottom current from trenches and canyons, as is observed, for example, at the estuary of the underwater Stolpe River in the Baltic Sea. Similar phenomena have also been observed in oceans where large sedimentary swells and drifts are formed (with a thickness of up to several hundred meters): strange deltas (see Part II.14).

Coast-Sea

Coasts are divided into abrasive and accumulative. Abrasive coasts mainly undergo destruction, while accumulative coasts increase at the expense of the accumulation of new portions of sedimentary matter supplied from abrasive and erosive sections of the coastal zone. The wider the shelf, the less sedimentary matter flux from the coast into the open sea. The flow of water through the coast–sea GBZ plays a significant role in processes such as the mechanical separation of terrigenous matter, the transportation of dissolved matter and the dispersion of organisms. Also, this flow is of great importance in controlling the mechanism, force and flow of energy (food) through food chains (Riedl and Ott, 1982). These aspects are of vital importance for porous meiofauna, i.e., for the animals inhabiting the pores of sediments between particles. The "biological contour of natural waters" (see part I.5) plays a very important role in the biological separation of chemical elements in the coast–sea GBZ.

Between the water line (usually in nontidal seas) and lower border of the second mechanical barrier, sandy sediments accumulate; at places with washout, residual (or relic) deposits occur, commonly represented by boulders, pebbles and gravel. In sediment sections, these deposits occur as sedimentary bodies (including bars) that extend along the coast. Parallel lamination, cross lamination, and ripple structures commonly characterize these sediments. Placers of titanomagnetite, rutile, monazite, zircon, garnet, and other heavy minerals are also characteristic features in such sediments. It has been suggested (Blazhchishin, 1994) that high hydrodynamic activity in the lower zone of the coast–sea GBZ in some cases results from an interaction the of seasonal thermocline with the sea bottom. At the depth where the thermocline touches the bottom (at a depth of 30–40 m in the Baltic Sea), internal waves break. These waves presumably give rise to jet streams; in the Baltic Sea they approach a velocity of about 30–50 cm $\times$ s^{-1}. Since the presence of a thermocline of internal waves and benthic jet streams operate almost constantly, the lower border of the coast–sea GBZ should be hydrodynamically very active. This contributes to the conservation of sand and gravel on the bottom surface.

In the ocean, the role of the thermocline is less clear. However, similar to the Baltic Sea, it can be assumed that on oceanic shelves there is a specific interconnection between the thermocline, internal waves, jet streams, and fields of relic sands (Emery,1968; Emelyanov et al.,1975).

Area of the Sambian Peninsula, Baltic Sea. The abrasive coast of the Sambian Peninsula in the Baltic Sea was subjected to detailed research of the coast–sea GBZ (Aibulatov et al., 1984, pp.62–68). On three transects perpendicular to the coast, samples of sediments (suspensions that accumulated in special traps) were taken at stations at a depth of about 3.5 m (zone of strong wave deformation before the breakwater), 7.0 m (zone of wave deformation) and 10 m (zone of weak wave deformation) (Figs. II.2.1, II.2.2, II.2.3). Moreover, at each of these stations samples of the suspension were taken at distances of about 0.2, 0.4, 1.0, 1.5, and 2.0 m from the bottom. Each of the selected samples was then exposed to granulometric and mineralogical analyses. On the coasts of the examined section of the sea, Quarternary moraine loam with sand and gravel lenses crop out. These sedimentary rocks are actively washed off and serve as sedimentary material sources for the adjacent section of the sea. Beyond the limits of the 5- to 10-m isobaths, sands, boulders, pebbles, or coarse-grained terrigenous sands are found almost everywhere in sediments (Fig. II.2.2). At the coast itself (and on the beach), and in the sea up to a depth of 10 m well-sorted medium- and fine-grained sands are distributed. The thickness of friable (Holocene) sediments is less than 1 m. Locally they are absent altogether. As a whole, distribution of sediments is very mosaic. Fields of fine-grained sands with an admixture of gravel alternate with small fields of coarse aleurite or fine-aleuritic muds.

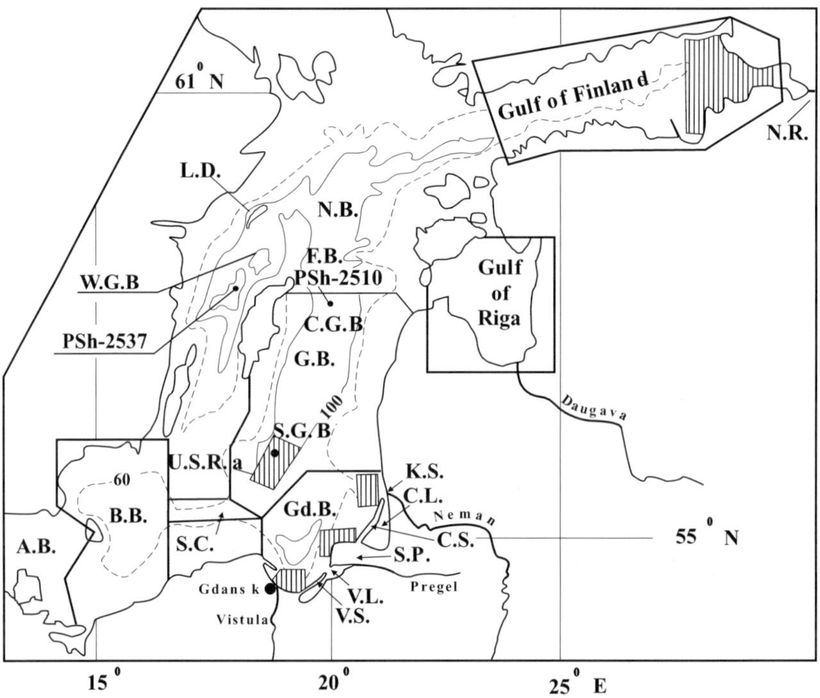

Fig. II.2.1. Map of studied areas of Baltic Sea.
A.B.—Arkona Basin; B.B.—Bornholm Basin; S.C.—Stolpe Channel; Gd.B.—Gdansk Basin; G.B.—Gotland Basin; S.G.B.—South Gotland Basin; C.G.B.—Central Gotland Basin; W.G.B.—Western Gotland Basin; N.B.—North Baltic Basin; F.D.—Farö Deep; V.L.—Vistula Lagoon; V. S.—Vistula Spit; C.L.—Curonian Lagoon; C.S.—Curonian Spit; S.P.—Sambian Peninsula; K.S.—Klaipeda Strait; L.B.—Landsort Deep; N.R.—Neva River; U.S.R.F.—underwater Stolpe river's foredelta. Isobaths 60 and 100 m are shown as well as locations of geological stations.

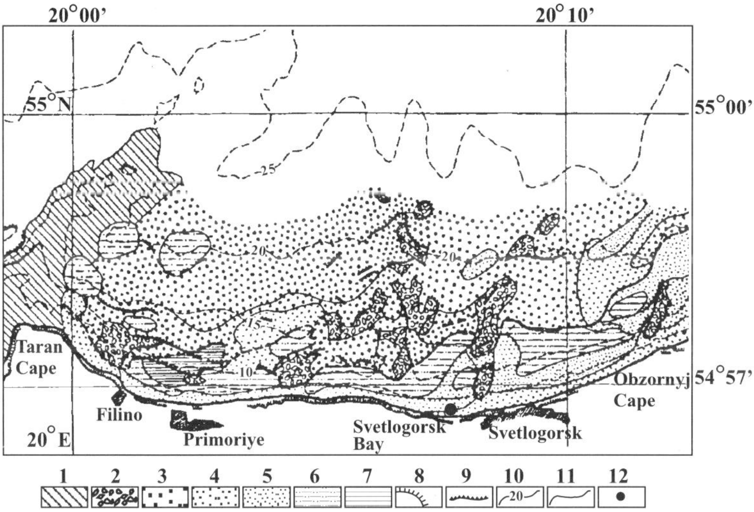

A. 1 - bedrocks; 2 - stone-pebbly pavements; 3 - coarse sand; 4 - medium sand; 5 - fine sand; 6 - coarse aleurites; 7 - fine-aleuritic mud; 8 - abrasional shores; 9 - submarine cliffs; 10 - isobaths; 11 - border between recent and relict sediments; 12 - test area of detailed studies (see Fig. II.2.3.).

B.

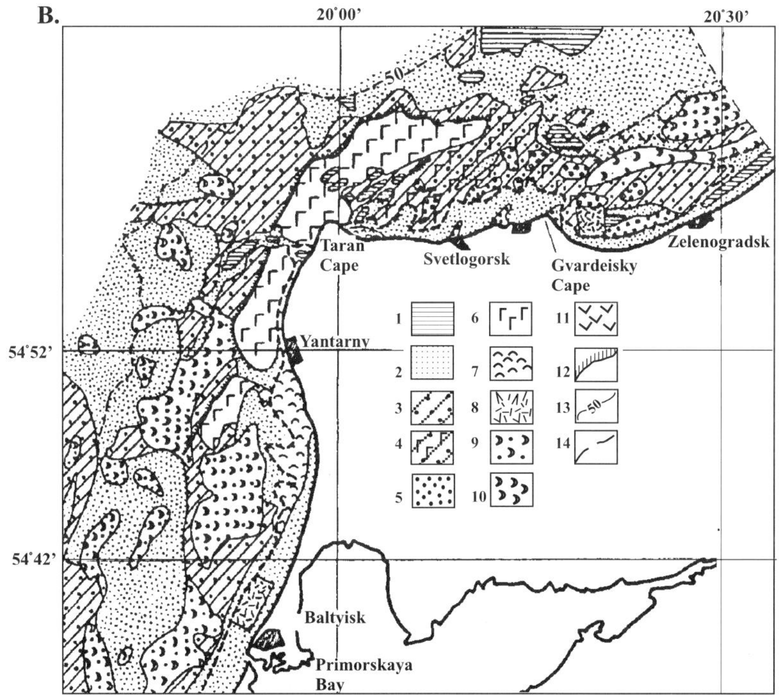

Caption see page 140

An increase in the area of a stony bottom leads to a general occurrence of thinner particulate matter and a reduction in the content of the sand fraction (Aibulatov et al.,1984, p. 6). A characteristic feature of the area studied (Fig. II.2.2) is that the amount of particulate matter in the water strata decreases with distance above the bottom. Consequently, two layers can be distinguished: the benthic layer (1–2 m above the bottom surface), rich in particulate matter; and the main layer, which is not so rich in PM.

Analysis of the results made it possible to discover the following regularities:

1 The quantity of the bulk suspension at the studied profiles is maximum at a layer 0.2–0.4 m from the bottom. As well, it is a universal regularity, confirmed by the data at all the stations of the three studied sections in the Baltic Sea; it has also been confirmed in other seas (see Fig. II.2.3);
2 Contents of fractions which are 0.1–0.25 mm, 0.25–0.5 mm, 0.5–1.0 mm and >1.0 mm in size are largest in the near-bottom water layer (0.2–0.4 m above the bottom); the 0.05- to 0.01- and <0.01 mm fractions are abundant at a layer higher than 1 m above the bottom. In Svetlogorsk Bay (Fig. II.2.2A), a regularity of somewhat different kind was observed: the concentration of particulate matter mostly decreases as depth increases. So, for example, the content of the >0.1 mm fraction in the suspension was maximum at a distance of about 1.5–2.0 m above the bottom, but not immediately over the bottom. The concentration of the suspension over a soft bottom was always 4–6 times greater than over a clumpy boulder bench (Jindarev, 1997, p.12).
3 The maximum quantity of heavy minerals in the 0.1- to 0.05-mm fraction was fixed in the layer 0.2–0.4 m from the bottom, above which it suddenly decreases.
4 The correlation of light and heavy minerals of the suspension in the samples from different horizons of the same station (at depths of 3.5, 7.0 and 10.0 m) remains practically the same. The insignifcant changes in content oberved by us appear to be more accidental than regular. A certain increase in black-ore minerals (ilmenite–magnetite), siderite, rutile and zircon at the very bottom can be observed only in some vertical sections.
5 In terms of the mineral composition and correlation of minerals, the studied suspension and bottom sediments (at these same stations) differ substantially.

Fig. II.2.2. Bottom deposits of submarine slope of northern part of Sambian Peninsula (A) and facies of submarine slope of Sambian Peninsula (B), Baltic Sea. After Blazhchishin, 1998. A. 1—bedrock; 2—stone-pebbly pavements; 3—coarse sand; 4—medium sand; 5—fine sand; 6—coarse aleurites; 7—fine-aleuritic mud; 8—abrasional shores; 9—submarine cliffs; 10—isobaths; 11—border between recent and relict sediments; 12—test area of detailed studies (see Fig. II.2.3). B. 1—coarse aleurites; 2—sand; 3—coarse deposits; 4—sandy-gravelly deposits with pebbly phosphorites; 5—stone-pebbly pavements on moraines; 6—bedrock; 7—anthropogenic deposits (mainly sand); 8—dumpings; 9—sandy waves; 10—sunken dunes; 11—peat; 12–submerged Ancylus–Litorina cliff; 13—50-m isobath; 14—border between recent (near-shore) and relict deposits.

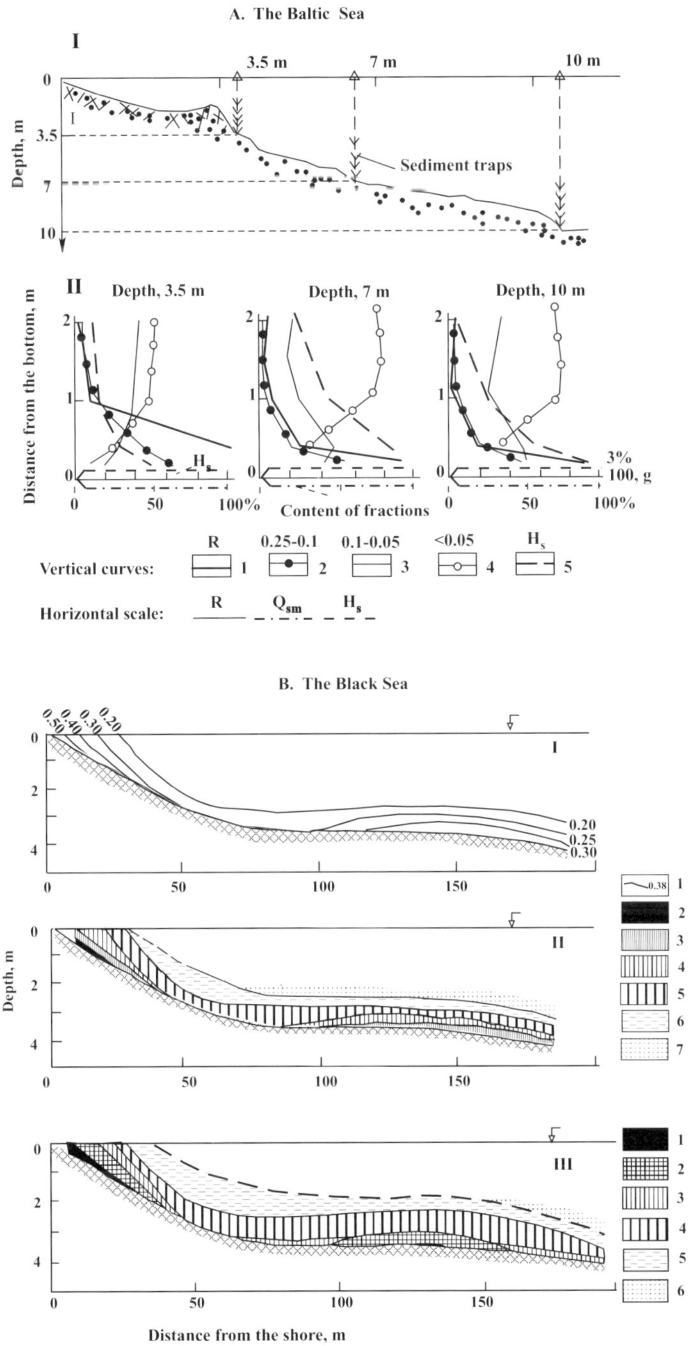

Caption see page 142

6 Increase or decrease in amounts of individual minerals on the three studied profiles is explained by different concentrations of these minerals in source areas (the nearest parts of the shore or bottom surface where washout of sedimentary rocks occurs).

From the above data and that of early publications (Emelyanov, 1968_1; Aibulatov et al., 1984; Jindarev, 1997) other conclusions have been drawn that are very important for the study sedimentation processes.

7 The major part (recalculated for water volume) of suspended matter is concentrated at the littoral zone up to a depth of about 5 m.

8 The main route of migration of sandy (0.1–0.5 mm fraction) and coarser fractions of sediments (>1.0 mm) falls within the benthic layer, the thickness of which is 0.5–2.0 m. At this same layer, migration of the main portion of heavy minerals of finer fraction (0.1–0.05 mm) occurs.

At depths of 3.5–15 m, sedimentary material moves mainly in the water layer, about 1.5 m from the bottom (Jindarev, 1997). The main direction of this movement is like rip currents: from the coast towards the open sea. In the littoral zone, which is 3.0–3.5 m deep, only a small amount of friable material migrates alongshore, and only in a 0.2-m-thick near-bottom water layer. This means that rip currents evacuate large volumes of beach-forming material from the waterline zone, leading thus to its exclusion from the sediment budget of the coastal zone. Consequently, within a sandbank coast with an irregular coastal line—just like the coast of Sambian Peninsula—there are no conditions for the extended, unidirectional alongshore movement of loose material. Alluvial evacuation of sand-sized material by rip currents from the waterline zone to greater depths is the dominant process there (Jindarev, 1997, p. 24). Another characteristic feature is that rip currents can be traced at a distance considerably farther than submarine bars, at depths of about 10–15 m or more. Extensive sandy fields, which are restricted to the central part of bays (see Fig. II.2.2A) are markers of rip currents routes. Benches and fields of coarse

Fig. II.2.3. Mechanical separation of sedimentary material at shore–sea GBZ near northern part of Sambian Peninsula (Svetlogorsk Bay), Baltic Sea (after Aibulatov et al., 1984, p. 64) and at Kamchia test area, Black Sea (after Kosyan and Pyhov, 1991).
A. Baltic Sea. I. Bottom relief on submarine profile (for location see Fig. II.2.2.A) and three bottom stations (at depths 3.5; 7.0 and 10.0 m) with sediment traps (located at 20, 40, 100, 150 and 200 cm from bottom).
II. Distribution of bulk suspended matter (in g per 6 days) (R), content of 0.25–0.1 fractions; 0.1–0.05 and <0.05 mm (in %) (Q_{sm}) and content of heavy subfraction in 0.1–0.05 mm fraction (in %) (H_s).
B. Black Sea. Average size of suspended particles (I), bulk concentration of suspended matter on shore slope during moderate roughness of sea (II) and expenditure (load) of bulk suspended material in lower 1.5-m layer of water column (III) in Kamchia test area.
I. 1—isolines of distribution of average size of suspended particles (fractions in mm).
II. 2—7—concentration of suspended matter, in g/l: 2—>5; 3—5–1; 4—1–0.5; 5—0.5–0.1; 6—0.1–0.05; 7—<0.05.
III. Load (flux), g/cm²/s: 1—>0.1; 2—0.1–0.05; 3—0.05–0.01; 4—0.01–0.005; 5—0.005–0.001; 6—<0.001.

and medium-grained sands in the vicinity of capes separating one bay from another form as a result of convection of strong alongshore currents and wave breaking (see Fig. I.4).

9 Pelitic (<0.01 mm) and fine-aleuritic (0.05–0.01 mm) materials in the zone up to a depth of about 10 m under the conditions studied (see Fig. II.2.3) hardly accumulate at the bottom (coarse aleuritic material in this zone has been discovered in the form of a small lens in the hydrodynamic "shadowy" section of the shelf). This material is transported by alongshore and rip currents at the edge of the littoral zone, 0–10 m. In the formation of an aleurite field on the shelf, the branches of the alongshore stream plays a large role: it is as if they are repelled from the capes (capes Taran, Gvardeiskiy and others in the Baltic Sea) and are directed to the open sea (Emelyanov, 1968). Where the velocity of the current decreases, practically at the edge of the coastal zone, aleuritic material precipitates from the current and forms lenses of aleuritic sediments on the bottom. Pelitic material is transported even farther than aleurite: mainly, it is carried out in the benthic muddy stream (the layer 2–5 m above the bottom surface).

10 At the bottom, muddy material actively starts to accumulate, not at a specific distance from the coast, but at a specific depth depending on distance (Fig. II.2.4, see also Fig II.1.6). This depth usually coincides with the salinity–halocline transition layer (or pycnocline). This is where the third (hydrodynamic) barrier is manifested. Irregularity of abrasive and accumulative processes is connected to the existence of various alongshore streams and currents, each of which has a different kinetic energy and velocity (Zenkovich, 1962; Longinov, 1973; Aibulatov, 1990; Kosyan, Pykhov, 1991). However, there are many situations when transverse currents (more frequently, rip currents) together with alongshore currents and wave processes combine to cause either very strong abrasive processes or formation of thick accumulative bodies such as bars. A feature of such a type is the Curonian spit, which originates on the northern coast of the Sambian Peninsula and extends northward to a distance of about 98 km in the form of a needle, separating the Curonian lagoon from the Baltic Sea (see Fig. II.2.1). In the body of the Curonian spit, about 6 billion m^3 of sandy material has accumulated (Aibulatov and Jindarev, 1992). The genetic spit is a coastal bar. This bar was fed by sedimentary materials, which were formed at the expense of erosion of regions contiguous to the Curonian-spit section of the sea bottom (Leontyev et al., 1985). The Vistula spit, which separates Vistula Bay from the Baltic Sea, to a greater degree is also a product more of transverse than longitudinal transporation of sand detritus (Aibulatov and Jindarev, 1992).

The Kamchia–Shkorpilovets area in the Black Sea. The area is situated at the estuary of the Kamchia River south of Varna, Bulgaria. Here have been observed strongly expressed alongshore flows of suspended matter. Detrital deposits are transported principally in the 1- to 1.5-m benthic layer. The average sizes of particles and their concentrations in the presence of strong ($>1\ m \times s^{-1}$) currents increased in the wave breaking zone (see Figs. I.2, II.2.3). The consumption rate of

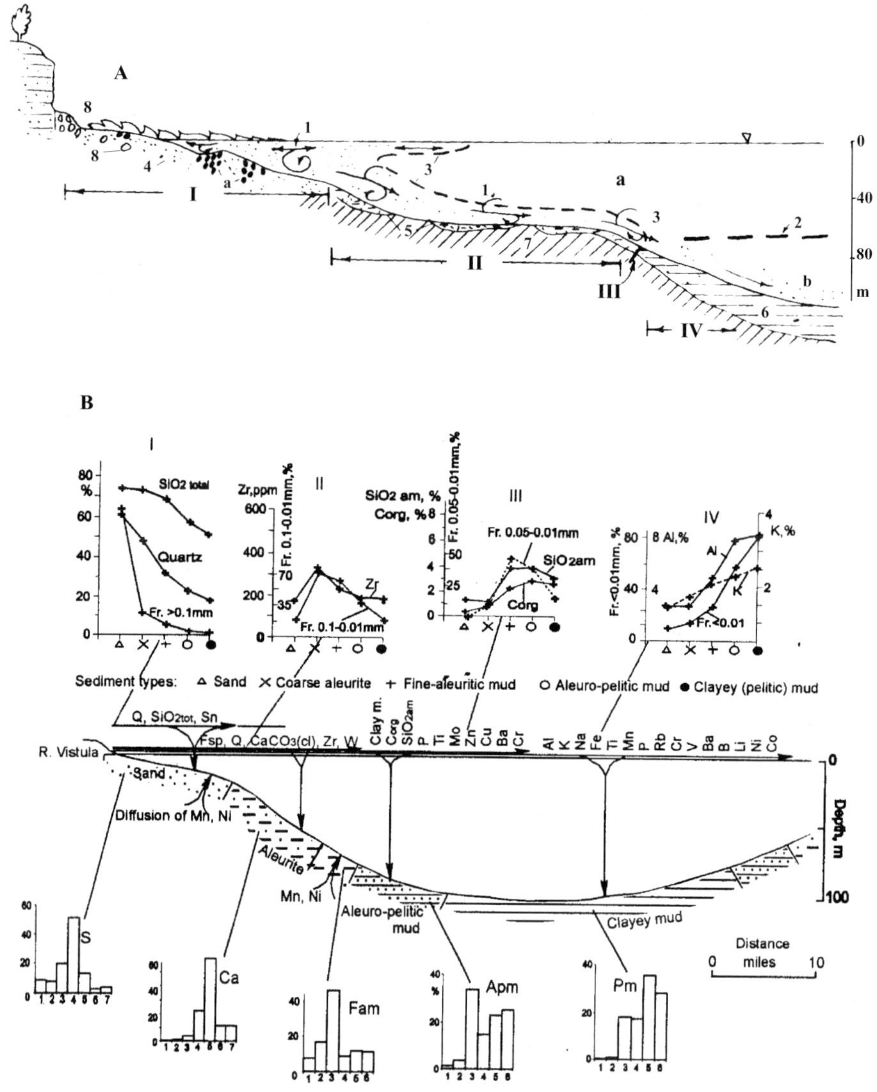

Fig. II.2.4. Position of mechanical (hydrodynamic) GBZs (after Emelyanov, 1981_1, 1982_1) and sedimentation on shore–sea cross profile in shallow sea (Baltic Sea).

A. I. GBZ shore–sea subzone (most intensive mechanical and mineralogical differentiation, formation of placers). Chemical components and elements distributed according rule of sandy fraction are accumulated.

II. Transit zone: strong near-bottom currents, sometimes erosion of bottom moraines (or clays), accumulation of aleuritic sediments, presence of relict sands. Accumulation of chemical components and elements distributed according rule of aleuritic fraction.

III. GBZ interface of clastic sediments replaced by clayey ones (third mechanical barrier). It usually coincides with a narrow area of fine-aleuritic mud and with the discontinuity layer (halocline, pycnocline).

IV. Area of clayey mud accumulation (hydrodinamically, a relatively calm zone). Accumulation of chemical components and elements distributed according rule of pelitic fraction.

sand detritus was maximum in this wave breaking zone (Kosyan and Pykhov, 1991, p. 208), and also in the surf zone, at depths of about 2 m: more than 0.12 gram × $cm^{-2} \times s^{-1}$.

The relief of the underwater slope here is complicated by submarine terraces of abrasive–landslide origin and swells (Kosyan and Pykhov, 1991, p. 220); outcrops of hard rocks are frequent. Bottom sediments are represented by sands which exist in the form of intermittent bands having thickness of 2–3 m. The predominant size of sand particles is 0.19–0.75 mm. The amount of transported sand is different for different parts of the underwater slope and ranges from 3.3×10^2 kg × s^{-1} (section IV) to 550 × 10^2 kg × s^{-1} (section I). The potential discharge of sediment is 4×10^3 kg × s^{-1}. It is precisely here where the fine sandy beaches of the Golden Sands resort were formed.

Sidra Bay (Mediterranean Sea) and the test area "Sirt"—the arid zone of the Eastern Mediterranean. Windstorms (Chili, Sirocco, Ghibi, Khamsin, Harmatan and others) carry millions of tons of aeolian dust from African deserts to the Mediterranean Sea. The concentration of dust in the air directly depends on the velocity of the wind and the altitude from the land surface. The maximum concentration of dust in the near-land layer (0–0.5 m) is up to 560 mg/1 m^3 and was observed during storms when the wind velocity was up to 20 m/sec^{-1}. In the area of test area II (Fig. II.2.5), above sandhills (barhans), the flux of dust in the layer 0.1–0.3 m above the land surface during a wind of 18 m/s was 560 mg/m^{-2}/sec^{-1}; from 1 to 6 m, only 5–10 mg/m^{-2}/sec^{-1}. In the foredune area (90 m from the coast) the concentration was for the most part the same (Aibulatov and Serova, 1983). The average flux at 0–0.5 m was 1389 mg/m^{-2}/sec^{-1}; at 0.5–10.0 m, 80 mg/m^{-2}/sec^{-1}.

There are about 40 stormy days per year on the Tunisian–Libyan coast (Aibulatov and Serova, 1983). The average flux of airborne dust at 0–10 m is up to 124.8 tons/km of coastline per day. The length of the Tunisian–Libyan coasts is about 1100 km. The total flux through this coastline in the 0- to 10-m layer is 5.5 million tons of dust per year.

Fig. II.2.4. *Continued*
1—strong waves, alongshore and near-bottom currents causing intensive mechanical differentiation of terrigenous matter; 2—layer of discontinuity (halocline, pycnocline); 3—interface of most intensive mechanical differentation of terrigenous matter; 4—8: 4—sand; 5—aleurite; 6—mud; 7—Pleistocene clay; 8—boulders, pebbles; in near-shore area, gravel and deluvium.
a—zone of intensive mechanical differentation of terrigenous matter and zone of phytoplankton development; b—zone of weakened mechanical differentiation of terrigenous matter and organic detritus dissolution.
B. Order of precipitation to bottom of minerals studied (Q—quartz, Fsp—feldspars) and chemical components and elements on river–sea profile (Gdansk Basin). I—IV rules of distribution of fractions (Fr.), minerals and chemical components and elements and order of precipitation: I—Rule 1 (rule of sandy fraction); II—Rule 2 (rule of aleuritic or silty fraction); III—Rule 3 (rule of aleuritic or silty fraction); IV—Rule 4 (rule of pelitic fraction). Diffusion of Mn, Ni and (in some cases) P, Mo, Fe, Zn, Co from interstitial water to near-bottom seawater. S, Ca, Fam, Apm and Pm—histograms of average granulometric composition of surface sediments: S—sand, Ca—coarse aleurite, Fam—fine-aleuritic mud, Apm—aleuro-pelitic mud, Pm—pelitic mud. Fractions on histograms for S and Ca: 1—>1; 2—1–0.5; 3—0.5–0.25; 4—0.25–0.1; 5—0.1–0.05; 6—0.05–0.01; 7—<0.01 mm; for Fam, Apm and Pm: 1—>0.1; 2 —0.1–0.05; 3—0.05–0.01; 4—0.01–0.005; 5—0.005–0.001; 6—<0.001 mm.

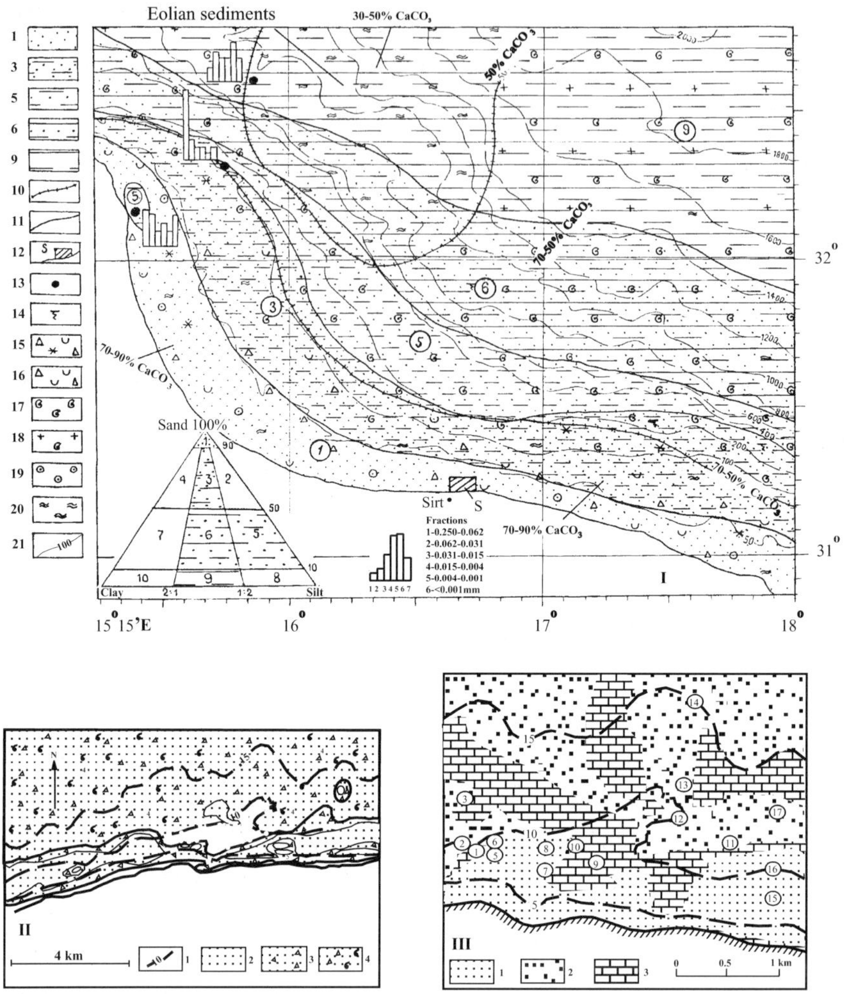

Fig. II.2.5. Recent sediment (0–5 cm) of region of Sidra Bay in Eastern Mediterranean Sea (I) (after Emelyanov et al. 1996.) and on two shallow test areas S near town of Sirt (II and III).
I. Types of bottom sediments (0–5 cm layer): 1—sand; 3—muddy sand; 5—sandy silt; 6—sandy mud; 9—mud (silt and clay); 10—isolines of 50 and 70 % $CaCO_3$ content; 11—borders of fields of sediment types; 12—test area S (see maps bellow); 13—stations, for which granulometric histograms (H) are given (fractions in mm); 14—chemogenic—diagenetic carbonate formations (magnesium–calcite concretions in mud); 15—sediments containing abundant bioclastic fragments (coquina, coral, algae); 16—sediments containing a considerable admixture of detritus (2—4 mm); 17—foraminiferal sand and silt; 18—coccolith–foraminiferal or foraminiferal–coccolith ooze; 19—oolithic, spherolitic sand; 20—aeolian sediments (pure or mixed with sediments); 21—isobaths.
II and III. Distribution of bottom sediments in two near-shore areas in desert environment with intensive flux of aeolian material (in area 100 km to east from Sirt in Sidra Bay) (for locations see test area S in map above). II—after Aibulatov and Serova, 1983, with some additions by the author; III—after Kosyan and Pyhov, 1991.
II. 1—isobaths, m; 2—terrigenous (aeolian) sand (>50% aeolian material, <30% $CaCO_3$); 3—carbonate shelly sand (50—70% $CaCO_3$) with admixture of aeolian material (30—50%); 4—carbonate shelly sand and shells (>70% of $CaCO_3$, 10–30% aeolian material).
III. 1—coarse aleurite and fine-grained sand; 2—medium and coarse sand; 3—calcareous plates (platforms) and crusts. Isolines are isobaths in m. Numbers in circles are search stations (see Table II.2.1).

The common grain-size distribution of airborne dust is pelitic, in which fractions >0.01 mm prevail (Fig. II.2.5). On the Libyan coasts in the 0.0- to 0.5-m layer, sandy (1–0.1 mm) fractions prevail; in the 0.5- to 1.4-m layer, sandy and muddy (0.1–0.05 mm) fractions prevail; in the 1.8- to 2.8-m layer, a coarse pelitic fraction (0.01–0.001 mm); in the 2.8- to 6.0-m layer fractions of 0.01–0.001 mm and less prevail. The clay fraction (<0.001 mm) makes up about 40% of total aeolian dust.

In the shelf area of the arid zone of the Eastern Mediterranean, biogenic carbonate sediments are widespread (Emelyanov, 1972; Emelyanov and Shimkus, 1986; Emelyanov et al., 1996): shelly (bioclastic) sands with an admixture of aragonitic pseudoolites and oolites (70–92% $CaCO_3$) (Fig. II.2.5).

Terrigenous aeolian sediments with an admixture of carbonates (30–50% $CaCO_3$) are distributed only in the near-coast areas of the Libyan coast (Fig. II.2.5). At test area III, between the waterline and 8- to 11-m isobaths, fields of sands were formed (Kosyan and Pykhov, 1991) (Table II.2.1). Fine-grained sands are usually well sorted and have a light-brown color. Deeper, mixed biogenic carbonate–terrigenous sediments with a content of 20–50% clastic aeolian components are distributed. At the shelf, deeper than 12–15 m, biogenic and mixed chemogenic–biogenic sediments with an admixture of 10–20% clastic aeolian components occur.

The main mineralogical airborne dust indicators are (Emelyanov and Shimkus, 1986):

Table II.2.1 The characteristics of the conditions of investigations in near-shore Region III in Sidra Bay, Mediterranean Sea (see Fig. II.2.5)

Site	Depth, m	Sediment type	Intervals of parameters of surface-disturbance state during investigation		
			h, m	λ, m	T, sec.
1	10.1	Rock	1.1–1.6	46–81	5.7–9.0
2	10.1	Medium sand	1.1–1.6	46–81	5.7–9.0
3	11.7	The same	1.1–1.4	46–59	5.7–6.8
4	14.0	Coarse sand	1.1–1.5	47–71	5.7–7.6
5	7.0	Medium sand	0.7–1.5	28–64	4.3–8.2
6	9.0	The same	1.1–1.6	46–77	5.7–9.0
7	8.0	Fine sand	0.8–1.9	18–64	3.2–8.2
8	9.0	Medium sand	0.8–1.9	18–64	3.2–8.2
9	7.0	Rock	0.8–1.9	18–64	3.2–8.2
10	8.5	Rock–medium sand barrier	0.8–1.9	18–64	3.2–8.2
11	10.5	Fine sand	1.1–1.6	46–81	3.7–9.0
12	11.7	Medium sand	1.1–1.6	46–81	5.7–9.0
13	10.4	The same	1.1–1.6	46–81	5.7–9.0
14	15.0	Coarse sand	1.1–1.6	47–82	5.7–9.0
15	7.0	Silty sand	0.9–1.5	39–64	5.4–8.2
16	10.0	The same	1.1–1.6	46–81	5.7–9.0
17	13.2	Medium sand	1.1–1.6	47–82	5.7–9.0

h – height of the wave, λ – length of the wave, T – time (duration) of the wave, sec.

1) carbonates: calcite, aragonite, dolomite;
2) clastic minerals: quartz, feldspars and mica (light minerals), iron hydroxides, ilmenite–magnetite, common horblende, rutile, tourmaline (heavy minerals);
3) clay minerals: palygorskite, kaolinite, illite;
4) biogenic compounds: spores, pollen grains, phytholites, diatoms (*Melosira granulata, M.ambiqua*, etc.). The Harmatan brings to the sea aeolian material that is mostly enriched in desert microflora.

The 0.1- to 0.0-mm fraction of aeolian dust above the Libyan coast (test area II, see Fig. II.2.5) is next: quartz (80–90%), feldspar (10–20%), biotite, organogenic carbonates (calcite, aragonite, Mg-calcite), dolomite; the <0.01 mm fraction, quartz, mica, feldspars, palygorskite and dolomite (Aibulatov and Serova, 1983). Aeolian sediments in the arid zone of the Eastern Mediterranean contain the lowest contents of aluminum and clay minerals.

Coast of Guyana, Atlantic Ocean. The shallow-water portion of the shelf from the estuary of the Amazon to the estuary of the Orinoco River is under the strong influence of the Amazon load (see Fig. II.1.1). In connection with the fact that the main mass of this load is sufficient enough to break the hydrodynamic barrier of the river–sea GBZ, the Guyana current stays close to the coast and is directed northwest; at the coast a superfast rate of sedimentation is observed. The kinetic energy of wave processes is not really enough for reworking such a large amount of sedimentary material; therefore, liquid muds are deposited spontaneously at the waterline ("semiliquid coast"). There are muddy banks (see Fig. II.1.6) which are moved along the coast by the current at a velocity of about 900 $m \times yr^{-1}$ (Froidefond et al., 1988). At least 16 banks composed of Holocene muds, with capes protruding out to sea, exist along the way from the mouth of Amazon to that of the Orinoco (a distance of about 600 km) (Allison et al., 1995). The tidal currents work over the muddy banks, transporting the fine sandy material into the mangroves. The length of these banks ranges from 10–100 km; the width towards the open sea is tens of kilometres. Banks migrate towards the direction of the Orinoco estuary and create a flat coastal landscape. They do not cover the estuaries of large rivers (Essekibo, Korantine, Marovitne, etc.). The banks and coast are constantly washed out by the alongcoast-stream suspension of the Amazon. The width of this stream is about 20–40 km. In one year it moves about 150 million km^3 of muds to a distance of 1600 km from the estuary of the Amazon. The oncentration of the suspension at the coast here is sometimes so great that it is difficult to determine where the coast ends and where the sea begins (Fig. II.2.6). The accumulation rate of sediments at the muddy banks from isobaths of 2 m to the coast is very high: 1.24–2.0 $cm \times y^{-1}$. The rise of the bank has a sporadic character: every few tens of years it is interrupted.

When, along its way, the littoral muddy stream meets deep trenches in the shelf, or a non-developed hydrodynamic barrier (Guyana current), a part of the current is directed to the edge of the shelf and enters the upper part of the slope and further—to the Guyana Basin (Emelyanov and Kharin, 1974) (see Fig. II.1.6).

Phenomena similar to that described at the Guyana coast are encountered in the estuary of other giant rivers in tropical zones of the globe, but such a powerful accumulation of mud drifts at the waterline is characteristic only for the Amazon–Orinoco area.

Sediment types on the Atlantic Ocean shelf. At the edges of the coast–sea GBZ many sedimentary bodies of various configurations are formed, varying in granulometric and mineral composition (Fig. II.2.7, insets) (Zenkovich, 1962; Longinov, 1973; Aibulatov, 1990; Kosyan and Peahov, 1991). One of the bodies is made of sand (swells). The majority of these types of ridges have been discovered at the Atlantic Shelf of the United States. They are distributed for the most part at a depth less than 20 m. Most of them have been formed as terminal deltas of tide or river deltas. Afterwards, during transgression, these deltas were reworked by wave processes and ocean currents and extend along the coast as elongated ridges. The second most important sedimentary form (body), which form mostly at the edges of coast–sea zones, are alluvial bodies. These bodies consist of sands accumulated by heavy minerals (Zenkovich, 1962; Shuisky, 1986; Dolotov, 1989; Aibulatov, 1990; Kosyan, Peahov, 1991). They are described in Part III.2.

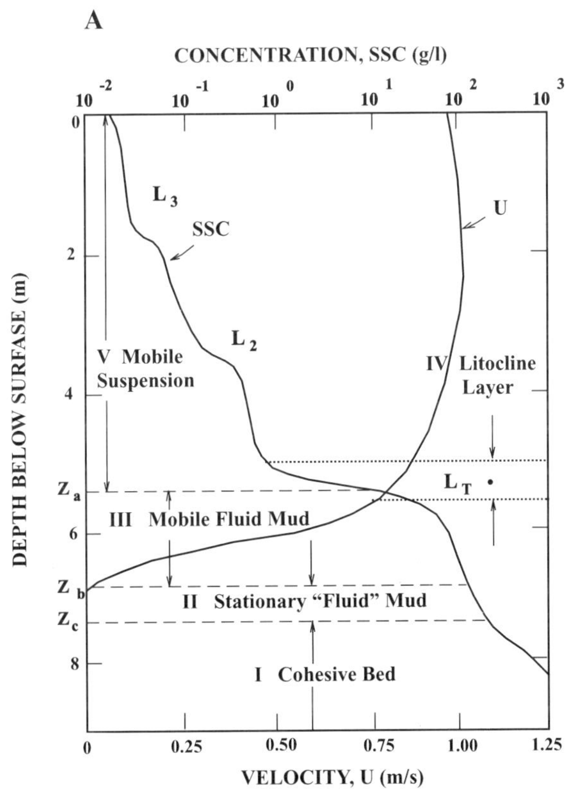

Fig. II.2.6. Sedimentation in near-shore area of South America between estuaries of the Amazon and Orinoco rivers.

A. Defining sketch of liquid-mud profile showing instaneous suspended–sediment concentration and velocity. Values are intended to be representative of field observations. After Kineke and Sternber, 1995. SSC—suspended matter–mud concentration.

Continues

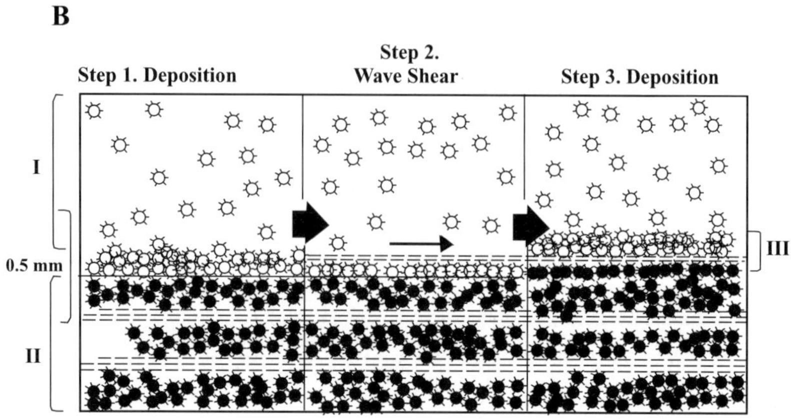

Fig. II.2.6. *Continued*
B. Idealized diagram for formation of plasmic fabric layers ~0.01–0.1 mm thick in surface sediment layer northwest of the mouth of the Amazon. Solitary waves have almost unidirectional flow oriented toward shore. Because of difference in viscosity, shear results at liquid mud–sediment boundary with passage of solitary wave crests. Shear is postulated to break surface floccules and orient plate-like mineral grains (plasmic layer) during wave passage. Between waves, floccule deposition is possible, creating unoriented interlaminations. After Allison et al., 1995.

Processes that occur at the coast–sea GBZ, and as a result of these processes, bottom sediments at various climatic zones of the World Ocean are not the same. We will illustrate this using types of shelf sediments of the Atlantic Ocean as an example.

On the shelf of World Ocean and at the coast–sea GBZ between glanulometric types of sediments, sands predominate (Fig. II.2.7). They cover about 90% of its area. In moderately humid and arid zones, sands cover almost the entire shelf; moreover, towards its edge, they are enlarged and their sorting worsens. At the equatorial humid zone, especially close to the estuaries of rivers, sands are pushed aside by muds in the central part of the shelf and to its edge. This is a general regularity for the entire World Ocean, connected with the geological history of shelves in the Late Quaternary (basically, with low sea level). In application to shelf areas, the earlier opinion that sediments tend to become thinner with increasing depth appears to be untrue.

There is very peculiar sediment formation at subestuary shelf zones, especially close to the estuaries of large rivers in equatorial regions. The inflow of muddy material is so intensive that terrigenous muds accumulate at the coastal portion of the open shelf (i.e., in the territory with the highest wave energy), which pushes out all other types of sediments. From the estuaries of rivers, muddy sediments are transported along the coast by the main alongshore currents and move to the open sea, where they form a massive accumulation of terrigenous material at the edge of the continental slope and at the foot of it.

In most regions of the shelf of the World Ocean, at the present time no accumulation of sedimentary material occurs. In these regions, usually characterized by high wave energy and strong currents, the occurrence of ancient relict sediments,

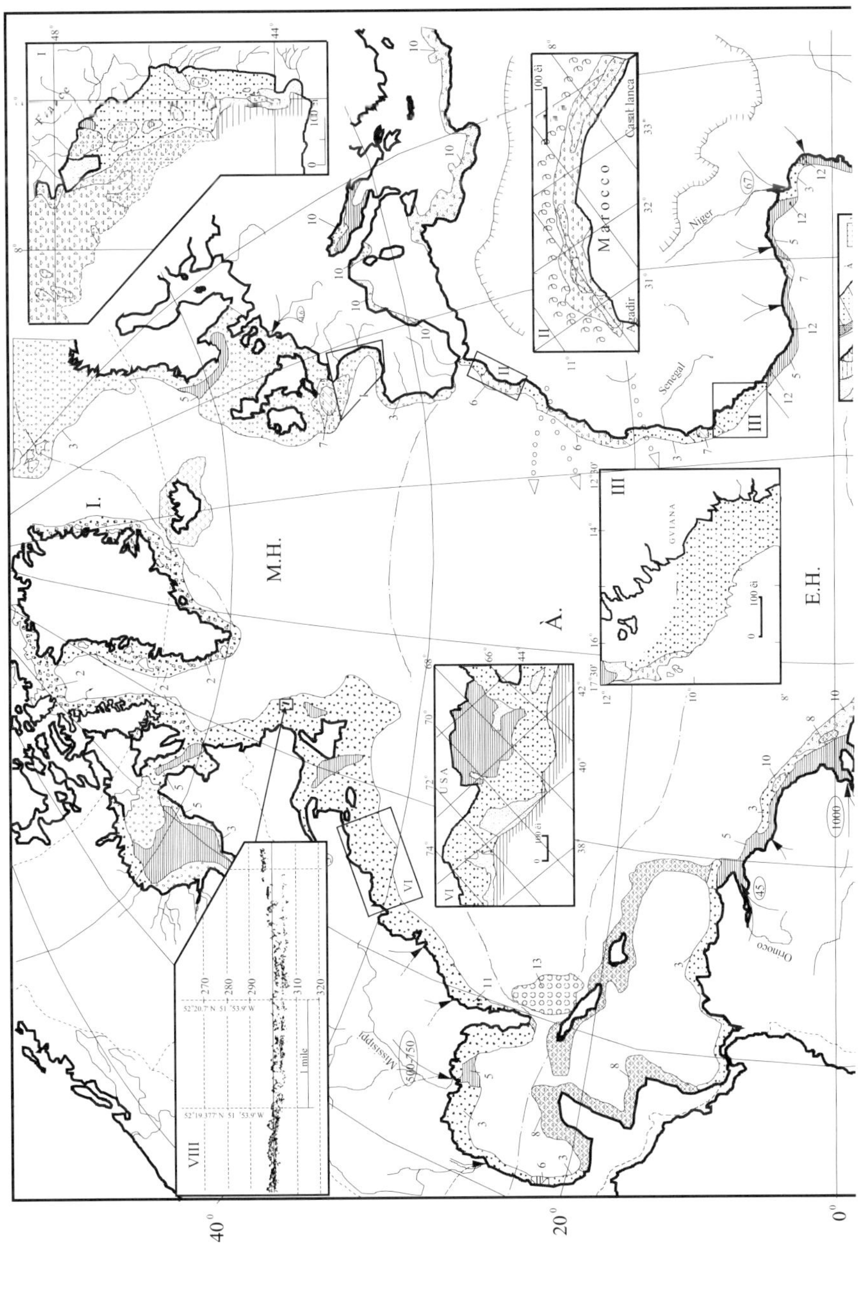

Caption see page 153

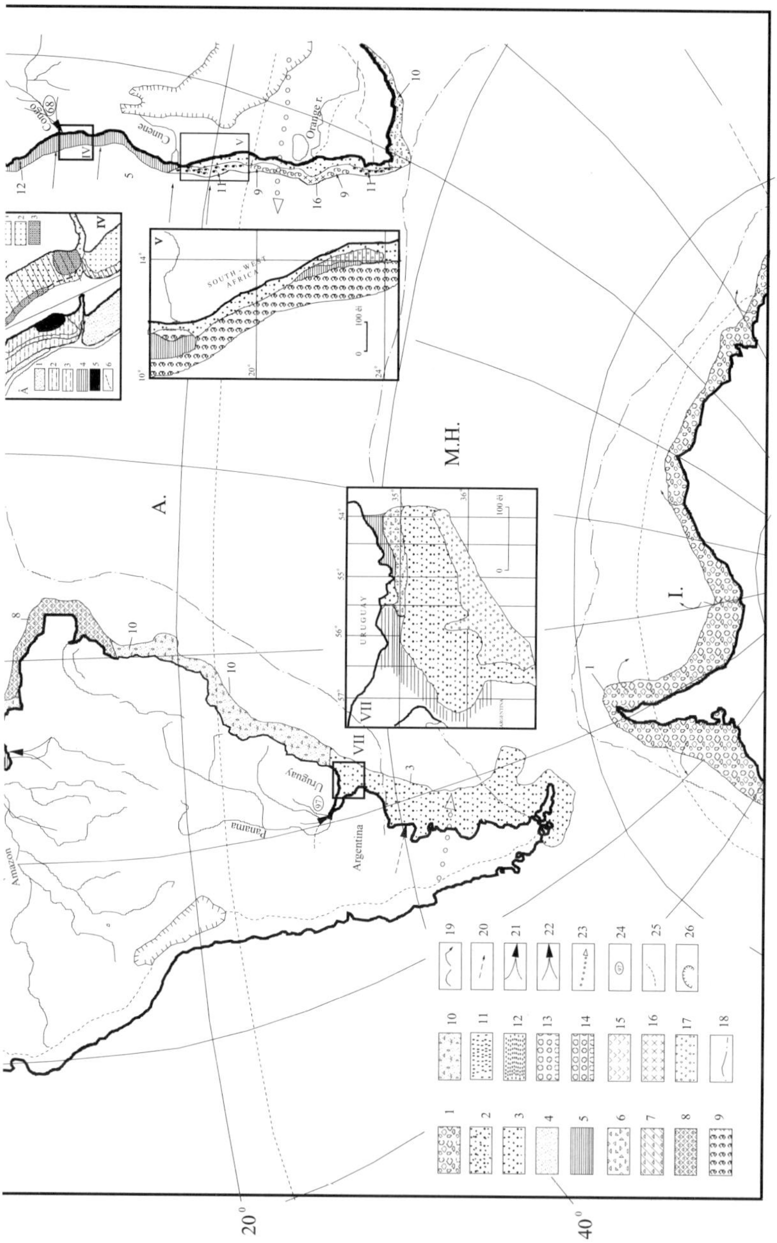
A.
M.H.
I.
Amazon
Uruguay
Argentina
Congo
Cunene
Orange r.
SOUTH-WEST AFRICA
URUGUAY
IV
V
VII
20°
40°
60°
80°
100°
120°

represented by terrigenous and calcareous shell sands, takes place. Their age in most cases varies from 50 to 30 thousand years. Relic sediments accumulated when the World Ocean was 100–120 m lower than its present level. A large portion of the present-day shelf was dry at that time, but the coastline lay either in the middle part of the shelf or along its outer zone. Below this line, coarser sediments accumulated. The following features are characteristic of relic sediments: deposits of sediments in moderately humid and arid zones occur at depths of 50–200 m (deeper than this, recent sands and aleurites have been deposited). In equatorial humid zones, they occur at the edge of the shelf; more rarely, at depths of 50–200 m (bottom regions with lower depths are usually covered with recent muds). There is an almost complete absence of well-preserved shells in shelly sands, good sorting, and an increase in the content of heavy minerals.

In the circumpolar area (ice zone of sedimentogenesis) iceberg-borne terrigenous sediments are distributed, represented by pebble–gravel deposits, sands and aleurites, poorly sorted with a noticeable admixture of spongy materials and very low content of carbonates (Fig. II.2.8). The presence of exits for moraine deposition to

Fig. II.2.7. Bottom sediments on shelf of Atlantic Ocean (0–5 cm). After Emelyanov, 1975_2, with additions.
1–5—terrigenous sediments: 1—iceberg sediments (<30% $CaCO_3$ and <30% SiO_{2am}); 2—iceberg and terrigenous (non-iceberg) (sand, gravel, pebble); 3—terrigenous sand and coarse aleurite (<30% $CaCO_3$) (in inset I–III and V–VIII—sand); 4—coarse aleurite (<30% $CaCO_3$) (only in inset I–III and V–VIII); 5—terrigenous mud (<30% $CaCO_3$). 6–9 and 14—biogenic sediments: 6—shelly sand and coarse aleurite (>40% $CaCO_3$); 7—shelly-bryozoan sand and coarse aleurite (>50% $CaCO_3$); 8—coral-bryozoan, coral-algae and coral sediments (>50% $CaCO_3$); 9—foraminiferal and shelly–foraminiferal sediments (>50% $CaCO_3$). 10—mixed shelly terrigenous sediments (30–50 % $CaCO_3$). 11–13 and 16—chemogenic and mixed sediments: 11—glauconitic (>10% glauconite, >5% Fe, <30% $CaCO_3$); 12—hydrogoethite–shamositic (>10% of ooids of hydrogoethite–shamosite, >5% Fe, <10% $CaCO_3$); 13—carbonate chemogenic and chemogenic–biogenic sediments (oolites, micrites, grapestones). 14—biogenic siliceous (diatomic) (only in inset V). 15—volcanoclastic sediments (<30% volcanoclastics and pyroclastics, >5% Fe, >0.7% Ti, <30% $CaCO_3$, <10% SiO_{2am}). 16—phosphatic (terrigenous and shelly) (>10% P_2O_5). 17—terrigenous carbonate sediments (>30% $CaCO_3$) (only Hudson Bay).
Inset IV: A—genetic types of sediments: 1—terrigenous (>30% $CaCO_3$); 2—terrigenous low-ferruginous (5–10% Fe); 3—terrigenous-ferruginous (>10% Fe); B—grain-size distribution: 4—sand; 5—coarse aleurite; 6—fine-aleuritic mud; 7—aleuro-pelitic mud; 8—pelitic mud.
Inset I, II, III, IV, V and VI, shown by dotted line:
200-m isobath or edge of shelf (inset VI), or 50-m isobath (inset II).
Inset I–VII sediment distribution shown according to (with our changes): I—after Blazhchishin, 1972; IV—after Nikolayeva et al., 1971; V—our data; VI—after Emery et al., 1965; VII—after Otman and Urien (1965); Etchichury and Remiro , 1960.
Inset VIII: roughness of bottom surface and structure of bottom sediments on glacial shelf of Labrador Sea (numbers 270–320—depths in m).
18–26—factors controlling deposition of sediment types on shelf: 18—borders of climatic zones (I—ice; MH—moderately humid; A—arid; EH—equatorial humid, after Emelyanov et al., 1975). 19—supply of glacier (ice) and iceberg sedimentary material; 20—main rivers which supply main quantity of sandy-silty material; 21—main rivers which supply main quantity of sandy-silty and pelitic material; 22—main rivers which supply main quantity of pelitic material; 23—main paths of aeolian material; 24—river loads of solid material, 10^6t/y; 25—border of Atlantic Basin drainage area; 26—deserts (without rivers).
For distribution of bottom sediments on shelf see also Figs II.1.6, II.1.7, II.1.11, II.1.11a, II.1.14, II.1.16, II.2.2, II.2.6, II.4.3, II.4.4, II.4.5, II.4.6, II.6.1, II.6.11.

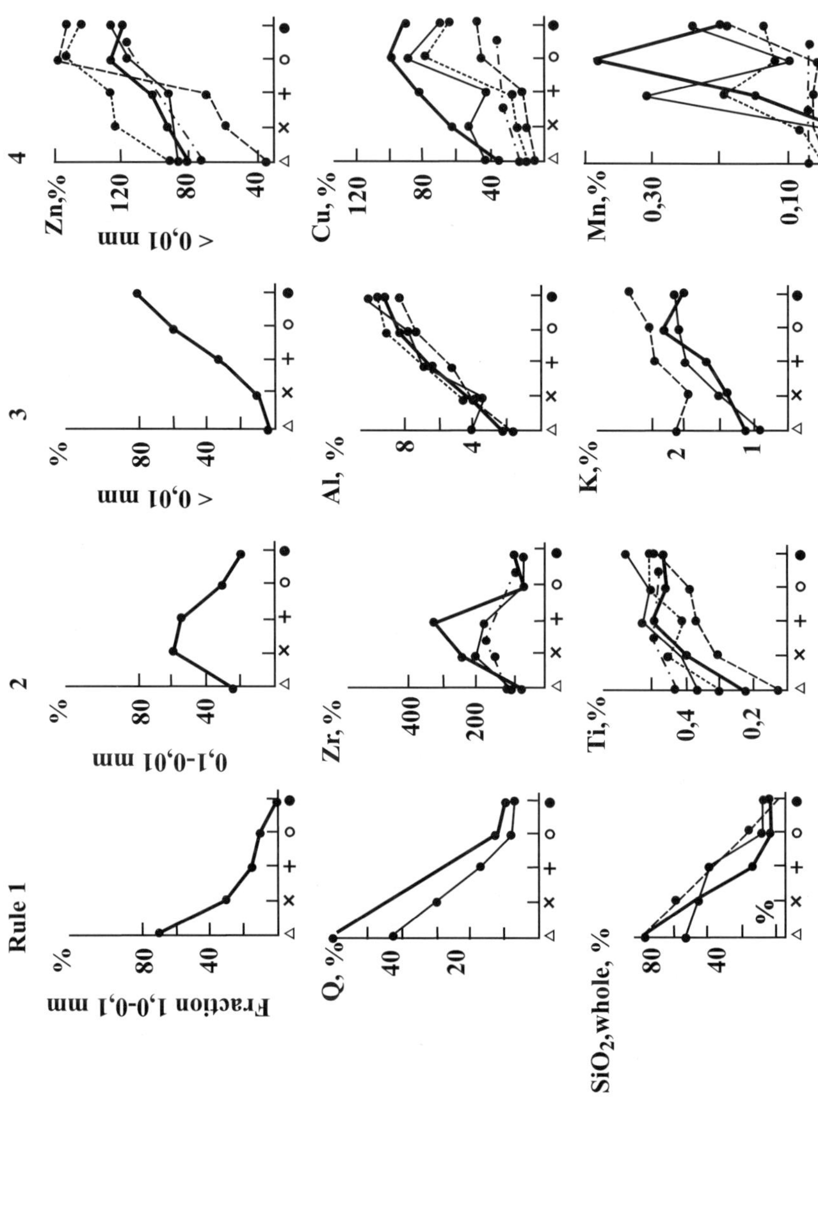
Rule 1
2
3
4
Fraction 1,0-0,1 mm
0,1-0,01 mm
< 0,01 mm
< 0,01 mm
%
Q, %
Zr, %
Al, %
Cu, %
Zn,%
SiO$_2$,whole, %
Ti,%
K,%
Mn,%
0,30
0,10

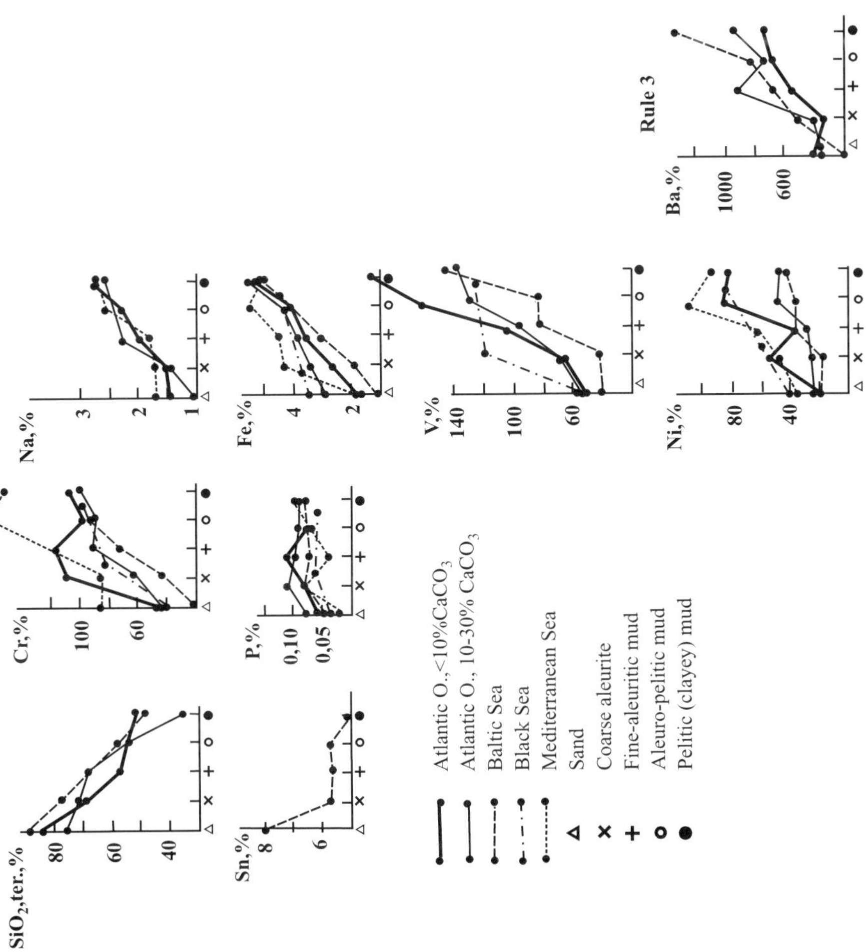

Caption see page 156

the surface of the shelf is characteristic of circumpolar zones, as well as of the northwest region of northern, moderately humid zones. Morraines are deposited here as a result of Late Quaternary glaciations and the low level of the ocean.

In moderately humid zones, terrigenous and, to a lesser degree, biogenic shelly and shelly-bryozoan deposits are distributed. They are represented by sand: muds accumulate only in basins and depressions. In arid zones, biogenic–clastic (shelly) and coral–bryozoan sediments form, and, to a lesser degree, terrigenous, oolitic, and mixed bio- and chemogenic sediments (micrite, grapestones). In the sediments there is a noticeable admixture of aeolian material, but in places they are entirely of aeolian material. In arid zones, there are also foraminiferal and diatomic oozes and glauconitic and biogenic sands containing phosphates. However, their accumulation is conditioned not by the aridity of the climate, but by the rising of deep water at the shelf zone.

In the equatorial humid zone, at the very waterline, mangroves have developed; at the coastal and the middle zones of the shelf, terrigenous muds have accumulated, but in the outer part and in the area of the edge, there are relict sands. In muds, especially close to the estuaries of African rivers, there occur brown hydrogoethite–chamosite ooids. In these cases, terrigenous muds change into chemogenic–diagenetic deposits with maximum contents of Fe (up to 42.0%).

Volcanoclastic sediments are azonal. They are distributed usually near volcanic islands. The shelf around these islands are covered by volcanoclastic (in some areas, pyroclastic) sands and coarse aleurites that contain from 5 to 12% Fe and from 0.7 to 2.0% Ti (Emelyanov and Kharin, 1987).

There are sharp differences in the sediment formation on the shelves of the eastern and western coasts not only of the Atlantic, but also of the Indian and Pacific oceans, which is conditioned by the presence of upwellings in the former, and by the presence of very large rivers in the latter.

In the coast–sea GBZ, important water-exchange processes occur. An especially active exchange occurs in zones of the ocean with intense tides. For example, in the shelf off the eastern coast of the United States, the exchange is $12 \times 2m^3/m^2$ of shelf per year or $33.2\ l \times m^{-2}$ of shelf per day (Riedl and Ott, 1982). This is sufficient to filter all the seawater on the Atlantic shelf of the United States in three years and four months. Water situated at the coast at depths less than 30 m are filtered through the shelf sediment in ten months. Extrapolating these data on the entire shelf of the tidal zone of the World Ocean, it is possible to approximate the volume of water exchange through the shelf zone. In one year, the "tidal pump" cranks 3480 thousand m^3 of water through the sediments of the shelf of the World Ocean. Apart from this, water is filtered through sediments and through the filter system between

Fig. II.2.8. Distribution of average contents of components and elements in different granulometric types of surface sediments (0–5 cm) of Atlantic Ocean and seas of its basin. Contents of Sn, Zr, Cr, V, Ni, Zn, Cu and Ba in ppm; ors, in %. (Q—quatrz)
1–5—ocean and seas: 1—Atlantic Ocean, <10% $CaCO_3$; 2—Atlantic Ocean, 10–30% $CaCO_3$; 3—Baltic Sea; 4—Black Sea; 5—Mediterranean Sea; 6—10—sediment types: 6—sand; 7—coarse aleurite; 8—fine-aleuritic mud; 9—aleuro-pelitic mud; 10—pelitic (clayey) mud.

tides. Taking into account this filtration, 96,900 km^3 of seawater pass through the sediments of the World Ocean per year. To thus filter all the water of the oceans would take 14 thousand years.

During tides (a rise in water level), large masses of seawater percolate through pores of beach and shelf sediments, transporting not only water, but also oxygen, food, organic matter, etc. "Seawater with suspended organic matter in it is left by the surf on the beach. Inasmuch as this water percolates down through sand as it returns to sea, organic detritus is filtered out in its uppermost layer, which has a thickness of several centimetres" (Neshyba, 1991, p. 239).

Organic matter and oxygen, which appear in the sediments of beach and shelf areas as a result of seepage of waters through sands, not only provide living conditions for populations of meiofauna, but also give rise to redox reactions. The upper sediment layer, which is especially well supplied with fresh seawater, is dominated by oxidation conditions, whereas the zone of reduction conditions is found at greater depths within the sediment column, below the zone of active influx of seawater. Of the total amount of organic matter infiltrating sediments together with water, only 25% is exposed to decomposition, whereas the remaining 75% of it remains in the sediments. The Eh barrier is the boundary between these two zones. The depth of the redox (Eh) barrier depends on such factors as tide level, type of sediment, water depth, the coastal and submarine bottom profile, etc. (Riedl and Ott, 1982, p. 52). Therefore, in some places, the redox (Eh) barrier is seen on the surface of the bottom (beach) and in other places it may be far below the sediment surface. In shelf sediments, the Eh barrier occurs at water depths of 38–50 m. At greater water depths, shelf sediments are reduced, and the redox Eh barrier approaches the water–bottom boundary, i.e., to the sediment surface. This narrow band of beach and shelf zone, extending up to the tidal zone, is called a "high-energy window" (Riedl and Ott, 1982).

Mechanical separation of sedimentary material and rules of fractions. In Part I.1, it was mentioned that the total amount of solid sediments delivered from land to the World Ocean annually is on the order of 25×10^9 tons, and from the coasts and benches, 9781×10^6 m^3 (or about 21×10^9 tons) (It is easy imagine the weight. Many geologists use m^3, not t) (Table II.2.2). When this sedimentary material occurs in the water medium, it is sorted out (fractionated) according to grain size and is distributed over the bottom.

As soon as the sedimentary material enters the sea (ocean), it is exposed to the effects of hydrodynamic processes. The zone of wave reworking (the first mechanical barrier zone) is responsible for the highest rates of mechanical fractionation of particles. There, sand-sized and partially silt-sized particles are deposited from water. At great distances from the coast, terrigenous suspended material is enriched with biogenic material, thus blurring the initial grain-size spectrum of the suspension. In open seas and oceans, suspended matter is often dominated (in weight) by particles, which comprise part of the aleuritic (0.1–0.01 mm) or coarse (0.01–0.005 mm) and medium-sized (0.005–0.001 mm) pelitic fractions. By excluding biogenic and chemogenic material, it is easy to see that in almost all known cases the average diameter of the terrigenous suspended particles tends to decrease with distance from

Table II.2.2 Volume of the clastic abrasional material rewashed to the World Ocean from various types of cliffs and benches. After Shuisky, 1986

Type of shore	Length of real abrasional shores 10^3 km	Production of cliffs		Production of benches		Total 10^6 $m^3 \times y^{-1}$
		$m^3/(m.y)$*)	10^6 $m^3 \times y^{-1}$	$m^3/(m.y.)$	$10^6 m^3 \times y^{-1}$	
Little modified by sea, including those of fjords, skerries, and the like	119.48	0.5	59.70	1	119.50	179.20
Denudation – abrasional, including those of the Riassic	55.50	3	166.50	3	166.50	333.00
Abrasional ingression and abrasional-bight	87.70	5	438.590	10	877.00	1315.50
Abrasional, smoothing and smoothed	30.60	15	459.00	30	918.00	1377.00
Abrasional-accumulative, smoothing and smoothed, including low shores	7.14	15	107.10	25	178.50	285.60
Abrasional-accumulative, dissected for the second time	30.56	20	611.20	45	1375.20	1986.40
Abrasional-accumulative, smoothed	14.10	20	282.00	40	564.00	846.00
Volcanogenic, rapidly retreating	6.42	40	256.80	80	513.60	770.40
Thermoabrasional	19.20	40	768.00	100	1920.00	2688.00
TOTAL:	370.70		3148.80		6632.30	9781.10

*) m.y. – meters/year

the coast. It should be said that the gradient of decreasing particle size is especially strong in the river–sea GBZ. Then, suspended matter begins to be dominated by the 0.1- to 0.001-mm fraction, and only then by pelitic fractions (<0.01 mm). Beyond the border of the shelf area, the range of grain sizes of terrigenous suspended matter along the whole length of the water column almost always corresponds to that of pelitic matter. Although sand- and silt-sized particles of organic matter are delivered by surface currents into open seas and oceans, the amount of these inputs is very small. Sedimentary material of this type migrates the most actively in the near-bottom layer. In addition, deposited sediments may experience erosion and resuspension by bottom currents and, at smaller depths (from 30 to 100 m), by wave processes, contributing thus to the transport of fine material to greater depths. In such a way, the process of mechanical fractionation of sediments continues even after particles settle on the seabed. The processes of near-bottom (or bottom) grain-size fractionation are especially active in shelf seas (the Baltic and North seas, etc), as well as in banks, on seamounts and in the open parts of shelves in seas and oceans.

According to the Stokes law, particles with a large specific weight are deposited together with light particles of much larger size (Strakhov et al., 1954). The result of this process is that small-sized particles of heavy minerals accumulate in coastal areas, whereas coarse but light-weight particles (feldspar and especially mica) are carried off far beyond the limits of areas where sand and aleuritic sediments are abundant. The combined effects of all factors and processes present are responsible for the accumulation of particular granulometric types of terrigenous sediments on shelves, as well as on the bottom of seas and oceans.

In epicontinental seas (e.g., in the Baltic Sea), where such events as permanent currents, temporal currents caused by strong storms, and wave processes play an essential role, mechanical sorting of sediments (mainly at depths no more than 100 m) is comparatively strong. As a result, the development of two sedimentary zones occurs (mostly on shelves) including (1) the coastal zone, where sands and coarse aleurite accumulate; and (2) the deep-sea zone, where fine-pelitic, aleuropelitic and pelitic muds develop. These sedimentary zones are separated by a non-deposition zone. Within the limits of this zone, pre-Quaternary deposits and products of their erosion (gravel–pebble sediments and sands) or relic sediments (commonly sands) are widely exposed.

Distribution of granulometric types in deep seas (in the Mediterranean Sea, for example) is of greater complexity than the pattern described above. However, in these areas the distribution of sediments also has a close relationship with such factors as the submarine topography, seawater hydrodynamics, and the hydrographic network of the drainage basin. Accumulation of fine sediments occurs in the deepest areas of a water basin. In contrast, fractions of 1–0.01 mm are characteristic of shelf areas, the upper parts of the continental slope, and submarine highs.

From the aforesaid it follows that terrigenous sediments are subdivided into two distinct sediment types, each recognized by a particular grain size and mineral composition, as well as by the mean contents of major and trace elements, as a result of mechanical fractionation: clastic (sands and aleurites) and clayey (pelite). Areas

polished skeletons of cowfish, shark's teeth, coprolites and weakly rounded, phosphoritized limestone debris (Gorsline, 1963).

On the Blake plateau, below the axis of the Gulf Stream, there are coral banks with heights of up to 20 m, and coral detritus (Jenkyns, 1986, p. 363). Due to the effects of strong bottom currents, the surface of nano-globigerina oozes is rough. The different velocities of bottom currents lead to the development of a variety of bed forms such as submarine dunes and sand waves. Ferromanganese crusts are very common on the surface of such sand waves (Jenkyns, 1963, p. 363). Factors such as erosion and low sedimentation rates lead to the establishment of a depositional environment favorable for the cementation of bottom sediments (see Fig. II.3.6). Magnesium-rich calcite is a cohesive substance in sediment. Developed crusts of sediments are often impregnated with manganese hydroxides (Milliman, 1971). Sand waves also appear to be locally cemented.

Farther to the north and northeastward, up to the Newfoundland Banks, the path of the Gulf Stream is well marked by an increased content of $CaCO_3$ in the upper layer of the sediment (1–30%), and also by the complex of planktonic foraminifera delivered by this current (see Fig. II.3.7). Also, in the distal part of this current, in the area extending up to Cape Hatteras, this current exerts a considerable influence on the ocean floor, leading to erosion, nondeposition or redeposition of the bottom sediments.

As the Gulf Stream moves northward, it supplies a great amount of water, heat and sedimentary material, which is characteristic of only Northwest Atlantic, to the northern part of the Atlantic Ocean. As a result, biogenic carbonaceous sediments are shifted towards the high latitudes of the Northern Hemisphere by tens of degrees (along the meridian) (Emelyanov et al., 1971). The melting of ice and icebergs in the zone of the NPF is clearly marked by a band of characteristic deposits within which

Fig. II.3.7. Gulf Stream and its traces in recent sediments (0–5 cm) of North Atlantic.
1—sedimentary bodies formed under influence of Gulf Stream; A—sedimentary ridge of Gulf Stream; B—layered sediments (according to acoustical data) of Corner Ridge.
2—zone of Gulf Stream's influence on bottom;
3—stripe of heightened content of $CaCO_3$ in sediments (30 and 50%);
4—borders of area in which fluctuations of position of main jet and front of Gulf Stream occured during 1946–1963;
5—area of distribution of planktonic foraminifera in sediments transported by Gulf Stream;
6—supposed trajectories of transportation of sedimentary material from Hudson (H) and Saint Lawrence (SL) rivers;
7—borders of heightened (>10%) content of SiO_{2am} (recalculated on cfb);
8—average long-term position of North Polar (subtropic) front (according to *Formation and Changes of Hydrophysical Fields...*, 1984);
9—position of section, on which flux of SPM was calculated;
10—average long-term position of north border of Gulf Stream;
11—subarctic convergence;
12—Labrador Current;
13—area with distribution of microlaminated diatomic oozes (with *Thalassiothrix*);
14—reversed Gulf Stream current and supposed position of recirculation front (eastern border of reversed current, where salinity is 35.34%). After Marchese and Gordon, 1996.

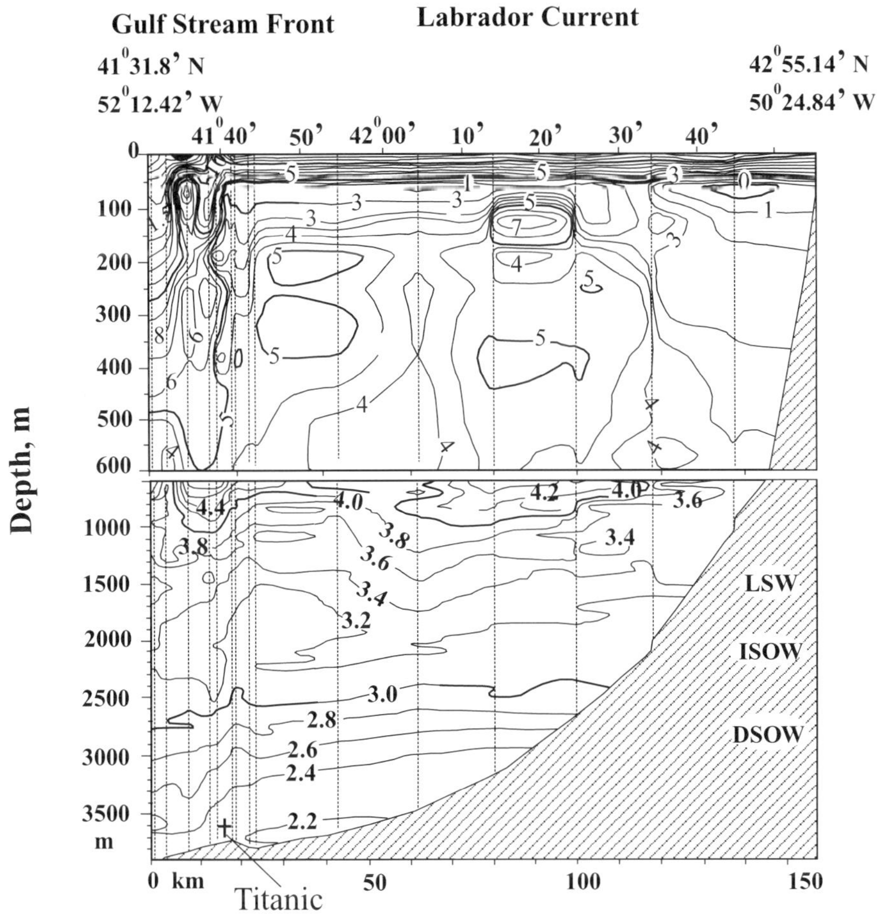

Fig. II.3.8. Vertical temperature of northwest Atlantic water strata across Gulf Stream and Labrador Current fronts in September 1999. After Aleinik, 2002. For location see Fig. II.3.9.

the deposition rates of marine-glacial and iceberg materials (cobblestones, boulders, pebbles, gravel, sand and aleurite) are maximal in the bottom sediments (Ruddiman, 1977). The enrichment of SiO_{2am} in sediments is attributed in large measure to the effect of diatoms, and partly of freshwater diatoms that inhabit ice. In places where the dynamics of the Gulf Stream is greatly diminished, a portion of sedimentary material (suspended matter) carried by this current settles to the bottom. The occurrence of such hydrodynamic traps leads to the development of sedimentary swells, such as the Gulf Stream Ridge (see Fig. II.3.7).

The cold Canary Current, which has spread as far north as 15°N along the northwestern coast of Africa during glaciation times of the Pleistocene, is marked by an

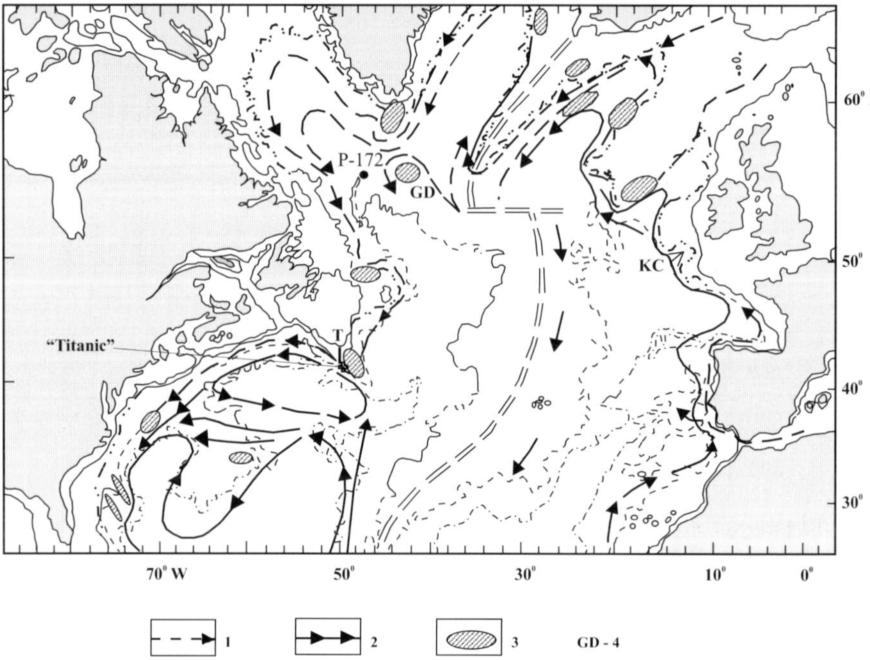

Fig. II.3.9. Deep-sea circulation in North Atlantic and location of gigantic conturites. After McCave and Tucholke, 1986.
1—NADW; 2—AABW; 3—sedimentary drifts: 4—GD—Gloria Drift; KC—King Arthur's canyon; T—temperature profile (see Fig. II.3.8.).

increased content of diatoms in sediments (Emelyanov et al., 1989). To the south, where this current appears to merge with the Atlantic North Equatorial Current, it can be traced by increased content of the giant diatom *Ethmodiscus rex* in sediments (Emelyanov and Kruglikova, 1990) (Figs. II.3.11, II.13.12).

In the Atlantic Ocean, the equatorial and Antarctic zones, where the rates of production of primary biogenic silica and quantity of diatoms are increased (>1000 cells/l in the 0- to 100-m layer in the equatorial zone; >10,000 cells/l in the Antarctic zone), leave their imprint on the ocean floor, where they can be indentified by increased contents of amorphous SiO_2 in sediments (Fig. II.3.11) (Emelyanov, 1975) and also by increased values of the mass of silica in sediments (Fig. II.3.12). Such distribution of materials is consistent with the pattern described for organic matter and biogenic $CaCO_3$ (Lisitzin et al., 1977; Emelyanov and Romankevitch, 1979). Thus, equatorial currents (and equatorial convergences) are marked by *E. rex*, as well as by increased sedimentation rates of biogenic $CaCO_3$, SiO_{2am}, and C_{org} (Emelyanov, 1982, p. 104). The Antarctic circumpolar current is well distinguishable by an increased mass accumulation rate of three biogenic components ($CaCO_3$, SiO_{2am}, and C_{org}). The distribution of diatoms related to the position of fronts also

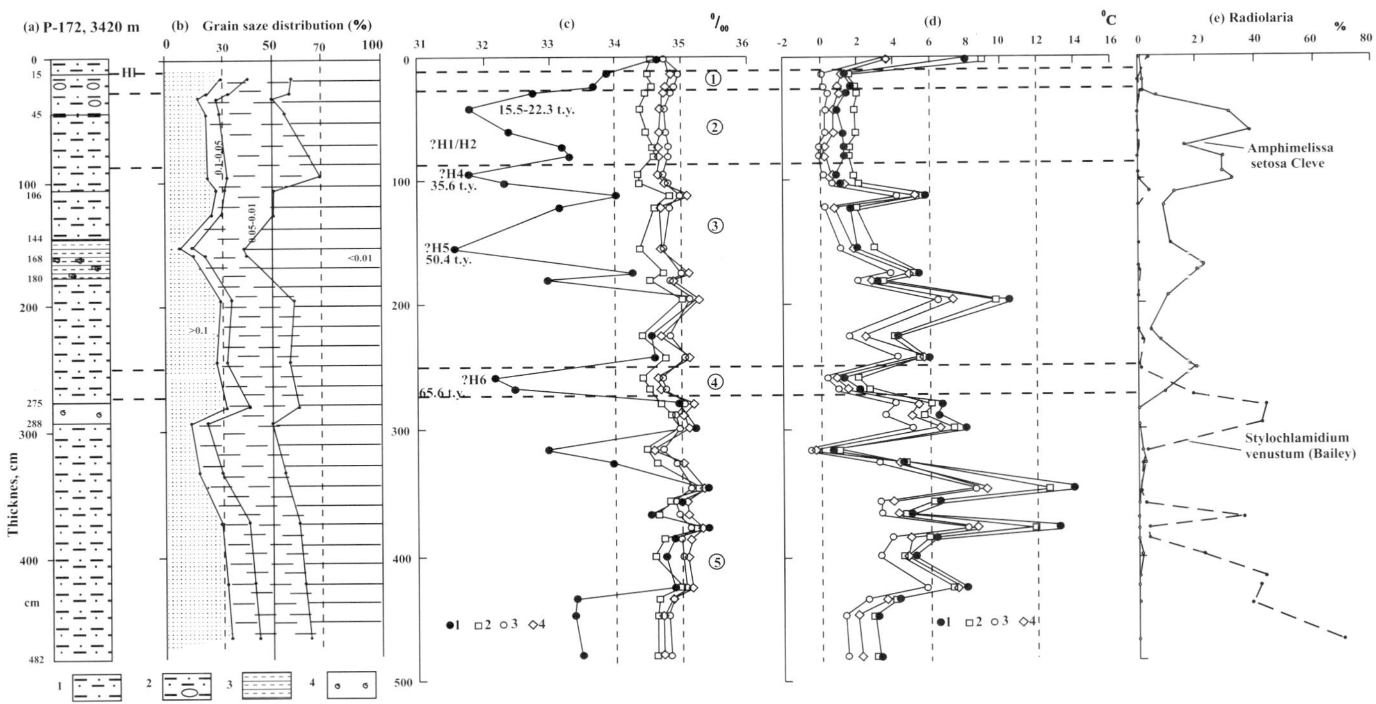

Fig. II.3.10. Lithology (a) and grain-size distribution (fraction in mm) (b) of core P–172 (depth 3420 m) in Labrador Sea (55°19′; 47°09′; O W) and paleoreconstructions.

a. Sediment types: 1—fine–aleuritic mud; 2—fine–aleuritic mud with pebbles; 3—aleuro–pelitic mud; 4—mud with foraminifera. 0–15 cm—fine-aleuritic mud, light gray, brownish (Hl, 0.015 mm/y); 15–106—fine-aleuritic mud, brownish gray, slightly greenish; 106–144 and 180–482—fine-aleuritic mud, greenish gray, at 145–163 and 288–300cm—with brownish shade; 144–168—same mud, but aleuro-pelitic.

c. Paleosalinity: 1—winter; 2—spring; 3—summer; 4—autumn.

d. Paleotemperature: 1—winter; 2—spring; 3—summer; 4—autumn. H—Heinrich events. Numbers in circles—oxygen stages c, d and g (in thousand years—t.y.).

e. Distribution in core radiolarians *Amphimelissa setosa Cleve* and *Stylochlamidium (Bailey)*. After Matul et al., 2001.

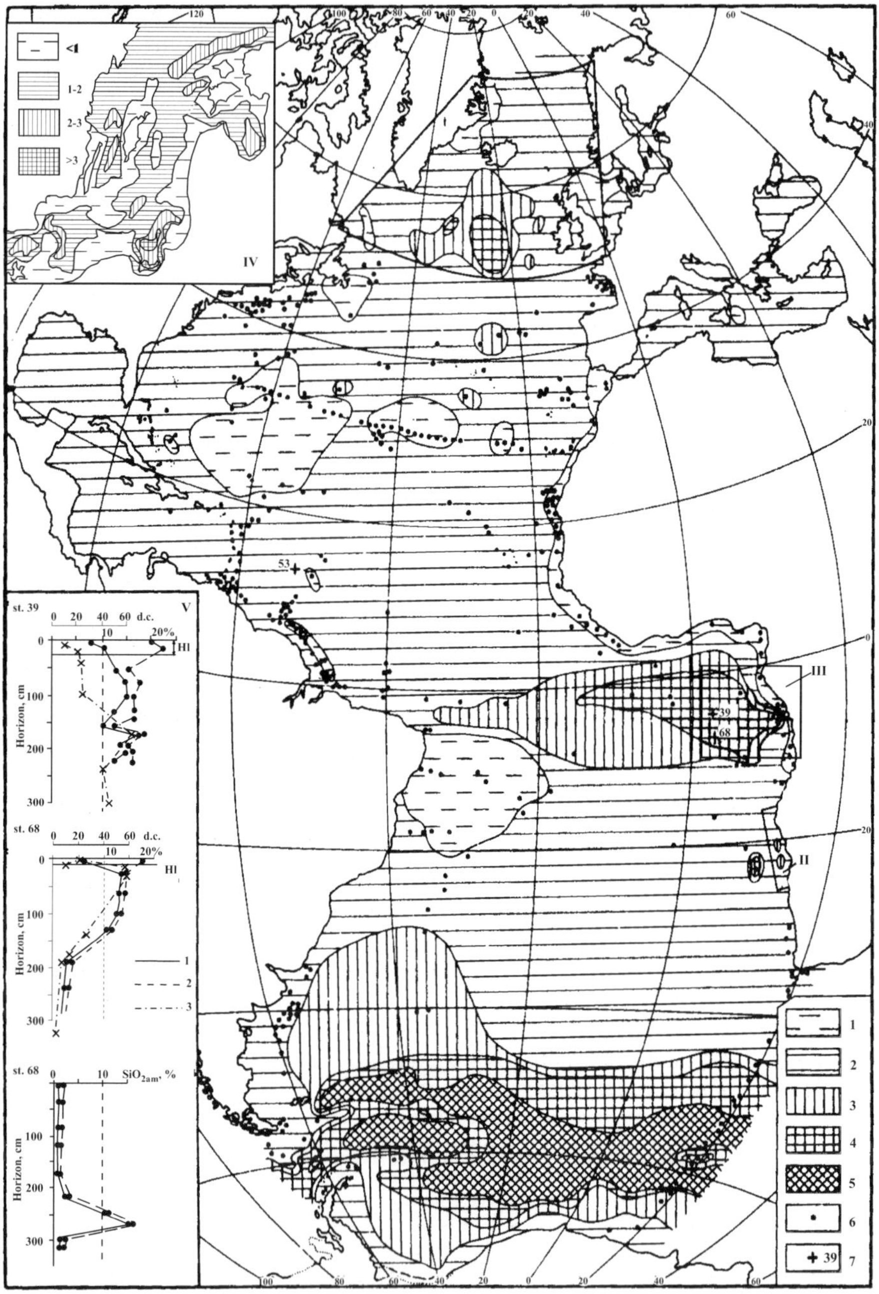

Fig. II.3.11. Distribution of SiO_{2am} in surficial bottom sediments (0–5 cm) of Atlantic Ocean (in %, recalculated on cfb). After Emelyanov, 1975, 1977.
1—<1; 2—1–5; 3—5–10; 4—10–30; 5—>30; 6—location of samples; 7—number of cited cores (see inset V).

can be readily traced. So, in the Antarctic sector of the Indian Ocean, the northern border of the area dominated by the Antarctic complex of diatoms (content >50%) coincides exactly with the position of the Antarctic divergence, while the northern border of sub-Antarctic species coincides with the Antarctic convergence (Figs. II.3.12, II.3.13, II.3.14). The distribution pattern of these complexes in waters is reproduced, without any distortions, on the ocean floor, where the skeletal remains of these diatoms are found in sediments. The limits of diatomic-glacial sediments and diatomic oozes also show good coincidence with the present-day front of sea ice (Emelyanov, 1982, 1998).

The near-bottom (benthic) front is more frequently found over sills, in straits and submarine canyons. In these areas, there is an intensive mixing of waters in the bottom layer due to effects such as frictional resistance over sills, in straits and in narrows (Fiodorov, 1983, p. 33). The near-bottom front in the southern part of the Pacific Ocean has been found at depths of 3300–3800 m. This front manifests itself as a strong drop in temperature, salinity, density and concentrations of O_2, Si, NO_3, PO_4, and Ra_{226} (Craig et al., 1972; Table II.3.1). It should be noted that this drop occurs within the narrow depth range of only 400–500 m. The front separates two water masses: Pacific deep waters (PDWs) and Pacific bottom waters (PBWs). This front is called a transition layer and the decrease in the water characteristics occurs just in this layer. The boundary between PDWs and the transition layer is the near-bottom front.

The depth of the near-bottom front may differ. This front may increase due to factors such as bottom topography or influx of PBWs. The near-bottom front occur in the ocean over areas of 10^7 km^2. This front spreads even into the Northern Hemisphere as far as to 20°N, intersects the East Pacific Rise, and can be traced as far as the continental slope of South America.

Along the boundary of the near-bottom front (at the border of the T–S transition layer), an increased horizontal exchange between water properties occurs. Sinking particles may be moved horizontally by these currents. In areas where the near-bottom front comes in contact with the ocean floor (slopes of seamounts and submarine rises), sediments must be exposed to the effects of mud resuspension, and in areas where currents play the role of screens (i.e., along fronts), the rate of their resuspension by bottom currents and the thickness of sediments should increase. The near-bottom front is probably the lower boundary of development of Mn crusts on the slopes of mountains and is the upper boundary of the distribution of biogenic siliceous oozes on the ocean floor.

Fig. II.3.11. *Continued*

I, II, III, IV—areas investigated in detail (for test areas see Figs. II.1.7. and II.4.4.) (for test area I see Lisitzin et al., 1977, Fig. 99).

Legend in inset V: Distribution of SiO_{2am} (in %) and diatomic remains in cores: 1—in natural dry sediment; 2—recalculated on cfb; 3—number of diatomic cells (d.c.), 10^6 cells in grams of sediment. See also Figs. II.11.8, II.12.3.

HL—Holocene, PL—Pleistocene.

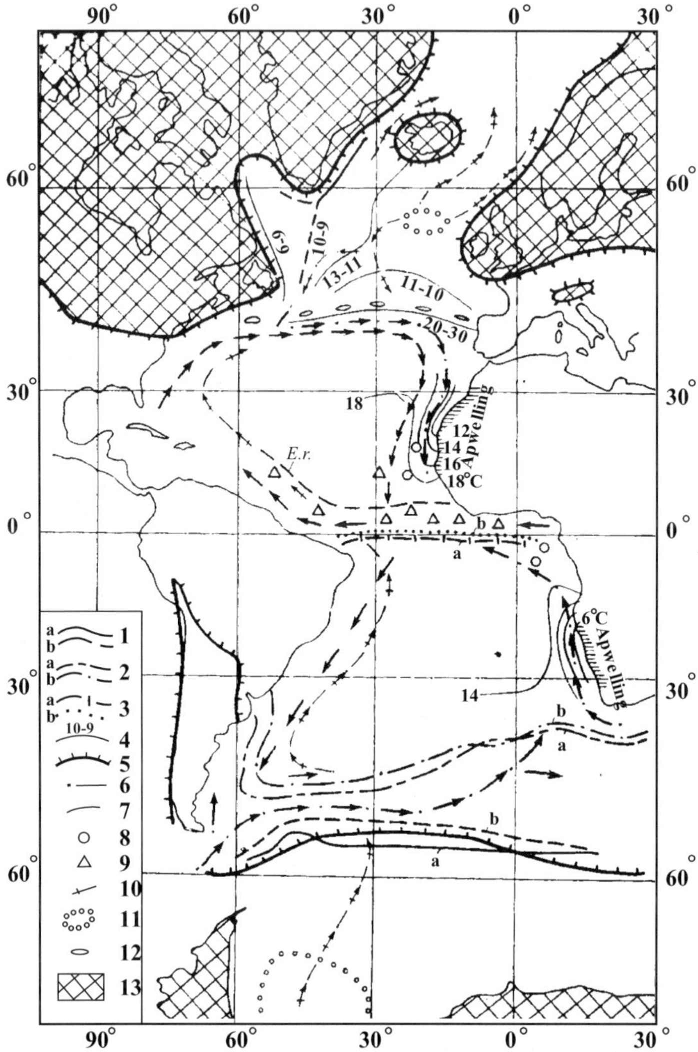

Fig. II.3.12. Position of polar front in North Atlantic and main types of convergences and divergences in South Atlantic during last 20,000 yr. After Emelyanov et al., 1989.
1–3—position of main types of divergences in recent time (a) and 18,000 yr B.P. (b): 1—Antarctic polar front; 2—subtropic convergence; 3—equatorial divergence; 4—position of retreat of polar front beginning from maximal glaciation 18,000 yr B.P.; 5—glaciers; 6–7—proposed sea currents: 6—cold; 7—warm; 8–9—presence of diatomic microlayers in upper Pleistocene deposits in equatorial Atlantic: 8—microlayers without *Ethmodiscus rex*.; 9—microlayers with *E. rex*. (symbol *E.r.* on scheme show north border of *E. rex*. distribution during glacial epochs—biozones Y and W); 10—cold near-bottom currents; 11—areas of cold deep-water formation; 12—north edge of compact ice (this edge in South Atlantic is not shown); 13—glacier on land.

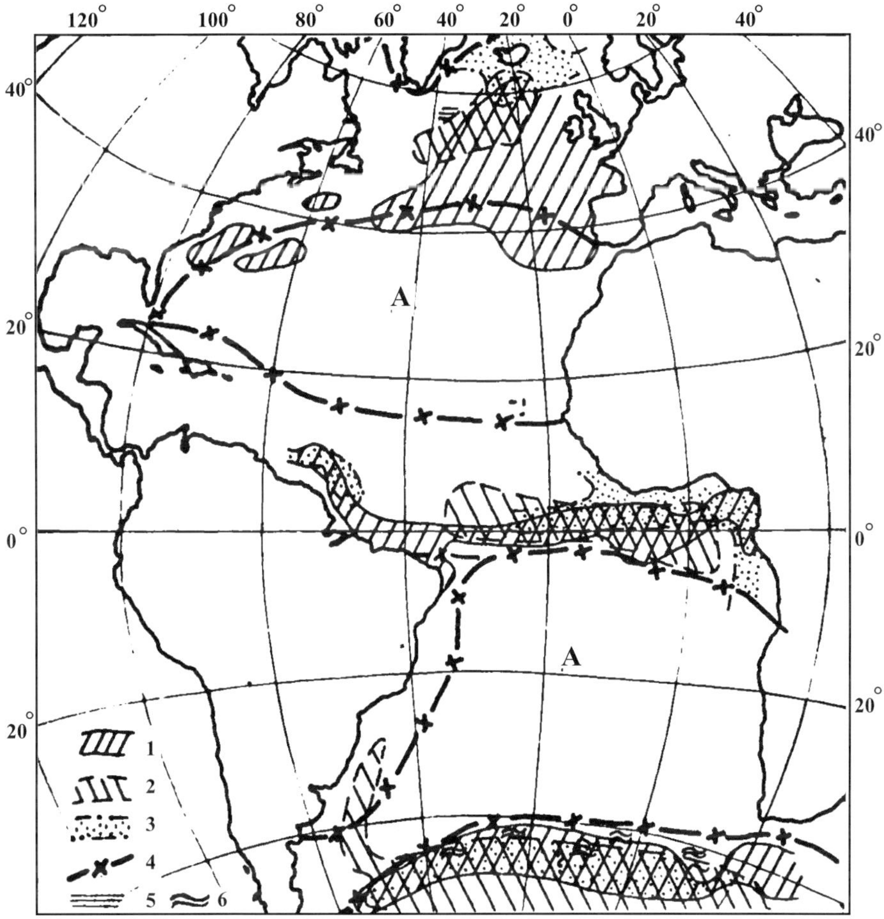

Fig. II.3.13. Location of zones of highest masses of biogenic group ($CaCO_3 + SiO_2 + C_{org}$) in pelagic areas of Atlantic Ocean. After Emelyanov, 1982.
1–3—Zones of highest masses: 1—$CaCO_3$; 2—SiO_{2am}; 3—C_{org} (most high masses are located in near-shore areas); 4—boundaries of climatic zones; 5—deposition of microlaminated diatomic oozes in Pleistocene; 6—proposed zone of deposition of microlaminated diatomic oozes; A—arid zone.

In the South Atlantic, the near-bottom front (which is here referred to as the "abyssal thermocline", Berger, 1968, 1981; Gardner, 1975; Johnson et al., 1977) is fairly consistent with the foraminiferal lysocline (depth 4050 m).

The depths of occurrence of the near-bottom front have varied in time. This front either moved to increasingly higher levels (during glaciations in the Atlantic Ocean) or fell (during the Holocene in the Atlantic Ocean). All this led, it is thought, to the vertical migration of the CCD level, and consequently, to distinguishable

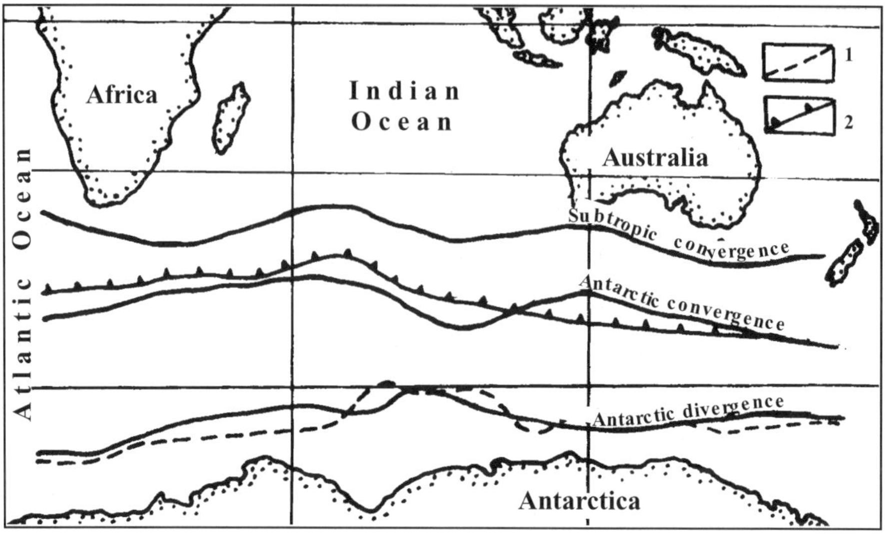

Fig. II.3.14. Zonality of bottom sediments in Indian Ocean according to diatomic complexes in bottom. After Juse et al., 1971.
1—north border of predomination of species of Antarctic complex (more than 50%);
2—north border of predomination of species of sub-Antarctic complex;
Continuous lines show position of main hydrological fronts (divergences and convergences).

interbedding of red pelagic clay with transitional low-calcareous mud and nanoforaminiferal oozes (Emelyanov et al., 1975; Emelyanov and Kruglikova, 1990).

Fronts and distribution of terrigenous material over the ocean floor. It is very common that fronts in the ocean are also borders where minerals of one type are substituted by other minerals in bottom sediments. This situation is clearly illustrated by the presence of distinct zonation in distribution of clay minerals (Griffin et al.,

Table II.3.1 Hydrochemical parameters of the near-bottom front GBZ (after H.Craig et al., 1972)

Hydrochemical indicator	Unit	Upper border of the front (depth 3380 m)	Lower border of the front (depth 3700–3800 m)
T°	°C	1.5	1.1
S	‰	34.65	34.71
δ_t	%	45.85	45.95
Si	$\mu g.kg^{-1}$	140	120
O_2	$\mu M.kg^{-1}$	160	240
NO_3	$\mu M.kg^{-1}$	37.5	34.5
PO_4	$\mu M.kg^{-1}$	2.7	2.2
Ra_{226}	10^{-4} g.kg^{-1}	11	8

μM – micromole; δ_t – density

1968) and of many authigenic minerals, especially in the shelf zone (glauconite, hydrogoethite–chamosite, and phosphates).

Fronts and borders of sediment types. The limits between sediments of different types in the World Ocean often coincide with the position of hydrological fronts. For example, semiliquid, sapropelic diatomic and low-diatomic mud (10–30% SiO_{2am}) (the middle part of the shelf) are separated from carbonate shelly phosphatic sands (the outer shelf and the shelf edge) and foraminiferal oozes (beyond the limits of the shelf edge) by the front in the zone of upwelling off Namibia (the Benguela current) (for details see Part II.4). The differentiation between sediments of different types is found also below the fronts of the Antarctic convergence (see Part II.5) and elsewhere (see Part II.17).

Geochemical studies have given grounds for speculation that lithogeochemical zonation in the Atlantic Ocean (lithogeochemical areas, regions, and provinces) (Fig. II.17.1) (Emelyanov, 1982_1) and, consequently, certain mineral resources are probably associated with ocean fronts.

The present-day zonation in distribution of sediments, their associations and types, is consistent with the pattern described for old sedimentary formations (Lisitzin, 1991, p. 218; Emelyanov et al., 1989). We believe that the study of sedimentary rocks will make it possible to solve the reverse problem: the reconstruction of palaeohydrodynamics and the geographical position of fronts in seas and oceans in the geological past. In application to the Atlantic Ocean, some steps to solve this problem have already been undertaken (Emelyanov et al., 1989).

Front of Coastal Upwelling

Upwelling is the rising of water toward the surface from deep layers. Upwelling is characterized by low horizontal gradients of temperature and density, which, together with the stress exerted by the wind upon the ocean surface, give rise to the development of geostrophic currents along shores (*Encyclopedia of Oceanography*, 1974, p. 21). A "two-cellular" circulation pattern is very common in zones of coastal upwelling (Figs. II.4.1, II.4.2). Due to the effect of wind blowing the surface water away from coast, "the water of the surface layer approaches a line called the frontal zone, where cold water of upwelling from one (land) side sinks under lighter water, contributing thus to the establishment of divergence and the rising of water on the other (seaward) side of the front" (Fiodorov, 1983, p. 134). Regions of upwelling waters are characterized by increased biological productivity. The frontal zones are a natural habitat for species such as crustaceans, plankton-feeders and predatory fish and squids; animals such as tuna and whales "forage" for food here. In the frontal zones, seawater sinks with a large velocity (Fiodorov, 1983, p. 273). Phytoplankton is also involved into the downward motion of water. In the zone of the coastal upwelling, water may be strongly driven downward also on the landward side of the front. The zone of water subsidence is characterized by rapid settling of phyto- and zooplankton onto the bottom (especially when they die). This produces a notable increase in biogenic matter in this region, with the result that the rates of sedimentation here and the content of organic matter in the bottom sediments are increased. This is why formation of terrigenous sapropelic muds with very high contents of C_{org} occurs precisely in this region, on the shelf floor. Seaward of this front, the thermocline is essentially found at the ocean surface. The result is wide differences in composition of bottom sediments, and an asymmetrical extension of living organisms, their composition, and biomass relative to the front.

In the areas of coastal upwelling, alongshore currents commonly flow along the shelf. The streamline of these currents divides waters into two water masses: cold coastal and warm oceanic. The difference in temperature of the surface waters is as much as 6°C, ranging from 12° to 18°C. Also, parameters such as saturation of waters with calcite and the value of the coefficient Alk/Cl $\times 10^{-4}$ tend to decrease considerably in coastal waters, from 300 to 70% and from >1340 to <1230, respectively (Emelyanov and Lyahin, 1973).

Areas of upwelling are characterized by the intense development of phytoplankton (the primary production of C_{org} is considered to be >200 mg/m^2 C per day) and fish. There are also calm zones and places of local sinking of surface water. H_2S occasionally occurs in sediments and near-bottom waters.

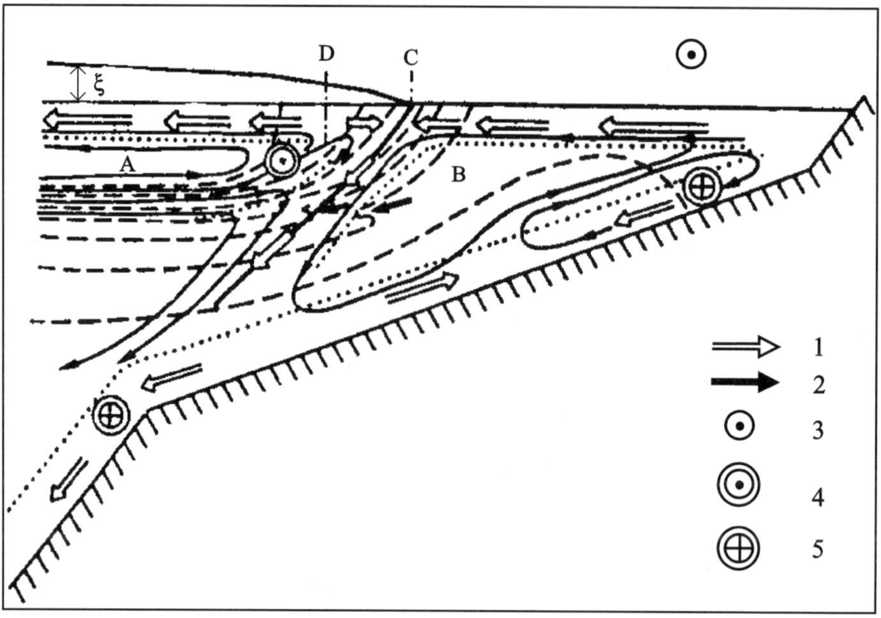

Fig. II.4.1. Scheme of bi-cell circulation in areas of coastal upwelling. After Fiodorov, 1983. Main cells of circulation in vertical plane, which is perpendicular to shore, are labeled A and B. Solid lines with arrows show general direction of water movement. Dashed lines are limits of Ekman boundary layers. Dotted lines are isotherms. Thick inclined line marks a frontal interface. Letters D and C show position of divergence and convergence, respectively.
Other notation: 1—direction of transport in Ekman layers; 2—direction of intrusive penetration of transformed upwelling waters across frontal interface; 3—winds having a direction outside the scheme; 4—axis of jet current having a direction outside the scheme; 5—axis of jet current winds having a direction outside the scheme; E—excess of observed level over undisturbed water surface.

In cold coastal waters, the following processes develop: (1) intensive extraction of major and trace elements (C, Si, P, N, Fe, Cu, Zn, Mn, Ni, Mo, Ba and others) from seawater by phytoplankton (mainly represented by diatoms) and settling of these elements together with organic detritus, diatom frustules, and fish bones; (2) accumulation on the bottom of sapropel-like and sapropelic diatomic oozes and terrigenous mud enriched in SiO_{2am}, C_{org}, P, Mo, Cu, Zn, Ni, Pb, and some other elements; (3) removal of Mn and Ba from this zone; (4) dissolution of carbonates; and (5) authigenic formation of phosphorites (Emelyanov, 1973, 1979_1).

Coastal upwellings are well developed in many regions of the World Ocean, but they mainly occur in the eastern margins of oceans. Among coastal upwellings, the most well known are the Oregon (Californian), Peru, West African (Benguela or Namibian), and Morocco (Fiodorov, 1983) and some other.

Oregon Upwelling. As revealed by satellite observations, a clearly visible front is present in the zone of the Oregon upwelling, limiting cold upwelled ocean water ocean to a distance of about 100 km from the coast. The most distinguishable

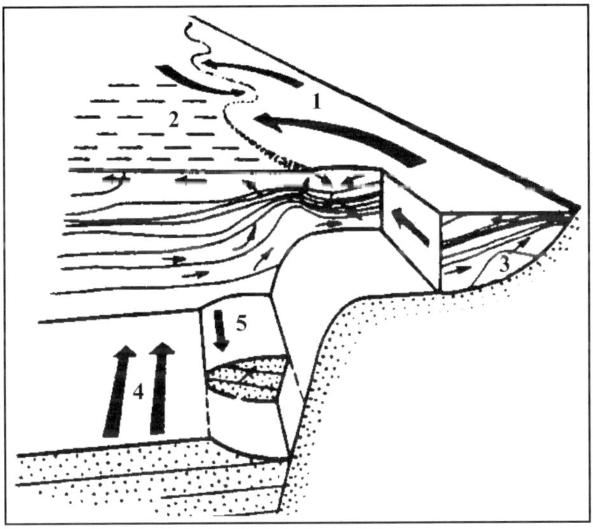

Fig. II.4.2. Scheme of coastal upwelling in Southern Hemisphere. After Hart and Currie, 1960.
1—surface alongshore current; 2—surface water of open ocean; 3—upwelling (rising) of subsurface water in near-shore zone; 4—deep compensation current.

sections of this front are characterized by cross-shelf extension, instead of an alongshelf pattern, and they extend some 200–300 km from the upwelling towards the open ocean. Naturally, the distribution of plankton and suspended matter in seawater, as well as distribution of organic matter in sediments, must follow this irregular (cellular) pattern.

Benguela (Namibian) upwelling. This upwelling is most distinct in the region of Walfish Bay off Namibia (Atlantic Ocean). The upwelling here is recognized by features such as sharp drops in temperature of the upper layer of seawater (from 11–12°C in the middle part of the shelf to 18°C on the upper part of the continental slope, with depths ranging from 1000 to 1500 m), increased concentrations of nutrients, increased primary production and other indicators. Concentrations of "a" chlorophyll near the coast increases to values as large as 5–15 mg $\times$ m^{-3} (in comparison with 0.3–1.0 mg $\times$ m^{-3} in the open ocean); the biomass of seston, to 500 mg $\times$ m^{-3} (in comparison with 25–50 mg $\times$ m^{-3} in the open ocean) (Romankevich, 1994, pp. 244–245).

The cold Benguela current (see Fig. I.20) is a very representative hydrological feature. The northern branches of this current between latitudes 24 and 18°S flow northward along the west coast of Africa. At about 18°S, this current meets the warm Angola current. The Benguela current is accompanied by rising of cold (15–16°C) intermediate waters towards the surface. Sea waters supersaturated by 250% or even higher over a normal level of calcium carbonate are found only in waters lying seaward of the Benguela current. In contrast, sea waters in areas between this current and coast are undersaturated with respect to $CaCO_3$ (Emelyanov and Lyahin, 1973). As a result, sediments of different types (Figs. II.4.3, II.4.4) and elemental composition are deposited on both sides of the main stream of the Benguela current (it flows over depths of 120–200 m). The bottom sediments present between the stream and coast are represented by sapropelic terrigenous muds and diatomic oozes containing up to 16.06% C_{org} (up to 25% organic matter), while seaward of the stream there are foraminiferal and shelly foraminiferal sediments (Table II.4.1).

In the transect along 23°S, at depths from 49 to 92 m, diatomic oozes contain from 5.02 to 6.25 C_{org}. The mass accumulation rates of muds here is estimated to be 20 g × cm^{-2} × 1000 y^{-1}. The rate of accumulation of C_{org} in the same area is about 900 mg × cm^{-2} × 1000 y^{-1} (Romankevitch, 1994, pp. 327–328). The sedimentation rates of calcareous–terrigenous aleuro-pelitic muds are markedly decreased on the outer side of the stream of the Benguela current (depths of 148–1980 m), where they are about 5–10 g × cm^{-2} × 1000 y^{-1}, while those of foraminiferal oozes (depth 3220 m) and C_{org} (which contribute 0.34% of muds) are even smaller, 2 g × cm^{-2} × 1000 y^{-1} and 4.4 mg × cm^{-2} × 1000 y^{-1}, respectively.

In the transect at 17°S, i.e., near the mouth of the Cunene River (depths of 62–155 m), the rates of accumulation of terrigenous and low-diatomic, aleuro-pelitic mud containing from 1.24 to 1.70 C_{org} (1.34% on average) are 21 g × cm^{-2} × 1000 y^{-1} or 200–240 mg C_{org} × cm^{-2} × 1000 y^{-1}. Recall that the rates of accumulation of terrigenous low-calcareous muds (contents of C_{org} are from 0.99 to 1.43) on the same transects but at depths of 4000–4100 m are 4–6 g × cm^{-2} × 1000 y^{-1} or 16–55 mg C_{org} × cm^{-2} × 1000 y^{-1}.

Diatomic oozes contain up to 78.58% of SiO_{2am} (Emelyanov, 1973) (Table II.4.2). These oozes occur at depths of 50–143 m and are 1–3 m thick. The age of these sediments dates back to the Holocene. Diatomic oozes are very moist (up to 75% water) and have dark green coloration. These strongly reduced muds (Eh is as low as −230 mV) contain also free H_2S. Amorphous silica is represented by diatoms. Mud is markedly enriched in phosphorus (0.17–0.60%P) and contain fish- and mammal bones, fish scales, and authigenic phosphate aggregates (Emelyanov, 1973_2; Baturin, 1978). They are also enriched in Mo (>0.01%), V (up to 0.054%), Cu (up to 0.041%), Ni (up to 0.107%), and Ba (to 0.058%) and are strongly depleted in elements associated with clay and coarse terrigenous material (Al, Ti, K, Li, Rb, and Zr). The content of Fe in muds is commonly <1–2%, Mn <0.01–0.02% (Table II.4.1, Figs. II.4.2, II.4.3).

At the boundary between cold shelf waters and a branch of the Benguela current, distribution of contents of $CaCO_3$ and SiO_{2am} (Fig. II.4.4), as well as of certain other components and elements (fraction <0.01 mm, C_{org}, some minor elements) differs: the higher the content of $CaCO_3$, the lower the content of SiO_{2am}.

The chemical-element composition of the sediments which occur just beneath the stream of the Benguela current is very different from the pattern described above. These sediments are commonly low-phosphatic (0.2–5% P_2O_5) or phosphatic cabonate (>5% P_2O_5), or terrigenous sands and aleurites containing from 5 to 70% $CaCO_3$, 0.1 to 2.0% SiO_{2am}, 0.2 to 10%P, and 1 to 4% C_{org} (Emelyanov, 1973_2). Calcareous

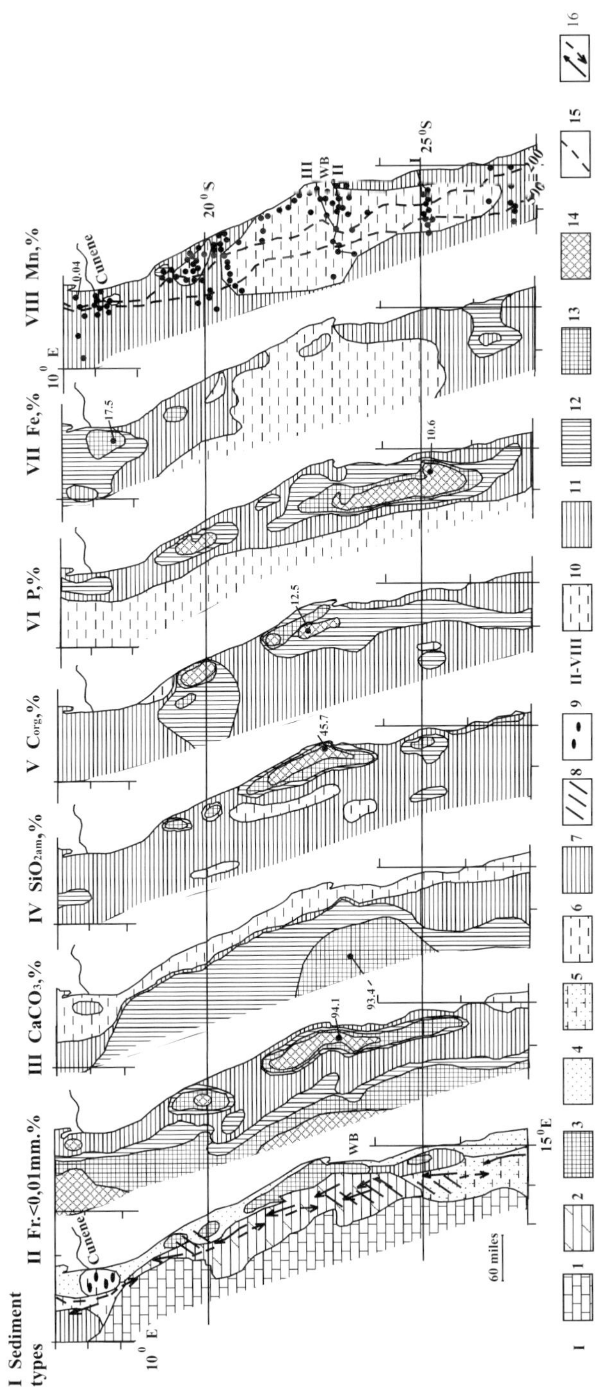

Caption see page 189

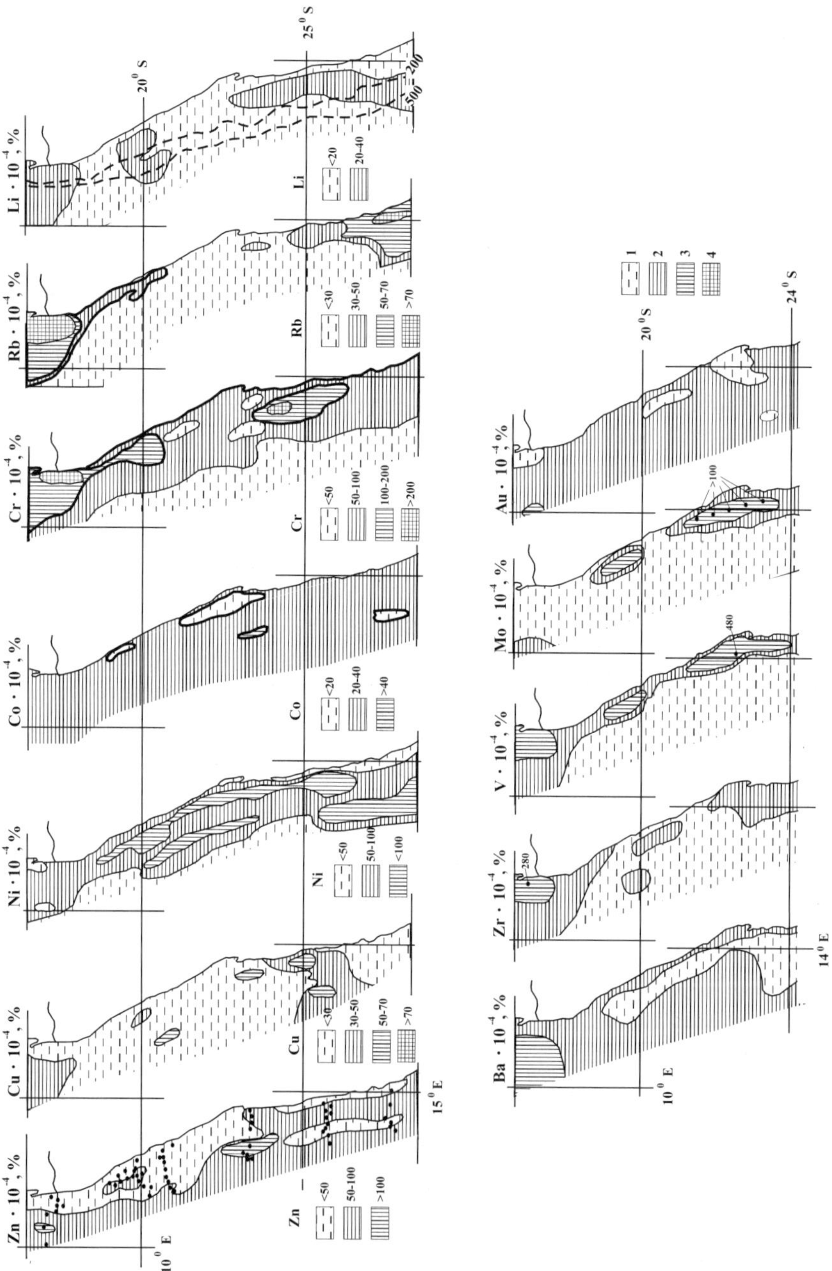
Zn · 10⁻⁴, %
Cu · 10⁻⁴, %
Ni · 10⁻⁴, %
Co · 10⁻⁴, %
Cr · 10⁻⁴, %
Rb · 10⁻⁴, %
Li · 10⁻⁴, %
20⁰ S
25⁰ S
10⁰ E
15⁰ E
200
500
Zn
<50
50-100
>100
Cu
<30
30-50
50-70
>70
Ni
<50
50-100
<100
Co
<20
20-40
>40
Cr
<50
50-100
100-200
>200
Rb
<30
30-50
50-70
>70
Li
<20
20-40
Ba · 10⁻⁴, %
Zr · 10⁻⁴, %
V · 10⁻⁴, %
Mo · 10⁻⁴, %
Au · 10⁻⁴, %
280
480
>100
10⁰ E
14⁰ E
20⁰ S
24⁰ S
1
2
3
4

material is more frequently represented by shell fragments of pelecipods and gastropods, and tests of benthic and planktonic foraminifers. Phosphate is represented by bones of fish and mammals, fish scales, and solid oval grains of phosphates (francolite, colofane). Seaward of the stream of the Benguela current, foraminiferal ooze commonly contains 50–80% $CaCO_3$, 1–3% SiO_{2am}, 1–3% C_{org}, 0.01–0.05% Mn, and 1–2% Fe; the content of Ba is somewhat increased (Emelyanov, $1973_{1,2}$; Emelyanov and Romankevitch, 1979; Lukashin et al., 1994) (Fig. II.4.4).

The area of the Benguela (Namibian) upwelling is a specific region in terms of lithology and geochemistry (we call it a carbon–phosphate–opal province). The sediments in this province are made up of those associations of chemical elements, authigenic minerals and faunal remains that are characteristic for this province only (Emelyanov, 1982). The characteristics of all three parameters in this province are notably different from the same parameters in lithogeochemical provinces in the deeps of the Baltic and Black seas, which are also characterized by accumulation of muds rich in C_{org} and saturated with free H_2S. Periodically, the content of free H_2S in muds of the shelf off Namibia is so high that anomalous concentrations of it may occur within several meters (or some tens of meters) above the seabed, due to its effective upward diffusion. It is a primary cause of death of fish. All this may not only lead to dramatic phenomena involving widescale extinction of all living

Fig. II.4.3. Sedimentation in region of South African (Namibian) upwelling.
I. Sediment types (0–5 cm layer)
1—biogenic carbonate (foraminiferal) sediments (sand, coarse aleurite, ooze, >50% $CaCO_3$); 2—biogenic carbonate (foraminiferal-shelly) sediments (sand, coarse aleurite, >50% $CaCO_3$); 3—diatomic ooze (>30% of SiO_{2am}); 4—terrigenous sand and coarse aleurite (<10% of $CaCO_3$ and <10% of SiO_{2am}); 5—mixed biogenic–terrigenous sand and coarse aleurite (10–50% $CaCO_3$ and <10% of SiO_{2am}); 6—terrigenous aleuro-pelitic mud (<30% of $CaCO_3$ and <10% of SiO_{2am}); 7—terrigenous pelitic mud (<30% of $CaCO_3$); 8—low phosphatic and phosphatic sediments (>10% P); 9—terrigenous mud, aleurite and sand with pseudooids of hydrogoethite–chamosite and glauconite (6–17% Fe).
II–VIII—content of components and elements in upper layer (0–5 cm) of sediments.
II—<0.01 mm fraction: 10—<10; 11—10–30; 12—30–50; 13—50–70; 14—>70%.
III—$CaCO_3$: 10—<10; 11—10–30; 12—30–50; 13—50–70; 14—>70%.
IV—SiO_{2am}: 10—<1; 11—1–5; 12—5–10; 13—10–30; 14—>30%.
V—C_{org}: 10—<1; 11—1–3; 12—3–5; 13—5–7; 14—>7%.
VI—P: 10—<0.1; 11—0.1–0.2; 12—0.2–0.5; 13—0.5–1; 14—>1%.
VII—Fe: 10—<1; 11—1–3; 12—3–5; 13—>5%.
VIII—Mn: 10—<0.01; 11—0.01–0.05; 12—0.05–0.1%.
15—200- and 500-m isobaths (scheme VIII only).
16—arrows show direction (to north) of Benguela Current; dotted-line arrows show countercurrent (to south) near bottom (aproximately above 200-m isobath). After Gordon et al., 1995.
Numbers I, II and III on scheme VIII show location of lithogeochemical profiles (see Fig. II.4.4). Points on Zn scheme shows location of studied samples of sediments for Zn, Cu, Ni, Co, Cr, Rb and Li. 200- and 500-m isobaths for the entire scheme are shown on the Li scheme.
Ba–Au contents in ppm on surficial sediments (0–5 cm, 32 samples):
Ba: 1—<500; 2—500–1000; 3—1000–2100;
Zr: 1—<50; 2—50–100; 3—>100 (maximum 280);
V: 1—<50; 2—50–100; 3—>100 (maximum 480);
Mo 1—<5; 2—5–10; 3—>10 (maximum more than 100);
Au: 1—<0.002; 2—0.002–0.005; 3—>0.005.

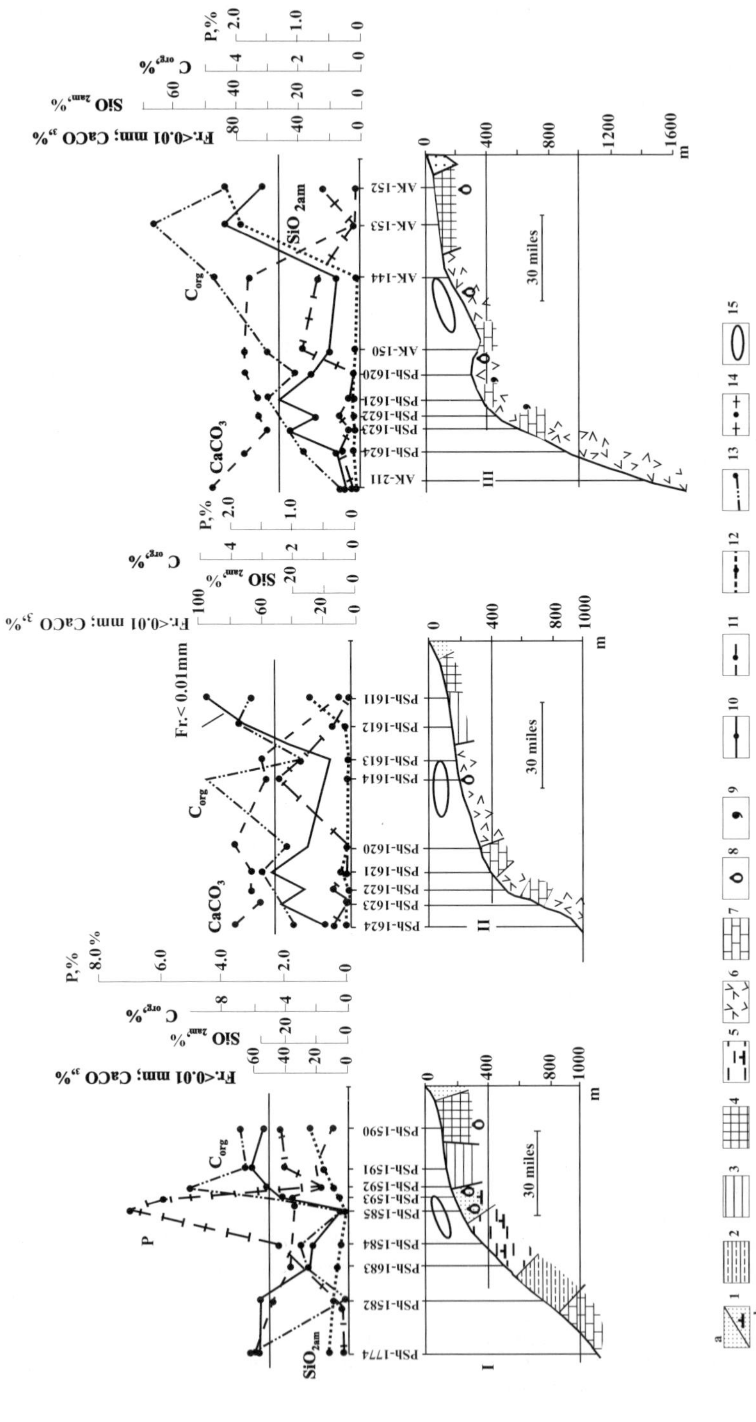

Caption see page 189

organisms on the seabed and in near-bottom waters, but may also contribute to better preservation of weakly modified organic matter and to intensive accumulation of remains of bottom fauna, fish and mammals on the sea bottom. This, in turn, is an additional stimulus for the development of fresh authigenic phosphates, and ultimately, phosphorites as mineral resources (see Part III).

Morocco Test Area. This area is part of the upwelling zone off northwestern Africa. It is situated between 25–27°N, i.e., somewhat north of the area (Figs. II.4.5, II.4.6).

The shelf within the survey area is 20–25 km wide and the shelf-break occurs at depths of 110–120 m. There are several near-bottom currents flowing over the continental slope. Cold Canary current flows southward. Within the shelf area, this current can be traced down to 35 m (Thiede, 1983, p. 185). The North Atlantic central waters (NACW) are clearly seen over the upper parts of the continental slope (400–800 m). The South Atlantic central waters, which flow northward, approach Cape Bakhador (25°N) from the south (Fig. II.4.6). Signs of the upwelling off northwest Africa are observed northward from 10° N up to the northern Portugal (Fütterer, 1983). Water over the continental slope (at a distance of about 200 km from land) has a temperature of about 20–21°C on average; over the shelf of the Morocco, 17–18°C. The primary productivity near Cape Bakhador is estimated at 200 $g \times cm^{-2} \times y^{-1}$ (Fütterer, 1983, p. 108). The oxygen minimum zone (1.5–2.0 $ml \times l^{-1}$) at 21°N is located at depths of 300–600 m (Suess and Thiede, 1983, p. 131).

Species such as diatoms, radiolarians, and planktonic foraminifers *Globigerina bulloides*, *Globigerinita glutinata, and Globoquadrina dutertrei* are good indicators of cold upwelled waters (Thiede, 1983, p. 184). Maximum (or increased) amounts of these planktonic organisms were observed in shelf waters (Fig. II.4.7). The sudden decrease in concentrations of such species in waters beyond the limits of the shelf is evidence that upwelling is most distinct only in shallow (coastal) areas.

The effects of rising water and the alongshore (Canary) current (see Fig. I.20) are markedly weaker than in areas off Namibia. In the Morocco test area, which embraces the shelf area with depths ranging from 20 to 500 m (Fig. II.4.6), the sediments are represented mostly by shelly sand and aleurite containing from 30 to 91% $CaCO_3$. Among biogenic remains, fragments of mollusk shells and corals and foraminiferal tests predominate. In >0.05 mm fractions, 50–96% of the total sediment is represented by biogenic remains. Foraminifera in the shelf sediments are dominated by benthic species. For >0.1 mm fractions, the amount of these species

Fig. II.4.4. Lithogeochemical profiles throughout Namibian upwelling area (see scheme VIII on Fig. II.4.3).

1–9—sediment types: 1—sand and aleurites terrigenous (a) and mixed biogenic–terrigenous (b); 2—aleuro-pelitic calcareous–terrigenous mud (30–50% $CaCO_3$); 3—pelitic carboniferous mud enriched in organic matter (3–10% of C_{org}); 4—diatomic ooze (>30% SiO_{2am}); 5—fine-aleuritic calcareous–terrigenous mud (30–50% $CaCO_3$); 6—biogenic carbonate sediments (foraminiferal and foraminiferal–shelly ooze, aleurites and sand, >50% $CaCO_3$): 7—foraminiferal ooze; 8—presence of phosphates; 9—presence of glauconite; 10–14—distribution of components and elements in upper layer of sediments (0–5 cm), %: 10—<0.01 mm fraction; 11—$CaCO_3$; 12—SiO_{2am}; 13—C_{org}; 14—P; 15—axis of alongshore Benguela current.

Table II.4.1 Comparative data of the average chemical composition and content of the <0.01 fraction mm in surficial sediments (0–5 cm) at the front of near-shore upwelling GBZ (near-shore area off Namibia, southwestern Africa) (after Emelyanov, 1973_1, with some additions)

Components, element	À 99–120*	B 50–129	C 75–120	D 130–345	AK-205 2400	AK-185 5470
Fraction <0,01 mm	30.8–50.4	40.4–79.3	62.5–88.1	9.2–23.9	-	77.4
$CaCo_3$	6.34–15.69	9.24	4.49	36.41–74.32	75.75	62.70
Sio_{2am} **	1.36–9.68	19.75	39.9	<0.1–0.86	2.02	1.0
$Sio_{2am} \times 1.6$ **	2.18–15.49	31.6	63.84	<0.16–1.38	3.23	1.6
C_{org}	1.21–8.80	7.01	5.35	2.51–4.71	1.18	0.30
Fe	1.13–3.96	2.02	0.92	0.57–0.85	0.93	2.20
Mn	0.01–0.03	0.01	<0.01	<0.01–0.02	<0.01	0.33
Ti	0.13–0.54	0.22	0.06	0.04–0.07	0.07	0.17
P	0.17–0.21	0.22	0.36	0.11–1.77	0.06	0. 8
Ba	520–740	220–740	570	<200–1260	17.30	-
Zr	90–280	<50–60	58	<20–70	40	-
V	60–68	117–122	256	<1–37	25	-
Ni	95–105	71–104	81	28–51	23	-
Cr	185–217	133–224	97	37–240	112	-
Cu	73	-	388	250–400	-	-
Mo	<5–36	72–>100	>100	<5–13	8,2	-

*) Sea depth, m

**) Chemical analytical data are 1.6 lower than real (Emelyanov, 1982_1). Data in this Table are multiplied by 1.6

<0,01 mm fraction; $CaCO_3$-P , in %; Ba-Mo, in 10^{-4}%

A – terrigenous mud, B – low diatomic ooze (containing up to 16.06% C_{org}), C – diatomic ooze; D – biogenic carbonate sand and coarse aleurite (containing up to 10.03% P)

The limits of contents are shown for A and D; average content for B and C.

A, B and C are located mainly toward the shore from the barrier "near-shore front", D – to the sea from the front (and occur on the edge of the shelf under the strong near-bottom current, see Fig. II.4.1). At the edge of the shelf are distributed mainly biogenic carbonates and phosphatic deposits (see Fig. II.3.4 also).

Station AK-205 – foraminiferal oozes; these occur under the OML.

Station AK-185 – typical pelagic nano-foraminiferal oozes (southern part of the Angola Basin). For the station locations see Fig. II.1.7.

Table II.4.2 Composition of Holocene diatomic oozes of the Namibian shelf (st. AK-163, depth 106 m, 21°28'7 S. 13°34'9 E.)

Horizon, cm	Sediment type	Fractions (mm). %			SiO_{2am}		C_{org}	P	Fe	Mn	Ti	Phos-phates **)	H.S.
		>0.1	<0.01	$CaCO_3$	a	b*)							
0–6	Diatomic ooze, pelitic, low-phosphatic, dark grey, semiliquid, H_2S smell	1.93	79.92	5.89	31.62	50.59	9.29	0.39	0.85	0.005	0.10	++	
10–20	The same	1.45	81.99	4.96	-	-	7.91	0.45	1.31	0.005	0.07	++	
38–49	Diatomic ooze, pelitic, low-phosphatic, dark grey, soft	1.14	80.30	7.14	35.61	57.00	5.94	0.24	1.03	0.02	0.04	++	I
68–80	Diatomic ooze, pelitic, dark-grey, soft	0.99	76.14	8.26	38.51	61.61	5.84	0.20	1.03	Tr.	0.06	+	
90–100	Diatomic ooze, aleuro-pelitic, dark grey, soft	9.58	62.56	6.57	41.48	66.37	4.98	0.19	1.31	0.01	0.07	-	I-II
105–113	Diatomic ooze, aleuro-pelitic, greenish grey, dense	20.10	41.23	6.50	42.43	67.89	4.92	0.13	1.31	0.008	0.07	-	
135–149	Diatomic ooze, aleuro-pelitic, greenish grey, dense	19.22	55.42	5.25	48.59	77.74	4.80	0.10	1.31	Tr.	0.05	-	
151–158	Diatomic ooze, aleuro-pelitic, greenish grey, dense	16.92	48.01	7.25	41.72	66.75	6.72	0.13	1.31	0.004	0.07	-	II
170–180	Diatomic ooze, fine-aleuritic, greenish grey, dense	25.35	41.96	7.50	49.1	78.58	5.12	0.13	1.21	Tr.	0.08	-	
190–201	The same as on horizon 151–158	19.12	49.28	7.82	51.53	82.49	5.34	0.13	1.40	Tr.	0.08	-	

*) $SiO_{2am} \times 1.6$

**) Phosphatic concretions in the >0.1 mm fraction (++ = many; + = present; – = absent)

H.S. – Holocene transgression stages: II – stage of weak transgression (weak rise in sea level); I – stage of intensive transgression (intensive rise in sea level); I-II – transition stage.

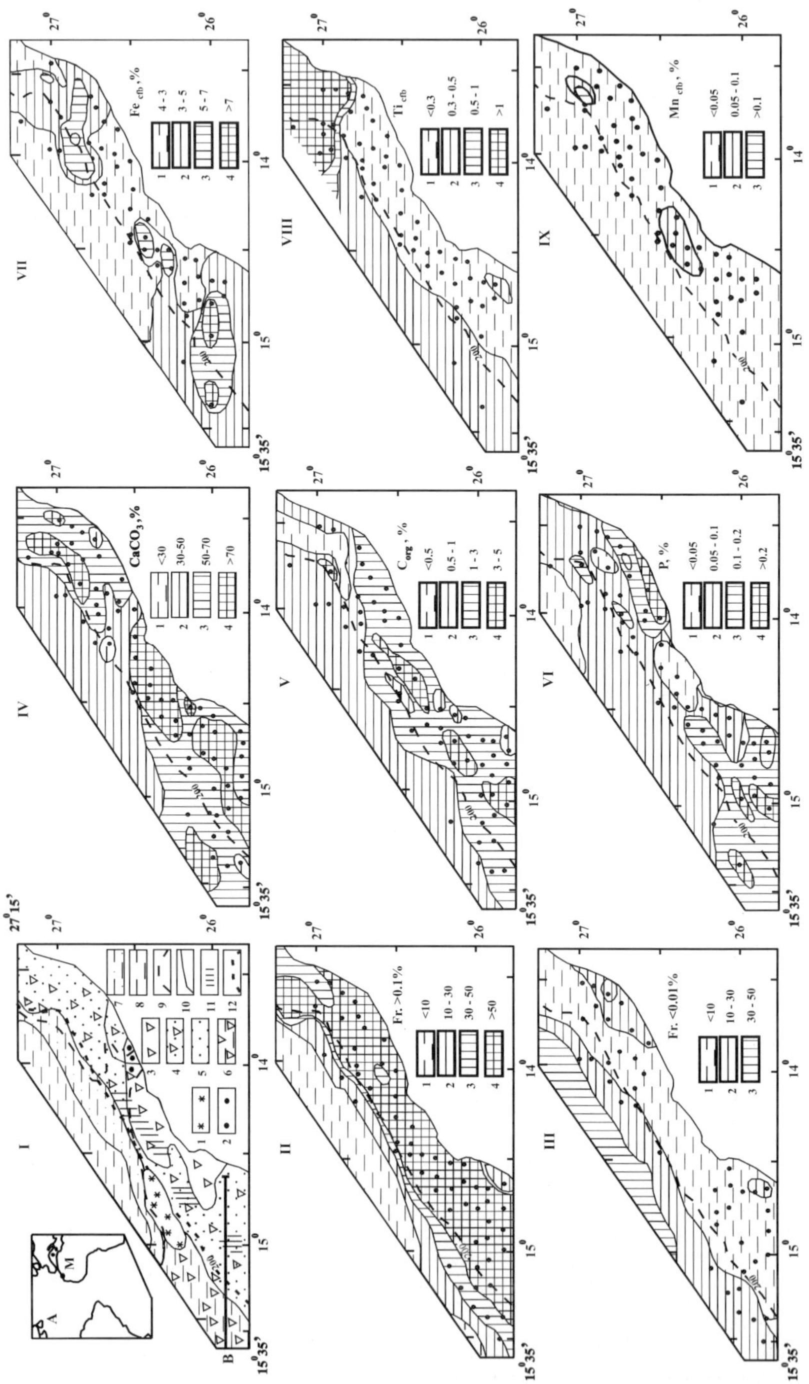
I
II
III
IV
V
VI
VII
VIII
IX
Fr. >0.1%
<10
10 - 30
30 - 50
>50
Fr. <0.01%
CaCO3, %
<30
30-50
50-70
>70
Corg, %
<0.5
0.5 - 1
1 - 3
3 - 5
P, %
<0.05
0.05 - 0.1
0.1 - 0.2
>0.2
Ticfb
<0.3
0.3 - 0.5
0.5 - 1
>1
Mn cfb, %
0.05 - 0.1
>0.1
Fe cfb, %
3 - 5
5 - 7
>7
27°15'
27°
26°
14°
15°
15°35'

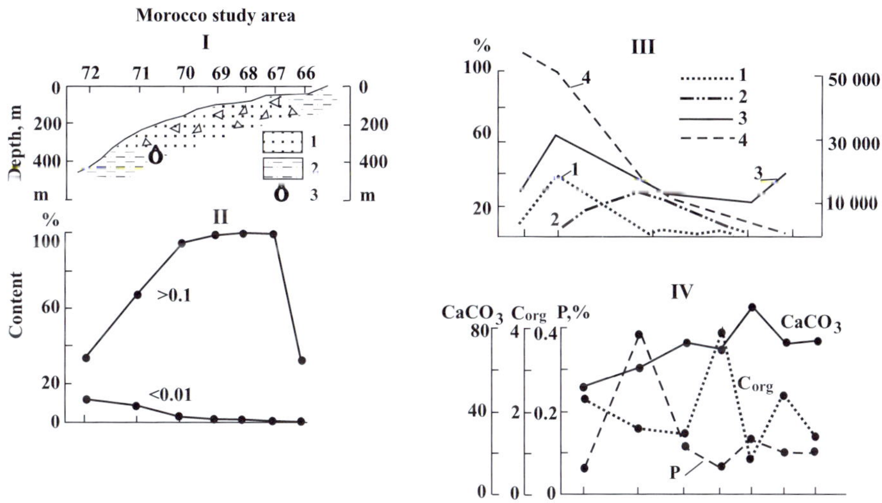

Fig. II.4.6. Distribution of (I) sediment types (0–5 cm layer); (II) fractions >0.1 and <0.01 mm content in sediments; (III) foraminifera; and (IV) $CaCO_3$, C_{org} and P in upper sediment layer over a transverse profile across shelf off northwest Africa, within Morocco study area (for position of profile see Fig. II.4.5(I)).

(I) Sediment types: (1) biogenic shelly sand; (2) biogenic foraminiferal–shelly aleurite; (3) occurrence of phosphate grains in sediments (0.2–0.4% P);

(III) foraminifera in >0.1 and <0.01 mm fractions: (1) *Cassidulina laevigata carinata*, % of total foraminifera; (2) *Textularia sagittula*, % of total foraminifera; (3) benthic foraminifera, number per gram of sediment; (4) planktonic foraminifera, number per gram of sediment.

in the upper layer of sediments (0–5 cm) is commonly found to be 11–50,000 specimens per gram of sediment. The amount of such species strongly increases (to values of 50–210,000 specimens per gram of sediment) only in the northern part of the test area (north of 26°40′N). The diversity of species is markedly higher in the southern part of the test area (20–38 species) than in its northern part (12–33

Fig. II.4.5. Types of bottom sediments (0–5 cm surface layer) within Morocco study area (shelf off northwest Africa), in Morocco upwelling zone, and distribution patterns of grain-size fractions and major chemical components.

(I) Sediment types: 1—coral buildups and limestone outcrops; 2—sand and gravel with shells (30–50% $CaCO_3$); 3—shelly detritus (>50% of fraction coarser than 1 mm, 50–90% $CaCO_3$); 4—carbonate shelly sand (50–90% $CaCO_3$); 5—mixed biogenic–terrigenous sand (>50% of fraction 1–0.1 mm, $CaCO_3$); 6—foraminiferal-shelly coarse aleurite (50–70% $CaCO_3$); 7—mixed biogenic–terrigenous coarse aleurite (30–50% $CaCO_3$); 8—mixed biogenic–terrigenous silty mud (30–50% $CaCO_3$); 9—isoline of 50% $CaCO_3$ content; 10—isoline of 0.2% P; 11—sediments enriched in C_{org} (3–5%); 12—200-m contour.

(A) location of Morocco study area; (B) position of lithogeochemical profile shown in Fig. II.4.5(I).

(II) >0.1 mm fraction content in sediments, %; (III) <0.01 mm fraction content in sediments, %; (IV) $CaCO_3$ content in sediments; (V) C_{org} content in sediments; (VI) P content in sediments (0–5 cm); (VII–IX) Fe, Ti and Mn content in sediments recalculated on cfb (0–5 cm), %; (VII) Fe; (VIII) Ti; (IX) Mn. Depth contour of 200 m is shown on all maps.

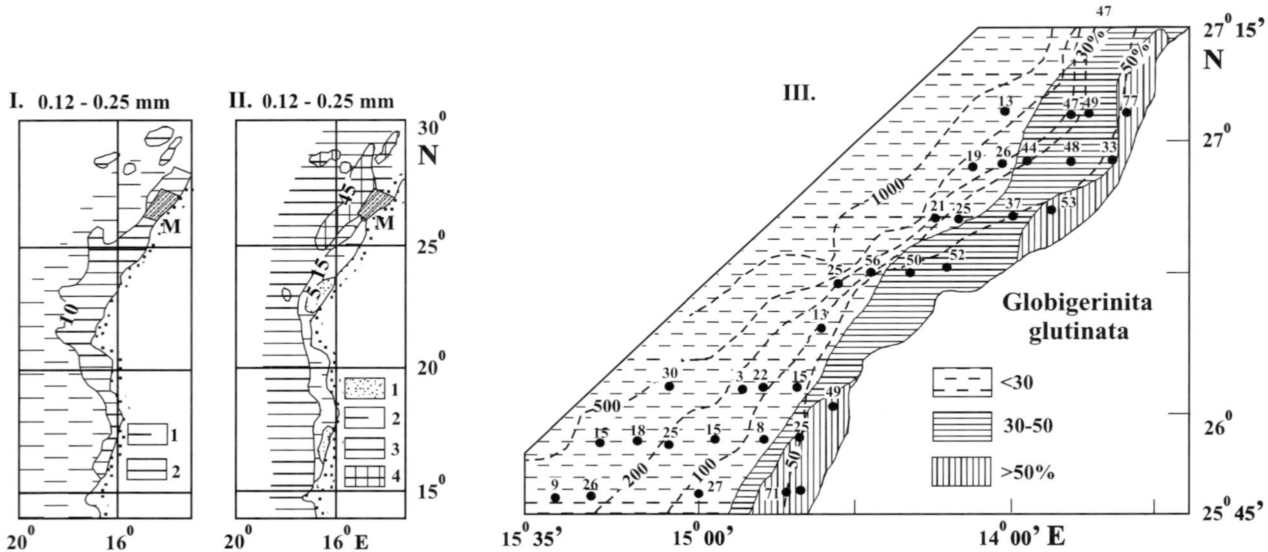

Fig. II.4.7. Region of coastal upwelling off North African coast, as revealed from planktonic foraminifera distribution in surface waters. After Thiede, 1983.

I. *Globigerina bulloides* in 0.12–0.25 mm fraction of suspended matter, % of total planktonic foraminifera: 1—<10; 2—>10.

II. *Globigerinoides ruber* in fraction 0.12–0.25 mm of suspended matter, % of total planktonic foraminifera: 1—<5; 2—5–15; 3—15–45; 4—>45. M denotes location of Morocco study area.

III. Distribution of planktonic species *Globigerinita glutinata* in sediments from Morocco study area (% of planktonic foraminiferal thanatocoenosis): 1—<30; 2—31–50; 3—>51.

species), and this is evidence that the intensity of upwelling is higher in the south than in the north of the test area. The species *Hoeglundina elegans (d'Orb,)* which is an indicator of upwelled waters, is also concentrated mostly in the southern parts of the test area (coastal areas) (Emelyanov and Lukashina, 1995).

Planktonic foraminifera concentrate in the sediments of the continental slope. At depths from 300 to 500 m, the amount of such species in fractions >0.1 mm is 50–205 thousand specimens per gram of sediment, while for the shelf this value does not exceed 10 thousand specimens per gram of sediment.

The mineralogical and geochemical data for the sediments from the shelf area off Namibia illustrate an important general point: upwelling is characterized by high concentrations of diatoms, organic detritus, authigenic phosphates and increased contents of phosphorus. Contents of all these components in the sediments of the Morocco test area are very small due to the effective near-bottom hydrodynamics, and evidently due to slow rising of water. The content of biogenic opal in >0.006 mm fractions in the vicinity of Cape Bakhador is very low: some 0.3–0.5% (of the total fraction) (Fütterer, 1983). The almost complete absence of pelitic fractions (<0.01 mm) and strong predominance of sand fractions (>0.1 mm) in the sediments are evidence that mobility of the near-bottom waters is very high. Increased contents of organic matter (1–5% C_{org}) and phosphorus (0.10–0.20%) in the sediments of the southern part of the test area is evidence that the upwelling is significantly developed in this area. Phosphorus is represented by dispersed phosphatic grains, the origin of which is not clear. A somewhat increased amount of coarse (0.06–0.25 mm) phosphatized particles (detritus of old phosphates) are found in some northern areas. There is no clear relationship between these particles and the content of P in the sediments.

The manganese content in sediments in the Morocco test area, like that in sediments of Namibia, is very low. The cause of sporadic distribution of Fe and Ti and high contents of these elements at some stations is due to the presence of minerals of basalts in the sediments, rather than to the effects of upwelling.

Peru Upwelling. In the surroundings of the Peru upwelling, the temperature of water decreases to values of 5–7°C (Stepanov, 1974). In these waters, increased contents of nutrients (N, P, Si and others), contributing to high biological productivity. Owing to low temperatures, the water is undersaturated with respect to calcium carbonate, with the result that processes of carbonate accumulation are substantially weakened here. The absence of large rivers in the adjacent land is not a contributing factor in the delivery of large amounts of terrigenous material into these regions: the largest input of sedimentary material entering this area appears to be related predominantly to abrasion. The shallow-water coastal zone is dominated by sands, and at certain greater depths (i.e., at a greater distance offshore), sediments are dominated by muds (Fig. II.4.8). Among fields of mud in shelf area, there are also diatomic oozes and diatomic–terrigenous mud, containing 15.5% SiO_{2am}. Moreover, muds enriched in diatoms extend from the shore down to a depth of 1500 m (Table II.4.3). At depths of 300–500 m, diatomic–terrigenous muds are markedly enriched in organic matter to as much as 11.63% C_{org}. The average content of C_{org} in sediments at these depths is 5.37%. At smaller and greater depths, values of organic

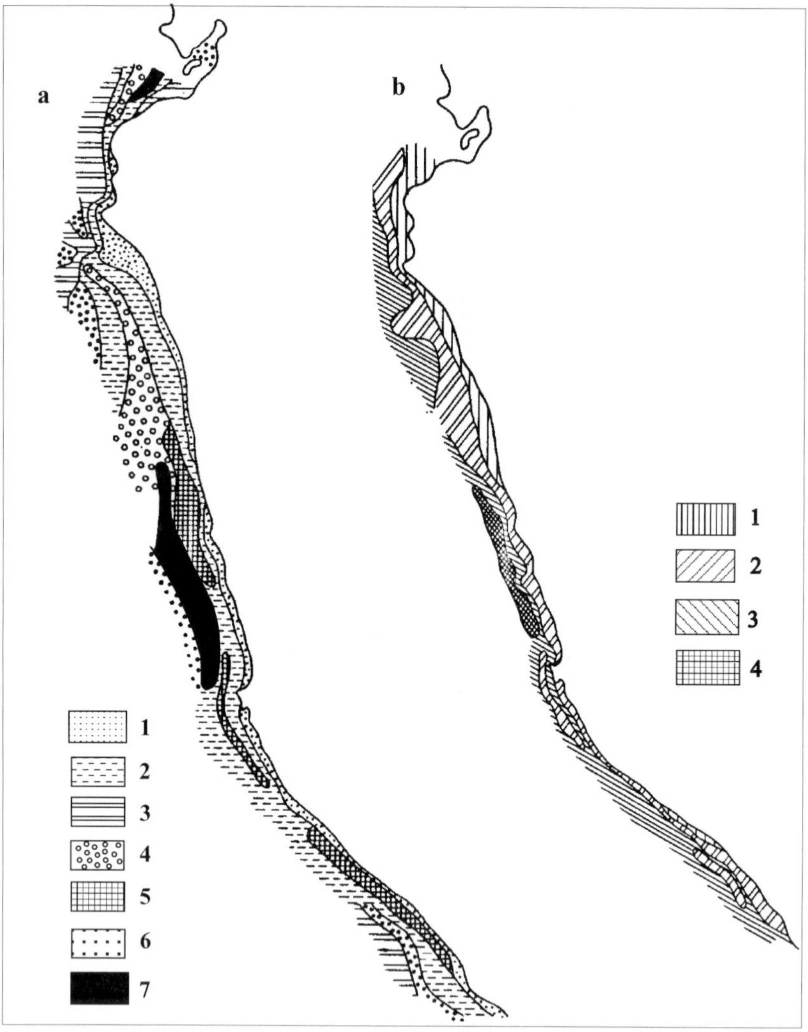

Fig. II.4.8. Distribution of surficial bottom sediments in Peru upwelling area (a) and content of C_{org} in sediments (0–5 cm layer), %. After Gershanovich, 1981.

a) Bottom sediments: 1–3—terrigenous sediments: 1—sand; 2—fine-aleuritic and aleuro-pelitic mud; 3—pelitic (clayey) mud; 4–5—biogenic sediments: 4—foraminiferal sand; 5—diatomic ooze; 6—glauconitic sand and mud; 7—outcrops of ancient deposits.

b) C_{org} (in %): 1—<1; 2—1–3; 3—3–5; 4—>5.

matter content are markedly lower. Foraminiferal and shell-foraminiferal sediments are abundant in areas lying deeper than diatomic oozes, and sediments within and even beyond the shelf-edge zone occur as bands of clayey-glauconite and sandy-glauconite extending along the coast.

Table II.4.3 Content of the main chemical components and elements in surficial sediments of the Peru upwelling zone, % (after Gershanovich, 1981)

Depth, m	$CaCO_3$	Fe	Mn	Ti	SiO_{2am}		P		C_{org}	
	a*)	a	a	a	a	m	a	m	a	m
0–150	15.7	2.56	0.04	0.25	7.76	15.5	0.26	0.72	1.29	7.82
150–300	28.8	2.15	0.02	0.22	6.62	8.68	0.66	3.65	4.50	9.91
300–500	3.1	2.45	0.03	0.25	11.50	-	0.76	1.32	5.37	11.63
500–1500	7.0	7.15	0.03	0.30	3.52	5.85	0.21	0.57	2.43	4.47
>1500	-	3.26	0.03	0.32	10.12	10.12	0.19	0.19	-	-
Average	11.2	3.34	0.03	0.25	7.29	-	0.31	-	2.36	-

*) a – average, m – maximum

The content of phosphorus in the sediments of the Peru upwelling are smaller, to a certain degree, than those in the sediments of Walfish Bay: the maximum content is 3.65%; the average content is 0.76% P.

Black Sea. The development of hydrofronts of coastal upwelling occurs not only in the open ocean but also in seas, including the Black Sea. The existence of a convergence over the upper part of the continental slope (depths from 100 to 500 m) predetermines the rising of cold and deep waters (from the intermediate layer) and this leads to a rapid drop in temperature in the coastal area. The frontal zone, which has formed in such a way near the continental slope, is a powerful hydrochemical (or geochemical, according to our terminology) barrier with maximum concentrations of dissolved forms of (Zn, Cu, Pb, Cd, Hg and Cr), as well as pesticides, polyaromatic hydrocarbon and other organic components (Sapozhnikov, 1991).

Ice–Water Boundary

In the transverse ice–water profile made in the Norwegian Sea in 1971–1972, the horizontal temperature gradient (T°) was 0.28°C per mile, and in some cases (near eastern Greenland, Fig. II.5.1, II.5.2) it was 2°C per mile. On average, a gradient of 0.1°C was found over a distance of 80 to 100 miles from the ice edge (Gatham, 1986). In the zone referred to as the ice edge, winds blow in different directions from the band of maximum gradient: in the north they blow from the −1°C isotherm; in the south, from the +5°C isotherm.

In the Nordic seas (North Sea, Greenland Sea, the western part of Barents Sea and waters around Iceland and the Faroe Islands), the existence of such a high T° gradient at the ice–water boundary is also caused by the effects of the Gulf Stream (more precisely, the North Atlantic current and its branches).

The sea state is moderate in the ice zone and is rather rough or heavy in the open water zone. The correlation coefficient between T°C and the sea state is 0.7. This correlation is very strong.

Moist warm air masses are transported by winds from ice-free waters. This factor contributes to the formation of fogs over ice. Fog is commonly observed under conditions when the temperature gradient is directed towards ice. In areas of warm water (i.e., beyond the ice–water boundary) the sky is clear and there is no fog.

The sound velocity in seawater covered by ice is low (1440–1450 m/s), while in regions lying beyond the ice–water boundary (i.e., in warm waters) it is high (1460–1470 m/s) (Wadhams, 1986, p. 69).

Upwelling is often observed near the ice edge (Wadhams, 1986, p. 71). In the Nordic seas, it can be traced over a distance of up to 8–10 miles from the ice edge.

The portion of light that penetrates the ice cover into seawater is less than 1% of its total energy on the surface and this imposes restrictions on the processes of photosynthesis. The primary production of water in ice conditions is very low—less than 1 $gC/cm^{-2} \times y^{-1}$ (Neshyba, 1991, p. 345). Only during the short summer season does the intensity of photosynthesis increase. The process of photosynthesis near the ice edge is so intensive that it can be called explosive.

It is very typical of ice conditions that there are some algae species (so-called epontic algae, i.e., developing out of water) capable of developing within ice (Neshyba, 1991, p. 346). Such species are represented mainly by pennate diatoms and microflagellants having dimensions of about 6 μm. A good deal of studies have confirmed that production of algae is dominated by pennate (*Pennatophycea*).

Furthermore, in regards to distribution of algae throughout the whole thickness of ice coating in Arctic areas, algae of this type develop mostly in the lower half

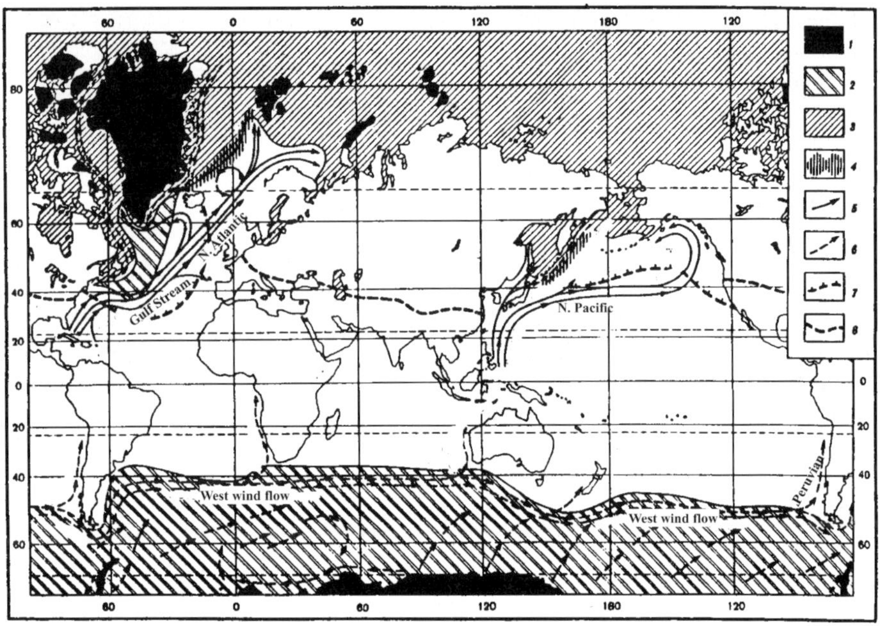

Fig. II.5.1. Scheme of sea ice distribution and recent bottom sediments connected with it. According to Lisitzin, 1994.

1—regions of freezing with excess ice masses to sea; 2—regions of iceberg distribution connected with contemporary precipitation; 3—regions of distribution of drifting sea ice connected with contemporary precipitation; 4—regions of linear unloading of ice at junction of cold and warm currents; 5–6—directions of basic currents transferring ice masses and icebergs; 7—extreme southern borderline of ice distribution and precipitation in Northern Hemisphere in period of maximum freezing; 8—contemporary borderline of water freezing on land.

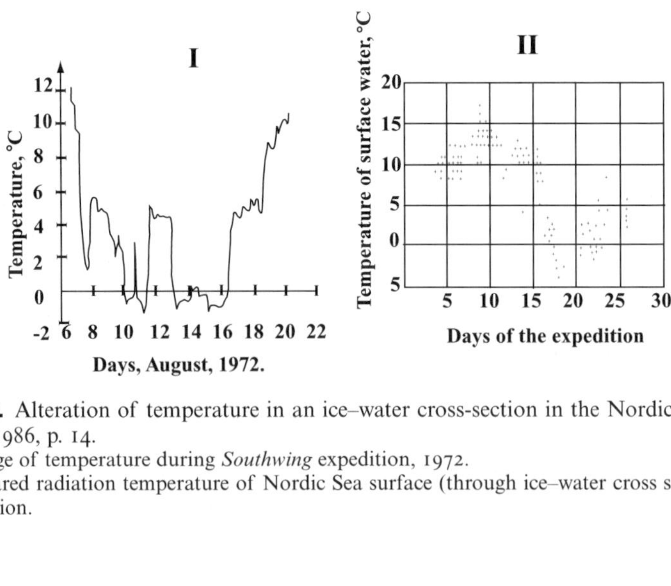

Fig. II.5.2. Alteration of temperature in an ice–water cross-section in the Nordic Sea. After Gatham, 1986, p. 14.

I—change of temperature during *Southwing* expedition, 1972.

II—infrared radiation temperature of Nordic Sea surface (through ice–water cross section) during expedition.

of these ices (Druzhkov et al., 2001, p. 386). The thick mass of ice as a whole can be subdivided into three structural zones according to biological characteristics: (1) zone of habitation of green algae; (2) transitive zone (combined freshwater–seawater composition of biota); and (3) zone of habitation of marine diatoms (Pennatophycea) (Melnikov, 1989; Druzhkov et al., 2001, p. 336).

In addition to diatoms and dinoflagellates, species such as infusorian and bacteria, have been found in ice.

In the course of melting of ice, these algae pass into the seawater.

The snow cover of ice, as well as ice itself, contains a great deal of various solid particles. Thus, the snow cover of ice in the Fram Strait was found to contain from 1.2 to 12.9 mg of suspended matter per liter of water (Lisitzin et al., 2001, p. 377). These solid particles are delivered to the Arctic predominantly by wind from the Asian continent. It should be mentioned that the total amount of aeolian input delivered to the Arctic Ocean basin is only on the order of 2.8–8.2 million tons annually, according to various estimates (Shevchenko et al, 2001, p. 388), whereas the largest input of sediments comes from rivers, making up 257 million tons per year (245 million in areas within the influence of the marginal filter, and 12 million in the open Arctic Ocean). Precipitation of aerosols in the Arctic occurs not onto the surface of water but onto ice. Suspended matter is released in areas of ice melting, i.e., predominantly in the northwestern part of the Greenland Sea, where all substances contained in ice (and lying on the ice surface) ultimately reach the seabed. Also, many chemical elements settle onto the bottom together with aeolian particles derived from ice. Aeolian material is a major source of such microelements as Pb, Sb, Se, and V. In general, aeolian input makes up about 10% of the total sediments delivered to the Arctic (Shevchenko et al., 2001, p. 390).

The intensity of development of biological communities in marine ice is especially large in dirty ice present in areas lying at the smallest distance from ice-free ocean waters, as well as in areas of ice melting.

In such a case, blooming occurs near the edge of melting ice. Suspended organic matter (including that in the form of large aggregates and floccules) may be deposited on the seabed or may be involved once again in the biological cycle and contribute to the blooming of phytoplankton (Ackley and Sullivan, 1994).

The phenomenon of "dead water" is observed in polar regions. This phenomenon is expressed as anomalously high resistance to ship motion due to the effects of internal waves generated at the interface between fresh and saline waters. Dead water is frequently observed during high water in the near-mouth areas of the Ob, Yenisey and Lena Rivers in the Arctic Ocean: in these rivers, the layer of water freshened due to the effects of flood is about 3–5 m thick (Fiodorov, 1983, p. 187). Furthermore, the seasonal thermocline most frequently coincides with the halocline here. Here, "the upper boundary of supercooled water is confined to the lower boundary of freshened surface water" (Golovin et al., 1999, p. 127, Fig. 2). "The supercooled layer itself at the weak inflow of river water occupies a small upper part of the pycnocline" (Golovin et al., 1999, p.139). A very typical feature of such areas is that internal waves are actively generated at the boundary between the pycnocline and upper, freshened layer of the water column. This is responsible for the intensification of processes such as mixing of waters and sinking of supercooled freshened waters to the lower layer of saline waters.

In the Arctic basin, lenses of freshened (river) water frequently depart the river–sea GBZ, with the result that they can be found in the open ocean a great distance from land. In the outer parts of such a lens, formation of both baroclinic and thermoclinic fronts takes place (Dmitrienko et al., 1999, p. 79, Fig. 5; see also Kassens et al., 1999). In the frontal part of the freshened water lens, the thermocline and pycnocline (halocline) have complex internal structure (Fig. 5, p. 79). "The discharge fronts of rivers in the Laptev Sea have a two-layer structure. In the depth interval from the sea surface to the depth of occurrence of penetrating river waters (seasonal thermocline), these fronts are baroclinic. In the lower layer, which extends down to the main pycnocline, the predominant mode of behavior of these fronts is thermoclinic" (Dmitrienko[3]). Dmitrienko et al. show in diagrams (Fig. 5) that there is intensive heat and salt exchange between the two types. The mechanism of this exchange is described in a special paper (Dmitrienko et al., 1999, pp.73–92).

Because glaciations in polar areas are of tremendous importance for geological processes, processes such as on-land destruction, transportation, modification alteration and deposition of sedimentary material have been united into a special type of lithogenesis which is called ice lithogenesis (Strakhov, 1960–1962). However, because not only land but also wide areas of the World Ocean (some 20 million km^2 or one-fourth its total surface) are covered by continuous fields of ice, ice floes and icebergs—and also because ice is the only agent responsible for preparation, transport and deposition of sedimentary material over large areas of the sea and ocean floors—A. Lisitzin (1994) identified and described an ice type of sedimentogenesis in the World Ocean. The very nature of the this type of sedimentogenesis in the World Ocean can be described as follows: "Due to the effect of relatively low temperatures, water as an agent which is responsible for preparation, transportation and deposition of sedimentary material is substituted by ice (i.e., water in the solid state), and this radically changes the course of all these processes" (Lisitzin, 1994, p. 5). Sea ice and icebergs are the main means whereby sedimentary material is transported from the source region into depositional basins. About 450 million tons of sedimentary material per year is delivered by each of the two agents. Deposits are made up of the sedimentary material from pelitic material ("glacial milk") to boulders about 1–10 m in size. In all, the amount of sedimentary material transported by ice and icebergs to the World Ocean is about 900 million tons per year (Lisitzin, 1994, p. 50).

Based on five year's worth of satellite observations, it has been found that in spring (November–December), the boundary of sea ice near the Antarctic (see Fig. II.5.1) corresponds to the boundary on the seafloor between typical low-siliceous glaciomarine muds (the south) and diatomic oozes (the north) (Burckle, 1984). In the Atlantic Ocean, this line is present in areas where sea ice covers 40–90% of the water surface (today, this boundary lies between 64 and 65°S). Glaciomarine, low-siliceous muds present to the south of this boundary are composed of glacier milk, aeolian particles (atmospheric dust), diatoms, and coarse ice-rafted debris. Diatomic oozes (>30% SiO_{2am}) have accumulated in areas lying to the north of the ice field boundary (40–90% of the total surface of the ocean) (see Fig. II.3.10). In the Antarctic, diatomic algae develop in areas north of polar front, but in spring and summer this zone extends farther south (Burckle., 1984; Burckle et al., 1992). It is the effect of ice melting which is responsible for the southward extension of this zone, rather than increased concentrations of nutrients. This is evidence for inferring that ice inhibits the development and reduces the productivity of diatomic algae in the ocean. Diatoms are the main component which finally forms diatomic oozes. Of course, diatomic algae also develop under the ice cover, but their productivity and role in the depositional processes are much smaller here.

In the course of geological evolution, this boundary shifted alternatively to the south or to the north. During the last glacial maximum (18,000 years ago), for example, the boundary between typical diatomic oozes and diatom-bearing glaciomarine sediments in the Southern Hemisphere was located within a band from 63 to 60° S, i.e. 2–2.5° closer to the equator than it is now (Burckle, 1982; Burckle et al., 1992).

The development of polar icecaps led to variations in climatic and oceanographic conditions not only in the vicinities of the poles but also everywhere in the World Ocean (Kennettt, 1987; Bleil and Thiede, 1990). The establishment of features such as cold bottom waters, their sinking along the slopes to deep ocean basins and their slow movement towards the equator was probably the chief process caused by glaciation in polar regions. Supercooling of seawater in polar regions results in an increase in their density. Such dense (heavy) water was transported down the continental slope to oceanic basins, and the process is still happening to this day. The strength of these bottom currents was so large that they led to the development of a giant river network (superrivers) on the ocean floor in polar regions (Fig. II.5.3). The valleys of these superrivers extend up to mid-latitudes, i.e., they can be traced as far south as 40–45°N. The result was that the deep-sea waters became well aerated and this, in turn, had an effect on (1) the biota (living conditions for cold-loving organisms became better); (2) the oxidation of organic matter; (3) the hydrodynamic conditions (increasingly high bottom currents velocities were responsible for intensive reworking of sediments), which led to the development of erosional and non-depositional hiatuses in some areas and intensive deposition of sediments (sedimentary ridges and so on) in other areas of the ocean (see Part II.14); (4) the activation of upwellings and transfer of nutrients into the photic zone, which, in turn, led to rapid development of phytoplankton; (5) an increase in pCO_2 and a rising of the CCD to higher levels, expansion of distribution areas of oxidized pelagic clays and active formation of pelagic FMNs (Kennett, 1987; Lisitzin, 1974, 1978, 1994; Emelyanov et al., 1991).

The edge of "eternal ice" is a BZ: in areas between such zones and poles, sedimentary material is prepared exclusively due to physical processes (Emelyanov et al., 1975; Lisitzin, 1994). Chemical elements in suspended matter are found in hard (clastic) forms or within their crystal lattices. Contents of organic matter, $CaCO_3$ and P present in moraine deposits and bottom sediments are also very small (Table II.5.1). Rapid formation of biota occurs only near the edges of glaciers (and sea ice) and within ice itself, and the rate of its production reaches maximum in the zone of Antarctic convergence (Voronina, 1977). The distribution of diatoms in the suspended matter of seawater and bottom sediments of the oceans in the Southern Hemisphere is consistent with the pattern of hydrologic fronts: the northern boundary

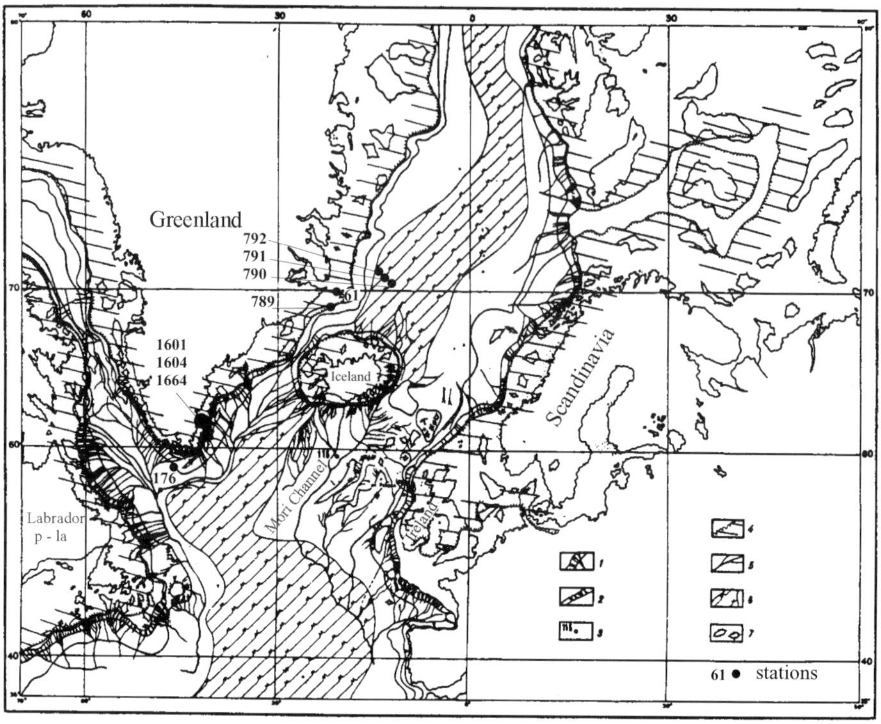

Fig. II.5.3. Drainage system on ocean bottom (North Atlantic) in connection with flow of heavy ice waters and glacial gravitites. According to Matishov, 1987.
1—underwater ravines and gullies; 2—underwater canyons; 4—ancient ice coverings; 5—abyssal channels; 6—zone of mid-oceanic ridge; 7—big submarine elevations and plateau.
Numbers on scheme—geological stations.

of the area characterized by maximum contents (>50%) of the Antarctic complex of diatomic algae corresponds to the boundary of the Antarctic divergence, and the northern boundary of the area dominated by species of the sub-Antarctic complex corresponds to the boundary of the Antarctic convergence (Jouse et al., 1963).

In the vicinity of the Antarctic, i.e., south of low-siliceous glaciomarine mud, there occur pelitic glaciomarine sediments with low contents of $CaCO_3$ (commonly <5–10%), C_{org} (commonly <1%) and SiO_{2am} (commonly <10%, but locally up to 20–30%, mainly at the expense of sponge spicules) (Table II.5.2). Because the edge of eternal ice migrates alternatively to the north or south depending on the season of the year (as was reported above), it could be expected that there is no distinct boundary between typical sediments and diatomic oozes on the ocean floor (see Fig. II.3.10), but this is not the case. The edge of the continuous glacier is well marked also in bottom sediments (Burckle, 1984).

Using the Weddel Sea as an example, models for describing the changes in the distribution of sediment types and foraminiferal complexes have been developed

Table II.5.1 The content of some chemical components and elements ($CaCO_3$-P in %, Ni-Y in 10^{-4}%) in moraine and marine-glacier (iceberg) deposits and in the recent (Holocene) diatomic oozes and terrigenous muds of the seas and oceans

Type of deposits and sediments	$CaCO_3$	C_{org}	Al	Fe	Mn	Ti	P	Ni	Cr	Co	Cu	V	Zr	Ga	Y
Tills of the Baltic Sea bottom (average)[1)]	11.49	0.3	-	3.12	0.06	0.05	0.05	62	59	22	25	-	-	-	-
The same Barents Sea [2)]	-	-	-	6.27	0.05	-	-	41	-	-	24	133	-	-	-
Varved clays, Baltic Sea	15.03	0.52	-	3.82	0.06	0.45	0.05	51	99	37	31	-	-	-	-
Marine iceberg deposits [4)] of the Antarctic	-	-	-	3.77-3.88	0.14-0.13	-	-	-	37-75	-	-	137-93	200-115	-	-
Terrigenous iceberg[5)], Indian sector	-	-	5.6 (6.4)	-	-	-	-	18 (20)	50 (54)	11 (12)	-	52 (59)	139 (156)	15 (14)	28 (31)
Diatomic oozes[3)] (30-50% SiO_{2am}), Indian sector	-	-	2.7 (4.3)	-	-	-	-	8 (26)	40-67 (50-72)	6 (26)	-	34 (88)	41 (71)	12 (16)	11 (18)
Terrigenous outside of the iceberg zone[2)]	4.78	0.78	-	4.86	0.11	0.55	0.06	65	82	25	38	-	-	-	-

1) Contains on average: 2.56% K; 0.64% Na; 0.0136% Li, 0.0076% Zn.
2) Cenozoic deposits, according to Kharin and Emelyanov, 1987. Contains on average: 2.35% K; 1.72% Na; 0.0113% Zn.
3) According to Biorlikue et al., 1978 (cited according to Lisitzin, 1994, p. 275).
4) According to Angino, 1966 and Frakes and Crowell, 1975, respectively (according to Lisitzin, 1994, p. 275).
5) According to Lisitzin, 1994, p. 275.
The values in brackets are given in carbonate-silica (amorphous)-free basis (csfb).

Table II.5.2 Average granulometric and chemical composition of iceberg deposits and diatomic oozes of the ocean zone near the Antarctic, %

Fractions (in mm) and components	Iceberg deposits		Diatomic oozes	
	Limits	Average	Limits	Average
Grain size distribution				
>0,1	0.01–78.99	-	Tr.39.1	7.4
0,1–0,05	0.76–64.00	-	0.60–26.17	6.12
0,05–0,01	0.29–45.27	-	4.82–61.15	21.51
0,01–0,005	0.52–82.80	-	1.75–86.52	37.00
0,005–0,001	0.17–28.40	-	0.15–23.08	10.75
<0,001	5.52–49.77	-	10.03–43.75	20.45
Md	0.0048–1.5	-	0.0055–0.01	0.0092
So	1.2–9.2	-	1.55–8.5	2.65
Content of chemical components and elements				
$SiO_{2am.}$	0.79–25.32	9.32	10.0–72.69	43.67
$CaCO_3$	0.11–15.37	2.39	0.04-40.32	13.88
Corg	0.04–1.16	0.39	0.08–0.82	0.28
Fe	0.85–8.04	3.36	0.01–5.04	1.66
Mn	0.02–0.15	0.06	cn.–0.57	0.11
P	0.013–0.23	0.06	cn.–0.12	0.035
TiO_2	0.16–2.80	0.58	cn.–0.76	0.29

Tr—traces

(Anderson, 1972; Kennett, 1986, p. 55). These models made it possible to distinguish two stages for a glacier with a dry base (i.e., grounded glacier) and two stages for a glacier with an aqueous base. In the first case, the main process is driven by the mechanical impact of ice upon the surface of the seafloor in the presence of a sedimentary environment characterized by essentially complete inertness of biological and chemical processes. In models of a glacier with an aqueous base, the geological process is much more complicated. In addition to the effects of mechanical (hydrodynamic) impact exerted on the bottom, the sedimentation rate is also increased here, first of all, near the edge of the glacier. Also, signs of vital activity of phytoplankton and benthos appear there. The edge of the glacier corresponds to the boundary of facial *Neogloboquadrina pachyderma* (Fig. II.5.4). This species is abundant in sediments on the ocean floor under ice floes and icebergs (*G. pachyderma* is absent in areas covered by continuous ice; deposits in these areas are inhabited only by shallow-water forms of agglutinated foraminifera). In the Arctic Basin, the edge of continuous ice is well marked by strong settling of agglutinated and non-agglutinated foraminiferal tests *Neogloboquadrina pachyderma* sin.). Under continuous ice, this type of foraminifera is absent and the amount of this species in open water with ice floes is insignificant. Facies of pelagic agglutinated foraminifera occur in areas between the zone of maximum density of icebergs and towards to the equator.

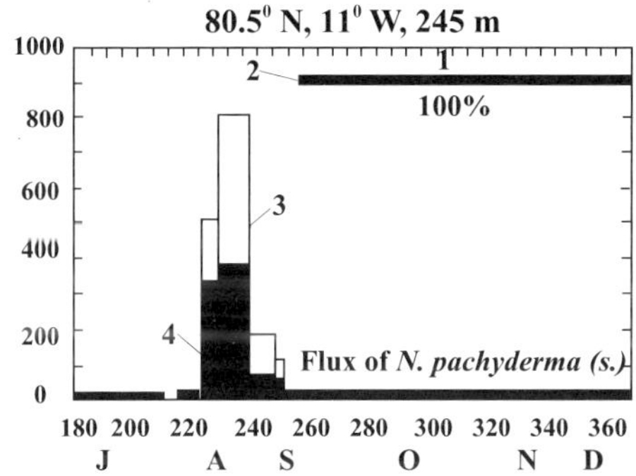

Fig. II.5.4. Flux of shells of *Neogloboquadrina pachyderma* (s) in region of New Polynya in Arctic ocean. After Kohfeld and Fairbanks, 1996.

Vertical scale—number of shells m^2/day; horizontal scale—months.

Main flux of encrusted and nonencrusted *N. Pachyderma* (s), which amount to about 50%, is deposited pulse-like in August. Fluxes of foraminifera on bottom are small.

Legend: 1—100% of compact ice; 2—edge of compact ice; left of it—free water with floating ice; 3—encrusted shells; 4—nonencrusted shells.

In addition to the models cited earlier, many other models of cryogenic and normal marine sedimentation in seas and oceans have been developed (Anderson, 1972; Kennett, 1987; Emelyanov, 1982_1; Matishov, 1984, 1987; Elverhoi, 1984; Domack, 1988). All these models indicate that the regime of sedimentogenesis is related to the location of the edge (barrier) of continuous glaciers and accumulations of ice floes and icebergs. We present two of these models, which are considered to be the most generalized (Figs. II.5.5, II.5.6).

As mentioned in Part I.6 (p. 57), in the ice zone of sedimentogenesis, the main agents for transporting sedimentary material to the sea are glaciers, and in the sea, ice floes and icebergs. This explains why there are large contents of coarse (>1 mm) clastic and silty-pelagic fractions having a grain size of 0.05–0.001 mm (glacier milk), and also very low contents of clay (<0.001mm) fractions. The ice zone is also characterized by very good preservation of minerals and rock fragments, which are transported within an "ice package" to distances of up to 1000 km from the ice basin. Owing to the presence of ice floes, wave abrasion in these areas is virtually absent. As a result, processes of mechanical differentiation are strongly diminished here, and grain-size differentiation of sediments is either poor or very poor (Emelyanov, 1982_1) (Table II.5.3, Fig. II.5.7).

Marine–glacial sediments of the Iceland–Greenland trench and Colbensay Ridge (the Greenland Sea) are represented by pelitic and aleuritic muds containing large amounts of coarse material. These sediments are dominated by glacial and iceberg material from northern Greenland. In sediment core AK-790 from this area (Tables II.5.4, II.5.5; Fig. II.5.8), which has been studied in more detail (a horizon of 20–217 cm), contents of minerals (in percent of the total core recovered) are as follows: quartz 10–20%; plagioclase, 9–24%; potassium feldspar, 3–9%; calcite, 0–7%;

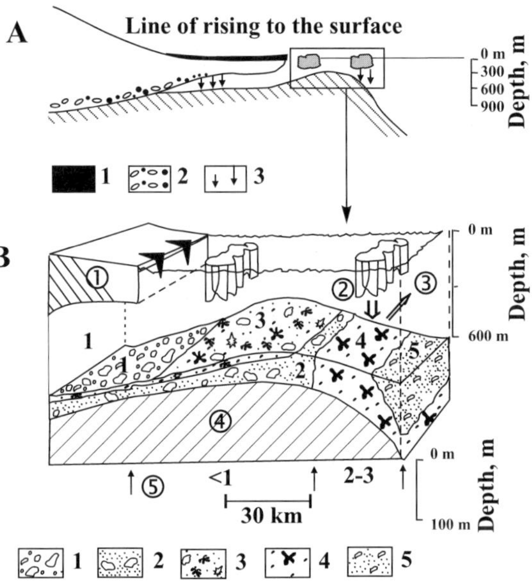

Fig. II.5.5. A model of marine cryosedimentation, which is based on studies of sediments in Weddel Sea and adjacent Riser-Larssen glacier. After Elverhoi (1984), with some additions by Lisitzin, 1993.

A—schematic diagram: 1—accretion of snow on shelf glacier; 2—moraine material at base of glacier (bottom moraine), which is about 30 m in thickness. Main zone of suspended matter deposition from water is found near zone where water moves towards surface; 3—scattered glacial material;

B—formation of ice deposits on outer parts of shelf (fragments): 1—homogenous mud with pebble and gravel; same material, of Wisconsin age; 3—sediment enriched in bioclastic material (more than 30%) (bryozoans, sponges); 4—soft mud with pebbles, containing from 10 to 30% bioclastic material; 5—soft mud with pebbles.

Numbers in circles: 1—shelf glacier, 2—deposition of glacial material; 3—evacuation of fine fractions; 4—floor of glacier; 5—rate of sedimentation, cm/1000 yr.

dolomite, 0–2%; amphiboles, 0–3%; illite, 27–39%, montmorillonite, 3–13%; chlorite, 4–7%; kaolinite, 4–6%; X-ray amorphous phase, 3–17%. The ratio of quartz to feldspar is 0.5 : 1.1. In glacial–marine muds (station AK-790), contents of potassium feldspar, plagioclase, illlite, K, and Na are distinctly higher than those in marine terrigenous muds (Table II.5.6). Contents of biogenic components in cores 791, 792 and 793 are low: 1.89–6.75% $CaCO_3$, 0.90–1.43% SiO_{2am}, and 0.13–0.73% C_{org} (Table II.5.7). Muds from all mentioned cores contain relatively large amounts of volcanoclastic (markedly weathered) material: in the heavy 0.1- to 0.05-mm subfraction, the content of ash particles ranges from 7 to 80%; monoclinic clinopyroxenes, 3–32%.

In glacial–marine sediments from the coastal area at the southern part of Greenland (from the Danish Strait to the south), contents of terrigenous SiO_2 are much higher and contents of $CaCO_3$ and Al are somewhat smaller than those in terrigenous clastic deposits from the Great Britain–Gulf of Biscay and

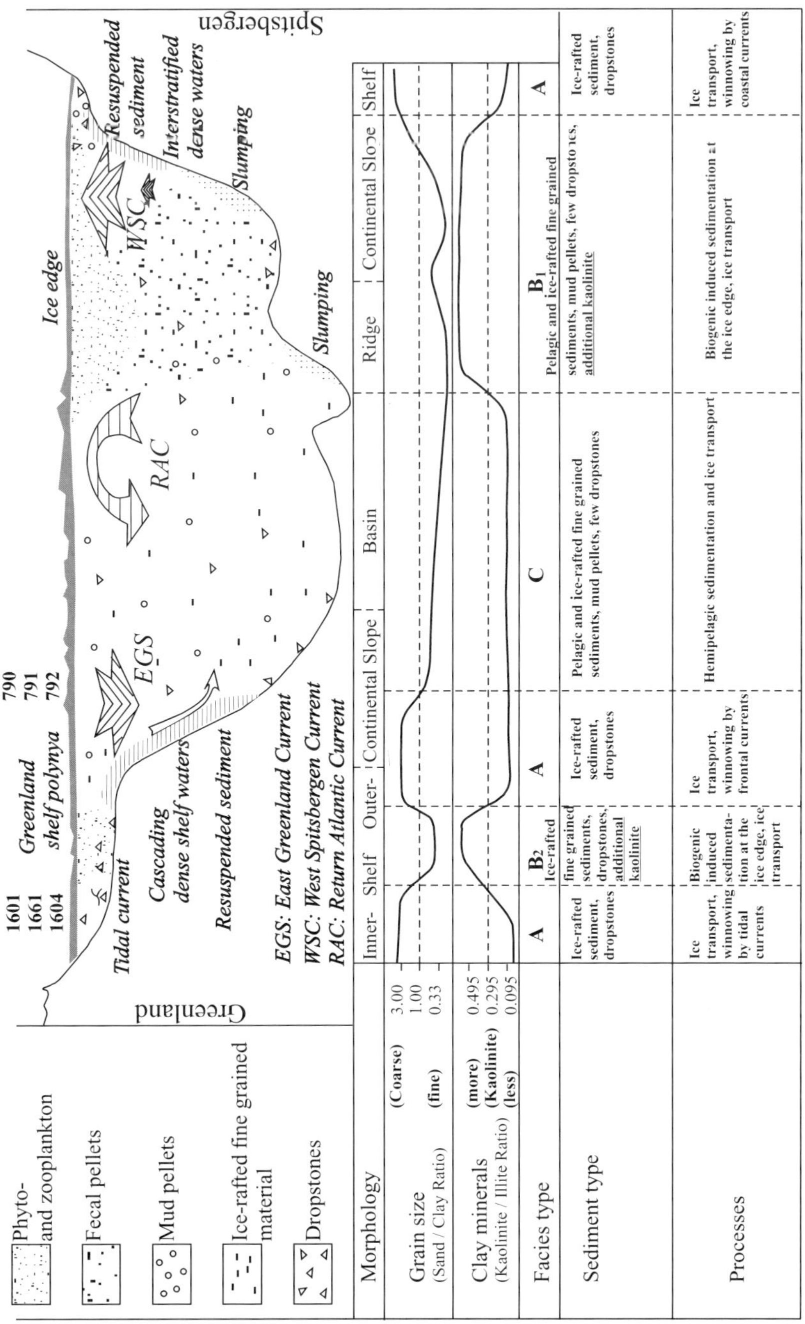

Caption see page 212

Table II.5.3 The composition (%) of the marine-glacial sands and fine-aleuriticmud off the eastern coast of Greenland

Fractions (mm), and chemical elements	Station and depth, m				
	A-1664	A-1601	P-61	A-1661	P-176
	210	250	340	550	3030
Grain size distribution, %					
>1.0	25.2	3.7	22.5	28.2	-
1.0–0.5	13.0	7.9	15.3	5.4	-
0.5–0.25	16.6	14.4	16.9	6.3	-
0.25–0.1	23.0	23.7	15.1	11.2	-
0.1–0.05	6.3	17.7	4.7	12.7	13.8
0.05–0.01	6.9	12.8	9.3	17.4	35.6
<0.01	9.0	19.8	17.2	18.8	29.0
>0.1	77.8	49.8	69.8	51.1	21.6
1.0–0.1	52.6	46.1	46.3	22.9	-
0.1–0.01	13.2	30.5	14.0	30.1	49.4
Md	0.30	0.10	0.31	0.11	0.03
So	3.2	3.8	2.4	9.0	3.7
Chemical components and elements (natural dry samples)					
$CaCO_3$	7.80	-	3.73	6.50	13.76
SiO_{2am}	2.36	-	1.69	3.01	0.65
C_{org}	0.60	-	0.42	0.80	0.15
Fe	5.24	-	9.15	4.05	4.35
Mn	0.05	-	0.16	0.06	0.08
Ti	-	-	0.62	-	0.65
P	-	-	0.09	-	0.10
Sediment type					
	Sand (mixed)	Fine sand	Sand (mixed)	Sand (mixed)	Fine-aleuritic mud

Newfoundland–Labrador provinces. As for the rest, features of terrigenous and glacial–marine clastic sediments in the aforementioned three provinces are fairly similar (Emelyanov and Kharin, 1987) (Table II.5.8).

The retreat of a shelf glacier with an aqueous base (subglacial water) due to melting processes is responsible for the appearance of fluxes of fresh water in the ocean. This fresh, mix of ice and water with seawater forms flows of low-salinity water which are directed from subglacial reservoirs into the open ocean. This low-salinity

Fig. II.5.6. Ice-edge-related sedimentation: Conceptual model. After Berner and Wefer, 1990. Numbers of geological stations, cited in tables, and approximate positions in model are shown above by arrows.

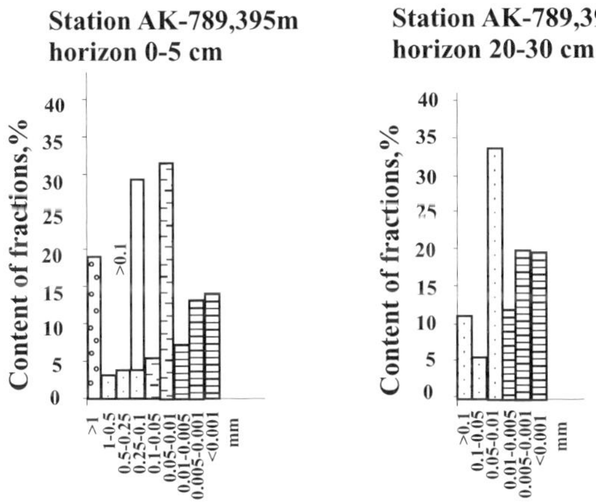

Fig. II.5.7. Histograms of grain-size distribution of surfical (0–5 cm) marine-glacial mud of Greenland Sea (station AK-789, depth 395 m, 68^{0}29'.8 N, 25^{0}08'.1 W, for station's location see Fig. II.5.3).

water forms the upper water mass; seawater forms the lower water mass. Owing to such a stratification pattern, water exchange is hindered in the deeps of the shelf (troughs) and this may lead to stagnation of near-bottom water (Kennett, 1987, p. 56). However, the reverse is also the case: subglacial waters, especially under the part of glacier which overlies shelf, tend to contain more dissolved oxygen. This process is facilitated by the effects of ebb tide phenomena. During ebb tide, water is evacuated from the subglacial space, while in time of high water, it is delivered into the subglacial space. All these processes lead to the development of strong subglacial (tidal) currents (see Fig. II.5.6). In addition, internal waves are generated at the interface between the upper mixed low-salinity water layer and lower layer of normal salinity (i.e., in a layer which may be considered as the halocline). All these factors, together with ebb and tidal currents, contribute to good ventilation in subglacial areas which are open only from the seaward side, as well as in ice-covered basins (Lisitzin, 1994, p. 33).

In the course of evolution, the geographical positions of the edge of eternal ice have changed. In a corresponding way, as evidenced by the geological record (stratigraphic section), two variants are possible: either some sediments have been substituted by sediments of other types, or the content of components (remains of diatoms, foraminifera, sand particles and other material which make up these sediments) have changed. For example, since the Early Miocene, a global increase in the level of the World Ocean took place. At the threshold of the Early and Middle Eocene, this level was higher than it is today, but at the very beginning of the Middle Eocene (approximately 13–14 million years ago), the development of the ice cap in the Antarctic began. The climatic changes from non-glacial to glacial regimes are clearly marked in the sediments of the peripheral Pacific (Ingle, 1981): at this stage, terrigenous sediments which were abundant in peripheral areas of the ocean were

Table II.5.4 Lithological description of the terrigenous marine-glacial mud of the Greenland Sea

Horizon, cm	Description
AK-790, 1735 m, 71°09′4 N, 17°49′9 W	
0–37	Aleuro-pelitic and pelitic mud, semiliquid (0-5 cm), viscous, brown, fat. Gradual contact
37–46	Pelitic mud, denser, soft, plastic
46–50	Pelitic mud, viscous, brown. Sharp horizontal, contact. Grain of quartzite (about 1 cm) in the 50-cm horizon
50–125	Pelitic mud, gray, plastic, viscous.
125–135	Aleuro-pelitic mud, plastic, dark gray. Gradual contact
135–240	Pelitic mud, viscous, plastic, gray. Debris of the sandstone (1–1.5 cm) in the 210- and 220-cm horizons
AK-791, 1550 m, 70°52′0 N, 16°22′0 W	
0–17	Aleuro-pelitic and pelitic mud, soft, plastic, brownish-gray. Fragment of the rock (1 cm) in the 5-cm horizon
17–206	Pelitic mud, viscous, soft, plastic, homogenous, brownish gray. 67.5–68 cm—interlayer of the aleuro-pelitic mud
206–210	Pelitic mud, viscous, fat, soft, plastic. The gravel and pebbles of hard rock in the 140-, 155- and 170-cm horizons
AK-792, 1420 m, 70°44′0 N, 15°42′0 W	
0–10	Aleuro-pelitic mud, viscous, fat, plastic, brownish gray
10–185	Aleuro-pelitic mud, viscous, with pebbles of metamorphic rocks. Brownish gray color with patchiness
185–208	Aleuro-pelitic mud, viscous, soft, plastic with the gravel and pebbles of metamorphic rocks, brownish gray
208–216	Aleuro-pelitic mud, consolidated, bluish gray
216–224	Aleuro-pelitic mud, viscous, dense, brownish gray.
	Gravelly-pebbles fragments on the 6-, 78—80-, 90-, 91-, 106-, 120- and 170-cm horizons

replaced almost everywhere by diatomic oozes. Later, these oozes were overlain by clastic glacial–marine sediments in the vicinity of centers of glaciation.

The time when sedimentary material derived from ice began to appear sporadically in the North Atlantic dates back 10.2 million years ago. However, a major increase in delivery of ice-borne sedimentary material took place 2.6 million years ago (Thiede

Table II.5.5 Grain size distribution (in %) of marine-glacial sediments of Greenland Sea (fractions in mm)

Horizon, cm	>0.1	0.1–0.05	0.05–0.01	0.01–0.005	0.005–0.001	<0.001	<0.01	0.1–0.01	Sed. type
Station AK-792, depth 1485 m., 70° 44′ 00N 15° 42′ 00W									
0–10	1.9	2.3	27.6	17.7	32.9	17.6	68.2	29.9	Apm *)
80–90	6.5	5.4	23.4	18.8	27.2	18.8	64.8	28.8	Apm
170–180	5.4	2.3	25.1	17.9	42.6	6.7	67.2	27.4	Apm
208–216	5.3	3.8	25.7	13.5	45.0	6.8	65.3	29.5	Apm
Station AK-791, depth 1585 m., 70° 52′ 00N 16° 22′ 10W									
0–8	2.7	3.3	26.1	13.1	39.2	15.7	68.0	29.4	Apm
8–17	1.7	2.4	25.0	14.7	46.5	9.8	71.0	27.4	Pm
90–100	2.9	3.0	21.9	6.2	47.5	18.5	72.2	24.9	Pm
170–180	1.6	2.2	20.5	10.8	54.4	10.8	75.6	22.7	Pm
206–210	6.7	4.9	23.7	7.2	39.5	18.0	64.7	28.6	Apm
Station AK-790, depth 1735 m., 71° 09′ 40N 17° 49′ 90W									
0–10	1.1	1.7	34.0	15.2	37.1	10.9	63.2	35.7	Apm
10–20	0.5	1.2	25.0	18.3	39.0	16.0	73.3	26.2	Pm
20–30	1.0	1.5	23.4	15.9	42.3	15.9	74.1	24.9	Pm
42–43	0.7	2.3	29.8	8.8	35.1	23.4	67.2	32.1	Apm
43–50	0.6	2.5	34.0	11.9	28.9	22.1	62.9	36.5	Apm
70–80	0.1	0.1	16.7	10.2	64.9	8.1	83.1	16.8	Pm
90–100	5.1	1.5	20.7	12.1	41.6	19.1	72.7	22.2	Pm
110–115	4.8	3.2	27.4	6.7	33.5	24.5	64.6	30.6	Apm
135–140	1.6	2.0	30.2	8.3	31.1	26.9	66.3	32.2	Apm
160–170	1.6	1.0	21.0	8.3	38.2	29.9	76.4	22.0	Pm
180–190	0.5	2.1	30.6	10.5	31.6	24.6	66.7	32.7	Apm
200–210	6.9	0.7	15.7	13.3	58.3	5.0	76.7	16.4	Pm
228–240	1.0	1.9	24.2	10.7	25.8	36.5	73.0	26.1	Pm

*) Sed. type–sediment type: Apm – aleuro-pelitic mud, Pm – pelitic mud

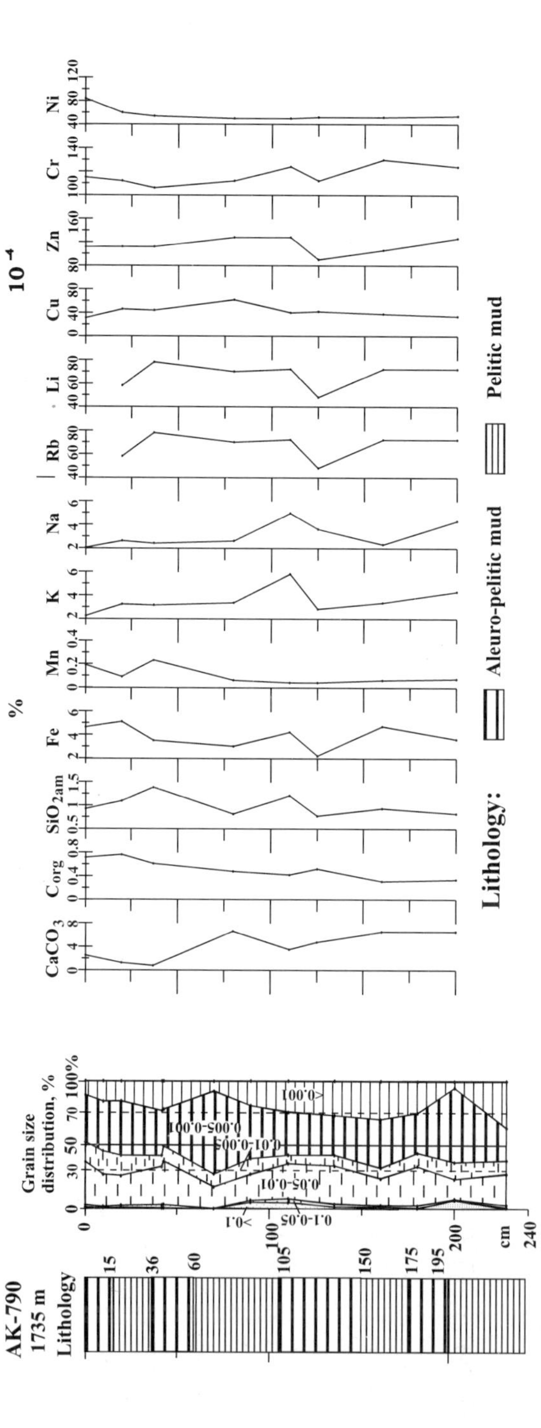

Fig. II.5.8. Lithology, grain-size distribution and content of some components and elements in sediment core AK-790 (depth 1735 m), Greenland Sea (for coordinates see Table II.5.7).

Table II.5.6 Mineral composition of the fraction 0.1-0.05 mm of the marine-glacial mud of the Greenland Sea

Station	Horizon, cm	Minerals													
		Q	Plag	Cal	Dol	V.g.	W.p.	Horb	Ore m	Lim	Zir	Gar	Bio	Gla	Foram
AK-791	0-3	xxx	xx	x	x	-	xx	xx	x	xx	x	x	-	-	x
	67.5-68	xxx	xx	x	-	-	xx	xx	x	xx	-	-	x	-	x
	140-141	xx	xx	x	-	-	xx	x	x	x	-	-	-	x	xx
	180-181	xxx	xx	x	-	-	x	xx	xx	-	-	-	x	x	xx
	206-210	xx	xx	-	-	-	x	xx	x	x	-	-	x	-	xxx
AK-792	0-5	xx	xxx	x	-	x	xxx	xx	x	-	x	x	-	-	xx
	150-160	xxx	xx	x	-	-	xx	xx	xx	-	-	-	-	-	x
	208-216	xx	xx	x	-	-	xx	xx	x	-	-	-	-	-	x

Minerals: Q – quartz; Plag – plagioclase; Cal – calcite; Dol – dolomite; V.g. – volcanogenic glass; W.p. – weathered particles; Horb – horblende; m – minerals; Lim – limonite; Zir-zircon; Gar – garnet; Bio – biotite; Gla – glauconite; Foram – foraminifera.

Content: xxx – abundant; xx – average content; x – present; – not detected.

Table II.6.3 Concentration of chemical elements in hydrothermal solutions. According to various data (from Lisitzin et al., 1993, p. 210)

Component, element	Unit of measurment	EPR, 21°N	EPR,13°N	Chuan-de Fuca Ridge	Explorer Ridge	Guyamas Deep	Galapagos rift	TAG, Atlantic ocean	Ocean water
Na	g/kg	9.9–11.7	12.9–15.1	11.1	-	10.9–11.8	-	13.432	10.67
K	"	0.91–1.01	1.16–1.31	-	1.7–1.8	1.18–1.92	0.73–0.74	968	0.383
Mg	"	0	0	0	0	0	0	0	0
Ca	"	469–834	2200–2750	260	1680–1730	1040–1660	985–1610	1040	409
SiO_2	µg/kg	937–1170	1320–1390	900	470–890	560–830	1320	1070–1100	9.6
Cl	"	17.3–20.5	26.2–32.4	18.8	20.4	20.6–22.6	-	23.36	19.18
H_2S	"	224–285	-	180	57–77	130–204	327	198–204	0
Br	"	-	93		-	-	-	-	-
Li	"	6.2–9.2	4.8–5.4	3.6	5.5–5.6	4.4–7.5	4.8–7.9	-	0.19
Rb	"	2.3–2.8	1.2–1.6	4.2	4.8	4.9–7.35	1.15–1.8	-	0.11
Sr	"	5.7–8.5	15.3–20.6	14.5	12.5	14.0–22.0	7.6	9	7.7
Ba	"	4.8–13	4.7	2.5	0.9	1–5.8	2.4–5.8	-	0.20
Fe	"	42–136	53–103	2.2 (872)	0.56–3.1	0.95–10	-	103–119	0.00006
Mn	"	38–55	44–86	24(195)	14–15	7.0–19	20–63	26.5–28	0.000012
B	"	5.4–5.9	4.9–5.3	49000	-	16.8–18.6	-	-	4.5
Zn	µg/kg	2600–6900	-	<100	5–72	6.5–2600	-	-	0.65
Cu	"	1,–2800	-	-	1–11	1.3–70	-	-	0.45
Al	"	110–140	-	-	-	0.11–2.5	-	-	1.0
Ag	"	0.11–4.1	-	-	-	1.1–5.2	-	-	0.002
Cd	"	1.9–20	-	-	-	4.1–135	-	-	0.1
Pb	"	38–74	12	186	-	-	-	-	0.002
Co	"	1.3–13	-	-	-	0.3	-	-	0.0028
As	"	2.2–34	-	-	-	21–80	-	-	2
Se	"	5.7	-	-	-	6.5	-	-	0.2
Be	"	0.09–0.33	-	-	-	0.11–0.82	0.10–0.33	-	0.00018
U	"	0	0	-	-	-	-	-	3.0
Ce	ng/kg	500–1700	1900–4000	-	-	-	-	-	1.36
Nd	"	140–500	1100–2100	-	-	-	-		5.52
Sm	"	30–140	180–620	-	-	-	-	-	1.02
Eu	"	50–280	400–900	-	-	-	-	-	0.28
Cd	"	40–90	150–340	-	-	-	-	-	1.56
Dy	"	30–70	80–240	-	-	-	-	-	1.75
Er	"	20–36	30–100	-	-	-	-	-	1.76
Yb	"	22–34	30–100	-	-	-	-	-	1.90
T °C		273–375	317	37	25–306	50–315	20	350	2–3
pH		33–3.8	3.3–7.6	6.18	4.62–7.26	5.85–5.90	-	3.6–5.02	7.8

EPR – East Pacific Rise

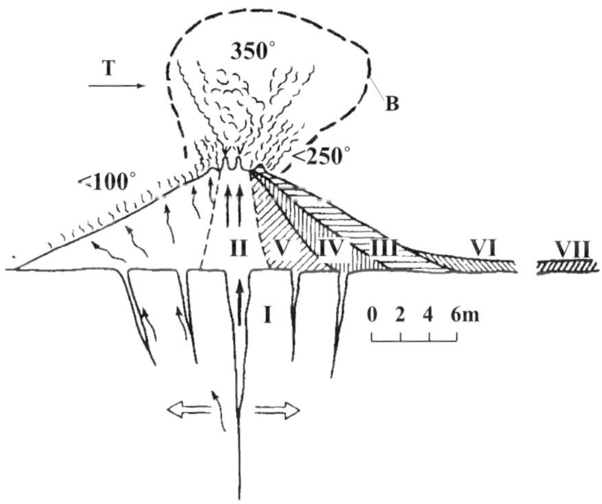

Fig. II.6.1. Generalized scheme of sulfide-mineral allocation in hydrothermal construction. After Hekinian and Fouquet, 1985, with additions.
I—sulfides precipitating in center (pyrite, copper sulfides, pyrrhotine and phases of silicon dioxide; these also meet in lodes in streams of changed basalts (stockworks), II—sulfides that have formed at formation of pipes (copper and zinc sulfides). III, IV and V—series of sulfide minerals that have formed at infiltration of hydrothermal solutions: III—low-temperature phases, such as associations of pyrite–marcasite–collomorphal pyrite–silicon dioxide; IV—medium-temperature phases, such as associations of pyrite–zinc sulfides; V—high-temperature phases, such as associations of pyrite–copper sulfides; VI—precipitations of high-temperature sulfide phases, which form oxidated precipitations around hydrotermal construction. Among them is an occasional association pyrrhotine ± pyrite ± zinc sulfides. Anhydrite is promptly dissolved; VII—metalliferous precipitations.
Areas of hydrothermal system: I—area of mobilization of metals and emergence of solutions (zone of leaching); II—area of solution motion.
Temperatures of hydrotermal solutions are shown above. B—approximate arrangement of a temperature barrier (it is shown outside of scale), T—sea current (see Fig. II.6.3).

A summary of data compiled by Y.A. Bogdanov and A.P. Lisitzin (1986, p. 200) in relation to the hydrothermal–sea water system indicates that:

—SiO_2 is intensively derived from rocks by solution (the reverse is periodically the case when SiO_2 is delivered into rock) during the development of metamorphic complexes enriched in epidote;
—CaO is intensively evacuated from rock;
—MgO and H_2O are intensively supplied into rock;
—Fe is derived from rocks in some cases and is accumulated in secondary minerals, in particular, and in other cases in sulfides;
—Na_2O, K_2O, MnO are most commonly derived from rocks;
—The mobility of TiO_2, P_2O_5 is very low.

Authigenic minerals and chemical elements precipitate at the temperature, oxidation, hydrodynamic and other barriers of the hydrothermal fluid–seawater system to form either ores (in close proximity to black smokers) or metalliferous sediments enriched in ore elements (Fig. II.6.2; Tables II.6.5, II.6.6). Essentially important is

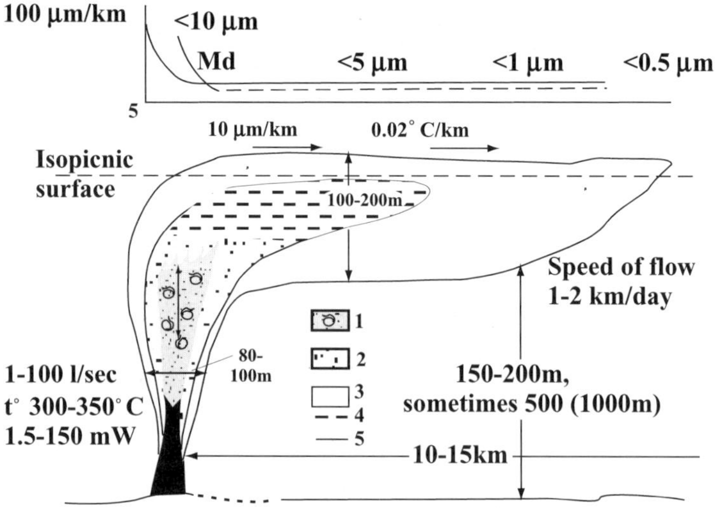

Fig. II.6.2. Mineralogical and geochemical differenciatin in middle hydrothermal torch. After Lisitzin et al., 1993.
1—a reducing part of torch, predominant elements Fe+S, precipitations of sulfides (pyrite, pyrrhotine, sphalerite, chalcopyrite), coprecipitation of microelements with sulfides, predominance of particulate matter over solutions, sulfo-bacterium, large amount of H_2S; 2—neutral part of torch—precipitation of sulfates and sulfides, predominance of suspended material; 3—oxidizing part of torch, predominant elements Fe, Mn, O_2, precipitation of oxyhydrates Fe and Mn, coprecipitation of micro-elements with oxyhydrates, predominance of solutions and most fine particulate matter, bacterium-metanotprophs, predominance of solutions and gases (CO_2, CH_4, $^3H_2$, ^3He) in external zone of oxidizing part of torch. Below formation of metalliferous deposits of two types: with positive anomaly of Eu (4) and with negative anomaly of Ce (5). Above—average median diameter of torch's particles (Md) and gradient of changes in sizes (μm/km).

the fact that the bottom sediments (which have an age similar to that of hydrothermal solutions) are enriched in hydrothermal elements only at the bathymetric level, which is found at depths where black smokers occur. So, with a hydrothermal solution at a depth of 3620 m (field TAG in the region of the Mid-Atlantic Ridge), sediments are enriched in "hydrothermal" metals only within the depth interval of 3500–3680 m (Table II.6.7; Figs. II.6.2, II.6.3, II.6.4; Lisitzin et al., 1993). If the hydrothermal plume happened to reach a higher level or was carried away and came in contact with the slope of the rift valley or an individual mountain, sediments deposited on the slope are considered to be enriched in oxide minerals of manganese and, evidently, in accompanying elements only at the level of this plume (see Fig. I.16 in Lisitzin et al., 1993, p. 70).

The regions of hydrothermal vents are often unique oases of life, which had been almost unknown until now and are found not everywhere on the surface of the earth, but only within the hydrothermal–sea water GBZ (Lonsdale, 1977; Edmond and Damm von, 1985).

The hydrothermal–sea water system is the basis of a chemosynthesis system (Lisitzin et al., 1993). Special areas form around hydrothermal features for the devel-

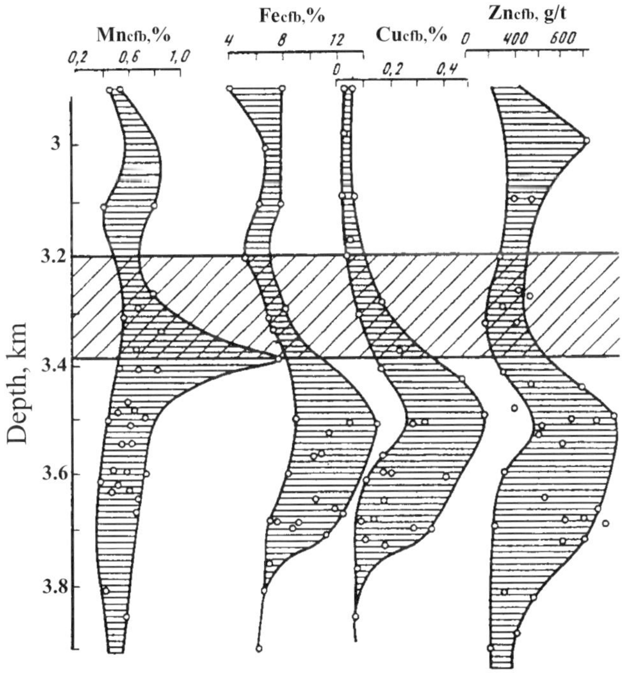

Fig. II.6.3. Content of main hydrothermal elements in surficial bottom sediments of TAG field, Atlantic Ocean (%, recalculated on carbonate free basis—cfb) depending on depth. After Lisitzin et al., 1993.

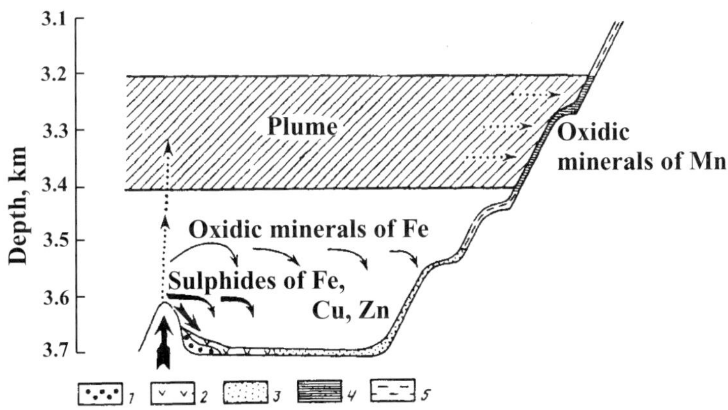

Fig. II.6.4. Scheme of a dispersion of hydrothermal sediment material. After Lisitzin et al., 1993.
1—deposition of stony deposits; 2—metalliferous sediments such as type I (see in text); 3—metalliferous sediments such as type II; 4—sediments enriched with manganese; 5—background sediments.

Table II.6.4 Flu x streams of hydrothermal matter into the ocean. After Corliss et al., 1979, p. 1073

Element	Hydrothermal flu x, mol/year	Global rivers flu x, mol/year
^{3}He	1.08×10^3	-
Li	-	13.5–10^9
high (max.)	130×10^9	-
low (min)	75×10^9	-
Mg	-9.3×10^{12}	5.5×10^{12}
Ca	-	12 × 1012
high	4.2×10^{12}	-
low	2.4×10^{12}	-
Ba	-	1.4×10^{10}
high	6.4×10^9	-
low	2.4×10^9	-
Si	2.9×10^{12}	7.1×10^{12}
CO_2	1.3×10^{12}	-
Ni	-25×10^6	-
Mn		
high	190×10^9	-
low	74×10^9	-
Alk	-1.3×10^{12}	30×10^{12}

opment of unique biocenoses: great amounts of planktonic and benthic organisms (values of biomasses are very large) living here are fed by organic matter formed from products of bacterial synthesis. Great accumulations of bacteria are observed in the surroundings of hydrothermal systems: some hydrothermal constructions (chimneys) are draped with continuous cover made up of mollusks, shrimps (*Remicaris sp.* and others), crabs, giant worms, and fish. In the vicinity of isothermal vents, the growth rate of specimens of such worm species as *vestimentifera Riftia pachyptilia* is enormously high—more than 35 cm per year; specimens of the species *Tevnia jerichonana* show a growth rate of more than 30 cm per year (Lutz et al., 1994). For invertebrates, such growth rates are among the largest ever measured in the ocean.

Chemo-autotrophic symbiotic bacteria are contained in special tissues—trophosomas—of *Vestimentiphera* (Cavanough et al., 1981). The main types of chemo-autotrophic bacteria, which are the primary producers, are *Beggiotoa, Thiothrix, Thiobacillus sp.* and methanotrophs. "White smokers", especially extinct ones, are inhabited by so-called Pompeian worms . The bodies of the worms are composed of silica, barite and various types of sulfides. After dying off, *Vestimentifera* are brought to a mineralized condition and become a constituent of sulfide ores.

In the course of studying hydrothermal sources of Guaimas and those in Baja, California, researchers paid special attention to the smell of the bottom sediments, which was very similar to that of diesel fuel. It was discovered that these sediments contained a large amount of wax balls. Further investigations showed that planktonic material which is very abundant in the surface waters of Baja California inten-

Table II.6.5 Mean chemical contents of elements in metalliferous sediments of various areas of the World Ocean. After Dekov, 1994, with additions by the author (Atlantic Ocean), $CaCO_3$–Mn in %; Cu–Y, in 10^{-4}%.

Element, component	11°–14°N	10°S	14°S	20°S	21°S	42°S	Red clay		
	EPR	EPR	EPR	EPR	EPR	EPR	Pacific Ocean	Indian Ocean	Atlantic Ocean
	1	2	3	4	5	6	7	8	9
$CaCO_3$	14.3	80.0	71.0	51.0	53.7	74.0	-	-	-
Si	16.83	7.10	4.38	3.94	3.60	4.11	-	-	-
Al	4.77	1.06	0.32	0.35	0.60	1.07	81	7.3	-
Mg	1.72	1.98	1.44	1.29	1.38	1.60	-	-	-
Ti	0.29	-	-	-	0.09	-	0.53	0.42	0.53
Fe	12.79	6.25	12.79	19.50	27.87	17.08	6.0	5.0	5.47
Mn	2.65	1.39	4.17	5.83	8.60	5.54	1.6	1.2	0.41
Cu	335	379	568	948	704	765	307	260	-
Zn	371	168	342	436	296	467	238	110	-
Pb	52	-	-	-	87	-	66	56	-
Co	110	80	66	73	73	68	185	73	-
Ni	245	261	328	499	532	372	412	205	140
Ba	4000	9000	5000	2300	2100	5800	-	-	710
V	238	-	-	-	741	-	126	97	96
Zr	-	-	-	-	-	-	174	139	140
Li	-	-	-	-	-	-	45	55	-
Au	0.02	0.008	0.012	-	-	-	-	-	-
Sb	5.9	1.9	1.9	5.0	13.8	-	-	-	-
As	49	142	249	266	306	-	-	62	-
Mo	146	-	-	-	125	-	56	11.5	9.1
Cd	1.73	-	-	-	47	-	-	-	-
Ga	-	-	-	-	-	-	18	21	-
Yb	-	-	-	-	-	-	-	5.3	-
Y	-	-	-	-	-	-	-	138	-
I	-	-	-	-	-	-	1.0	0.85	-
II	-	-	-	-	-	-	46	15	10

*) The data have been recalculated on a carbonate-silica–free basis (csfb). 1–8 – literature data: 1 – Cherkashov, 1990. 2, 3, 6 – Kunzendorf et al., 1984; Walter and Stoffers, 1985. 4 – Walter and Stoffers, 1985; 5 – Dekov, 1994; 7–8 – Lisitzin et al., 1987; 9 – Emelyanov, 1992 and new data.

Table II.6.6 Contents of components and elements in the upper sediment layer (0–3 cm) of the northeast equatorial part of the Pacific Ocean and Guatemala Basin

Component, elements	Polygon[1]				EPR[2]	Whole area (Fig. II.6.6)
	I	II	IV	V		
$CaCO_3$	0–3	0–86	0–1	0–13.7	46.6–65.0	0–71.3
SiO_{2am}	4.25–5.62	3.75–6.0	1–3	4.90–5.60	2.32–6.50	1–7.70
C_{org}	0.23–0.46	0.10–0.30	0.11– 0.18	0.22–0.88	0.28–3.45	0.11–3.45
P	0.24–0.42	0.11–0.21	–	0.06–0.30	0.05–0.17	0.03–0.44
Fe	5.62–7.50	3.93–7.41	4.40–5.84	4.54–9.75	5.97–13.55	3.93–13.55
Mn	0.22–0.49	0.20–0.53	0.58–0.69	2.50–8.50	1.60–5.67	0.20–8.50
Cu	362–632	295–510	307–351	338– 625	481–879	210–879
Zn	205–713	104–204	–	307–651	288–446	104–713
Ni	148–505	89–400	197–351	463–995	261–692	89–995
Co	81–392	69–198	85–101	56–142	29–154	29–392

Contents: $CaCO_3$–Mn, in %; Cu–Co, -in 10^{-4}%

[1] For the location of the polygons see Fig. II.6.6.

[2] Without Gulf of California (only part of the East Pacific Rise, which is shown in Fig. II.6.6).

Table II.6.7 Content of certain components and elements in Quaternary pelagic oozes, test area G in the Guatemala Basin, Pacific Ocean , % (st. DM-3883, depth 3610 m, 6°30′1 N, 93°00′1 W, brown, miopelagic, radiolarian, manganiferous mud). ($CaCO_3$–Mn in %; Cu–Rb in 10^{-4}%). After Emelyanov, 1991.

Horizon, cm	$CaCO_3$	C_{org}	Fe	Mn	Cu	Zn	Ni	Co	Li	Rb
5–15	12.1	0.25	3.85	4.60	423	456	360	55	42	35
18–23	13.8	0.75	4.24	2.22	316	379	249	59	40	38
30–35	19.8	0.30	3.30	1.64	310	316	245	44	40	48
44–49	<10	0.17	3.01	1.23	169	183	146	22	34	64
49–54	<10	0.11	3.36	3.68	410	346	369	33	42	64
54–58	<1.0	0.11	4.70	3.27	447	355	391	52	40	50
63–67	<1.0	0.18	5.35	3.03	441	316	408	55	42	40
84–87	<1.0	0.24	6.73	2.03	419	331	360	59	50	42
92–96	<1.0	0.26	6.73	1.73	374	336	373	52	60	44
108–112	5.4	0.41	5.61	3.35	569	413	627	81	48	40
130–135	14.8	0.78	4.50	0.65	267	357	249	37	44	35
315–320	19.8	0.18	3.39	1.76	243	357	369	48	42	40
325–330	27.6	0.70	3.39	0.38	163	360	215	55	48	35
395–400	11.8	0.38	4.90	-	377	360	339	66	50	37

sively precipitates to the ocean floor. Planktogenic material under the effect of heat in a high-temperature hydrothermal zone transforms into hydrocarbons. Bacteria living here in great numbers feed on these hydrocarbons (Edmond and von Damm, 1983). Crabs feed on *vestimentifera*.

Near the outlets of hydrotherms, the formation of specific type of ores occurs—hydrothermal sulfide ores of copper, zinc and some other elements. At a distance from the hydrothermal vents, specific facies of sediments are formed, which are commonly enriched in Fe, Mn, Cu, Ni, Co and certain other components (Tables II.6.5, II.6.6).

In the surroundings of the Mid-Atlantic Ridge, such minerals as gypsum, halite, and quartz are derived from hydrothermal solutions (Drever et al., 1970; Epstein, 1980; citation is after Jenkyns, 1986, p. 353).

In a distance of 10–1000 km away from hydrothermal fields, accumulation of metalliferous sediments occurs, which are characterized by specific properties, including markedly increased contents of Fe and Mn (usually >10%) and some minor elements. The East-Pacific Rise (EPR) is an area which is well known for the wide occurrence of such sediments (Skornyakova, 1964; Boström, 1970, 1973; Lisitzin et al., 1972; Lisitzin et al., 1990; Dekov, 1994; Krasnov et al., 1992). Studies of the northernmost part of the EPR and Guatemala basin, which is separated by this rise from the open ocean, have been undertaken recently (Emelyanov, 1991, 1992) (Table II.6.7, Figs. II.6.5, II.6.6). Up to 7.48% Fe (up to 13.55% if calculated for carbonate free basis (cfb)) is found in the upper (0–3 cm) layer of bottom sediments in these areas, while in the Guatemala basin these sediments contain up to 9.75% Fe and 8.50% Mn. Thus, muds in the crest part of the EPR and in some parts of the Guatemala basin are distinguished by high contents of Fe and Mn (Fig. II.6.5). Contents of Fe and Mn in sediment cores collected from the Oligocene–Quaternary layer (the 3- to 500-cm layer) are almost the same as those in the 0- to 3-cm layer (Emelyanov, 1991, 1992). In general, in muds, there is a clear correlation between Cu and Ni, and the correlation is weak between Ni and Mn.

Data from our studies have confirmed the fact, discovered earlier, that pelagic muds are markedly enriched in the triad of elements Cu–Ni–Co in the field of the Clarion–Clipperton fracture zones in the Pacific. FMNs in which contents of this triad of elements are the largest were found precisely in this zone, mainly over pelagic clays of the Oligocene and Miocene (Horne's zone, according to G. Greenslate et al., 1973). On the other hand, some local areas characterized by high contents of ore elements were found on the EPR and also in the Guatemala basin (Fig. II.6.6). Hydrothermal solutions are considered to be the main source of Mn and probably of Cu and Ni as well (Emelyanov et al., 2002). It is precisely the EPR [21°N, 109°E (Hekinian et al., 1987)], Galapagos rise [85–86°W, 0–45°N (Lalou et al., 1984)] and the Hess deep (*Metalliferous Sediments*, 1976) where features such as hydrothermal springs—black and white "smokers", which supply a great amount of ore components (Fe, Mn, Cu, Ni, and Zn) and other elements (Li, Si, and Ba) into the near-bottom layer—have been observed. Sulfides of such elements as zinc (which periodically contain up to 49.7% Zn), copper and iron accumulate directly in the zone of the action of smokers (Greenslate et al., 1973). As for Mn, Ni, and, to a lesser degree, Fe, Zn, Co, and Cu, elements are carried away from smokers. Fe is

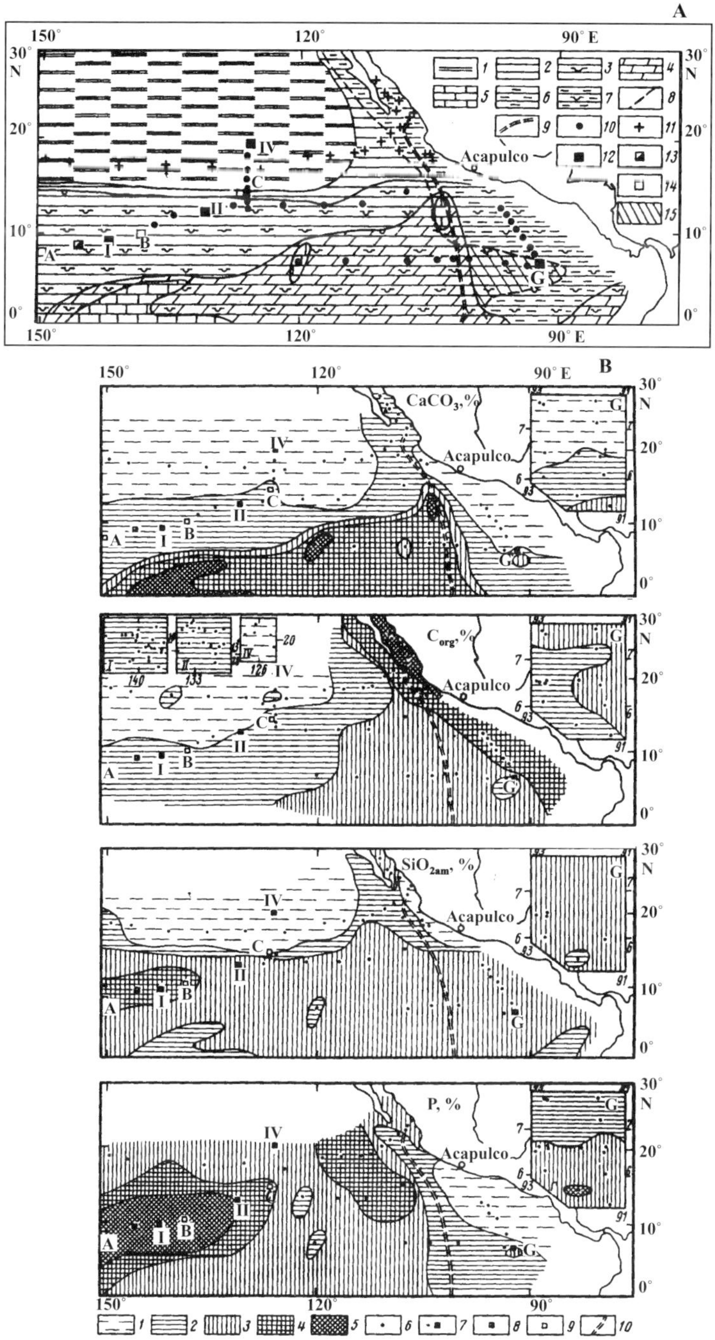

Caption see page 234

deposited at the smallest distance from them; and the opposite case for Mn . This is why areas displaying the highest Fe contents occur on the crest of the EPR; with respect to Mn, such areas are found at a great distance from the EPR, mainly in pelagic muds of deeps. The Guatemala basin, which is separated from the open part of the Pacific Ocean by such features as the EPR, Galapagos rise and Cocos Ridge is a good trap for hydrothermal manganese and associated elements (Ni, Cu). It is likely that a great portion of these elements which come from the EPR is delivered also into the Clarion-Clipperton FMN province.

The formation of specific types of sediments and rocks takes place under conditions when hydrothermal solutions are supplied to the oceans not from hydrothermal vents but seep through pelagic clays on planar areas of the ocean floor. The two main types of events involving the expulsion of volcanogenic material to the ocean can be distinguished: (1) when basaltic lava is expelled onto the surface of pelagic clay; (2) when hydrothermal solutions penetrate into the upper layer of clay due to the effects of seepage (Skornyakova et al., 1973). In the first case, thin (about 2 cm) glassy basaltic plates are formed. A thin layer of these deposits covers the bottom, which is made up of pelagic red clay. Wavy and planar areas may alternate on the surface of these plates, while on their bottom side there are "fingers" or "icicles", which penetrate the clay to a depth of 1–10 cm.

Liquid lava, it is thought, reached the bottom surface of the ocean (depths of 4800–4900 m) through fissures, where it rapidly spread as a thin layer which covers some tens of square kilometers of the ocean floor. In the second case, saturation of clays with hydrothermal solutions led to its alteration. Solidified pelagic red clays of

Fig. II.6.5. Sedimentation in eastern part of eastern equatorial zone of Pacific Ocean. After Emelyanov, 1991_2.
A. Map of distribution of bottom sediments (0—3-cm layer).
(1) Miocene pelagic clay, brown, sometimes with zeolites, often with ferromanganese concretions ($CaCO_3$ = 1%, SiO_{2am} = 1%, C_{org} 0.25%); (2) Miocene pelagic clay, brown and reddish brown, with higher content of $CaCO_3$ (1–10%) and SiO_{2am} (2–5%, diatoms, radiolarians and sillicoflagellates), C_{org} 0.25–0.40%; (3) low-siliceous clayey mud (or miopelagic clay) (SiO_{2am} 5–10% or 10–15% diatoms, radiolarians and silicoflagellates), $CaCO_3$ 1–10%; 4) low-siliceous, marly nano-foraminiferal ooze ($CaCO_3$ 50-70%, SiO_{2am} 5–10% or 10–15% diatoms, radiolarians and silicoflagellates); (5) nano-foraminiferal carbonate ooze ($CaCO_3$ 70–85%); (6) terrigenous, brown hemipelagic mud with higher C_{org} content (1–7%); (7) same, but slightly siliceous (SiO_{2am} 5–10% or 10–15% diatoms, radiolarians and silicoflagellates), C_{org} 1–2%;(8) contact between oxidized (brown to reddish brown) and reduced (gray) muds; (9) rift zone of East Pacific Rise (EPR); (10, 11) stations of expeditions DM-41 and DM-9, respectively; (12–14) profiles (test areas) of expeditions DM-41, DM-2483 and DOMES, respectively; (15) miopelagic manganese (Mn 1–8%) and siliceous (SiO_{2am} 5–10%) clay with higher C_{org} content (0.5–1%); Oligocene and Miocene coccolith ooze and pelagic clay found in profiles I and II.
B. Maps showing distribution of biogenic components: $CaCO_3$ (a), C_{org} (b), SiO_{2am} (c) and P (d) in upper layer (0–3 cm) of bottom sediments.
$CaCO_3$ content, %: (1) <1; (2) 1–10; (3) 10–50; (4) 50–70; (5) 70–85.
C_{org} content, %: (1) <0.25; (2) 0.25–0.50; 3) 0.50–1.0; (4) 1–2; (5) >2.
SiO_{2am} content, %: (1) <2; (2) 2–5; (3) 5–10; (4) >10.
P content, %: (1) <0.05; (2) 0.05–0.1; (3) 0.1–0.15; (4) 0.15–0.20; (5) 0.2–0.3; (6) Station; (7–9) profiles (test areas) of expeditions DM-41, DOMES and DM-2483, respectively; (10) rift zone EPR; G, profile in Guatemala Basin; I, II, IV and G, profiles of expedition DM-41.

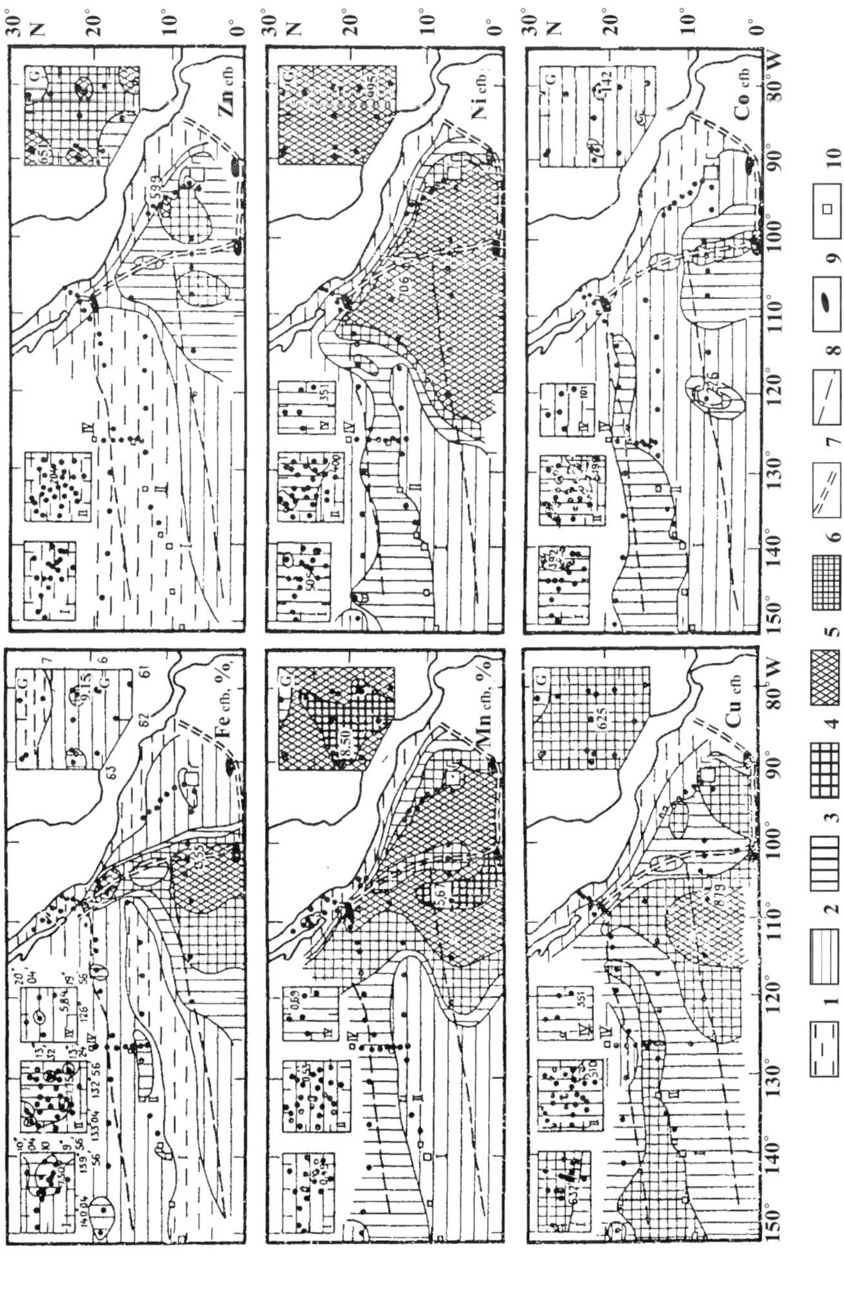

Caption see page 236

the Late Quaternary containing authigenic minerals were discovered (Skornyakova et al., 1973). Large round ferromanganese concretions which fall within the influence of such hydrothermal fields turned out to be very fragile features (Emelyanov et al., 2002). They fell to pieces in the hands of researchers; the internal part of such features is black in color.

The total number of hydrothermal sources in the Atlantic Ocean is much less than that in the area of the EPR. Nevertheless, active smokers situated predominantly in the rift zone of the North-Atlantic Ridge (fields TAG, FAMOUS, SNAKE PIT and many others), were also found in these regions (Scott et al., 1974; Rona et al., 1975; Almuhamedov et al., 1990; Lisitzin, 1992; Lisitzin et al., 1993). Metalliferous sediments are present in many test areas of the Atlantic, just like in the Pacific and Indian oceans, and away from areas of active hydrothermal activity. In the Reikjanes test area in the Atlantic Ocean, features such as hydrothermally altered lava and high concentrations of certain chemical elements in sediments have been found. Here, average contents of Fe and Mn in the upper layer of sediments are 6–7% and 0.1–0.2%, respectively (Fig. II.6.7), and these values are 20–30% in excess of those found in sediments beyond the limits of the North Atlantic Ridge. Sometimes, maximums of Fe are as large as 21–22%; Mn, 1%; copper, 378×10^{-4}%; Zn, 215×10^{-4}%. Formations having such considerable concentrations of elements exist commonly in the form of crusts which cover basalts and muds. In areas where such crusts occur, it is very typical of hydrothermally modified basalts to form ocherous deposits.

Hydrothermal activity in the TAG field in the Atlantic (rift zone of the North Atlantic Ridge, 26°N) has been reported. This field is situated at a depth of 3400–3650 m. Temperatures of hydrothermal solutions are 321–290°C; the pH is 3.66–5.02. Here, the hydrothermal solution is enriched in Si, Cl, Na, Li, Se, Fe, Mn, and CH_4 (Campbell et al., 1988) and contains H_2S. Five types of hydrothermal ore formations have been determined within the TAG field (Lisitzin, 1989). They contain up to 35.8% Fe, up to 62.0% Cu, up to 26.4% Zn, up to 0.05% Pb, up to 0.038% Cd, up to 0.100%Co, and are enriched in Ag and Au. Sediments around the hydrothermal field are also enriched in metals. Sinters, crusts and concretions contain up to 55.12% Mn, up to 0.02% Cu, up to 0.02% Ni, and up to 0.02% Co (Thompson et al., 1988).

Fig. II.6.6. Maps showing distribution of Fe, Mn, Zn, Cu, Ni and Co in upper layer (0–3 cm) of bottom sediments in eastern equatorial zone of Pacific Ocean (area of Clarion–Clipperton fracture zones and Guatemala Basin). After Emelyanov, 1991_3.
Fe, %: 1—<5; 2—5–6; 3—6–7; 4—7–10; 5—>10.
Mn, %: 1—<0.2; 2—0.2–0.5; 3—0.5–1.4; 4—1–3; 5—3–5; 6—5–10.
Cu, 10^{-4}%: 1—<200; 2—200–300; 3—300–400; 4—400–600; 5—>500.
Zn, 10^{-4}%: 1—100–200; 2—200–300; 3—300–400; 4—400–500; 5—>500.
Ni, 10^{-4}%: 1—<100; 2—100–200; 3—200–300; 4—300–400; 5—>400.
Co, 10^{-4}%: 1—<50; 2—50–100; 3—100–150; 4—150–200; 5—>200;
7—ridges and border of Guatemala Basin; 8—Clarion and Cliperton fractures zones; 9—hydrothermal fields; 10—test areas (shown in insets I, II, IV and G). Llegend in insets is same as on base map.
Numbers 879 and others—maximal contents of element in corresponding region; points on maps—geological stations.

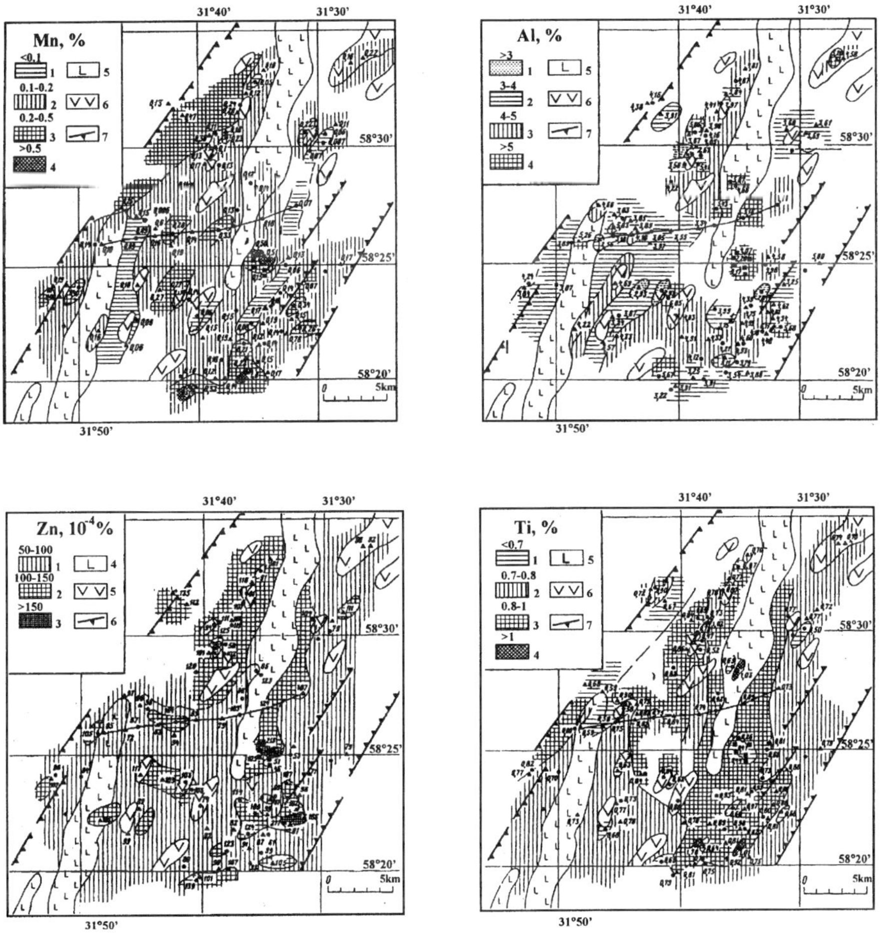

Fig. II.6.7. Distribution of Mn, Al, Zn and Ti in upper sediment layer (0–5 cm) of Reykyanes test area, North Atlantic. After Almouhamedov et al., 1990.
For location of test area see Fig. II.6.11c.
Legend: 1–4—content of elements, in %; 5—bazalts of Holocene; 6—Weishelian (Würm) basalts; 7—escarpments.

At present (in the Holocene), there are several low-temperature hydrothermal sources in the Tyrrhenian Sea. Ferruginous hydrothermal sediments were found near Vulcano volcano; atop the volcanic Marsilli seamount (Emelyanov, 1988_3) (Fig. II.6.8); and on Polynuro, Vercelli and certain other seamounts (see Table II.6.8). Ore hydrothermal material accumulates predominantly in the form of ferruginous deposits near hydrothermal vents, while such elements as silica, phosphorus, manganese and other associated minor elements are dispersed over the whole aquatic area of the sea and increase the background contents of these elements

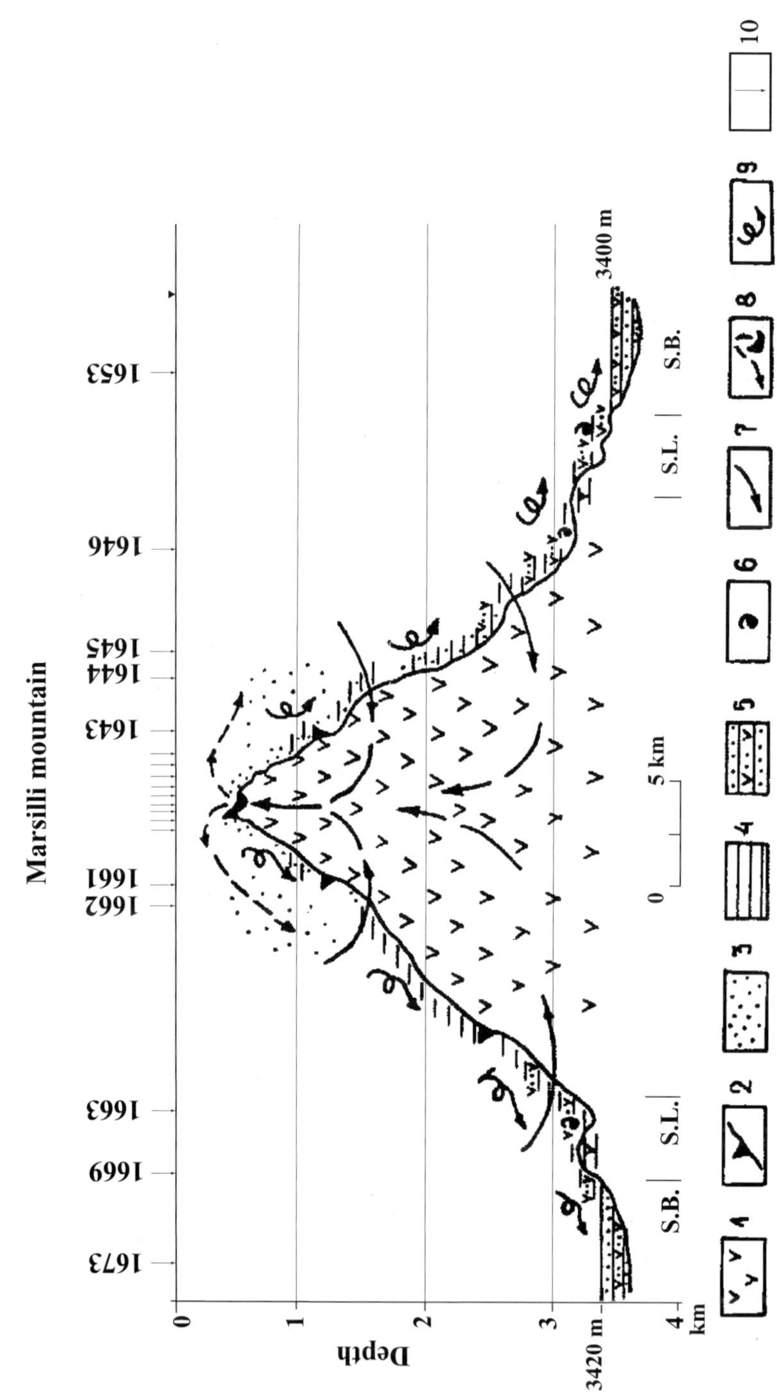

Caption see page 239

either in seawater or in bottom sediments. In the latter case, these elements tend, in contrast to Fe, to the central part of the Tyrrhenian Sea (Fig. II.6.9).

A distinguishing feature indicating the presence of an admixture of hydrothermal material in pelagic sediments is the module (Fe+Mn)/Ti (Strakhov, 1976). When this ratio is higher than 25, it means that the amount of the admixture in the hydrothermal material is considerable and such sediments most frequently become metalliferous. In the Atlantic, such sediments are most commonly confined to the North Atlantic Ridge in arid climatic zones, where the rates of sedimentation are minimal. Contents of Fe+Mn in such sediments (if calculated for cfb) commonly exceed 7% and even reach 20% (see Fig. 38 in Emelyanov, 1982); the contents of Fe are usually more than 7% (Fig. II.6.10), and the contents of minor ore elements, including Cu, Ni and Co, are markedly increased (Figs. II.6.11, II.6.12, II.6.13). Chromium (Fig. II.6.14) is a more clastic than mobile (hydrothermal) microelement. Increased contents of this element are located near the outcrops of ultrabasic (serpentinites) rocks, i.e., near shores and in certain places in the transform fracture zones rift valley.

Differentiation of material in the hydrotherm–seawater GBZ. The process of differentiation of material commences as a result of variations in each of the main parameters of the hydrothermal system: pH, temperature, concentration of H_2S or O_2, salinity, etc. The influence the process has on the system is found in sediments as follows: contents and compositions of gases decrease with distance from the hydrothermal vent; the total amount of suspended matter decreases and the transparency of the plume increases; Fe, Mn and accompanying elements are separated. Sharp chemical differentiation of elements occurs in the plume. Three zones of reactions are clearly recognized within this plume: sulfate reducing (sulfides of Cu, Zn, Fe; sulfates of Ba and Ca), neutral (Fe+Si—nontronite), and oxidizing (Lisitzin et al., 1993, p. 223).

Among barriers giving rise to the mineral and chemical composition of hydrothermal solutions, plumes, ore formations and bottom sediments, there are three major barriers: temperature, redox and acidic–alkaline. This has an effect on the chemical and elemental composition of the hydrothermal systems themselves

Fig. II.6.8. Sedimentation on Marsilli seamountain. After Emelyanov, 1988_3, 1989.
1—basalts and lavas; 2—blocks of lava (outcrops); 3—redistribution of hydrothermal iron and manganese hydroxides in Holocene; 4—mixed volcanoclastic–terrigenous low calcareous and calcareous (foraminiferal) mud; 5—volcanoclastic terrigenous and foraminiferal turbidites; 6—foraminiferal ooze; 7—circulation of sea and hydrothermal waters; 8—hydrothermal iron deposits (cap about 1 km^2) and main directions of distribution of hydrothermal material in Late Pleistocene; 9—slumps and turbidities.
Numbers and arrows on top of picture—geological stations (sifters, gravity corers, dredges and dives on Argus and Mir submersibles).
S.B.—hanging sedimentary basins with ideally horizontal surface; Sl.–hills, the origin of which are slumps.

Table II.6.8 Chemical composition of manganese, ferrugineous and manganese-ferrugineous sediments of the Mediterranean Sea and Atlantic Ocean

	Mediterranean Sea										Atlantic ocean				
	M		Str	E+E	San		ST	ET	P		Er	Str	TAG	NM	NS
	1	2			1	2			M	S					
Fe	32.40	29.50	14.30	4.0	31.38	32.60	0.26	31.82	12.9	32.0	4.97	6.51	0.13	9.99	12.92
Mn	2.59	0.42	22.40	≤8.3	0.05	0.03	48.02	3.26	-	-	9.10	2.08	41.00	2.31	17.36
Al	-	-	-	-	0.30	1.10	0.41	-	-	-	7.43	8.47	0.07	-	-
Ca	-	-	0.50	-	0.91	1.09	2.33	-	-	-	22.24	4.62	-	-	-
Mg	-	-	0.74	-	0.63	0.95	1.52	-	-	-	3.25	2.65	-	-	-
K	0.30	0.89	0.78	-	0.58	0.69	0.38	-	-	-	1.34	1.12	-	1.50	-
Na	3.53	2.20	2.20	-	4.30	-	2.61	-	-	-	3.13	2.82	-	2.95	-
Ti	0.09	0.18	0.006	-	-	0.04	0.02	-	-	-	0.58	0.60	-	0.25	0.50
P	0.62	0.70	0.30	-	0.06	-	-	-	-	-	-	-	-	0.72	0.16
$CaCO_3$	3.55	15.51	-	-	2.64	-	-	-	-	-	47.64	-	-	5.91	0.00
$C_{org.}$	0618	1607	-	-	1622	-	-	-	-	-	0.001	-	-	0.47	0.00
Sio_{2am}	-	-	-	-	-	-	-	-	-	-	-	-	-	1,26	-
Fe/Mn	12.50	70.00	0.60	-	628	1087	0,01	-	-	-	0.55	3.13	315	4.30	0.70
Cu	42	30	420	400–900	27	34	8200	40	2700	22400	167	37	104	308	440
Zn	54	56	-	≤1000	139	63	83	-	1400	-	157	115	-	320	525
Cr	10	10	5	-	-	5	-	-	-	-	-	-	-	-	31
Ni	115	130	15	≤3000	-	5	249	230	-	-	135	82	440	314	208
Co	66	80	30	-	-	1	223	250	-	-	52	43	6	124	338
V	78	84	55	-	-	66	84	-	-	-	180	-	-	538	230
Mo	7	9	-	-	-	29	557	-	-	-	89	17	-	730	100
Ba	200	200	1070	-	-	103	3200	-	-	6	7862	226	400	-	660
Li	147	139	-	-	-	3	-	-	-	-	178	-	-	-	-
Rb	93	92	-	-	-	-	22	-	-	-	54	-	-	-	-
Sr	-	-	-	-	-	400	600	-	-	-	1384	168	-	-	-
Cd	-	-	-	-	-	-	-	-	-	-	2,4	-	-	-	-
Age	Plt	Plt	Plt(?)	-	Hl	Hl	Plt(?)	Plt	-	-	Hl-Plt[3]	Plt(?)	Plt(?)	Plt	Mio

M – Marsilli; Str – Strombolichio, Tyrrhenian Sea (Bonatti et al., 1972); E.E. -Enarette and Eolo (Rossi et al., 1980); San – Santorini Lagoon (1 – Butuzova, 1969; 2 – Smith a. Cronan, 1983); ST – Southern Tyrrhenian (Lametino) (Rossi et al., 1980); ET – Eastern slope of the Tyrrhenian Sea (Castelarin a. Sartori, 1978); P – Palinuro, Tyrrhenian Sea (M – ore deposits: Sb – 1000, Bi – 500, As – 860 ppm); S – sulphides from the ore deposits: Sb – 9100, As – 2400, Ag – 220, Hg – 400 ppm ; Er – Eratosphen seamountain (Varnavas et al., 1988); Str – Strabo Trench; TAG – hydrothermal area (Toth, 1980); NM – Brasilian basin, nameless seamountain (Emelyanov, 1986_4); NS – Norwegian Sea (Emelyanov et al., 1976).

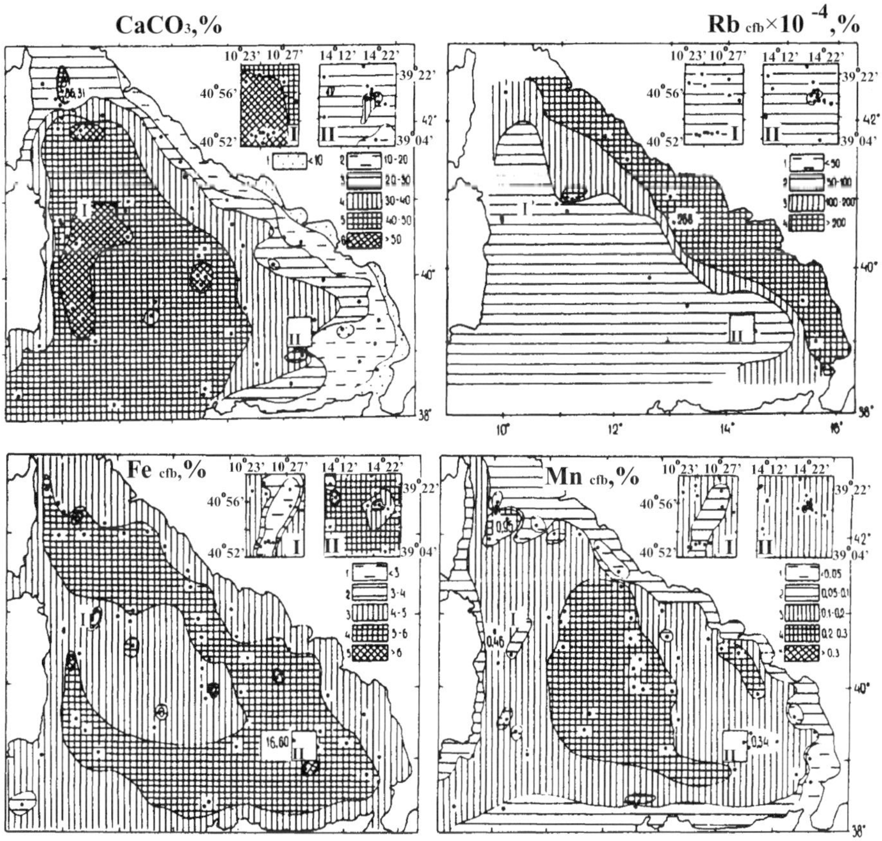

Fig. II.6.9. Distribution of $CaCO_3$, Rb, Fe and Mn in surficial bottom sediments (0–5 cm) of Tyrrhenian Sea. $CaCO_3$, Fe and Mn in %, Rb in 10^{-4}% (contents for Fe, Mn and Rb are recalculated on cfb).
I and II—test areas Baroni (I) and Marsili (II) investigated in detail (legend for test areas is same as on base map).

and on the sequence of precipitating minerals (Table II.6.9). Such minerals as pyrite, pyrrhotite, copper sulfides, and silica are also commonly deposited in the lower central zone (see Fig. II.6.1). Pyrrhotite begins to precipitate at a temperature of 400°C and the process is terminated at 260°C. Chalkopyrite and stevensite eventually begins to precipitate as well. Among lower-temperature elements are Zn- and Cu sulfides, which are followed by anhydrite, halenite, aragonite, and calcite. Barite, markasite, and amorphous silica precipitate at about 250–190°C. Pyrite is the last element in this series (Fig. II.6.15). In the presence of a sufficient amount of oxygen, Fe- and especially Mn hydroxides precipitate at much lower temperatures under almost natural or even low-alkaline conditions.

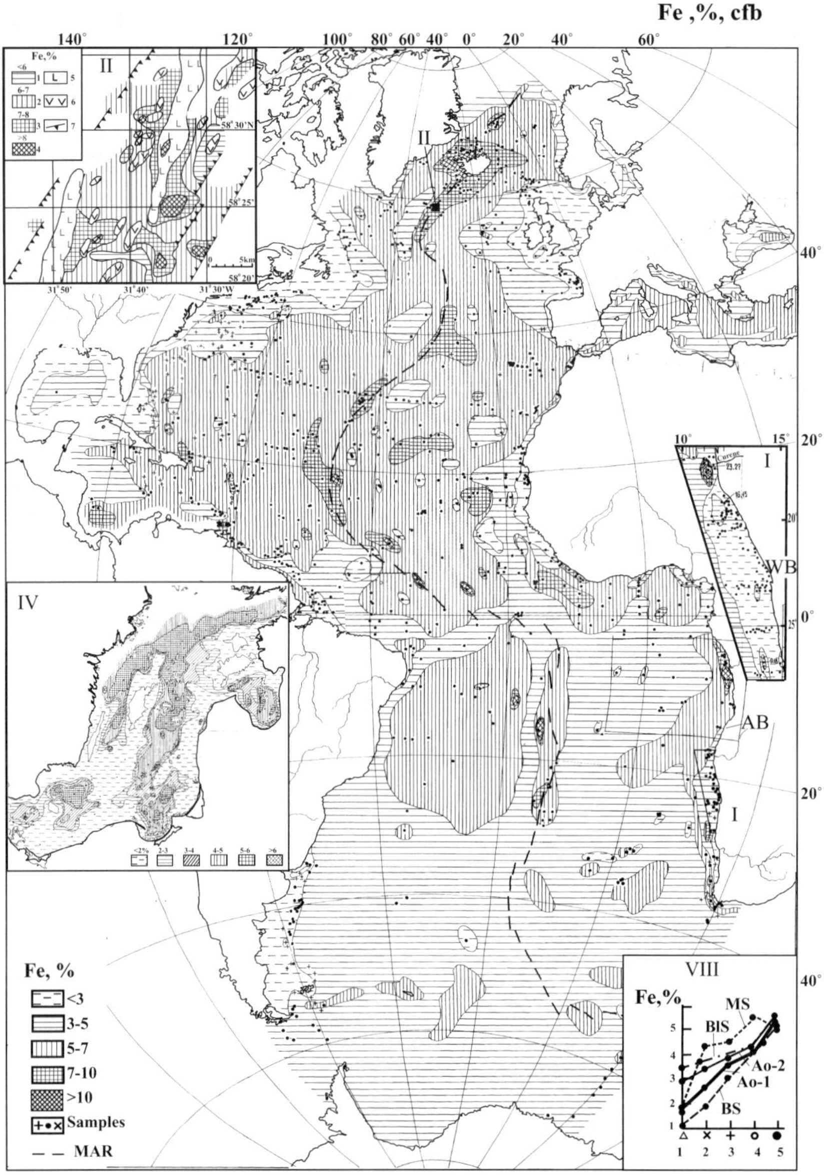

Fig. II.6.10. Fe distribution in upper (0–5 cm) layer of Atlantic Ocean bottom sediments, % (recalculated to carbonate and silica free basis—cfsb).

1–5—content in %; 6—samples; Mid-Atlantic Ridge—position of rift valley of Mid-Atlantic Ridge. I, II, IV, AB—areas (test areas, studied in detail): I—Walvis Bay; II—Reykjanes Ridge; IV—Baltic Sea; AB –Angola Basin.

Legend for inset II: 1–4—Fe (cfb) content, in %: 1—<6; 2—6–7: 3—7–8; 4—>8; 5–6—outcrops of basalts (5—Holocene; 6—Würm); 7—escarpments and fractures.

VIII. Distribution of average Fe content in granulometric sediment types of Atlantic Ocean Basin. 1–5—sediment types (0–5 cm): 1—sand, 2—coarse aleurite, 3—fine aleuritic mud, 4—aleuro-pelitic mud, 5—pelitic (clayey) mud. Ao-1—terrigenous sediments of Atlantic Ocean with $CaCO_3$ content <10%; Ao-2—same, with 10–30% $CaCO_3$; BS—Baltic Sea; BlS—Black Sea; MS—Mediterranean Sea.

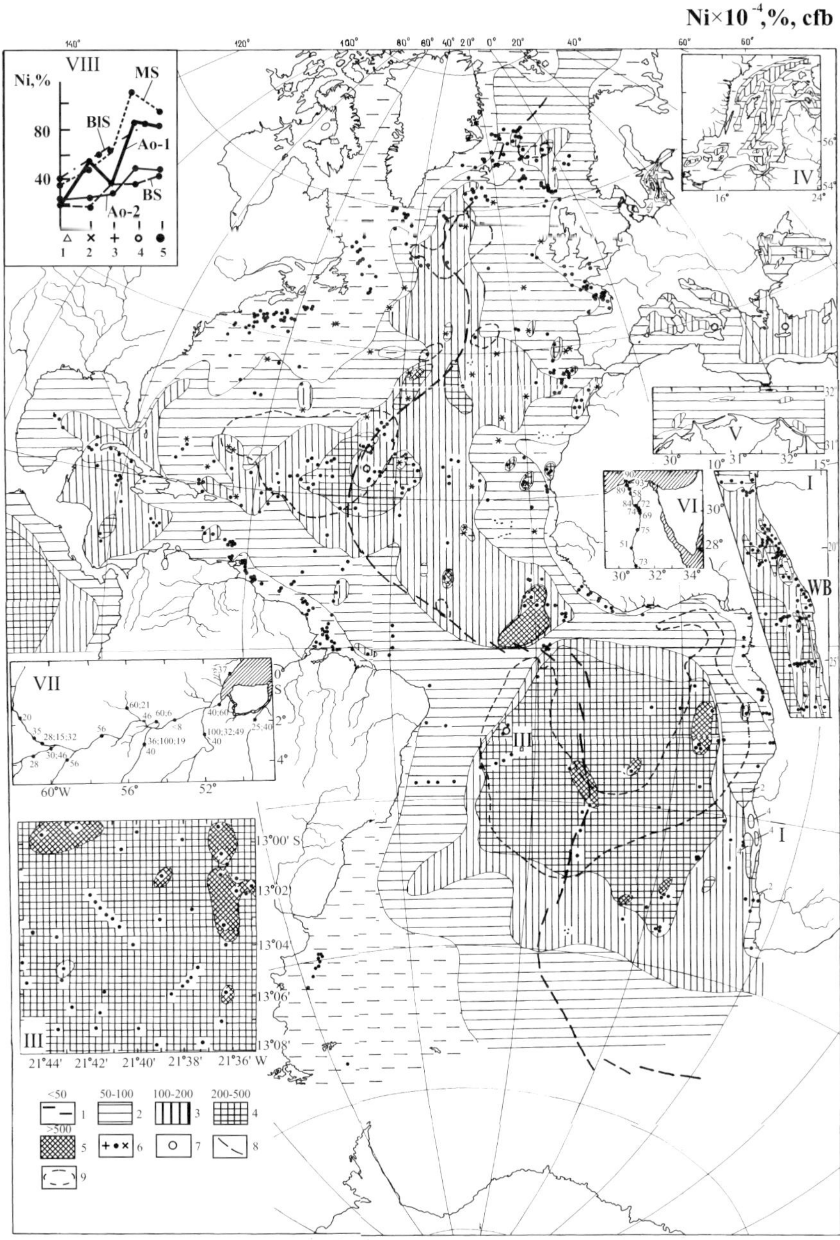

Fig. II.6.11. Ni distribution in upper (0–5 cm) layer of Atlantic Ocean bottom sediments, 10^{-4}% (recalculated to cfsb). After Emelyanov, 1992.

1—<50; 2—50–100; 3—100–200; 4—200–500; 5 >500; 6—samples; 7—main hydrothermal sediment distribution areas; 8—Mid-Atlantic Ridge axis; 9—0.5% Mn isolines (recalculated to cfsb). Inset maps: (I) Upwelling area (WB—Walvis Bay); (III) Brazil Basin; (IV) Baltic Sea; (V) Nile foredelta; (VI) Nile River alluvium; (VII) Amazon River alluvium. Inset maps I, III and V—legend symbols identical to those of base map. IV: (1) <25; (2) 25–30; (3) >50. Figures in inset map VI and VII—Ni content, in ppm. Legend of inset VIII is the same as on Fig.II.6.10.

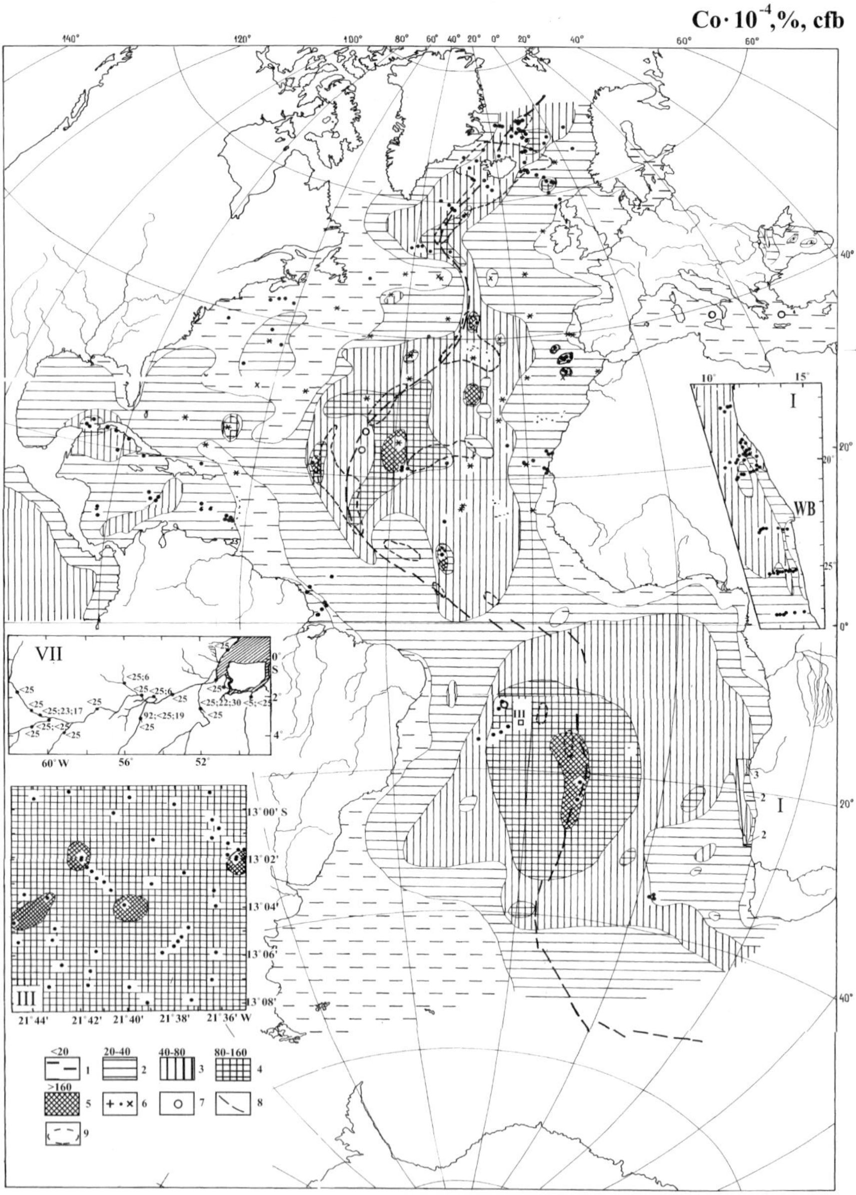

Fig. II.6.12. Co distribution in upper layer (0–5 cm) of Atlantic Ocean bottom sediments, 10^{-4}%, (recalculated on cfb).
1—<20; 2—20–40; 3—40–80; 4—80–160; 5 >160; 6—samples; 7—hydrothermal ore deposit distribution zones; 8—Mid-Atlantic Ridge rift zone; 9—sediment distribution area with Fe content >7% (recalculated to cfsb). Insets: (I) upwelling zone (WB—Walvis Bay); (III) Brazil Basin; (VII) Amazon alluvium, natural sediment. Legend in insets I and III same as in base map. Numbers in inset VII: Co content in 10^{-4} %.

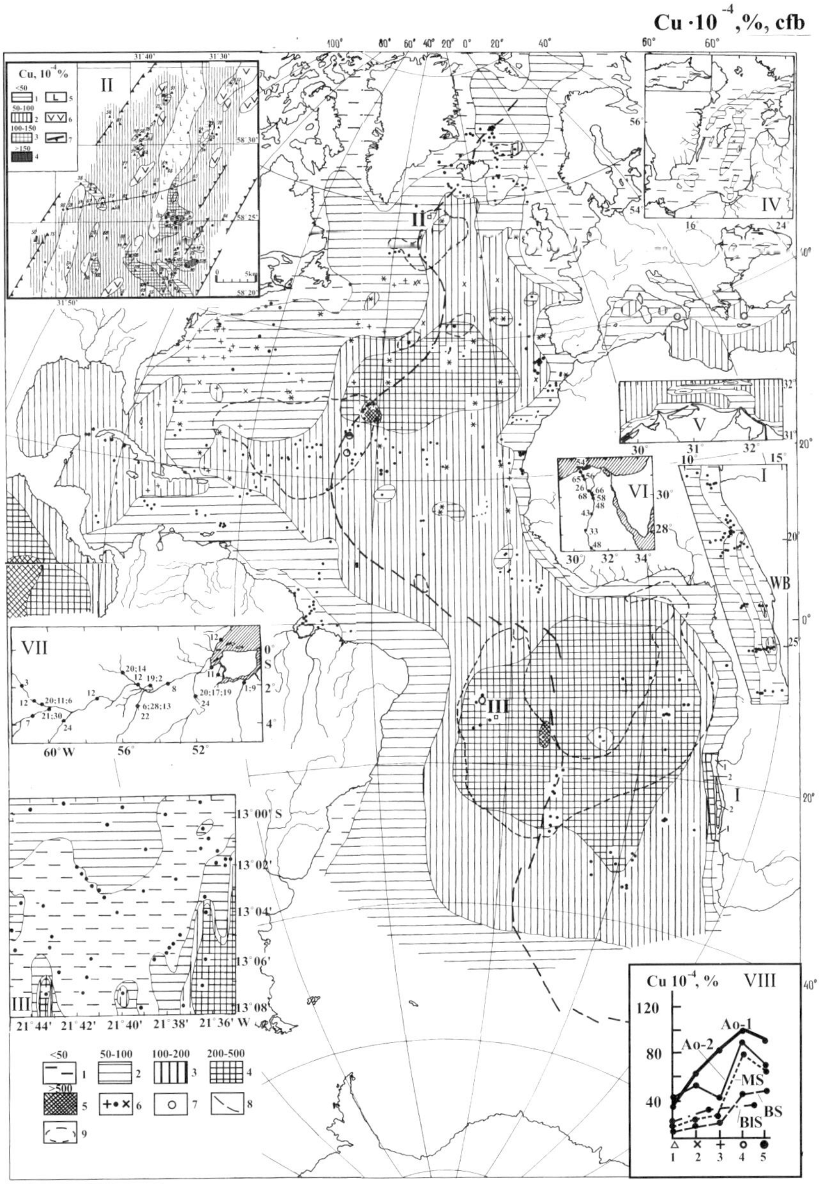

Fig. II.6.13. Cu distribution in upper layer (0–5cm) of Atlantic Ocean Basin bottom sediments 10^{-4}%, (recalculated on cfb).

1—<50; 2—50–100; 3—100–200; 4—200–500; 5—>500; 6—samples; 7—area of widely distributed hydrothermal sediments; 8—Mid-Atlantic Ridge rift zone; 9—contour of >0.5% Mn area (recalculated on cfb) in upper sediment layer. Insets: (I) Upwelling area (WB—Walvis Bay); (II) Reykjanes Ridge; (III) Brasil Basin; (IV) Baltic Sea; (V) Nile foredelta; (VI) Nile River alluvium; (VII) Amazon River alluvium, air-dried sediment. Symbols in inset I same as in base map; inset II: 1—<50; 2—50–100; 3—100–150; 4—>150; 5—location of studied samples; 6–7—outcrops of Holocenic (6) and Late Pleistocenic (7) bazalts. Maximal contents of Cu (388 and 378 ppm) are located in areas of hydrotherms. Inset III: 1—100–200; 2—200–250; 3—250–300; 4—<300; Insets IV–V: 1—<40; 2—40–60; 3—>60.

Numbers on inset maps VI and VII: Cu content, 10^{-4} %. Legend of inset VIII is the same as on Fig.II.6.10.

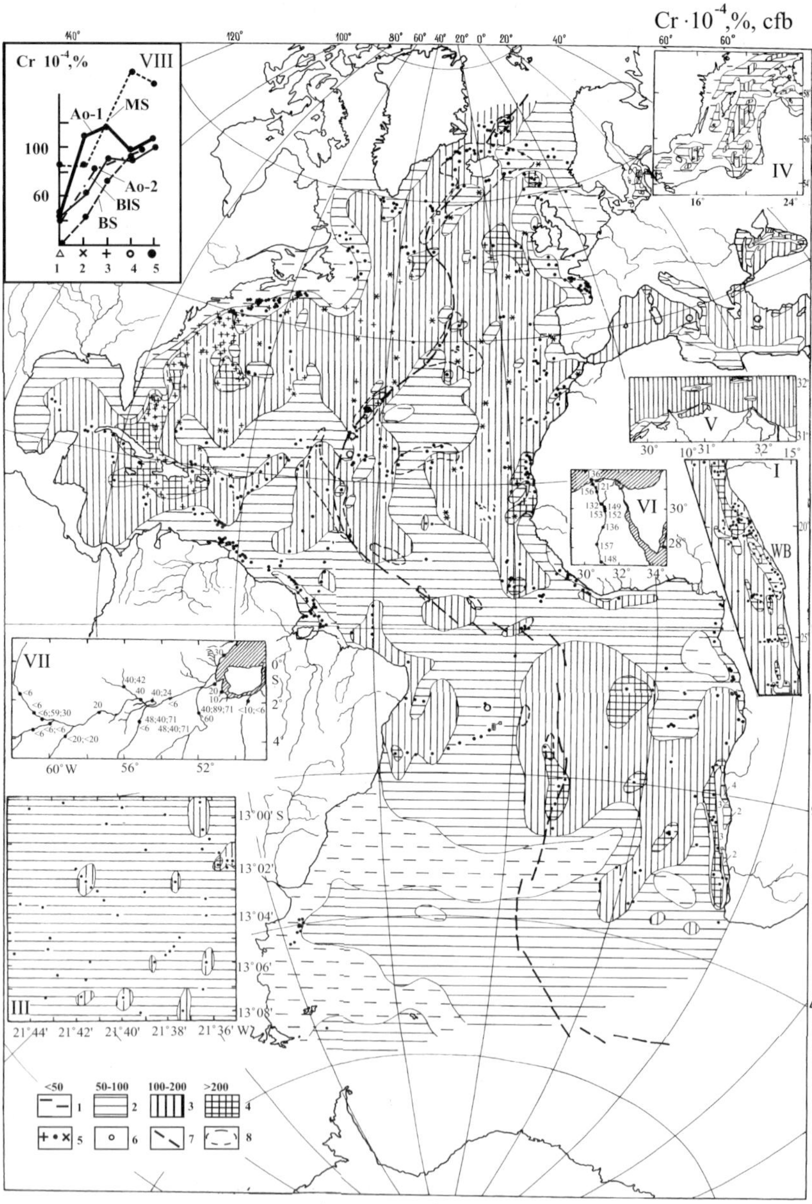

Fig. II.6.14. Cr distribution in upper layer (0–5 cm) of Atlantic Ocean bottom sediments, 10^{-4} %
(recalculated on cfb). 1–4—content in 10^{-4} %: 1—<50; 2—50–100; 3—100–200; 4—>200; 5—samples; 6—area of widely distributed hydrothermal sediments; 7—Mid-Atlantic Ridge rift zone; 8—sediment distribution area with
Fe>7% content (recalculated on cfb). Legend in insets I an III same as in base map. Numbers in inset VII—content of Co in 10^{-4} %.

Table II.6.9 The consecutive stages of hydrotermal mineral formation in the TAG area, Atlantic Ocean. After Lisitzin et al., 1993, p. 136.

The stages of mineral formation	Mineral parageneses
The early stage:	
Of low temperature	Anhydrite + pyrite + ephalerite
Of high temperature	Anhydrite + pyrite (marcasite) + halcopyrite
Of main ore formation	Pyrite (marcasite) + halcopyrite (cubanite)
Of late stage	Pyrite + halcopyrite (cubanite) + minerals of silica + nontronite
Of hypergenic change (crusts on the ore)	Minerals of silica + Fe-hydrooxide ++ native sulfur

Based on the temperature of ore solutions and fluids, A. Lisitzin (Lisitzin et al., 1993, p. 178) recognized the following main forms of hydrothermal formations:

I. Concentrated (massive)
   1. High-temperature (above 300°C).
      a. Sulfide formations in the sequence of the ocean crust (stockwork deposits, disseminations, veins).
      b. Sulfide constructions on the ocean floor are massive.
   2. Mid-temperature (50–300°C).
      a. Buildups consisting of Ca- and Ba sulfates (150–300°C).
   3. Low-temperature (below 50°C).
      a. Constructions (hills) of nontronite (20–30°C),

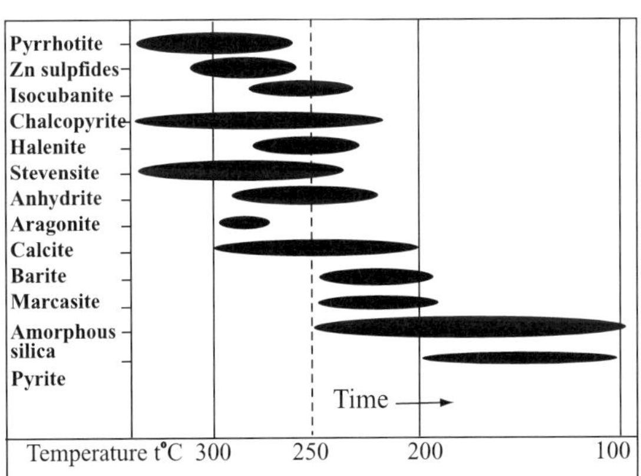

Fig. II.6.15. Successive precipitation of minerals in hydrothermal sulfidic chimney. After Lisitzin et al., 1993.

b. Silica and ferrosiliceous constructions,
c. Ferruginous and ferromanganese constructions,
d. Manganese constructions and crusts.

II. Dispersed
1. Metalliferous sediments (in the reducing environment of the brine-filled deeps of the Red Sea),
2. Oxidized metalliferous sediments.

The pattern of elemental and mineral differentiation is, in many respects, different from what it is found in low-temperature hydrothermal systems of the type of the Santorin lagoon (Butuzova, 1969) or the deeps of the Red Sea (Bogdanov et al., 1986). In the Atlantis-II deep, the highest-temperature formations (+60°C) are anhydride, Zn- and Cu sulfides, Fe–silicate (nontronite), and Fe–smectite. These minerals form in the absence of oxygen. At +50°C, even in the presence of small amounts of oxygen, the development of Fe- and Mn carbonates, oxyhydroxides of these elements (including manganite), takes place. Among compounds formed in normal seawater, the temperature of which is, however, considerably increased (up to 15–20°C), carbonates and Fe- and Mn hydroxides are predominant.

At a considerable distance beyond the limits of this hydrothermal system (10–500 km), the chemical differentiation of single minerals and chemical elements also occurs, but it is much more weakly pronounced compared with hydrothermal solutions and is found in sediments as small variations in the contents of iron, manganese, phosphorus, silica and minor elements.

Ocean-Atmosphere

Mass and energy exchange across the ocean–atmosphere boundary. The SML is a boundary across which the process of global exchange occurs between the atmosphere and hydrosphere by different gaseous, liquid and solid matter. Of special importance for the earth as a planet in controlling its climate are processes involving exchange of carbon dioxide, oxygen, salts, heat, and moisture (Figs. II.7.1, II.7.2).

Intensity and direction of the CO_2 flux are specified by equilibrium of the carbonate system, which controls the content of CO_2 in seawater. Carbon dioxide combines with other components of the carbonate system via the following series of chemical reactions (Bezborodov and Eremeev, 1984, p. 27):

$$CO_{2g} = CO_{2f}$$
$$H_2O + CO_{2f} = H_2CO_3$$
$$H_2CO_3 = H^+ + HCO_3^-$$
$$HCO_3 = H + CO_2^3,$$

where g is gas and f is fluid.

Many factors, including latitude, influence the direction and intensity of the CO_2 flux across the ocean–atmosphere interface (Fig. II.7.1). Active extraction of carbon dioxide from the atmosphere occurs only in the Atlantic, mainly in its northern and southern parts.

In other regions, on the contrary, carbon dioxide is released from the ocean into the atmosphere. Total annual fluxes of carbon dioxide across the ocean–atmosphere interface are – $+11.6 \times 10^9$ t in the Atlantic Ocean, -4.3×10^9 t in the Pacific Ocean, and -0.7×10^9 t in the Indian Ocean (+ means flow into the ocean; – means flow out of the ocean). The amount of CO_2 absorbed by the World Ocean over the area extending from 60° N to 60° S is 6.6×10^9 tons (Bezborodov and Eremeev, 1984, p. 143). The ocean is the reservoir of our planet responsible for absorption of excess CO_2 from the atmosphere. This excess makes up 39% of the total CO_2 released into the atmosphere annually (Macintyre, 1980). It should be noted that 5% of CO_2 is absorbed by the surface layer of the ocean, 26% by the intermediate layer, and 8% by the deep layer.

The ocean controls not only the balance of natural, but also of anthropogenic, carbon dioxide. The natural processes of formation and exchange of CO_2 are disturbed by a number of anthropogenic factors such as burning of fuel, steelworks, cutting of forests, plowing of lands, and oil slicks on the water surface. The amount of CO_2 in the atmosphere continuously increases and it has been predicted that the

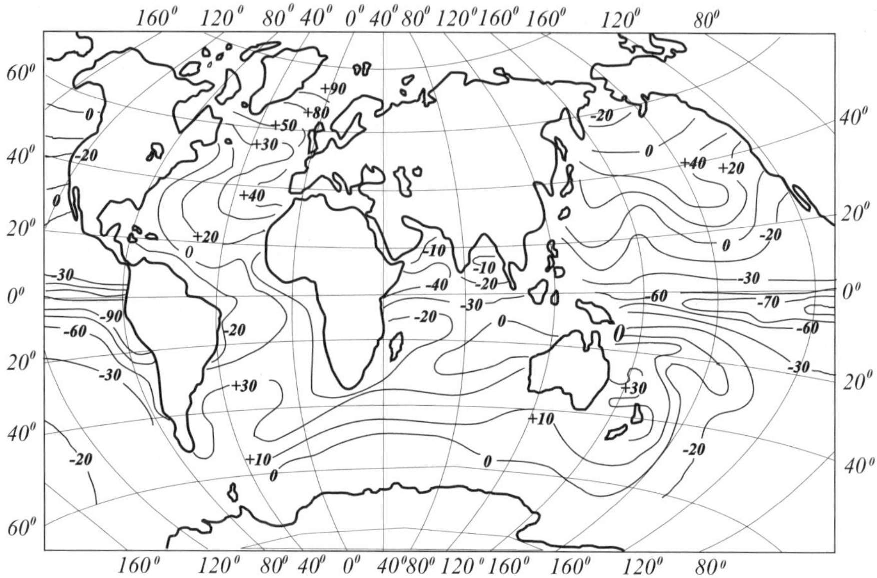

Fig. II.7.1. Average value of ΔP_{CO2} for World Ocean, n · 10 Pa. After Bezborodov and Eremeev, 1984. Direction: "+"—into ocean; "–"—from ocean.

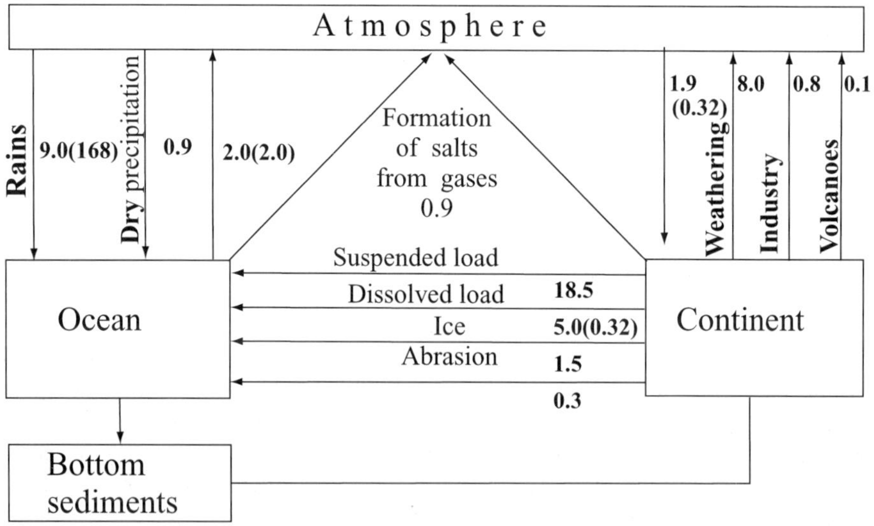

Fig. II.7.2. Balance of salts in ocean–atmosphere–continent system, 10^9 t/yr. In brackets—salts of oceanic origin. After Bezborodov and Eremeev, 1984.

partial pressure of CO_2 in the atmosphere will become three times as large as the present-day level by the year 2050. According to Kagan et al. (1990), the concentration of atmospheric CO_2 has increased by 65–70 million tons over the last 200 years, and this has already led to increase in temperature of the near-surface atmosphere by several tenths of a degree centigrade. Thus, variations in the CO_2 content in the atmosphere strongly influence the earth's climate. It is difficult to predict the extent of such changes in the future if the fluxes of anthropogenic CO_2 into the atmosphere are not controlled.

In the ocean, CO_2 is in its final form of $CaCO_3$ which settles to the bottom as biogenic calcite or aragonite. Today, still no danger exists that the degree of supersaturation of ocean water with respect to calcite and the rates of natural processes of precipitation of $CaCO_3$ to the ocean floor will be strongly decreased due to increasingly large fluxes of CO_2 from the atmosphere.

The oxygen flux and directions by which it migrates through the ocean–atmosphere GBZ depends on the following factors: (1) geographical latitude; (2) season of the year; (3) occurrence of oil film and its thickness (Bezborodov and Eremeev, 1984, p. 161). In the Southern Hemisphere, during the summer season of the Northern Hemisphere, the oxygen flux is from the atmosphere to the ocean, and in the Northern Hemisphere, vice versa, from the ocean to the atmosphere. In total, the net oxygen flux is directed from atmosphere to ocean. In northern latitudes, during the winter season in the Northern Hemisphere, oxygen flux from ocean to atmosphere still exists, but its intensity is strongly reduced; in the southern latitudes, the opposite occurs, from ocean to atmosphere. For this season, the net oxygen flux is from oceans to atmosphere. In the presence of oil film, the oxygen flux becomes weaker: the thicker this film, the weaker the oxygen flux. When the thickness of such a film is as large as 1 mm, oxygen flux eventually ceases. Factors such as increased sea strength, temperature variations, and increased intensity of photosynthesis increase the intensity of oxygen exchange.

The flux of salts is essentially unidirectional: from ocean to atmosphere. As much as 0.6×10^9 tons of sodium are delivered to the atmosphere every year (Fig. II.7.2). The total amount of salts delivered to the earth's surface is estimated at 11.8×10^9 tons per year, including an input of 8×10^9 tons per year, which is the flux of products of terrestrial weathering to the atmosphere (Bezborodov and Eremeev, 1984, p. 21). The remaining 3.8×10^9 tons per year are delivered from the surface of the ocean (some 1.9×10^9 tons per year or 17% of the total amount of delivered material) including the products of volcanic eruptions and fuel burning (about 1.8×10^9 tons per year). In conditions of wind force up to 1.5 m/s, the amount of salts delivered to the atmosphere is estimated to be 3.34–8.71 µg on average (Prospero, 1979). The ratio of salt to the terrigenous component in the atmosphere varies within the limits of 3 to 770, but values of 0.5–5.6 are the most typical (Miklishansky, 1983, p. 83). This ratio is minimum in tropical areas of the Atlantic. At a height of 15 m above the ocean surface, from 10 to 40 µg (periodically more) of salts are found per cubic meter of air. According to other estimations (Ericson and Duce, 1988), the amount of salts that escaped from the ocean into the atmosphere is $1–3 \times 10^{16}$ g per year. It should be noted that content of salts in air in high latitudes is two to four times as large as that in low latitudes

(Erickson et al., 1986). Such a phenomenon is facilitated by strong winds in high latitudes.

Part of the salts migrates from oceans to the atmosphere to finally precipitate on land. This is estimated to be 0.32×10^9 tons per year, which makes up 16% of the total amount of oceanic salts delivered to the atmosphere (Fig. II.7.2).

The ocean–atmosphere system is non-equilibrium, because the processes carried out at this boundary are irreversible. Solar radiation brings this system to a thermally non-equilibrium state (Karavaeva et al., 1990): the temperature of the upper 100-m layer of seawater is 0.1–1.4°C higher than that of the troposphere, and this gives rise to continuous heat exchange between the atmosphere and hydrosphere at the ocean–atmosphere interface. These processes are responsible for cooling of the upper microlayer: a so-called cold film of the ocean, or cold skin layer, is formed. The average thickness of the skin layer is 3–5 mm, and the temperature gradient is 2–5°C/cm (Hunarzhua et al., 1977). The skin layer is found even when the water temperature is lower that that of air, i.e., when the heat flux is directed from hydrosphere to atmosphere.

Heat transfer in the laminar sublayer is produced by thermal conductivity (or molecular diffusion), and temperature decreases linearly. In the ocean, the temperature of the SML is commonly lower (by 1°C or more) than that of the subsurface waters. The uppermost film, the SML, has a salinity higher than that of the subsurface waters (the drop in salinity is about 0.32‰).

The surface of the ocean is almost everywhere covered by organic films of natural (biogenic and abiogenic) origin, as well as by anthropogenic films. The surface film is commonly a monomolecular layer of surface-active matter. In addition to oil films, they are represented also by films of biological origin (high-boiling hydrocarbons), films from detergents and so on. The development of abiogenic films appears to be related to events involving "Gibson's absorption of surface-active matter from the surface volume of suspended matter and its delivery to the ocean surface by air bubbles" (Bezborodov and Eremeev, 1984, p. 37). The presence of a layer of surface-active matter on the ocean surface leads to weakened capillary waves and is a factor decelerating the formation of waves of greater amplitude. Due to the effect of organic films, the evaporation of water slowing down, and fluxes of gases across the ocean–atmosphere boundary are also diminished (organic films represent a diffusion barrier). Thus, for example, due to the effects of a film of paraffin oil and surface-active matter 5 μm in thickness, the rate of evaporation is 15% less than that for a clean water surface (Garret, 1972). Under calm weather conditions, evaporation from the water surface can be reduced by up to 90% due to the effects of oil films.

The natural exchange between substances such as solid and liquid matter, air and water, is strongly hindered by surface-active matter. This is because of the shorter lifetime of air bubbles. A film several molecules in thickness appears to reduce the rate of fluxes of material from ocean to atmosphere, in contrast to a monomolecular layer, which increases this rate (Bezborodov and Eremeev, 1984, p. 42). In the upper side of the microlayer, in the course of exchange of matter at the ocean–atmosphere boundary, instability of the interface (wind-induced waves, presence of water spray in the air just above the water surface, the character of temperature stratification in this layer, etc.) plays the main role.

Processes occurring on the ocean–atmosphere GBZ tend to intensify during formation of splashes and air bubbles (Fig. II.7.3). Splashes and air bubbles considerably increase the surface area for interaction between the hydrosphere and atmosphere. In the ocean, air bubbles may be carried downward to a depth of up to 20 m. The probability of air bubbles forming in seawater increases with increasing diameter of drops. A drop having a diameter of 0.4 mm is able to produce two to three air bubbles, while a drop 4.4 mm in diameter produces 200–400 air bubbles. "One drop 5 mm in diameter can produce up to 1000 drops 0.2 mm in diameter, which in turn will produce about 5000 air bubbles. Air bubbles produced by rain drops sink to a depths of 4 cm^2" (Bezborodov and Eremeev, 1984, p. 43). Waves, rain, atmospheric fallout of dust and snow, decay of sedimentary material on the seabed, hydrothermal activity, rapid increase in temperature of seawater, etc., are the reasons for the origin of bubbles. In a rough sea (when wind velocity is 6–8 $m \times s^{-1}$), the total surface area of air bubbles, which is an internal air–seawater interface at a depth of 1–2 m in the ocean, amounts to about 250 $cm^2 \times m^{-3}$. The conventional surface area of the internal air–seawater interface in the ocean is about 2.10^3 $cm^2 \times m^{-3}$ (Bezborodov and Eremeev, 1984, p. 46), and the flux across the internal interface is 1–10 $cm^2 \times m^{-2} \times s^{-1}$. When the wind speed in the surface layer is 8 $m \times s^{-1}$, 4% of the total ocean surface will be covered by foam.

A great amount of salts, organic matter and microelements is delivered to the atmosphere together with air bubbles. The amount of salts delivered to the atmosphere by air bubbles which then burst there range from 10^{-12} to 10^{-6}g. Filmy

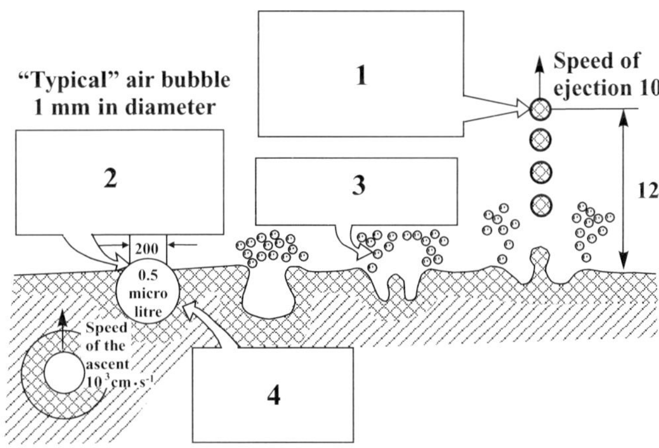

Fig. II.7.3. Scheme of bursting of air bubble. After Macintyre, 1965 (citation after Horne, 1972).
1—higher jet drop/diameter 100 μm, volume 1.5 nanoliters, on surface there are $3 \cdot 10^{11}$ molecules of water, 200 positive charges, 30 nanograms of salts, 0.5 ergs of kinetic energy;
2—film cap, thickness—2 μm, or 4000 monomolecular layers, radius of cap—one-half the bubble radius, area $3 \cdot 10^{-4}$ μm^2, or $3 \cdot 10^{12}$ molecules of water on surface;
3—film drops forming, numbers 0–20, 1–20 μm in diameter, rising up to 1 km above surface;
4—surface, area 3 mm^2, $3 \cdot 10^{13}$ molecules of water, 2 ergs of surface energy.

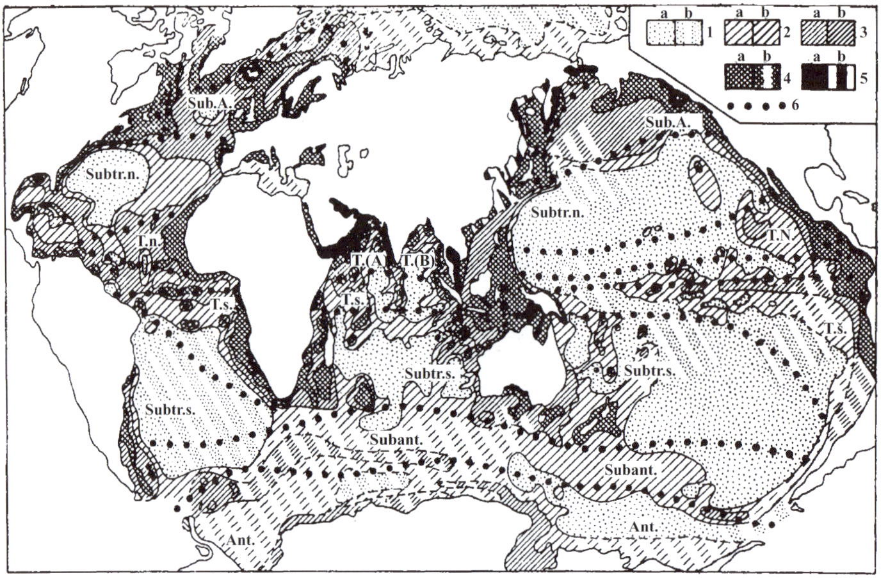

Fig. II.8.1. Distribution of average annual primary production (mg C/m²/day) in seawater area of World Ocean. After Koblents-Mishke, 1985.
a—according to results of definition of primary production by radiocarbon method; b—according to indirect data: 1—<100, 2—100–150, 3—150–250, 4—250–500, 5—>500, 6—circulation borders; Sub. A.—subarctic circulation; Subtr.n.—subarctic north; Subtr.s.—subarctic South; T.n.—tropical north; T.s.—tropical south; T (A)—tropical (Arabian Sea); T(B)—tropical (Gulf of Bengal); Subant.—Subantarctic; Ant.—antarctic.

In the course of photosynthesis, the process of biomechanical fractionation of dissolved matter, which is related to the appearance of the solid phase due to dissolved forms of elements (Ca, Si, O, C, N, H, and microelements), takes place: $CaCO_3$, SiO_2, and organic matter form. This is class 1 of biomechanical fractionation of material in the ocean (Lisitzin, 1986, p. 7). Class 2 comprises biofiltration: separation of fine particles of suspended matter, passage of these through the digestive systems of living organisms, partial transformation of this solid material in digestive systems, and its use in the normal vital processes of all organisms—packaging of excess organic matter and inorganic components into fecal pellets and entry of these pellets into the water.

Planktonic organisms contain more than 70 chemical elements (Fig. II.8.4). In addition to 12 major elements (see Table II.8.1) and 18 minor elements (Table II.8.3, II.8.4, II.8.5, II.8.6), these elements are F, Be, V, Ga, As, Br, Mo, Sn, J and some others (Mertz, 1981; Vinogradov, 1969). All of them, i.e., both minor elements and 12 major macrocomponents, are organized at the molecular level (Lisitzin, 1986, p. 47). At the same time, some of these elements are concentrated in organic matter, some elements are incorporated into biogenic $CaCO_3$, and some of them into amorphous

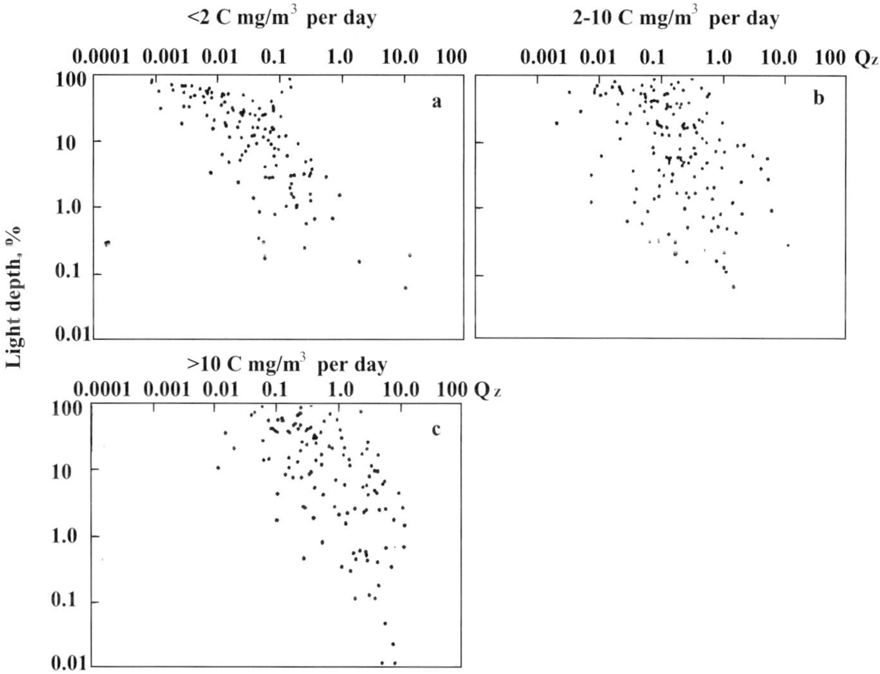

Fig. II.8.2. Energetic efficiency of photosynthesis (Q_z) in areas with different levels of primary production on the surface (mg C/m^3/day). After Koblents-Mishke and Vedernikov, 1977. a—<2; b—2–10; c >10. Axis of ordinates—depth of light (in % of penetrating radiation).

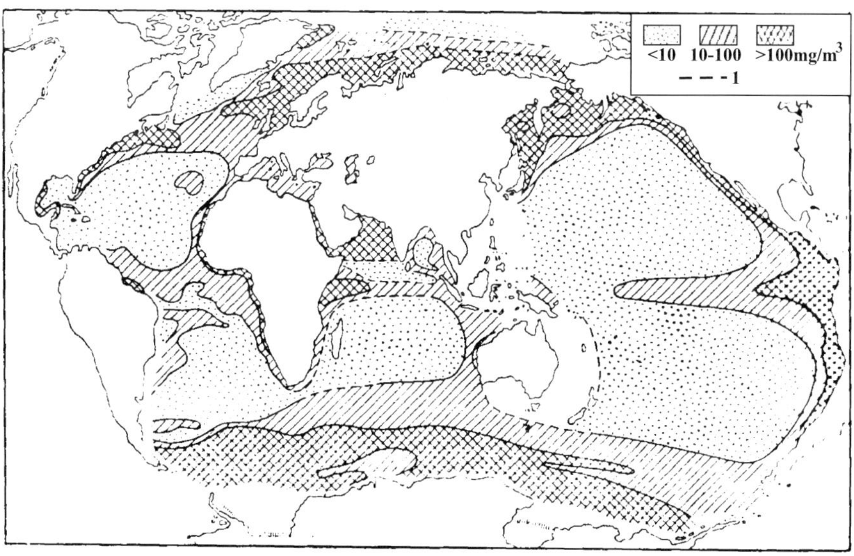

Fig. II.8.3. Distribution of phytoplankton biomass (mg/m^3) in 0–100 m layer in World Ocean. After Zernova, 1985. I—according to indirect data.

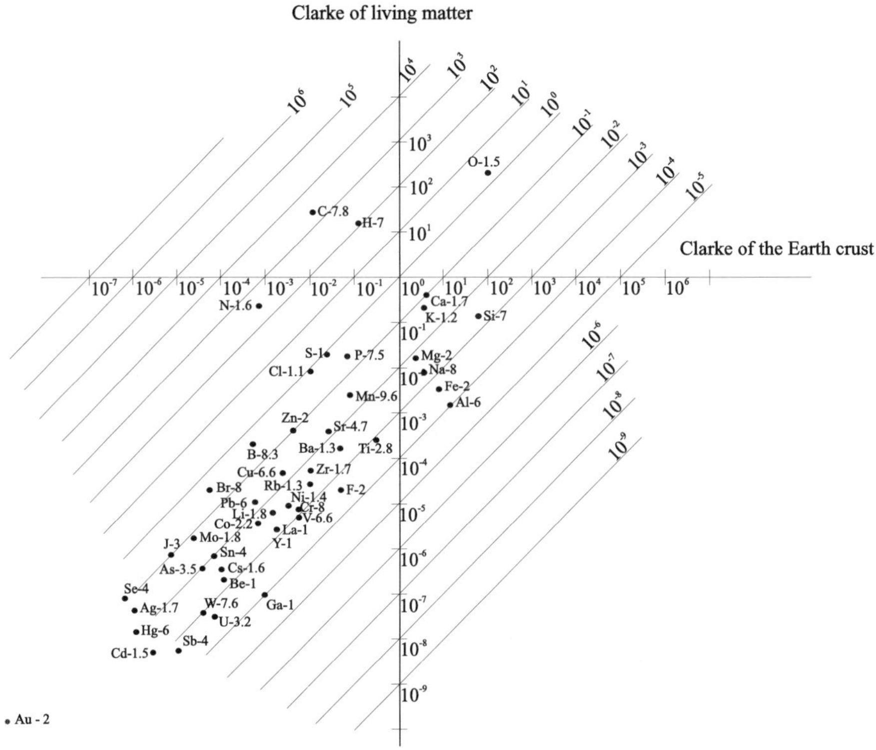

Fig. II.8.4. Biological activity of elements. After Perelman, 1989, p. 249.

SiO_2 (Table II.8.7). From this point on, the fate of these elements depends on the fate of that which contains these materials—matrices-carriers. Depending on when and where these matrices are dissolved or recrystallized, minor elements are released or reorganized in another bonds. In such a way, other authigenic minerals are formed. But these processes are controlled by other physical and chemical conditions are typical of the environment in which these elements came in contact with the matrices-carriers.

Migration of chemical elements in the ocean takes place either with the participation of living matter (biogenic migration) or it proceeds in an environment the geochemical features of which (O_2, CO_2, H_2S and others) are conditioned by both that which inhabits this system now or that which existed in the biosphere in the geological past (Perelman, 1979, p. 215).

All these elements circulate in the ocean via specific routes from an external medium into organisms and back, thus forming more or less closed biochemical cycles (Lisitzin, 1986, p. 29).

These elements either enter an organism (i.e., become organized) or they escape from an organism (i.e., become mineralized).

Table II.8.3. The content of chemical elements and components in plankton and suspensions of the World Ocean ($CaCO_3$, SiO_{2am}, C_{org}, Fe and Mn in %; Cu-Cr 10^{-4}%)

Area of the ocean	$CaCO_3$	SiO_{2am}	C_{org}	Cu	Ni	Co	Cr	Fe	Mn
Phytoplankton (0–1 m layer)									
BS	2.6	-	37.26	30	60	23	30	0.30	0.07
BS	-	-	-	5	50	17	20	0.25	0.01
BS	0.52	-	38.79	27	60	30	20	0.46	0.14
Separated suspension (5–7 m layer)									
BS	4.25	-	42.64	60	50	10	-	0.60	0.32
BS	2.14	-	44.17	170	50	20	70	0.30	0.10
AO	-	-	-	90	70	10	207	0.70	0.03
AO	-	-	-	35	30	15	135	2.50	0.06
AO	1.34	-	1.50	-	80	5	313	2.10	0.03
PO	5.25	5.1	21.1	6	250	30	1420	2.24	0.01
PO	35.8	6.0	11.9	50	220	30	340	0.48	0.01
PO	25.7	2.0	16.6	40	100	30	280	0.32	0.01
PO	26.7	1.3	17.8	40	120	20	240	0.54	0.01
PO	42.2	2.3	10.8	40	80	20	152	0.72	0.01
PO	-	1.6	-	20	50	30	128	0.30	0.01
Aeolian suspension									
AO	3.00	-	0.93	40	50	27	100	5.10	0.01

* BS – Baltic Sea, AO – Atlantic Ocean, PO – Pacific Ocean.

Table II.8.4. The average contents of chemical components and elements in plankton (bulk samples) of the upper layer (0-1 m) of the Baltic Sea. C_{org}–Mn, in % of dry weight; Zn–Li, in 10^{-4}%

C_{org}	$CaCO_3$	Ca	Mg	P	K	Na	Al	Ti	Fe	Mn	Zn	Cu	Cr	Ni	Rb	Li
35.3	5.9	0.26	0.76	0.66	0.76	2.97	0.10	0.02	0.12	0.02	266	34	17	28	6	7

Table II.8.5. Concentration of elements in euphausids and in their living products, mg/1 of dry matter. After Lisitzin, 1986, p.18

Sample	Ratio net weight/ dry weight	Ag	Cd	Co	Cr	Cu	Fe	Mn	Ni	Pb	Zn	Ce	Cs	Eu	Hg	Sb	Sc	Se	Sr
Pellets	4.4	2.1	9.6	3.5	38.0	226	24000	243.0	20.0	34	950	200.0	6.0	0.66	0.34	71	2.8	6.6	78
Fur (skin)	4.6	2.9	2.1	0.80	5.3	35	232	11.7	6.7	22	146	1.2	0.019	0.0077	0.17	0.80	0.030	1.9	350
Euphasids	4.7	0.71	0.74	0.18	0.85	48	64	4.2	0.66	1.1	62	0.21	0.062	0.0023	0.35	0.071	0.090	4.4	117
Micro plankton	0.7	0.67	2.1	0.87	4.9	39	570	17.9	8.4	11	483	0.30	0.080	0.013	0.05	0.22	0.13	2.7	520

Table II.8.6. Chemical composition of the water suspension (PSM) of the World ocean and its potential sources*

Element	1	2	3	Σ1–3	Σ2+3	4	5	6	7	8	9	Σ4–9	Σ5–9	10	11	12	13	14
Microelements, %																		
Si	5.04	12.54	18.12	11.9	15.33	8.84	7.49	5.48	0.87	1.49	1.19	4.2	3.3	4.5	6.0	-	22.25	23.9
Al	0.20	3.78	2.55	2.15	3.16	0.08	0.02	0.02	0.25	0.02	0.93	0.07	0.06	0.6	0.01	0.48	7.59	9.47
Ti	0.03	0.1	(0.1)	0.09	0.158	0.040	0.034	0.026	0.053	0.023	0.024	0.034	0.032	0.02	0.005	0.085	0.344	0.42
Na	7.9	0.38	1.3	3.2	0.84	4.2	6.2	9.1	6.8	12.0	9.8	8.2	8.8	4.0	3.5	23	0.55	0.66
K	0.56	1.12	1.0	0.86	1.06	0.23	0.31	0.30	0.43	0.48	0.53	0.38	0.41	0.80	1.0	1.1	1.70	2.29
Mg	1.80	1.00	(1.2)	1.4	1.00	1.79	1.75	2.04	1.69	1.98	2.14	1.90	1.92	0.5	0.8	3.1	1.46	1.38
Ca	1.17	8.73	11.9	7.2	10.31	5.51	6.49	5.04	4.97	11.80	8.30	7.0	7.3	4.0	1.5	2.9	1.62	2.50
Fe	0.82	3.16	2.4	2.13	2.78	2.24	0.28	0.37	0.36	0.56	0.26	0.68	0.37	0.8	0.08	0.54	4.2	3.32
Mn	0.033	0.065	0.030	0.033	0.048	0.009	0.006	0.004	0.41	0.006	0.003	0.012	0.013	0.020	0.001	0.009	0.004	0.075
P	(0.1)	0.20	0.06	(0.12)	0.13	0.26	0.33	0.25	(0.30)	(0.30)	0.42	0.1	0.33	0.25	0.9	0.25	0.07	0.07
Sr	0.018	0.048	0.053	0.04	0.05	0.071	0.247	0.207	0.357	0.438	0.168	0.248	0.283	0.03	0.03	0.03	0.019	0.030
Ba	0.09	0.28	0.03	0.13	0.15	0.25	0.20	0.20	0.22	0.30	0.24	0.23	0.23	0.04	0.01	0.005	0.11	0.08
Br	0.151	0.056	0.017	0.075	0.036	0.087	0.163	0.293	0.370	0.202	0.211	0.188	0.208	0.035	0.05	0.1	0.0006	(0.07)
C_{org}	-	-	1.5	-	-	21.1	11.9	16.6	17.8	10.8	-	15.6	14.3	20.0	40.0	-	0.93	1.0

Continues

Table II.8.6. Chemical composition of the water suspension (PSM) of the World ocean and its potential sources*—*cont'd*

Ele-ment	1	2	3	Σ1–3	Σ2+3	4	5	6	7	8	9	Σ4–9	Σ5–9	10	11	12	13	14
Microelements, g/t																		
Ni	70	30	87	60	58	250	70	50	90	70	30	93	62	70	10	20	40	68
Co	16	20	6	14	13	27	19	28	29	27	27	26	26	5	1,5	4	27	20
Cu	92	35	223	117	130	55	100	61	88	57	48	67	70	300	40	70	43	45
Zn	130	450	280	293	375	210	160	110	150	100	110	130	125	600	300	250	100	95
Pb	30	50	(50)	(43)	50	30	60	70	70	60	50	57	62	200	20	180	50	20
Mo	18	19	11	16	15	41	23	25	25	19	10	23	20	10	1	0,4	12	2
Cr	207	135	313	218	223	1420	667	262	309	199	128	490	305	100	10	35	107	100
V	11	55	(50)	35	50	4	41	61	21	12	20	26	31	50	4	30	82	130
Li	6	29	38	24	34	5	9	8	6	5	4	6	6	20	40	-	46	60
Rb	32	74	50	52	62	4	4	3	8	3,5	3	4	4	50	3	12	123	140
Cs	0.1	5.7	2.9	2.9	4.3	0.07	0.24	0.26	0.27	0.23	0.19	0.21	0.24	2	0.04	0.3	4.5	5.5
Yf	0,12	1.8	1.0	4.0	5.9	0.07	0.06	0.07	0.15	0.11	0.07	0.09	0.09	-	-	0.4	7.8	6
Sc	0.43	7.0	6.1	4.5	6.5	0.13	0.15	0.22	1.0	0.27	0.30	0.35	0.39	0.4	0.2	1.5	1.6	13
As	7	14	13	11	13	7	6	9,5	18	29	15	14	15,5	-	10	-	7,4	10
Sb	6,2	2,0	15	3,2	1,7	1,0	0,66	1,1	0,85	0,97	0,47	0,83	0,81	5	0,1	6	2,7	1,5
Se	3.1	4.6	2.8	3.5	3.7	1.8	1.8	1.4	4.4	5.4	2.7	3.0	3.2	8	4	15	3.4	(0.17)
Te	2,1	1,9	2,8	2,3	2,3	1,5	1,6	3,3	4,2	1,8	1,3	2,3	2,4	-	-	-	1,0	-
Cd	5	6	4	5	5	10	19	15	22	20	11	14	15	6	3	-	5	0.8
Hg	1.4	0.56	1.0	1.0	0.8	0.09	0.58	0.32	0.96	0.96	0.44	0.57	0.65	5	0.2	20	0.20	0.9
U	1.5	1.2	1.8	1.5	1.5	6	1.1	1.3	2.4	0.8	1.0	1.2	1.3	1	0.6	-	2.2	3.2
Th	0.27	5.7	8.3	4.8	7	0.13	0.07	0.08	0.28	0.20	0.26	0.17	0.18	0.3	0.1	1.3	14	12
La	1.5	22	32	18	27	1.2	1.0	0.5	2.7	2.3	2.0	1.6	1.7	1.2	0.8	4.8	58	35.5
Ce	2.6	37	58	33	48	2.1	1.6	1.5	5.6	3.1	3.0	2.8	3.1	2.5	1.2	8	104	67
Nd	2.3	15	27	15	21	0.9	1.1		2.	3.4	2.4	1.7	1.9	1.6	0.7	2.5	60	33
Sm	0.39	3.2	5.0	2.9	4.1	0.19	0.26	0.34	0.61	0.38	0.32	0.40	0.46	0.4	0.07	0.4	9.0	6.7
Eu	0.05	0.9	0.85	0.60	0.87	0.03	0.10	0.06	0.09	0.06	0.07	0.07	0.08	0.06	0.02	0.09	2.2	1.2
Tb	0.03	0.43	0.87	0.41	0.65	0.02	0.03	0,04	0.07	0.04	0.03	0.04	0.04	(0.03)	0.03	0.15	1.4	1.0
Yb	0.13	1.20	2.90	1.4	2.05	0.06	0.06	0.08	0.27	0.12	0.06	0.09	0.10	0.08	0.07	0.3	4.6	2.9
Lu	0.02	0.19	0.38	0.20	0.28	0.015	0.025	0.016	0.05	0.035	0.023	0.020	0.021	0.02	0.015	0.07	0.48	0.45
Ag	0.12	0.1	1.1	0.4	0.6	0.19	0.13	0.12	0.93	0.38	0.32	0.35	0.38	4	0.4	1.5	0.65	0.07

*) 1-3 – Atlantic Ocean suspension; 4-9 –Pacific Ocean suspension; 10 – the average ocean suspension composition (Savenko, 1988_1); 11– the average ocean plankton composition (Savenko, 1988); 12 – the average ocean aerosol composition (Savenko, 1988); 13 – aeolian dust according to the our data (station PSh-7-6E); 14 – the average continents clay sedimentary rocks structure (Vinogradov, 1962).

Table II.8.7. The order of biological absorption of elements. After Perelman, 1989.

		Biological absorption firures				
		100.n	10.n	n	0.n	0.0n–0.00n
Elements of biological accumulation	Active	P,S,Cl,Br,L,I				
	Strong		Ca,Na,R,Mg, Sr,Zn,B,Se			
Elements of biological capture	Medium			Mn,F,Ba,Ni,Cu, Ca,Co,Pb,Sn, As,Mo,Hg,Ag,Ra		
	Weak and very weak				Si,Al,Fe,Ti,Zr,Rb,V,Cr, Li,V,Nb,Th,Se,Be, Cs,Ta,U,W,Sb,Cd	

Based on the reports of several authors (Vinogradov, 1935–1944; Patin, 1979; Morozov, 1983), some of the most important regularities with respect to biodifferentiation on the elemental level have been formulated:

(1) the content of minor and major elements in seawater and in newly formed suspensions (plankton) decreases as the atomic number increases;
(2) for elements with an atomic number less than 30 (12 macrocomponents, see Table II.8.1, and also a series from K, Ca, to Cu and Zn), there is a pair linear correlation (0.7–0.8) between the coefficient of accumulation and the atomic number;
(3) the concentration of elements by plankton correlates positively with the value of ion potential (relation of valence to ion radius) with 95% confidence;
(4) the elemental composition of organisms is strongly dominated by elements of initial numbers in the periodic system (from 1 to 20), and 12 of them are of basic elemental composition (see Table II.8.1) and make up 10^{-3}% of the total material. To provide for the buildup of skeletons, organisms use those salts which have a solubility of not less than 10^{-3} (Ca, Si, Mg, Al, Fe and others);
(5) ions are transported across cell membranes in the form of helates;
(6) the total amount of elements used by organisms currently exceeds 70. The ability of organisms to concentrate chemical elements is qualitatively determined by the coefficient of accumulation (CA): the ratio of elemental contents in organisms to their contents in seawater.

Four groups of elements can be identified: (a) elements which are parts of living organisms, but their concentrations in marine organisms are not in excess of the values found in seawater (Na, (Mg), K, (F), Cl—high contents of which, however, are found in oceanic waters); (b) elements which are relatively almost not accumulated by organisms (Lg with CA < 2)—K, Pb, Cs, Ca, Sr, Ba, B, Mo, W, F, Br, and I;(c) elements which are, to a great extent, accumulated by marine organisms (Lg with CA = 2–4)—Cu, Ag, Au, Zn, Cd, Hg, Al, Ca, C, Si, Ti, Sn, V, As, Sb, S, Se, Cr, Mn, (I), Fe, Co, and Ni; (d) elements which are concentrated intensively by one or several groups of the studied organisms (Lg with CA >4)—Al, C, (Sn), N, P, Mn, Fe, and Co). Consequently, elements can be subdivided into the following groups and rows by biofiltration (i.e. according to the clarkes of these elements in living matter) (Table II.8.7, Fig. II.8.4).

In such a way, a great variety of compounds are formed in the course of photosynthesis and this leads to an increasingly large amount of various chemical information. A new and more complicated type of information, so-called "biological information", comes into being. By this is meant that the order, complexity and organization of nature becomes higher; negative entropy increases and informational and thermodynamic entropy decreases" (Perelman, 1989, p. 246).

Biofiltration of seawater. Organisms of zooplankton pass a great amount of water when filtering seawater in order to obtain food: over a 182-day period, zooplankton is capable of passing through themselves a volume of water equal to that of the whole World Ocean (Bogorov, 1974). However, because 60–65% of zooplankton live in the upper water layer, it takes this zooplankton only 20 day to pass the whole volume of this upper layer through themselves (Lisitzin, 1983_2, p. 226). Zooplankton pass through its digestive tract not only biogenic but also terrigenous material to transform waste products into fecal pellets.

Over 1.5- to 2-day period, zooplankton must filter and use for nourishment an amount of suspended particles equal to its own weight. On the whole, this value for zooplankton is estimated to be about 30 billion tons of raw suspension, or 2 billion tons of C_{org} per day. Because the surface suspension contains about 15–20% C_{org} on average (Tables II.8.3, II,8.6.), the amount of suspension which must be filtered by zooplankton to provide nourishment comprises 75–100 billion tons of raw suspension or 7.5–10 billion tons in dry weight. To obtain such a quantity of food by filtration, zooplankton must filter about 50,000 km^3 of seawater (Lisitzin a. Vinogradov, 1983). Zooplankton do not filter the water volume of the whole World Ocean but only the water of the layers they inhabit (Fig. II.8.5), i.e., a layer from 0 to 500 m, which is repeatedly reprocessed; in some seasons and in certain latitudes the lower level of this layer is located at depths of up to 1500–3000 m (Fig. II.8.6).

The second circumstance to be taken into account is that the terrigenous suspension from river mouths and abrasive coasts extends into the interior areas of the ocean (sea) and travels mainly not in the upper (photic) layer but in the near-bottom turbid layer. Depending on the current velocity, the suspension passes over the 100-km-wide shelf zone in a matter of tens of hours to several days. This period of time is naturally insufficient for the river suspended load to be completely filtrated either by plankton or benthos. This suspended matter, almost unchanged, is carried over the shelf edge and in several days passes beyond the limits of the zone of mass habitation of filter-feeders (Fig. II.8.5, II.8.6), extending as the near-bottom nepheloid layer. This suspension is also inaccessible to bottom filter-feeders. Thus, river suspended matter in the near-bottom layers, which is little affected if at all by benthos, reaches ocean basins, where it is deposited on the bottom. Because of the

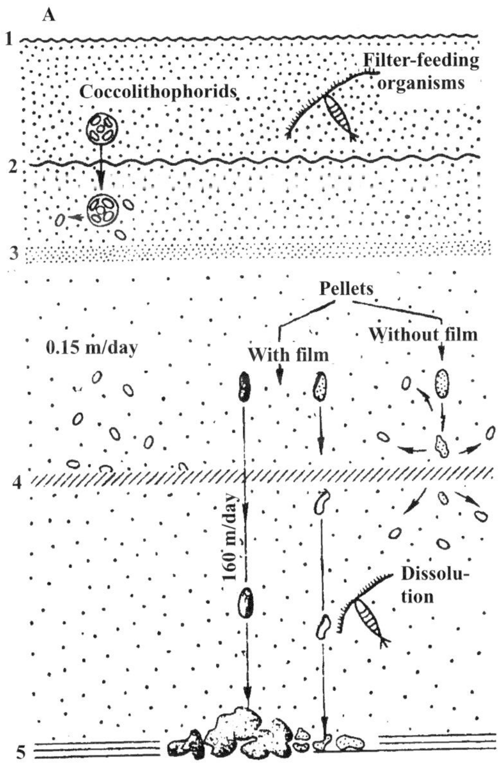

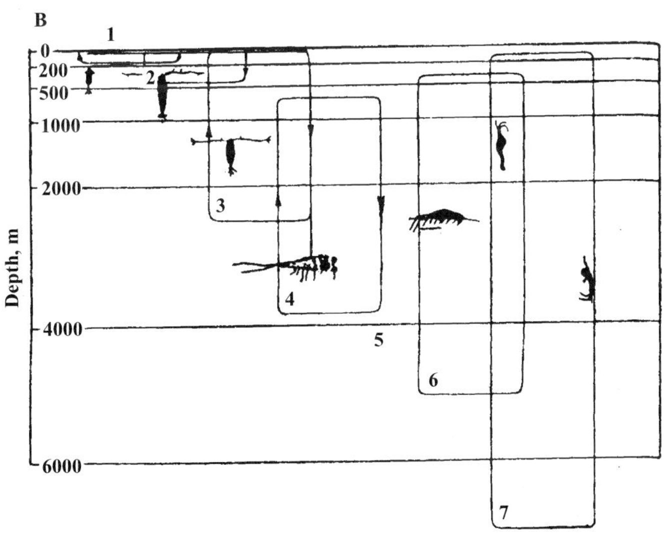

Caption see page 276

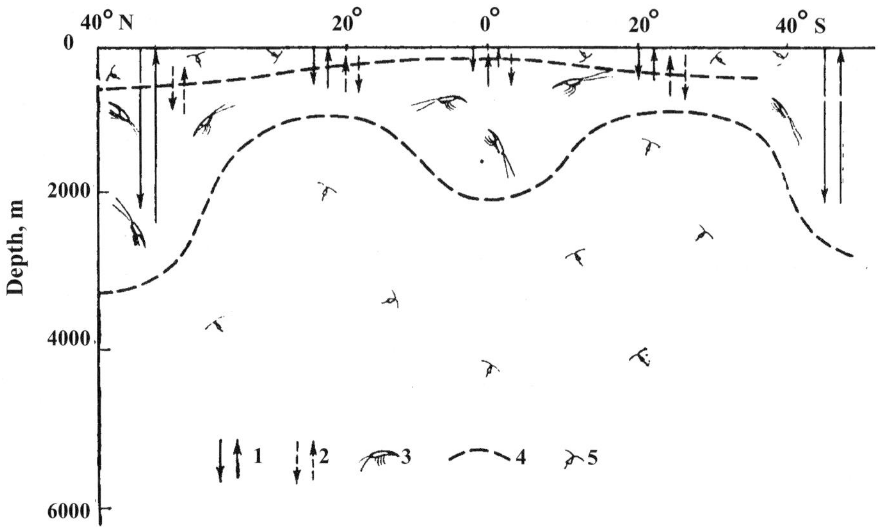

Fig. II.8.6. Distribution of meso- and macroplankton in meridional profile of ocean. After Vinogradov, 1977.
1—amplitude of daily migration of mesoplankton; 2—amplitude of migration of macroplankton (according to echo sounding data); 3—zone of concentration of macroplankton; 4—border of concentration of macroplankton; 5—zone of prevailing mesoplankton.

extremely small density of benthos in these basins, the consumption of benthos by bottom filter-feeders is also small. However, some other biotic factors come into effect: detritus feeders and burrowing organisms (see Section II.14).

Suspended organic matter in water. The main producer of organic matter in a suspension (70%) is diatomic phytoplankton, and in tropical waters, diatoms and coccolithophores. High contents of C_{org} in suspended matter of the oceans have been found both in temperate humid and arid climatic zones (Lisitzin, 1964; Lisitzin et al., 1975; Emelyanov and Romankevich, 1979; Baturin et al., 1995). This is because predominant constituents of plankton in these zones are peridinic algae and copepods, i.e. species which have no skeletons and contain on average 33.49 and 45.52% C_{org}, respectively (Table II.8.1). High contents of C_{org} in suspended matter (Table

Fig. II.8.5. Mechanism of biological sedimentation in ocean. After Lisitzin, 1983.
A—pellet mechanism:
1—layer of minimal distribution of phytoplankton and detritus in ocean, zone of distribution of coccolithophorides and filtrators–copepods.
2—depth at which usually occurs destruction of coccolithophorides on separate plates after death of organisms (mean velocity of sinking of coccoliths—0.15 m/day);
3—area of maximal concentration of detritus—pycnocline (detritus is shown by points);
4—maximal depth of distribution of coccoliths during particle-by-particle sedimentation ;
5—bottom sediments.
B—stages (bold) of migration of zooplankton. After Vinogradov, 1968.

II.8.3, II.8.6) in arid zones are also conditioned by the weak dilution of suspended matter by terrigenous matter: areas with maximum (>20%) contents of C_{org} (Fig. II.8.7) correspond approximately to areas where the content of terrigenous particles in suspension is minimum. Also, they correspond to areas of high $CaCO_3$ contents in suspension. The content of C_{org} is markedly decreased in those regions where diatoms predominate in the phytoplankton. In the Antarctic zone of the ocean, increased content of C_{org} is found only in summer; in winter this value is very low. On the whole, there is a direct relationship between the concentration of C_{org} in the surface particulate matter and the production of C_{org}, as determined by the C^{14} method (Lisitzin et al., 1975). In connection with the seasonal variations in this production, contents of C_{org} also show the same fluctuations.

The C_{org} content in suspended matter decreases with increasing depth. At depths greater than 500–750 m, the content of C_{org} makes up only 10–20% of the amound in the surface layer. In the Atlantic Ocean, at a depth of about 3500 m, the content of C_{org} makes up 10–15% of its amount in the surface layer (Menzel, 1967; Emelyanov and Romankevich, 1979).

The distribution of organic detritus—the main form of particulate organic matter (fragments of tissues of phytoplankton), which was studied under a microscope—give a general picture of the C_{org} concentrations in the deep-sea suspended matter (Emelyanov and Romankevich, 1979).

In the upper photic layer, particles of detritus having a diameter less than 0.1 mm make up 56–84% of the total mass of detritus (Fig. II.8.7); particles 0.1–1 mm in diameter make up 14–33%. More than 80% of their mass is represented by the fecal pellets of animals. Particles larger than 2.5 mm make up 0.2–2% of the total mass of detritus (Vinogradov, 1986, p. 66).

More than 90% of coarse detritus (larger than 0.1 mm) is evacuated from the photic layer due to the effects of gravitational precipitation. Gravitational precipitation is responsible for removal of only 5% of detritus less than 0.1 mm in diameter. The remainder of detritus is eaten by phytoplankton and packaged into fecal pellets (Vinogradov, 1986, p. 87). Filter-feeders are the major consumers of detritus. The waters of the 200 m-thick water layer are filtered in 3–20 days. As a result of such intensive filtration, only from 0.5 to 3% of daily primary production extends beyond the lower photic zone (120 m) in waters of various trophicity (Emelyanov and Romankevich, 1979).

In the Baltic Sea, organic detritus makes up from 2.5 to 53.6% of the total weight of suspended matter, or 18.8% on average (Blazchishin and Pustelnikov, 1975). The maximum average content of detritus (about 28%) is found in open-sea waters, where the organic type of suspended matter is abundant, whereas the minimum content (10.3%) is found in coastal waters, i.e., in suspended matter of terrigenous type.

In the Mediterranean Sea, organic detritus makes up from 29.5 to 91.2% of the total suspension by weight (Emelyanov and Shimkus, 1972). The highest average content of organic detritus was found in a suspension contained in the upper 2-m-thick water layer (78.8%). In the upper part of the water column, the concentration of detritus tends to increase with depth, whereas in deeper water layers there is a tendency for detritus to decrease.

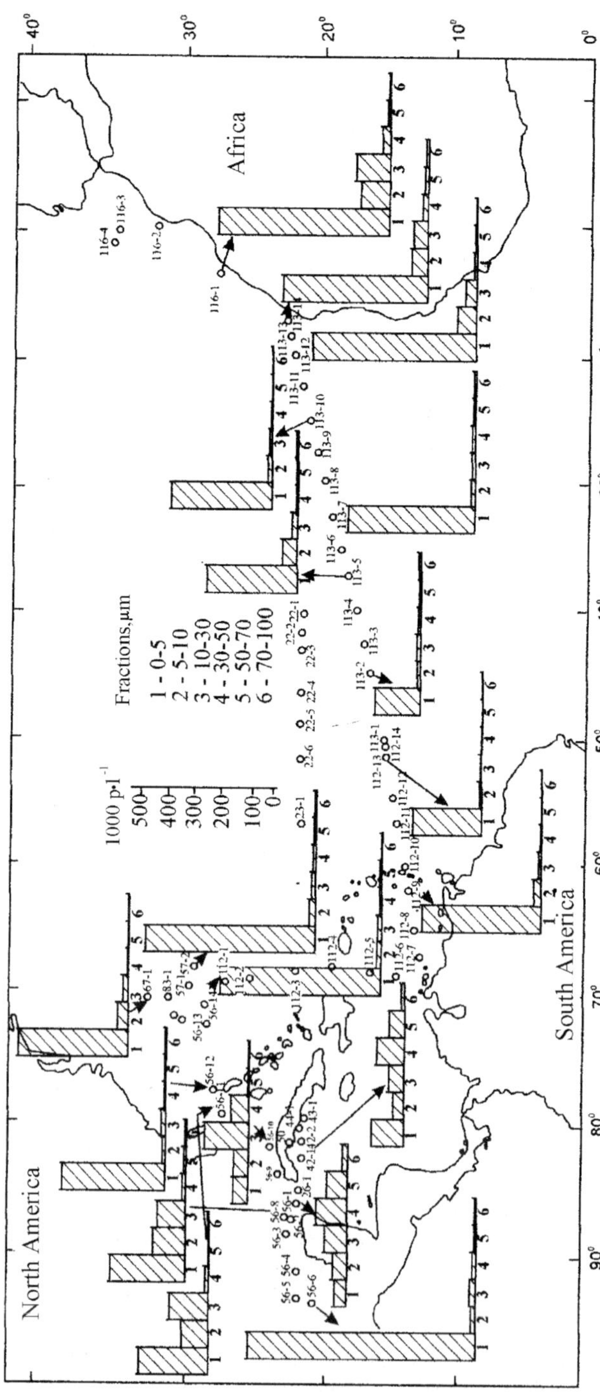

Fig. II.8.7. Histograms of grain-size distribution of water suspended particular matter (layer 2 m) in North Equatorial Atlantic according to Coulter counter data, 1981. Scale: particles per liter.

The content of organic detritus in the Atlantic Ocean varies from 0 to 100% of the total volume of suspended matter (larger than 0.7µm). It should be said that the maximum number of particles of organic detritus was found in coastal waters, primarily in areas of coastal upwelling (up to 200 thousand specimens per liter of water). The size of organic detritus particles here commonly range from 5 to 50 µm in diameter. The amount of organic detritus in the open ocean is very small (less than 10,000 specimens per liter of water), and the sizes of its particles here are much smaller than those found near coasts (Emelyanov and Romankevich, 1979). Wide fluctuations in the number of organic matter particles are characteristic not only of surface waters but also deep waters. This is evidence that some particles have time to "pass rapidly" through a layer 3 km thick.

Suspended amorphous SiO_2 in water. From trace amounts to 334.6 µg SiO_2 per liter were found in the surface waters of the Atlantic Ocean (Lisitzin and Emelyanov, 1977). The concentration distribution of this component in the surface waters of the ocean is consistent with the pattern for contents of SiO_2 in suspensions: in both cases, latitudinal zonation of the component distribution is evident. Concentrations of SiO_2 in coastal waters are markedly higher than those in the waters of the open ocean. Concentrations of SiO_2 decrease with increasing water depth.

The amount of diatomic cells in the upper 100-m layer is controlled by climatic zonation of the ocean (Zernova, 1985). The quantity of these cells is abundant in temperate humid zones, as well as in areas adjacent to zones of coastal upwellings. The role of diatoms in phytocene sharply increases in frontal zones and near coastal upwellings. This phenomenon can be explained by the redistribution of diatoms by ocean currents. These currents are believed to be responsible for the accumulation of great amounts of diatoms in sediments from the Angola basin (Emelyanov, 1973, 1982).

A map of annual primary production and SiO_2/C ratios has been used to calculate the annual primary production of SiO_2 in the surface waters of the Atlantic Ocean (Lisitzin et al., 1975). It was found that the maximum production is characteristic of the Antarctic zone of the ocean, increased production is characteristic of high-latitude regions in both hemispheres, and low production is typical for arid zones of the open ocean. Also, the primary production in the near-shore areas off Europe was found to be markedly higher than that near the coasts of the North and South America.

Therefore, the primary production of SiO_2, as well as the distribution of cells, are closely connected with climatic zonation and also with upwelling. Such a distribution of the primary production is well consistent with the distribution of SiO_2 in suspended matter and in bottom sediments (Emelyanov, 1977, Fig. II.8.8). Consequently, some elements of zonal distribution of silica on the ocean floor originated as early as in the upper layer of the water column, i.e., in the photic zone.

The content of amorphous silica in ocean suspended matter markedly decreases with depth. An especially sudden decrease in this element is fixed below the active layer (100–200 m). At depths greater than 2000 m, the content of SiO_2 in the suspended matter is commonly less than 1–3%; this is illustrated by the calculated numbers of biogenic siliceous particles in the suspended matter (Fig. II.8.9).

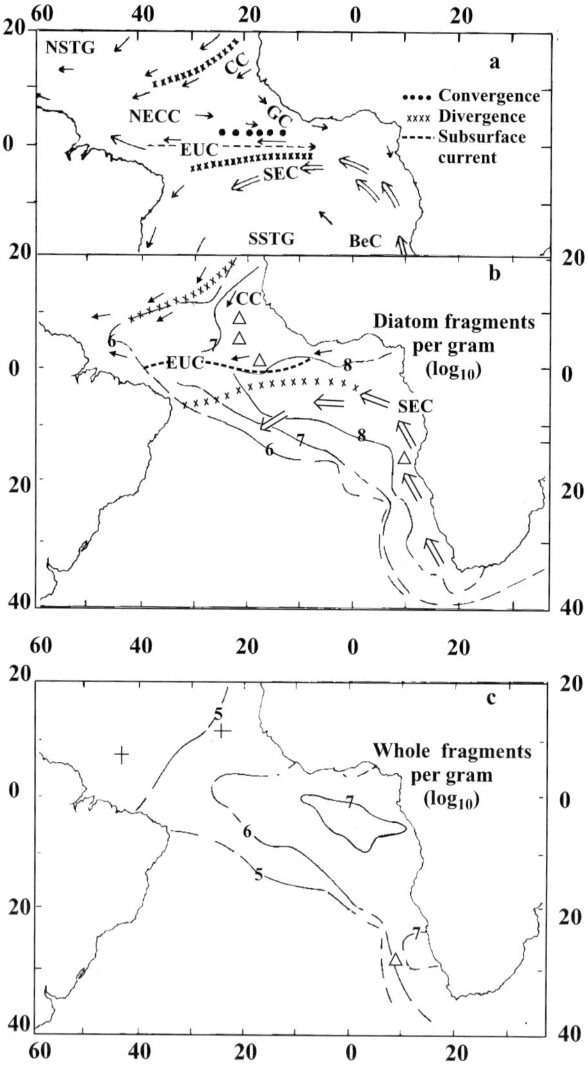

Fig. II.8.8. Geographic distribution of diatomic abundance in 81 surface sediment samples. After Pokras and Molfino, 1986.

Contours are $\log_{10}$ of numbers per gram. Triangles indicate below contour; crosses indicate above contour. Surface circulation: BeC = Benguela Current; CC = Canary Current; EUC = Equatorial Undercurrent; GC = Guinea Current; NECC = North Equatorial Countercurrent; NSTG = North Subtropical Gyre; SEC = South Equatorial Current; SSTC = South Subtropical Gyre.

Calcium carbonate. The amount of carbonate ($CaCO_3$) delivered to the ocean from land is estimated at 1.36 billion tons per year, and this value is three times as large as that for silica (Lisitzin, 1978). However, silica is captured mainly in the course of photosynthesis (the first link in food chains), while calcium carbonate is captured by

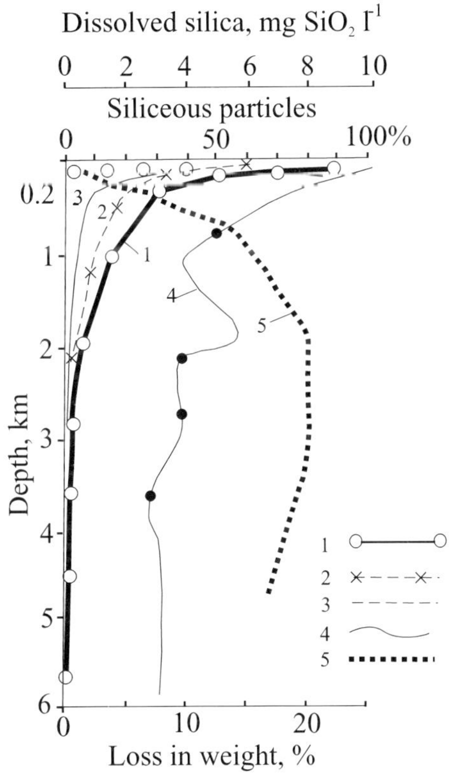

Fig. II.8.9. Distribution of biogenic siliceous particles in suspended matter of Atlantic Ocean (1), Mediterranean sea (2) (in volume %) and preservation of diatomic skeletal remains (3) (in %, upper scale according to Kozlova's data) and skeletal radiolarian remains (4) (loss of weight in %, lower scale) in water strata; 5—mean concentration of dissolved SiO_2 in Hawaii area of Pacific Ocean. After Emelyanov, 1982; Berger, 1968 (radiolarians and SiO_2)

organisms of the second and subsequent links (Lisitzin, 1986, p. 43). In the first link, $CaCO_3$ is trapped only by coccolithophores, and on the bottom, by algal-macrophytes (lime algae). Of the total 1.08 billion tons of $CaCO_3$ annually trapped mostly in the upper active layer of the World Ocean (foraminifera, heteropodes, coccolithophores, crustaceous), 79.4% settles onto the bottom, and the remaining 20.6% remains suspended in the water column (Lisitzin, 1986, p. 43).

Distribution of P in suspended matter. The distribution of the P contents in samples of the surface suspension collected in the Atlantic Ocean is consistent with the pattern described for C_{org} (Lisitzin et al., 1975; Emelyanov, 1979_2, Emelyanov and Romankevich, 1979; Baturin et al., 1995). Various correlation profiles show that P and C_{org} have a directly proportional relation: P increases proportionally as C and larger particles to be bound then into large and relatively heavy aggregates (Fig. II.8.10). Dimensions of pelletal particles are directly proportional to the size of crustaceans, while the settling rate of pellets is directly proportional to their size (Pattenhoter and Kaonles, 1979). Pellets are commonly formed by means of passing through the digestive system of zooplankton: reworked food is ejected into the water as fecal pellets. Gravity causes the downward movement of pellets out of the photic zone, and it takes them a short time to reach the ocean bottom (Fig. II.8.11). In general terms, such settling of small-sized biogenic particles had been suggested as early as the beginning of 20th century (Lohmann, 1902), but this suggestion was proved only 20–25 years ago, when researchers started to use sediment traps to determine the flux of suspended matter to the seabed from the upper layers of the ocean (Hanjo, 1976). Having studied the available information, the author concludes that as much as 92% of coccoliths produced in the photic layer in pelagic regions of the Pacific settle to the bottom (depths of 2–3 km) by means of pelletal transport. According to reports of other authors (Komar et al., 1981), the sinking velocity of pellets with dimensions of 250 μm is from 40 to 440 m per day. As mentioned earlier, the settling rate of single coccoliths is only 0.14 m per day, or 1000 times less.

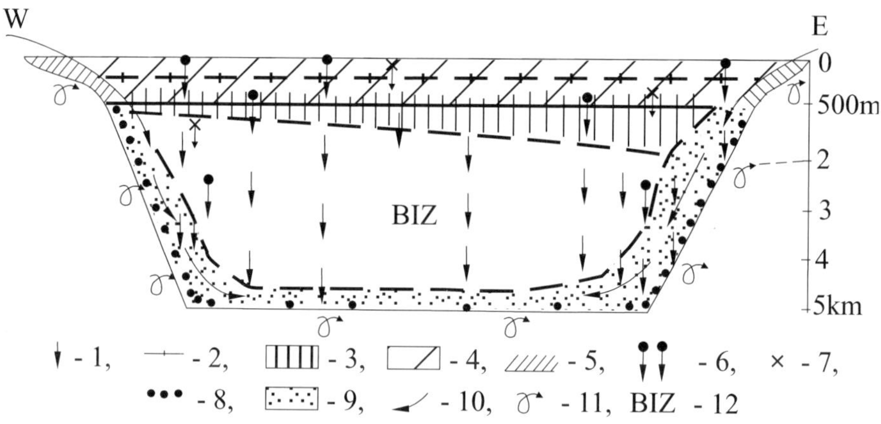

Fig. II.8.10. Principal scheme of biogeochemical transformation (processing) of sedimentary material in ocean.

1—photic layer, extraction from water of dissolved matter, filtration of water and transformation of particulate suspended matter;

2—layer of discontinuity (thethermocline, halocline, pycnocline);

3—oxygen minimum layer (OML), dissolution of organic detritus;

4—zone of most active filtration of water by zooplankton, formation of fecal pellets;

5—zone of active filtration of water by benthic filtrators;

6—pellet transport;

7—sea snow (aggregates or flakes of organic matter, etc.), which sinks with a velocity of tens of meters per day);

8—zone of weak filtration by benthic filtrators;

9—transit zone of terrigenous suspended matter from river mouth to pelagic area;

10—areas of fastest load of terrigenous suspended matter to pelagic areas;

11—transformation (processing) of bottom sediments by burrowing benthic organisms and mud-eaters;

12— most biologically inert zone.

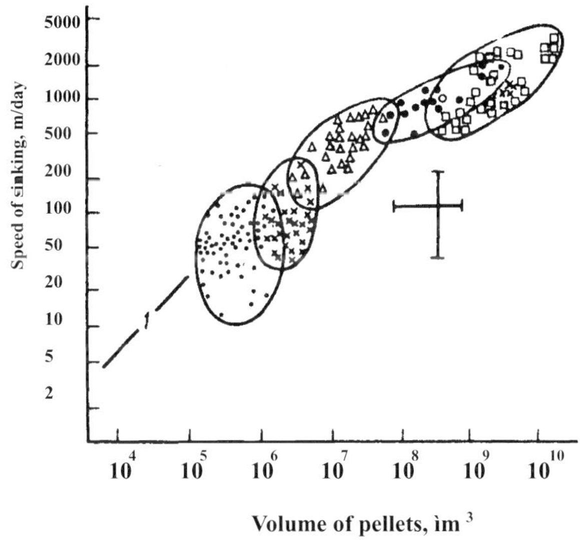

Fig. II.8.11. The speed of sinking of pellets in relation with their dimension (volume). After Lisitzin, 1991.

Thus, it takes pellets and, consequently, fragile diatoms, coccoliths and organic amorphous flakes, some 20–30 days to sink to the bottom of basins (5 km depth). Thus, pelletal transport is a means whereby sediments, including organic matter, reach the ocean bottom without considerable changes. Only the smallest particles which have broken up in the water column fail to reach the ocean bottom (Fig. II.8.9).

Many chemical elements settle onto the bottom with organic detritus, pellets, remains of skeletons and terrigenous particles. The sedimentation conditions here are as follows: the higher the intensity of the sinking particle flux from the source region and the larger the contents of elements in pellets, the stronger the flux of these elements to the bottom. The fluxes of organic sedimentary material are maximum where hydrothermal springs occur. (the Azores and around Iceland, Table II.8.8) or near river mouths in the ocean, as well as in arid climatic zones and especially in areas of coastal upwellings (Romankevich, 1994; Emelyanov and Romankevich, 1979).

The maximum organic matter accumulation rates in general (see Fig.II.3.8) and C_{org} in particular in the Atlantic Ocean occur also in the near-shore areas (circum-continental zonality) and in humid climatic zones, especially in the equatorial divergence zone.

The descending organic matter consists predominantly dead phytoplankton and organic aggregates. Some 95% of organic matter is exposed to the effects of oxidation and decomposition, and this leads to regeneration of nutrients (re-mineralization), which return to the photic layer, nitrogen and phosphorus being the first elements among them, as a result of decomposition of organic matter.

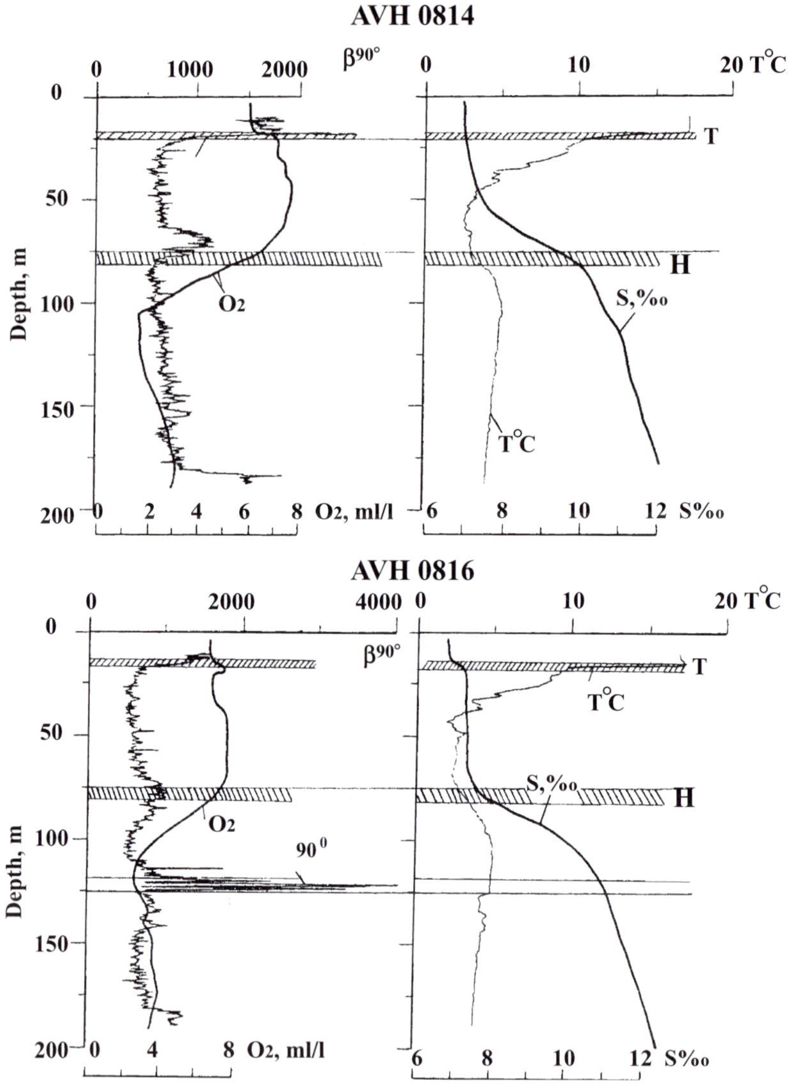

Fig. II.9.1. Vertical profiles of temperature (T°C), salinity (S ‰), oxygen (O_2) and indicator of light dispersion $\beta^{90°}$ on two stations of Gotland Basin, Baltic Sea (Summer, 1994).
H—halocline.

In the Black Sea, the thermocline is found at depths of 18–27 m (eastern part) and 30 m (western part). The maximum temperature gradient here has been as large as 3–6°C/m of water column (Vinogradov and Shushkina, 1980, pp. 181–182). At the same place, the density gradient was observed to be maximum and turbidity was also at its highest. In the lower part of the gradient layer there is a sudden drop in O2 concentration, but after that it decreases slowly.

In the Black Sea, the maximum bacterial biomass is present at the boundary of the thermocline (depth 25–30 m) (Fig. II.9.2) (Sorokin and Kovalevskaya, 1980, p. 165). The amounts of infusoria and nauplius are maximum at the level of the thermocline or slightly below it; as well, a major mass of zooflagellates (representatives of microzooplankton) is present there (Mamayeva, 1980). The infusoria *Ponarium* inhabits a water layer which is only 10–20 m in thickness, and this layer lies 5–9 m from the upper O_2–H_2S boundary. This layer is depleted in oxygen (0.13–0.17 μl × l^{-1}), but salinity and temperature here are slightly increased (Mamayeva, 1980).

The distribution of plankton is very often controlled by the density stratification of seawater (Fig. II.9.3). The maximum zooplankton concentration (euriphags and nanophags, Vinogradov and Shushkina, 1980, p. 185) always coincides with the layer of maximum temperature and density gradients. The more pronounced the thermocline and the stronger the gradients of properties present in this thermocline, the narrower and more pronounced the maximums of euriphags and nanophags (Vinogradov and Shushkina, 1980, p. 186). This is explained by the fact that the occurrence of animals at the density boundaries requires minimum energy losses (because these animals are either in a passive position or move upwards and downwards in conditions close to neutral buoyancy). Layers where there are accumulations of low-active animals correspond essentially to specific isopycnals. This is especially true with respect to *Calanus*. In the Black Sea, in June–September, *Calanus* were restricted to a horizon of δ_t = 15.61 ± 0.02, and the lower maximum ("hibernating fund"), to a horizon of δ_t = 16.02 ± 0.03. In general, the maximum *Calanus* biomass is represented by a thin layer extending along the density boundaries where density values vary within the range of only 1% (Vinogradov et al., 1992).

The thermal transitive layer in the Black Sea is the interface between two ecosystems (Petipa et al., 1970). Among mass species of zooplankton, the most notable are

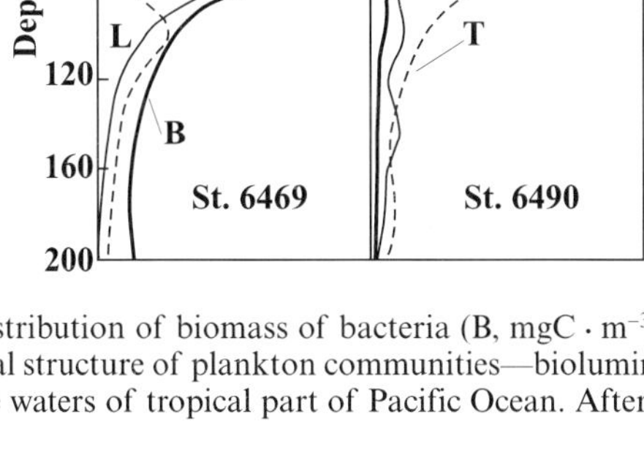

Fig. II.9.2. Vertical distribution of biomass of bacteria (B, mgC · m^{-3}) and profiles of physical indicators of vertical structure of plankton communities—bioluminiscence (L, %) and turbidity (T, %) in surface waters of tropical part of Pacific Ocean. After Sorokin, 1977.

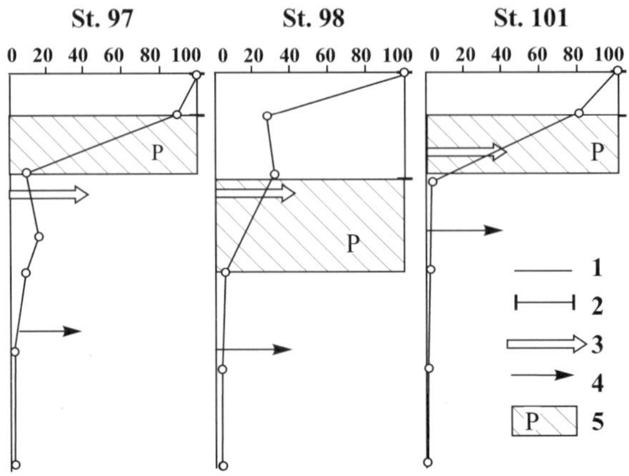

Fig. II.9.3. Distribution of phytoplankton in southeastern part of tropical zone of Atlantic Ocean in relation to halocline (H) position. After Semina, 1977.

Maximum amount of cells for each station is 100 percent.

1—phytoplankton; 2—thickness of trophogenic layer; 3—depth of isolume with value which is 10% of that on surface (taken for 100%); 4—depth of compensation point (isolume for illumination making up 1% of that on surface). P—pycnocline.

species inhabiting only warm waters of the upper layer, i.e., above the thermocline (for example, *Penillia avirostris*, and partly, *Paracalanus parvus*). Another group of species is found only in cold waters below the upper boundary of the gradient (transitive) layer (*Pseudocalanus elongatus, Noctiluca miliaris*) (Vinogradov and Shushkina, 1980, p. 182). However, there is also a group of zooplanktonic species which live throughout the whole length of the water column (to a depth of 120 m), independently of the position of the thermocline (*Saggita setosa, Aicopleura dioica*, etc.).

In seas such as the Baltic and Black seas, which communicate with the ocean through shallow-water straits, the pycnocline acts as a tight lid to cover and isolate deep waters, which make up 80–90% of the total volume of nourished seawater. However, this lid is not capable of preventing water movement and component exchange. In the winter season, severe storms are responsible for cooling and mixing of the upper layers of water, with the result that the pycnocline may either disappear (in shallow-water areas of seas) or effectively reach a considerably higher level. For example, domes of major cyclonic gyres in the Black Sea appear to be zones of specific Black Sea upwelling (Vinogradov et al., 1992). Under such conditions, the pycnocline may extend up into the photic layer, i.e., at depths of 20–40 m. In such periods, the water exchange across the barrier of the pycnocline increases to values as large as one to two orders of magnitude. Welled-up deep water enriches the euphotic layer in biogenic elements, which results in a rapid increase in biologic productivity.

The density barrier is indistinct in high-latitude regions of the ocean, whereas this barrier is strongly pronounced in the equatorial climatic zone (Fig. II.9.4). The lower

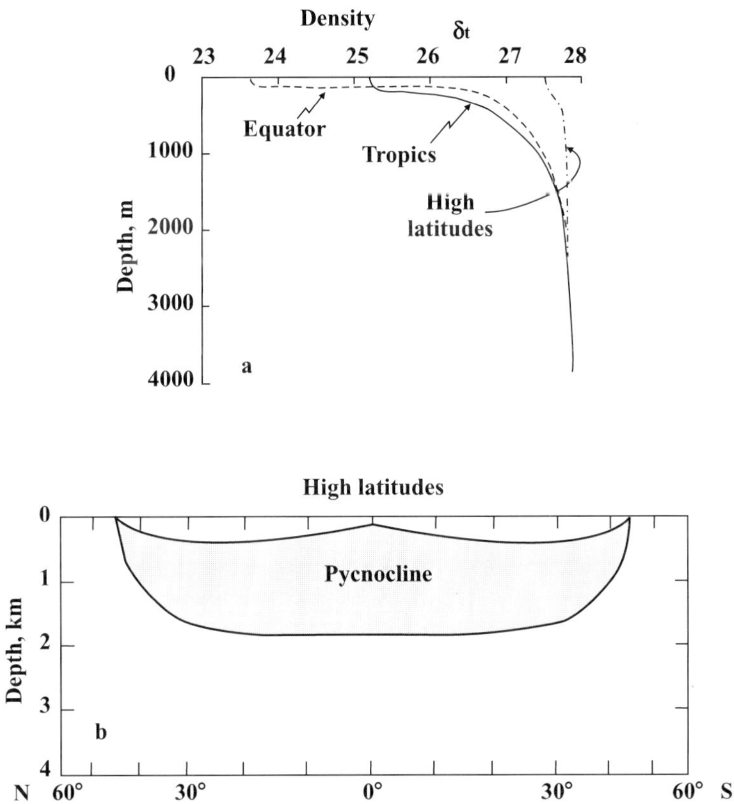

Fig. II.9.4. Depth of occurrence of density barrier (pycnocline) in ocean. After Gross, 1977.

boundary of the pycnocline in the ocean reaches depths of 1.5–2 km. In the equatorial zone of the Pacific, the pycnocline is the most pronounced at depths of 75–120 m (Fig. II.9.5). The thermocline is found at depths of 60–120 m: here there is a tendency for the temperature to fall from 28°C at a depth of 60 m to 15°C at a depth of 120 m. Thus, the upper mixed waters and waters of the oxygen minimum are effectively separated from each other by the thermocline and halocline. Initially, the number of species of phytoplankton, as well as the amount of chlorophyll in the thermocline, increases, and then, towards the base of the thermocline, decreases (El-Sayed and Taguchi, 1979). Consequently, the thermocline is a specific habitat for phytoplankton, as well as for bacterioplankton.

In the ocean, species composition of phytoplankton is commonly dominated by diatoms. They are followed by coccolithophorides, and then by dinoflagellates. It is typical that the maximum primary production in the ocean is observed not in the top sunlit layer, but near the uppermost boundary of the thermocline. This is evidently due to the effective upward diffusion of nutrients (phosphates, nitrates, silica) from the thermocline to the upper mixed layer.

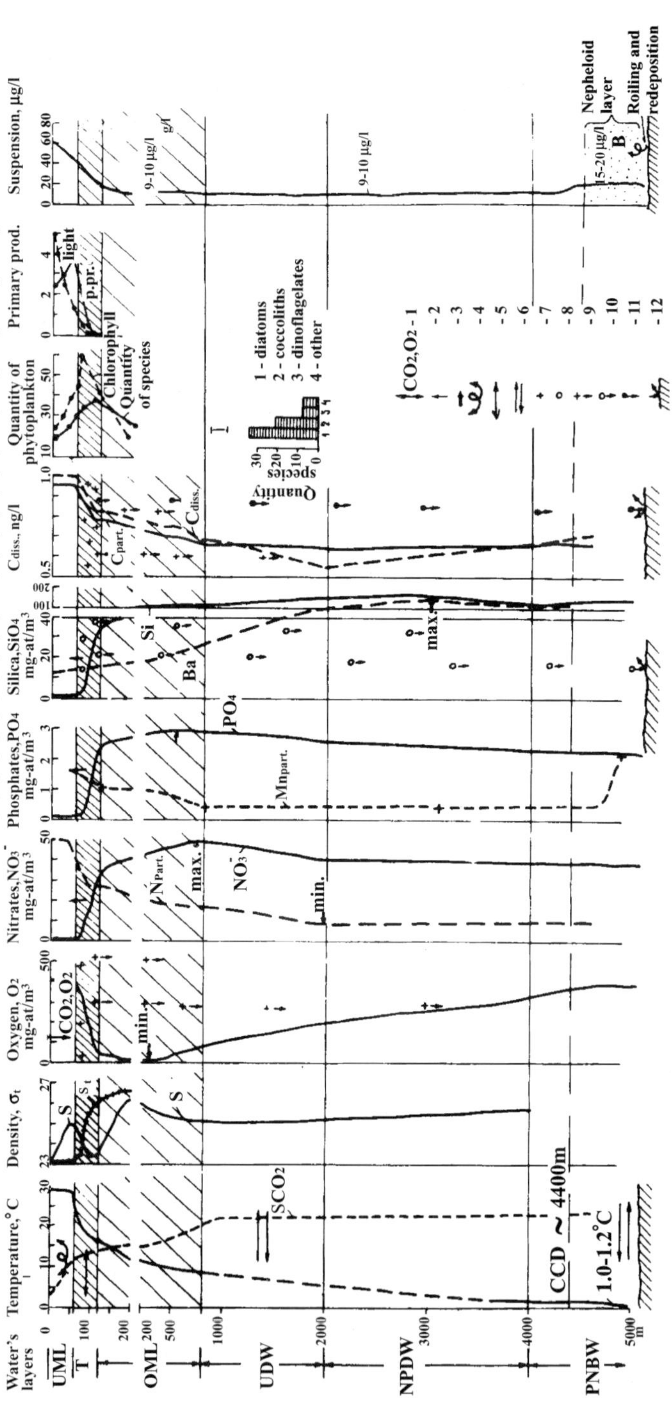

Water's layers: UML - upper mixed layer; T - thermocline; OML - oxygen minimum layer; UDW - upper deep water; NPDW - North Pacific deep water; PNBW - Pacific near-bottom water.

Caption see page 293

The horizon of light scattering coincides with the chlorophyll-A maximum (Baker et al., 1979). This maximum almost always occurs in a layer of the thermocline. The distribution of concentrations of particulate material follows the curve of light scattering most frequently. So, the average concentration of suspended matter in the upper layer of the water column is 47 µg/l; in the thermocline, 50 70 µg/l; below the thermocline (depth 200 m), 10–20 µg/l; in the layer of seawater between 450 and 4600 m, 9–10 µg/l; in the near-bottom waters (turbid layer, i.e., at a distance of 400–500m above the bottom surface), 14–15 µg/l (Baker et al., 1979, p. 172).

The reason for increased concentrations of suspended matter in the upper mixed layer and in the thermocline is accumulation of phytoplankton, and in the turbid layer, it is the disturbance and resuspension of the uppermost layer of sediments. The latter is proved by an abrupt decrease in values of the K/AL, Ni/Al, Fe/Al ratios for the near-bottom suspension, and also by direct observations of near-bottom currents (Baker et al., 1979).

In highly productive areas of oceans, where manifestations of coastal upwelling are the most active (near Peru; in the Walfish Bay region, Namibia; the northern part of the Arabian Sea), a complex, vertical, fine structure of the water column has been revealed (see Figs. 7 and 10 in Ivanenkov, 1979, p. 238):

The 0- to 25-m water layer is homogeneous in all parameters; the 35- to 45-m layer represents the upper layer of the pycnocline (caused by decreasing temperature as a result of the rising of intermediate waters). Above this pycnocline, maximums

Fig. II.9.5. GBZ thermocline (T) and OML in eastern equatorial zone of Pacific Ocean (test area B of project DOMES, 11°42′N, 138°24′W, depth 5100 m). According to various sources, after Bischoff and Piper, 1979.

Legend: UML—upper mixed layer; UDW—upper deep water; NPDW—North Pacific deep water; PNBW—Pacific near-bottom water; C—organic carbon; p—particulate; s—dissolved; Pp—primary production.

Legend for numbers 1–12: 1—CO_2 and O_2 exchange between atmosphere and UML; 2—diffusion of NO_3, PO_4 and SiO_4 from layer T into UML; 3—minimum (or maximum) concentration; 4—active mixing of water; 5—horizontal movements (current) in thermocline (Emelyanov's supposition); 6—deep and near-bottom currents (3–10 cm/s); 7—concentration of organic detritus; 8—concentration of diatomic algae; 9—rain of diatomic remains and skeletal remains; 10—flux of biogenic carbonate material on bottom (foraminifera, coccoliths); 11—flux of pellets to bottom; 12—material deposition on bottom and its dissolution (partly or fully).

On SiO_2 scheme is shown dissolved Ba distribution at GEOSECS station near test area M (9°12′N, 119°18′W), after Bacon and Edmond, 1972. There is a positive relation between Si and Ba. Ba associates with diatomic and radiolarians remains.

Histogram I shows the quantity of species of phytoplankton: 1—diatoms; 2—coccolithoforids; 3—dinoflagellates; 4—others.

CCD—carbonate compensation depth in test area M.

The profile "Light" depicts a sharp decrease in illumination (light) and the distribution of primary production (Pp), mg/C/day.

Dashed line—thermocline and OML.

The graph "suspension" (or particulate suspended matter—PSM) depicts the near-bottom nepheloid layer. B—resuspension of upper layer of sediments and delivery of suspended particles into nepheloid layer. Concentration of PSM is indicated by numbers, in µg/l.

of phytoplankton and primary production occur, which gives rise to maximums of particulate suspended matter and dissolved forms of C_{org}, N, P, ammonium nitrate and O_2.

The 60- to 70-m water layer is another layer with a maximum vertical density gradient (which is related to such factors as a sudden decrease in temperature and an increase in salinity). In this layer there are maximums of zooplankton and bacteria; a second maximum of particulate organic matter; dissolved forms of C_{org}, N, P; an upper maximum of nitrites and nitrates; and a second maximum of ammonium nitrate. The 100- to 250-m layer is the layer of the maximum density (due to advection of the waters with increased salinity). Also, there are accumulations of zooplankton and nekton, particulate organic matter, and dissolved forms of C_{org}, N, P, ammonium nitrate and O_2.

At a depth of about 1500 m, there is small density maximum (at the boundary between the intermediate and deep waters).

Up to now, the question as to what role is played by GBZs remains unanswered. As for the physical essence of these phenomena, the pycnocline is a means whereby the upper mixed layer is separated from the lower weakly mixed layer. The lower boundary of the halocline is in reality the boundary limiting the downward sinking of fresh water (Horne, 1972, p. 125).

In the thermocline there is a tendency for some properties to decrease abruptly with depth, including the primary production, dissolved oxygen content, content of particulate and dissolved forms of organic carbon, and particulate nitrogen. At the same time, the amount of phosphates, nitrates and dissolved silica tend to increase in the thermocline (see Fig. I.11). Below the pycnocline, the waters are markedly enriched in nitrates, phosphates, silica and other elements of vital importance. A strong (well-pronounced) pycnocline favors the accumulation of these substances, which are thus preserved here before being exported to the photic layer. Components such as carbon dioxide and solar radiation accumulate in deep waters. In high latitudes, where deep waters come to the surface, this energy enters the air to thereby increase the temperature of the atmosphere.

The closer the decrease in temperature and salinity to the ocean surface, the larger the productivity of these waters.

On the boundary of the pycnocline, where organic detritus sinking from surface waters is not very fast for a certain period of time, accumulation of phytoplankton and bacterioplankton occurs (see Figs. II.1.15, II.9.5). There are several active processes occurring within the pycnocline, including (1) further consumption of suspended matter by zooplankton, aggregation of particles into larger fecal pellets and rapid (within 5–15 days) settling of them onto the seafloor; (2) intensive decomposition of organic particles and release of elements trapped by protoplasm, which then pass to solution in the form of soluble metalloorganic compounds (the process is most active within the first 10–20 days (Gromov and Starodubzev, 1974; Gromov and Spitzin, 1975); (3) partial hydrolysis of these compounds and transition of some metals into oxidized compounds (for example, $Fe(OH)_3$, MnO_2 etc.); (4) sorption of microelements from seawater by these hydroxides and organic detritus; (5) consumption of dissolved (or suspended) forms of elements by phytoplankton and bacterioplankton.

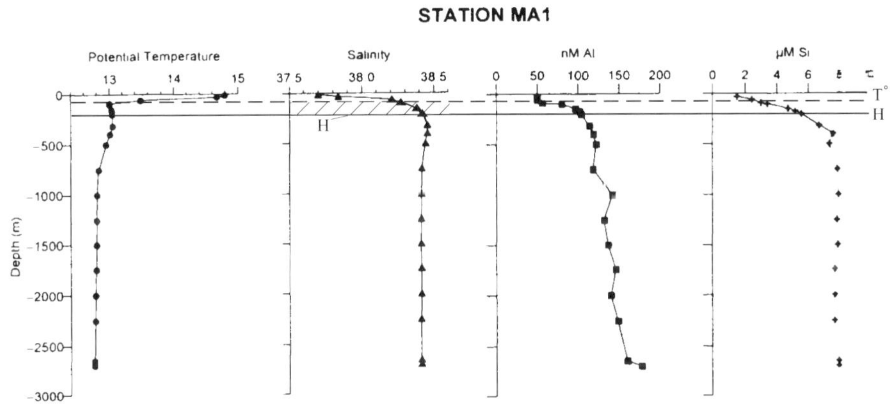

Fig. II.9.6. Vertical profiles of potential temperature, salinity and dissolved Al and Si in Provence Basin of Mediterranean Sea. After Chou and Wollast, 1997, p. 748.
T°—thermocline, H—halocline.

The halocline plays an substantial role in the distribution of both suspended and dissolved elements, not only in application to biogenic materials but also to chemically inert elements, such as Al and Ti. There is evidence, for example, that waters in the western part of the Mediterranean Sea contain very high concentrations of dissolved Al (Chou and Wollast, 1977, Fig II.9.6). A major source of this element is aeolian dust from the Sahara. ("The high concentrations of events", Chou and Wollast, 1977, p. 765). When aeolian particles fall into water, the water begins to dissolve these particles. As a result, the water becomes considerably enriched with respect to dissolved Al. Enrichment in Al strongly increases with depth, especially within the zone extending from higher levels to the depth of the halocline. Below the halocline, the concentration of dissolved Al is two to three times as large as that in the water column above the halocline. There is a strong correlation between concentrations of dissolved Al and Si. Such correlation can be attributed to the effects of biochemical processes in the photic layer. "The very good correlation between dissolved Al and Si suggests that the uptake of Al is due to the incorporation of this element in the siliceous forms of diatoms. Organic debris (including diatoms) are dissolved in deep waters and return the Si and Al incorporated in diatoms. As a result, concentration of dissolved Al increases in deep waters" (Chou and Wollast, 1977, p. 765).

Peak values in concentrations of suspended matter are very often observed both above the upper boundary of the pycnocline or within the pycnocline itself (station 0816) with the result that transparency of the water rapidly declines (see Figs. II.1.15, II.9.1). The sound-scattering layer is commonly recorded using echo-sounding records. The distribution of sea snow (particles with a diameter >0.5 mm) is in full acordance with the vertical density stratification (87% of peaks, according McIntyre et al., 1995). In the pycnocline, the number of aggregates was found to have increased: the amount of aggregates in the Pacific off California was estimated to be in the range from 38.6 ± 18.3 specimens per liter of water above the halocline, to 59.3 ± 26.0 specimens per liter within the halocline itself. The size of sea-snow aggregates range from 1 to 60 mm. Small snowflakes consist of individual diatomic frustules or fecal pellets, while the larger snowflakes are made up of aggregates of pellets, diatomic frustules, and unidentified (amorphous) particles (Kilps et al., 1994). Diatomic aggregates have the largest dimensions.

The occurrence of sea-snow aggregates is easily detected not only by optical imagery but even by visual observations. Confirmation of their increased concentrations in the water column lying on the thermocline or halocline has been obtained from direct observations by divers, including the present author during submersible operations above the Baroni Ridge in the Tyrrhenian Sea and in the northern Baltic Sea (Emelyanov et al., 1994).

In the pycnocline, where accumulation of zooplankton feeding on phytoplankton and organic detritus occurs, the process of fecal pellet formation becomes more active and this considerably accelerates the downward transport of chemical elements by these pellets to the interior of the sea, down to the seafloor (this depth interval is covered in 5–15 days). Intensive decomposition of organic detritus within the pycnocline and somewhat below it (at depths of up to 500–800 m) results in decrease of O_2 in seawater and increase in concentrations of dissolved forms of C, Si, P, Fe, Mn and certain trace elements.

Let us note that the amount of C_{org} mineralized (to CO_2) in the water column of the World Ocean is estimated at 18–20 billion tons C_{org} per year (Romankevich, 1977). A major portion of organic matter is mineralized in the upper part of the water column (to a depth of 200–500 m), including the pycnocline. The process which occurs here is opposite to that in the photic layer. This process leads to decomposition of organic detritus and its transformation into other forms of matter, such as dissolved, colloid, and sorbed forms. As a result, in the pycnocline (and below it) there is an increase in concentrations of dissolved forms of biogenic elements, CO2 (Fig. II.9.5) and, locally, Fe and Mn [within the pycnocline itself, concentrations of dissolved reactionable forms and Mn are periodically much less (Emelyanov, 1979_3)]. Thus, a major mass of both macro- and microelements which have been extracted from seawater by phytoplankton in photoic layer and in the pycnocline return to solution.

In the Kiel bight, the pycnocline provides for a distinct subdivision between sediment facies, communities of benthic organisms and their biomasses. Above the pycnocline, i.e., in hydrodynamically active conditions, sand and muddy sand is deposited; below this layer, mud with H_2S occurs. Above the pycnocline, 150 species of macrobenthos have been observed and their biomass (wet) is 730 g $\times$ m^{-2} (Cyprina predominates), whereas below the pycnocline there are only 10 species and their biomass is less than 2 g $\times$ m^{-2} (Theede, 1981). The pycnocline (17–20 m) in the Kiel bight provides a distinct subdivision between the main types of bottom foraminifera. *Elphidium excavatum* predominates in the layer above the pycnocline (0–17 m); *Ammotium casis* is abundant in the upper part of the pycnocline (17–20 m); *Elphidium incertum*, in its bottom part; and just below the pycnocline (20–25 m), *Elphidium clavatum* predominates in the deeper layer and up to the bottom (on a mud substrate at depths 25–28 m). Saturation of seawater with oxygen abruptly decreases within the pycnocline: from 110–60% above the pycnocline to 60–40% within the pycnocline, and up to 5–20% below this layer.

The content of soluble Mn^{2+} was 0–20 μg × l^{-1} in the oxygenated layer (above the pycnocline), 30–180 μg × l^{-1} within the pycnocline, and 200–350 μg × l^{-1} below this layer (Djafari, 1977; Balzer et al., 1987). Formation of ferromanganese concretions on this profile occurred in mixed sediments of the pycnocline zone (18–22 m).

In the Bornholm deep, like the Gdansk basin, Gotland and other deeps of the Baltic Sea (Emelyanov, 1995), the lower boundary of the halocline is the upper limit of abundance of pelitic and aleuro-pelitic muds and is the lower limit of sand and coarse aleurite (Fig. II.9.7). Because the contents of many chemical components, C_{org} and elements (Fe, Al, K, Ti, P, Li, Rb, Cu, Zn, Ni, Cr) are in a close, direct relationship with the proportion of the <0.01 mm fraction in the sediments, the occurrence of deeps at a particular site of the seafloor and the depth of the pycnocline are factors believed to be responsible for the abundance of these components in the sea, rather than geographical position: maximums of aforementioned elements are contained in the sediments beneath the pycnocline in deeps, while their minimums are in areas shallower than the pycnocline. In seas such as the Baltic and Black seas, at depths greater than the pycnocline and down to the seafloor, deposition of reduced

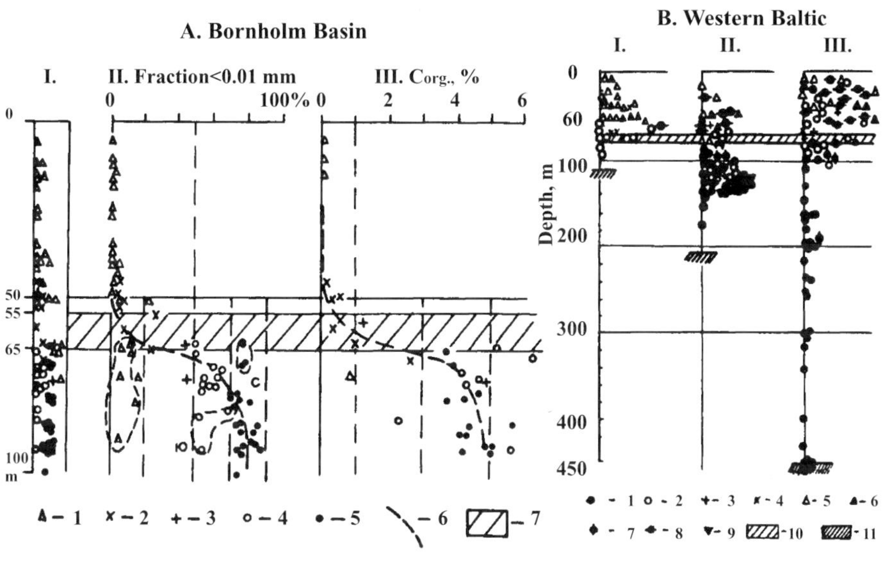

Fig. II.9.7. Distribution of sediment types, <0.01 mm fraction and C_{org} in sediments in Bornholm Basin (Baltic Sea) (A) and in Western Baltic (B) versus sea depth.
A. I—sediment types (0–5 cm layer); II—<0.01 fraction mm content in surficial sediments (0–5 cm); III—C_{org} content in surficial sediments. 1—5—sediment types: 1—sand; 2—coarse aleurite; 3—fine-aleuritic mud; 4—aleuro-pelitic mud; 5—pelitic (clayey) mud; 6—average curve; 7—halocline.
B. Western Baltic: I—region of fjords; II—Western Gotland Basin; III—Landsort Basin.
1–9—sediment types: 1—pelitic (clayey) mud; 2—aleuro-pelitic mud; 3—fine-aleuritic mud; 4—coarse aleurite; 5—sand; 6—gravel; 7 and 8—clay of lakes; 9—till; 10—halocline; 11—bottom.

sapropelic mud occurs, which is terrigenous. In areas of coastal upwellings, diatomic oozes are often also sapropelic and contain phosphates.

In oceanic surface layers, seawater density tends to increase gradually from the equator towards the poles (Figs. II.9.8, II.9.9); i.e., the pycnocline attains increasingly higher levels with decreasing distance to the poles. This phenomenon plays a vital role in geological development. Heavy waters formed in high latitudes sink from the surface and are transported by deep cold currents towards the equator. As these waters move equatorward, they perform a job of tremendous importance: (1) they ventilate the near-bottom layers of the ocean, and this finally gives rise to processes which are responsible for either preservation or dissolution of biogenic, especially, carbonate material; (2) due to the activity of bottom currents, great amounts of sediments remain suspended in the near-bottom waters (turbid layers); (3) sediments experience erosion and resuspension by bottom currents in some places, whereas in other places sedimentary material is deposited to form sedimentary drifts of large thickness or to give rise to non-depositional and erosive hiatuses within the sediment section.

It is important to stress that the halocline (pycnocline) is also an important means whereby sedimentary material is transported laterally. According to the model obtained at Atlantic Branch of the Shirshov Institute of Oceanology, RAS (Fig. II.9.10), suspended matter carried down the slope by turbidity currents "rests" on the pycnocline: one portion of it (periodically, a greater one) extends over the surface of the pycnocline to the open sea; the remainder (which overcomes density bar-

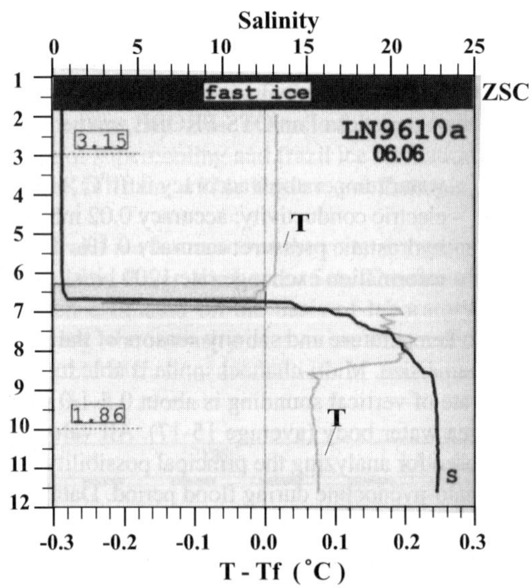

Fig. II.9.8. Vertical salinity distribution and differences between in situ temperature and freezing temperature (T) at a given salinity (S) for station LN9610a, Arctic Ocean. After Golovin et al., 1999, p.128. Fast ice—zone of supercooling (zsc).

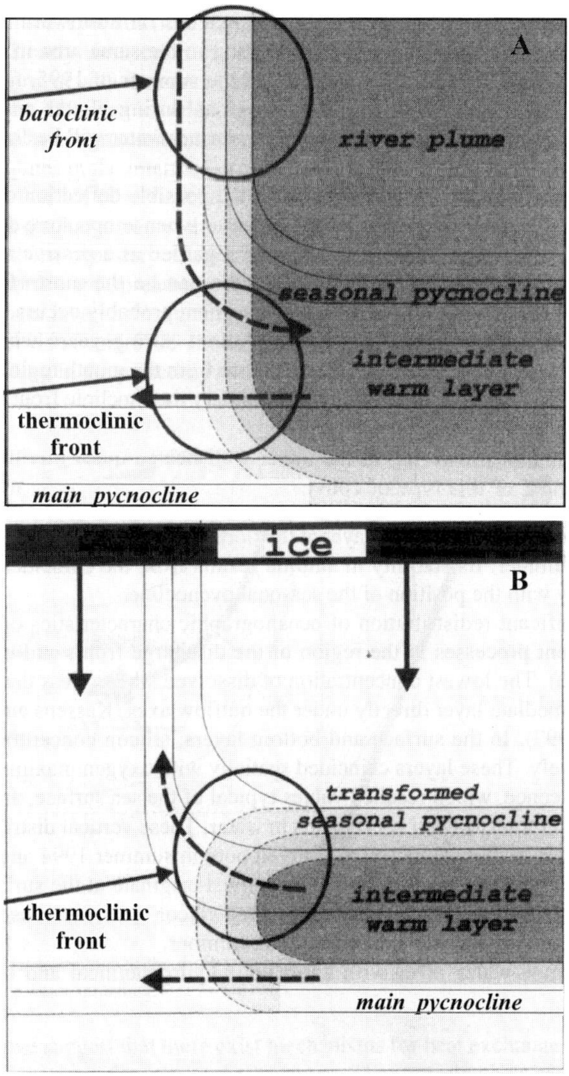

——— isohalines, ----------- isotherms, ——▶ salt flux, — — —▶ heat flux.

Fig. II.9.9. Structure of river discharge front and its evolution from summer (A) to winter (B). After Dmitrienko et al., 1999, p. 79.

rier) continues to move to the deep sea and ocean basins following the law of mechanical differentiation.

In the ocean, well-sorted sand-aleurite sediments are found at the ocean floor at a level above the pycnocline; transient sediments which are strongly enriched in organic matter, phosphates, various organic remains (skeletons of diatomic algae,

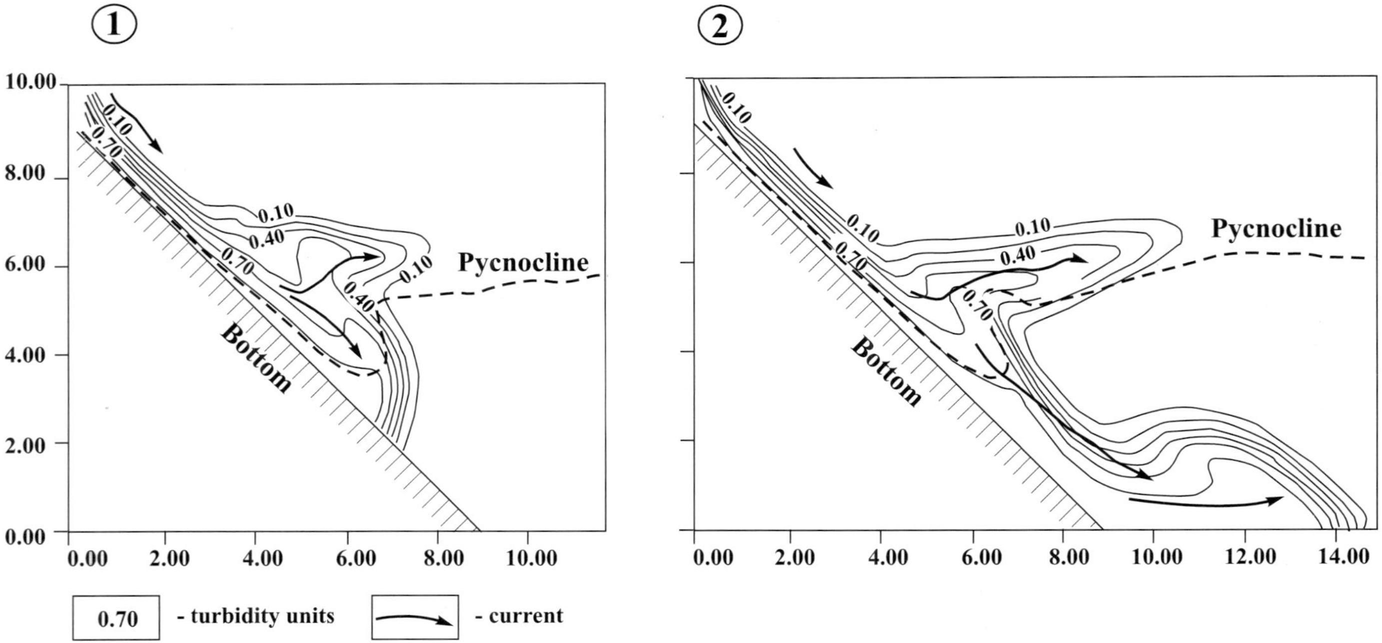

Fig. II.9.10. Model of two phases of bottom gravity current spreading along bottom slope in two-layer density fluid. Model is calculated by A.V. Gritsenko. After Emelyanov and Gritsenko, 1998.

bones and scales of fish, etc.) lie in zones of the pycnocline. Below the pycnocline, muddy sediments commonly accumulate.

In ocean regions of high productivity, primarily in areas of coastal upwellings, the importance of biogenic and hydrogenic migration of metals increases in sediments below the level of the pycnocline, whereas the role of their silica forms becomes smaller (see Part II.10). In the zone of the Peruvian upwelling, for example, hydrogenous Fe makes up about 50% of total Fe, and biogenic Fe makes up 10–29%. Hydrogenous forms of Mn and Cu contribute about 90–95% of the total abundance of these elements, and their siliceous forms make up 1–2% (Gordeyev and Lisitzin, 1979). This circumstance gives rise to high mobility of metals in sediments. Because of extremely unfavorable conditions, almost all of the Mn and 50–80% of Fe are evacuated beyond the limits of the shelf area.

It is known that the symmetry of fine-structure temperature profiles are broken over the summits (or slopes) of seamounts, whereas the intensity of turbulent velocity fluctuations tend to increase there. For example, over the Empire seamount in the Atlantic Ocean, these phenomena occur at depths of 40–80 m (Monin and Ozmidov, 1986) (the summit of this mountain is 43 m below the sea level). These fluctuations and currents are responsible for fully non-depositional conditions for fine-grained (aleuritic and pelitic) material on top of this seamount, with the result that outcrops of rocks, gravel, sand and "hardgrounds" predominate here.

In the deeps of the Baltic Sea, in areas where the halocline comes in contact with the slope, turbulent events (internal waves, seiches) are very typical phenomena which prevent mud deposition in these regions, with the result that the surface of the seabed continues to be covered by old lacustrine clays or Late Pleistocene moraines (Emelyanov et al., 1994).

The important role of the thermocline in geological development has been found to be essential not only for saltwater basins, but also for freshwater lakes. So, in Brienz Lake in Switzerland, in summer, the density of river water is less than that of lake water, with the result that the influx of river water into the lake occurs in the upper water layer (Allen and Kollinson, 1996, pp. 93–94). If the river water is very turbid, it has a density which is higher than that of lake water in the epilimnion, but is low compared to cold water in the hypolimnion, and this is why it flows above the thermocline. A considerable proportion of sand-aleuritic material settles onto the bottom mainly from such turbidity currents. Fine (pelitic) material precipitates to the bottom only when the water column is fully mixed throughout its entire length. This fine suspension forms a light-colored layer on the bottom (Allen and Kollinson, 1996, pp. 93–94).

Turbidity currents build up a succession of sediments with a grain-size gradation and seasonal variability. This is the way by which varved clays are formed in lakes (Sturm and Matter, 1978).

Owing to the temperature barrier, the seawater of basins in arid climatic zones is suddenly supersaturated with calcite and this leads to precipitation of $CaCO_3$ from seawater and the formation of aragonite ooliths, calcite and magnesium–calcite crusts on the seabed (see Fig. I.5). According to the principles of Les-Schatelier, systems of solutions show a pattern of behavior which is reactive to external impacts. Consequently, heating of seawater and its supersaturation with calcite lead to removal of minerals (calcite, aragonite), which absorbs a great amount of heat. In lagoons and shallow seas, where the temperature of water and air depends on the season of the year (the thermal barrier is shifted in the water column in a particular way), the temperature of bottom sediments varies within a wide range. So, in the shallow-water Arkona deep in the Baltic Sea (depth, up to 40 m), the water temperature is 1°C in winter and 9°C in autumn (see Fig. I.6). Consequently, even within the bottom sediments conditions may be created that give rise, for example, in winter, to those minerals the formation of which is accompanied by heat release . In summer, on the contrary, these minerals tend to dissolve again as temperature increases. In such a way, heat is released.

Due to the influence of the thermodynamic barrier, there are certain types of thermal springs on the surface of the seabed. In connection, precipitation of calcite and aragonite occurring in fractures (crevasses) of ultrabazites (serpentinites) of the Mid-Atlantic Ridge (Emelyanov and Kharin, 1974), is associated with the sudden decrease in CO_2 pressure at the thermodynamic barrier in the hydrotherm–water system and should be accompanied by the following reaction:

$$\underset{\text{solution}}{Ca(HCO_3)_2} = \underset{\text{solid}}{CaCO_3} + H_2O + \underset{\text{gas}}{CO_2}$$

In the hydrotherm–water system, the thermodynamic barrier is of extreme importance for the distribution of life, chemical elements, bottom sediments and ores. Here, over a distance of several hundreds or even tens of meters, the temperature of seawater falls from 400–350°C to −1°C (see Part II.6).

Oxygen Minimum Layer

In the Pacific Ocean, north of the equator (from approximately 8–9°N to 20°N), off Central America, the oxygen minimum layer (OML) is totally depleted of oxygen (with values below 0.2 ml/l), giving rise to anaerobic conditions over large areas (Anderson, 1979). This water area has the form of a sharp wedge. The base of this wedge is in the near-shore regions off Central America, and the edge of this wedge points westward (along 12–15°N latitude) up to 135°W. The position of this edge corresponds to the zone of maximum bioproductivity in the upper ocean layer (Koblentz-Mishke, 1977). The existence of a well-pronounced OML to the north of the equator in the Pacific Ocean, off Central America, is conditioned by (1) a sharp thermocline, which limits the diffusion of O_2 into the OML; and (2) the absence of considerable currents there (Anderson, 1979, p. 146).

The OML is absent in areas where high-salinity Arctic, North Atlantic and Antarctic waters are formed (Ivanenkov and Chernyakova, 1979).

In the World Ocean, the OML commonly occurs at the depths from 300–400 m to 2400–2500 m (Fig. II.10.1). The higher the productivity of the upper ocean layer, the closer to the ocean surface the oxygen minimum lies (Bay of Bengal and Persian Gulf; the eastern parts of equatorial zones in the Atlantic and Pacific oceans). The depth of the OML depends on the circulation pattern. The OML is deepest in the southern anticyclonal gyre of the Pacific (1500–2500 m).

In the Pacific ocean, off Central America, the OML is especially distinct and lies at a depth of 100 m. Here, at a depth of 200–300 m, the O_2 decreases to values as low as 0.10 or even 0.005 ml $\times$ l^{-1} (Brinton, 1979).

The core of the OML coincides with the 9°C isotherm (Alekin and Lyakhin, 1984). Both the CO_2 partial pressure and the concentrations of dissolved organic matter reach maximum levels in the OML.

In the eastern part of the Pacific Ocean, the OML (0.1–0.2 ml $\times$ l^{-1} O_2) occurs at a depth of 170–500 m, and its thickness is 100–150 m near the continental slope (Fig. II.10.2).

A layer containing 1 ml $\times$ l^{-1} O_2, rises up to a depth of 70 m in the open ocean (100–120 miles away from the coast) and occurs at a depth of 200–220 m near the continental slope. The lower boundary of the OML (1.0 ml $\times$ l^{-1} O_2) appears at a depth of 1000 m and has a horizontal setting.

pH values are markedly decreased in the OML (<0.2 ml $\times$ l^{-1}) of the Peruvian upwelling. To be more precise, the minimum pH value (<7.7) was found in the lower part of the OML or just beneath the OML. Thus, pH values fall from 8.1 to 8.2 in the upper ocean layer to 7.6 below the OML, and then increase again. Immediately

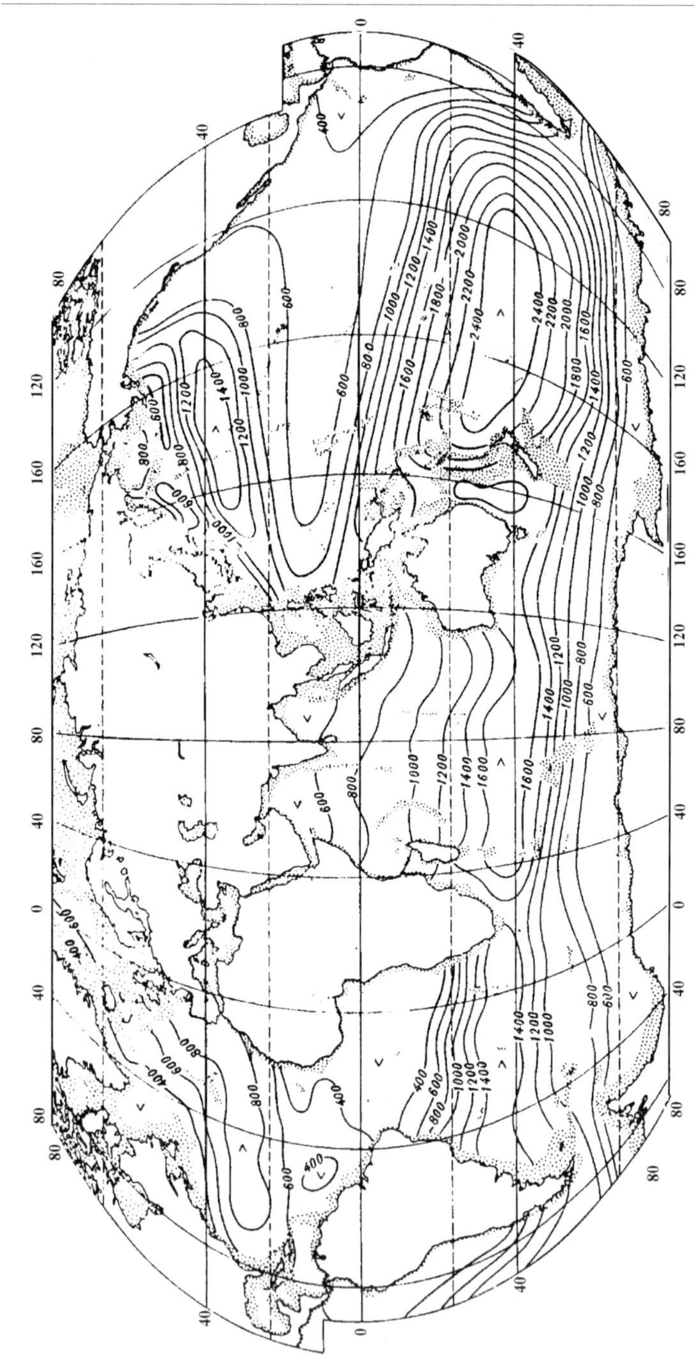

Fig. II.10.1. Depth of OML in World Ocean. After Ivanenkov a. Cherniakova, 1979.

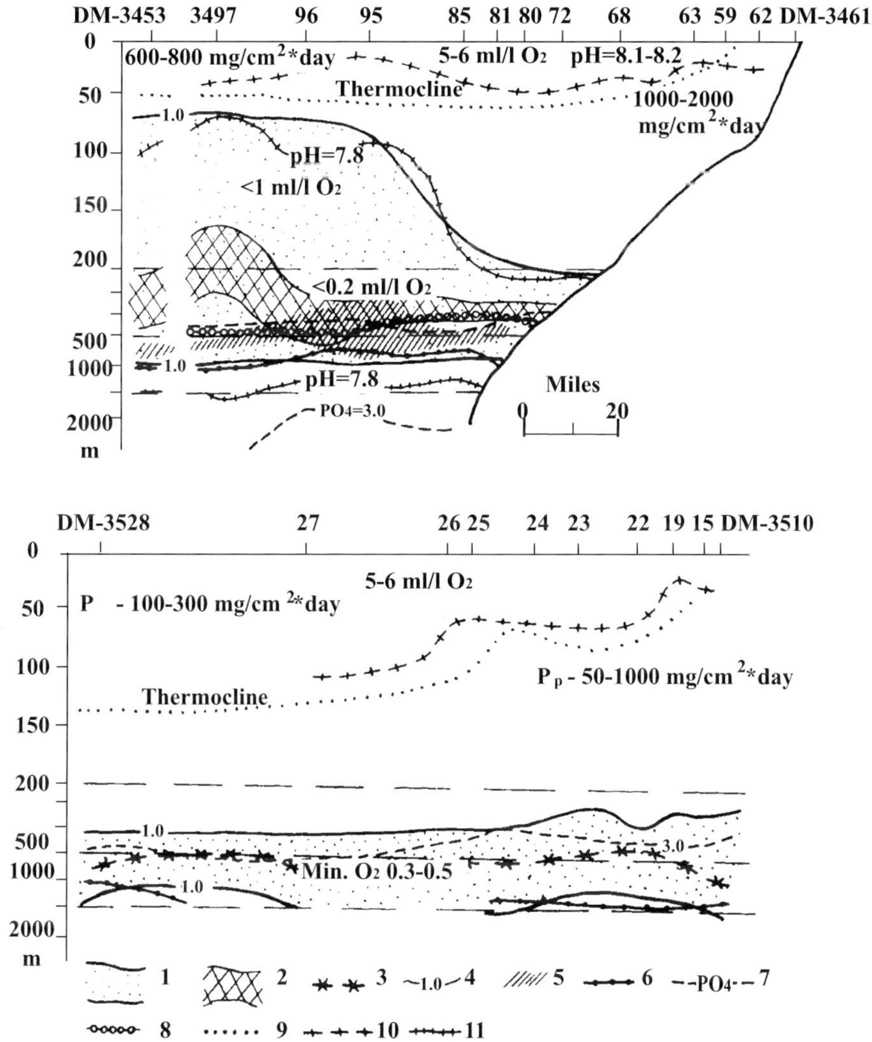

Fig. II.10.2. GBZ OML at Peru upwelling area (profile I along 8°S) and in California area (profile II, along 25–30°N). After Vinogradov and Mamaeva, 1990.
1–4—OML: 1–concentration of O_2 is <0.1 ml/l; 2—concentration of O_2 is <0.2 ml/l; 3—minimal concentration of O_2 on profile (0.3–0.5 ml/l); 4—isoline of concentration O_2 = 1.0 ml/l; 5—minimal value of pH (<7.7); 6—maximum nitrates + nitrates; 7—concentration of PO_4 is >3.0 mg/l; 8—maximum of summary biomass of mesoplankton (>2 mm) (100–400 gr/m^3); 9—thermocline; 10—maximum of intensity of bioluminescence; 11—isoline of pH = 7.8; Pp—primary production for given part of profiles.
Numbers of stations of RV *Dmitry Mendeleyev* (DM)—on upper part of profiles.

below this layer, there is a maximum in the overall concentration of nitrites and nitrates and a maximum in the concentration of phosphates (>3.0 μg-at × l^{-1}).

The distribution of mesoplankton in relation to the OML is very illustrative: the maximum for mesoplankton (>250 μg × m^{-3}) occurs at the thermocline or just beneath it, and the maximum overall biomass of mesozooplankton with a particle size >2 mm (100–400 g × m^{-3}) occurs within the OML.

Total productivity of mesoplankton in the OML, in the eastern part of the Pacific Ocean, is increased (Vinogradov and Mamayeva, 1990). Small fish accumulate in this area. In one way or another, a considerable portion of pelagic inhabitants are in active contact with the OML, and some types of organisms live within this layer. In the area of the Peruvian upwelling, for example, *Calanus australis* thrive between the 0.20 to 0.25-ml × l^{-1} isooxylines, and *Eucalanus inermis*, in the 0.10- to 0.15-ml × l^{-1} layer (Brinton, 1979). *Euphasiidae* and the copepod *E. inermis* accumulate between the 0.30- to 0.2-ml × l^{-1} isooxylines. It has been found that the habitat areas of these organisms are very distinct in the layer between the 0.11–0.15 ml × l^{-1} isooxylines (Fig. II.10.3). The biomass of crustaceans in such layers is three times as large, or even larger, as that of mesoplankton. In the Black Sea, *Calanus helgolandicus* accumulate in a water layer outlined by the 0.25- to 0.50-ml × l^{-1} isooxyline (Flint, 1994). Another layer inhabited by *Calanus australis* lies between the 0.4- to 0.5-ml × l^{-1} isooxylines. Information obtained from direct observations by during submersible operations made it possible to understand that *Calanus pacificus* and *Euphausia pacificus* inhabit thin interlayers of the OML (where their concentrations may reach very high values) at depths of 400–500 m (Aldredge et al., 1984).

It is evident that this regularity in faunal distribution is true for the entire World Ocean. Copepods consume a large amount of food, which is then quickly released into seawater in the form of fecal pellets. Usually, copepods migrate upwards into the photic layer to feed, but periodically they feed on organic detritus and fecal pellets sinking from the thermocline (pycnocline).

As mentioned above, some types of organisms actively choose conditions optimal for them in terms of oxygen, and the result of such conditioning is that the rates of "exchange" become smaller due to decreased consumption of oxygen. In the OML, mesoplankton occurs as "homeostasis". Accumulations of mesoplankton in the OML is a "forage reserve" for plankton-feeders able to live in oxygen-deficient waters for a short period of time (Timonin and Flint, 1985).

The living conditions for mesoplankton lead to an important conclusion in terms of geological development: pellets are released by zooplankton into the water not only from the photic layer and thermocline (pycnocline) but also from the OML. The most abundant species of zooplankton (such as *Calanus pacificus*, *Calanus australis*, *Eucalanus subtenius*, and *Eucalanus elongatus*) (Flint and Kolosova, 1990) empty their digestive tracts in 30–40 min, whereas for *E. inermis* it takes 3–4 h to complete the process.

In different areas of the Pacific, the organic matter flux at depths of 100–500 m is 0.01–20 g of raw material over 1m²/day on average (Sazhin, 1986, p. 103). The greater portion of fecal pellets rarely reach depths as great as 1500 m; i.e., they do not go beyond the limits of the OML and are dissolved within it. Detritus begins to loss P and, to a lesser degree N, in the first hours of its existence (Sapozhnikov

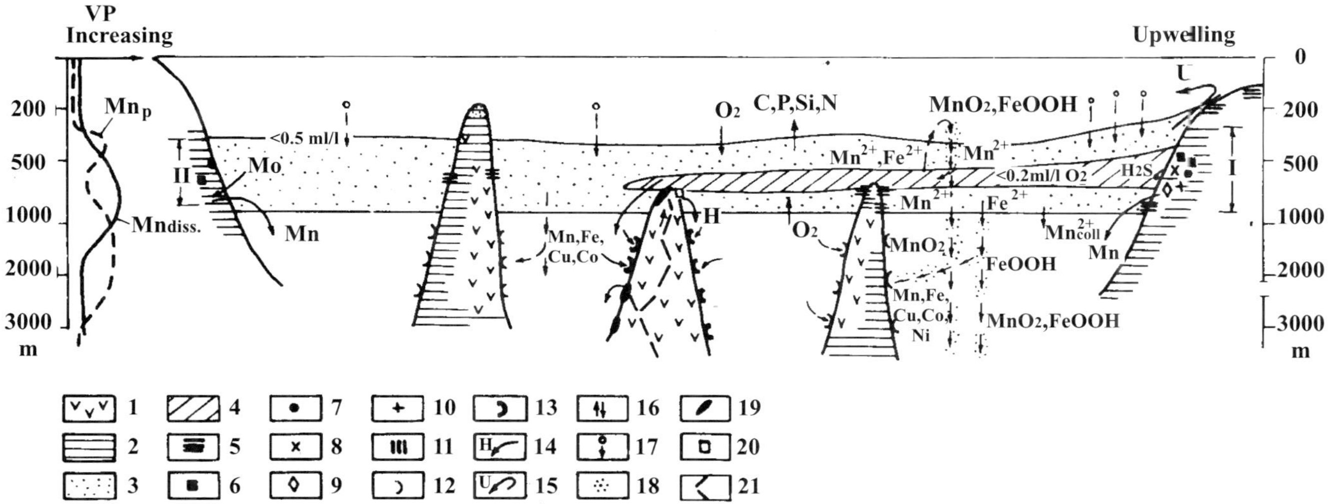

Fig. II.10.3. Processes in GBZ OML in ocean.

1—basalts; 2—sediments; 3—OML with concentrations of O_2=0.5–0.2 ml/l (sometimes 1.0–0.5 ml/l); 4—central part of OML with <0.2 ml/l of O_2 (or sometimes under anaerobic conditions); 5—formation of phosphorites and enrichment of sediments by phosphates during diagenetic processes; 6—iron sulfides; 7—glauconite; 8—vivianite; 9—barite; 10—magnesium carbonates (?); 11—phosphorites redeposited; 12—ferromanganese crusts on mountain slopes; 13—same, but under influence of hydrotherms (H); 14—hydrotherms (H); 15—uppwelling (U); 16—direction of elemental exchange; 17—flux of fresh biogenic material (pellets, foraminifera, pteropods, coccoliths, diatoms, silicoflagellates and others) from photic layer and from halocline; 18—Fe- and Mn oxide formation; 19—ferruginous hydrothermal deposits; 20—possible formation of sulfides in hydrothermal deposits; 21—fractures.
VP—vertical profile, which shows a tendency to particulate (Mn_p) and dissolved (Mn_{dis}) manganese distribution in sea water: I—facies of reduced mud in upper part of continental slope in upwelling zone; II—same, but outside upwelling zone. Coll—colloidal.

et al., 1984). This is the reason for rapid accumulation of these elements in the OML.

Within the limits of the OML, the process of dissolution of organic detritus follows the formulas

$$(CH_2O)_{106} (NH_3)_{16} (H_3PO_4) + 138\ O_2$$
$$106\ CO_2 + 122\ H_2O + 16\ HNO_3 + H_3PO_4 \quad \text{(Richards, 1965).}$$

Also, the regeneration of biogenic components in this layer is, in most cases, accompanied by the regeneration of trace elements. Increased concentrations of Mn in OML waters were discovered for the first time in 1970 (Table II.10.1). Water samples collected at station AK-130 from depths of 0–1 m and 100 m contained only 0.5 $\mu g \times l^{-1}$ Mn whereas at the 1000–3000 m the Mn contents were 5.0 and 8.8 $\mu g \times l^{-1}$, respectively. At station AK-182, the Mn concentrations were at increased the 500–3000 m. Increased concentrations of Cu have also been found there. We observed the highest amounts of suspended particulate Mn in the Mediterranean Sea and Atlantic Ocean at depths of 50–500 m, i.e., where the most intensive decomposition of suspended biogenic material (organic detritus) occurs. During decomposition of biogenic material, manganese seemingly passes into solution or into colloidal form (Emelyanov, 1977, p. 50). We reported in previous publications that increased concentrations of chemical elements (Fe, Mn, Cu) are commonly confined to the 200–1000 m levels. In the equatorial zone, markedly increased concentrations of Cd were observed in the OML (Yeats et al., 1995). Nickel is also regenerated similarly to biogenic components, whereas the Cu concentrations in the water column tend to increase gradually, from ocean surface to bottom, indicating a bottom source for this element.

Table II.10.1. Concentration of $Mn_{dis.}$ ($\mu g.l^{-1}$) in seawater of the Angola Basin. After Emelyanov and Vlasenko, 1972, p. 1274.

Horizon of sampling, m	$Mn_{dis.}$
Station AK-30, 5450 m	
0–1	0.5
100	0.5
1000	5.0
3000	8.8
Station AK-182, 5200 m	
0–1	3.6
50	2.4
100	2.4
500	3.5
3000	4.0

In the Sargasso Sea, there is a tendency for concentrations of dissolved Mn to decrease abruptly with depth in the photic layer (200–250 m) from 2.0 to 0.5 nmole/kg of water, whereas concentrations of suspended Mn increase (from 0.1 to 1.8 nmole/kg) (Sholkowitz et al., 1994). In the OML, the concentration of Mn_{part} becomes somewhat smaller, and the concentration of dissolved Mn increases (see Fig. 4 in Yeats et al., 1995, p. 1570).

The total amount of organic matter in the ocean is estimated at 21.1 billion tons C, 95% of which is produced by phytoplankton in the ocean itself, and 5% is supplied from on-land areas. The process of decomposition of this matter commences as early as in the photic layer, but the process is most intensive at the boundary of the pycnocline and at depths of 500–1000 m (in the OML).

Role of the OML in geology. In zones of coastal upwelling, the OML comes in contact with the shelf bottom. Beyond the limits of coastal upwellings, this layer comes in contact with the bottom of the upper parts of the continental slope, and in the open ocean, with the tops and slopes of high seamounts. Weak reducing conditions are not rare in areas where the OML comes in contact with the bottom, and periodically even stagnation conditions are observed there (Fig. II.10.3). All these factors combine to cause the death of fauna, accumulation of their remains (bones, scales) on the bottom, and strong enrichment of bottom sediments in organic matter. In interstitial waters of such regions, there is an abrupt increase in concentrations of phosphates, nitrates, dissolved forms of organic matter, and Mn. The sediments are dominated by two types of diagenetic reactions: aerobic–anaerobic and, sporadically, anaerobic. Mn and Fe are mobilized from sediments into seawater, whereas Mo passes from seawater into sediment.

Several researchers (Aplin and Cronan, 1985) believe that the increase in concentrations of dissolved Mn in the OML is due to its effective diffusion from muds of the continental slope (from depths of 500–2000 m), which come in contact with the OML. Manganese in this layer is transported laterally by advective-diffusion processes. Dissolved cobalt in the water column shows a distribution generally similar to that of dissolved manganese: this element is abundant in the OML, but beneath this layer, contents of it are small (Aplin and Cronan, 1985; Knauer and Martin, 1980).

In sediments in contact with the OML, accumulation of barium sulfate ($BaSO_4$) occurs, and pyrite and glauconite are formed. The sediments are generally enriched in C, SiO_2, P, Ba, Mo, and probably in Se and As. In areas of upwelling, there are active processes of diagenetic phosphate production at the level of the upper boundary of the OML (Baturin, 1969; Emelyanov, 1973). According to some authors, authigenic accumulation of phosphates occurs also in sediments of the continental slopes, in zones where there is contact between the lower boundary of the OML and sediments (Jenkyns, 1986).

So, in areas where the continental slopes are washed by water of the OML, there are facies conditions which lead to accumulation of specific types of muds, the constituents of which include faunal remains (Table II.10.2), chemical elements and authigenic minerals, each characteristic of a particular facies (Emelyanov, 1982_1;

Table II.10.2. Benthic foraminifera as indicators of the contact of the OML with bottom sediments. After Hermelin and Shimmield, 1990.

Environment	Facial zone	Bottom sediments	Forams-indicators
OML-1 (upper part)	Shelf	Sediments enriched in OM	Bolivina pygmaea, Bulimina sp. 1, Lenticulina iota
OML-II (lower part)	Upper part of continental slope	Sediments enriched in $CaCO_3$	Ehrenbergina trigona, Hyalinea balthica, Tritaxia sp.1, Uvigerina, Peregrina
Beneath OML	Lower part of continental slope	-	Bulimina aculeata, Uvigerina hispida
Pelagic area	Abyssal area	-	Bulimina sculeata, Oridorsalis Umboratus, Uvigerina spinicostata

Jenkyns, 1986). Under such facies conditions, in zones of coastal upwelling, accumulation of phosphorites also occurs.

In the northwestern equatorial zone of the Pacific, the O_2 content in the OML decreases to 0.10–0.05 ml × l^{-1}, pH values decrease to 7.6–7.7, the Alk/Cl ratio increases to 0.128–0.129, and pCO_2 increases to (10–12) × 10^{-4} atm (Ivanenkov, 1966). The waters within this layer (200–1000 m) become undersaturated with respect to calcite [the degree of saturation is only 60–80% (Alekin and Lyakhin et al., 1979)] and, consequently, aggressive to carbonates. In places where the OML rests on the continental slope, carbonates dissolve and sapropel-like muds commonly predominate on the ocean floor. These muds give rise to favorable conditions for diagenetic formation of barite, precipitation of phosphates from interstitial waters, and formation of phosphorites on fish bones and scales. Sediments are enriched in Fe sulfides (Volkov et al., 1980). Thus, the sediments of the continental slopes assume all the signs of so-called black clays or black shales.

The formation of sapropel-like and sapropelic sediments (present-day black shales) requires at least two conditions: intensive supply of organic matter (for example, in the Sea of Azov, the Gulf of California and Walfish Bay) and good preservation (the Black Sea). Factors of secondary importance are low oxygen contents in the intermediate and near-bottom waters, a fine grain-size spectrum of sediments and high rates of their accumulation.

The C_{org} content in muds on the continental slope in the Gulf of California, at water depths of 0.5–1.5 km, showed an increase of up to 5–6%. This slope is washed by the intermediate waters with the OML (Thiede and Van Andel, 1977). According to Rozanov et al. (1976) muds of the Gulf of California (station DM-669, depth

Table II.10.3. The content of chemical components and elements in the surficial (0–5 cm) sediments of the continental slope in the Cape Basin. The profile of 23°S. The OML (with O_2 = 1–2 ml/l) occur at depths of 200–800 m. CaCO3–Mn in %; Ba–V in 10^{-4}% After Emelyanov, 1973, and Lukashin et al., 1994.

Station	Depth, m	Horizon, cm	Sediment type	$CaCO_3$	$C_{org.}$	SiO_{2am}	P	Si	Al	Ti	Fe	Mn	Ba	Sr	Cu	Zn	Ni	Co	Cr	V
AK-210	200	0–5	Sh. sand	68.48	3.59	0.12	1.77	-	-	0.05	0.75	Traces	<200	-	250	-	51	-	240(?)	28
B-3226	240	0–5	Terr a-pel.m	27.2	2.45	3.3	4.69	4.56	0.26	0.10	1.45	0.002	1300	3000	30	47	48	6	97	54
AK-150	345	0–5	Sh.sand	72.94	2.91	0.23	0.66	-		0.07	0.94	traces	1260	-	400	-	28	-	37	22
B-3218	418	0–5	Foram. ooze	72.8	2.14	2.7	0.39	5.48	1.15	0.12	1.19	0.030	260	1320	20	78	22	3	64	49
B-3228	565	0–5	Terr-calc. a-pel.m	39.3	3.49	9.3	1.79	15.11	1.25	0.16	1.68	0.007	260	1040	40	33	44	5	79	61
B-3229	1100	0–5	Foram. ooze	56.0	5.96	2.7	0.12	6.95	1.49	0.13	1.15	0.006	700	1370	65	187	73	7	71	61
B-3210	1101	0–5	-"-	71.7	3.50	0.8	0.08	5.25	1.16	0.14	0.05	0.006	670	1640	35	60	46	10	53	60
AK-205	2400	0–5	-"-	75.75	1.18	2.07	0.06	-	-	0.07	0.93	<0.01	17.30	-	-	-	23	112(?)	-	25
B-3221	3220	0–5	-"-.	86.9	0.34	1.0	0.04	3.14	0.75	0.09	0.61	0.032	900	2030	20	30	27	5	26	78

Sediment type: Sh – shelly, Terr – terrigenous, Terr-calc. – terrigenous-calcareous, a-pel.m.– aleuro-pelitic mud.

1000 m) contain up to 8.70% C_{org}. These hemipelagic (Holocene) muds are colored in shades from green to gray to greenish olive, Eh values vary from 200 to 260 mW, pH amounts to 7.6–7.8 (pH 7.3–7.5 in the water 1–3 m above the seafloor), and muds contain free H_2S, 2.94–2.99% Fe_{total}, 0.37–0.40% $Fe^{2+}{}_{H_2SO_4}$, 0.02–0.12% $Fe^{3+}{}_{H_2SO_4}$, 0.37% $Fe^{3+}{}_{sulf}$ and very small amounts of Mn (0.02%). In the same area, C_{org} contents in sediments which were taken from different depths fall within the following range: 140 m, 2.70–3.62%; 1450 m, 6.21–6.79%; 2650 m, 2.24–2.38%; 3280 m, 0.51–1.38%. FMNs occur in semiliquid reddish brown muds, at a depth of 3280 m.

On the continental slope off Namibia (Cape Basin, Atlantic Ocean), which is strongly influenced by the Benguela upwelling (2 ml × l^{-1} isooxyline), the OML commences from a depth of about 800 m and extends to increasingly shallower parts of the shelf. Sediments deposited beyond the shelf area and within the influence of the OML (200–800 m deep) or immediately beneath the OML (to a depth of 1100 m) are represented either by terrigenous, low calcareous or calcareous oozes, or by foraminiferal oozes (Table II.10.3). Oozes are markedly enriched in amorphous silica, strontium (0.104–0.203%), and in one sample, in Ba (0.130%), and contain 2.14% C_{org} and 0.08–4.69% P. It is interesting to note that very low contents of Mn and decreased values of Ni, Cu, Zn, and Co are representative of these oozes. Foraminiferal oozes occurring much deeper than the OML (at about 3220 m) were found to contain 0.34% C_{org}, 0.04% P, and 0.04% Mn. It is thus evident that muds and oozes occurring within the influence of the OML are characterized by the removal of a large portion of Mn due to diffusive processes, as well as small amounts of U, Co, Cu, Zn and Ni.

Sediments with high contents of organic matter (1–3% C_{org}) do not accumulate in pelagic areas of the World Ocean (deeper than 4.5–5.0 km). The only exception is the Angola Basin in the Atlantic Ocean (Emelyanov and Romankevich, 1979). This circumstance makes the present-day oceans different from some aquatic basins of the geological past, within which thick layers of black clays had accumulated.

It has been found that only strictly predetermined periods of time both in the Paleozoic and Mesozoic were responsible for accumulation of black clays enriched in organic matter (Fisher and Arthur, 1977; Jenkyns, 1980). These stages correspond to periods of a very low OML level. These global phenomena led to accumulation of black shale in the Middle Cambrian, Early and Middle Ordovician, Lower Silurian, Late Devonian, Early Carboniferous, Early and Late Jurassic, and Late Cretaceous (Arthur, 1983). Because ice caps were still absent, weak renewal of near-bottom waters was the main reason for accumulations of black shale in these periods. Warm ocean waters contained smaller amounts of oxygen, and the surface waters were more productive.

There is a good correlation between stagnancy conditions in the ocean and transgressions (Jenkyns, 1986, pp. 392–395). Submergence of wide shelf areas gave rise to the formation of many epicontinental seas, delivery of great amounts of nutrients to the ocean, an explosion of biological productivity, and deepening of the OML.

Redox (Eh) Barrier in Seawater

Basins with the constant hydrogen sulfide contamination. In nature, some water basins are characterized by constant contamination of deep waters with hydrogen sulfide, but in other basins its occurrence is random. The Black Sea represents a characteristic example of the first type of basins. In the Black Sea, oxygen almost completely disappears at depths as shallow as 120 m (120–200 m in nearshore areas), and hydrogen sulfide appears below this level (Figs. II. 11. 1, II. 11. 2). In the sea, hydrogen sulfide can be observed throughout the whole length of the water column, down to 2200 m. Hydrosulfuric contamination in this basin has started 7,000 years ago, when it was connected to the Sea of Marmara by the Bosphorus Strait. The Black Sea is the world's largest marine basin where processes of hydrosulfuric contamination occur: here, the volume of waters containing H_2S is 5470 km^3 or 90% of the total water volume.

The Cariaco Trench (depth 1350 m) in the Caribbean Basin is another example of basins where constant hydrogen sulfide contamination of the near-bottom waters occurs. The boundary between the upper (oxygen-rich) and lower (hydrogen sulfide) layers is situated at a depth of 350–400m. The hydrogen sulfide layer is 900 m thick and the volume of waters contaminated by hydrogen sulfide is estimated at 5200 km^3 (Richards, 1975). This basin is second in size among all basins with hydrosulfuric contamination. Bottom sediments (upper 6 m) of the Cariaco Trench are represented by greenish gray mud. The mud contains 10–30 % $CaCO_3$ and 2–5% C_{org} (Blazchishin and Lukashina, 1994). The <0.002 fraction consists of quartz, mica and kaolinite with an admixture of montmorillonite, chlorite and feldspars; authigenic pyrite and gypsum are fixed. Low Eh values (−307 to −409 mV) are measured in the mud (100- to 500-cm interval) from the western part of the Cariaco Basin. Some layers of this mud were found to contain microlaminations. Some microlaminations consist of grey terrigenous matter, whereas another are composed of brown sapropels or diatomic detritus. The biogenic remnants are represented by planktonic foraminifera: *Globorotalia menardii* and *Globigerina bulloides*.

The boundary between the Late Pleistocene and Early Holocene is very distinct: the transition layer is enriched in iron sulfides, whereas deposits of the Late Pleistocene are characterized by the abundance of benthic foraminifera (*Planorbulina mediterranensis*, *Cibicides kullenbergi* and others) (Blazchishin and Lukashina, 1994). The benthic foraminifera occur also in the upper, Holocene sediments. This is evidence that conditions of stagnancy in the Cariaco Trench during the Holocene were not continuous: this trench was periodically ventilated by oxygen-rich waters.

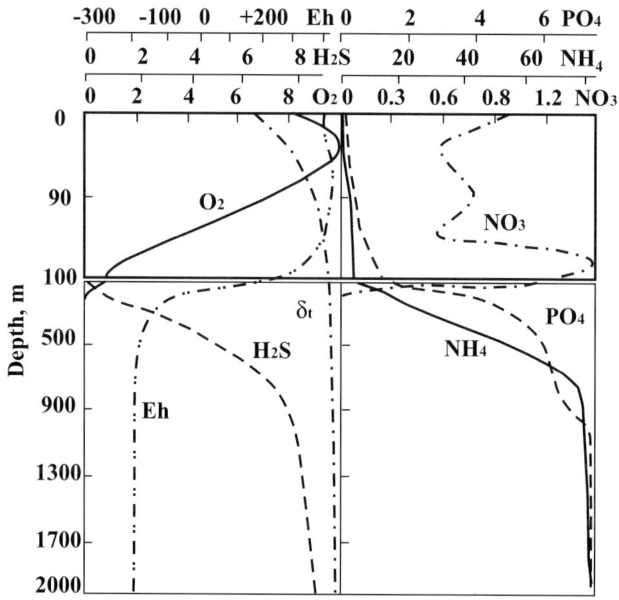

Fig. II.11.1. Vertical distribution of parameters in water strata of Black Sea which characterize redox condition. After Skopintsev, 1975.
O_2 and H_2S—in ml/l; NN_4–N, PO_4–P and NO_3–N—in µg-at/l; δ_t—density.

Deep waters of the basins with hydrosulfuric contamination (the Black Sea, Cariaco Trench) do not mix with adjacent deep waters of other water basins because of topographic barriers to mixing (sills). Nevertheless, even in basins of such type, hydrosulfuric waters are not completely stagnant. Water exchange in them occurs (although very slowly) between the upper oxygenated layer and lower hydrosulfuric layer.

In addition to the Black Sea, many shallower basins with constant hydrogen sulfide contamination exist. Among them are Orke Deep in the Mexican Gulf, some small depressions in deeps of the Red and Baltic Seas, Norwegian fjords, Drams, Skiomen, the Saanish gulf-fjord, Nitinat lake-fjord in British Columbia, many areas of the Arabian Sea, the Gulf of Oman, and in the vicinity of the Bermuda Islands (often called the Devil's Hole). Hydrosulfuric zones periodically appear in the near-bottom layer of river estuaries and in bays of islands. Also, such zones arise periodically in the near-bottom water layer in regions of coastal upwelling, primarily off southwestern Africa, Peru and Chile.

In areas around the Bermuda Islands, stratified seawater is divided into three layers (Table II.11.1). In each of these water layers, sediments contain different types of carbonates (Wefer, 1979), as well as their concentrations. In layer I, seawater is supersaturated with respect to calcite and aragonite, whereas the supersaturation of magnesium calcite is insignificant (the content of $MgCO_3$ in the calcite lattice is 12 mol %). With undersaturation of magnesium calcite of only 17 mol % $MgCO_3$ can

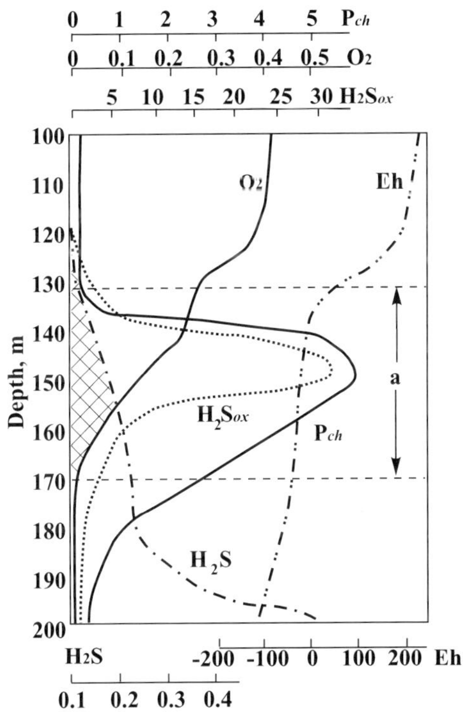

Fig. II.11.2. Vertical distribution of parameters which characterize redox conditions, and microbiological processes in O_2–H_2S transition layer (redox barrier) of water strata of Black Sea. After Sorokin, 1982.
O_2 and H_2S—in ml/l; Eh—in mV; H_2S_{ox}—intensity of H_2S oxidation; P_{ch}—intensity of bacterial chemosynthesis, mg/C/m^3/day^{-1}; a—transition zone of coexistence of O_2 and H_2S.

restore saturation. This is the reason why in layer I there occurs active production of biogenous carbonates, especially calcite and aragonite (organic detritus, *Strombus* shells, and *Diploria* corrals). Magnesium calcite is in the state of equilibrium: production of this substance equals to its dissolution. In layer III there occurs dissolution not only of magnesium calcite, but also of aragonite and calcite.

Table II.11.1. The layering of the water in the Devil's Hole of Bermuda Island. After Balzer and Wefer, 1981.

Layer	Depth, m	O_2, ml.1^{-1}	T°C	Alk mg-eq.l^{-1}	P, ppm
I. Oxygenated	0–18	4–5	28	2.3	480
II. Transition	18–21	3–1	25–24	2.4	500–1000
III. With hydrogen sulfide	21–24	0	27–23	2.5	1200–1500

In the Black Sea, besides the thermocline and halocline, the (O_2–H_2S) transition layer and deep hydrogen sulfide layer can also be clearly distinguished.

In seas containing hydrogen sulfide in water, an abrupt stratification between zones of oxidizing and reducing reactions exists (Rozanov, 1995). This stratification is a result of a succession of biochemical reactions (Table II.11.2), each of which is controlled by thermodynamic processes. Reactions of each type are characteristic of a certain water layer: reactions 1 and 2 occur in the photic layer; reactions 3, 2 and 5 occur in water layers 10–40 m above redox (Eh) barrier/O_2–H_2S layer; reactions 4, 7 and 9 occur in layers 5–10 m above the redox (Eh) barrier; reaction 10 occurs within the O_2–H_2S zone; reactions 6 and 8 occur in the hydrosulfuric water layer 50–250 m below the redox (Eh) barrier; and reactions 10 and 12 occur in the bottom water layer. All these reactions were described by Rozanov (1995). In this paper, we only will describe the behavior (in terms of geochemistry) of metals with changeable valence, mainly within the redox (Eh) barrier.

In the Black Sea, the contents of dissolved Mn and, to a lesser degree, Ni rapidly increase both at the boundary of the redox (Eh) barrier and below it. At the same time, increased concentrations of Mn (up to 400–500 µg per 1 liter) continue to exist throughout the whole water layer with hydrosulfuric contamination (Mokievskaya, 1961; Skopintsev and Popova, 1963; Brewer and Spencer, 1974). Although concentrations of dissolved Fe and Co slightly increase in the water layer below the redox (Eh) barrier and then slightly decrease within the zone of reducing (H_2S) reactions, their concentrations remain to be still higher than in the oxygenation zone. Concentrations of dissolved Cu, Zn, Mo, Se and As quickly decrease on the redox (Eh) barrier and then remain to be low throughout the whole length of hydrosulfuric water layer. The particulate forms of these elements are distributed in the water strata in the following way: (1) an increase in elemental concentration is characteristic of Mn, and partially of Fe and Al at the redox (Eh) barrier, whereas decrease in concentrations and, consequently, in relative contents of these elements in suspended matter occurs below this barrier zone; (2) a characteristic feature for Cu and Zn is increase in their concentrations and relative contents at the redox (Eh) barrier, and decrease in contents of these elements below this barrier; (3) the behavior of Ti and Al at the redox (Eh) barrier is almost nil. Such a distribution of these elements in particulate matter can be explained by the input of terrigenous material (mostly due to abrasion or from the mouth of rivers).

According to the data of Rozanov and Volkov (2002, p. 193), very low concentrations of dissolved manganese occur in the surface water of the open part of the Black Sea (less than 0.2–0.5 µM, Fig. II.11.3). Its concentration sharply increases in the redox (Eh) layer and especially in the first ten meters of hydrosulfuric water. Below the redox (Eh) layer and down to the bottom, the $Mn^{2+}{}_{diss}$ concentration becomes smaller (up to 4–5 µM) and remains constant throughout this depth interval. The maximal concentration of particulate manganese occurs in the depth interval between the redox (Eh) layer and NO_3^- maximum (Fig. II.11.3).

According to Rozanov and Volkov (2002, p. 149), successive reactions occur within the transitive O_2–H_2S water layer, which is enriched in Fe- and Mn hydroxides: $CO_2 \rightarrow O_2$, $NH_4^+ \rightarrow NH_3^+$, $Mn^{2+} \rightarrow MnO_2$, $Fe^{2+} \rightarrow FeOOH$, $S_2 \rightarrow SO_4^{2-}$. Each reaction is smoothly replaced by another.

Table II.11.2 Redox stratification in natural waters. After Rozanov, 1995.

No	Major processes	Reaction	Horizon characteristics	Eh, V	E_{pt}, V	Microorganisms
1.	Photosynthesis	$CO_2 + H_2O = CH_2O + O_2$	O_2 maximum			Phytoplankton
2.	O_2 consumption	$O_2 + C = CO_2$	O_2 decrease	0.74	0.4	Heterotrophs (saprophytes)
3.	Nitrification	$2NH_4^+ + 3O_2 = 2NO_2^- + 4H^+ + 2\,H_2O$ $2NO_2^- + O_2 = 2\,NO_3^-$	NO_2^- и NO_3^- maxima O_2 minimum			Chemoautotrophs (nitrificators)
4.	Denitrification	$3NO_3^- + C = 2N0_2 + CO_2$ $2NO_2^- + C = N + 2CO_2$	NO_2^- maximum, and NO_3^- decreae	0.59	0.3	Heterotrophs (denitrificators), Chemoautotrophs (theodenitrificators)
5	Mn^{2+} oxidation	$2Mn^{2+} + O_2 + 2H_2O = 2MnO_2 + 4H^+$	MnO_2 maximum			Chemoautotrophs (iron bacteria)
6.	MnO_2 reduction	$2MnO_2 + C + 4H^+ = 2Mn^{2+} + CO_2 + 2H_2O$	Mn^{2+} maximum	0.47	0.2	Heterotrophs (manganese reducing)
7.	Fe^{2+} oxidation	$2Fe^{2+} + MnO_2 + 2H_2O = 2FeOOH + Mn^{2+} + 2H^+$	FeOOH maximum			Chemoautotrophs (iron bacteria)
8.	FeOOH reduction	$4FeOOH + C + 8H^+ = 4Fe^{2+} + CO_2 + 6H_2O$	Fe^{2+} maximum	0.24	0	Heterotrophs (iron reducing)
9.	S^{2-} oxidatiom	$S^{2-} + 2FeOOH = SO_4^{2-} + 2Fe^{2+} + 2H^+$	SO_4^{2-} maximum			Chemoautotrophs (theobacteria)

Continues

Table II.11.2 Redox stratification in natural waters. After Rozanov, 1995.—*cont'd*

No	Major processes	Reaction	Horizon characteristics	Eh, V	E_{pt}, V	Microorganisms
10.	SO_4^{2-} reduction	$SO_4^{2-} + 2C + 2H^+ = H_2S + 2CO_2$	ΣH_2S increase, SO_4^{2-} decline	–0.24	–0.2	Heterotrophs (sulfate reducing)
11.	CH_4 oxidation	$CH_4 + CO_2 = 2CH_2O$; $CH_4 + CO_2 = CH_3COOH$	CO_2 decrease			Chemoautotrophs (methane oxidizing)
12.	CH_4 formation	$CO_2 + 4H_2 = CH_4 + 2H_2O$ mainly $CH_3COOH = CH_4 + CO_2$ partly	CH_4 increase	–0.35	–0.2	Chemoauto-heterotrophs (methane producing)
13.	$12H_2$ oxidation	$12H_2 + 6CO_2 = C_6H_{12}O_6 + 6H_2O$				Chemoautotrophs (hydrogen)
14.	H_2 formation	$C_6H_{12}O_6 + 6H_2O = 6CO_2 + 12H_2$				Chemoautotrophs (fermentation)

Notes: 1. To simplify stoicheometry, the OM empirical formulae (CH_2O) is sometimes reduced to a carbon atom (C); 2. The composition of solid phases of MnO_2 and FeOOH are relative, these phases are polyvalent and change oxygen content under natural marine conditions, ions of Mn^{2+} and Fe^{2+} can have organic and inorganic legends; 3. Fermentation, as well as H_2 oxidation can occur both under anaerobic, and aerobic conditions; 4. The rates of pure chemical interaction, which can occur in natural systems, are significantly lower (by 1–6 orders of magnitude for reactions 5–9) than the rates of bacterial processes; 5. Microbiological processes can involve both the components from the bordering horizons, and the "more remote" oxidants and reductants (reactions 4–9).

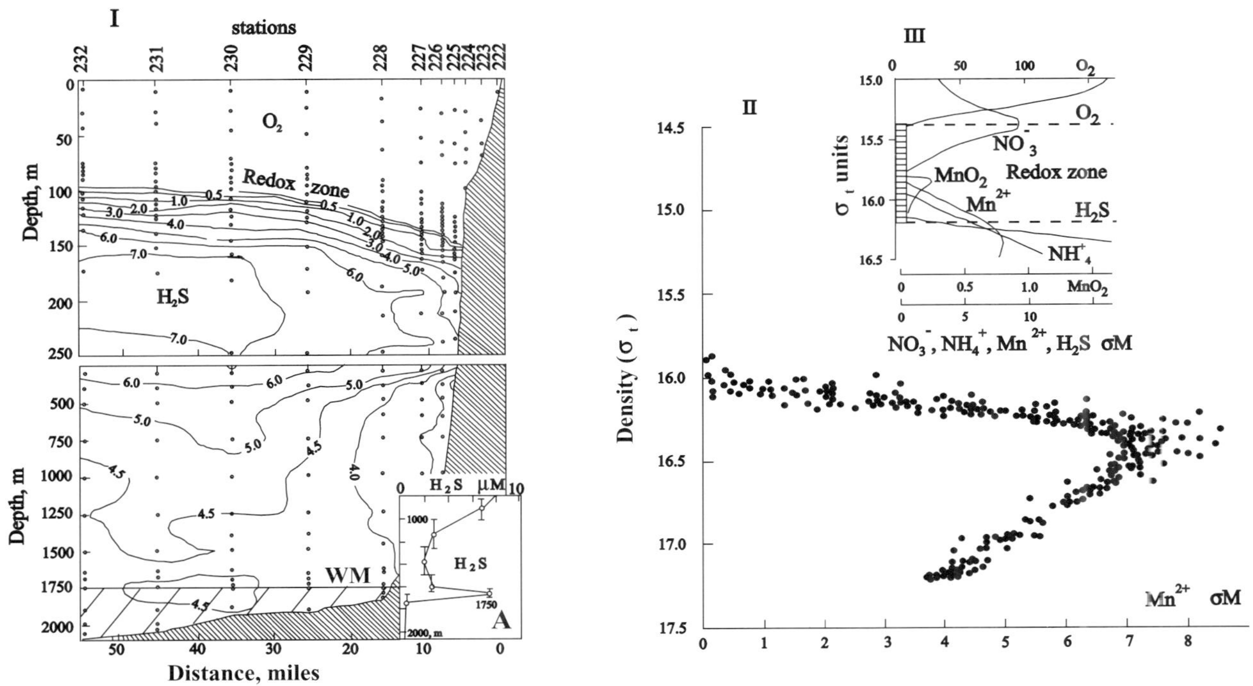

Fig. II.11.3. Manganese concentration (μM) in water strata and at O_2–H_2S barrier of Black Sea (1997). After Rozanov and Volkov, 2002, pp. 192–193, with some changes by the author.

Profile I—concentration of dissolved Mn^{+2} in water strata of the Black Sea near the slope of the Caucasus (profile from Gelendgik to SSW). There is a sharp increase in concentration at the redox zone, but the maximum concentration is located 50–200 m lower.

WM—border (at 1670–1750) between two water masses (upper mixed and near-bottom homogenous in salinity and density, but stratified in temperature).

A. Inset A: gradient (jump) of H_2S shown between two water masses (μM).

II. Position of maximal concentration of Mn_{diss} in density field of profile I (this figure).

III. Distribution of redox components in redox zone (σ_t 15.4–16.2 units) of Black Sea.

Factors such as the presence of a continuous and sharp redox (Eh) zone and active formation of Fe- and Mn hydroxides (and sulfides) are responsible for the accumulation of very fine gel-like particles within the redox (Eh) layer of the Black Sea. As a result, the percentage of Fe and Mn in suspended matter increase up to 8–18% and 30–45%, respectively (Emelyanov and Aibulatov, 1989; Emelyanov et. al., 1979, Fig. II.11.4). At the same time, concentrations of Fe and Mn as large as these have not been found in bottom sediments. In muds of the hydrosulfuric zone, the Fe and Mn contents are either no more than the concentrations of these

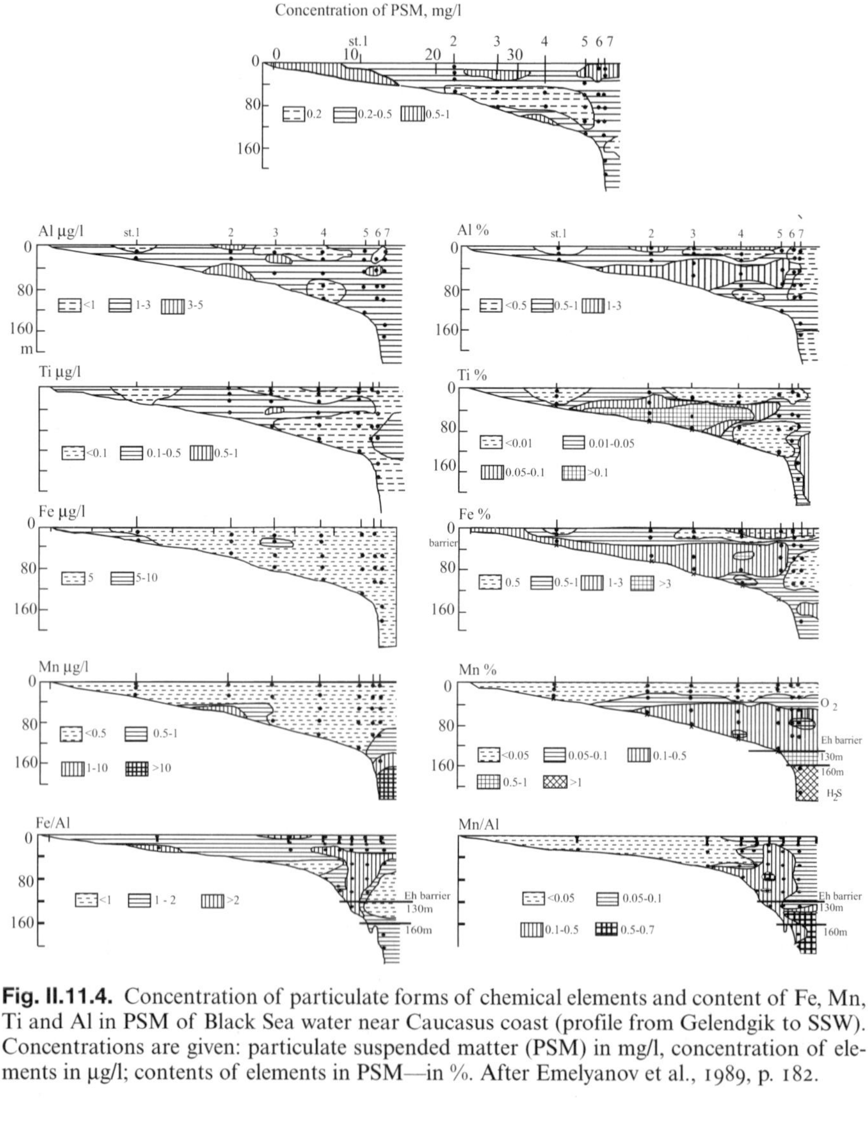

Fig. II.11.4. Concentration of particulate forms of chemical elements and content of Fe, Mn, Ti and Al in PSM of Black Sea water near Caucasus coast (profile from Gelendgik to SSW). Concentrations are given: particulate suspended matter (PSM) in mg/l, concentration of elements in µg/l; contents of elements in PSM—in %. After Emelyanov et al., 1989, p. 182.

elements in muds of seas under normal gas conditions (Fe) or even smaller (Mn). However, a characteristic feature of Mn is the occurrence of well-expressed pelagic maximums of this element. The reason for such maximums is the accumulation of very fine particles of suspended matter (evidently, terrigenous) with which this element associates in rivers. The proportion of mobile Mn in these pelagic muds is thought to be small. It is likely that Mn exists here in the form of chemogenic–diagenetic carbonates or sulfides, similar to its behavior in sediments of the Baltic Sea [or in Late Cenosoic sediments of the Black Sea (Emelyanov et al., 1982)]. Carbonates, sulfides and phosphates of Mn (as in the Baltic Sea), as well as siderite and vivianite, can be form in the muds of the upper part of the continental slope, where physiochemical conditions can be defined as transitional, and are characterized by disappearance of O_2 and appearance of H_2S. The Black Sea, like all deep (basin) seas with constant hydrosulfuric contamination, is characterized by a unidirectional process (in contrast to deeps of the Baltic Sea). This process provides for: (1) a good preservation (and, consequently, high contents) of organic matter; (2) reduction of authigenic minerals (to their extreme forms) during diagenesis, which leads to the formation of iron sulfides in sediments; (3) the accumulation of large amounts of Mn and, to a lesser degree, of Fe, Ni and Co in sea and interstitial waters; (4) rapid removal from seawater of elements able to exist in the form of iron sulfides or to precipitate together with these minerals (mainly, Mo, Se, As, and, to a lesser degree, Zn, Cu, Ni, Co and others); (5) the enrichment of these elements in the sediments; (6) considerable enrichment of trace elements in sulfides of the stagnant zone in contrast with sulfides of the oxidation zone; (7) depletion of manganese in sediments (Emelyanov, 1982_1).

Periodic hydrosulfuric contamination of the Baltic Sea and physiochemical processes. Periodic contamination of the bottom waters by hydrogen sulfide is typical for the Baltic Sea. The process began 7,800 ago, when the Baltic Sea united with the Northern Sea. In contrast to the Black Sea, the Baltic Sea is a platform (shallow-water) basin. Here, against the background of intermediate depths (80 m), shallow deeps separated by sills are clearly seen, including: Bornholm (108 m deep), Gdansk (118 m deep), Gotland (249 m deep), Faröe (205 m deep), North Baltic Sea (219 m deep), and Landsort (459 m deep) deeps. This depth is maximal for the whole Baltic Sea). Many other (shallower) deeps exist in the Gulfs of Finland and Bothnia. The situation in most of these deeps is characterized by periodic disappearance of oxygen and appearance of hydrogen sulfide.

A complicated structure of the water strata was identified not only in the Black Sea but also in the Baltic Sea, as well as in fjords (Syvitski et al., 1987). The transition redox (Eh) layer here usually consists of several other layers, which could be clearly seen during continuous profiling, especially with such instruments as a nephelometer (Figs. II.11.5–II.11.8; cf. Fig. II.9.1). The thermocline is distinct in the Baltic Sea during the summer season: it is commonly found at a depth of 30–40 m and is overlain by a well-mixed water layer. The halocline occurs at a depth of 80–90 m in the central Baltic Sea and at a depth of 50–70 m in the Bornholm Deep. The halocline in the central areas of the Baltic Sea is the place where the occurrence of the intermediate water layer with a low (1–3 ml/l) content of oxygen takes place. In the Bornholm Deep, this layer is compacted to 5–10 m or absent altogether. During

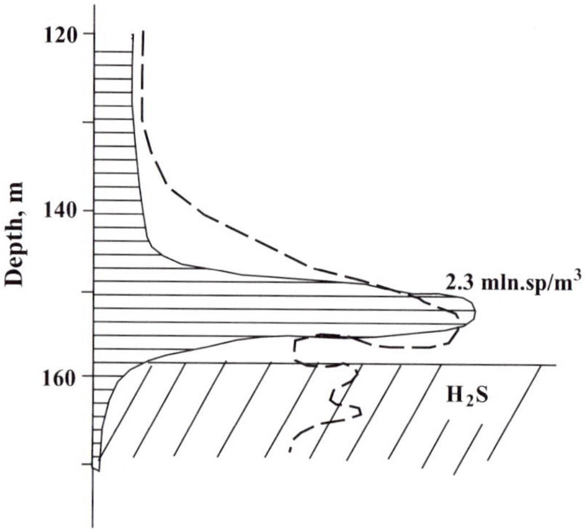

Fig. II.11.9. Number of *Infusoria* (dashed) above redox (Eh) barrier (O_2–H_2S interface) at station AK-4212 in Baltic Sea, $2.3 \cdot 10^6$ specimens · m^{-3}. After Mamaeva, 1987.
Dotted line—transparency profile.

length, specimens from the suborder *Hymenostomata*. The quantity of these infusoria amounts to 400,000 specimens per liter, which is several times more than in the upper water layer. The biomass of this infusoria is 8 mg × m^{-3}.

Infusoria feed on bacteria and possibly colorless flagella. Infusoria is, in turn, good food for crustaceans (*Calanus*), which migrate to zone in 5–10 m above the redox (Eh) layer from the upper water layer during daytime.

In the water layer above the redox (Eh) layer, there occurs accumulation not only of infusoria but also of colorless flagella (Moiseev, 1980, p. 178). The quantity of their biomass in the Black Sea sometimes amounts to 30 mg × m^{-3}. Because zooflagellates feed on bacteria, they concentrate in places where infusoria are abundant. In turn, zooflagellite is a food for infusoria.

In the daytime in the Black Sea, *Calanus helgolandicus* crustaceans accumulate 5–8 m above the redox (Eh) barrier, where the content of oxygen is 0.23–0.34 ml/l (Vinogradov and Shushkina, 1980, p. 183). The concentration of *Calanus* within this layer amounts to 2.5 g/m^3, which is the maximum for the whole oxygen layer.

Infusoria of the Baltic Sea are mostly represented by such species as *Metopus*, *Litonotus* and *Pleuronema*. In the water layer above the turbidity layer (i.e., noticeably above the boundary of the O_2–H_2S layer), the most abundant species is *strombidium*, which is very sensitive to oxygen concentrations, whereas within the turbidity layer (i.e., just above the redox (Eh) barrier), the abundance is characteristic of only saprobionts.

In this layer, "infusoria is an intermediate link providing for consumption of energy of the biomass of so called chemosynthetic bacteria, which accumulate near the upper boundary of the hydrosulfuric zone and thus involve the biomass again in the biological cycle" (Mamaeva, 1987, p. 158). Thus, infusoria is an important agent for the self-cleaning eutrophic waters of the Baltic Sea.

Role of the redox layer in the geological development. Because of periodic hydrosulfuric contamination of the bottom waters in deeps of the Baltic Sea, the accumulation of sediments is dominated by microlaminated sapropelic–clayey (pelitic) muds containing from 5.0 to 9.5% C_{org} (Emelyanov, 1981_2, $1986_{1,2}$). In the margins of deeps, these muds alternate to sapropelitic muds (2–5% C_{org}).

The most distinguishing feature of muds from these deeps is the high content of Mn (up to 13%) (or up to 26% $MgCO_3$), which may be as large as 26.5% Mn in some microlayers. For example, average contents of elements in microlayers in sediment core G-184 from the Gotland Deep are as follows: 26.5% Mn; 7.07% Ca; 0.45% Mg; trace elements (in 10^{-4}%): Zn, 38; Ba, 181; Ni, 8.6 (Huckriede and Meischner, 1996). The chemical formula of rhodochrosite in microlayers can be written as $Mn_{0.74}Ca_{0.24}Mg_{0.03}CO_3$.

The formation of manganese muds in deeps with periodic hydrosulfuric contamination of bottom waters is conditioned by the inflow of saline, oxygen-rich North Sea water to these deeps (Emelyanov, 1979_3, 1981_2, 1982_2, 1986_1, 1988_1). A manganese (rhodochrosite) microlayer is characteristic of the stage when waters in these deeps were well-oxygenated. Another confirmatory fact is the presence of benthic foraminifera *Elphidium excavatum* in rhodochrosite microlayers, which may develop under oxygen-deficient conditions. This species of foraminifera appeared in these deeps precisely at periods of time when oxygen was available in bottom waters (Saidova, 1981; Huckriede and Meischner, 1996). Sapropelic interlayers that accumulated during stages of stagnancy and are free of rhodochrosite represent a medium where habitation of benthic foraminifera is impossible.

A characteristic feature of marine Holocene mud in the Central Gotland Basin is that the sediment has a microlayered structure not throughout the whole length of the sediment section, but only in some parts of this section (Figs. II.11.10, II.11.11). Moreover, the total thickness of microlaminated mud appears to be largest in the deepest part of the basin (220–249 m), where it amounts to approximately 1.5–2 m. In the margins of the Central Gotland Basin, as well as in the Faröe and North Baltic deeps, microlayers are not continuous (see the lithological section in Emelyanov, 1986_2, 1995_1, Figs. 18-2 and 7). In the Faröe Deep, there are only two distinct microlayered beds, and in the margins of the Central Gotland Deep (at depths of about 160–200m), four interlayers exist (see Fig. 6 in Emelyanov, 1995_1). Because microlayered muds formed only under stagnant conditions, it should be assumed that non-microlayered (homogeneous) muds accumulated at the stage of a weakly oxygenated medium, i.e., when burrowing organisms lived on the sea floor and muds were disturbed by their activity. One organism of such type is *Mesodesma entomon*, which, as evidenced by our direct observations from Mir submersibles, can live even in regions where the oxygen content in bottom water is no more than 0.5-1.0 ml/l (Emelyanov et.al., 1994). The number of specimens of *Mesodesma entomon* living on the seafloor (depth 190 m) was considerable (up to 10 specimens/m^2), so that their activity led to disturbance and resuspension of black (hydrotroilitic) mud.

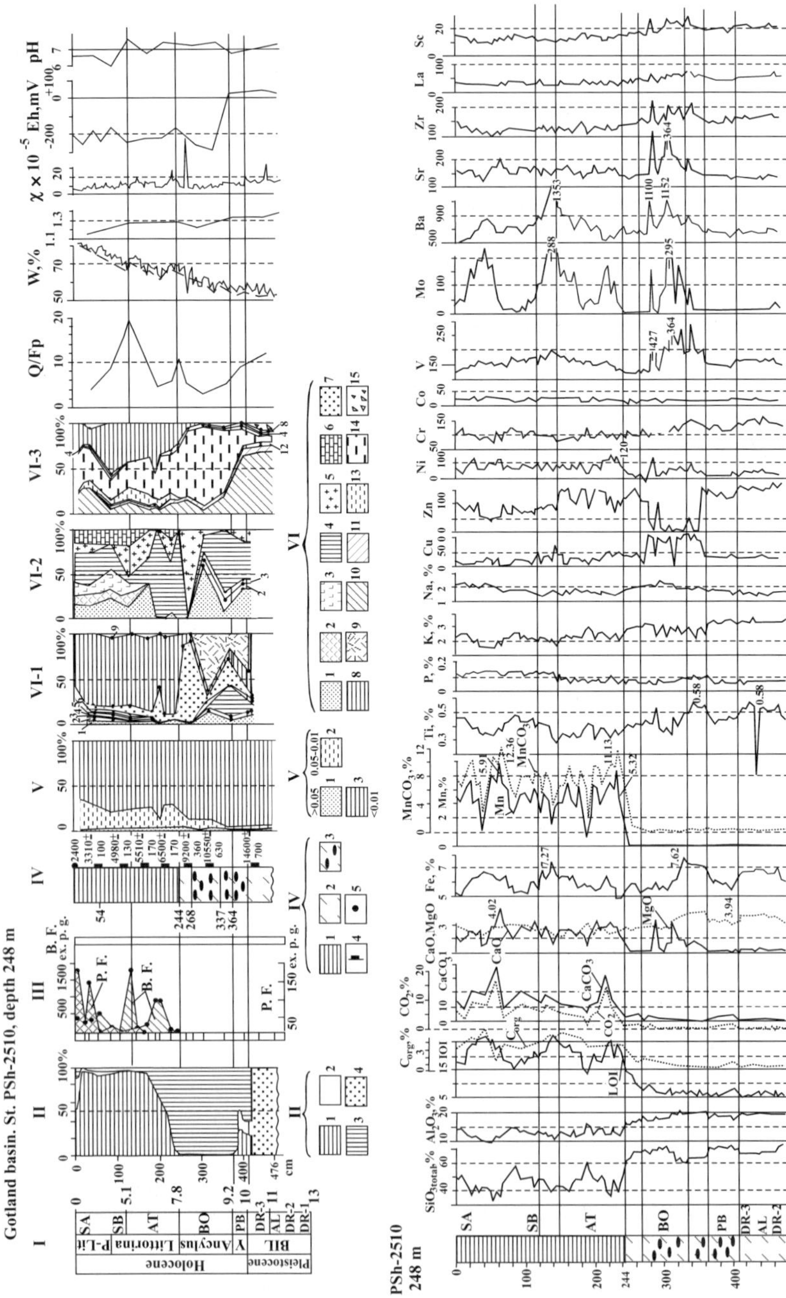

Caption see page 331

Manganese combines with other substances and exists in the form of carbonates with complex composition, which changes to weakly crystallized rhodochrosite with time (Jacobson and Postma, 1989). The upper layer of these muds is commonly represented by semi-liquid material made up of black hydrotroilite, with high concentrations of Mn, Fe, Cu, Zn, Cd, and P in interstitial waters. This mud contains a large amount of reactionable iron (4.46%, or up to 6.80% in total). In addition to large concentrations of CO_2, C_{org}, Mn and Fe in mud of the Gotland Basin, a characteristic feature of this mud is high contents of Mo, Ni, Cr, Zn, and Cu, as well as authigenic minerals: rhodochrosite, greigite; in peripheral areas of the deep, barite and vivianite (Blazhchishin, 1976).

Sedimentary conditions and composition of muds in the Faröe Deep are close to that in the Gotland Basin. Hydrogen sulfide periodically appears in this region as well. The muds are black (hydrotroilitic) in color; they contain up to 7.02% C_{org} and, thus, belong to sapropel-like and sapropelic muds, which often display microlayered structure. At the same time, the manganese content in the muds of the Faröe Deep usually do not exceed 0.50%. Only in the deepest part (depth 190–210m) of this deep, the muds contain 1.00–4.48% Mn (Emelyanov, 1995_1). In addition to Mn and C_{org}, the muds are also enriched in CO_2, Fe, Mo, Ni, Zn and Cu; i.e., chemical elemental composition of the muds from the Faröe Deep is almost similar to that in the muds from Gotland Deep.

To the north of the Faröe Deep is the North Baltic Deep, which is also a good settling basin for sedimentary material (Emelyanov, $1988_{1,2}$). The depth of this deep is about 180–200m. The deep is filled up with Holocene sapropel-like and sapropelic muds black (hydrotroilitic) or grayish black in color. The upper layer of these muds is a liquid substance, containing up to 84.28% water. In such liquid, grayish black

Fig. II.11.10. Stratigraphy and composition of Holocene mud and clay and varved clay of BIL in core PSh-2510 from Gotland Deep (depth 248 m). After Emelyanov et al., 1995.

I—stratigraphic scale. Abbreviations: BIL—Baltic Ice lake; Y—Yoldia; P–Lit—Post-Litorina; DR—dryas; Al—alleräd; PB—preboreal; SB—subboreal; SA—subatlantic; numbers from right—age in 10^3 ago.

II. Content of diatomic remains: 1—marine; 2—brackish; 3—lacustrine; 4—no diatoms.

III. Foraminifera, samples per gram (ex · p · g) in 1 g of sediment: BF—benthic; PF—planktonic.

IV. Lithology: 1—terrigenous–sapropelic mud (4–6% C_{org}); 2—gray clay; 3—sulfidic (back) clay; 4—samples in which C^{14} was analyzed (numbers—age); 5—grab sample.

V. Grain-size distribution (fractions in mm), %.

VI. Mineral composition of 0.1–0.05 mm fraction.

VI-1—heavy subfraction;

VI-2—only terrigenous minerals of heavy subfraction;

VI-3—light subfraction;

1–14—minerals: 1—opaque ore minerals (ilmenite–magnetite–leucoxene); 2—amphiboles (mainly green hornblende); 3—epidote–clynozoisite; 4—mica; 5—garnet, zircon, tourmaline and others; 6—clastic carbonates; 7—micronodules of iron sulfides; 8—clayey–carbonate aggregates; 7—micronodules of iron sulfides; 8—clayey-carbonate aggregates; 9—Fe hydroxides; 10—quartz; 11—feldspars; 12—plant remains, diatoms, foraminifera, others.; 13—clayey aggregates; 14—glauconite; 15—debris of rocks. Q/Fp—quartz (Q)—feldspar (Fsp) ratio; W—moisture (short lines mean jumps in moisture); ρ—density; $\chi^* 10^{-5}$Si—magnetic susceptibility.

Numbers of chemical schemes designate maximal content.

Cu–Sc—in 10^{-4}%; SiO_2–Na—in % (in natural dry sample).

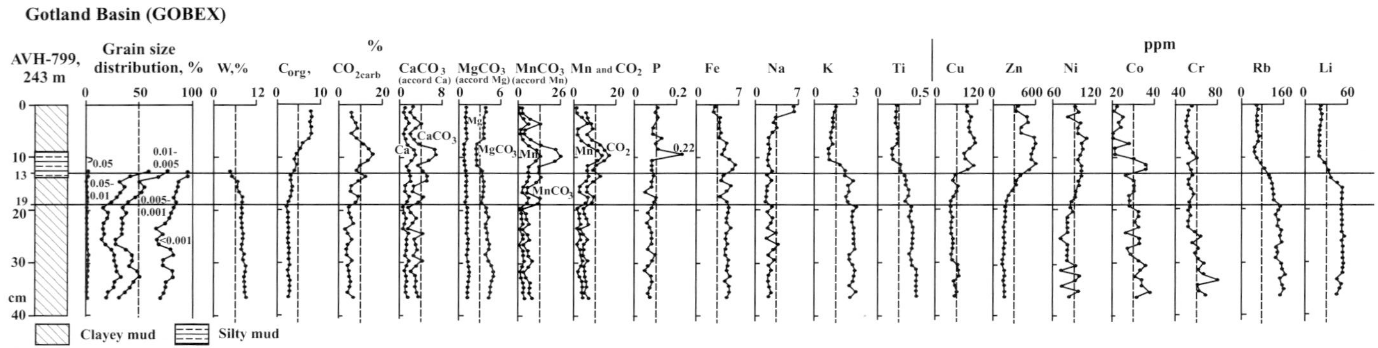

Fig. II.11.11. Grain-size distribution (%) and chemical composition (in %, Cu–Li in 10^{-4}%) of sapropelic–manganese mud of Gotland Basin, Baltic Sea (GOBEX station AVH-799, depth 243 m). W—moisture, %. Core obtained by K. Emeis and V. Sivkov.
Grain-size distribution—fractions in mm.

mud, the content of reactionable Fe^{2+} (7.26%) is highest throughout the World Ocean, and contents of $Fe^{3+}_{react.}$ and iron sulfides amount to 1.42% and 0.71%, respectively. Dry samples of these muds contain 8.68% Fe (this value is maximal for reduced muds of the Baltic Sea); 0.13% Mn; 0.00% $CaCO_3$; 3.21% C_{org}; 0.06% P; 2.72% K; 170 ppm Rb; 333 ppm Zn; 136 ppm Cu; 57ppm Cr; 57 ppm Ni.

Hydrodynamic and geomorphological traps of the Gulf of Finland are features where accumulation occurs of very loose hydrotroilitic muds black in color with a large content of gases. When the sediment core is removed from the metallic corer, the structure of muds becomes porous and cavernous. In addition to methane, these muds contain free hydrogen sulfide. A characteristic feature of these muds is their latent microlaminated structure (layering is indistinct). However, if one inclines a sediment core composed of this mud, the sediment will fragments into very small pieces or thin sheets. These sheets are marks of annual sediment deposition. The muds consist of clayey, well-sorted material with a relatively low content of $CaCO_3$ (less than 3%, clastic calcite) and C_{org} (1–3%). Contents of $Fe^{2+}_{react.}$ and $Fe^{3+}_{react.}$ are usually not more than 1% and 0.5%, respectively.

Black (hydrotroilitic) and grayish black muds of the Bornholm Deep contain up to 5.67% C_{org} and up to 0.40% Mn.

The formation of increased concentration of Mn and Fe in the sediment of the Baltic Sea does not require preliminary separation of Mn and Fe within the drainage area and fixation of Fe in the weathering crust (Strakhov et al., 1968): both elements may be transported into the water basin in the same ratios as in their rocks of origin. For example, the value of the Fe/Mn ratio in different rivers of the Baltic Sea Basin varies within the limits of 5–120, or 38 on average (Blazhchishin and Emelyanov, 1977, p. 91). At the same time, this ratio in sediments equals 35, i.e., almost similar to that in rivers. Mn more frequently separates from Fe as early as in the water basin: Fe accumulates in the form of concretions and the clarke values of this element are found everywhere in water basin sediments, whereas the concentration of Mn occurs both in nodules and muds, but only under very favorable facial conditions. Supply of additional portions of Mn into this region occurs at the cost of depletion of this element in sediments, which are abundant in weakly reducing conditions (diffusion of Mn from sediments into bottom waters). According to the estimates of Blazhchishin (1982), the compositional ratio of Fe to Mn being supplied to the Baltic Sea Basin is nearly the same as in the sediments of their origin, i.e., in continental rocks or soils.

The sapropelic and sapropel-like muds described above began to accumulate in deeps after the inflow of saline North Sea water to the Baltic Sea, i.e. 7,800 ago (beginning of the Litorina transgression). A reflection of this moment in the geological history is the change in species composition of diatoms; the appearance of foraminifera; the sudden decrease of SiO_{2total}, Al and Ti in muds; and the increase of content of C_{org}, CO_2 and especially Mn (in the form of $MnCO_3$), Mo and some other microelements. Because the frequency of stagnant conditions in deeps have changed throughout the history of oceans, the influence of these events is seen as a nonuniform distribution of these elements in the sediment column, especially when thin layers of sediments 1 cm in thickness or less are being studied (Figs. II.11.10, II.11.11). Analysis of peaks in the distribution of Mn, C_{org} and microelements enables stages of strongest stagnation or oxygen-rich bottom water to be reproduced. Such an analysis bears fruit if the coefficient of stagnation $\{(Mo + 10Se) \div Mn\}^{*}100$ is employed (Emelyanov, 1978, $1982_{1,2}$). For muds that formed when H_2S was present in near-bottom waters, the coefficient is usually more than 5, and this coefficient is less than 5 when muds develop in the presence of oxygen in the near-bottom water layer.

A considerable number of isolated and small areas of ore shows are a characteristic feature of the Baltic Sea. Increased contents of Mn (>0.20%) were discovered in many areas of the Baltic Sea. Pure-manganese ore manifestation in the Baltic Sea is restricted to deeps which are filled up with fine pelitic mud, whereas iron manganese ore manifestations are confined by sand, aleurite or moraine outcrops. The greatest masses of Mn have accumulated in deeps (basins), where deposition of these muds occurs. Stagnant muds of marine Holocene (7800–0 years ago) and Baltic Sea deeps contain 12.5 million tons of Mn, or 831 tons/km^2, which exist in the form of carbonates and other authigenic minerals (Blazhchishin, 1982).

Authigenic deposition of Mn in the Baltic Sea occurs also in an oxidizing environment, where formation of FMNs and crusts takes place (Glasby et al., 1996). The content of Mn in nodules and crusts amounts to 8.3 million tons, or 10.8 million tons including Mn that exists in the form of carbonates and other divalent forms. The total amount of Mn which is removed from seawater annually to become a constituent of crusts and nodules is 830 tons, and 1353 tons of this element are used for the formation of carbonate minerals. In all, 2183 tons of Mn are required for the production of authigenic compounds. Annual supply of Mn by rivers amounts to 2585 tons in addition to this. As well, 6500 tons of Mn are supplied to the sea as a result of shore abrasion and bottom erosion (Blazhchishin, 1982). Thus, river discharge provides a sufficient amount of sedimentary material for authigenic processes in the Baltic Sea. Only 20–25% of the total influx of Mn is used for development of authigenic sediments, whereas the remaining 75–80% are dispersed in non-carbonaceous manganese deposits, where the Mn content is the same as the clarke value of this element.

Chapter 12

Salinity Barrier

Deeps with brines in the Mediterranean Sea. Cold brines the salinity of which is ten times as high as that of seawater have been discovered (Cita et al., 1989) in three deeps of the eastern Mediterranean Sea (Bannok, Tyro, Poseidon), at depths of 3276 and 3328 m (Fig. II.12.1, Table II.12.1). The density of these brines is 1.21 g/cm^3. Therefore, the fractionation of sedimentary material that sinks from the upper water layer occurs at this density barrier: heavy particles, the specific weight of which is more than 1.21 g/cm^3, sink through the barrier to be deposited on the seabed. At the same time, light particles, which are mainly represented by organic detritus (1.06–1.15 g/cm^3) and also by skeletons of diatoms, accumulate at this barrier, i.e., on the surface of the brines. Therefore, the concentration of suspended matter increases several times there, whereas the water transparency decreases. In the water layer several meters below the density barrier, water transparency again becomes higher (95–97%) and the content of suspended matter sharply decreases.

The thickness of Holocene muds sampled by us (Emelyanov, 1992$_2$) at the five stations in Libeccio subbasin is 20–40 cm. In the central part of the Libeccio subbasin, Holocene muds are represented by greenish gray calcareous-clayey oozes with a very high water content: 89–90%. These are almost liquid muds. They contain free hydrogen sulphide. At station AMK-2128, at a depth of 15–25 cm from the bottom surface, the sediments contain about 200 g of transparent, colorless, euhedral, flat-prismatic gypsum crystals 2–4 cm long, 2–3 cm wide, and 3–5 mm thick. Some of these features have the form of "swallow-tail" intergrowths. Particles were present as inclusions in different parts of the crystals. In all, 26 crystals were found in the sample, or 104 crystals per square meter of bottom surface. No gypsum crystals were found in the upper (01–10 cm) layer of the bottom sediments.

A similar liquid greenish gray ooze was also found in sediment cores AMK-2129 (0–20 cm) and AMK-2132 (0–24 cm). However, no gypsum crystals were discovered in Holocene muds of these cores.

Beneath the greenish gray ooze (at stations AMK-2127, AMK-2129, and AMK-2132) lie semi-liquid Pleistocene, homogeneous, bluish gray clayey-calcareous oozes with a 50–60% content of $CaCO_3$. At the 70- to 80-cm level below the sediment surface, the semi-liquid ooze takes the form of very soft, soft, and plastic ooze. The series of Pleistocene sediments exposed (20–432 cm) corresponds to the *Pseudoeunotia doliolus* zone or the *Emiliania huxleyi* nanozone. Radioisotopic dating confirms this (Fig. II.12.2). In the 95- to 45-cm (Upper Würm) interval, ooze accumulated at a rate of 14 mm/1000 yr. If sedimentation took place at such a rate earlier as well, the whole Pleistocene part of the AMK-2129 sediment core (40–532

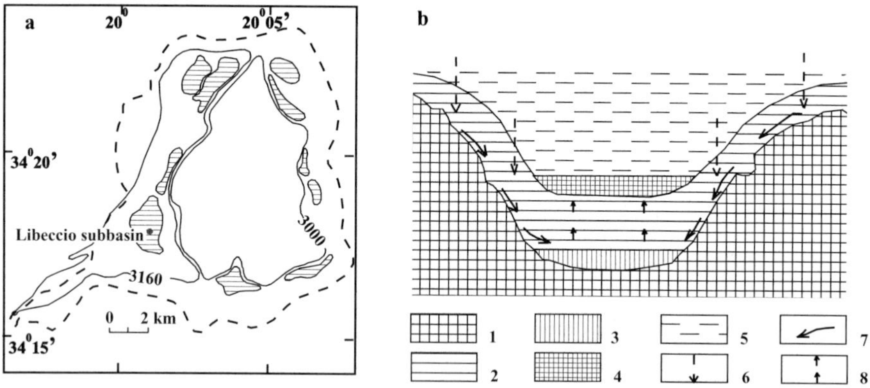

Fig. II.12.1. Sedimentation in anoxic hypersaline basins of Eastern Mediterranean.
a) Bannock basin: locations of subbasins (cross-hatched) in Bannock basin (outlined by dashed line). After Cita et al., 1989.
3000-m isobath is shown; 3160-m isobath marks edge of strongest, most concentrated brines. The black dot shows the main area where samples were recovered in Libeccio subbasin (station AMK-2129).
b) Inferred scheme of influx of brines into depressions of subbasins (after Cita et al., 1989):
1—Messinian evaporates; 2—sedimentary series formed after Messinian time, 3—accumulation of brines in depression of halogenic deposits; 4—brines at surface of sea bottom, 5—seawater; 6—osmotic movement,;7—solution of salts; 8—diffusion of salts.

cm) accumulated in 34,000 yr. In the Holocene, the sedimentation rate was considerably higher: about 20 mm/1000 yr.

The Pleistocene sediments are represented by pelitic, and the Holocene by aleuro-pelitic, ooze.

At the brines–seawater boundary, the Eh redox potential falls from +240 mV to +30mV (De Lange et al., 1989), and the pH value decreases from 8.20 to 6.3–6.6 (Table II.12.1). The water column becomes almost completely devoid of oxygen, and H_2S appears. In addition, brines differ from seawater by higher concentrations of SO_4, PO_4, NH_3, Mn^{2+}, Mg, Ba, As^{3+}, Sb^{3+} and certain other elements (Catalano et al., 1989; De Lange et al., 1989). Brines are characterized by growth in concentrations of Fe (from 1–3 to 105 nanomoles/kg), Mn (from 0 to 5300 nanomoles/kg), and dissolved forms of REE (La, Nd, Yb, Ce, Sm, Sr, and Lu). So, concentrations of these elements in brines reach values as large as those found in interstitial waters of reduced sediments (see Table II.12.1).

The activity of anaerobic microbial flora is very high at the brines–seawater boundary (Domenico and Domenico, 1989). The total amount of bacteria in the marine environment of the Mediterranean Sea, at a depth of 400 m, is 3.9×10^5 cells/ml, whereas at a depth of 3331 dbar, or 3377 m (that is, 1 m below the surface of brines), their amount is already 1.1×10^6 cells/ml (Ferla and Crisafi, 1989, p. 103). Accumulation of bacteria was found to be maximum within a water layer the upper limit of which is about 3 m above the brines–seawater density barrier and the lower limit is 10 m below this barrier. It is evident that the trapping of particles within this

Table II.12.1. Average chemical characteristics of brines in the Libeccio subbasin of Bannock Basin and in Tyro Basin, Eastern Mediterranean. After Catalano et al., 1989; de Lange et al., 1989

Sampling depth, m	pH	Alk*	H_2S,	R-SH,	SO_4,	F	$P0_4$,	$Si0_2$,	NH_3,	Ca,	Mg,	Ba
		$mM.l^{-1}$				$\mu M.l^{-1}$			$mM.l^{-1}$			
Seawater												
–	8.20	2.67	0	0	31.1	78.3	0.2	8.7	0	11.7	60–70	8
Brines of Libeccio and Bannock basins												
Upper layer (3275–3390)	6.46	4.22	1.89	0.23	100.6	106.9	13.1	201.5	3.39	21.0	600	7
Lower layer (3400–3500	6.38	4.10	2.69	0.28	131.9	138.9	13.7	258.0	3.31	16.3	650	–
Brines of Tyro basin												
–	–	–	2.63	0.26	49.2	40.9	9.9	267.5	1.26	34.0	70–80	30

*) Alk – total alkalinity

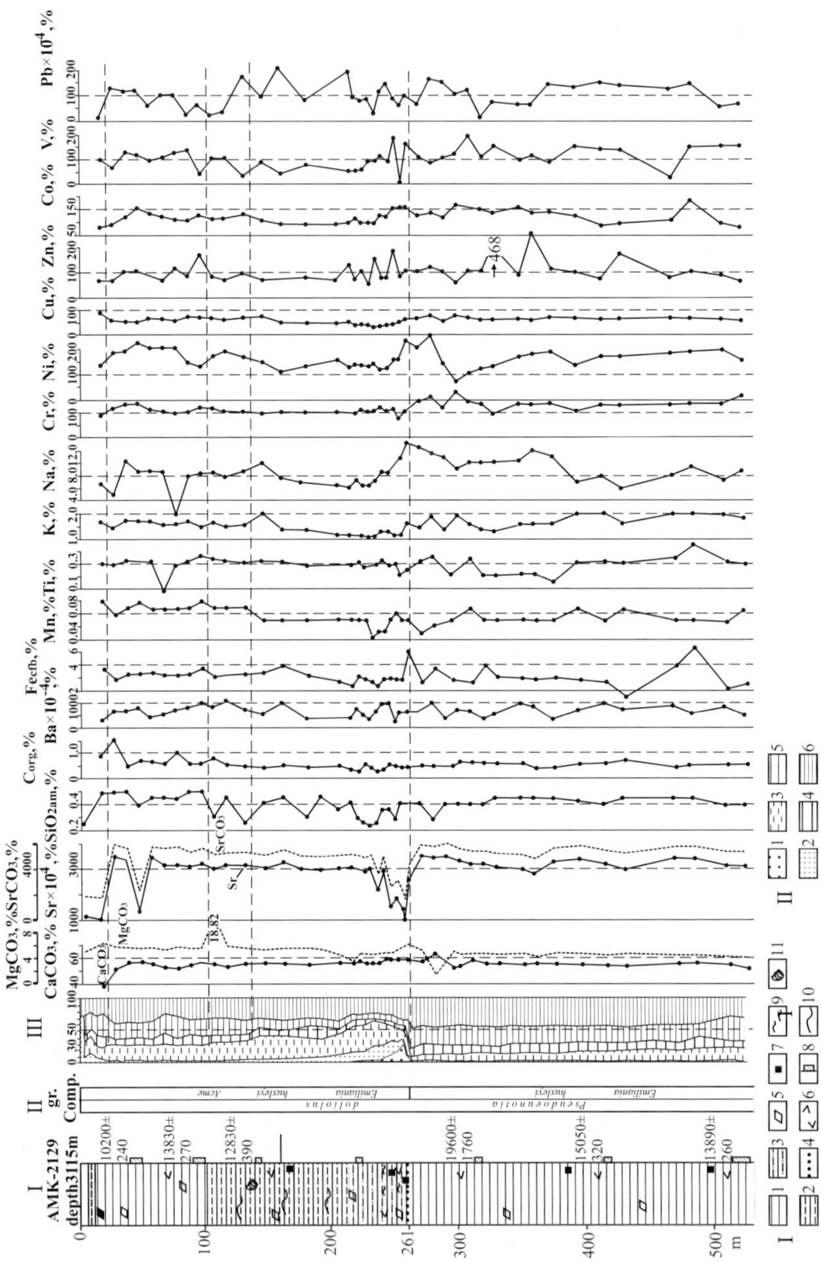

Caption see page 339

layer is the main reason for high concentrations of bacteria. The amount of bacteria tends to fall downward from this layer, but it increases near the bottom.

Gelatin-like films, "pellicles", begin to form at the brines–seawater boundary, under the influence of bacteria. Pellicles are composed of organic coal-like material, the remains of siliceous organisms (predominantly diatoms), and, rarely, calcareous shells. They are of bacterial origin (Erba, 1989, p. 45). Pellicles are thought to form at the brines–seawater barrier, but evidence of this process is lacking.

What are the mechanisms that cause the light particles to sink to the bottom together with brines? In the case under consideration, the traditional gravity mechanism is of no avail. Two mechanisms are believed to be responsible for delivering light material to the bottom of deeps (Corselli and McCoy, 1989, p. 50): (1) pelletal transport, i.e., the process when light particles are ingested by zooplankton and nekton, digested, and aggregated into larger fecal pellets, which rapidly sink towards the bottom. Because the density of fecal pellets is larger than that of brines, pellets succeed in overcoming the barrier and settle to the bottom to replenish reserves of organic matter in muds; (2) density (turbidity) currents, which are caused mainly by tectonic movements. The uppermost layers of brines are disturbed by the effects of these currents, which trap light particles, including pellicles, and carry them downwards until they settle on the bottom. Traces of a powerful turbidity current were found in the sediment core collected at station AMK-2129 (Fig. II.12.2) as a turbidite sequence (261- to 200-cm horizon), which is thought to have been formed about 13–15,000 years ago.

It is supposed that a certain portion of pellicles becomes heavier as a result of biological aggregation. Therefore, the flow of sedimentary material sinking from the surface can catch these particles to carry them to the seabed.

The consequences of the brines–seawater density barrier, in terms of geological development, are fairly considerable. Below brines, accumulation of fine muds occurs, which consist predominantly of clay particles, clastic minerals (quartz, feldspar), planktonic foraminifera tests, coccoliths, authigenic and biogenic carbonates, salts, and pellicles (Tables II.12.2, II.12.3; Fig. II.12.2, Emelyanov, 1992). Mud is commonly homogeneous, gray, and contain iron sulphides (chains 1–2 cm long).

The sand-sized (more than 0.1 mm) and coarse aleurite [or finest sand] (0.1–0.05 mm) sized fractions together amount to no more than 1% of the sediment. These fractions predominantly consist of Mediterranean planktonic foraminifera tests.

Fig. II.12.2. Lithological composition and distribution of cemical components and elements in sediment core AMK-2129. After Emelyanov, 1992.

I) Types of bottom sediments: 1—4) granulometric types: 1) pelitic oozes, 2) silty-pelitic oozes, 3) fine-silty oozes, 4) coarse silts and sands; 5) authigenic carbonates (strontianite, calcite, dolomite); 6) gypsum crystals; 7) iron sulfides (including vermiform concretions); 8) radiocarbon age of sediments; 9) pellicles (gelatinous membranes); 10) signs of sediment slides; 11) clumps of allothigenic ooze: HL—Holocene.

II) Biozones (diatom and nanozones).

III) Grain size composition of sediments (fractions in mm): 1) >0.1; 2) 0.1—0.05; 3) 0.05—0.01; 4) 0.01—0.005; 5) 0.005—0.001; 6) <0.001. $CaCO_3$ contents calculated from CO_2, $MgCO_3$ from Mg and $SrCO_3$ from Sr. Contents of Ba, Fe and all or elements recalculated on cfb. All minor elements (Ba, Cr etc.) given in 10^{-4} % (or ppm).

Table II.12.2 Decription of sediments taken in Libeccio subbasin, Bannock Basin. After Emelyanov, 1992

Station	Depth, m	Core thickness, cm	Age	Sediment type	Presence of gypsum crystals
AMK-2127	3410	26	Hl-Pl	Clayey-calcareous oozes, greenish gray (clay), with gypsum crystals	+++
AMK-2128	3512	25	Hl	Clayey-calcareous ooze, greenish gray, semiliquid, with gypsum crystals	+++
AMK-2129	3515	532	Hl-Pl	Clayey-calcareous ooze, greenish gray, with gypsum crystals and layer of turbidite	++
AMK-2130	3150	15	Pl	Clay, bluish gray, soft	–
AMK-2132	3442	465	Hl-Pl	Clayey-calcareous ooze, greenish gray, with gypsum crystals, and layer of turbidite	+++

Note: +++ – much; ++ – average amount, – – none found.
Age: Hl – Holocene, PL – Pleistocene

Table II.12.3. Mineral composition of sediments (in % of overall sample) of core AMK-2129, according to results of diffractometric analysis. After Emelyanov, 1992

Horizon, m	Q	P	KFS	I	K	M	Ca	A	D	Ph+Ha
0–3	21	0	6	17	6	6	21	3	13	7
75–80	19	4	9	5	5	6	22	5	9	16
130–135	26	10	6	10	6	5	18	4	8	7
250–256	22	24	11	4	0	4	17	3	7	8
261–265	13	5	23	7	4	5	21	3	8	11
300–306	16	3	0	8	6	7	31	0	2	20
375–380	18	4	0	9	6	4	31	6	4	18
430–435	18	4	0	10	7	8	28	6	5	14
525–530	20	5	3	10	6	5	28	5	4	15

*) Minerals: Q – quartz, P – plagioclases, KFS – potassic feldspars, I – illite, K – kaolinite, M – montmorillonite, Ca – calcite, A – aragonite, D – dolomite, APh + Ha – amorphous phases + halite.

The >0.05-mm fraction shows elevated contents only in the Late Holocene (0–20 cm)—up to 15% of the sediment. Benthic foraminifera were also found: 80–100 specimens per gram of sediment in the 0- to 20-cm layer, and a few specimens in Pleistocene ooze (260–530 cm layer): *Quinqueloculina juw, Spirillina wrightii* and *Astrononion stelligerum* (out of 58 species in all).

In addition to foraminefera, there were shelly particles and also numerous coccoliths (*Emiliania huxleyi, Gephyrocapsa carribeanica, G. oceanica, Calcidiscus leptoporus, Coccolithus pelagicus, Rhabdosphaera clavigera,* etc.) and reworked discoasters of Neogene age (*Discoaster broweri, D. challangeri, D. surculus,* etc.).

The amorphous silica in the oozes amounts to an average of 0.5% of the sediment (Table II.12.4). It consists of rare diatoms (*Pseudoeunotia doliolus*, etc.; 39 species in

Table II.12.4 Average contents of chemical components and elements in clayey-calcareous oozes of the anoxic Bannock Basin (st. AMK-2129, depth 3515 m) in comparison with composition of the same oozes, but outside of anoxic basin

Components, elements	Bannock Basin		E. Mediterranean*	
	a	b	a	b
Components and elements, %				
$CaCO_3$	56	–	59	–
$SiO_{2am.}$	0.5	–	0.8	–
C_{org}	0.6	–	0.6	–
Fe	1.50	3.50	2.98	5.33
Mn	0.04	0.09	0.05	0.17
Ti	0.20	0.47	0.31	0.52
P	0.03	0.07	0.04	0.09
K	0.70	1.63	(1.34)	(2.24)
Na	4.00	9.32	(1.46)	(2.88)
Sr	0.33	0.77	(0.10)	(0.22)
Elements, 10^{-4}%				
Li	76	177	(51)	(96)
Rb	6	14	(82)	(158)
Zn	35	82	(85)	(156)
Cu	25	58	19	37
Ni	80	186	41	76
Co	<20	<47	<6	<13
Cr	<20	<47	34	65
V	40	93	63	108
Pb	<20	<47	(20)	(50)
Ba	250	583	(<200)	(<400)
Zr	85	195	59	118

Note: Mo, Be and Ge not found in sediments by spectroscopic method (less than 5×10^{-4}%).

*After Emelyanov and Shimkus, 1986, for recent nano-foraminiferal aleuro-pelitic oozes of the Eastern Mediterranean (for the entire Mediterranean Sea in parentheses): (a) natural dry sediment; (b) recalculated, carbonates and silica free basis (cfsb).

all), silicoflagellates (the genera *Dictyocha, Distephanus, Mesocena*), and radiolarians.

Authigenic carbonates (calcite, dolomite, and strontionite) are concentrated mainly in the 0.01- to 0.001-mm size fractions.

The oozes contained 0.4–0.5% of organic carbon, and in one sample of Holocene ooze, up to 1.54%.

The contents of Fe, Mn, Ti, K, Rb, Cr, Zn, Cu, Co, V, Pb in the oozes are small; the content of phosphorus is 0.03%. Against the background of their average contents in the calcareous-clayey ooze of the eastern Mediterranean, contents of Zr, Li, Ni, and Ba in the sediments of the Libeccio subbasin are slightly elevated, and Na and Sr are strongly elevated. Low contents of Sr were found only in Holocene ooze (Table II.1.1b).

The "normal" pelagic sedimentation in the Libeccio subbasin was interrupted by a turbidity current 14,000 years ago. The occurrence of the slide can be inferred from the inclusions of oval clumps of allothigenic ooze (Fig. II.12.2), and the turbidity flow, by the presence of turbidite 161 cm thick (100- to 161-cm interval). The turbidite consists (from the bottom) of sands (fragments of pelecypod shells, planktonic foraminifera tests, quartz, feldspars, micas, iron sulphides, and gypsum) (Fig, II.12.2). Distinguishing features of this interrupting bed are gray mottled color, grain-size composition (upward gradation into finer sediment), lower Sr contents (showing a downward decrease in the seabed), the uneven distribution of other chemical elements (Zn, V, Pb) in the layer, and a number of remains of flora and fauna. In the 110- to 265-cm interval of sediments, for example, the diatomic assemblage is characterized by the presence of benthic marine species (*Diphloneis crabro, D. smithii, Cocconeis scutellum, Bidulphia alternans,* etc.), freshwater bottom foraminifera (*Anlacosira sp., Stephanodiscus astraea, Epithemia zerba*, etc.), and redeposited benthic foraminifera (*Numnulites sp., Nodosaria sp.*, and *Planulina sp.*), which have been transported into the anoxic Bannock basin from areas with a normal oxygen regime.

In comparison with "normal" clayey-carbonate (nano-foraminiferal) oozes, oozes of anoxic basins are also slightly enriched—in addition to Sr and Na—in Cu, Li and Ni, and depleted in Fe, Mn, Ti, K, Rb, Zn, and Cr. The concentrations of these elements in sediments do not depend on the presence of brines and are practically the same as in the biogenic carbonate sediments outside the brines area .

At the base of the turbidite layer (261–225 cm), sediments contain iron sulphides, gypsum crystals up to 1–2 cm in size and—the most interesting—gray translucent gelatinous membranes (pellicles). The thickness of the pellicles extracted from the sediments was 1–3 mm, and the area of their individual pieces, 2–5 cm^2. These membranes, or pellicles, lie horizontally, along the bedding planes, in the sediments. Upon being dried (at 100°C) they become cloudy and opaque white, and can be seen to consist of very small crystallites of sulphates and salts: gypsum, halite, and anhydrite. In a dry state the pellicles retain their forms, but at even the slightest touch they turn into powder.

Brines in the Bannok basin (Fig.II.12.1) represent a double-layer system. Their upper layer, which is 123 m thick, has a temperature of 14.25–14.40°C, and the lower layer is 123 m thick and has a temperature of 15.06–15.14°C. There is a sharp boundary between these two layers (Rabitti and Boldron, 1989, p. 67). The upper layer of these brines is enriched in Ca^{2+}, and the lower layer, in SO_4^{2-} (Table II.11.1). Also, both layers are nearly on the verge of supersaturation by gypsum. As a result of mixing of brines at the boundary between these two layers, water becomes supersaturated with respect to gypsum, and, consequently, the gypsum precipitates from water (de Lange et al., 1989, p. 83). This is evidenced by the Sr/Ca ratios in water. However, gypsum precipitates predominantly at the boundary between these two solutions. Gypsum formation is facilitated by *Desulphuvibrio* bacteria. Large accumulations of gypsum—up to 100 crystals/m^2 and even more—have been found along the margins of the Bannok deep (Emelyanov, 1992_2). Some crystals are as long as 35 cm. Amounts of gypsum are so large in spots that these places are called "gypsum gardens" (Cifuentes et al., 1989).

Brines are supersaturated with respect to SO_4^{2-}, but undersaturated in barium. As evidenced by the saturation curve of Ba concentration versus depth, values of Ba in the water column tend to decrease with depth at the brines–water boundary in the Bannok basin (in contrast to the Tyro basin, where these values tend to increase). Therefore, barite is absent in sediments lying below the surface of brines.

Gypsum precipitates from brines to the extent that there is a decrease in the relative contents of Ca^{2+} in solution and an increase in the Sr/Ca ratio. As a result, concentrations of Sr are in excess of coexisting Ca. This is probably the reason why Sr is so abundant (up to 0.30–0.40% Sr) in the sediments of the Bannok basin (Emelyanov, 1992_1). Sr is found in sediments in the form of small-sized chemogenic–diagenetic crystals of strontianite (Fig.II.12.2).

The degree to which the margins of deeps filled with brines have been studied up to now has been insufficient. We believe that precisely where brines–water boundary comes in contact with the bottom, accumulation of elements occurs, such as vivianite, barite, Mn carbonates (rhodochrosite), and enrichment in P, Mn, Ba, Cu, Zn and other trace elements in the sediments is especially considerable.

The role of these barriers, in terms of geological development, is given further consideration in other sections of this book (see Parts II.1, II.6 and II.9).

Chapter 13

Calcium Carbonate Compensation Depth (CCD)

In low latitudes of the World Ocean, the level of the CCD is the shallowest in the Guatemala deep in the eastern part of the Pacific Ocean, where it occur at 3400 m (Figs. II.13.1, II.13.2), and in the Brazilian deep of the Atlantic Ocean (the CCD is 4700 m) (Lisitzin et al., 1977), which is well supplied with cold and aggressive (with respect to $CaCO_3$) Antarctic waters. The deepest level of the CCD is up to 5900–6000 m in the southern part of the North American and Cabo Verde basins (Emelyanov et al., 1971; Lisitzin et al., 1977; Emelyanov and Kruglikova, 1992). At depths greater than 6020 m (down to 8000 m), contents of $CaCO_3$ in the upper layer of sediments are less than 10%. In other basins of the Atlantic Ocean, the depth of the CCD ranges from 4700 to 5900 m. At the same time, the position of the CCD in basins of the equatorial zone (Guinea, Angola, and Cabo Verde basins) is always deeper than in moderate or high-latitudinal basins.

The depth of the lower boundary of the CCD at which saturation of ocean water in respect to calcite occurs, depends on latitude. In areas adjacent to polar areas, the depth of the CCD is shallowest: 200–150 m for calcite and not more than 100 m for aragonite.

In polar areas, the CCD for calcite is on the ocean surface. In arid zones, the depth of saturation is much lower: 500–600 m for calcite and 300–400 m for aragonite (Bordovsky and Ivanenkov, 1979, p. 90). In the Atlantic Ocean, the level of the CCD for calcite and aragonite is 500–800 m deeper than that in the Pacific Ocean (Fig. II.13.3). The main reason for this difference is that Atlantic deep waters are well ventilated. This is because strong deep and bottom currents of the Arctic Ocean supply this region with a large amount of cold water relatively depleted in carbon dioxide, in contrast to the circulation pattern in the northern areas of the Pacific and Indian oceans, where such inflows of cold polar waters are absent (this is the reason for relatively high rates of CO_2 accumulation in deep waters). In the southern areas of all three oceans, regimes of water ventilation and the distribution of carbonate systems are nearly the same (Bordovsky and Ivanenkov, 1979).

Therefore, the positions of the CCD in these oceans are very similar.

In mid- and high latitudes of the Atlantic Ocean, the level of the CCD varies within the range of 4700–6000 m, and the curve corresponding to 100% saturation of waters with respect to calcite occurs at a depth of about 3000 m (Lisitzin et al., 1977). The difference in depths is 1700–3000 m. This is because the depth of the CCD is controlled by several factors including the flux of calcareous particles to deep layers, the speed of sinking particles and the rate of their dissolution. Because the rate of dissolution generally lags behind the rate at which particles sink,

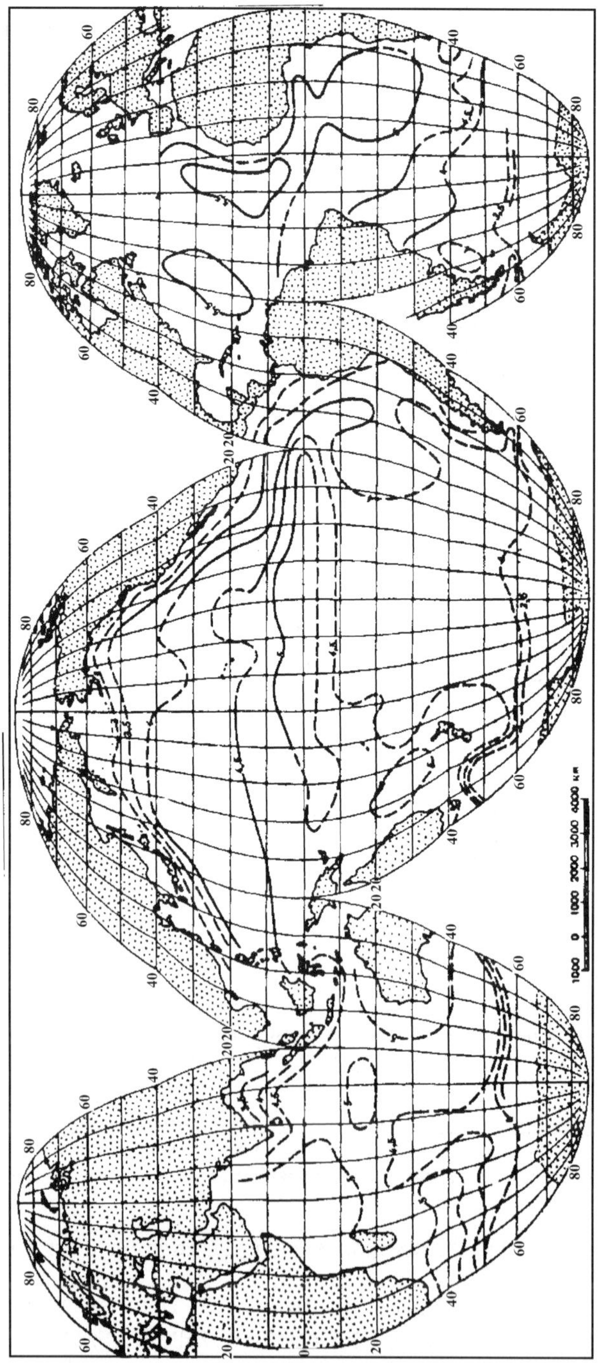

Fig. II.13.1. Topography of calcium carbonate compensation depth (CCD) defined by interpolating local fading boundaries between calcareous sediments and sediments with no or only a low percentage of carbonate. Numbers on contours denote kilometers below sea surface. After Berger and Winterer, 1974.

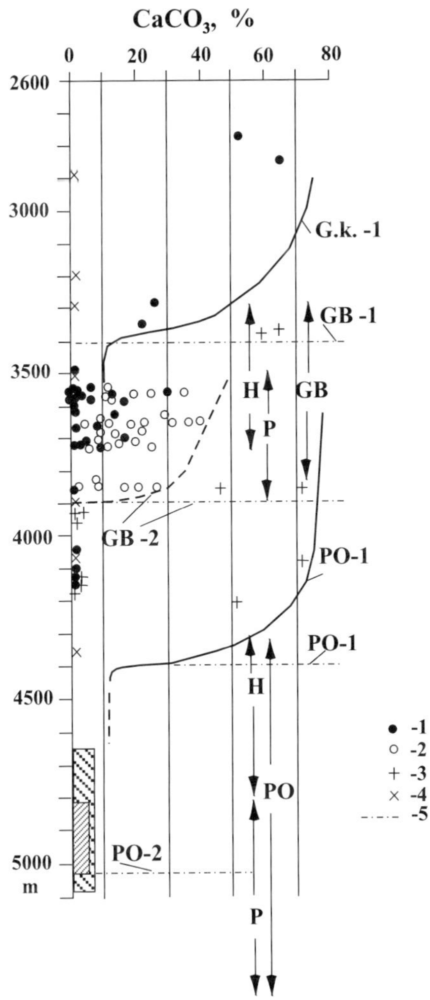

Fig. II.13.2. CCD in north part of equatorial Pacific. After Emelyanov, 1991.
Samples: 1—Guatemala Basin (layer 0–3 cm, mainly late Holocene); 2—same, 25–400 cm layer (Pleistocene); 3—open Pacific (west of EPR); 4—samples from expedition DM-9 (see Fig. II.6.5), 0–3 cm layer; 5—CCD: GB, Guatemala Basin (GB-1, Holocene; GB-2, individual stages of Pleistocene); PO, Pacific Ocean west of EPR (PO-1, Holocene; PO-2, Oligocene); I and II, samples from profiles I and II. Vertical arrows indicate depths most favorable for development of FMNs: GB, in Guatemala Basin (H, Holocene; P, Pleistocene); PO, in open Pacific west of EPR (H, Holocene; P, Oligocene). Sample areas outlined by curved lines.

carbonate particles have time to penetrate through an additional 1700–3000 m of water column.

There is a close correlation between the depth of the CCD and biological productivity: the higher the biological productivity, the shallower the CCD, and vice versa (Berger and Winterer, 1974). This is why in the eastern part of the equatorial

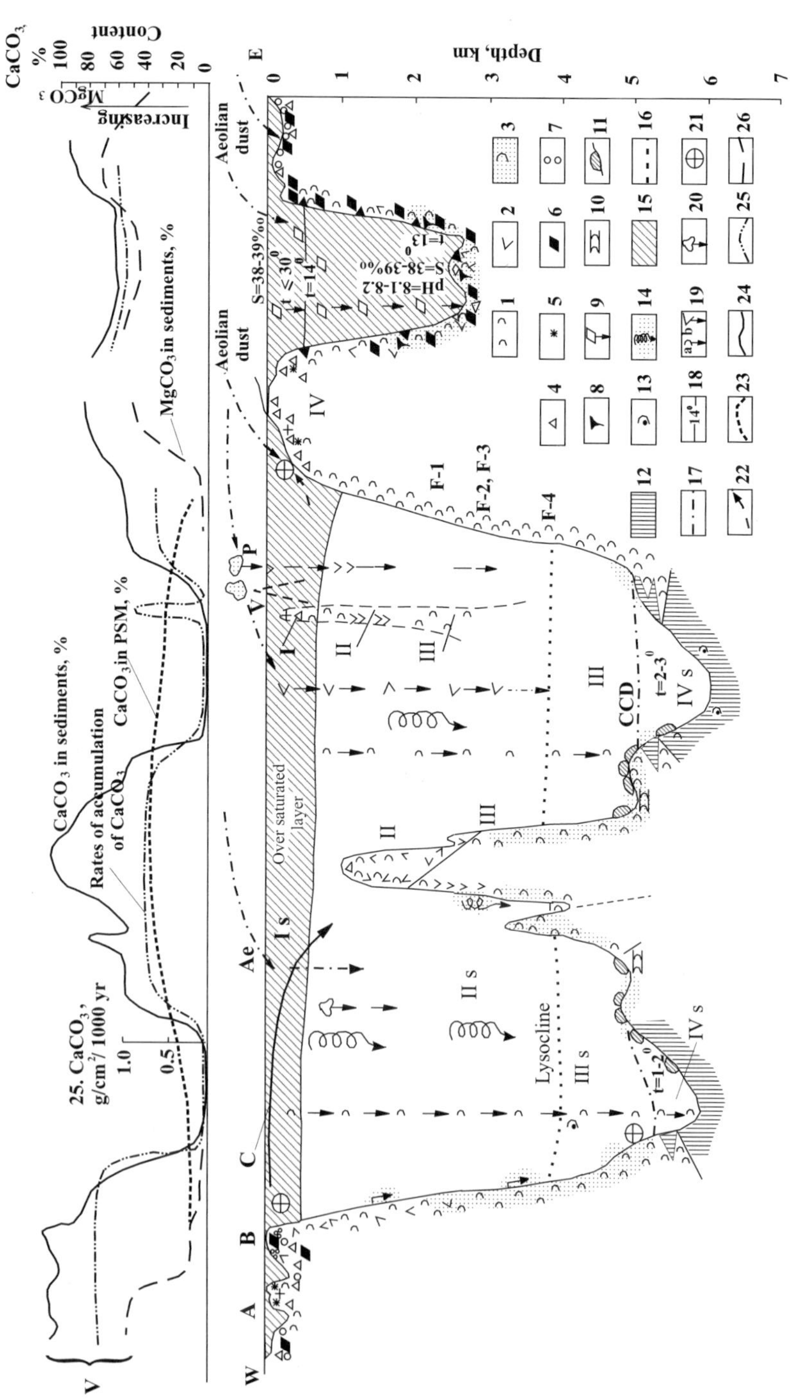

Caption see page 349

Pacific (Guatemala basin), where biological productivity very high, the CCD is located at a depth of 3400 m, which is an extremely shallow level throughout most of the equatorial zone of the World Ocean (Emelyanov, 1992).

In the eastern zone of the Atlantic Ocean, the CCD does not rise to the higher levels of the water column. In this region, the depth of the CCD is 5200 m for the Cape basin (Fig. II.13.4), 5600 m for the Angola Basin (Fig. II.13.5, II.13.6), and 5700 m for the Cabo Verde and Canary basins, whereas it lies at 4700–4800 m in the Brazilian Basin and at 5500–5600 m in the North American basin deepening up to 5900–6000 m in its southern part (Emelyanov et al., 1971; Lisitzin et al., 1977; Emelyanov, 1982_2).

The eastern area of the Atlantic Ocean as a whole and North American basin in particular are isolated geomorphologically from the Brazilian and Cape basins: although aggressive Antarctic bottom waters do penetrate in these areas through the

Fig. II.13.3. Scheme of sedimentation and transformation of carbonates in north arid climatic zone of Atlantic (Florida passage is in the West; the Mediterranean and Sea of Azov are in the East).

1—foraminifera (foraminiferal oozes); 2—pteropods; 3—coccoliths and foraminifera; 4—mollusks and bryozoans; 5—calcareous algae, corals; 6—chemogenic calcite (Magnesium calcite, ankerite, dolomite in Mediterranean Sea); 7—calcareous oolites and pseudoolites; 8—calcareous (magnesium) clayey formations (concretions) (mainly in Mediterranean Sea); 9—origin of chemogenic calcite in water strata; 10—foraminiferal turbidities; 11—FMNs; 12—pelagic red clay; 13—relicts of foraminiferal shells; 14—slumps; 15—water oversaturated with $CaCO_3$; 16—lysocline level; 17—CCD; 18—14°C isotherm; 19—flux of foraminifera (a) and pteropods (b); 20—intensive dissolution of siliceous remains; 21—strong near shore and deep currents; 22—upwelling; 23–26—Distribution of lithological components in bottom sediments: 23—content of $CaCO_3$ in PSM of sea surface layer (in % of total PSM-weight); 24—$CaCO_3$ content in bottom sediments; 25—rate of accumulation of $CaCO_3$ in Holocene sediments (g/cm²/1000 yr); 26—$MgCO_3$ content in sediments.

I–IV—prevailing types of bottom sediments, which are composed mainly of: I—bryozoan remains, mollusk shells and benthic foraminifera (consisting of magnesium calcite and aragonite); II—pteropodal remains and planktonic foraminifera (consisting of aragonite and calcite); III—planktonic foraminifera and coccoliths (consisting of calcite); IV—coral-algae remains, corals, mollusks, bryozoa, benthic foraminifera (consisting of Magnesium calcite and calcite).

I_s–IV_s—layers of seawater of different saturation with $CaCO_3$, with different temperature (t°C) and salinity (S‰): I_s—oversaturation (>100% $CaCO_3$), t—25–27°C; S—37‰; II_s—100%—saturation, S—35‰; III_s—saturation is <100% (layer between lysocline and CCD); IV_s—water is not saturated (layer between CCD), t—+1 to +3°C.

V—lines show contents of $CaCO_3$ in PSM and bottom sediments; $MgCO_3$ in bottom sediments (0–5 m layer) and rate of $CaCO_3$ accumulation in Holocene sediments.

A—coral reefs, deposition of clayey-calcareous clots, pelletoids, aggregates, coral (algae *Halimeda*) and mollusks remains,; minerals aragonite and magnesium calcite predominate.

B—shallow banks with warm water (up to 29–30°C). Deposition of aragonitic oolites and grapestones; sediments contain 90–99% $CaCO_3$; Sr content is hightened. V—volcanogenic mountains and rain of pyroclastics (P).

Arrows show: Ae—intensive flux of aaeolian material (quartz, feldspars, illlite, calcite, dolomite); C—water current from shore to open sea; F—deposition of foraminifera: F-1—foraminifera of first group of stability predominate (Berger, 1971)—*Globigerinoides ruber, Hastigerina pelagica, Globigerina bulloides*; F-2, F-3—foraminifera of second and third group of stability predominate: *Globigerinoides sacculifer, G. Conglobatus, Orbulina universa, Globigerina aeqilateralis*; F-4—foraminifera of fourth group of stability predominate: *Globorotalia truncatulinoides, Globoquadrina dutertrei, G. menardii, Pulleniatina obliquiloculata, Sphaeroidinella dehiscens.*

The Water-Bottom Interface

The hydrodynamic conditions at the seawater–bottom interface comprise a complicated interaction of different processes which are controlled by a number of factors. Here we can examine only some of the most important aspects.

Water exchange at the seawater–bottom interface. Water exchange at this interface occurs everywhere in the ocean. The discharge and subbottom recirculation of fresh groundwater occur in many regions along the continental margins (Schüter, 2002, p. 213): in the Mediterranean Sea, the Gulf of Mexico, the continental margin off Florida, the Baltic and North seas, and in many others regions of the World Ocean.

In some cases, cold groundwaters penetrate to the seabed with the result that the near-bottom water layer is freshened. In other cases, it is seawater which penetrates into the sedimentary sequence (through fractures and pores of igneous rocks) to salinize groundwaters so that they become unsuitable for drinking (Schüter, 2002). In zones of transform faults and in rift zones of the ocean, seawater penetrates to the earth's interior to a depth of several kilometers below the bottom surface, where it is heated and then appears at the bottom surface once again as hot thermal springs. In specific environmental conditions where there are salt-bearing deposits (evaporites) enclosed within the sedimentary sequence, hypersaline groundwaters may reach the surface of the ocean bottom, where they may flow down the slopes of small depressions (troughs) on the ocean floor and form stagnant basins with brines similar to those described in Part II.2.

According to Djamalov (1991), input of groundwaters to the ocean is estimated at 2.4×10^{18} g/yr, including inputs due to deposition of sediments and their densification, 18×10^{15} g/yr; and juvenile waters entering the ocean, 0.5×10^{15} g/yr. The volume of seawater penetrating into sediments (rocks) is only 0.5×10^{15} g/yr. Thus, it is clear that ocean water reserves are being replenished (across the water–bottom boundary) by underground flows, mainly due to the input of groundwaters from continents (Fig. II.14.1).

Hydrodynamics, transport of sedimentary material and its deposition. The main process which occurs at the water–bottom interface in the World Ocean is the deposition of sedimentary material and accumulation of bottom sediments. This is a complex but one-way process, which is responsible for delivering suspension particles from land to sea and from sea surface to bottom. The vertical fluxes of sedimentary material in the deep ocean ultimately result in accumulation of sediments of two different types: hemipelagites (in the near-continental areas of the ocean) and

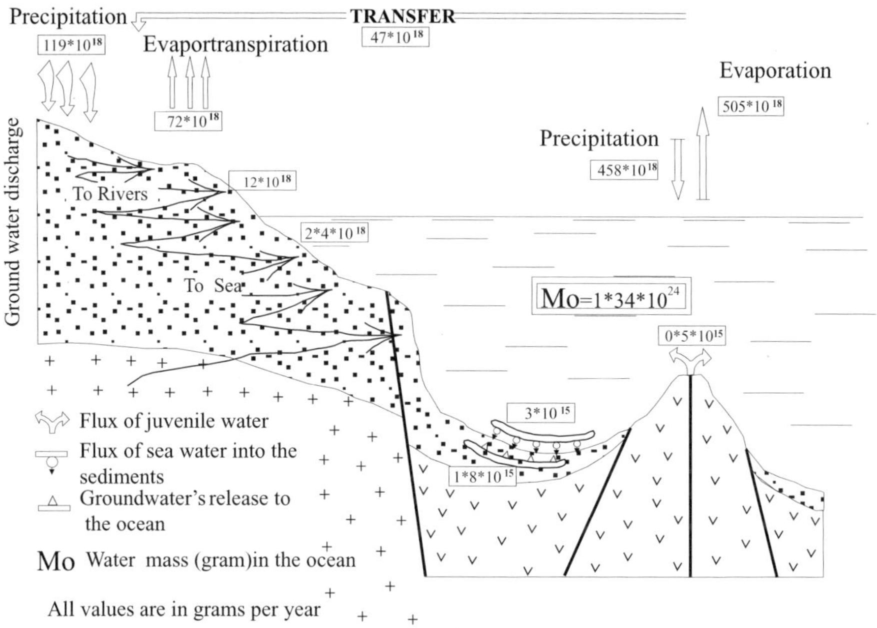

Fig. II.14.1. General representation of submarine ground water generation: Water exchange on atwater–bottom interface. After Djamalov, 1991.

pelagites (in pelagic areas of the ocean). The vertical fluxes of sediments create a universal pelagic background against which the formation of other types of sediments occurs (Murdmaa, 1987, p. 95). Hemipelagites are predominantly gray (reduced) sediments enriched to a certain degree in organic matter, the rate of accumulation of which is 5–100 times higher than that for pelagites. Pelagites, for the most part, are oxidized sediments accumulating with rates less than 1–10 mm/kyr.

The main mechanism that causes the deposition of hemipelagites is the settling of particles onto the seabed from the near-bottom water layer. These particles appear to enter the near-bottom waters, including those at depths of 4.0–10.0 km, predominantly not from the upper active water layer, but as the near-bottom nepheloid layer enriched in suspended matter. This layer forms mainly at the river–sea barrier and over the shelf edge. The near-bottom layer forces its way through hydrodynamic barrier present on the outer edge of shelf, crosses its border, and then moves down the continental slope. Only upon reaching the base of the continental slope, where this layer is already within the pelagic areas of the ocean, i.e., under more or less quiet hydrodynamic conditions, it extends from hundreds to thousands of kilometers away from the base of the continental slope, most commonly as the nepheloid layer. Such a mechanism of exporting fine muddy material from the river mouth to pelagic zone, without passing through the upper photic layer of the ocean, was observed in the World Ocean, including, for example, the Bay of Bengal (Gordeev, 1979), the Amazon (Svirenko, 1970, Emelyanov, Kharin, 1974; Monin and Gordeev, 1988), Congo (Romankevich, 1994), and Nile (Emelyanov, 1994_1) rivers. In many areas of the World Ocean, including those far from the mouths of large rivers, "clouds" of suspension have been observed almost everywhere above the continental slope (Lisitzin, 1964, 1979). From the mouths of small rivers or from abrasive coasts in shallow seas, where there are no submarine valleys and canyons, the near-bottom waters enriched in suspended matter appear to spread as a laminar flow. Even at small angles of bottom inclination (less than 1°), these waters cross a "transition zone" (at depths above the pycnocline) and fall to deeps, which play the role of settling basins (Emelyanov, 1968).

As revealed by the international expedition onboard the *Amas Seds* (Kineke and Sternberg, 1995; Alison et al., 1995) at the Amazonian River mouth there is a huge mass flow (near-bottom, several meters thick) which transports fluid mud suspensions (the concentration of suspended matter is >10 g/l) across the interior and middle parts of shelf areas to the northwest of the mouth of the Amazon River. The area of shelf covered by such a flow is 5700 km^2 when this flow is small, and it reaches 10000 km^2 when this flow is strong (Kineke and Sternberg, 1995). The near-bottom mud-rich flow coincides with the near-bottom salinity front. The mud flux is entrained by saline near-bottom waters, which thus play the role of a sediment trap. This flux moves not only along the coast, but also across the shelf. When the intensity of this flux is high, the volume of mud transport over the shelf is equal to the annual input of solid matter from the Amazon River (1.2 billion tons per year), and more than 90% of this solid matter flux is transported across the shelf as fluid mud (Kineke and Sternberg, 1995). Above the salinity front, the sedimentary material is transported as a suspension, and near the bottom, as a mud flux. Thus, for example, at a depth of 20 m, the upper 18 m are turbid waters strongly enriched in suspended matter. Below this level, the concentration of suspended matter rapidly increases (this is a so-called lutocline), and the lower 1.5–2 m is underwater mud flow. Below this mud flow, there is stationary fluid mud (or interbedding of plasmatic mud about 0.5 mm thick), which is underlain by viscous sediment (Allison et al., 1995) (Fig. II.1.6).

Clouds of suspended matter are characteristic not only of the continental slope but also of deep basins. It is known as nepheloid layer (Biscaye and Eittreim, 1977). This is why "pelletic transport" is not a necessary means of delivering fine terrigenous suspension to deep oceans (4–6 km), as is now widely accepted (Lisitzin, 1983, 1986). Terrigenous material reaches the near-bottom layer without passing through the photic layer, zooplanktonic biofiltering of this material and bottom filter-feeders also are not capable of completely filtering this material because the density (and biomass) of these organisms rapidly decreases toward pelagic zones.

Recent observations have shown that in oceans, in the near-bottom water layer, the velocity of bottom currents is most frequently within the range from 2 to 10 cm $\times s^{-1}$, but periodically the velocity increases up to 20–30 cm $\times s^{-1}$ or even more. Due to the effects of "bottom friction", the velocity of near-bottom currents is diminished and turbulence becomes stronger (than that in free or mean currents) (Craig, 1979, p. 547). Such a near-bottom current is called the near-bottom boundary current. It is 5–30 m thick (Wimbush and Munk, 1970; Craig, 1979). Within 1 m above the bottom surface, the current velocity decreases logarithmically, from the average current velocity (U) to the friction velocity at the bottom–seawater boundary (Figs. II.14.2, II.14.3).

Depending on roughness of the bottom topography, the hydrodynamic conditions near the bottom can be divided into three main subdivisions: "quiet", "transition", or "rough". Under quiet conditions, a thin (about 2 cm thick) viscous layer (where the average velocity of flow is 3 cm × s^{-1}) does not allow the edge of a turbulent eddy to touch the bottom. Under "transition" and "rough" conditions, when the scale of bottom features become comparable to the thickness of the viscous layer, the turbulent eddy comes in contact with the bottom, with the result that the sediments become exposed to an increased shear stress. In the equatorial part of the Pacific, the typical roughness of the bottom surface is 2–3 cm. This is due to the presence of FMNs which cover the ocean floor there. Due to these factors, the velocities of the bottom currents are 0.7–10.2 cm × s^{-1}, on average (Craig, 1979, p. 547).

Under conditions of rough bottom surface (clay and FMNs), erosion of sediments begins to take place at V = 48.2 cm × s^{-1} (Lonsdale and Southard, 1974). According to Craig (1979, p.548), velocities of current ranging within V = 18.5–37.0 cm × s^{-1} are quite sufficient to start the process of erosion of a rough surface. These velocities are commonly higher than those recorded by direct measurements. In the equatorial part of the Pacific, velocities of the bottom currents vary commonly within 2 to 3 cm × s^{-1}, and only during "bottom storms" do they reach 10–25 cm × s^{-1} (the cause of these storms are phenomena such as ebbs, tides, tsunamis and other events). Settling of clay particles on the bottom takes place even in the presence of current velocities as large as 7.8 cm × s^{-1}; therefore, a decrease in current velocities to 2–3 cm × s^{-1} leads to accumulation of clay onto the bottom. A proper understanding of processes of sedimentation and redeposition in ocean pelagic areas are described by Noorang, (1972), Murdmaa (1987), J.Kennett (1987), V.Svalnov (1991) and many other scientists.

Cold bottom currents (typically contour currents) flowing from high latitudes towards the equator keep a great amount of sediment in a suspended state (or even rework the upper sediment layer). Nevertheless, when a current meets a topographic barrier on its way, it loses kinetic energy and "excess" sediment load is deposited. In such places, rapid deposition of sedimentary material onto the bottom occurs. This leads to thickening of sediment layers of the same age. It is this mechanism which accounts for the development of contourites, which form huge sedimentary bodies called as "outer ridges" of oceans (Heezen and Hollister, 1971; Hollister and Heezen, 1972). These ridges extend hundreds of kilometers, and their thickness is up to 1–2 km. These ridges are distinct in the western part of the North Atlantic (the Blake Bahamas and the Great Antilles outer ridges), where the Labrador Current meets the Antarctic near-bottom current.

Marks on the ocean floor left by the action of bottom currents are also found in pelagic areas of oceans, i.e., in areas at great distances from continental slopes. Most commonly these features are "tails" of sediments redeposited from the slopes of seamounts, island slopes, and bottom elevations. Here, the rates of accumulation of these sediments are somewhat higher than those in areas well away from the feet of seamounts and hills on the ocean floor. Increased contents of biogenic carbonates (and, in some cases, of volcanoclastics) also occur in the sediments of such tails.

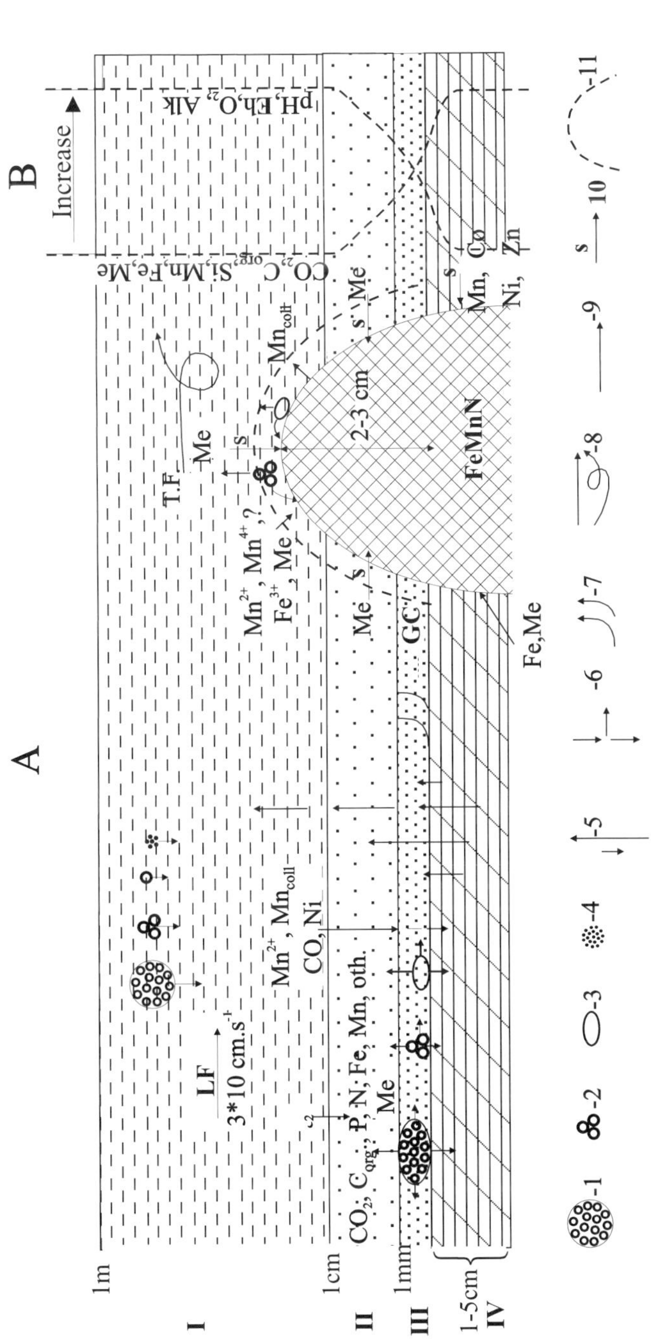

Caption see page 368

In many areas of the open ocean where terrigenous muds cover the bottom, waves are observed on the ocean floor; the length of these waves is some tens of kilometers, the height of their crests is up to 50–60 m. The amount of such waves is largest where there are strong near-bottom currents, primarily, the southern part of the Argentine basin (Flood et al., 1993), Cape basin and areas in the western equatorial Atlantic between the equator and 10°S and the Amazon River fan (Droz et al., 1996).

The main exchange between the bottom waters of the North and South Atlantic (Fig. II.14.4) occurs at the boundary between the Brazil and Guyana basins (Warren, 1981; Molinari et al., 1992; Rhein et al., 1995; Hall et al., 1997; Zangenberg and Siedler, 1998), or, more precisely, in the bottleneck between the submarine slope of northeast Brazil, near the port of Recifi, and the mountains of the western flank of the Mid-Atlantic Ridge, i.e., on a profile stretching approximately from Recifi (M mountain)–Fernando-de-Noronha Island–RM mountain on the eastern flank of the Mid-Atlantic Ridge (05°00′S, 35°00′ W–01°33′S, 30′00′W) (Fig. II.14.5).

Three main channels have been identified on the submeridional (M–RM) profile: (1) the Equatorial Mid-Ocean Channel (EMOC) (Fig. II.14.6), the relative depth of channel floor being 149 m (depth 4753 m) (Fig. II.14.7); (2) the Akademik Ioffe (AI) channel situated in the southern part of the profile and having a relative depth of 116 m (132 m relative to the northern wall;depth 4892 m); and (3) the Brazil channel (B) in the southernmost part of the profile, situated somewhat north of point 587 and having a relative depth of 130 m (water depth 4855 m) (Fig. II.14.8). The remaining channels are significantly smaller and have relative depths of 30–40 m. Thus, channel AI (4892 m) is the deepest in the study area. Its depth is very similar to that of trench in the north NT.

EMOC lies in the highest part of the ocean floor on the M–RM profile. This elevated part of the ocean bottom on the M–RM profile represents the Central

Fig. II.14.2. Processes at water–bottom GBZ in pelagic area of ocean, below CCD level (sediments containing FMNs). According to data of Bischoff and Piper (1979), Boudrau and Guinasso (1982), and Emelyanov (1982, 1986), with additions and corrections of author.
A—Bedding of a 1-m-thick near-bottom water layer and processes of elemental exchanges; I—a 1-m-thick near-bottom water layer; II—a 1-cm-thick near-bottom viscous layer; III—a 1-mm-thick near-bottom diffusion (turbulent) layer; IV—uppermost 1–5 cm of muds (clay).
1—pellets; 2—shells of planktonic foraminifera; 3—skeletons of diatomic algae and/or organisms with siliceous skeletons (radiolarians, dino- and silicoflagellates); 4—coccoliths, terrigenous (clayey) particles; 5—direction of prevalent sediment flux; 6—flux of chemical elements as a result of dissolution of biogenic material; 7—interference of light at a between layer interface (at V = 1 cm/s); 8—flows of near-bottom water (speeds are about 3–10 cm/s); L.F.—laminar flow; T.F.—turbulent flow; 9—diffusive elemental exchange at boundary; 10—sorption of microelements (Me) from seawater and interstitial water; 11—upper (diffusional) water microlayer around FMNs.
Mn_{coll}—colloidal Mn occurs in seawater (according to Tikhomirov, 1986). All major chemical elements participating in exchange at water–bottom boundary are represented.
GC—gradient of concentration of matter (a 0.15-mm-thick near-bottom water layer accounts for 90% of this gradient).
B—schematic distribution of hydrochemical characteristics and chemical elements at water-bottom boundary (Me—microelements).

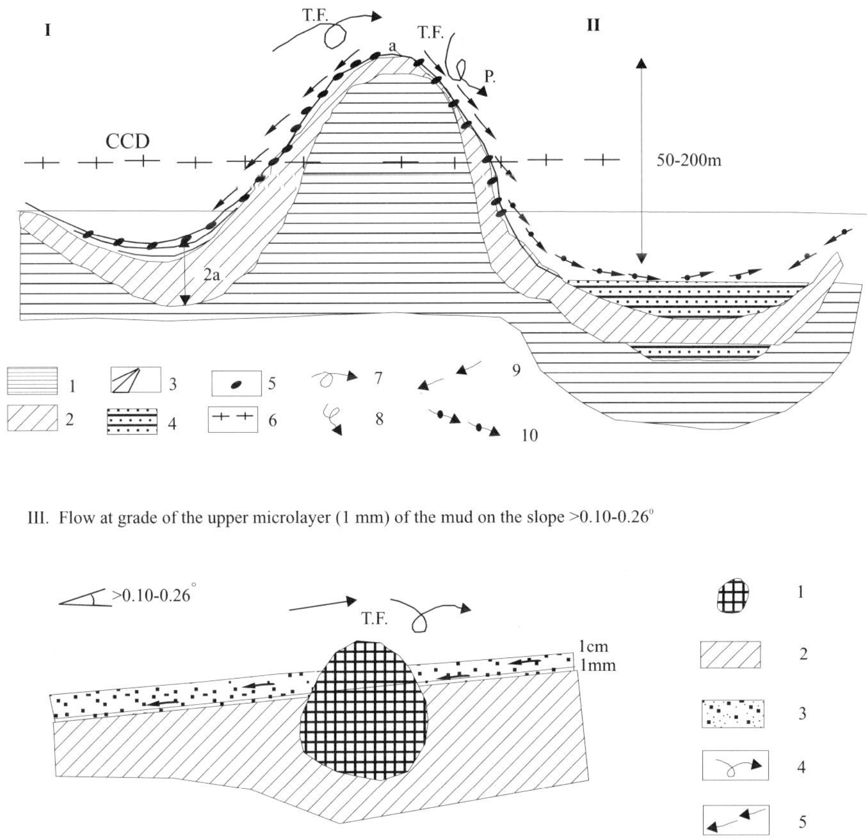

Fig. II.14.3. Facial distribution of FMNs in some pelagic oceanic deeps and downward motion of uppermost microlayer of muds.
I—situation characteristic of Cape Basin and/or basins of Atlantic Ocean; II—situation characteristic of Cabo Verde Basin; III—situation characteristic of World Ocean basins with an inclined bottom topography (angle of inclination is more than 0.10–0.26°) (below CCD).
For key to I and II, see below:
1—old sediment; 2—relatively recent sediment; 3—recent sediment formed as a result of downward flux of uppermost paste-like film of mud (<1 mm, see facial situation III); 4—turbidites and muds formed by downflowing mud (see conventional sign 3); 5—FMNs (below surficial FMNs, small FMNs are buried); 6—CCD; 7—near-bottom turbulent flow (speed is 3–13 cm/s); 8—strong near-bottom turbulent flow (TF) resulting in dilution (D) of bottom sediment (bottom storms); 9—downslope laminar flow of uppermost (<1 mm) semi-liquid (paste-like) mud; 10—flows of turbidites.
(a—thickness of recent sedimentary layer on hill; 2a—same in between hill space).
III—downflowing of uppermost microlayer of mud in area where bottom is a plane, gently sloping feature, with slope angle more than 0.10–0.26^0. 1—FMNs; 2—bottom (mud); 3—diffusive turbulent near-bottom water interlayer (thickness is 1 cm and 1 mm); 4—turbulent flow of near-bottom waters (TF); 5—downflowing of uppermost (paste-like) mud.

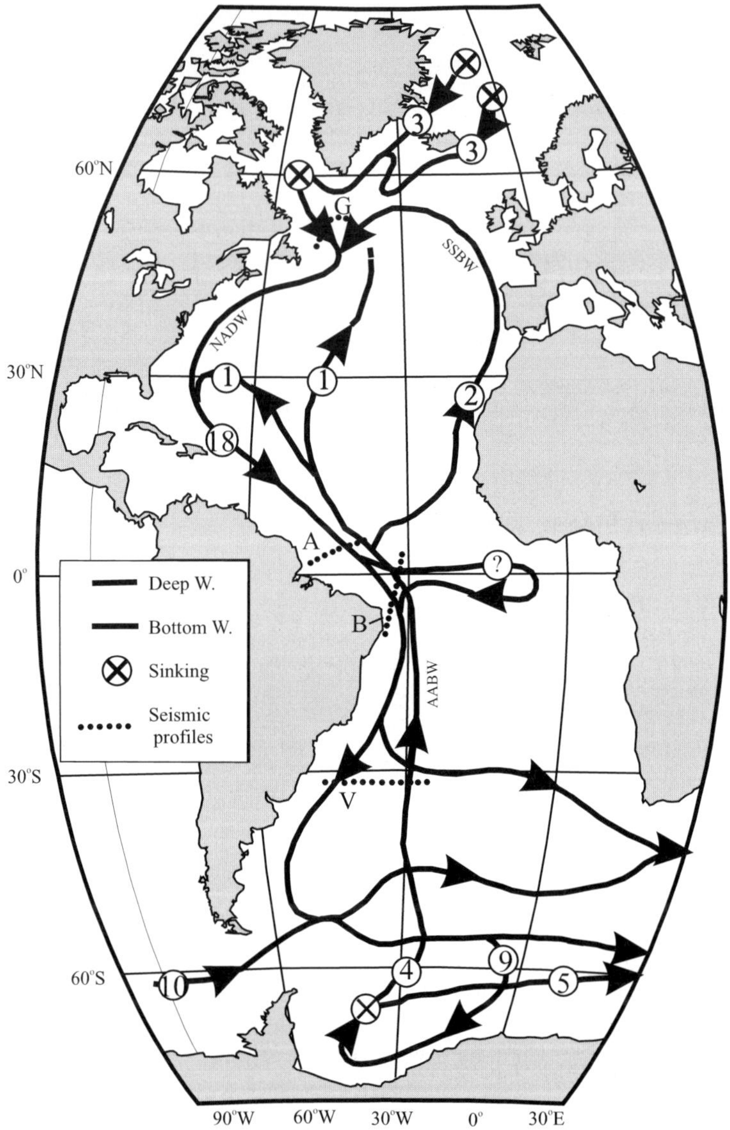

Fig. II.14.4. Modern Atlantic thermohaline circulation of deep and bottom water. After Sarnthein et al., 2001, p.367, with some modifications by the author.
AABW—Antarctic Bottom Water; NADW—North Atlantic Deep Water; SSBW—Southern Source Bottom Water.
Numbers denote water transport in Sverdrups (1 million m^3 per second)
Location of main seismic (Parasound) profiles are shown (see text) (G—Gloria, A—Amazon, B—Brazil, V—Vima)

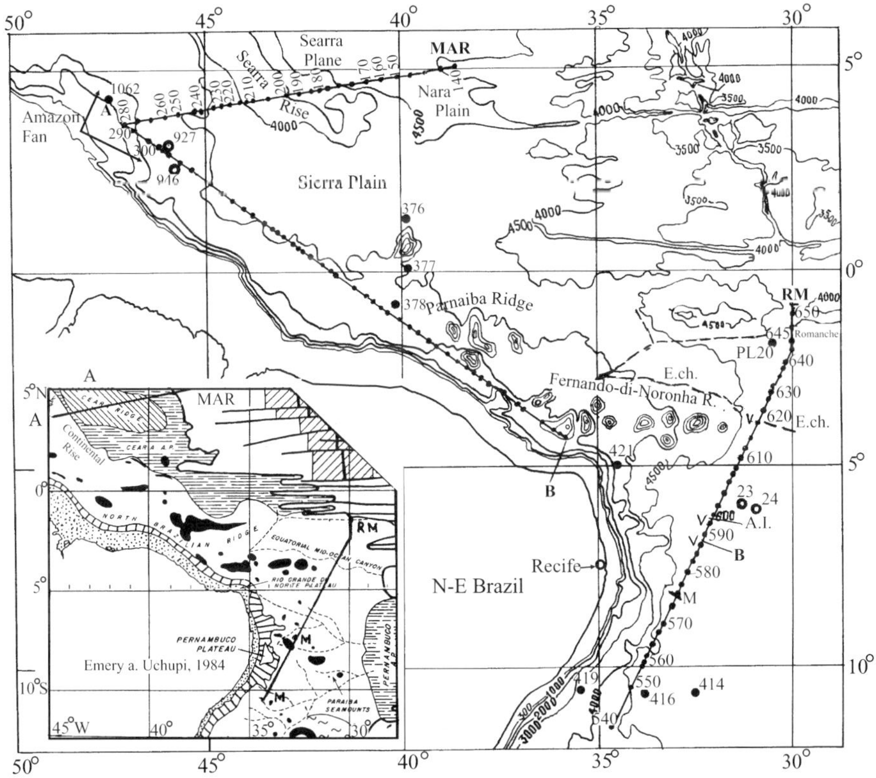

Fig. II.14.5. The location the parasound profiles, drill holes and geological stations in the western part of the the Equatorial Atlantic. Bathymetry (in m) after A. V. Yliyn, 1976.
A – MAR; A – B and M – RM-parasound profiles;

● 376 -geological stations;
○ 23 -drillholes

E. Ch.-Equatorial Mid -Ocean Channel;
number of points of parasound profiles

Cut-in A: Morphotectonics of the investigated region after Emery and Uchupi, 1984. The location of profiles A-MAR and M-RM are shown.

Sedimentary Swell (CSS) (Emery and Uchupi, 1984). The CSS is bissected by the EMOC. One can see on the M–RM profile that the CSS is 180–190 m in height. The EMOC is asymmetric: its northern wall is 59 m higher than the southern and has a depth of 4564 m (Fig. II.14.7). The northern slope of the channel gradually drops northward from its highest spot to 4940–4950 m (01°31′:947S; 29°58′:810W). The factor of the EMOC extending over strong swells clearly distinguishes it from channels AI, B and others, which have no such swells in their territory.

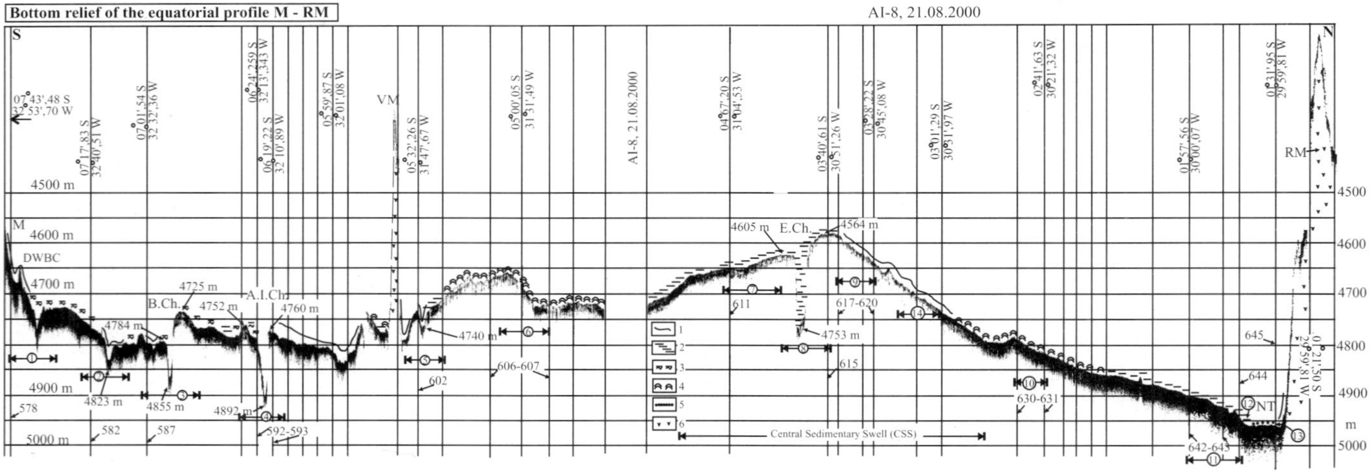

Fig. II.14.6. Bottom relief of Western Equatorial Atlantic on submeridional profile M (mountain)—RM (Ridge mountain). Depth in m. VM—volcanogenic mountain; E.Ch.— Equatorial Mid-Ocean Channel (EMOC); A.I. Ch.—AI (Akademic Ioffe) Channel; B.Ch.—Brazilian Channel; Numbers in circles—sites on profile for which sedimentary strata are shown on separate figures; numbers without circles—sites for which coordinates are given.

1–6—types of bottom surface: 1—bottom surface is rough, bottom strata hard; 2—bottom surface is smooth, bottom strata banded; 3—intermediate bottom surface; 4—sedimentary waves on bottom; 5—layered sedimentary strata with turbidities; 6—volcanogenic mountains.

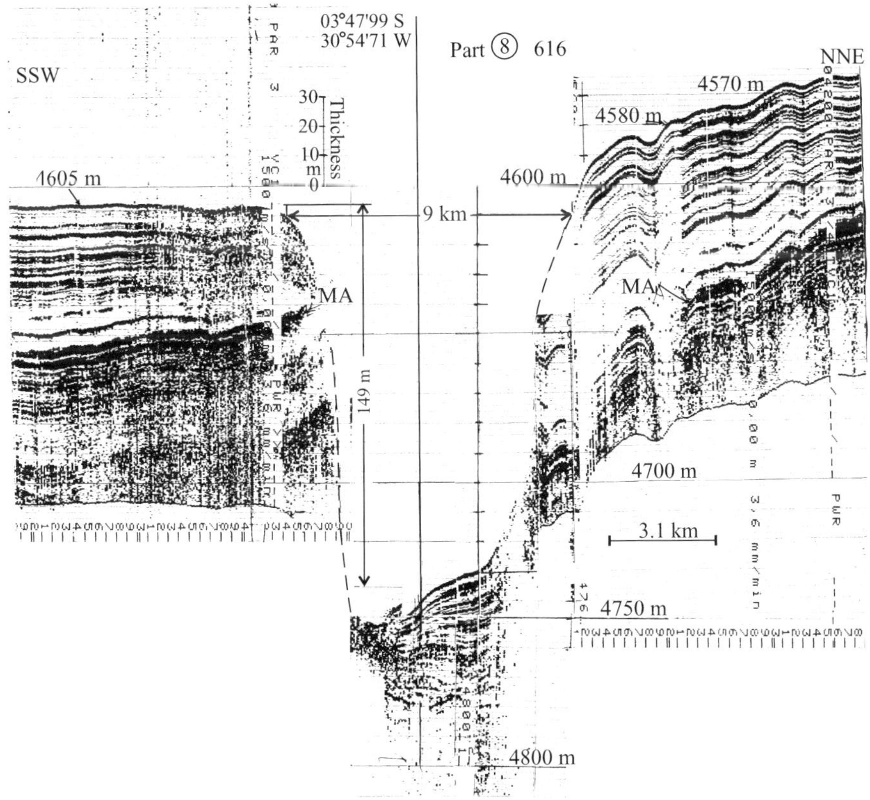

Fig. II.14.7. Equatorial Mid-Ocean Channel (EMOC) between North and South Atlantic at part 8 (points 615 and 617) of M–RM profile (see Fig. II.14.6.). MA—proposed border of Amazon Messinian. Area Q of cross-section of Channel is 0.780 km^2.

Deep channels (or abyssal canyons) have not been found on the sublatitudinal A–MAR profile (Fig. II.14.9). The deepest point of the profile (4765 m) lies between volcanic mountains (05°13′N, 38°45′W). In general, the Nara plain lies at depths of 4660–4640 m (Fig. II.14.10), i.e., about 100 m shallower than those in the southern part of the M–RM profile, and 300 m shallower than the passage near the base of the RM seamount in the northern part of this profile. The depth of Searra plain ranges from 4300 to 4500 m. On the A–MAR profile the maximum depth of this plain is 4423 m (see point 1540 in Fig. II.14.9). To the southwest of the Searra rise, which is a natural obstacle to the distal part of the Amazon River fan, depths as large as 4200–4150 m (Fig. II.14.9) exist on the A–MAR profile, and these depths are 400–450 m shallower than those for the Nara plain.

The almost horizontal flat bottom surface is typical only for restricted areas in the northernmost part of the M–MR profile (site 11), and also for the southern slope of the CSS (site 7, see Fig. II.14.6), where the EMOC is present. One can see

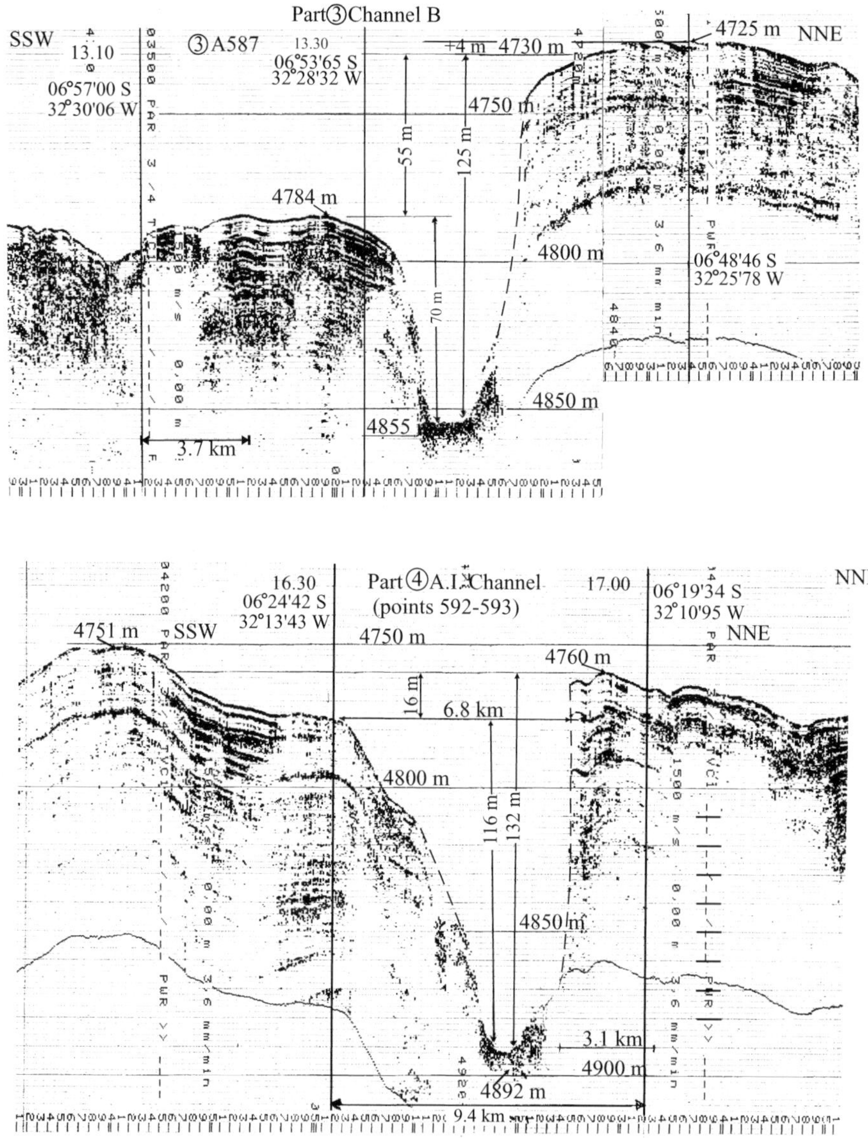

Fig. II.14.8. Profiles of B and A.I. Channels at parts 3 and 4 of M–RM profile (see Fig. II.14.6.).

that the floor of the northern trough (NT) is almost horizontal at a depth of 4958 m) (Fig. II.14.11). The floor of the trough is about 12 miles wide (sites 12 and 13) and a 2- to 3-m deep erosion channel occurs near the foot of the RM seamount. The southern slope of the CSS is also nearly horizontal. The bottom surface of the southern slope of the CSS is flat, with slight undulations.

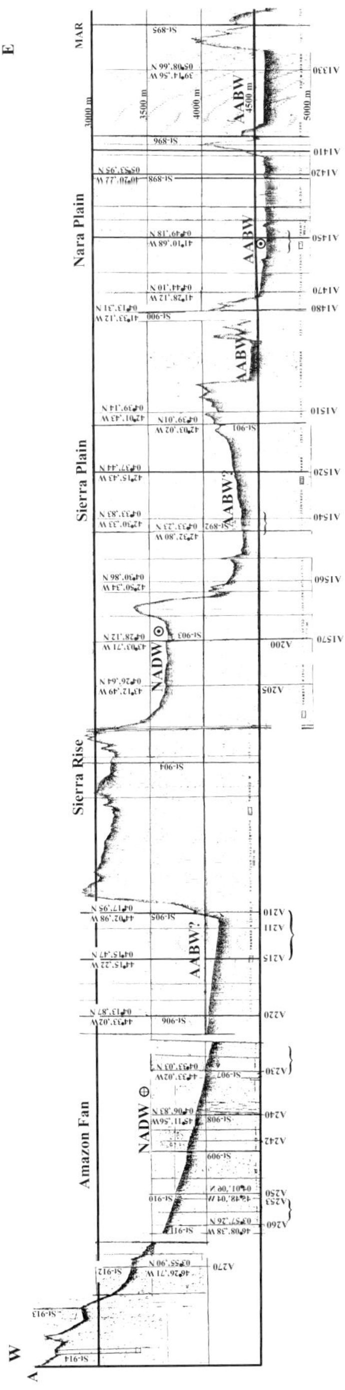

Fig. II.14.9. Bottom relief on longitudinal A–Mid-Atlantic Ridge profile along 5–4°N

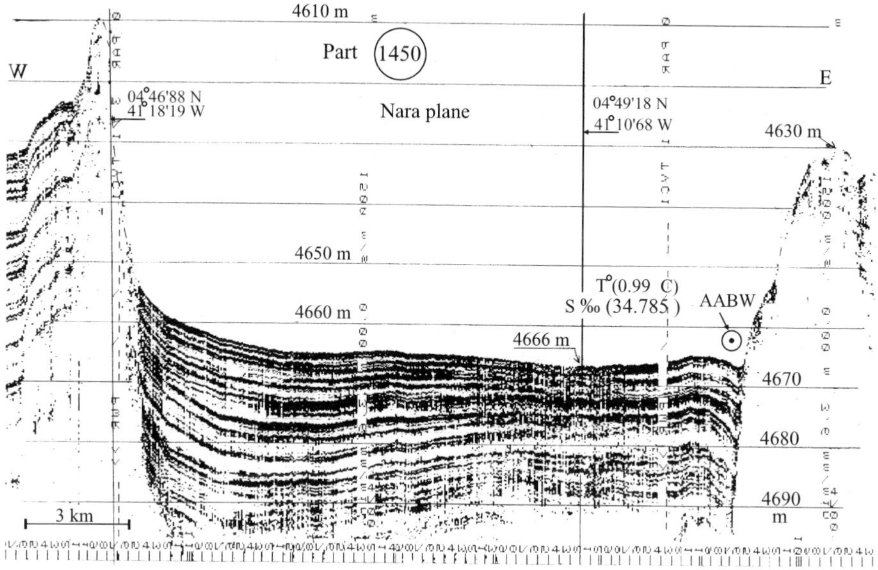

Fig. II.14.10. Structure of sedimentary strata in Nara Plain (point 1450, see Fig. II.14.9.). There is sedimentary evidence of AABW stream with lowest temperature and salinity in left-hand part of Nara Plain. Surface of Sierra Plain bottom is rough; layering of sediments is well expressed.

Two areas on the seafloor in the profile have slightly undulating surface: one on the northern slope of the CSS at medium depths from 4750 to 4850 m (site 10, Fig. II.14.6), and another on the hill in the central part of the profile (site 6). A single wave is 1-2 m high, and about 1 km wide.

The sedimentary sequence recorded at site 7 comprises 25–30 reflectors. The stratified structure is especially distinct in the uppermost 45-m layer (16 boundaries). The thickness of individual interbeds varies from 1 to 3 m. At depths greater than 45 m below the seafloor, reflecting boundaries are not continuous: in places they are quite distinct, but at other sites they are broken or completely indistinct.

A very pronounced acoustic boundary was recorded at 80–85 m. This boundary consists of several thinner reflecting boundaries, which are not always visible in records.

The structure of sedimentary strata described for site 7 is the same for site 11 (Fig. II.14.6), for elevated walls of channels B, EMOC, and also for the northern part of site 5.

At sites 6 and 10 (Fig. II.14.6), the bottom topography of which is represented by undulating areas, acoustics signal extended to a depth of 30 m. These sites also show signs of layering, and wave features are found throughout the entire 30 m of the sediment. There are 12–14 reflectors in these sediments. We suggest that this stratified 30-m-thick sedimentary sequence present at sites 6 and 10 stratigraphically

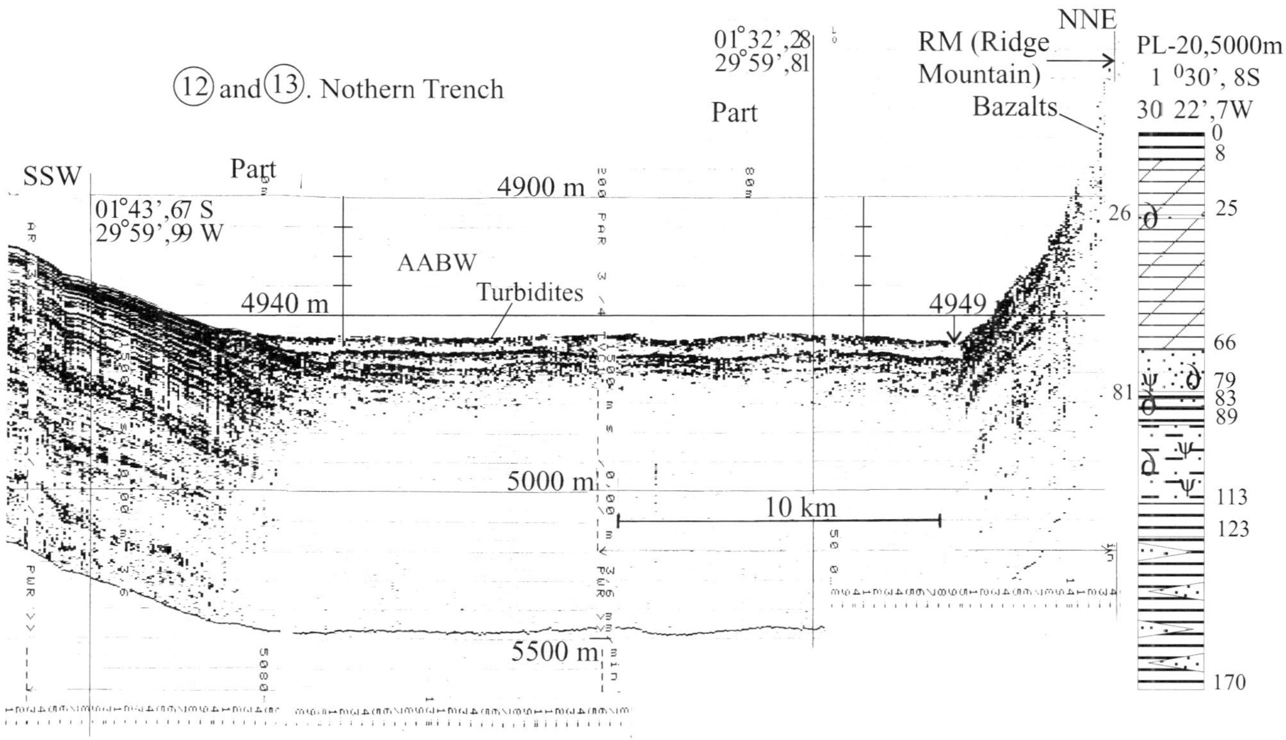

Fig. II.14.11. Structure of sedimentary strata in western part of Romanche Trench (or Northern Trench—NT) at parts 12 and 13 (see Fig. II.14.6). Core PL-20 was taken in the trench at a depth of 5000 m about 23 miles to west of point 645 (see Fig. II.14.5) or part 13 (see Fig. II.14.13). The core consists of terrigeneous turbidites with plant remains. For legend see Fig. II.14.13

Fig. II.14.20. Distribution of biomass of benthos (g/m^2) in World Ocean. After Zenkevich et al., 1971.

Table II.14.1. Accumulation of products of some filtrators. After Lee and Schwartz, 1980)

Species	Rate of accumulation	
	Individual, mg/ind/day	Bulk sedimentation, g/m²/yr
Cardium edule	648	(fecal)
Crassostrea virainica	231/560*)	2.8 kg (fecals and pseudofecals)
Geukensia dimissus	235	549 (fecals pseudofecals)
Pylynesolaea roliniana	30	1.1 (fecals pseudofecals)

*) Numerator – average; denominator – maximum depths; ind – individual

Carnivorous benthos-eaters, which feed on living organisms, are also responsible for disturbance and mixing of the upper several centimeters of sediments. This group of benthophagues comprises some species of starfish and polychaetes, squids, crabs, crustaceans, and ophiuroids. When they are in danger, they bury themselves into the bottom sediment, which is thus disturbed.

Many types of fish also act in a manner that disturbs sediment, with their fins, tails, and, occasionally, teeth. Some predatory species prefer to lie in wait for victims, burying themselves in sediment. Some other types of fish and octopi transport particles of sediments with sizes ranging up to pebbles, in order to construct their dwelling places.

Holothuria also intensively rework bottom sediments (for example, *Cucumaria abysoorum*). As a result of their biological activity, there are deep furrows and coils of holothurian feces on the sediment surface. Some types of sea urchin do the same. The width of a furrow produced by a species corresponds to the width of the

Table II.14.2. Velocity and depth of bioturbation and reworking of sediments by some species of benthic organisms. After Lee and Schwartz, 1980.

Species	Velocity of the reworking, gr/cm²/yr	Depth of reworking
Annelidae *Clymenenella torguata*	73.062	20 cm
Pelecypoda *Macoma baltica*	415	Surface
Gastropoda *Hydrobia minuta*	26–8871	2 mm
Crustaela *Paraphoxus ipinosus*	54.4–2.159 (kg)	-
Echinodermata *Liptosynapta tenuis*	585–2.924	1.15 cm

and (2.2–6.4) × 10^{-12} g × cm^{-2} × s^{-1}, respectively. Maximum flows are in shelf area and minimum flows in the basin.

Along with organic matter, SiO_{2am} and $CaCO_4$, trace elements also precipitate onto the ocean floor. The flux of these elements is related to the primary production and dissolution processes, which in turn depend on sedimentation rates. The release of metals captured by biogenic matter in the photic zone (organic detritus, fecal pellets, calcareous material) occurs predominantly at the bottom, at depths below the CCD level. These are the mechanisms (elements which are captured by sedimentary particles in the photic layer are released at a depth below the CCD level) whereby much Fe and trace elements are delivered to the bottom. Much of the Mn mass is supplied in the same manner. Then this element may be released in the form of colloidal Mn (Tikhomirov, 1986) into the bottom water microlayer to be catalytically oxidized and finally deposited on the active surfaces of hydroxide phases.

In shallow-water areas of seas (Kiel Bay in the Baltic Sea), out of the total amount of C_{org}, N, and P that reaches the bottom, only 22% C_{org}, 13%N and 15% P appear to be buried in bottom sediments at depths greater than 10 cm below the seabed (Balzer, 1984) (Table II.14.4). At the same time, 75% C_{org} and 66% P return from sediments into the bottom water. As for nitrogen, about 50% of the total amount of this element comes back into the bottom water, and the remaining 50%, it is thought, is lost due to denitrification. In Kiel Bay, a rapid decrease (20–80 times) of biogenic components in interstitial waters in mud at a water depth of 20 m (the pycnocline layer in this bay is at a depth of 17–23 m), commences commonly at a depth of 20 cm below the sediment surface (Balzer, 1984). As for SiO_4, the strongest gradient of this constituent is found within the 0- to 10-cm layer of sediments. In Kiel Bay, at depths of 20–28 m, the concentration of SiO_4 in interstitial water decreased from 120–900 μM at a depth of 10 cm to 30–50 μM at a depth of 1 cm. The concentration of SiO_4 in bottom waters was essentially similar to that in waters at a depth of 1 cm within the sediment (Balzer et al., 1988).

If sediments are rich in organic matter (if they contain more than 0.5% C_{org}), then under reducing conditions there are gas fluxes: CO_2, H_2, N_2, H_2S, which pass from entrails to bottom waters; and CH_4, CO_2, H_2, N_2, H_2S, etc., from the upper layer of bottom sediments (Fig. I.24).

Table II.14.4. Decrease in concentration (μM) of biogenic components in interstitial waters of the upper (20 cm) layer of mud in Kiel Bight (average data). After Balzer, 1984

Component	Sediments (layer, cm)		Near-bottom water
	20	5	
PO^{3-}_4-P	25	15	3
NH^+_4-N	180	50	5
N : P	8	5	4
SiO_4 -Si	450	400	50

Under reducing conditions, within the zone where sulfate-reducing reactions with precipitation of FeS occur, the Fe flux from sediments into near-bottom water amounts to about 733 μmol $\times$ m^{-2} $\times$ day^{-1} (Balzer, 1982, p. 1159). This flux is considered to be very intensive. For example, in the sediments of Puget Sound, such a flux was only 12.8 μmol Fe $\times$ m^{-2} $\times$ day^{-1}. If the sediment is overlain by an oxidized layer, then Fe flux from this sediment becomes impossible.

The Mn flux from sediment to water is 362 μmol $\times$ m^{-2} $\times$ day^{-1}, on average (Balzer, 1982, p. 1159). However, this value depends on the Mn content in the sediment, oxygen concentrations, and variations in pH values. Under reduced conditions, the Mn flux increases to 1127 μmol $\times$ m^{-2} $\times$ day^{-1}. As reported by Graham et al. (1976), the Mn flux from sediments to bottom water (oxygen zone) is 364 μmol Mn $\times$ m^{-2} $\times$ day^{-1}.

In Chesapeake Bay, this flux amounts to 300 μmol Mn $\times$ m^{-2} $\times$ day^{-1} (Eaton, 1979). Under stagnant conditions, the Mn flux from sediment to water varies from −600 to +400 μmol $\times$ m^{-2} $\times$ day^{-1}, or even up to 7000 μmol $\times$ m^{-2} $\times$ day^{-1} (Eaton, 1979).

In the shelf areas near the mouth of the Congo River, the Mn^{2+} flux was directed from the bottom waters to sediment and was (111–115) $\times$ 10^{-14} g $\times$ cm^{-2} $\times$ s^{-1}. Conversely, in the deep areas of the Angola basin, the Mn^{2+} flux occurred from sediments to bottom water. The rate of the Mn^{2+} flux from the sediments was (9.5–89.2) $\times$ 10^{-14} g $\times$ cm^{-2} $\times$ s^{-1}. (Romankevich, 1994).

Although the rates of Fe and Mn reduction are very considerable, only small portions of Fe^{2+} and Mn^{2+} formed during early diagenesis are buried in the sediments (Cappellen and Wang, 1996). A major portion of dissolved Fe^{2+} and Mn^{2+} is again oxidized in the upper film of the sediments or is diffused to the bottom water.

Fluxes of elements are especially variable in deep basins periodically exposed to hydrogen sulfide contamination (Baltic Sea). Here, up to one million tons of dissolved Mn appear to move from sediments to the bottom water in a matter of months (or several days or several tens of days?) (Blazhchishin and Emelyanov, 1977; Emelyanov, 1981_2, $1982_{1,2}$).

Under reducing conditions (at Eh <+0.1V), As (VI) changes to As (IV), resulting in precipitation of As from the porous solution (Volkov, 1980). Arsenic is commonly found in iron sulfides (pyrite and greigite).

In waters with H_2S, the molybdenum ion MoO_4^{2-}, the form in which it occurs in normal seawater, passes to tioxymolibdenum. In reduced sediments, coprecipitation of molybdenum is described by the formulas

$$FeS.nH_2S + MoS^{2-}_4 \rightarrow 6\ FeS.MoS_3 \times nH_2S \qquad \text{(Volkov, 1980, p. 155).}$$

In muds of the Black Sea, molybdenum is found as $MoSO_3$. The uranium, like Se, As and Mo, also accumulates in reduced muds. In the bottom water, U (IV) exists as a stable three-carbonate complex. Precipitation of U from solution into the solid phase is described by the reaction

$$U\ (VI) \rightarrow 6\ U\ (IV) \qquad \text{(Volkov, 1980, p. 156).}$$

The cited elements (for example, in deeps of the Baltic Sea), V and Cr also accumulate in reduced muds. This is because V and Cr compounds become less soluble, as a result of the changes in their valence state under oxidation and reduction conditions, which are as follows:

$$V (V) \rightarrow V (IV);\ Cr (VI) \rightarrow Cr (III)$$

The main reason for diffusion of As, Se, Mo, U, V and Cr from bottom water to sediments is the reduction and fixation of these elements in minerals of lower solubility [sulfides for Se and As, oxides for Cr, authigenic phosphates for U (IV) (Volkov, 1980, p. 156)]. Within an oxidizing environment, the sorption of elements onto terrigenous, biogenic and chemogenic particles prevail—in contrast to the situation in reducing environments, which is characterized by fluxes of Se, As, Mo, U, V and Cr from the bottom water to sediments.

Under conditions where the redox (Eh) barrier is slightly above the water–bottom interface (sediments are reduced), selenium (as selenate SeO_4^{2-}) diffuses from the bottom water to sediments because concentrations of this element in the bottom and interstitial waters are different. In the uppermost film, selenate enters selenite (SeO_3^{2-}), which is then rapidly reduced to elementary selenium; selenium passes to the solid phase, and the concentration of Se in interstitial waters decreases. Fixation of Se in the solid phase occurs under redox conditions that are "softer" for this element than for sulfur; this is why selenium accumulates in sediments earlier than sulfur (sulfides). As sulfide formation becomes more intensive, Se enters iron sulfides, isomorphically substituting sulfur in pyrite (Volkov, 1980, p. 154).

Se does not accumulate in oxidized sediments: it freely goes across the redox barrier from the side where positive values of Eh are high, then this element is reduced from Se (VI) to Se (IV) and combines with hydroxide as selenite. Thus, the enrichment of Se in muds occurs on the side of Eh barrier that is rich in Fe (III) compounds, being able to form combinations with selenite ions (Volkov, 1980, p. 155).

Research conducted on waters and sediments from Alaskan fjords (depth 90–280 m) made it clear that Cu is removed from seawater by being adsorbed by suspended matter, and then this element is transported by sinking particles to the bottom. In the upper layer of reduced sediments (0–7 cm), Cu becomes remobilized. About 20% of the Cu that settles onto the bottom is recycled again by being evacuated across seawater–bottom GBZ to bottom waters. Cu occurring in sediments below 7 cm is likely to be removed from interstitial waters as Cu sulfides.

Processes on the bottom surface at depths below the CCD level (strongly oxidizing conditions). Not only large-scale topographic features such as abyssal hills (Fig. II.14.3), but also the smallest irregularities on the bottom play an important role in the distribution of manganese nodules. The effect of small features on the seafloor (bottom roughness) has been studied by Boudreau and Guinasso (1982, p. 127). These irregularities include ripples, furrows of bottom organisms, manganese nodules and the grain size of sediments. If the bottom roughness is smaller than the "laminar" part of a viscous sublayer (see Fig. II.14.2), then features present on the bottom will have only a negative effect on bottom currents and the characteristics of sediment transport. When these irregularities are within the viscous sublayer, they are not able to generate turbulence strong enough to break a diffusive interlayer (Boudreau and Guinasso, 1982, p. 127). The thickness of the laminar layer typical for pelagic conditions in the ocean is about 0.9 cm (Figs. II.14.2, II.14.3). Because pelagic sediments are composed of fine-grained material, the grain size of the sediments involved do not actually affect turbulence. When both bottom irregularities and a laminar sublayer are present on the seafloor, turbulence appears to be random.

Dissolution of sediment at the bottom surface cannot be faster than the evacuation of dissolved products away (upward) from the seawater–bottom boundary; i.e., the rate of reaction cannot be faster than the rate of material transport across the diffusive sublayer. Boudreau and Scott (1978) suggested that the growth of manganese (hydrogenetic) nodules is caused by diffusion of metal ions from seawater to the surface of these nodules. These authors predicted the velocities with which metals diffuse from seawater and compared them with the growth rates of manganese nodules calculated from radiometric data. They proposed that not only Mn but also all trace elements can be delivered and precipitated on manganese nodules by means of diffusion from seawater.

As has been experimentally demonstrated (Tikhomirov, 1986), bottom waters contain not only dissolved Mn, but also its colloidal form. This colloidal form makes up about 30% of the total amount of Mn dissolved in seawater. The bottom sediments actively absorb such elements as Mn, Co and Ni from the bottom water, and to a lesser degree, Cu and Zn. Also, the same elements are sorbed from water and interstitial water by nodules.

Mn in the interstitial waters of reduced sediments accumulates in the course of the following reaction:

$$2MnO_2 + C_{org} + 2H_2O = 2Mn^{2+} + CO_2 + 4\,OH^- \text{ (reaction 1)}$$

(Savenko and Baturin, 1981; Baturin, 1986, p. 110)

In oxidized sediments (which experience exposure to oxygen), the course of reaction is as follows:

$$2Mn^{2+} + O_2 + 4OH^- = 2MnO_2 + 2H_2O \text{ (reaction 2)}$$

As a result, Mn precipitates from interstitial waters. At the same time, reaction (1) also takes place in oxidized pelagic clays, because in these muds (in their interstitial waters) a positive correlation between dissolved C_{org} and dissolved Mn has been revealed (Hartmann and Müller, 1974).

Fluxes of various metals from interstitial waters into bottom water have been the subject of research by many scientists (Manheim, 1976; Elderfield, 1976; Hartmann and Muller, 1982; Calendar and Bouser, 1980; Klinkhammer et al., 1982; Baturin et al., 1986). Baturin et al. (1986) describes the intensity of elements fluxes from sediments of different types.

The flux of element Q can be calculated by using the following formulas (Manheim, 1976):

$$Q = DDC/DX \times ts,$$

where D is the diffusion coefficient, DC is the difference in concentrations of interstitial and bottom waters; DX is the thickness of the sediment layer under consideration, t is time, and S is the surface area of interface.

In seas and oceans, variations in the gradient of an element concentration are within several hundreds to a few hundreds of thousands of times. For Mn, this gradient is the highest in the Baltic Sea: the Mn concentration here, at the upper boundary of the O_2–H_2S barrier, is 3–10 μg/l; below this barrier, 1000–2000 μg/l; in interstitial water, up to 30,000 μg/l. Tables II.14.5 and II.14.6 are good illustrations of the difference between concentrations of elements in deep waters of oceans and interstitial waters within sediments for the respective facial zones.

For mud and ooze of the pelagic (equatorial) zone of the Pacific, fluxes of elements from sediment to bottom water were estimated to be as follows (Table II.14.7).

In red clays, almost all Mn input remains deposited in these sediments, whereas in calcareous ooze, 40% of Mn input diffuses to bottom water. Only from 9 to 18% of Cu is fixed in the sediments, whereas the remaining 82–91% of this element returns to bottom water. Consequently, Cu migrates more intensively from the sediments into bottom waters than Mn. Mn diffusion increases with the transition from red clays to calcareous oozes. The explanation advanced for this is as follows. Calcareous oozes occur above the CCD, and consequently, their sedimentation rate is five to ten times higher than those for red clays. This causes calcareous oozes to become less oxidized and more porous, which enables Mn (and Cu) to accumulate more intensively in the interstitial waters of low-calcareous and calcareous oozes (occurring between the CCD and the calcite lysocline). The Cu flux from sediments into water is even more intensive than that from water into sediments (Boström et al., 1973).

As evidenced by Tsunogai and Kusakabe (1982), in open (pelagic) ocean areas Mn diffuses from sediment into bottom water. In general, the distribution of Mn in the ocean is balanced in terms of geochemistry, because its contents in the upper layers of sediments are higher than those in the lower layers. Calculations suggest that in the northwestern part of the Pacific, within the upper layer of red clays, the amount of Mn diffusing from sediments (namely, from the interstitial waters of the upper oxidized layer of sediments) to the bottom water layer is 1.5×10^{-5} μg $\times$ cm^{-2} $\times$ s^{-1}, and the Mn flux from underlying sediments (also within red clays) is 5.4×10^{-5} μg $\times$ cm^{-2} $\times$ s^{-1} (Tsunogai and Kusakabe, 1982).

In the equatorial radiolarian belt of the Pacific, manganese nodules are additionally fed by manganese coming to the water–bottom barrier from interstitial waters during early diagenesis. As a result, manganese nodules are markedly enriched in Mn: the Mn/Fe ratio for such nodules exceeds 2 and may reach 10 (Skornyakova et al., 1983).

At the water–bottom GBZ, Cu, Ni, Zn and Ag, which were delivered to the bottom by biogenic material, pass partially into gels. Because there are vast amounts of phosphates on the bottom and because their surface area is very large, sediments are become very good sorbents (Greenslate et al., 1973). REE and U are effectively absorbed by microcrystalline apatite.

Cu, Ni and Ag are fixed predominantly by proteins, and Zn appears to form salts with fatty acids in organic fractions of skeletal remains (Greenslate et al., 1973). On the basis of data published by the same authors, in addition to REE, bone phosphates actively accumulate Mo, Cr, Fe, Ni, Cu, Zn, Sn, Pb, Co, Zr, and Ti; according to other authors (see, e.g., Baturin, 1993) they also accumulate U and certain other elements.

Table II.14.5. The metals content in the near-bottom waters of the radiolarian zone of the ocean, $\mu g.l^{-1}$. According to varioss authors*) (after Baturin et al., 1986)

Sediments type	Mn	Fe	Cu	Ni	Zn	Ca	Source*)
Siliceous ooze	0.07	-	0.34	-	-	-	1
The same	0.02	-	0.3–0.4	0.4–0.5	-	0.07	2
Siliceous ooze and red clay	0.1–0.4	-	0.1–0.3	0.3–0.5	1.3–7.76	-	3
Clayey mud	0.07–0.20	0.09–0.70	0.05–0.70	0.3–0.9	-	-	4
Pelagic mud	0.05	-	0.30	-	-	-	1
Calcareous ooze	0.025	-	0.3–0.4	0.4	-	0.08	2
The same	-	-	1.6	0.5	1.3	-	3
Siliceous, clayey and calcareous oozes and muds	0.05–0.20	0.9	0.28–0.82	0.5	-	-	5
The zone at all (average)	0.2	0.9	0.3	-	-	-	-

*)1 – Callender and Bowser, 1980; 2 – Klinkhammer et al., 1982; 3 – Hartmann and Müller, 1982; 4 – Raab, 1972; 5 – after V. Gordeev (unpublished).

Table II.14.6. Comparison of metals content (μg.l^{-1}) in the ocean, near-bottom and interstitial waters (average, in brackets). According to various authors*) (after Baturin et al., 1986)

The water type	Mn	Fe	Cu	Ni	Zn	Co	Cd
Ocean	0.02	0.05	0.2	0.3	0.4	0.001	0.03
Near-bottom	0.02–0,4 (0.1)	0.09–12 (3)	0.05–1.6 (0.5)	0.3–0.9 (0.6)	1.3–7.7 (5)	-	0.07–0.08 (0.07)
Interstitial	0.04–50 (2)	1–100 (20)	1–30 (3)	0.5–30 (3)	2–100 (18)	0.05–3.6 (0.18)	0.19–0.27 (0.20)
Ratio near-bottom/ocean	1-20 (5)	2–200 (10)	0.2-8 (2.5)	1–3 (2)	1.6–10 (6)	-	1–10 (2)
Ratio interstitial/ near-bottom	0.4–500 (20)	0.3–30 (7)	2–60 (6)	0.9–50 (5)	0.4–20 (4)	-	2–4 (3)

*)Gordeev and Lisitzin, 1979; Boyle et al., 1981; Bruland, 1980, 1983; Knauer et al., 1982; Bewers and Windom, 1983; Mart et al., 1983; Moore, 1983; Wong, 1983.

According to calculations for pelagic areas of the Pacific (Hartmann and Müller, 1982, p. 297), if the sedimentation rates are about 1 mm/1000 yr, then the accumulation of Cu, Zn and Ni in the upper 2-cm layer of oxidized ooze lags behind the diffusion of these elements from this layer to the bottom water.

Table II.14.7. Concentration of metals in the interstitial waters of the pelagic sediments of the Pacific ocean. After Baturin, 1993, p. 68

Sediment type	Mn	Fe	Cu	Ni	Zn	Co
Concentration in the porous water, μg/kg						
Siliceous oozes	0.04–56.2	11–785	0.4–132	0.6–369	4.8–138	0.05–3.6
Siliceous-clayey mud	0.6–62	0.2–19	0.1–12	0.1–22	4.5–22	-
Clayey muds	80–100	5–90	10–60	5–20	-	-
Calcareous oozes	0.35–192	0.2–17.6	0.4–15	0.3–7.6	2.3–58	-
Average	2–10	1–20	3–15	1–3	5–18	0.18
Diffusional flux, μg/cm²/1000 y						
Siliceous oozes	20	-	140	3.5	-	0.3
Siliceous-clayey muds	100	-	270	80	580	-
Clayey muds	124	-	784	-	-	-
Average	40–70	8	14–260	0.4–20	9–290	0.3

The rates of diffusion for heavy and transition metals are comparable to the rates of precipitation of these metals onto manganese nodules. For instance, the rate of Mn diffusion from sediment is 180 μg × cm^{-2} × 1000 yr, and the rate of its precipitation onto manganese nodules is 190 μg × cm^{-2} × 1000 yr. Thus, the amount of Mn delivered to the ocean due to diffusion may be sufficient for oceanic nodules to grow (Hartmann and Müller, 1982, p. 298). The rate of diffusion for Cu, Ni and Zn from sediment to seawater is more intensive than their precipitation onto manganese nodules: input of these elements is characterized by redundancy.

The cited data indicate that the most important geochemical processes occur in the upper 1–2 mm of sediment (Hartmann and Müller, 1982, p. 298).

As mentioned earlier, the downward flux of elements to the ocean floor is nonuniform and depends on facies conditions; i.e., it is controlled by interactions of various GBZs within the upper part of the water column and on the bottom surface. In pelagic areas of the Pacific, the Mn flux from the water column to the ocean floor was found to be 100–2000 μg/cm^2/1000 yr (Boström et al., 1973; Elderfield, 1976; Krishnaswamy, 1982; Callender, 1980), whereas input of this element from sediments to the near-bottom waters is 1–80 μg/cm^2/1000 yr, i.e., 10–100 times less (Table II.14.8). So, much of the Mn remains in bottom sediments, constituting a reserve for subsequent (diagenetic) redistribution. Earlier, it was also believed (Manheim, 1976) that, on a global scale, the diagenetic flux of many elements from interstitial waters to bottom waters is comparable to their input coming from rivers.

The same can be said about diffusion of K and Mn from interstitial to bottom waters (Sayles, 1979). For Na, Ca, SO_4^{2-}, HCO_3^-, input due to this diffusion is 50% of that from rivers.

Drawing on the literature data, Baturin (1986) estimated the Mn flux from interstitial waters to be in the range 50–70 μg × cm^{-2}/1000 yr for pelagic areas of the WorldOocean, and from 5–10 to 100–400 μg × cm^{-2}/1000 yr for its peripheral areas. If we assume that the area of the pelagic World Ocean is 272.6 million km^2 and the area of its peripheral part is 87.6 million km^2 (Lisitzin, 1974), annual maximum fluxes of Mn from these zones would be 0.13 and 4.35 million tons, respectively (Baturin, 1986, p. 112).

It is evident that only part of the Mn diffusing from hemipelagic sediments of the World Ocean into bottom water passes to pelagic parts of the ocean (Sundby et al., 1981; see Baturin, 1986, p. 112). It is possible that a major portion of this element is finally buried in hemipelagic sediments or dispersed within the sedimentary sequence.

As a result of convection and advection in the bottom waters of the World Ocean, chemical elements diffusing from sediments to bottom waters may rise to a considerable height above the bottom and could thus provide for the enrichment of biogenic components in deep waters (Berner, 1980; Lerman a. Lietzke, 1977).

Sediments enriched in iron may accumulate at the water–bottom interface in pelagic areas of the ocean. The accumulation of Fe is going on under conditions where the Eh barrier is far below the bottom surface and the dissolution of biogenic particles on the bottom surface is intensive, especially in radiolarian–clay sediments. At very low sedimentation (<1 mm × 1000 yr), sediments containing up to 10–12% Fe may accumulate (Hein et al., 1980, p. 388). Similar iron-rich sediments that

Table II.14.8. Diffusional flux of metals from pelagic sediments into near-bottom water. After Baturin et al., 1986

Element	Region	Sediment type	Layer of sediment, cm	Porous/near-bottom water		Diffusional flux, µg.cm^{-2}/1000 yr	Sources*
				Gradient of concentration	Coefficient of enrichment		
	Ocean (in total)	-	100	1000	1000	44	1
	The same	-	20	-	-	70	2
	Radiolarian zone of Pacific Ocean	Siliceous-clayey oozes	2	3.5	19	180	3
Mn	The same	The same	10	2.6	40	11–32	4
	"	Siliceous oozes	12	1.8	30	9–40	4
	"	The same	8	0.02	2	0,8	4
	"	Calcareous oozes	1	1.4	23	124	4
Fe	Ocean (in total)		100	17	6	8	1
	Ocean (in total)	-	100	4	5.5	14	1
	Radiolarian zone	Siliceous oozes	2	4.6	28	230	3
Cu	The same	The same	1	3.9	13	124–492	4
	The same	Siliceous ooze	1	1.7	6	76–254	4
	The same	The same	2	2.7	9	114	5
	The same	Calcareous oozes	1	4.9	16	784	4
	Ocean (in total)	-	100	1	1.2	0.4	1
	Radiolarian zone	Siliceous-clayey oozes	2	1.7	4.6	80	3
	The same	Siliceous oozes	2	0.1	1.3	23–4.6	5
Zn	Ocean (in total)	-	100	20	5	9	1
	Radiolarian zone	Siliceous-clayey oozes	2	12	3.2	580	3
Cd	The same	Siliceous oozes	?	0.04	3	2	5
Co	Ocean (in total)	-	100	0.9	10	0.3	1

*) 1 – Manheim, 1976; 2 – Elderfield, 1976; 3 – Hartmann and Müller, 1982; 4 – Callender and Bowser, 1980; 5 – Klinkhammer et al., 1982.

accumulated under the influence of active dissolution of SiO_{2am} were discovered in core no. 159 of DSDP (Pacific Ocean). The formation of such sediments requires long hiatuses (not less than 1–2 million years).

Erosion, hiatuses and lithification of bottom sediments. Outcrops of ancient (commonly pre-Quaternary or even Cretaceous) basalts, serpentinites or sedimentary rocks on the tops of seamounts, including guyots, are evidence that the time span during which these areas are exposed to the effects of strong hydrodynamic activity is on the order of several hundreds of thousands or even a few tens of millions of years. The existence of strong bottom currents hugging the tops of seamounts is evidenced by such features as (a) relict, commonly coarse, sediments; (b) ripple marks on the sediment surface; (c) recent or ancient coral buildups; d) other evidence, including current velocities measured above the tops of seamounts. As a result, thick sedimentary strata develop around mountains, whereas their tops and slopes frequently remain bare.

If the top of a seamount is in the upper active layer, it is densely populated with algae and living organisms, and sediments enriched in organic matter accumulate in small bottom depressions. When such a top of a seamount is in an arid climatic zone where waters are supersaturated with respect to calcite, crusts are formed on its slopes and sediments are lithified to form hardgrounds (see Fig. I.13). Examples of such seamounts are Empire and Coral Patch mountains (Fig. II.14.21), Gettysberg and Ormond summits, and Gorringe Ridge in the Atlantic.

If the top of a seamount is located in the OML, the processes in these environments are substantially different. There, very small oxygen contents present difficulties for habitation of bottom organisms and plankton (plankton lives mainly well above the OML). Nevertheless, the environmental conditions are more favorable for conservation of organic matter and the deposition of phosphates and carbonates. In the photic zone, there are such examples as the Cobb seamount (Pacific Ocean; the top of this seamount lies 34 m below the ocean surface), Plantagenet bank (Atlantic Ocean, the top of which lies 54 m below the ocean surface), and Vema seamount (Cape basin, the top of which lies 120 m below the ocean surface) (Jenkyns, 1986, p. 357). On these seamounts (first of all, Cobb and Plantagenet), in addition to a large amount of biogenic calcareous remains (algae, corals, foraminifera, echinodermata, bryozoa, serpulid, gastropods, brachiopods, polichaetes, and especially the bivalve *Mytilus*), calcareous aggregates developing around calcareous algae have also been found. The same aggregates, cementing remains of red algae, were found on the Vema seamount summit.

The formation of hardground crusts of cementation on the bottom occurs under conditions of very small rates of sedimentation or hiatuses, or in the presence of precipitation of cement from solutions due to chemical reactions. Most frequently, calcite; magnesium calcite (Miliman, 1971), aragonite, and more rarely, dolomite, gypsum, anhydrite, and Fe and Mn hydroxides, play the role of cement.

Calcareous cement is frequently replaced by phosphatic cement, especially on the summits of seamounts in the Pacific (Baturin, 1978). The author believes that the time when the tops of seamounts (guyots) were within the OML may also be responsible for the accumulation of phosphorites.

filled with cold brines [depth 3270–3500 m (Adhib et al., 1989, p. 62)]. They are white to brownish gray in color. Lithification of these crusts increases upwardly, from the base to the top. The cement of the crusts, which are composed mostly of biogenic calcareous material, consists of high-magnesian calcite with a small amount of low-magnesian calcite and dolomite. These crusts periodically incorporate large crystals of gypsum, whereas small crystals of this mineral are dispersed throughout the cement.

Some areas of the ocean floor are occasionally covered by ferruginous crusts. We mention here only two types: hydrogen–diagenetic ferruginous and hydrothermal silica–ferruginous. Crusts of type 1, for example, formed during the late Pleistocene on the erosive (or non-depositional) surface of mud, where they exist as interbeds, each up to a few centimeters thick. These interbeds consist mostly of Fe hydroxides. The areals of these crusts extend a few thousands of kilometers. These crusts are abundant both in the upper (see Fig. II.1.6, overleaf II) and distal parts of the Amazon River fan (Damuth and Kumar, 1975). Fe used for the formation of these crusts reach the bottom from two sources—seawater and interstitial water. The reason for their development is an abrupt decrease in sedimentation rate (deficit in sediments derived from the Amazon River) as a result of Holocene transgression and that a major portion of the sediment load delivered to the ocean by the Amazon River was deposited on the continental shelf.

We discovered the second type of crusts on the abyssal plain of the Central basin in the Indian Ocean, in the distal zone of the fans of the Ganges and Brahmaputra rivers (Emelyanov and Kruglikova, 1990). It is evident that the appearance of these crusts was synchronous with eustatic rising of the ocean level during interglacial periods.

Miocene mineralization on the Blake plateau (the Atlantic ocean) was phosphatic. The productivity of seawaters at that time was very high, whereas the present-day productivity is responsible for the precipitation of Fe- and Mn hydroxides there, and this is a reflection of strongly oxidizing conditions in the area (Jenkyns, 1986, p. 393). Phosphatization of Miocene calcareous ooze occurred during the Early and Middle Miocene. Phosphatic pellets, grains of gravel, and individual boulders and crusts were observed there (Manheim et al., 1980). Deposits show signs of erosion and redeposition. Some phosphatic aggregates were repeatedly exposed to phosphatization. Grains of glauconite have also been found there.

On the calcareous–phosphatic bed of the Blake plateau, there are deposits of FMNs. The surface area covered by these deposits is about 5000 km^2. In the north of this plateau, these deposits occur at depths of about 400 m, and in the south, at depths of about 800 m. In the west of this plateau, where the ocean floor is exposed to the erosive effects of the Gulf Stream, Miocenic phosphorites are abundant. Manganese nodules actually form a pavement about 7 cm thick.

Manganese nodules from the Blake plateau are clearly not pelagic: they are enriched in carbonates (which occur as veins of aragonite) and in silica, and are depleted in Mn and minor elements. In terms of mineralogy, they are composed of crystallized Fe- and Mn oxyhydroxides. These nodules were formed either originally or due to replacement of primary phosphatic material (Jenkyns, 1986, p. 363).

Microrelief on the bottom surface. In seas such as the Red, Mediterranean, and Caribbean and in some areas of ocean, there are a variety of bottom surface microforms, including centimeter-sized hillocks, holes with a cone crater—so-called microvolcanoes—p to 20–25 cm in height (holes of the first type) and various holes a few centimeters in diameter and up to 25 cm deep. Holes of the first type are often localized in their occurrence: 1 m^2 of the bottom may contain up to 80–100 such holes. In close vicinity to these bottom features, there are small swells or, vice versa, small lows on the sediment surface. In some cases, features like microvolcanoes with distinct craters and without holes, or, in contrast, volcanoes without craters but with holes, exist on the bottom surface. In some places, holes are arranged in a manner closely resembling a horseshoe or bird's claw in shape.

Some of these holes are elongated in shape or have the form of cuts. In sediments with holes of both types, there are no signs of any specific minerals, the formation of which could lead to the development of microforms on the bottom similar to those described earlier.

With reference to genesis of microforms, some scientists suggest that the microforms may be of biological origin. They believe that holes are habitats of bottom organisms: worms, mollusks, crustaceans, fish, eels, etc. Indeed, some of these holes were previously occupied by bottom organisms. Nevertheless, although attempts were made to reveal organisms capable of digging through sediments, all of them failed to show positive results. For the first time in oceanography, we investigated microforms on the bottom (in the eastern Mediterranean) by using TV deep-sea camera, designed in 1961 (Emelyanov and Marakuyev, 1962).

Although many microforms of the bottom microrelief are recognized as being of biological origin, it may also be expected that some of these microforms, primarily holes of second type, are of diagenetic (gaseous) origin (Emelyanov and Marakyev, 1962). If the release of gas occurs within coarse sediments, gas escapes the sediment through numerous pores and does not form holes in it. In cases when the processes of gas release occur within semi-fluid or fluid mud, channels formed due to gas fluxes immediately fill up with sediment and disappear without leaving a trace on the sediment surface. Such phenomena can be observed in swamps, lakes, craters of mud volcanoes, etc. The picture is quite different in areas where consolidated and viscous plastic oozes are widespread. The structure of these sediments is such that more or less considerable seepage of gases is hindered. Under such conditions, gases accumulated on a particular horizon tend to release instantly, bursting and disturbing the surface of viscous compacted ooze. If ooze is characterized by moderate density and viscosity, and the strength of the gaseous "burst" is small, then a microvolcano with oval edges and small (3–10 cm) crater develops on the bottom surface. If the process of gas infiltration through the canal formed is continuous, then this canal continues to exist as a hole. When ooze is fairly compact and there are bottom currents, then disturbed pieces of ooze are rewashed and redeposited, and a canal with an opening for gas release continues to exist. A cone around the hole does not develop in such circumstances. If the strength of a gaseous burst is considerable, the diameter of the burst crater is also considerable. Pieces of ejected ooze fall back to bury the crater. Subsequent gas vents again produce holes, which are arranged in a line or in a ring-like manner, or form a bird's claw structure.

Why are holes of both types, hummocks and swellings found most frequently on the bottom of arid climatic zones (the Red, Mediterranean and Caribbean Seas)? The answer is simple: the seafloor is covered by biogenic foraminiferal and pteropodal–foraminiferal oozes. These oozes are very compacted and viscous compared to clay. Bricks made up of these oozes are saved from destruction for a long time, existing in their original form. The higher the $CaCO_3$content in these oozes, the lower the content of water and the higher their compaction (Emelyanov, $1965_{1,2}$). Consequently, microforms that develop in these oozes do not disappear (do not become "swollen") and preserve their shape for a long time.

Is the abundance of these microforms the same for all three of the seas mentioned above? It has been found that their abundance is quite different. For example, these features are completely absent in shelf areas and on the tops of seamounts. The explanation advanced for this is that these bed forms are dominated by coarse sediments (sands, aleurites), which are porous and have low plasticity and compaction. Even if holes appear in such sediments, they finally disappear without leaving traces on the sediment surface. A contributing factor in such a situation is bottom currents, which are commonly strong enough to strip the sediment surface of features. Fine-grained muds in which content of interstitial water is higher than that in shallow water biogenic oozes are widespread in deep areas of seas and oceans. This fine mud is characterized by an increased water content (fluidity). Muds are considered to have higher fluidity: the higher the content of clay minerals in them, the higher their sedimentation rate. As for areas where fine non-carbonaceous muds (commonly clay) are abundant (the Sea of Azov, Black and Baltic seas, high-latitudinal areas of oceans), even though holes develop, they immediately swell and the bottom surface remains featureless.

Because biogenic carbonate oozes are abundant in low latitudes (mainly in areas of arid climate) (Fig. II.14.22) of the World ocean, the bottom microforms described earlier also exist mainly in the same latitudinal belts.

Hydrothermal processes are also a factor in development of holes on the bottom surface. When hydrothermal solutions become exposed on the bottom surface, a certain portion of these solutions passes into the gaseous state due to decrease in pressure, and, as a result, sediments are disturbed by the effects of gas release. Formation of gases leads to accelerated heating of sediments: from below by hydrothermal solutions, and from above by seawater and brines having temperatures up to 22–25 and 62°, respectively (for example, in Red Sea deeps). At increased temperatures, the rate of biochemical reactions that decompose organic matter become more intensive.

Because biogenic carbonate oozes are compact and viscous, they do not slump even in the case of steep slopes. This is why phenomena such as sediment sliding and turbidity currents are essentially absent on the slopes of the continental margin, seamounts, and heights in aridic climatic zones of oceans, which are mostly covered by foraminiferal oozes.

In the oceans, small irregularities on the bottom surface occur predominantly at depths of 200–4500 m. This is because the amount of organic matter in ocean bottom sediments (see Fig. I.8), which is the main source of gases, decreases with distance from land, i.e., its distribution is circumcontinental. Moreover, the content of

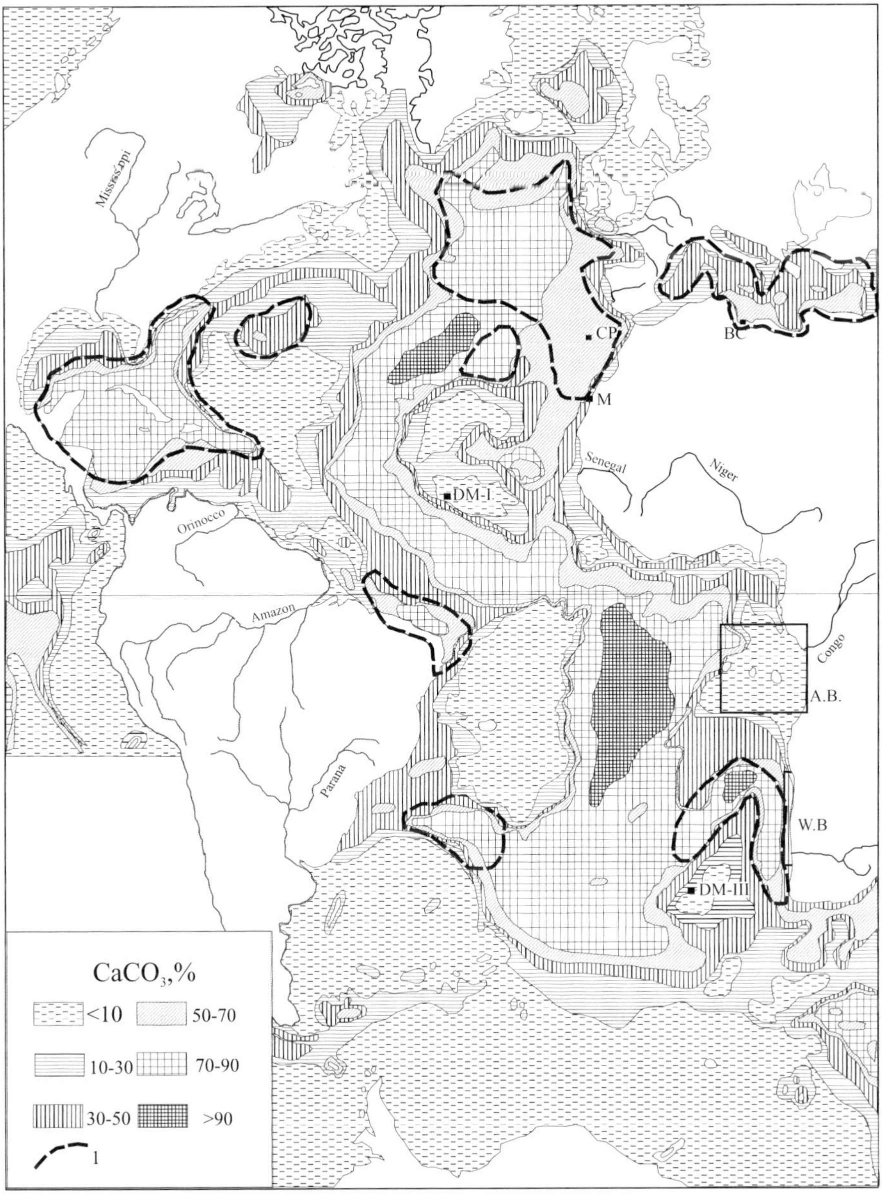

Fig. II.14.22. Distribution of $CaCO_3$ in surficial (0–5cm) sediments of Atlantic Ocean, % (after Emelyanov et al., 1971) and location of test areas investigated in detail.
Contours by dotted lines (1) show areas where microforms of relief are distributed on the bottom (microvolcanoes, holes, microhills, etc.).
Test areas: CP—Coral Patch mountain (see Fig. II.14.21); BC—Baroni crest (see Fig. II.14.19); M—Morocco test area (see Fig. II.4.5); DM-I—Cabo Verde Basin test area; DM-III—Cape Basin test area (see Fig. II. 13.6); W.B.—Walfish Bay (see Fig. II.4.3); A.B.—Angola Basin (see Fig. II.1.7.)

organic matter decreases as with increasing depth. In a broad sense, mass accumulation rates of sedimentary material are generally similar to the pattern described for organic matter (Emelyanov and Romankevich, 1979). Consequently, the greater the distance from coast and the greater the ocean depth, the smaller the amount of organic material buried in sediments and the lower the possibility for gases. This explains why holes of both types have not been observed in oceans at large distances from coasts, for example, in areas of mid-oceanic ridges, where high-carbonaceous, compacted and viscous muds, very similar to those present in the Red, Mediterranean and Caribbean seas, are abundant. In red clays, holes are also commonly absent. The clays themselves contain only 0.2–0.4% organic matter, and this is 10–20 times less than in sediments in the upper parts of the continental slope. In sediments of the Mediterranean Sea, there are layers of sapropelic oozes containing up to 5–10% organic matter (Emelyanov and Shimkus, 1986). In the Red Sea, where the number of holes in sediments was found to be maximum, formation of gases is probably facilitated by the presence of carbonate-rich oozes, which are heated from below by hydrothermal solutions.

One of the most interesting phenomena in the World Ocean is "non-sinking" manganese nodules (i.e. combined effects keeping them at the sediment surface) (see Fig. II.14.2). Although the nodules are considerably older than the sediments (the age of the nodules is a few millions years), they lie on the sediment surface, despite the fact that sediments accumulate 2–3 orders of magnitude faster (a few millimeters per 1000 yr) than the nodules themselves grow (a few millimeters per million years). The problem was seemingly solved by studying the physical properties of muds

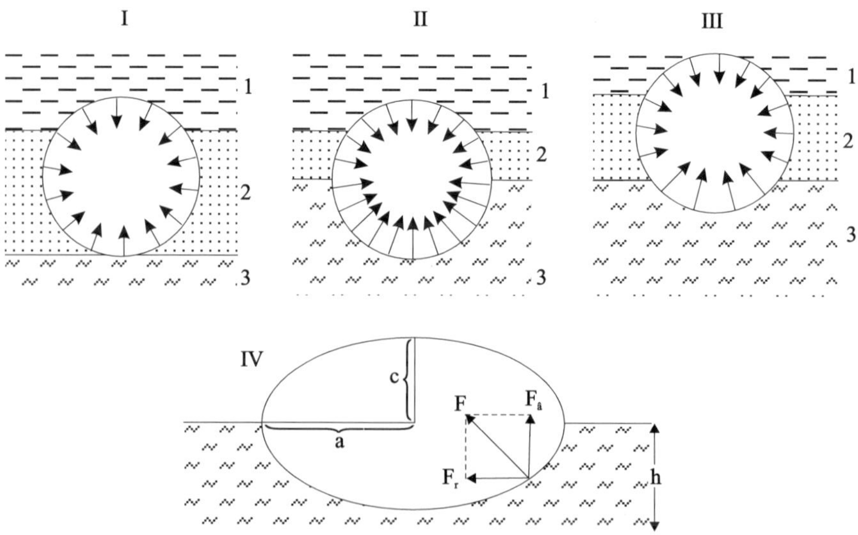

Fig. II.14.23. Scheme of force pushing nodules out of sediments. After Goryainov and Goryainova, 1983.
1—seawater; 2—near-bottom suspension; 3—sediment. Other symbols are explained in text.

(Piper and Fowler, 1980; Goryainov and Goryainova, 1983). Studies led researches to infer that the lifting ability of the sediment present at a depth of 7 cm below the sediment surface is about 100 times higher than that at the sediment surface itself. When the nodule is within the clay, the lower part of this nodule is pressed much more than its upper part, which is exposed only to hydrostatic pressure (Fig. II.14.23). If the force acting vertically upward is in excess of the attachment force, the nodule will be pushed out from the sediment to the surface. In my opinion, this mechanism is disputable. According to another scenario, burrowing organisms may also be responsible for positioning nodules at the sediment surface: the nodules are being nudged by mud-feeders, which roll the nodules over and keep them at the sediment surface (Glasby, 1977). However, this mechanism is also arguable. Skornyakova and Bezrukov (1976) showed that they are not rolled). Radiometric methods have confirmed the presence of these events in the "life" of the nodules.

Redox (Eh) Barrier in Sediments

Structure of the redox (Eh) barrier and behavior of Mn and Fe in sediments. The oxidized (commonly, the top) layer of sediments is sometimes called the oxygen (o_2) layer. Reaction No.4 (Table II.15.1, Fig. II.15.1) is the main reaction in this layer. Owing to this reaction, stable compounds of Mn^{4+} develop in sediments (Fig. II.15.2). In a sediment column below the Eh barrier, sediments are dominated by reaction No. 5, resulting in the decomposition of stable compounds of Mn^{4+}. An Mn^{2+} ion generated by the previous reaction migrates upward, towards the Eh barrier, where reaction No. 4 occurs again. Therefore, the lower part of the oxidation layer, where the change in the Eh potential is virtually strongest (Eh barrier), is markedly enriched in compounds of Mn^{4+} to as much as 6–8%Mn, in contrast to the background values of 0.04–0.10% Mn (Figs. II.15.3, II.15.4). This manganese-enriched interbed is called the Mn (IV) interbed (Fig. II.15.1) (Rozanov, 1988).

This interbed is an indicator showing the depth within the sediment column to which oxygen has penetrated. If the process of sedimentation continues for a long time under the same physicochemical conditions, the thickness of the Mn interbed gradually increases, reaching tens of centimeters (Fig. II.15.3), and in pelagic areas of oceans, even several meters or a few tens of meters (in red clays) (see Figs. II.15.4, II.15.5). In contrast, if both physicochemical conditions and the process of sedimentation are variable, the Mn (IV) interbed show signs of microlamination. Microlaminations vary from fractions of a millimeter to several centimeters in thickness. The degree of oxidation of these microlaminations is different: some strong oxidized, others less so, or even reduced.

As for element mobility within sediments, the largest flux from reduced muds to oxidized muds is that for Mn^{2+} (because manganese is more soluble in its reduced form, the Mn contents in oxidized clays are 10–15 times as large as those in reduced muds) (Table II.15.2), which is followed by Fe^{2+}, P and other elements. Moreover, a characteristic feature of element behavior is that the intensity of such flux increases with distance from shore (in Fig. II.15.2, station DM-668 is closer to shore, while station DM-674 is closer to the center of the ocean). Manifestation of these features is especially clear for Mo and Mn, because the proportion of mobile forms of these elements in sediments is much higher than those for other elements. For each particular element, the proportion of mobile forms of this element in sediments increases with distance from shore (Volkov, 1979). Not only Mn^{2+} but also Fe^{2+} migrates from interstitial waters, contained in the underlying (reduced) layer, upward to the Eh barrier. This barrier is dominated by chemical reaction No. 6 (Table II.15.1), owing to which iron precipitates from seawater to then exist in

Table II.15.1. Redox stratification in sea bottom sediments.
Compiled by Rozanov (1988) on the basis of 4 – Fralich et al., 1979; 5 – Margin, 1982; 6 – Thomson et al., 1984; 7 – Belyaev et al., 1981

Horizon	Main processes	Reactions	Redox largers characteristic	ΔG°, kg.joules/ mole	Eh, V	Ep_t,V
1	Expenditure of O_2	$0_2 + C = CO_2$	Decreasing $[0_2]$ to the depth of sediment	–3190	0.74	0.5–0.6
2	Nitrification	$2NH_3 + 30_2 = 2NO_2^- + 4H^+ + 2H_20$; $2N0_2^- + 0_2 = 2N0_3^-$	Maximum $[N0_3^-]$	-	-	-"-
3	Denifrication	$3N0_3^- + C = 2N0_2^- + C0_2$; $2N0_2^- + 2C = N_2 + 2C0_2$	Decreasing $[NO_3^-]$	–3030	0.59	-"-
4	Oxidation of Mn^{2+}	$2Mn^{2+} + 0_2 + 2H_20 = 2Mn0_2 + 4H^+$	Maximum of $Mn0_2$	-	-	0.6
5	Reduction of $Mn0_2$	$2Mn0_2 + C + 4H^+ = 2Mn^2 + C0_2 + 2H_20$	Maximum of $[Mn^{2+}]$	–2920 (pyrolusite)	0.47	0.4–0.5
6	Oxidation of Fe^{2+}	$2Fe^{2+} + Mn0_2 + 2H_20 = 2Fe00H + Mn^{2+} + 2H^+$	Maximum of FeOOH	-	-	-"-
7	Reduction of Fe00H	$4FeOOH + C + 4H^+ = 4Fe^2 + CO_2 + 4H_20$	Maximum of $[Fe^{2+}]$	–1330 (geothite)	0.24	0.2–0.4
8	Oxidation of S^{2-}	$S^{2-} + 2FeOOH = SO_4^{2-} + 2Fe^{2+} + 2H^+$	Maximum of $[S0_4^{2-}]$	-	-	-"-

9	Reduction of SO_4^{2-}	$SO_4^{2} + 2C + 2H^+ =$ $H_2S + 2CO_2$	Increasing iron sulfides	–380	–0.24	<0.2
10	Oxidation of CH_4	$CH_4 + SO_4^{2-} \rightarrow HS + HCO_3^-$ $+ H_2O$; $CH_4 + CO_2 =$ CH_3COOH	Decreasing $[SO_4^{2-}]$	-	-	-"-
11	Formation of CH_4	$CO_2+4H_2=CH_4+2H_2O$ (mainly) $CH_3COOH=CH_4+CO_2$ (partly)	Increasing CH_4	–350	–0.35	-"-
12	Oxidation of H_2	$12H_2+6CO_2 =$ $C_6H_{12}O_6+6H_2O$	-	-	-	-"-
13	Formation of H_2	$CH_{12}O_6+6HO_2 =$ $6HO_2+12H_2$	-	-	-	-"-

Remarks:

(1) The composition of solid phases of MnO_2 and FeOOH is shown conditionally; these phases have variable valences in the real sediments; ions Mn^{2+} and Fe^{2+} can have organic or inorganic ligands.

(2) Process of reduction of NO_3 begins with decreasing oxygen content to 0.1 ml/l (Nehring, 1982); according to other data, the reduction of MnO_2 can be before nitrification (Frolich et al., 1979; Thomson et al., 1984).

(3) Microbiological formation of H_2 (fermentation) can occur in all sedimentary strata located below horizon 8 (Belyaev et al., 1981).

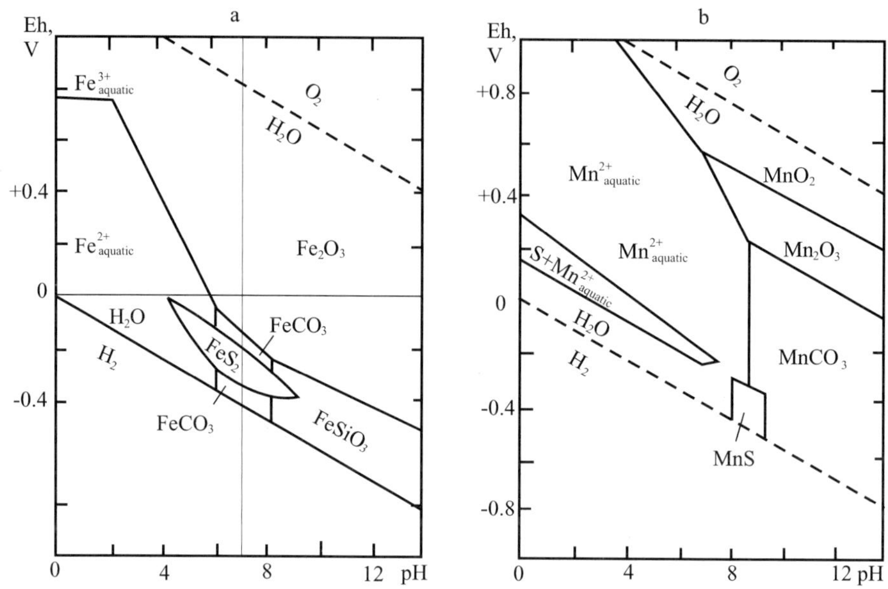

Fig. II.15.1. Diagram of formation of iron and manganese minerals in early diagentic stage (Eh and pH barriers).

particulate form as FeOOH. Because the mobility of Fe is lower than that of Mn [when the pH value is 7, the rate of oxidation of Mn^{2+} is 50 times smaller than that of Fe^{2+} (Stumm and Morgan, 1981)], the ability of Mn^{2+} to precipitate from seawater lags behind precipitation of Fe^{2+}. The result of this process is a thin bed with authigenic compounds of Fe^{3+} called the Fe (III) interlayer bed. The thickness of this bed is smaller than that of the Mn (IV) bed. As a result, the sequence of beds (from bottom to top) that built up the sediment column that developed on the Eh barrier is as follows: reduced greenish gray bed–oxidized yellowish brown bed (compounds of Fe^{2+})–strongly oxidized brown bed (compounds of Mn^{2+}). Sometimes, beds of Fe (IV) and Mn (IV) merge to form one bed; in this case, increased contents of both Fe and Mn are found in this bed.

Diagenetic processes depend upon a number of factors such as the sedimentation rate, bioturbation, porosity of sediments, and the intensity at which organic matter accumulates in sediments. For example, the flux of manganese in bedded sediments (displaying variations in the Eh values) precipitates either near the redox (Eh) barrier or within the oxidized sediment layer, depending on the depth of manganese oxidation and bioturbation zones (Dhakar and Burdige, 1996). Sediments containing an oxidized layer having a thickness <5cm are dominated by the flux of Mn (which is carried by interstitial waters) moving upwards to the upper oxidized layer, in contrast to its behavior within the redox (Eh) barrier, where Mn precipitates. In such a case, the majority of Mn precipitates in the upper oxidation layer, which is actively

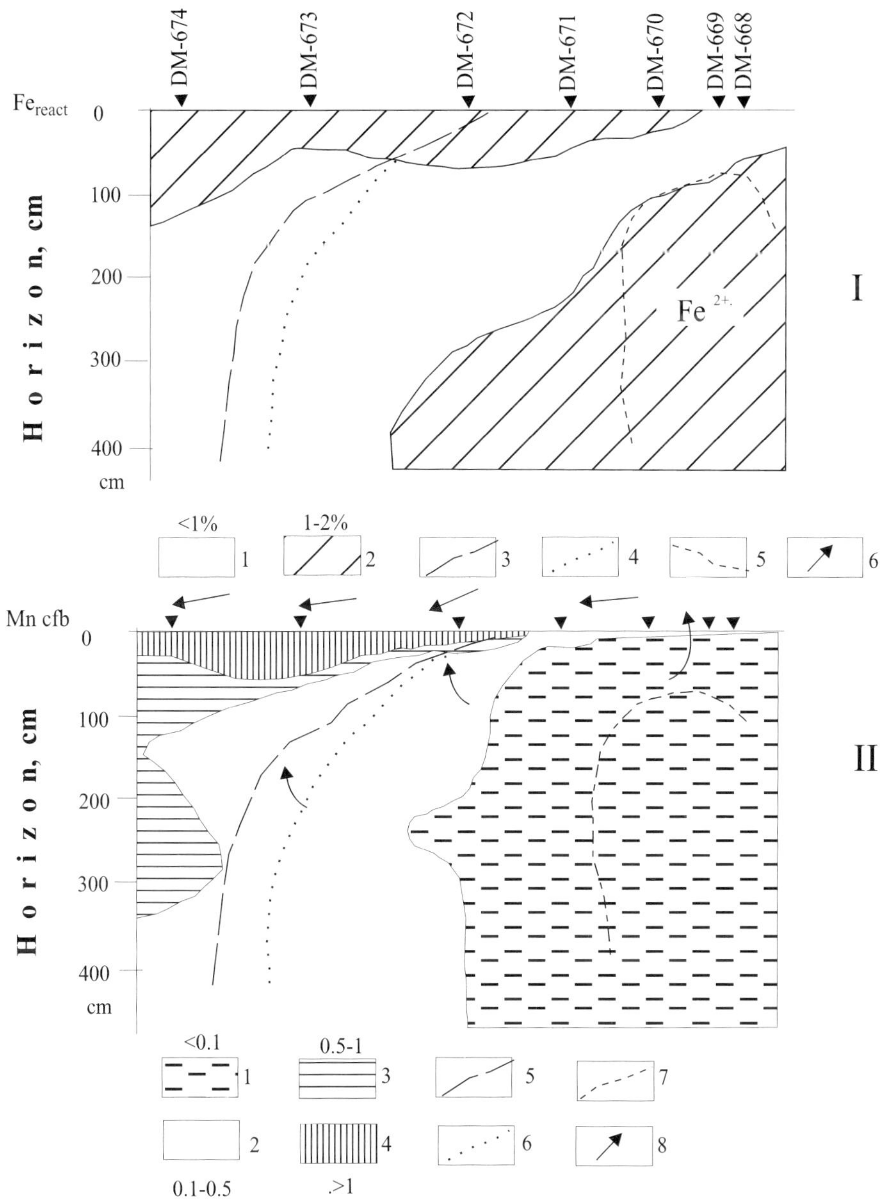

Fig. II.15.2. Eh and content of Fe_{react} and Mn (recalculated on cfb) in bottom sediments of Gulf of California (stations DM-668–DM-672, depth 140–2650 m) and continental slope of Pacific Ocean (stations DM-672–DM-674, depth 2900–3470 m). According to data of A.G. Rozanov et al., 1976, with additions by the author.
Profile I: 1–2 Fe_{react} content, %: 1—<1; 2—1–2; 3—isoline of Eh = +400 mV (mud above this isoline is oxidized, Eh = +600 mV; beneath it, weakly reduced and reduced, Eh = +400–350 mV); 4—border of distribution of sulfate-reduction processes; 5—border of distribution of free H_2S in sediments; 6—direction of iron flux. Prevailing forms of reactionable Fe are shown (Fe^{2+} and Fe^{3+}).
Profile II: 1–4—Mn content, in %: 1—<0.1; 2—0.1–0.5; 3—0.5–1; 4—>1; 5–7—same borders as on profile I; 8—direction of Mn flux (in sediments and in near-bottom water).

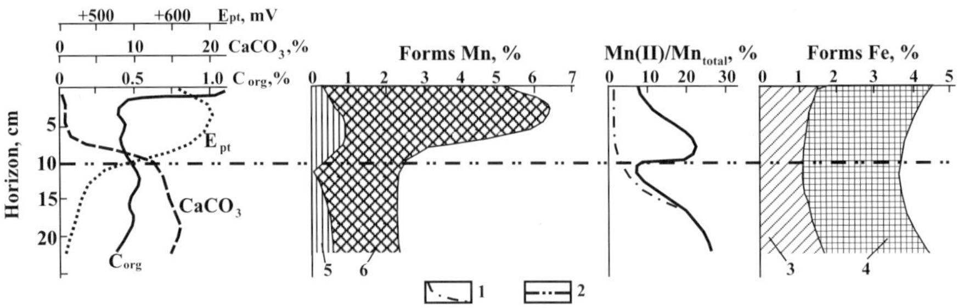

Fig. II.15.3. Distribution of $CaCO_3$, C_{org}, E_{pt} and forms of Mn and Fe in pelagic mud of ocean. After A.A. Morozov, 1986 (unpublished).
1—supposed distribution of ratio of $Mn(II)/Mn_{total}$ during mobility phase and active aeration of sediment; 2—condition border of mobile layer of sediment; 3—$Fe_{clastic}$; 4—Fe^{3+}_{react}; 5—Mn^{2+}; 6—Mn^{4+}.

bioturbated. If the thickness of the oxidized layer is more than 10 cm, the predominant pattern (mode) of Mn behavior is precipitation of this element mostly near the redox (Eh) barrier, rather than Mn diffusion into the oxidized layer. In this case, peak (maximum) values of Mn will be found near the redox (Eh) barrier, below the bioturbation zone (8–10 cm of the sediment column) (Dhakar and Burdige, 1996).

Along with Fe and Mn, microelements such as Cu, Ni, Cr, V, and Mo can also be more or less enriched in sediments of the oxidized layer (within the Eh barrier).

Within a reducing environment (Eh value less than +0.2 V), reduced forms of sulfur undergo reoxidation, as a result of reaction No. 8. In this case, sulfate sulfur SO_4^{2-} accumulates in sediments, where this precipitated phase forms the SO_4^{2-} bed (Rozanov, 1988; Rozanov, et al., 1971). This bed is commonly found below the Fe (III) bed.

So, in zones of transition from oxidizing to reducing conditions in bottom sediments, there are several successive redox horizons of interstitial waters, each of which is dominated by a particular type of chemical reactions, which are listed in Table II.15.1 (Fig. II.15.3). Consequently, in sediment cores collected from pelagic muds, there are several layers enriched in Mn and other microelements (Emelyanov et al., 1975; Emelyanov, 1992). Such layers are characteristic of peripheral zones of deep basins in the Atlantic Ocean, where red clays are changed to mixed coccolithic–foraminiferal oozes or terrigenous (hemipelagic) muds. The thickness of such oxidized layers ranges from 10–20 to 80–100 cm, two to three times smaller than that of the reducing (grayish brown or gray) muds containing these beds. The reverse is the case in the Mediterranean Sea, where low-oxidized muds contain thin (1–10 cm) beds of reduced (dark grayish green) sapropelic sediments (Shimkus, 1981; Emelyanov and Shimkus, 1986).

Although organic matter contents in some shelf areas of oceans, as well as in some seas and bays, are higher than those in surrounding areas (more than 1% C_{org}), there is a thin (a few millimeters or centimeters in thickness) layer of oxidized sediments. This bed, which is commonly found at the water–bottom boundary, is also a

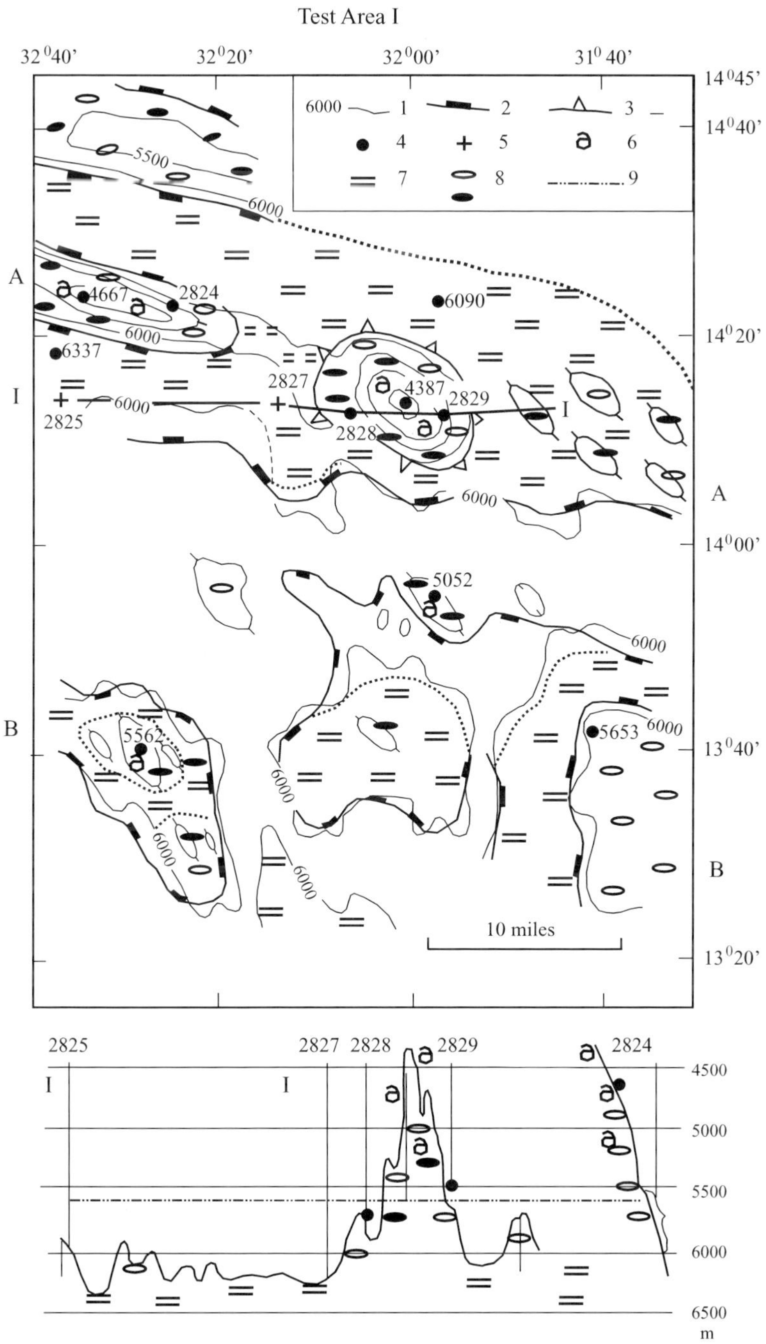

Caption see page 420

boundary between the zone of oxidizing reactions present in the water column and zone of strongly reducing conditions found in the sediment strata below a thin oxidized layer. In seas and bays, flat ferromanganese concretions commonly develop in the uppermost thin layer of oxidized muds (see Part III).

Even when pelagic muds were constantly found in oxidizing conditions (Eh value more than +300 mV) and below the CCD level, but Eh values in these muds ranged from +300 to +600mV, manganese beds also developed (see Fig. II.15.5). They were confined to environments having the highest Eh values. Moreover, the abundant Mn (when the Mn content was more than 1.0%) was nearly always used for the growth of manganese micronodules (see Fig. II.13.8).

Such micronodules contain from 22.41 to 40.2% Mn and from 0.83 to 4.95% Fe (Svalnov, 1991).

In seas and oceans, sediments found on both sides of the Eh barrier are, in general, often of the same type, but the properties of these sediments are different. The first striking thing is the coloration of the bottom sediments. Above the Eh barrier, i.e., in oxidizing conditions, sediments are generally colored in shades of brown. Due to compounds of Mn^{4+}, the color of sediments is from brown to brownish black, whereas Fe^{3+} gives them a pronounced dark yellow coloration. On the other side of the barrier, where compounds of Fe^{2+} (iron sulfides) are dominant, sediments have a pronounced greenish gray coloration. A considerable admixture of organic matter only increases this effect: the coloration of sediments becomes a dark, dirty green (for example, in reduced sapropelic muds). In reducing environments, where hydrotroilite is present, sediments are grayish black (the Black and Baltic seas).

The color of reduced sediments is conditioned not only by sulfide compounds, organic matter, and compounds of Mn^{4+}, Fe^{3+} and Fe^{2+}, but also by silicates. For example, owing to the presence of nontronite in sediments, they have yellowish or greenish hues, whereas smectite gives them a white or grayish color. As a result of chemical reactions in the reduction of iron, which provide a component of silicates, many silicates change their color. At the Eh barrier, oxidation or reduction of minerals occurs, or more precisely, iron, which comprises a part of their elemental composition. This is another factor contributing to change in sediment color. Thus, the color of muds depends on the contribution of a particular type of silicate.

The Eh barrier is a place where processes of initial ore formation in sediments or even accumulation of ferromanganese concretions or ferromanganese crusts are very active.

Fig. II.15.4. Relief, bottom sediments and nodules in test area DM-1 in Cabo Verde Basin, Atlantic Ocean. For location see Fig. II.14.9. After Emelyanov et al., 1982.

1—isobaths, m; 2—foot of large ridges and block highlands; 3—foot of volcanic mountain BVG; 4—seabed locations where collected samples of sediments and bedrock have been found to contain FMNs; 5—seabed locations where collected samples of sediments and bedrock do not contain FMNs; coccolith-foraminiferal sediments; 7—plains of maximum aggradation; 8—fields which are the most likely the best for FMN prospecting (a—Holocene, b—Holocene and Pleistocene); 9—position of CCD for Holocene; TP—trough-like plain.

I—Profile of bottom topography made across a test area, together with location of stations.

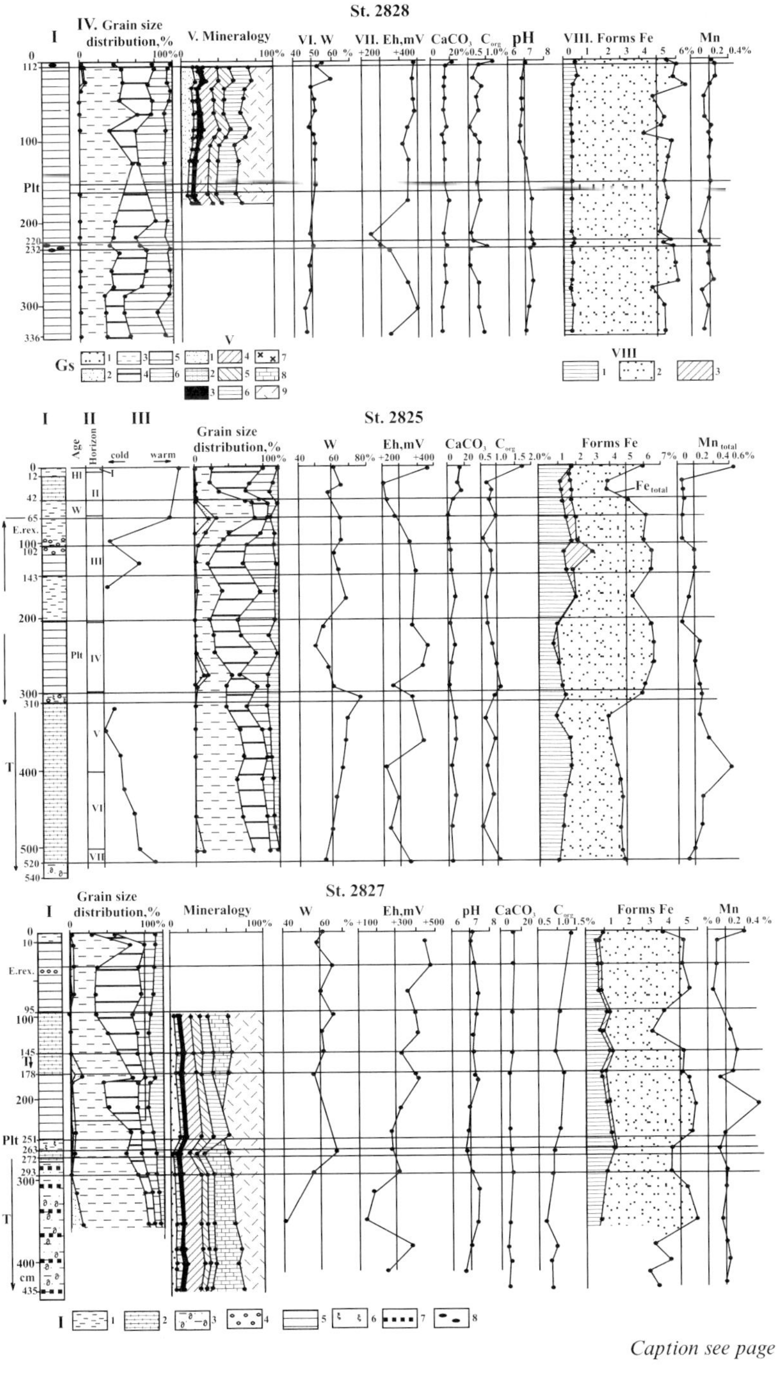

Caption see page 423

Table II.15.2. Content of Mn and other elements and <0.01 mm (in %) fraction in the oxidized (O) and reduced (R) pelagic muds of the Angola (stations AK-39 and AK-68) and Brazil (station AK-414) basins. After Emelyanov et al., 1975; Emelyanov, 1992.

Station	Depth, m	Horizon, cm	Layer	Fraction <0,01 MM	$CaCO_3$	SiO_{2am}	C_{org}	Fe	Mn	Ti	P
AK-39	5250	13–14	R	76.4	3.57	-	2.21	2.82	0.11	0.22	0.05
		33–35	0	73.0	15.39	13.04	1.97	2.63	1.51	0.22	0.05
		75–81	R	75.5	14.28	14.70	1.64	2.45	0.12	0.19	0.05
AK-68	5500	94–97	R	66.6	0.0	13.00	0.44	3.77	0.12	0.31	0.08
		129–133	0	68.2	0.0	11.42	0.19	3.77	1.28	0.35	0.10
AK-414	4880	91–96	0	78.18	0.0	1.25	0.21	5.28	1.43	0.53	0.06
		146–152	R	81.62	0.0	1.40	0.21	5.47	0.12	0.54	0.06

Fig.II.15.5. Lithological composition of sediment cores from Cabo Verde Basin (test area DM-I) (see Fig. II.15.4) and distribution in cores of certain chemical components and elements.
I—1–8—sediment types: 1—terrigenous (gray) mud; 2—low-calcareous mud with coccoliths; 3—low-calcareous mud with foraminifera; 4—mud with significant amount of diatomic remains; 5—pelagic red clay (brown); 6—holes of mud-eaters; 7—sandy-aleuritic interlayers (mainly graded beddings); 8 FMNs.
II—age and numbers of horizons.
III—paleotemperature curve.
IV—grain-size (gs) distribution (fractions in mm), %. 1—>0.1; 2—0.1–0.05; 3—0.05–0.01; 4—0.01–0.005; 5—0.005–0.001; 6—<0.001.
V—x-ray mineralogy of bulk sediment samples, % from bulk sample: 1—quartz; 2–K feldspars; 3—plagioclases; 4—illite; 5—caolinite; 6—montmorillonite; 7—chlorite; 8—calcite; 9—x-ray amorphous phase.
VI—W—moisture, %.
VII—Eh—redox potential, mV.
VIII—forms of Fe: 1—Fe^{3+} reactionable; 2—Fe^{2+} reactionable; 3—clastic and pyritic Fe.
E.rex—considerable amount of *Ethmodiscus rex* in sediment;
T—turbidites (with foraminifera).

Diagenetic distribution of sedimentary material is active not only within the redox (Eh) barrier, but also within the oxidized and, especially, in the reduced sediment layer at a considerable distance from the barrier. This process is given further consideration in Part II.16.

Chapter 16

The Upper Active Sediment Layer

Anaerobic-type diagenesis. This type of sedimentary material alteration characterizes sediments in the Baltic and Black seas, Cariaco trench (Caribbean Sea) and some other basins depleted in oxygen. Figures II.16.1 and II.16.2 represent concentration–depth profiles for methane (CH_4), total sulfuric dioxide (ΣCO_2), sulfates (SO_4^{2-}) and compositional ratios of oxygen isotopes versus carbon dioxide ($\delta^{13}C$ of CO_2) in interstitial waters of marine sediments from the Cariaco trench. The "zone of low concentrations of methane" (ZLCM) (Reeburgh, 1982) is clearly distinguishable against the surrounding conditions. One can see that the content of sulfates sharply decreases downward from the sediment surface and at the ZLCM they are absent. The CH_4 concentrates beneath this zone. The ΣCO_2 values gradually increase with depth, and contents of $\delta^{13}CO_2$ are minimal in the ZLCM.

Broadly speaking, the ZLCM is a specific geochemical barrier, where concentrations of sulfate ion ($SO^{2-}{}_4$), CH_4, ΣCO_2, δ^{13} CO_2 and some other components of interstitial water on one side of this barrier are different from those on the other side (Reeburgh, 1982). Methane is used up completely in the sediment column above the ZLCM, and sulfates (SO_4^{2-}) are almost completely used up within this zone. A downward increase in CO_2 is evidence that the flux of this component to higher levels is increasing. Figure II.16.2 shows the direction and intensity of fluxes of various components in the Cariaco trench (Reeburgh, 1976, 1982). Methane migrates from the upper 45 cm of the sediments into the bottom water and concentrates in the anaerobic zone. Then, methane leaves this zone to enter the mixed-water layer, where oxidation of methane occurs. CO_2 migrates from the sediment upward, crosses the ZLCM, and is conveyed to the water–bottom interface. CO_2 and SO_4^{2-} migrate in the opposite direction.

Most surprising is that methane usage (oxidation) is very distinct: the ZLCM is located within a very thin layer, which is marked by sharp minimums in $\delta^{13}CO_2$ concentrations. From this it follows that the distribution of microbes, for example, sulfate-reducing bacteria, as well as the pattern of authigenic minerals formation, should also change strongly here.

Above the ZLCM, oxidation proceeds at the cost of sulfates, whereas below this zone, sulfates disappear and cannot be oxidizing agents.

The thickness of the sedimentary layer above the ZLCM is different for various seas and oceans. It depends on the sedimentation rate and on the flux of C to the seabed from overlying waters.

Thus, the ZLCM in reduced sediments is also the lower boundary of the upper active sediment layer barrier.

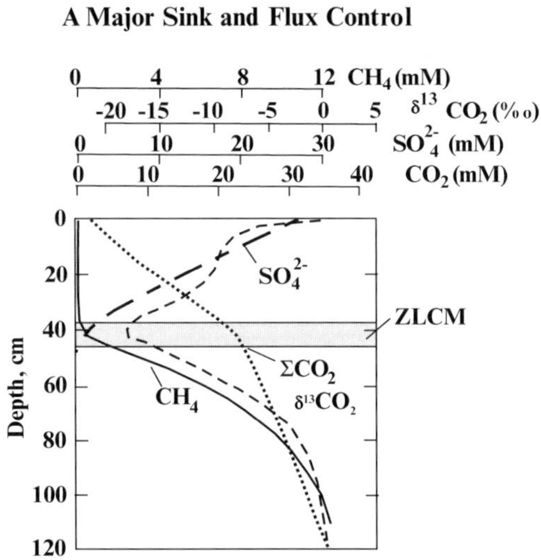

Fig. II.16.1. Schematic diagram showing depth distribution of methane, sulfate, total carbon dioxide, and carbon isotope ratio of carbon dioxide in interstitial waters of marine sediments. All distributions show breaks or slope changes in stippled area, which represents zone of anaerobic methane oxidation. After Reerburgh, 1982.

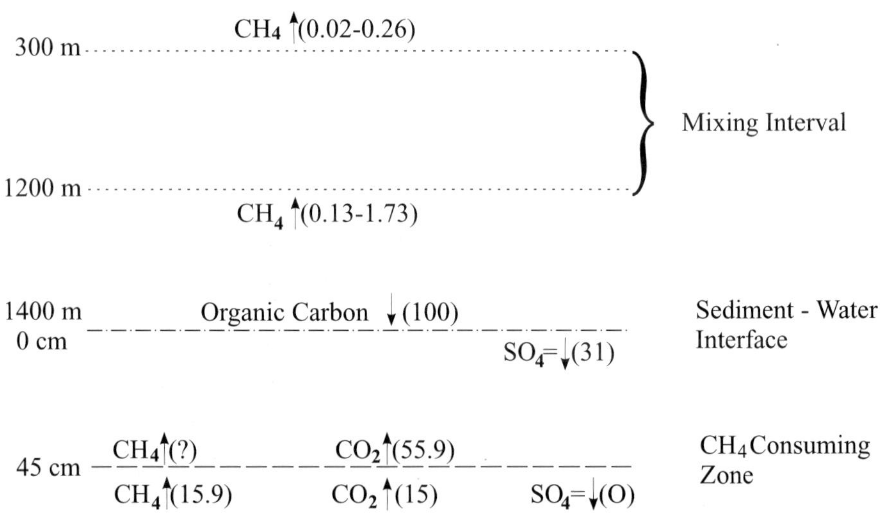

Fig. II.16.2. Summary of methane fluxes in Cariaco Trench water column and methane, carbon dioxide, sulfate, and organic carbon fluxes in sediments [in μmol/(cm^2 · yr)]. Reerburgh, 1982.

K.U. Hinrichs and A. Boetius (2002) reviewed data collected over the course of thirty years. Their new insights into the quantitative significance of anaerobic oxidation of methane were briefly described (Hinrichs and Boetius, 2002, p. 474) and are presented in Fig. II.16.3.

The distinct boundary in sediments can also be identified using other geochemical parameters, including sulfur isotopes (Lein et al., 1986), pH, alkalinity, and forms of silica (Fig. II.16.4). It is evident that this boundary coincides with that of the ZLCM.

The content of H_2S in muds (Fig. II.16.5) occurs as result of two opposite processes: (1) bacterial synthesis of organic matter during diagenesis and (2) removal of H_2S (gas phase) from solution to be aggregated with various compounds. In muds of Baltic deeps, sulfur is represented by elementary, pyritic, and organic forms (Emelyanov et al., 1986; Fig. II.16.6). Moreover, among all these forms, the pyritic form strongly predominates. A major portion of reactionable iron is represented in the form of pyrite.

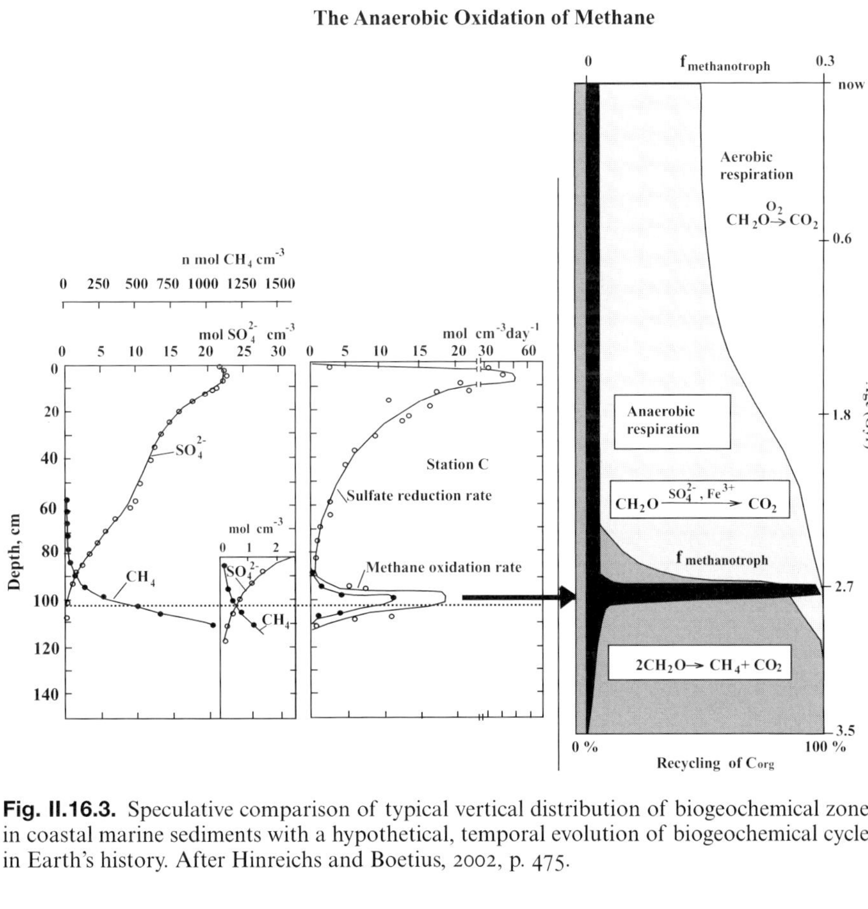

Fig. II.16.3. Speculative comparison of typical vertical distribution of biogeochemical zones in coastal marine sediments with a hypothetical, temporal evolution of biogeochemical cycles in Earth's history. After Hinreichs and Boetius, 2002, p. 475.

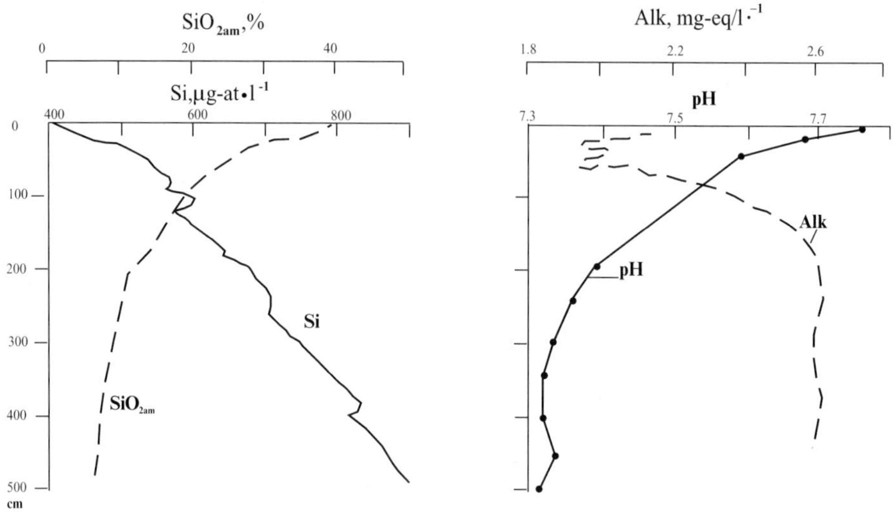

Fig. II.16.4. Distribution of biogenic siliceous particles (SiO_{2am}), dissolved silica (Si), alkalinity (Alk) and pH in interstitial water of pelagic red clay of eastern equatorial part of Pacific Ocean (see Fig. II.6.6). According to data of Hartmann et al., 1973.

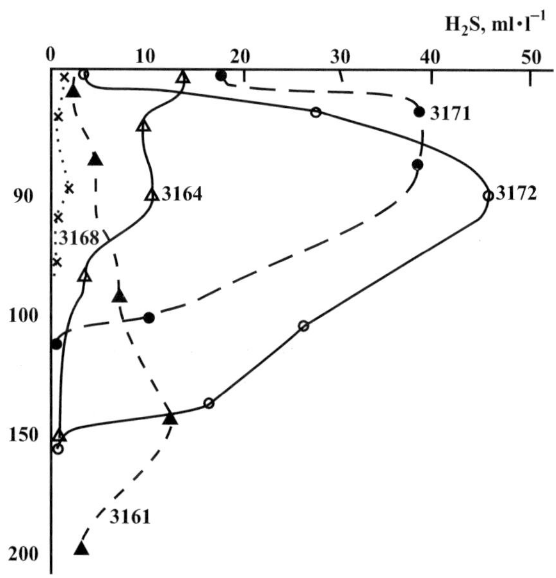

Fig. II.16.5. Distribution of free H_2S in Holocene mud of Baltic Sea. Numbers on scheme designate stations. After Emelyanov et al., 1983.

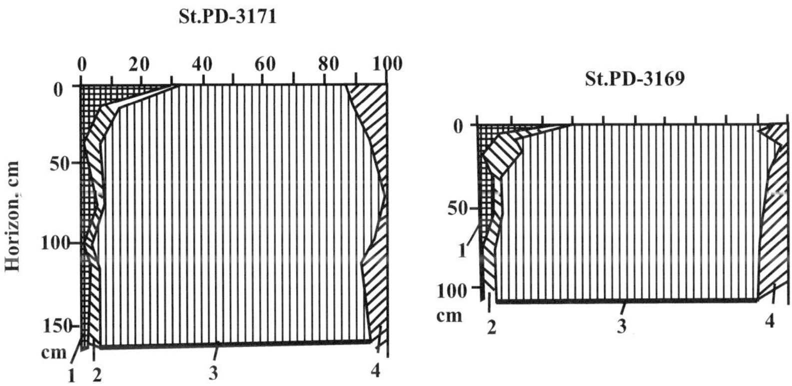

Fig. II.16.6. Forms of S in mud of Baltic Sea (station PD-3171, depth 162 m, and station PD-3169, depth 200 m). After Emelyanov et al., 1983.
Forms of S : 1—sulfidic; 2—elemental; 3—pyritic; 4—organic.

In peripheral areas of the World Ocean, containing about 88% of the total organic carbon in sediments (Table II.16.1), the processes of anaerobic diagenesis predominate: there occur rapid burial of organic matter in sediments; diffusion of organic components and metals into the bottom waters; and development of (a) bacterial methane CH_4, (b) chemogenic Ca(Mg) CO_3, and (c) authigenic phosphates.

The early stage of diagenetic transformation of sedimentary material, which was accompanied by intensive development of authigenic minerals (Fig. II.16.7), is believed to be responsible for considerable variations in the isotope composition of various forms of sulfur (Volkov, 1979_1).

A very characteristic feature of water basins (and individual deeps) in which bottom waters are polluted with hydrogen sulfide is the presence of yellowish white or yellow film (1–2 mm thick) undergoing transition between anoxic and oxic zones. This film covers black hydrotroilite muds like a continuous carpet and consists

Table II.16.1 The deposition of C_{org} in the Holocene sediments of the World Ocean. After Gershanovich et al., 1974 (citation by Volkov, 1979)

Geomorphological zone of the ocean	Square (10^6 km^2)	Average thickness of Holocene layer, m	Average content of C_{org}, %	Mass of C_{org} in the sediments, 10^{10}t
Shelf	26.7	0.8	0.7	22.4
Continental slope and its foot	76.5	1.5	1.3	195.0
Pelagic area	257.0	0.05	0.3	5.4

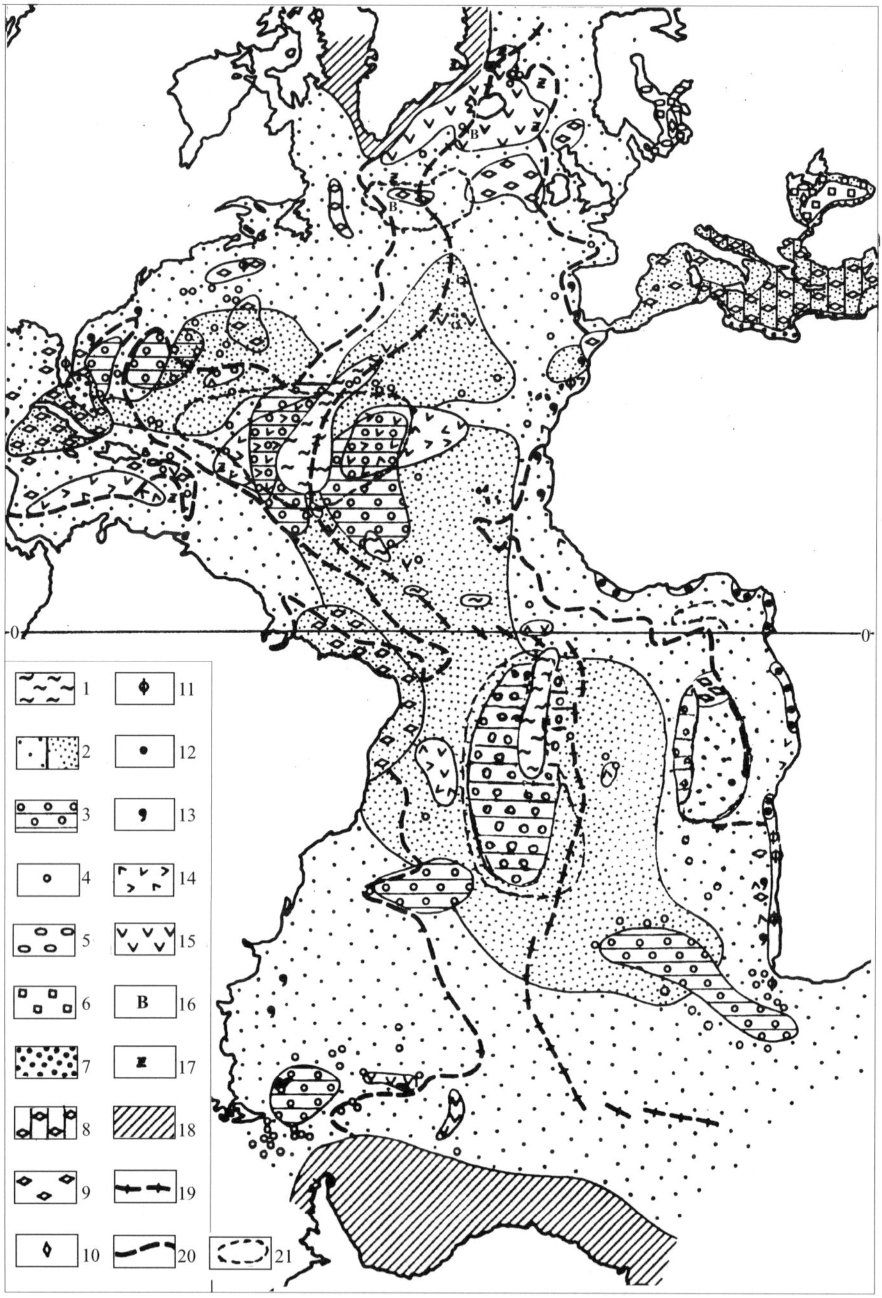

Caption see pg. 431

predominantly of sulfate-reducing bacteria. The same "bacterial carpets" were observed by the author in Baltic deeps aboard a Mir submersible (Emelyanov et al., 1996).

A large number of mud-volcano "craters" have been revealed in the Baltic Sea, North Sea and in other places of the World Ocean (see Fig. I.24). They are situated either in Quaternary deposits or in the fissures (or fractures) of pre-Quaternary hard rocks. Gases, primarily methane, escape from these fissures as well as from overlying Holocene muds (see Fig. II.16.1). In areas where these craters are especially numerous, concentrations of free H_2S in muds were highest. In general, there is no clear relationship between contents of elements in the muds of mud craters (pockmarks) and the positions of craters and acoustic anomalies (Romankevich et al., 1991).

Processes of sedimentogenesis occurring in extremely oxidizing conditions (i.e., in conditions of aerobic diagenesis) can be subdivided into several types. We examine here the processes of only two types occurring in (1) biogenic siliceous sediments, with their insoluble remains represented by pelagic clay; and (2) sediments represented by typical pelagic (red) clays. In the radiolarian zone of the Pacific Ocean, north of the equator, processes of sedimentation belong to the first type (Hein et al., 1979; Emelyanov, 1991_2; Svalnov, 1994). Here, sediments are represented by clay and biogenic silica (radiolarians, diatoms). Concentrations of Si in interstitial waters depend on the amount of biogenic siliceous remains in the clay. Water is undersaturated with respect to amorphous silica, and therefore the biogenic siliceous remains dissolve rapidly. The bottom water contains 8–10 mg/l SiO_2, whereas the solubility limit for SiO_2 at 1°C is about 90–25 mg/l (Pushkina, 1980). The content of SiO_2 in interstitial waters is generally the same as that in bottom waters. The amount of PO_4^{3-} content in interstitial solutions depends on the decomposition of phosphate-containing organic matter.

In the equatorial part of the Pacific, about 92–99% of the total amount of biogenic siliceous particles dissolve in the water column before reaching the bottom (Johnson, 1976). The remaining 8–10% of siliceous material which does reach the bottom continues to dissolve on the bottom surface and also in the top few tens of centimeters of sediment. The amount of siliceous remains and their grain size become markedly smaller with increasing depth below the sediment surface. The reason for such behavior is that dissolution of siliceous particles continues in the sediment. Therefore, only particles displaying the highest degree of resistance to

Fig. II.16.7. Distribution of authigenic minerals in bottom sediments of Atlantic Ocean (0–5 cm layer).
1—abundance of Fe- and Mn hydroxides (>7–10% of Fe_{cfb}+Mn_{cfb}); 2—dispersed Fe hydroxides: a—a few; b—many; 3–5—distribution of FMNs: 3—wide distribution of round nodules; 4—separate nodules (round); 5—flat concretions and iron crusts; 6–13—prevalence of authigenic formations: 6—iron sulfides; 7—calcareous oolites, aragonitic mud; 8—calcite, ankerite, dolomite; muddy-calcareous concretions; 9—calcite; 10—manganese carbonates, iron and manganese sulfides, vivianite, barite; 11—phosphates (recent and ancient); 12—hydrogethites–shamosites; 13—glauconite (recent and ancient); 14—phosphatic remains of fishes; 15–17—separate grains: 15—palagonite; 16—barite; 17—zeolites; 18—areas without authigenic minerals (marine-glacial sediments); 19—rift zone of Mid-Atlantic Ridge; 20—border of maximal distribution of grayish hemipelagic (terrigenous) mud with grains of iron sulfides; 21—isoline of 0.5% distribution of Mn (in cfb).

solution are preserved in the sediment sequence: spicules, large radiolarians and diatoms. Buried biogenic particles undergo additional solution within the top 20- to 40-cm layer of sediment. The grain size of particles tends to become thinner with increasing depth below the sediment surface. Several tens of thousands of years may be required to dissolve these particles (Hein et al., 1979, p. 385). The dissolution of siliceous particles shows a strong cyclical character. These cycles fluctuate on time scales of about 20–80,000 years. There are several hypotheses explaining these cycles (Hein et al., 1979, p. 386), but the most probable reason for cyclic recurrence was glacial to interglacial climatic oscillations during the Quaternary (dissolution of SiO_2 tended to become stronger during interglacial and weaker during glacial epochs). Also, cyclic recurrence in the primary productivity of organisms with silica skeletons occurred at that time.

About 50% of siliceous remains contained in sediments dissolve at the water–bottom GBZ (i.e., on the bottom surface). This factor facilitates structural jumps in sediment properties and leads to release of Cu, Fe, Si and other metals which had been absorbed by biogenic siliceous particles (Bischoff and Piper, 1979, p. 386). Amounts of elements passing into solution are so large that only siliceous remains in the bottom sediments are able to supply the major portion of Cu, Fe and Zn, which are preserved in the sediments of the equatorial part of the Pacific ocean. It was also found that Al in interstitial waters associates with SiO_2. This is because Al and Si are released into interstitial waters as a result of dissolution of SiO_{2am}. Subsequently, Al and Si are used up in the diagenetic development of aluminosilicates.

It is evident that biogenic activity is responsible for the concentration of Al in seawater (Mackenzie et al., 1978; see Bischoff and Piper, 1979). Ni and Mn may also be delivered to the seafloor with biogenic siliceous material (Leinen and Stakes, 1981).

The time for metals to be released during dissolution of SiO_{2am} is relatively short: only a few tens of thousands of years. Such an interval is necessary for the accumulation of 40–50 cm of sediment in which the amount of siliceous remains decreases sharply downwards.

Among authigenic minerals, which accumulate in the upper active sediment layer GBZ in open-ocean conditions, are ferruginous smectites $(Ca_{0.14}, K_{0.22}, Na_{0.4})$ $(Ni_{0.004}, Cu_{0.01}, Mn_{0.03}, Ti_{0.04}, Mg_{0.59}, Fe^{3+}{}_{0.53}, Al_{0.75}) \times (Al_{0.52}, Si_{3.48})\ O_{10}\ (OH)_2$ (Heinen et al., 1979, p. 369). Ferruginous smectites are not the product of weathering of volcanic remains, as was adopted earlier, but appear as a result of reactions with Si- and Fe-hydroxides. All this led to the inference that two types of material had been formed authigenically during the early stage of diagenesis in pelagic areas of the Pacific Ocean: Fe smectites, and FMNs and micronodules. Smectite is the most abundant mineral in clays in the Pacific Ocean (Svalnov, 1991).

The chief raw materials for production of smectites are Fe hydroxides delivered from the EPR; Si that accumulates in interstitial waters as the result of dissolution of SiO_{2am}; and Al, the accumulation mechanism of which in interstitial waters is the same as that for Si and includes the sorption of Fe- and Mn hydroxides (Heath and Dymond, 1977). Furthermore, ferruginous nontronite forms near hydrothermal sources, whereas Fe montmorillonite forms at a greater distance from them (Aoki et al., 1974). Cu, Zn, Mn, and Ni are constituents of smectites.

The presence of dissolved Si in interstitial waters in the radiolarian zone is a good for possible formation of smectites using Fe. Therefore, areas where such formation occurs are characterized by the separation of Fe and Mn: Fe is used for smectites, whereas Mn is used to build up manganese nodules. This is why the content of Mn in manganese nodules is higher for areas within the radiolarian zone than that for areas beyond the limits of this zone (Heinen et al., 1979, 1989). In red clays, the separation of Fe and Mn is weaker and the content of smectites is much lower, with the result that the enrichment of Fe in manganese nodules is greater. Manganese nodules in the radiolarian zone of the Pacific Ocean are characterized by increased Mn/Fe ratios. Moreover, they are also enriched in Cu, Ni and Zn. The reason for such a distribution of these elements is probably smectite formation.

The enrichment of Mn, Cu, Ni and Zn in manganese nodules in near-equatorial areas of the Pacific Ocean, as interpreted by the author, may be bolstered by two factors: high biological productivity and the OML, which are very distinct in this region (Emelyanov, 1991_3).

In the surface sediments of the radiolarian zone of the Pacific Ocean, smectites are the most important component of sediments. The total volume of this component is greater than that of manganese nodules. In smectites, the Zn contents are greater, but contents of Cu and Ni are smaller, than in manganese nodules. It is quite possible that in the future the importance of such deposits for commercial production of metals may become greater than that of manganese nodules.

Interstitial waters of ocean sediments are markedly richer in Fe, Mn, Cu, Ni, and V than seawater (Pushkina, 1980). This enrichment is closely related to the lithological–facial type of sediments.

Because the rate of dissolution of organic matter and SiO_{2am} rapidly decreases with depth below the sediment surface (Hurd, 1973), the transition of trace elements from solid phase to interstitial solution is maximum precisely in the uppermost (0–10, 0–20 cm) layers of pelagic muds (Hartmann and Müller, 1982, p. 295). There is evidence that in lenses composed of light ooze, supplied by mud-eaters and other benthic organisms, the contents of Mn, Cu, and Ni are markedly larger than those in dark-colored muds. The contents of Fe and Zn were the same in both types of muds. Biogeochemical microconditions of such or similar type are responsible for remobilization of Mn and associated Cu and Ni (Hartmann and Müller, 1982), leading to an increasingly higher release of these elements into interstitial waters. The reason for such a pattern of behavior is that the contents of organic matter and SiO_{2am} are markedly higher in the upper layer of sediments than in their lower layers. In sediments, oxidized compounds of Mn are in equilibrium with interstitial waters, when Eh values fall within the range of +400 to +500 mV (Lyle, 1978). The solubility of Mn increases by about an order of magnitude for a drop in Eh value from +500 to +450 mV, whereas the solubility of Fe remains unchanged (Borchert, 1970).

In the upper layer of clay, not only Mn (this element undergoes solution on the surface of particles) but also trace elements, which are structurally incorporated by Mn hydroxides, experience remobilization during early diagenesis. When interstitial waters become enriched in Mn, Cu, Ni, Zn and other trace elements, the concentration gradient formed will result in these elements diffusing along this gradient from interstitial waters into the bottom water. Oxidation of Mn^{2+} occurs in the lower microlayer of these water (see Fig. II.14.1). Cu, Ni, Zn and probably Mn^{2+} are adsorbed onto newly formed Mn hydroxides, which in turn are adsorbed onto the surface of manganese nodules.

This is evidence that concentrations of Cu in interstitial waters of the upper layer of pelagic sediments in the Pacific Ocean are higher than those in the lower layers of these sediments. The concentration of Cu in interstitial waters tends to decrease with depth below the sediment surface. This leads to the inference that there is no diffusion of Cu (at least from sediments at depths greater than 2 m below the sediment surface) from interstitial to bottom waters (Hartmann and Müller, 1982). At the same time, concentrations of Cu and some other trace elements in interstitial waters are different from those in bottom waters, which suggests that diffusion should occur there; elements diffuse into near-bottom waters from the uppermost 2 cm of sediment. Moreover, it is evident that remobilization of elements occurs in the upper 2 mm of ooze. If such is the case, then less than 1% of the total amount of elements that settled onto the seabed are buried in sediments. The remaining 99% dissolve or escape to seawater as a result of diffusion. The diffusion mechanism of metals from interstitial waters to bottom waters is especially effective when the rate of diffusion exceeds that of sedimentation, i.e., when the rate of sedimentation is less than 1 mm/1000 yr, or when the uppermost layers of sediments are disturbed (mixed) by the activity of benthic organisms (bioturbation) or by the effects of strong bottom currents, including erosion and resuspension. In the radiolarian zone of the Pacific Ocean, effective diffusion is aided by exceedingly high porosity of siliceous oozes (their water content is 80–85%).

The oxidation rate of organic matter on the seafloor, which appears to be the main driving force of diagenetic redistribution of matter, is greatest in the top 10–20 cm of sediment. After oxidation, the system is in the steady state. The increase of gradients in the concentration of various elements, including O_2, at the water–bottom boundary provide for greater efficiency of diffusion there. In the equatorial zone of the Pacific Ocean, i.e., within the zone of anaerobic reactions, decomposition of organic matter is maximum in the uppermost 10 cm of the sediment. About 150–450,000 years are required to dissolve this material completely, and this yields a thickness of the order of 40–50 cm for the diagenetically active layer (the upper active sediment layer GBZ) in the radiolarian zone. If organic-rich layers become buried within the sediment sequence, including zones dominated by oxidizing reactions, the formation of Fe sulfides and the ensuing cementation of these beds may occur there as a result of diagenetic processes.

Such pyrite-cemented layers have been found, for example, in the North American Basin, not far from Cape Hatteras, at depths of 4770–4950 m (Field and Pilkey, 1970), in Pleistocene–Eocene sediments. When such beds become exposed at the bottom surface (where the environment remains oxidizing), they undergo transformation into sediments with Fe hydroxide cementation (Rona et al., 1982).

In the upper active sediment layer, carbonaceous sediments are dominated by early diagenetic reactions of a particular type. In many respects these processes depend primarily on climate (in shallow-water seas, shallow-water areas of basin seas, and oceans) or the temperature of near-bottom waters. As mentioned earlier, the temperature barrier plays an important role in accumulation of various types of carbonates in the shallow areas (see Part I, Fig. I.8). Processes of accumulation of carbonate sediments, as well as their reformation, are especially strong in seas of arid climatic zones such as the Red, Arabian, Caspian, Mediterranean, and Caribbean seas. There, at the stage of sedimentogenesis–early diagenesis, sediments are dominated by processes involving precipitation of such minerals as aragonite, calcite and dolomite, the formation of magnesium-clayey aggregations (concretions) on the bottom surface (the top 0–25 cm) (Emelyanov, 1972), cementation of shallow-water sediments, development of hard carbonate rocks (beachrock) and hard-rock surface of the bottom (hard-ground).

Some of these cementation processes were described in Chapter II.14, while some of the background work on the subject is contained in papers by J. Chillingar et al., (1970) ("Carbonate Rocks"), Lisitzin et al., (1977), Logvinenko (1979), and Sellwood (1990). For a better understanding of what role the facial environments play in carbonate accumulation and in early regeneration of carbonates in sediments, we represent here only one transverse profile across arid zones of the Atlantic Ocean (see Fig. II.13.3). This profile shows zones of chemogenic–diagenetic production of carbonate minerals, including calcite, dolomite, and ankerite; as well as areas where modern magnesium-clayey aggregations (concretions) are formed during early diagenesis. To see these processes in more detail, see previous works on the subject (Emelyanov, 1972; Emelyanov, 1982_1; Lisitzin et al., 1977).

Chapter 17

Lithogeochemical Areas, Regions and Provinces

Differences in geotectonic, hydrodynamic and hydrochemical conditions under which sedimentary material enters and develops in the ocean and the fractionation of sediments present in the water column and caused by the effects of GBZs, influence the type and composition of sediments that accumulate in a particular region (Emelyanov, 1982). Ten distinct lithogeochemical areas (LGAs) can be recognized in the Atlantic Ocean by variations in chemical and mineral composition of sediments, various types of sedimentation, and the abundance of particular mineral forms of chemical elements in sediments (Fig. II.17). Sediments in these areas are characterized by difference in genesis, bulk chemical elemental composition, compositional ratios of the main sediment-forming elements, and granulometric and mineralogical characteristics. Whereas some lithogeochemical areas are restricted to certain climatic zones, others extend beyond their limits. Within the limits of terrigenous-clastic and terrrigenous-clayey LGAs (i.e. in near-continental hemipelagic areas of the ocean), processes of mechanical fractionation of denudation products of land rocks are the main factors affecting the formation of chemical composition of sediments. The process of mechanical fractionation also takes place beyond the limits of these areas, but there it is of secondary importance. The chemical composition of suspended matter and sediments is mainly formed here by biochemical and, in some cases, chemical differentiation. The distribution of chemical elements in pyroclastic, hydrothermal and edaphogenic LGAs is conditioned by the composition and amount of pyroclastics, hydrothermal, and edaphogenic (products of bedrock destruction) materials, respectively. Due to the existence of a variety of complex factors (interactions) involving various types of zonation; different hydrodynamics; geotectonic and physicochemical regimes; and the effects of GBZs—particular types of LGAs or provinces (LGPs) with various types of sediments become localized in occurrence within the limits of each lithogeochemical region (LGR). There are two types of zonation—climatic and vertical—the contribution of which is especially evident in LGR or LGP sediment composition. For each climatic zone, there are specific (indicatory) LGRs which cannot be found in other latitudinal belts (Emelyanov, 1982).

Sediments from much of the known LGAs, LGRs and LGPs show signs of specific lithogeochemical indicators (associations of minerals and chemical elements, their relative proportions, and geochemical coefficients) and biogenic remains, which are a reflection of the past sedimentary environment and therefore can be used for paleoreconstructions (Emelyanov, 1982; Emelyanov et al., 1989).

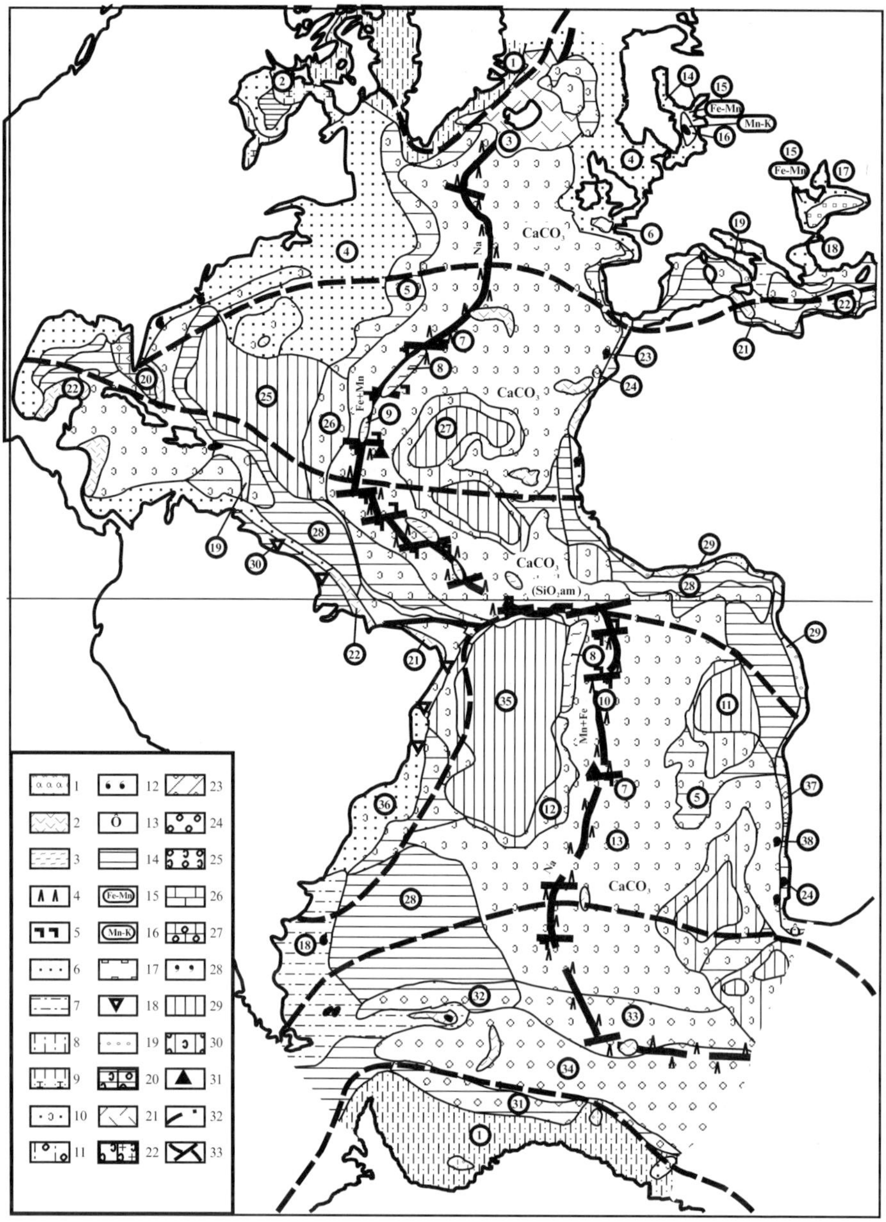

Fig. II.17.1. Scheme of lithogeochemical areas (LGA), regions (LGR) and provinces (LGP) in surface bottom sediments of Atlantic Ocean.

Legend: 1–2—pyroclastic LGA: 1—andesite-clastic LGR, 2—basaltic LGR; 3—endogenic–hydrothermal LGA with ore and metal-bearing manganese–iron LGR; 4–5—endogenic–volcanogenic LGA: 4—basalt LGR; 5—ultrabasic LGR;

6–13—terrigenous–clastic LGA: 6—quartzy-silicitic LGR; 7—feldspar–quartzy–silicitic LGR; 8—quartzy–feldspar–silicitic LGR; 9—quartzy–feldspar carbonate–siliceous LGP; 10—quartzy–feldspar–carbonate LGP; 11—quartzy–feldspar silicitic–siliceous LGP; 12—feldspa–quartzy LGP with glauconite; 13—feldspar–quartzy LGP with phosphates;

Fig. II.17.1. ***cont'd*** 14–20—terrigenous-clayey LGA: 14—alumosilicate LGR; 15—shallow alumosilicate manganese-concretional LGR; 16—alumosilicate manganese–carbonate LGR; 17—alumosilicate iron–sulfidic LGR; 18—carbonate–organic-clayey (mangrove) LGR; 18—iron–silicate hydrogoethite–shamositic LGR; 19–20—mixed alumosilicate LGR with provinces: calcite–alumosilicate (a) and opal–silicate (b);

21–22—biogenic–carbonate LGA: 21—shallow aragonite–calcitic LGR with provinces: shallow aragonite–calcitic, shallow dolomite–magnesium–calcitic and shallow calcite–aragonitic; 22—deep-sea calcitic LGR with provinces: deep-sea foraminiferal–calcitic (a), deep-sea aragonite–calcitic and deep-sea chemogenic–biogenic (magnesium) (b);

23–25—biogenic siliceous LGA. 23—shallow carbon phosphate–opal LGR; 24—deep-sea diatomic–opal LGR; 25—deep-sea mixed calcite-opal LGR;

26–27—chemogenic–aragonitic LGA: 26—oolite–aragonitic LGR; 27—aragonite–agregate LGR;

28—chemogenic–glauconitic LGA (with glauconitic LGP);

29–30—pelagic oxide–clayey LGA: 29—pelagic oxide–clayey manganese LGR; 30—pelagic mixed calcite–clayey LGR;

31—ferromanganese crusts; 32—borders of climatic zones; 33—rift zone of Mid-Atlantic Ridge with transform fractures.

Numbers in circles show the most characteristic chemical components and elements for sediments.

1—$SiO_{2\ silicate,\ quartzy}$, (Al, K, Na, Zr);
2—$SiO_{2\ amorph,\ quartzy}$, $CaCO_3$ (clastic);
3—Ti, Fe, V, (Mn, P, Cr, Ni);
4—$SiO_{2\ bulk}$, Zr, Sn;
5—$SiO_{2\ silicate}$, $CaCO_3$, Al;
6—$CaCO_3$, (C_{org});
7—Cr, Ni, V (for edaphogenic ultrabasic sediments);
8—Mn, Fe, P, Pb, B, Zn, Mo, Ba, Cu, Ni, Cd (for metalliferous sediments);
9—Mn, Fe, Ni, Co, Cu (mainly for hydrothermal manganese crusts);
10—Pb, Zn, Mg, Ag, Co, Cr, Ni (for rocks of rift zone of Mid-Atlantic Ridge; sometimes, for micro-layers of rift-valley sediments);
11—$SiO_{2\ silicate}$, Ba, Al, Fe, Mn, Mo, Ni, Co, Cu, C_{org} (oxidized clays in Angola Basin occur in small areas; the main pelagic area in this basin is occupied by reduced gray (clayey) mud;
12—$SiO_{2\ silicate}$, $CaCO_3$, (Fe, Mn);
13—Fe, Ti, V, (Cr, Ni—for edaphogenic basaltic sediments distributed in rift valleys);
14—Fe (crusts), Mn, P;
15—FMNs, $SiO_{2\ silicate}$, Fe^{3+}, Mn^{4+}, Co, Mo;
16—Mn^{2+}, C_{org}, P, Mo, Se, AS, Ba, Zn, Fe^{2+}, SiO_2 amorph;
17—$Fe_{sulfidic}$, Mo, Se, As, Cu, C_{org};
18—$SiO_{2\ quartzy,\ silicate}$, Al, Zr, Sn;
19—Na, P, Zr, B, Be;
20—$CaCO_3$, $SrCO_3$, (C_{org});
21—$CaCO_3$, $SrCO_3$;
22—$CaCO_3$, $MgCO_3$;
23—$SiO_{2\ silicate,\ quartzy}$, Al, P, Fe, K, Na, $CaCO_3$;
24—$SiO_{2\ quartzy,\ silicate}$, P;
25—$SiO_{2\ silicate}$, Al, Mn, Fe, Ni, Co, Cu;
26—$SiO_{2\ silicate}$, $CaCO_3$, Al, (Mn, Fe);
27—$SiO_{2\ silicate}$, Al, Mn, Fe, Zr, Cr, V;
28—$SiO_{2\ silicate}$, Al, Fe, K, Na, (Ni, Co, Cu, Zn);
29—Fe, (C_{org});
30—$CaCO_{3\ chemogenic}$, C_{org}, ($MgCO_3$, Mo, Se, As, Ni);
31—SiO_2 silicate, amorph, quartzy;
32—$SiO_{2\ silicate,\ amorph}$, Al;
33—$CaCO_3$, $SiO_{2\ amorph}$;
34—$SiO_{2\ amorph}$;
35—$SiO_{2\ silicate}$, Al, Mn, Fe, Ni, Co, Cu, Mo;
36—$SiO_{2\ quatzy,\ silicate}$, $CaCO_3$, $MgCO_3$;
37—$SiO_{2\ amorph}$, C_{org}, P, Mo, Se, As, U, Au;
38—$SiO_{2\ quartzy,\ silicate}$, Al, P, Fe.

For detailed characteristics of chemical zonation of ocean bottom sediments see Emelyanov, 1982_1.

PART III

LITHOGEOCHEMICAL BARRIERS AS LEADING FACTORS IN ORE FORMATION

General

There are two major principles which provide the basis for using regularities in hypergenic processes in geochemistry: (1) heterogeneity of ore provinces (according to conditions of geochemical prospecting) and (2) a historical approach to the problem (historical geochemistry, that is, a field of science on the development of migratory processes in the course of geological history) (A.I. Perelman, 1989, pp. 28–29). We believe that these principles can be considered as a basis for geochemical predictions and prospecting of solid mineral resources both in the present-day and ancient seas and oceans.

We can identify three accumulation stages of mineral resources: (1) The stage of sedimentogenesis, when sediments enriched in ore elements or embryonal ores accumulate. This stage is controlled by facies conditions. The principle of mechanical, chemical and biological differentiation is fully realized at the stage of sedimentogenesis. (2) The stage of secondary reworking (rewashing of sediments and removal of non-ore material) and diagenetic transformation of sediments enriched in ore components into ore deposits. (3) The stage of catagenetic transformation of ore-bearing sediments into ore deposits. Description of environmental conditions under which mineral resources—geochemical relics—accumulated in the geological past is the subject of paleoceanography. Nowadays, paleoreconstruction of a sedimentary environment responsible for sediment and ore accumulation in the geological past can be performed by using different criteria, based on deep-sea drilling data collected by the R/V *Glomar Challenger* (see *Initial Reports of DSDP* and *JOIDES Resolution*), which have been summarized in the Russian literature by A.P. Lisitzin and co-authors [Monin and Lisitzin, 1980, Emelyanov et al., 1989, and many other Russian scientists]. These works on the subject will make it possible to appreciate the geochemical evolution of seas and oceans and to determine more exactly sedimentary conditions under which a particular geochemical relic formed in the past.

The combination of a number of revealed barriers forms a complete system, displaying the infrastructure of the ocean. The integral contribution of this infrastructure (including the entire complex of its functions) to lithogenesis and ore formation is, at least, no smaller than the contribution from processes of distribution of material due to tectonic, climatic, or circumcontinental and vertical zonations in the ocean (Lisitzin, 1974, 1978, 1991, 1994). These processes, which are complicated by interaction between barriers and GBZs, combine to form a picture of the present-day distribution of sedimentary features in the ocean. These implications occur not only as the present-day sediment distribution (Bezrukov, 1979) and composition of sediments, but also as the distribution of lithogeochemical facies [for the first time in application to the Atlantic Ocean (Emelyanov, 1982)]. Now it is evident that the combination of sedimentation processes occurring on different scales, on the one hand, each characteristic of particular climatic and depth-related zones, and the effects of barrier zones, on the other hand, forms a framework for the geochemical processes in oceans. Some of these factors have a global, transregional influence on depositional processes, whereas others are represented only locally, in specific regions. This circumstance is not grounds for underestimating the role of barriers. On the contrary, it draws attention to them, because they have functions very important for ore formation: mobilization and concentration of ore material in certain, relatively small areas of the seafloor.

With these two functions in view, all lithogenetic barriers related to ore formation can be subdivided into two parts: mobilizing and concentrating. Naturally, such a functional division of barriers is rather conventional, and it is valid only in application to certain types of mineral resources. In pelagic areas of the ocean these include FMNs, metalliferous sediments, and deep-sea polymetallic sulfides. The relativity of such an approach is very clear when we begin to discuss the mobilizing function of such a barrier as basalt–water, which is only one variety of the previously described water–bottom and hydrothermal solution–seawater GBZs. Waters of hot springs are responsible for leaching all chemical elements from basalt, including iron and manganese. However, as soon as a mixed solution becomes neutral or weakly alkaline (pH on the order of 7.0–7.8) as a result of mixing of hydrothermal solutions with oceanic water, iron becomes immobile and this makes its occurrence in solution impossible, whereas Mn, which continues to be present in solution and separates from iron, moves away from the hydrothermal spring to accumulate at great distances from its place of origin. Thus, under conditions of this type, the mobilizing function of a barrier changes to a concentrating one. However, division of barriers into two principal types, according to their main functions, is correct, because this allows the reader to concentrate on the conventional aspects of ore material. On the other hand, if such division is correct, it points to the generality of ore formation processes for all mentioned types of mineral (hydrothermal) resources in pelagic areas of the ocean. The principal similarity of ore formation processes is that reaction zones lie at relatively small distances from the sources of material participating in the formation of ore deposits. By passing through mobilizing barriers, mobile forms of metals, both in metalliferous sediments and sulfides, which entered the ocean from various sources, become depersonalized. The major mass of these metals, which exists in hydrogenous forms, pass into solid ore material at concentrating barriers.

Geologists engaged in the studies of concretions and their local variability often point out a variety of factors contributing to formation of concretions. Also, many scientists (Cronan, 1980; Rona, 1986, Lisitzin et al., 1990; Lisitzin et al., 1993) maintain that formation of sulfides and metalliferous sediments is conditioned by a number of factors. Thus, they are mainly talking about the influence of oceanographic factors, which, however, are only indirectly related to the formation mechanisms for high ore concentrations. Among these are factors such as the thickness of a transparent water layer; the occurrence of rich deposits of FMNs mainly on slopes of submarine hills, highs and seamounts; a large amount of total heat flux in rift zones of the North Atlantic Ridge, etc. At the same time, some important events involving mechanisms, environmental conditions, and the initial and final parameters of ore-forming processes remain hidden in the background.

It is believed that a more reasonable approach to the problem will make it possible to increase the efficiency of studies, including prospecting works: development of models of ore formation should become the first step of each research program. This is why the study of barriers where the concentration of ore material occurs has become a subject of special importance.

Among the major distinguishing features of concentration barriers such as interstitial water–sediment, bottom water–sediment and water–basalt barriers is the proportion of solid and liquid phases being in contact, which controls the initial and final parameters of nodule and hydrothermal ore formation. This is because processes involving the adsorption of hydrogenic forms onto the solid phase (sediment particle) from seawater or, vice versa, leaching of elements from rocks and the passage of their mobile forms to solution, are possible only under particular environmental conditions, for a concrete phase-to-phase ratio and redox (pH) potential. As we know, FMNs are very enriched in transitive elements. The main mechanism that causes migration of these elements from interstitial and bottom solutions and accumulation of these elements on the FMNs is the adsorption of metal ions onto particle surfaces. Consequently, the formation of FMNs requires such sediments which provide for an optimal ratio between liquid and solid phases. Belonging to them are, first of all, sediments which adsorb only a certain portion of mobile metals derived from interstitial and bottom waters, whereas the other part of these metals should be adsorbed onto the nodules. On the other hand, sediments should contain considerable amount of material having increased adsorption ability, first of all, Fe and Mn. Nodules are more abundant in eupelagic clay facies, but there they are less enriched in Cu and Ni than the diagenetic nodules of the radiolarian belt.

The water content of sediments may by considered as an indicator for liquid-to-solid-phase ratio in the interstitial water–sediment system. As revealed by geochemical observations in the Pacific Ocean and the Mediterranean Sea (Emelyanov, 1965), the pH value is a function of the water content in sediment, which is why this parameter can be used to characterize the phase-to-phase ratio. When the water content of sediments is low, the pH value is generally similar to that in ocean water. At the same time, the pH value becomes substantially smaller as the water content of sediment increases.

Similarly, the optimal water-to-rock ratio at the water–basalt barrier is possible only in the presence of intensive water fluxes circulating through bottom rock and has a close relationship with anomalously high permeability and heat transfer of the ocean floor within local structures. Experiments (Bischoff and Jackson, 1975; Mottl and Seyfried, 1980) and studies of hydrothermal systems at individual bottom locations of the East Pacific Rise (Rona, 1986) have shown that when the water flux is greater than the ability of this water to react with rocks (high water-to-rock ratio), the acidity of solutions increases due to extraction of Mn^{2+} and OH^- ions from ocean water, as well as due to the increasing ability of solutions to solve metals. In contrast, when the velocity of water is lower than its ability to react with rock (low water-to-rock ratio), the acidity of solutions decreases, up to neutral and low-alkaline, and the concentrations of metals in them decrease.

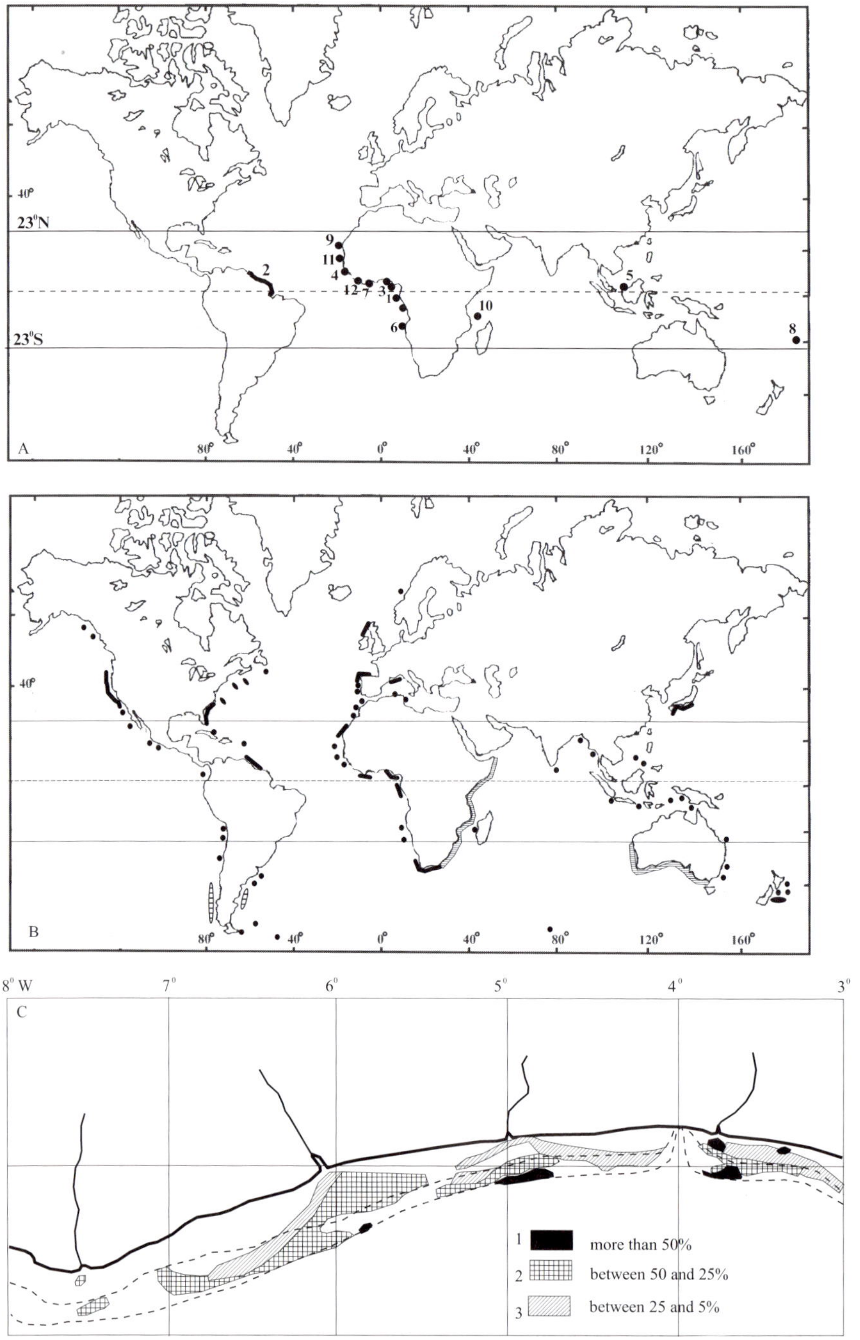

Caption see page 459

The uppermost layer of these hydrogoethite–chamosite sediments (several centimeters or in some places even a few tens of centimeters) is weakly oxidized and has a brownish gray or even brown coloration (Nikolayev and Oskina, 1994). The pellet size is variable: from 0.4 to 0.7 mm. They can be divided into three main subdivisions: black (large), brown and yellow (small). Yellow pellets are the most friable. Lithified pellets (as well as lithified sediments in general) fall within the areas of stations 3186, 3190, 3195, 3196, 3197, and 3199 (Fig. III.4.2), i.e., at depths from 100 to 150 m. Lithification of sediments is the most distinct in the upper layer of sediments (0–15 cm). In sediments below this layer, destruction of pellets occurs. Pellet-rich sediments are represented by kaolinite, quartz, potassium feldspar, calcite, and glauconite, as well as Fe hydroxides, gypsum and goethite (Table III.4.3).

Many glauconite grains are distorted by edgings, which suggests that this is chamosite. Chamosite was identified as one of the main minerals of pellets by Giresse (1965), Nikolayeva et al. (1971), and later by many other researchers (Odin, 1985, 1988). Among biogenic remains in sediments, there are numerous planktonic foraminiferal tests, diatomic frustules, sponge spicules, and chitine remains of crabs. Biogenic remains are seen to be markedly filled up with (or replaced by) glauconite–chamosite and pyrite (Nikolayev and Oskina, 1994).

Sediments at test area V-20-III (Table III.4.4) are rather young: Kuptsov (1994) dated them at 2.5±0.6 thousand years using the carbon-14 method. This is evidence that both sediments up to 1.5–1.9 m in thickness and numerous pellets in these sediments formed during the Holocene, so that their age is no more than 3.3 Ka. The Holocene age of the sediments has also been confirmed by foraminiferal analysis. The average sedimentation rates are very high: 29–50 cm/Ka (or 0.7–6.2 mm/yr).

Core B2-3197 is considered to be very characteristic of sedimentary environments (for locations see Fig.III.4.2). The uppermost 105 cm of the sediment core are represented by pelitic and aleuro-pelitic muds ranging in age from 3600 to 3080 years. Below this layer, at depths of 170–105 cm, there are sands the age of which dates back 8280–5010 years. Such abrupt changes in the ages of sediments are probably related to events involving considerable variations in the hydrodynamic regime and sediment rewash which occurred in the recent geological past (during an early Holocene transgression).

Fig. III.4.1. Scheme of the principal distribution of the green facies and the glauconite. After Odin, 1988; Odin et al., 1988; Emelyanov, 1972, with some additions by the author.

A. General distribution of te verdine facies in present-day oceans. 1—Ogooue; 2—Amazon to Orinoco; 3—Niger; 4—Koukoure; 5—Sarawak; 6—Congo + N'Dogo; 7—Ivory Coast; 8—New Caledonia; 9—Senegal; 10—Mayotte; 11—Casamance.

B. Presence of glaucony on present-day sea-bottom. After Odin and Fullagar, 1988, p. 325. Shaded areas indicate presumed glaucony for which no mineralogical details are available. Areas shown are generally larger than are the actual deposits in order to facilitate legibility.

C. Content of fecal pellets rich in iron in the bottom sediments of the Ivory Coast (in % from the sediment). After Martin, 1972.

Table III.4.1. The content of chemical elements in ferruginous (hydrogoethite–shamosite) sediments (0–3 cm) in test area V-20-III (for location see Fig. III.4.2). After Lukashin et al., 1994.

Station	Depth, m	C_{org}	Si	Fe	Al	P	Ti	Ba	Mn	Ni	Co	Cu	Zn	Cr	V	Pb	Sr
		%								10^{-4}%							
V 2-3180	60	2.57	17.49	11.74	13.7	0.17	0.73	0.16	0.05	52	28	40	121	143	212	32	193
V 2-3181	70	2.40	-	12.24	13.0	-	0.65	0.045	0.059	47	36	50	130	121	185	-	165
V 2-3182	63	2.28	-	15.2	10.9	-	0.65	0.064	0.09	54	37	50	131	130	223	30	292
V 2-3186	175	0.87	13.30	34.6	6.0	0.43	0.42	0.12	0.11	57	87	30	188	169	333	-	323
V 2-3190	102	1.48	14.43	31.1	7.5	0.43	0.49	0.13	0.08	47	60	40	226	151	329	13	288
V 2-3191	100	2.47	-	11.67	12.7	-	0.65	0.079	0.04	53	26	40	157	128	228	18	246
V 2-3192	180	2.35	-	10.33	12.1	-	0.64	0.059	0.04	53	28	50	106	115	202	17	265
V 2-3196	175	0.72	11.45	42.2	5.2	0.47	0.36	0.13	0.12	67	96	30	290	183	458	-	308
V 2-3197	495	1.72	15.40	20.2	8.9	0.23	0.53	0.11	0.06	58	40	45	179	154	229	-	254
V 2-3199	260	2.18	-	12.1	10.8	-	0.64	0.065	0.04	62	29	40	137	134	202	17	230

Table III.4.2. Granulometric and chemical compositions of hydroghetite–shamosite sands, aleurites Aleuro-pelitic oozes and monomineral fractions of shelf pellets off Liberia and the mouth of the Congo (sections 12 and 6 in the Fig. III.4.1A). After Emelyanov, 1972.

Fractions in mm and chemical components	Shelf of Liberia (12)*					Area near the mouth of the Congo (6)*					Pellets from sector 6		
	729*	739	744	750	737	780	789	54	Atl-796	56*	51	54	56
	55**	55	50	60	69	27	42	51	65	162	90	51	162
	1. Granulometric composition, %												
<0.1	14.5	10.6	45.2	47.9	53.4	1.2	1.0	-	10.9	54.5	-	-	-
0.1-0.05	11.9	59.0	18.6	4.7	10.6	55.1	17.7	-	7.5	18.7	-	-	-
0.05-0.01	26.6	31.5	12.2	7.1	9.1	21.9	25.5	-	21.8	13.4	-	-	-
<0.01	47.5	19.1	24.1	40.5	26.7	41.9	55.8	-	59.7	13.4	-	-	-
	2. Chemical composition, %												
SiO_{2tot}	39.25	38.24	46.92	28.71	16.00	41.16	58.51	64.40	36.92	-	40.24	29.38	29.72
TiO_2	0.99	0.74	0.66	0.82	0.37	1.15	1.10	0.24	1.04	0.45	0.36	0.40	0.32
Al_2O_3	20.28	9.64	9.42	13.11	7.12	17.57	10.74	4.75	21.41	-	19.31	10.06	9.49
Fe_2O_3	8.95	8.80	9.62	16.72	7.29	16.43	15.18	16.27	15.00	18.60	20.36	37.04	55.02
MnO	0.05	0.03	0.02	0.15	0.01	0.13	0.08	0.14	0.03	0.02	0.02	0.14	traces
Na_2O	3.30	1.39	1.48	1.87	1.36	2.63	2.90	0.42	2.31	-	1.40	1.07	1.20
K_2O	1.01	0.70	0.58	0.91	0.31	1.20	1.20	0.34	1.15	-	0.87	0.54	1.71
P_2O_5	0.28	0.50	0.45	0.65	0.48	0.17	0.13	0.41	0.35	0.38	0.23	0.90	1.01
CO_2	2.74	11.84	8.09	10.99	23.47	0.61	0.94	1.54	0.79	9.02	-	-	-
$CaCO_3$	6.23	26.93	18.39	24.99	53.97	1.59	2.14	3.03	1.80	20.51	-	-	-
SiO_{2am}	cn.	0.56	cn.	0.60	0.58	cn.	0.58	0.77	0.53	1.11	-	-	-
Corg	3.02	-	1.40	0.78	1.20	2.62	2.58	0.29	2.95	0.89	-	-	-
H_2O	3.82	20.1	2.14	5.44	2.25	4.10	4.50	1.86	4.02	-	5.40	5.08	4.24

Continued

Table III.4.2. Granulometric and chemical compositions of hydroghetite–shamosite sands, aleurites Aleuro-pelitic oozes and monomineral fractions of shelf pellets off Liberia and the mouth of the Congo (sections 12 and 6 in the Fig. III.4.1A). After Emelyanov, 1972.—*continued*

Fractions in mm and chemical components	Shelf of Liberia (12)*					Area near the mouth of the Congo (6)*					Pellets from sector 6		
3. Content, 10^{-4}%													
Ba.10^{-4}	<2.0	-	<2.0	<2.0	<2.0	-	<2.0	<2.0	-	<2.0	-	-	-
Zr.10^{-4}	1.7	-	2.8	0.2	3.6	-	1.8	9.3	-	5.5	-	-	-
Cr.10^{-4}	1.3	-	11.9	6.4	9.4	-	8.2	5.2	-	25.4	-	-	-
Ni.10^{-4}	6.0	-	2.9	-	2.0	-	-	-	-	-	-	-	-
V.10^{-4}	18.2	-	3.2	2.6	0.8	-	3.0	7.8	-	-	-	-	-
Mo.10^{-4}	<0.5	-	-	<0.5	-	-	<0.5	<0.5	-	<0.5	-	-	-
Ce.10^{-4}	<0.5	-	<0.5	<0.5	<0.5	-	<0.5	<0.5	-	<0.5	-	-	-
Be.10^{-4}	<1.0	-	<1.0	<1.0	<1.0	-	<1.0	<1.0	-	<1.0	-	-	-

*)Number of sector in Fig. III.4.1.
**)Number of section and depth in m.

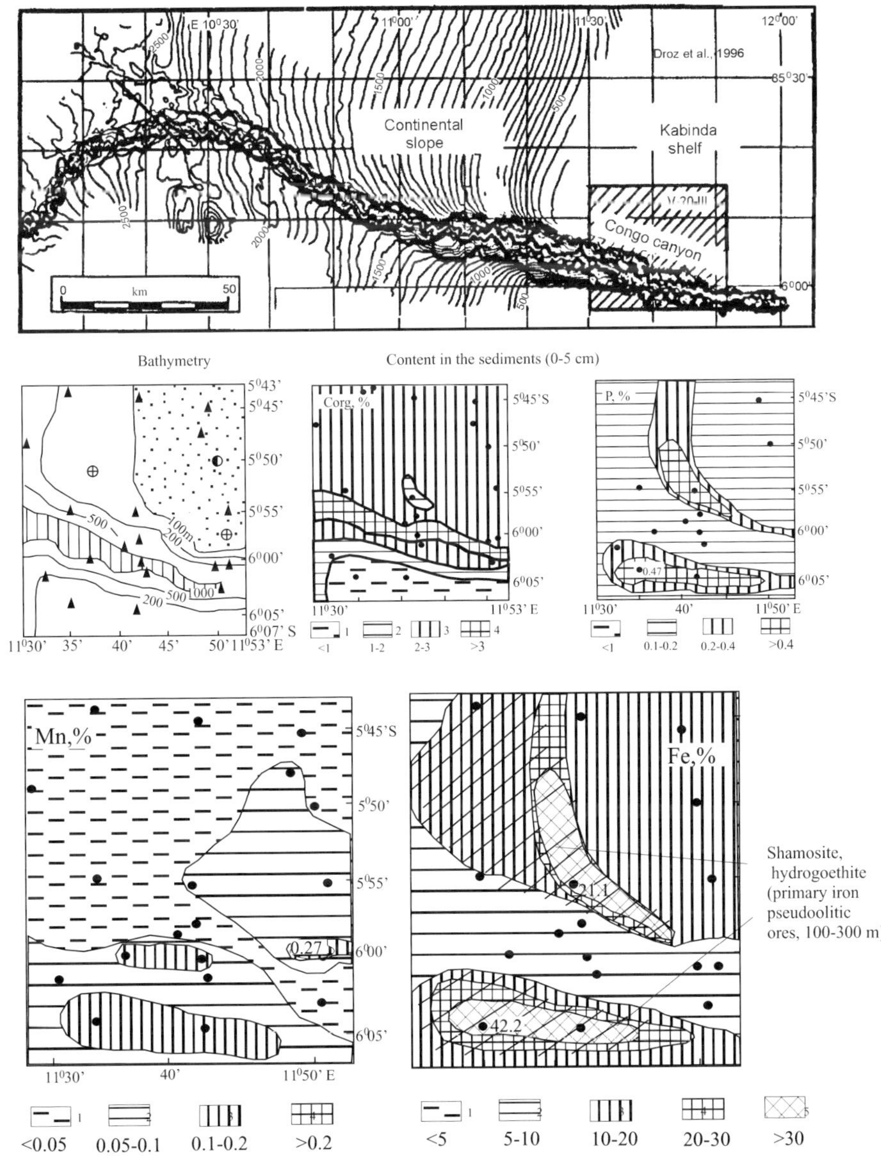

Fig. III.4.2. Congo (Zaire) canyon (after Droz et al., 1996, isobaths in m), position of test area V-20-III and geological stations, surficial sediment types, and distribution of C_{org}, P, Mn, Fe, and Ti in surficial sediments (0–5 cm), in %. Sidelong line on the Fe scheme show the areas where surficial sediments are slightly lithified (the rates of sedimentation are low). scheme has been completed by author and according to Lukashins et al. (1994).

Table III.4.3 Mineral content (in %) of the surface layer of ferruginous hydrogoethite sediments in test area V-20-III at the mouth of the Congo, Atlantic Ocean (see Fig. III.4.2) (x-ray analysis data). According to Lukashin et al., 1994; age, according to Kuptsov, 1994)

Station	Depth, M	Quartz	K-feldspars	Kaolinite	Calcite	Halite	Gypsum	Goethite*)	Other minerals	X-ray amorphous phase*)	Age of bottom sediments (years)	
											a	b
V 2-3180	60	5.1	1.2	8.8	0.7	3.6	0.3	-	-	80.3	10–155	103–1680
V 2-3181	70	4.5	1.2	8.6	0.6	2.5	0.5	-	-	82.1	-	-
V 2-3182	63	4.8	2.1	7.9	0.6	3.2	0.5	-	-	80.9	15–80	230–2200
V 2-3186	175	2.5	-	5.3	2.4	0.3	0.4	17.8	Phillipsite 6.6; Aragonite 1.8	62.9	-	-
V 2-3190	102	3.0	-	6.6	1.5	1.6	-	12.3	Aragonite 2.2	72.8	-	-
V 2-3191	100	4.3	1.7	7.5	1.7	4.2	-	-	-	80.6	-	-
V 2-3192	180	4.1	2.2	7.2	2.4	4.7	-	-	Aragonite 2.9	76.5	10–90	1390–2340
V 2-3196	175	2.3	-	5.3	2.1	-	-	25,4	Aragonite 2.1; Magnetite 1.4	61.4	-	-
V 2-3197	495	3.3	-	6.8	3.0	2.0	0.5	1.6	Aragonite 2.2	80.6	-	-
V 2-3199	260	4.7	1.7	8.1		3.0	2.4	-	-	80.1	-	-

*)Mineral shamosite was not found by the authors; a – upper layer (0–5 cm), b – lower layer (10–200 cm).

Table III.4.4. Rates of sedimentation of iron (hydrogoethite–shamosite) in sediments at the mouth of the Congo, test area V-20-III. Defined by ^{210}Pb. After Kuptsov, 1994.

Station (see Fig. III.4.3)	Sediment interval (horizon, cm)*	Rates of sedimentation, mm/yr
V2-3180	1–5	0.67
V2-3186*	1–5	6.20
V2-3190*	0–5	3.95
V2-3191	0–5	0.72
V2-3192	0–1	-
V2-3197*	3–5	1.69
V2-3199*	0–5	1.06

*⁾Pellets are noticeably lithified in the upper (0–15 cm) layer. For depths in the stations see Table III.4.3.

Hydrogoethite–chamosite sediments, both recent (Late Holocene) and pre-Holocene, have also described for other areas of the World Ocean. Pre-Holocene sediments are especially characteristic of the continental shelf off Surinam/Guyana (site 2 in Fig. III.4.1A). They are thought to have formed at depths of 30–50 m in times when the sea level was low, approximately 18–13 thousand years ago. Owing to a global sea level rise of some 120–150 m during the Holocene transgression, these sediments occurred at the shelf edge, at depths as great as 150–200 m (Odin, 1988).

Facies of green marine clays (or hydrogoethite–chamosite sediments) are localized in their occurrence in the river mouths of the tropical zone, which supply a large amount of products of chemical weathering from land to oceans, which are markedly enriched in Fe, both in dissolved and particulate forms (Fig. III.4.3). The largest contributor of iron at the mouth of the Congo are the waters of this river itself (Eisma and Van Bennekom, 1978). Dissolved iron is almost completely removed from water solution before reaching the shoreline. This leads to the development of a "seal", which is strongly enriched in Fe hydroxides, a short distance from the mouth (Fig. III.4.3B). Nowadays, formation of this seal occurs in the area between the coastal plain and shallow-water shelf. At the salinity barrier, iron is removed from solution to accumulate as Fe hydroxide flakes in hydrodynamic (or geomorphological) traps in the shelf area. Another source of iron, which is responsible for the enrichment of this element both in pellets and sediments, comes from interstitial waters, which are very rich in dissolved iron. Contents of Fe^{2+}_{diss} in interstitial waters of sediments (0- to 5-cm sediment layer) at test area V-20-III were as follows: from 20 to 290 μg/l quite near the bottom of the shelf area, and from 180 to 1780μg/l at the bottom of the Congo canyon (see Fig. III.4.2) (Vershinin et al., 1994, p. 209). Thus, contents of Fe^{2+}_{diss} in ferruginous sediments of the shelf area were found to be much smaller than those in reduced muds on the canyon floor in the shelf area or in gray sediments (muds of the continental slope; see station 3174 in Table III.4.5). It is evident that much of the dissolved Fe^{2+} in ferruginous sediments has aggregated to goethite. Fe^{2+} migrates upward from reduced sediments of the underlying layer (Eh = –80 mV) into a thin (2–5 cm) weakly oxidized layer (Eh = +50 mV), where it becomes oxidized at the Eh barrier before finally being deposited as goethite–hydrogoethite.

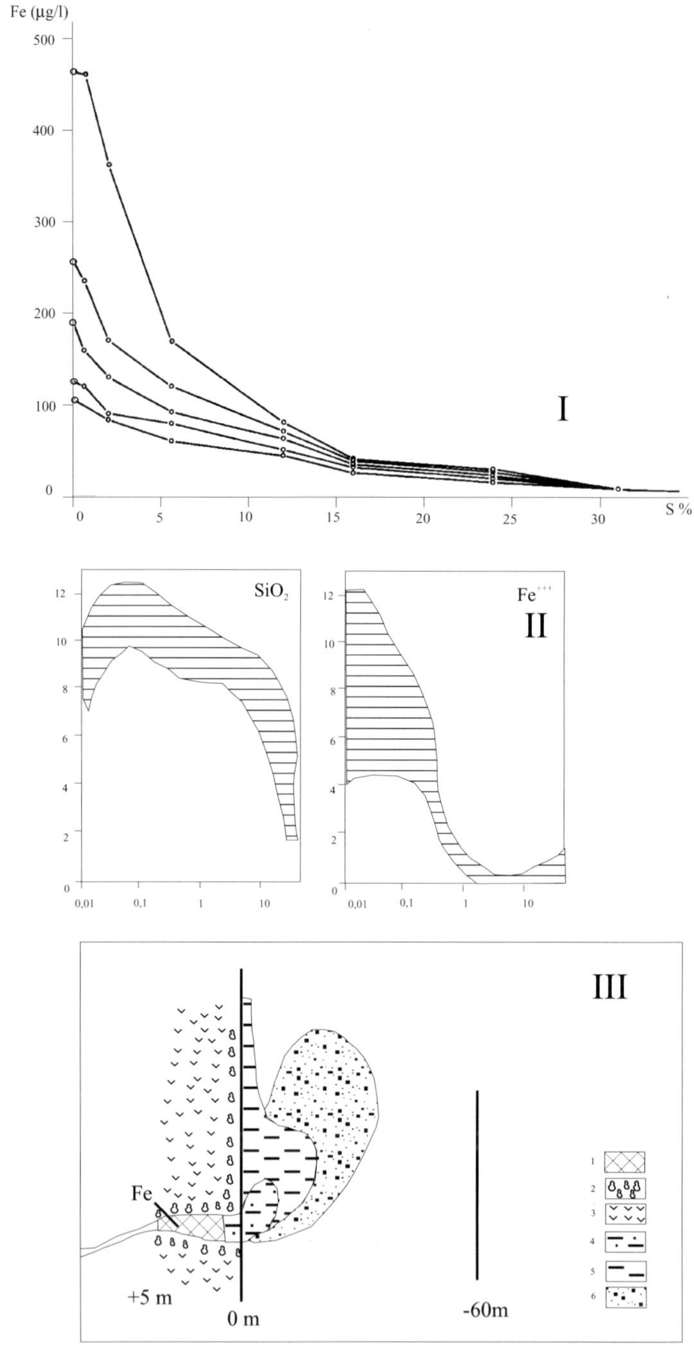

Fe (µg/l)
500
400
300
200
100
0
0
5
10
15
20
25
30
S %
I
SiO2
12
10
8
6
4
2
0
0,01
0,1
1
10
Fe+++
II
III
Fe
+5 m
0 m
-60m
1
2
3
4
5
6

Table III.4.5 Limits of pH values and content of dissolved forms of chemical elements in interstitial water of sediments in the area of influence of the Congo River (test area V-20-III) of Angola Basin. After Vershinin et al., 1997.

Form elements, pH	Unit of measure-mont	Station of test area V20-III[x]			Station V-3174	Station V-20-I
		V-3180	V-3192	V-3197		
SO_4^{2-}	g/l	1.72–2.43	2.33–2.53	2.13–3.73	2.03–2.73	2.23–2.54
H_2S	mg.l^{-1}	n.d.	n.d.	n.d.	n.d.	1.7–5.0
NH_4^+	µg-at.l^{-1}	220–2300	150–500	390–2100	220–740	70–440
Alk	µg-at.l^{-1}	2.25–15.31	2.10–3.65	3.40–6.46	4.17–9.32	3.35–7.46
Mn^{2+}	µg.l^{-1}	310–630	160–440	250–470	3530–7450	350–500
Fe^{2+}	µg.l^{-1}	40–190	30–140	40–140	430–2350	80–180
Cl^-	g.l^{-1}	17.51–18.44	-	-	17.79–19.08	17.43–17.83
pH	-	7.55–8.10	7.75–7.85	7.55–7.75	7.30–7.45	7.95–8.05

[x] For position of test areas and stations see Figs. II.1.7 and III.4.2. n.d. – not determined.

The gravity "flow" of interstitial waters along the sloping surface of the bottom at the shelf edge is thought to be the mechanism that delivers Fe^{2+} (by means of interstitial waters) (Lukashin et al., 1994). There, muddy waters reach the surface of the bottom and are deposited at the redox (Eh) barrier.

Ferruginous pellets are commonly lithified near the mouth of the Congo River, where they are represented by goethite–hydrogoethite. Glauconite, siderite, and Fe sulfides are found in the sediments (Nikolayeva et al, 1971; Emelyanov, 1972; Odin, 1988).

Several hypotheses explaining the accumulation of hydrogoethite–chamosite (ferruginous oolitic and pseudo-oolitic) facies have been proposed (Odin, 1988). According to them, oolitic and pseudo-oolitic sediments might have been formed in different sedimentary environments, as that of (1) freshwater and marine lagoons in the equatorial zone, which are traps for fluvial sediments; (2) outer river cones on the shelf of the ocean; (3) shallow bays; (4) areas of sand bars; and (5) open shelf areas (Figs. III.4.4, III.4.5). The formation of ferruginous oolite and pseudo-oolite occurs at depths of not more than 80–90 m.

Fig. III.4.3. Iron geochemistry at the river–sea GBZ.
I. Filtered iron versus salinity in the Zaire estuary (November 1976) depending on filter interstitial sizes used; from top to bottom, plots for interstitial sizes of 0.1, 0.45, 0.22, 0.05 and 0.025 µm. After Figueres et al., 1978, p. 334.
II. Concentration of SiO_2 and Fe^{3+} in the estuary of the tropical Kuru River in relation to the electrical conductivity (salinity), in M hos/cm. After Roche, 1977 (citation after Odin and Sen Gupta, 1988).
III. Sedimentological model of the deposition of the green facies in the estuary of the tropical river. After Odin and Sen Gupta, 1988.
1—zone of the formation of iron oxides; 2—mangroves; 3—swamps; 4—sandy, delta deposits; 5—terrigenous mud; 6—zone of the formation of green facies.

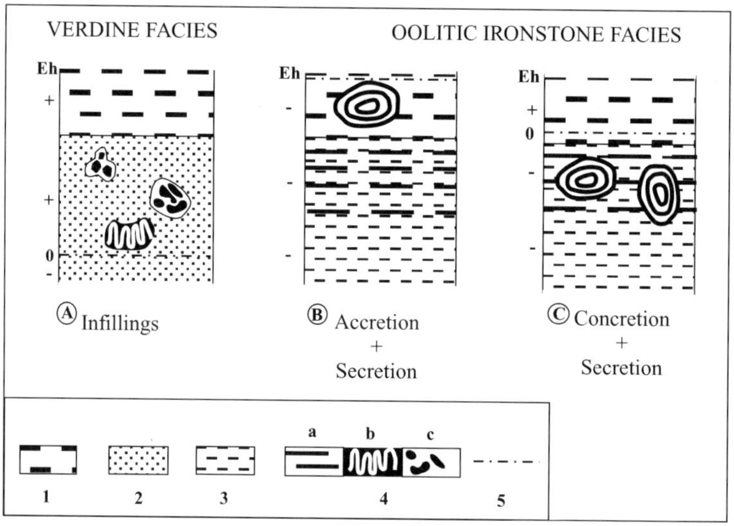

Fig. III.4.6. Location of the zero Eh plane in relation to the sediment surface. In recent verdine facies (A), it is located well below the sediment surface and a sheltered microenvironment is necessary for the growth of ferriferous green clays in the first decimeter of sediment. In oolitic ironstone facies, the zero Eh plane has been postulated to be above the sediment surface if the ooids form by accretion (B), and at the sediment surface if the ooids grow by an intrasedimentary concretionary process or derive from an accretion-replacement process (C). After Odin and Sen Gupta, 1988 (from Odin, 1988).
1—sea; 2—sand; 3—mud; 4—green clay (a, b, c-different kinds of clay); 5—0 Eh plane (position of Eh).

Oolites consist mostly of varying proportions of hematite, hydrogoethite, siderite, and chamosite. In addition to oolites, also found there are mineralized ferruginate fragments of shells and whole shells of mollusks, as well as burrows. Periodically, organic remains are so abundant in oolitic ores that they are referred as "fossil ores", especially in the United States (Strakhov, 1947, p. 308). Ores from Kerch (Crimea peninsula) contain large amount of fragments and whole shells of *Dressensia* and *Didacna*. Organic remains are cemented by various ferruginous minerals, which are commonly represented by oxides and hydroxides.

According to Strakhov (1947, p. 310), "the conditions of a very shallow coastal sea were responsible for the development of oolitic hematite–chamosite–siderite ores, which are represented by deposits in bays, bights, lagoons and in the upper part of the shelf area; moreover, such factors as submarine topography influence the type of sediment that accumulates in a particular region of the coastal zone". Thus, Strakhov's hypothesis, formulated 50 years ago, was confirmed by studies of several authors (10–25 years ago; for illustration see Fig. III.4.3), as well as by the author's ideas on genesis of ferruginous oolitic sediments (future ores) (Emelyanov, 1972).

As noted by Strakhov (1947, pp. 324–325), the deposition of oolitic ores occurred in geomorphological traps, generally as thin (clayey, sandy-clayey and sandy) layers of sediments. The sedimentation conditions most favorable for "the formation of marine oolitic iron ores were those of an inland sea or transverse coast, which are typical for submerging mountain land" (p.327).

According to Strakhov, transgressions and regressions are of crucial importance for the development of shelf ores ("there is a strong correlation between the behavior of marine ores and the oscillatory motions of the earth"). Our data completely confirm this conclusion. We suggested a scheme displaying three stages of development of ferruginous pseudo-oolitic and oolitic ores (Fig. III.4.4). Stage I is transgressive. On the shelf, not far from the mouth of a river (some distance away from the mouth, in the direction of the coastal current), there is an area where accumulation of green marine clays occurs. These clays contain high contents of Fe hydroxides (up to 10–30% and, rarely, to 40–42% of the total sediment sample) (Fig. III.4.5). Sediments, which are represented mainly by sands and muds, are intensely reworked by mud-feeders and burrowing organisms. The location of the layer with the largest difference in Eh values is shown in Fig. III.4.6. Manganese is transported to the ocean beyond the limits of the shelf area, and at the shelf edge, pellets are converted into hydrogoethite pellets (pseudoolites). During a global drop in sea level of some 100–150 m (stage II, regressive), the reworking of green marine clays occurs, resulting in the oxidation of ferruginous pellets, so that they become enriched in Fe, and then muddy (non-ore) material is evacuated beyond the limits of the ore bed. At stage III (transgressive), the ore bed is buried and sediment undergoes genetic reworking (with the formation of siderite and other carbonaceous Fe sulfides, etc.). If the transgression–regression–transgression cycles are repetitive, this leads to the formation of a multilayer ore bed (showing alternation of ore, poor ore and non-ore layers), as can be seen in many old ferruginous pseudo-oolitic and oolitic ores (Strakhov, 1947). As a result of further evolution of the ocean, a fully developed ore bed ultimately ends up within an on-land sedimentary sequence, where it may be exposed to further diagenetic or even catagenetic processing, to finally become ore with high metal concentrations, making it attractive for commercial use.

Glauconite. Glauconite occurs in weakly reduced (gray) sediments, in peripheral areas of oceans (Fig. III.4.1). A weakly reducing environment, providing a sufficient reserve of reactionable iron, causes, in combination with two other two important factors, the diagenetic formation of glauconite. These factors are as follows: (1) the presence of montmorillonite in mud and mixed-layer mica-montmorillonite; and (2) the occurrence of a sandy-aleuritic layer with an increased water resistance, providing for an active interaction of mud with potassium-rich interstitial waters (Lisitzina and Butuzova, 1979, p. 71). According to the theory of these authors, reactionable Fe in muds forms collomorphic aggregations. Small lumps of greenish gray coloration are initially formed in mud, without any visible changes in structure and composition of the mud. Later, Fe and other elements are concentrated around these lumps to finally form clearly outlined globules. These globules have green coloration, and their X-ray patterns are consistent with those of glauconite. At this stage, such globules are Fe-rich mica-montmorillonite layered features, where montmorillonite packets dominate over co-existing mica packets. It is evident that Fe is a component of bedded minerals with degraded structure, where the type of compositional ratio is 2 : 1 (Burst, 1958; Hower, 1961), and Al in octahedral positions is substituted for this element. Ions of potassium, the amount of which is sufficient for reaction, are delivered from interstitial waters. Potassium becomes trapped among the layers of partially transformed mineral. P participates in a variety of complex chemical reactions, with the result that the structure of glauconite becomes more ordered. The regularity of structure of this mineral increases as the amount of micaceous packets and K_2O contents increase.

In sediments of the World Ocean, there are several distinct features that can be related to the authigenic origin of glauconite (Lisitzina and Butuzova, 1979, p. 72). These features are as follows: "the distribution of glauconite in the sediment column is irregular and patchy, unrelated to the bedded structure of the sediments; the concentrations of globules is maximum in coarse, sandy-aleurite muds; the terrigenous fragments, remains of organisms, ash particles and other elements occur in glauconite–rich mud; mineral grains and cementing mass are replaced for glauconite; participation of glauconite in construction of internal cavities of shells".

In a fresh approach to this problem, we believe that facial conditions most favorable for formation of glauconite appear in those areas where two GBZs intersect: the OML and the water–sediment interface. In contrast to chamosite–hydrogoethite, the formation of glauconite is not as closely related to the river–sea GBZ. This probably explains why glauconite is so abundant in aleurite sediments of the continental slopes of oceans, including areas at large distances from the mouths of rivers.

The distribution of glauconite and pyrite in reduced sediments (at least in the sediments of the eastern part of the Pacific Ocean) is very different. Glauconite accumulates predominantly in sandy-aleuritic and aleurite-clayey muds, whereas pyrite is found in clayey mud. Also, in clayey mud there are maximum C_{org} contents. In pelagic areas of the ocean, beyond the limits of the reducing zone, glauconite and pyrite are essentially absent.

In individual sediment strata, the process of mineral formation begins with the appearance of glauconite, which is followed by pyrite (Lisitzina and Butuzova, 1979). The formation of glauconite requires a weaker reducing environment than that of pyrite, so that the development of glauconite slightly lags behind that of pyrite. The reason for such a division between glauconite and pyrite is probably that although these muds in sandy-aleuritic sediments are slightly depleted in C_{org} (relative to clayey mud), their porosity is much greater. The intensity of sulfate-reducing processes in these sediments is weaker than that in muds. However, sandy-aleuritic sediments are probably dominated by decomposition processes of silicate material, resulting in the release of components required for formation of glauconite. The decomposition process is evidently favored by the presence of CO_3^{2-} and HCO_3^- ions. Their concentrations are increased in the sediments of the reducing zone.

According to the model by Odin (1988, p.329), glauconite forms at a somewhat shallower depth than is generally accepted (Lisitzina and Butuzova, 1979). As follows from this model (Fig. III.4.7), the depths most favorable for glauconitization of recent sediments are shelf edges, that is, depths ranging from 60–70 to 450–500 m. Moreover, glauconitization begins with biogenic–planktonogenic (foraminiferal) sediments, and only then the process begins in bioclastic shelly sediments, containing large amount of pellets. A condition required for glauconitization to be successful, especially in the case of bioclastic sediments, initially accumulating mainly in the coastal part of shelf area (from 0 to 5–10 m), is transgression, when bioclastic sedi-

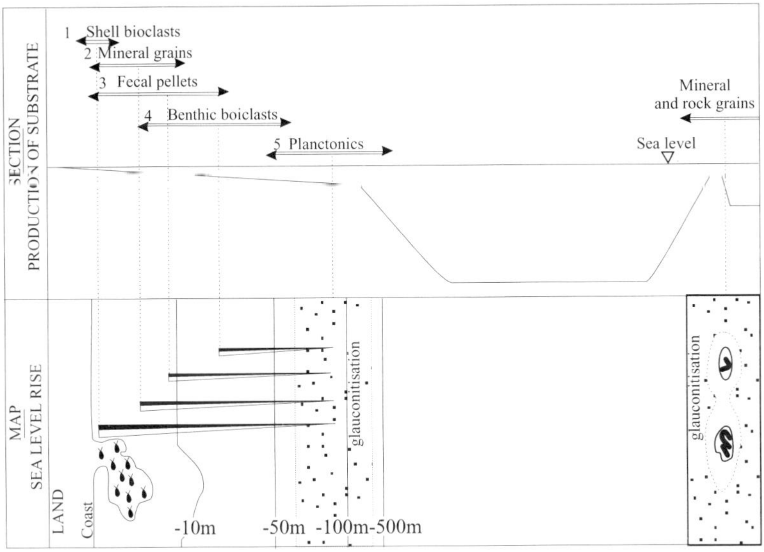

Fig. III.4.7. Distance between the zone of production of substrates and the zone of glauconitization. Larger the distance, the larger the transgression needed before glauconitization is possible. Glauconitized shell bioclasts will denote an important transgressive phase; glauconitized planktonic tests will not.
After Odin and Fullagar, 1988, p. 329.

ments should become available within the depth range from 60 to 450 m, as a result of this transgression. Therefore, glauconitization of biogenic-shelly sediments occurs during the transgression stage, whereas for foraminiferal sediments, accumulating initially at moderate depths (60–500 m), this stage is not necessary.

It is significant that the existence of a warm, wet climate is not a necessary condition for the formation of glauconite: this mineral, in contrast to chamosite-hydrogoethite, forms both in arid and moderate humid climatic conditions (Fig. III.4.1).

Glauconite and glauconite-containing sediments (like a chamosite) are markedly enriched in Fe (Nikolayeva et al., 1971). However, the process of enrichment of these sediments is somewhat different from that described above: in this case, such a strong source of Fe as river water or riverborne suspended matter of tropical rivers is not required. Because glauconite is formed diagenetically in the uppermost part of the sediment strata, the main input of Fe comes from its dissolved form (Fe^{2+}), increased concentrations of which occur in interstitial waters of reduced sediments.

Front of Coastal Upwelling and Accumulation of Phosphorites

Phospate accumulation and enrichment of sediments in phosphorus occur predominantly in coastal upwelling zones (Baturin, 1969, 1978; Emelyanov, $1973_{1,2}$; Jenkyns, 1986) (Fig.III.5.1). Recent accumulation of slightly phosphatic sediments (0.5–10% P or 0.2–5% P_2O_5) takes place in two different facies environments, at both sides of the front of coastal upwelling (Fig. III.5.2, see also Fig. II.4.3, II.4.4): slightly phosphatic sapropelic diatomic ooze accumulates landward of the front, at depths of 50–140 m; whereas slightly phosphatic or phosphatic shell–foraminiferal and mixed carbonate–terrigenous sediments accumulate seaward of this front (within the depth range from 100 to 340 m). Thus, the front of coastal upwelling, especially in areas where it is well pronounced, represents an irresistible barrier for both sediment facies.

New phosphates originate in facies of diatomic, mixed diatomic–terrigenous and terrigenous muds, accumulating in the middle part of shelf zone. However, phosphorus content in these sediments is commonly rather low, from 1 to 2% (Emelyanov, 1973_2). Shelf diatomic oozes are the most characteristic type of slightly phosphatic sediments (Tables III.5.1, III.5.2). These oozes are very common in oceanic areas, where the coldest, more nutrient-rich deep waters reach the ocean surface. Development of this type of sediments in the arid zone of the African shelf is the most characteristic feature of modern sedimentation in the Atlantic Ocean.

The intensity of upwelling is highest on the continental shelves off Peru and Namibia. Intermediate water off the Namibian coast is characterized by low temperature, increased phosphate and CO_2 contents, and decreased values of pH, Alk and O_2. On the continental shelf, these waters move upward and come to the surface in areas where water depths are not more than 50–150 m (Emelyanov and Lyakhin, 1973). The above-mentioned hydrological and chemical parameters experience rapid variations with distance offshore. The result of such a situation is a strong physicochemical boundary responsible for the occurrence of the biogeochemical boundary in seawater and sediments. In shelf areas between this boundary and the shore, environmental conditions spur the production of phytoplankton. In these regions, dead phytoplankton results in the development of a huge mass of organic detritus, much of which sinks through the water column without dissolving, to be finally buried in the bottom sediments. A vivid reflection of this process is the distribution and composition of recent sediments in shelf areas, which are represented there by diatomic and slightly diatomic sapropel-like semiliquid oozes (at depths of 50–143 m), bearing the smell of H_2S, and low Eh values (values no more than -230 mV). These sediments contain up to 67% SiO_{2am} diatomic frustules. Oozes are strongly enriched

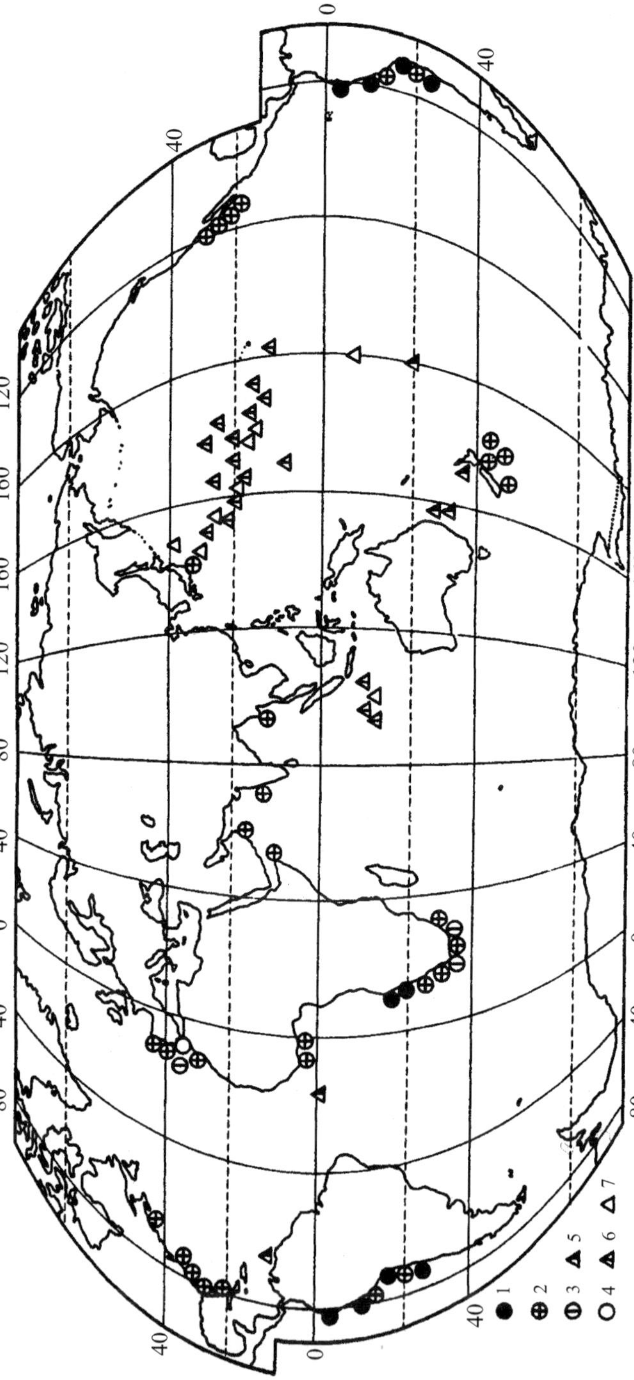

Fig. III.5.1. Distribution of phosphorites on the ocean bottom. After Baturin, 1978. 1–4—phosphorites on continental margins: 5–7—phosphorites on ocean mountains (1—Holocene; 2 and 5—Neogene; 3 and 6—Paleocene; 4 and 7—Cretaceous).

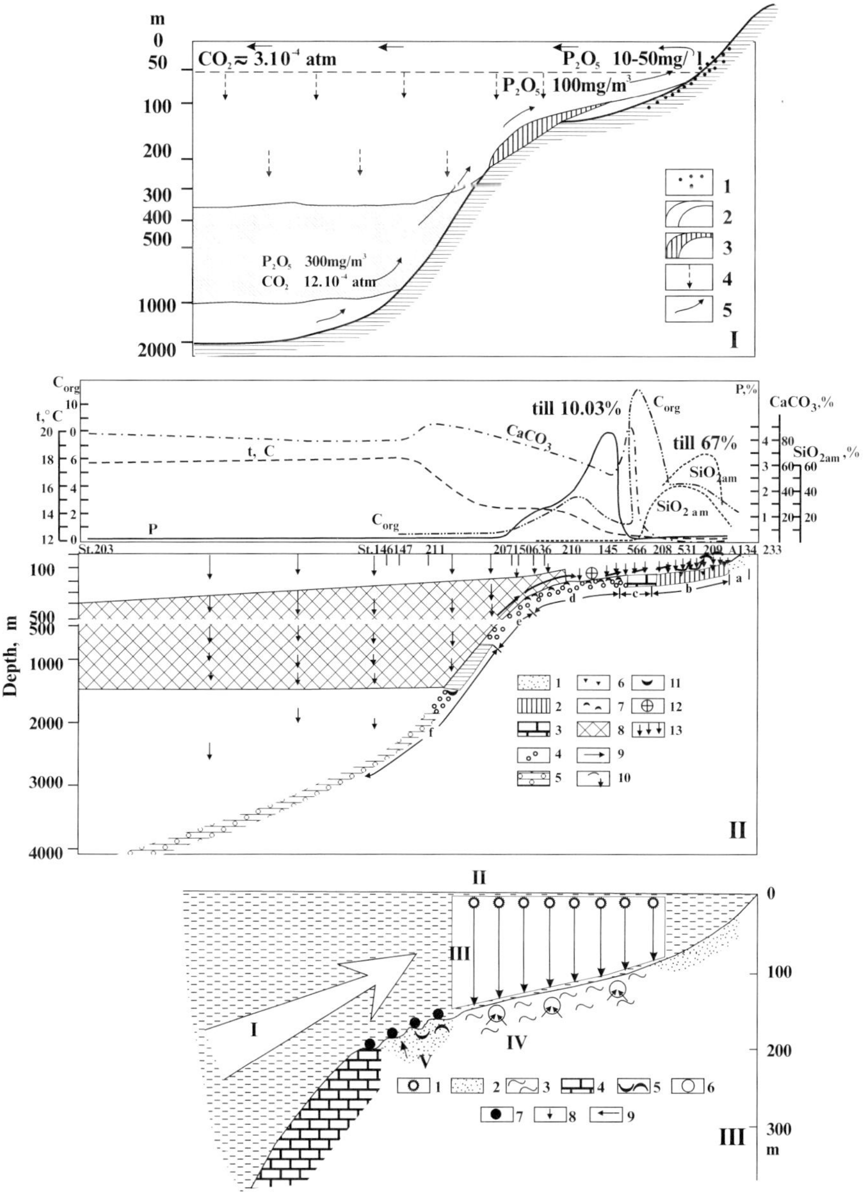

caption see page 478

in C_{org}, P, Mo and other elements, but are considerably depleted in Fe, Ti, Ba, as well as Al_2O_3 and K_2O. There are many scales and skeletal remains of fish, and also authigenic phosphorites, but low contents of $CaCO_3$. The field of diatomic oozes is surrounded on all sides by terrigenous muds and aleurites with low contents of $CaCO_3$. As for the whole Atlantic Ocean, maximum contents of organic matter are found in terrigenous (up to 16.06% C_{org}) rather than diatomic sediments (Emelyanov, 1973_2).

Accumulation of phosphorus in shelf sediments had been associated with accumulation of silica and organic matter, as well as the remains of fish and mammals. Organic detritus is a major supplier of phosphorus to interstitial waters, which are source of phosphatic aggregates (nodules). The contribution of other types of biogenic material to the process is of minor importance. As a result of secondary rewashing of diatomic and terrigenous sapropel-like oozes, the phosphorus and biogenic opal is partly lost.

A transition occurs from muds either to biogenic carbonate oozes or (rarely) to terrigenous sands and aleurites. In most sediment samples collected in the coastal zone off Africa, phosphatic sediments are of biogenic, carbonate type (30.78–78.7% $CaCO_3$). These sediments are characterized by low contents of amorphous silica and high contents of organic carbon.

Fig. III.5.2. Models and schemes of the formation of phosphorites on the continental margin in upwelling areas of the ocean: I—model after Kazakov, 1939; II—scheme (profile) after Emelyanov, 1979; III—model after Baturin, 1986.

I. 1–3—facies: 1—near-shore gravelly and sandy; 2—phosphoritic; 3—carbonate sediments; 4—sinking of plankton remains; 5—direction of currents.

II. Facial-geochemical profile I-I through the shelf and continental slope of the West Africa and formation of phosphorites: 1—terrigenous sand and coarse aleurite; 2—diatomic ooze; 3—aleuro-pelitic mud or ooze of mixed composition (terrigenous-low siliceous-low calcareous, sometimes calcareous, foraminiferal, enriched on C_{org}); 4—shelly sand, foraminiferal or mixed foraminiferal-shelly, high calcareous (>50% $CaCO_3$); 5—foraminiferal ooze (>50% $CaCO_3$); 6—recent phosphates; 7—ancient phosphorites; 8—oxygen minimum layer (OML), enriched in PO_4, CO_2 and others; 9—upwelling (rising water from OML to the shelf); 10—areas of fast sinking of surface water; 11—phosphorites forming at the lower border of the OML on the continental slope; 12—the main along-surface current; 13—the sinking organic detritus (Od) and pellets (P).
Facial zones of sedimentation: a—zone of near-shore terrigenous sedimentation (depth 0–50 m) and reworking of the upper layer of sediments by waves and near-shore currents; b—zone of intensive biogenic sedimentation (diatomic plankton and organic detritus and early diagenetic formation of soft phosphates (depth 50–120 m); c—transition zone between b and d (depth 120–150 m); d—zone of weak or zero sedimentation rewashing of and redeposition of sediments by strong near-bottom currents and zone of enrichment by phosphorites that formed before in zone b (depth 150–350 m); e—zone of low or zero sedimentation (depth 350–600 m); f—zone of normal (particle-by-particle) sedimentation on the continental slope (depth 600–4000 m).

III. 1—upwelling and delivery of phosphorus into surface waters; II—biological productivity; III—biosedimentation; IV—diagenesis and formation of phosphoritic concretions; V—rewashing of sediments during changes in sea level and relict concentration of phosphatic concretions.

Legend: 1—plankton; 2–5—bottom sediments; 2—sand; 3—biogenic siliceous ooze; 4—biogenic calcareous ooze; 5—shelly; 6—forming phosphatic concretions; 7—lithified phosphatic concretions of earlier generations; 8—deposition of phosphorus from seawater on the bottom in the composition of organic detritus; 9—migration of phosphorus in sediments during diagenesis.
re are curves of the contents in the surface bottom sediments (0-5 cm) of $CaCO_3$, SiO_{2am} and C_{org}, and temperature (t°C) in the surface seawater at top of profile II.

A major part of phosphatic material present in the sediments of the African coast is concentrated in various phosphate concretions and aggregates. In diatomic and low-silica oozes, these features are represented by recent (Holocene) sediment, and in all other cases they are old (pre-Holocene) sediments.

Considerable amounts of organic detritus are contained in all genetic types of sediments of the coastal area of Western Africa. This form of phosphatic compounds is characteristic only of recent low-phosphatic and phosphatic sediments. All types of shelf sediments in the area of Walfish Bay contain large amounts of various friable (unconsolidated) aggregates. They can be divided into five main groups: (1) organic matter (phytodetritus); (2) diatoms algae, which are clamped by organic filaments; (3) foraminiferal tests with organic "cement"; (4) terrigenous particles and foraminiferal shells with organic "cement"; and (5) terrigenous clayey material (Emelyanov, 1973_2, 1979_2).

The sediments also contain large amount of fish remains (skeletons, teeth, scales), which make up about 1–10% of the total sediment. The size of bones is variable, more frequently from 0.1–2.0 to 10–20 mm or more. This material is represented by vertebrae, ribs, fragments of fish heads and so on. Among all fish remains, teeth are the most resistant to dissolution and therefore the most widespread. However, accumulations of teeth are nowhere near as large as those of fish bones. The majority of fish remains are clean and have gray or brownish gray coloration. Only some of them are replaced or covered by phosphatic material. The fish teeth are most frequently replaced by phosphate (Emelyanov, 1979_2).

Slightly phosphatic and phosphatic sediments contain only small amounts of Fe sulfides. In oozes, sulfides are commonly represented by hydrotroillite.

Phosphorus in the sediments is represented not only by biogenic remnants, but also by authigenic grains. Among black grains, oolite-like and oval grains are predominant. The angular and elongated grains or phosphatized remains of fish are rarer. Both types of grains are composed of very dense and glassy material, with a smooth, shiny surface. Oolite-like grains vary in size from 0.5 to 0.005 mm (mostly, from 0.25 to 0.05 mm).

They are commonly represented by collophane; rarely, by francolite. Gradual transitions between these minerals are common. The occurrence of fluorapatite is also possible.

Four types of dark, brown and yellow phosphate aggregations have been identified: (1) phosphatized clots, which are typically 5–10 cm across; (2) friable phosphate aggregations with dimensions of up to 1–4 cm; (3) granular dense pieces up to 3–4 cm in size; and (4) dense massive pieces of dark brown color up to a few centimeters in size (Baturin et al., 1970; Emelyanov, 1973_2, 1979). Phosphates show signs of transition from clot-like and friable structure of light brown color to dense brownish and finally to very dense (glassy) black phosphates. In sediments which fall within this sequence, the parameter which most often increases is the refractory index: from 1.579–1.591 for friable, light phosphates to 1.591–1.630 for glassy, black phosphates (which are spherical or irregular in shape).

As revealed by geochemical studies (Baturin, 1969; Baturin et al., 1970; Emelyanov, $1973_{1,2}$; Emelyanov et al., 1975; Emelyanov, 1979_2), accumulation of phosphate material and development of phosphates in the shelf area off Africa (Fig. III.5.3)

Table III.5.1. Chemical composition of low-phosphatic and phosphatic sediments of the shelf of West Africa (layer 0-5, sometimes 0-10 CM) and content of <0.01 mm fraction in them (fraction and elements SiO_{2tot}.–C_{org} in %; F–Mo, in 10^{-4}%)

Station	Depth, m	Fraction <0.01, mm	$SiO_{2\,total}$	TiO_2	Al_2O_3	Fe_2O_3	MnO	P_2O_5	P	Na_2Or	K_2O	CO_2	$CaCO_3$	SiO_{2am}	C_{org}	F	Ba	Cr	Zr	Ni	V	Mo
Diatomic oozes																						
AK-157	75	41.4	-	0.14	-	1.48	0.01	1.22	0.53	-	-	3.74	8.50	42.44	3.66	-	-	41	-	8	-	8
AK-152	76	64.3	-	0.12	-	1.62	tr.	1.38	0.60	-	-	1.93	4.39	40.40	4.28	-	1040	75	70	53	156	38
AK-159	80	44.1	-	0.17	-	1.21	0.01	2.44	1.6	-	-	1.46	3.32	35.11	4.46	-	820	95	130	71	570	62
Shelly-foraminiferal sands																						
Atl-581	130	15.2	40.74	0.10	4.03	1.23	tr.	1.60	0.70	1.95	1.06	16.06	36.41	0.86	2.69	-	-	-	-	-	-	-
AK-144	148	15.6	-	0.12	-	0.94	0.005	1.63	0.71	-	-	30.93	70.34	0.53	4.71	-	210	72	70	44	37	14
AK-145	160	9.7	3.78	0.09	10.04	2.08	tr.	11.52	5.03	1.37	0.38	22.60	51.40	0.28	2.87	-	200	131	70	42	29	13
Atl-602	182	22.2	7.38	0.17	1.43	1.36	tr.	1.23	0.54	1.60	0.42	33.26	75.53	0.30	2.60	1910	<200	64	110	30	<10	-
AK-210	200	9.9	4.06	0.08	-	1.07	tr.	4.06	1.77	-	-	30.11	68.8	0.12	3.59	-	2200	250	70	51	28	10
Terrigenous sands, auleurites																						
Atl-611	230	12.6	50.42	0.90	7.01	8.02	0.02	2.19	0.96	1.73	2.10	7.26	16.51	1.46	1.29	-	-	-	-	-	-	-
Atl-654	250	6.7	43.92	0.27	8.14	16.90	tr.	4.25	1.85	0.89	3.60	2.81	6.41	1.10	0.76	6440	-	-	-	-	-	-
Shelly-foraminiferal sands																						
Atl-636	300	3.9	4.57	0.07	3.59	0.81	tr.	2.91	1.27	1.60	0.34	32.68	74.32	tr.	3.57	4050	<200	85	<20	39	<10	-
Atl-657	305	13.2	20.33	0.17	2.40	2.74	tr.	1.28	0.56	1.20	0.92	28.39	64.56	0.32	1.15	-	<200	65	<20	<20	14	-
AK-150	345	18.6	-	0.11	-	1.35	tr.	1.96	0.86	-	-	32.07	72.93	0.29	2.01	-	-	-	-	-	-	-

*) For location of stations see Fig. II.1.7.

Table III.5.2. Chemical composition of low-phosphatic sediments (0–5 cm layer) of the shelf of Namibia ($CaCO_3$–Mn in %; Ba–V, in 10^{-4}%). After Lukashin et al., 1994

Station	Depth, m	Sediment type	$CaCO_3$	SiO_{2am}	C_{org}	P	Si_{tot}	Al	Ti	Fe	Mn	Ba	Sr	Cu	Zn	Ni	Co	Cr	V
Natural dry sediment																			
V-3224	103	Tm, d, s	15.2	8.5	8.83	2.04	13.62	1.78	0.17	1.60	0.01	120	630	25	117	91	8	112	86
V-3226	240	Tm	27.2	3.3	2.45	4.69	4.56	0.26	0.10	1.48	0.002	1300	3000	30	47	48	6	97	54
V-3215	265	Fo	66.5	1.5	4.04	2.07	2.78	0.53	0.11	0.74	0.004	170	1900	45	54	53	5	67	53
V-3228	565	Tm, d	39.3	9.3	3.49	1.79	15.11	1.25	0.16	1.68	0.007	260	1040	40	33	44	5	79	61
Recalculated on clastic basis																			
V-3224	103	Tm, d, s	-	10.0	10.41	2.41	16.06	2.10	0.20	1.89	0.01	740	740	29	138	106	9	132	101
V-3226	240	Tm	-	45	3.32	6.44	6.26	0.36	0.13	2.00	0.003	1800	4060	41	64	65	8	531	73
V-3215	265	Fo, s	-	4.5	12.06	6.18	8.30	1.58	0.33	2.19	0.01	500	5690	134	161	158	15	200	158
V-3228	565	Tm, d	-	15.3	5.75	2.95	24.89	2.06	0.26	2.77	0.01	430	1710	66	54	72	8	130	100

Sediment type: Tm – terrigenous mud; Fo – foraminiferal ooze; Tm, d – mixed terrigenous-calcareous with large amount of diatoms; s – sapropelic

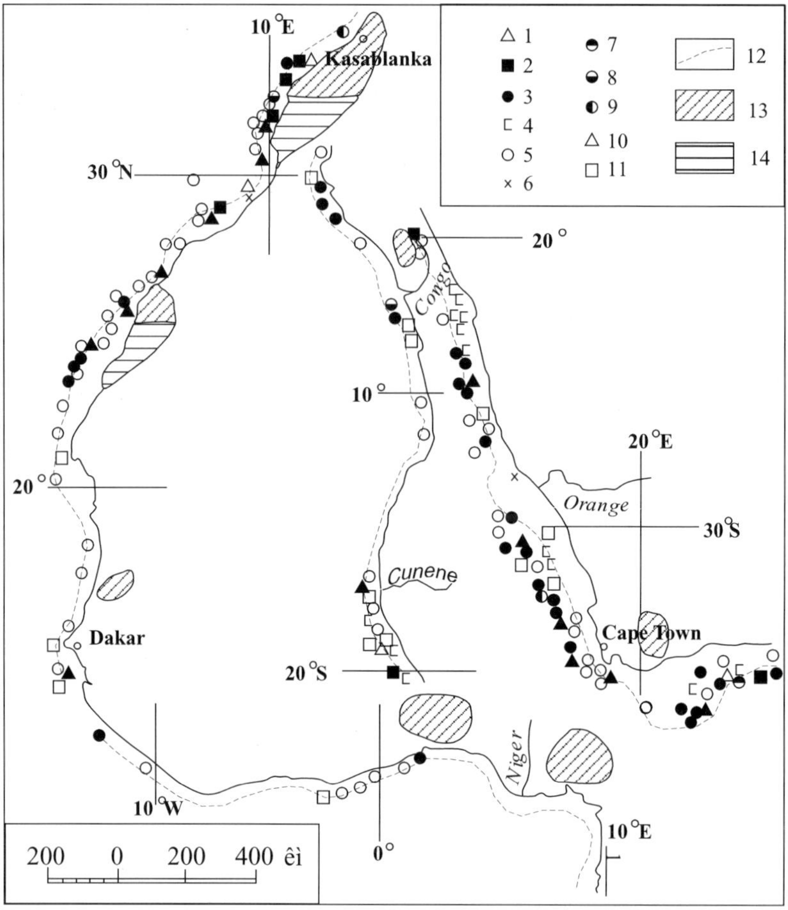

Fig. III.5.3. Distribution of phosphorites and phosphatic deposits on the shelf of Western Africa. After Kharin and Soldatov, 1975.

Phosphorites: 1—breccialike; 2—conglomeratelike; 3—fine grained; 4—bones of mammalia and fishes. Phosphatized rocks: 5—carbonates; 6—terrigenous; 7—limestones; 8—marls; 9—dolomites; 10—aleurolites and argilates; 11—sandstones; 12—200-m isobath. Phosphatic basins near Africa: 13—Paleogene; 14—Upper Cretaceous.

occur in accordance with the biogeochemical hypothesis suggested by J. Murray (1891), which has been described in detail by G.N. Bushinsky (1963, 1966). The basic statement of this theory says that planktonic organisms play a decisive role in removing phosphorus from water. The coastal waters off Namibia are rich in plankton. The biomass of phytoplankton in water is estimated at 2–3 mg/m^3, and this is 10 times as large (periodically even more) as that of the open ocean

(Khovansky, 1962). In the environs of Walfish Bay, the primary biological production is often very high—3.8 gC/m^2 per day (Steeman-Nielsen and Jensen, 1957). In the coastal areas off southwest Africa, the value of primary production is commonly no more than 0.2–1 gC/m^2 per day, but periodically it is as large as 235.5 gC/m^2 per half-day (Koblents-Mishke et al., 1973).

High concentrations of fish bones, teeth, and scales are characteristic of phosphatic sediments. In other genetic types of sediments, the proportion of fish remains is very insignificant. Therefore, the role they play in precipitation of phosphorus is of minor importance. Consequently, the theory that fragments of fish scales, bones and teeth are the main forms of phosphorus has not been confirmed in relation to phosphatic deposits of the African shelf. In most cases, these remains are either partly phosphatized or partly dissolved, providing for the enrichment of phosphorus in near-bottom sea- and interstitial waters, or they transform into phosphatic aggregations, subrounded or rounded in shape. Phosphatized skeletal remains are most commonly clumped (bound) together by phosphatic matter into clots, nodules and concretions.

Phosphatic sediments contain small amounts of terrigenous apatite. This is evidence that role of this mineral in the total phosphorus budget is insignificant.

Carbonates of chemical origin, which could be used as evidence that phosphates are chemically precipitated from seawater, have not been found in phosphatic sediments.

In glauconite sediments, P is represented by glauconite itself and by phosphatic aggregations. Phosphorus occurs in hydrogoethite–chamosite grains (coprolites and ooids), which are common in the sediments.

The mechanism that causes the present-day accumulation of phosphates in bottom sediments of the shelf area and accumulation of phosphorites in oceans (Figs. III.5.1; III.5.3) was described in detail by A.V. Kazakov (1939), G.B. Baturin (1969, 1978, 1986), and details related to the shelf area off Western Africa are also given in papers of E.M. Emelyanov (1973_2, 1979_2, 1982).

Phosphorus contents are higher than the clarke of this element at various areas of coastal upwelling, first of all in sediments of the coastal areas off Chile, California, eastern Australia, Northwest Africa, Portugal, and so on. Nevertheless, in many of these regions, upwelling, and, consequently, processes of accumulation of phosphatic sediments, are not as distinct as in the shelf area off Namibia (see Figs. II.4.3, II.4.4). For instance, although the phosphorus content in biogenic carbonate sediments of the continental shelf off Morocco is increased, it does not exceed 1%, in general (see Figs. II.4.5, II.4.6).

Jenkyns (1986, p. 348) proposed that authigenic accumulation of phosphates in areas of coastal upwelling occurs not only in sediments of the continental shelf but also in sediments of the continental slope, at the lower boundary between the OML and the bottom (Fig. III.5.2-II). Analysis of data for the continental shelf off Namibia (see description of the effect of barriers in Part II.4) led us to the conclusion that the zone of the so-called "phosphatic shelf" should be expanded from 100±50 m, according to A.V. Kazakov (1939), to 200±50 m according to our data (Emelyanov, 1979_2, 1982). As can easily be seen, phosphatic sediments are restricted to the outer zone of the shelf area and to the shelf edge; i.e., they are localized within the zone of the highest hydrodynamic activity. In this region, diatomic oozes and terrigenous muds are repeatedly rewashed. Another processes occurring there are transport of clayey minerals and biogenic opal, mineralization of organic matter, further cementation of phosphate material, and enrichment of biogenic carbonates in sediments.

Hydrofronts and Ore Formation in Pelagic Areas of the Ocean

In the World Ocean, a clear relationship occurs between the distribution of FMNs and oceanic zonation, which in turn can be subdivided into three main types: climatic (latitudinal), circumcontinental and vertical. Climatic (latitudinal) zonation in the ocean is a result of processes related to preparation of sedimentary material in catchment areas of continents, the hydrodynamics of the ocean and its biological productivity. Zones of high productivity of FMNs are northern-equatorial zones and also three zones in the Southern Hemisphere (Fig. III.6.1, Table III.6.1).

The area of highest bioproductivity is the northern-equatorial zone (Arrhenius, 1952; Bender et al., 1966; Finney et al., 1984; Bezrukov, 1976, 1979; Skornyakova and Murdmaa, 1989; Baturin, 1986), which is located in the zone of northern tropical convergence (see Fig. I.21), or in the subequatorial zone with distinct fronts (see Fig. II.3.4) where biological productivity is very high (see Fig. II.3.4). As evidenced by Horn (Horn et al. 1973), in the equatorial zone of the Pacific Ocean (which is a productive region in application to FMNs), between the Clipperton and Clarion faults, FMNs with the highest contents of Cu and Ni overlie the Miocene substrate (radiolarian red clay of light-brown color), which is situated between a zone of carbonate sediments of the Pliocene–Quaternary (near the equator) and a zone of siliceous (radiolarian) oozes. These oozes (up to 1–3 m) overlie the Oligocene and Eocene substrates and lie on the northern side of the productive zone (Horn et al., 1973). This productive zone is a narrow strip FMNs which are the most enriched in Cu and Ni. The strip is about 150 miles in width and extends from 8° N, 145° W to 10° N, 131° 30′W (Fig. III. 6.1). This strip is called the "Horn zone" (Greenslate et al., 1973). The geographical demarcation (position) of this Horn zone corresponds roughly to the most pronounced part of the OML, which spreads from Central America precisely in the same direction, along the same latitude (see Figs. II.10.1, II.10.3). In the same "a line in the sea," a frontal zone represented by a dark green strip, spreads on the surface of the ocean with a "carpet" of *Rhizosolenia* (Yoder et al., 1994; see Figs. II.3.4 and III.6.1).

Radiolarian clays are characterized by very high porosity. Biogenic material, first of all SiO_{2am}, dissolves in sediments, leading to increased concentrations of Si in interstitial waters. It is evident that, in addition to biogenic components (Si, Ca, P, C_{org}), elements such as Cu and Ni also enter interstitial waters.

The high porosity of sediments provensures effective upward diffusion of elements, up to the water–bottom boundary. This is a mechanism for delivering Cu and Ni from interstitial waters to FMNs (the content of Cu and Ni is from 0.3 to 0.8%), i.e., to the surface of the ocean floor. The amount of Cu and Ni delivered during the

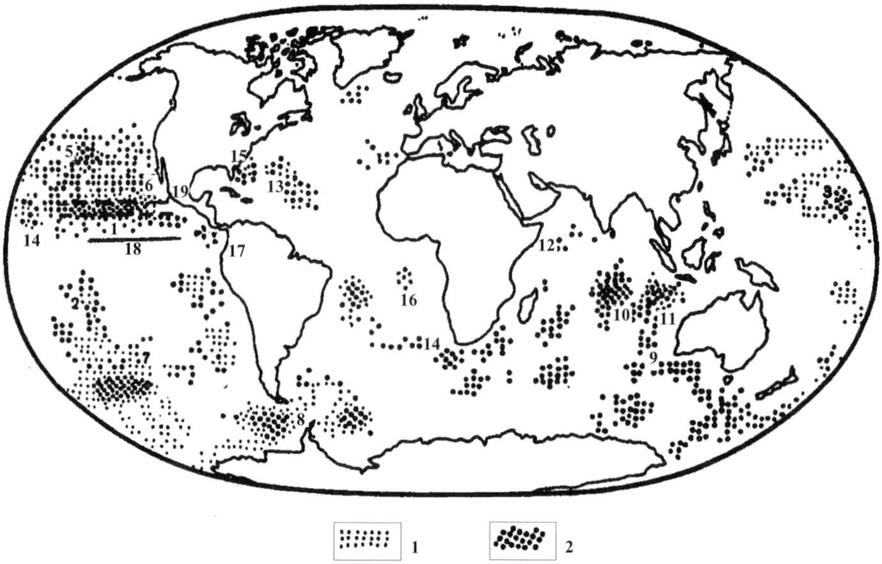

Fig. III.6.1. Distribution of FMNs in the World Ocean. After Cronan, 1982, with additions by the author (symbol 3, regions 16, 17, 18, 19, 20).

1—moderate productivity; 2—high productivity; 3—Horne belt, where nodules richest in Cu and Ni occur.

Fields (regions) of concretions: 1—Clarion-Klipperton; 2—South Pacific; 3—Wake-Nekker; 4—Central Pacific; 5—Guinea; 6—California; 7—Menard; 8—Drake Strait and Skotia Sea; 9—Diamantina; 10—Central Indian; 11—West Australian; 12—Somali; 13—North American-Guyana; 14—Cape Agulhas; 15—Plato Blake; 16—Angolan; 17—Guatemala; 18—Brazilian; 19—Baltic; 20—Black Sea; 21—frontal line on the ocean water surface: dark green line (width 2–10 km, length 200 km, to the north of the convergence line, about 2°N, see Fig. II.3.4).

last 1–10 million years (the age of FMNs in this zone) is sufficient for the enrichment of these elements in FMNs. In contrast, sediments in this productive zone (the Horn zone) are depleted in Cu and Ni (Greenslate et al., 1973, p. 59).

From the above, we may conclude that the main properties of FMNs and their productivity are controlled by the situation in the upper active water layer (photic

Table III.6.1 Distribution of the ferromanganese nodules according to latitude. After Andreev et al., 1984.

Ocean	Northern Hemisphere	Southern Hemisphere.
Pacific	4–30°	20–46° 55–63°
Atlantic	18–23°	30–40° 50–60°
Indian	0–11°	15–20° 32–40°

layer) and depend on a number of factors, such as climatic zonation of the ocean (position of hydrofronts, biological productivity of ocean waters), the CCD, etc. (Arrhenius, 1952; Greenslate et al., 1973; Bezrukov, 1976, 1979; Skornyakova and Murdmaa, 1986; Baturin, 1986).

The Halocline (Pycnocline) and Accumulation of Shallow-water Ferromanganese Crusts and Nodules

The process of active accumulation of ferromanganese crusts and nodules occurs in some lakes, bays and shallow-water seas. The Baltic Sea is a very illustrative area, the study of which makes it possible to outline the facial sequence and sediments and situations in which accumulation of increased amounts of Mn and Fe occurs (Emelyanov, 1981, 1986; Lisitsin and Emelyanov, 1986):

(1) Muds at the mouths of gulfs (Riga, Finland and certain others), at depths of 30–60 m (Fig. III.7.1). In these areas, the halocline is indistinct (as a result of general freshening of a gulf's waters by river waters) and is found at depths of 30–40 m in the Gulf of Riga and 40–50 m in the eastern part of the Gulf of Finland. A thin oxidized layer of semiliquid muds is the place where accumulation of flat (most frequently coin-like in shape) FMNs occurs (under conditions characterized by intensive input of terrigenous pellitic material, including Fe and Mn, where the hydrodynamic activity is moderate). In the Gulf of Finland, nodules are abundant mainly in its eastern part (Butylin et al., 1985), i.e., in the area well supplied with suspended matter from the Neva River, rich in heavy metals and P, Fe and Mn (Emelyanov, 1995_2). Crusts and nodules are found predominantly at depths of 10–20 and 35–50 m (Butylin et al., 1985), and their contents reach 18–24 kg/m^2 (Glasby et al., 1997). The surface area of the seabed where the production of nodules is very high is about 300 km^2, and the total mass of nodules here is estimated at about 6×10^6 tons. Such a considerable amount of nodules might be of commercial interest.

In the Gulf of Riga, nodules actively accumulate in the muds of its marginal southeastern part, which is the inflow area of the Daugava River. The amount of nodules is somewhat smaller in other peripheral parts of this gulf (depths of 19–37 m). Concretions are absent in the central, deepest part of this area (50–60 m), the bottom of which is covered by gray muds.

(2) Clastic sandy-aleurite terrigenous sediments around islands and bays (the environmental situation in zones above the halocline and seasonal thermocline is characterized by increased hydrodynamic activity) (Fig. III.7.2). Development of round, Fe-rich nodules (which closely resemble small-size fractions in shape) occurs under such conditions.

(3) Shallow-water areas of the open sea, below the zone of wave reworking, but above the upper boundary of the occurrence of mud (at the level of the halocline, where the environmental situation is dominated by strong bottom currents and zero

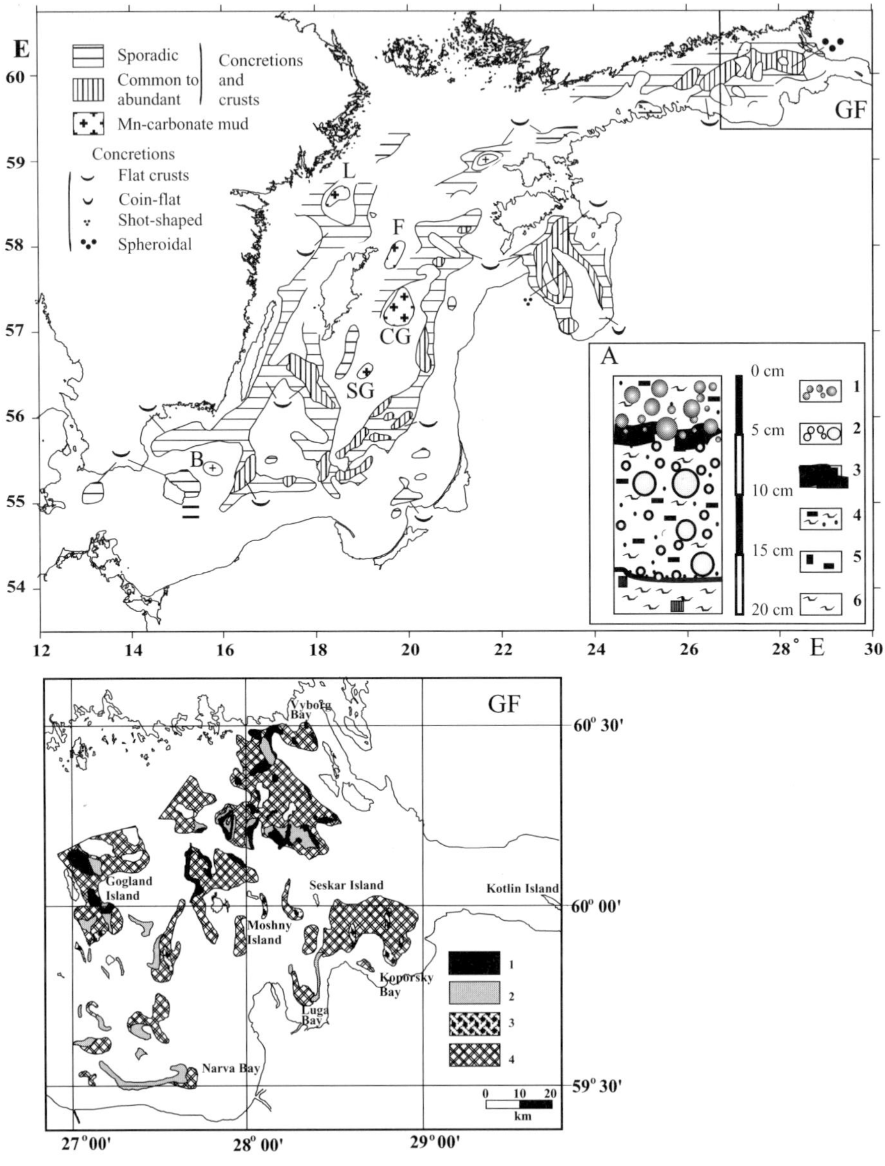

Fig. III.7.1. Schematic map showing distribution of ferromanganese crusts and concretions in Baltic basins (region 19 on the Fig. III.6.1): B—Bornholm; SG—Southern Gotland; CG—Central Gotland; F—Faröe; L—Landsort Trench; GF—Gulf of Finland (see Fig. GF).

GF. Concretion fields in the eastern Gulf of Finland (after Zhamoida et al., 2002): 1–2—relatively "deep" fields of Mn-rich spheroidal concretions; the potential abundance of concretions varies: 1—from 5 to 40 kg/m^2; 2—from 1 to 10 kg/m^2; 3–4—relatively shallow fields of Fe-rich discoidal concretions and rings of ferromanganese oxides around erratic nuclei; the potential abundance of concretions is: 3—up to 5 kg/m^2; 4—not more than 1 kg/m^2.

A. Example of the occurrence of concretions in the sediment column of a relatively deep field in the Gulf of Finland (after Zhamoida et al., 2002): 1—spheroidal concretions displaying granular surface texture; 2—spheroidal concretions displaying smooth surface texture; 3—iron oxyhydroxide cementation; 4—sandy-clayey silt; 5—amorphous iron sulfides; 6—lacustrine clays.

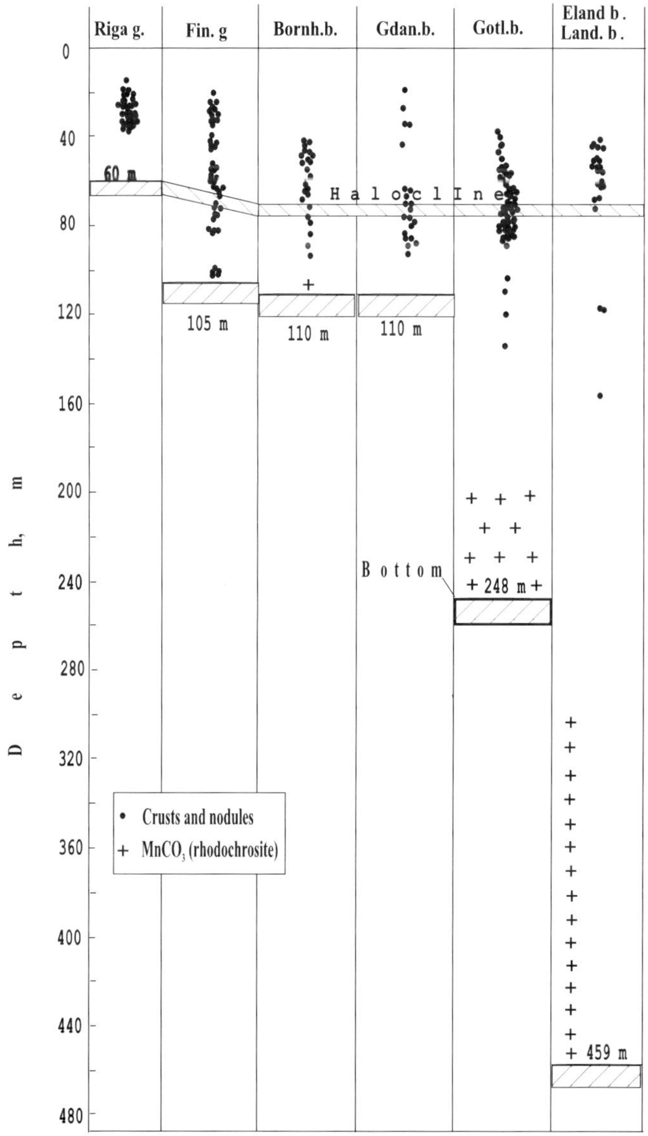

Fig. III.7.2. Distribution of the ferromanganese crusts and nodules in different basins of the Baltic versus sea depth.

sedimentation, including events of bottom erosion). Nodules and crusts in the central Baltic are also found mainly around the margins of deep basins at depths of 40–100 m (Figs. III.7.1, III.7.2), where strong near-bottom currents occur. The number of crusts is high only on the Gdansk–Gotland Sill. The quantity of them in places reaches 5–16 kg/m^2 there.

(4) The deepest parts of basins, where bottom waters are periodically dominated by stagnant conditions (much below the level of the halocline). The depositional environment of these regions is almost completely stagnant, where the hydrodynamic activity of waters is very low. Accumulation of primary carbonate–manganese muds occurs there (Fig. III.4.4; see also Fig. III.9.2).

This type of sedimentary environment is very characteristic of the Baltic Sea. The reason for the large productivity of seawater within the given depth interval is most likely the intensive delivery of ore elements from overlying waters, from the photic layer (predominantly, together with organic detritus) and from underlying waters, from environments with reducing conditions (dissolved forms of Mn^{2+} and Fe^{2+}, and hydroxides of these elements).

In the lateral extension (that is, along a bed composed of sediments of the same age), sediments of the above-mentioned four facial situations do not border each other, so that there is no gradual transition from one type of sediment into another (moreover, it is impossible): the space between these sediment facies is always covered by other sediments, containing clarke (or even lower) values of Fe and Mn.

Flat FMNs, as well as Fe crusts, cannot be preserved after their burial in sediments. In contrast, carbonate–manganese muds tend to become richer in Mn as a result of removal of this element from interstitial waters.

The contents of Fe, Mn and other microelements in the nodules of the Baltic Sea vary within a wide range. With the exception of four samples, Fe is the most abundant element in these nodules (Table III.7.1). The chemical composition of nodules in the Gulf of Riga is very similar to those in the Gulf of Finland (on average, the Mn/Fe ratios are 0.22 and 0.43, respectively). However, Fe and Mn contents in these nodules vary considerably from place to place, and the occurrence of these nodules is frequently sparse. In other areas of the Baltic Sea (the Gotland, Bornholm and Eland deeps), Fe nodules were found mostly at large depths, in contrast to Mn-rich features, which were found generally in shallow water. These Mn-bearing features, as interpreted from previous studies (Varentsov and Blazhchishin, 1976), are thought to be Mn crusts.

The contents of minor elements in Baltic nodules are by 1–2 orders less than in oceanic nodules. In researching nodules of the Gulf of Finland, some differences were found in the composition of minor elements between the samples from shallow (30–70 m) and deeper (102 m) sites. Shallow nodules are enriched in Cr, but have a very low Li content. Deep nodules, on the contrary, have a low Cr content and are almost 10 times richer in Li. Together with Li, the Ca + Mg (carbonate) contents increase. Nodules in the Gulf of Riga are also rich in Li, but they lie at a depth of only about 30 m. A possible explanation for these variations can be found in the composition of sediments containing nodules. The bottom of the Gulf of Finland is very rough. On the steep slopes of many islands and hills, granites and gneisses are exposed, whereas *in* depressions the bottom is covered mainly by fine terrigenous muds. Cr, a less mobile element, was probably delivered to nodules along with the fine debris of crystalline rocks. Nodules enriched in Li developed in clayey deposits and have evidently inherited some of the geochemical features of granitoids and moraines. The increase in the Ca^{+} and Mn^{+} contents in the mud (and, consequently, in nodules) could be connected with the erosion of Ordovician carbonate rocks occurring at the southern coast of the Gulf of Finland.

Table III.7.1 Chemical composition of ferromanganese concretions and crusts and sapropelic carbonate–manganese mud (only the last column) of the Baltic Sea

Element	Gulf of Riga			Gulf of Finland			Eland Basin		Bornholm Basin		Gotland Basin	
	PD-3158 (36 m)	AK-2578 (30 m)	Mean*)	Sh.a. 27-53m)	D.a. (61-102m)	Mean*)	PSh-2525 (61 m)	Mean (118-155m)	PSh-2551*) (73 m)	Mean (79-93m)	PSh-2526 (60 m)	Mud***) (210-248m) (mean)
Fe	11.00	21.55	30.27	29.38	17.12	20.57	10.50	33.93	24.90	21.43	11.28	5.10
Mn	1.50	4.75	5.50	5.27	9.55	8.71	22.57	0.05	9.75	0.04	0.65	2.40
Si	-	-	8.37	-	-	8.14	-	-	7.93	-	24.64	-
Al	-	-	1.75	-	-	2.64	2.35	2.92	1.04	3.32	6.85	7.10
Na	0.67	1.02	0.50	0.36	1.02	0.83	0.80	0.69	0.76	1.05	0.78	2.14
K	2.46	1.28	0.91	0.81	1.26	1.21	1.38	2.36	0.80	2.37	2.14	2.95
Mg	-	0.74	0.90	0.50	0.86	1.23	1.26	1.70	1.02	1.65	1.30	1.31
Ca	-	1.10	1.33	0.26	1.00	1.85	1.46	1.20	2.04	0.74	0.97	1.20
Ti	-	0.25	0.11	0.08	0.19	0.11	0.05	0.26	0.03	0.27	0.15	0.45
P	-	-	1.49	1.81	1.86	1.20	0.93	0.06	2.98	0.04	0.55	0.09
Li	-	146	-	17	118	15	13	47	10	50	37	64
Rb	28	48	-	75	50	70	-	-	20	-	60	137
Sc	-	-	3.9	-	-	4.8	-	-	2.1	-	13.7	(14.2)
V	-	-	(68)**)	-	-	100	13	90	13	90	11	-
Cr	100	37	23	254	70	37	31	33	34	53	80	72
Co	-	-	114	106	93	70	185	98	60	54	25	42
Ni	55	102	(35)	116	131	130	764	148	104	93	88	79
Cu	14	37	(9)	27	28	22	80	35	13	28	16	113
Zn	135	394	255	209	215	115	480	124	210	63	130	174
As	-	-	334	-	-	-	180	325	653	121	96	(22)
Sr	-	-	-	-	-	-	630	280	1110	150	250	-
Mo	-	-	-	-	-	-	480(?)	60	20	60	12	-
Rd	-	-	0.003	-	-	-	-	-	(0.002)	-	(0.006)	-
Pd	-	-	0.002	-	-	-	-	-	0.001	-	(0.002)	-
Cd	-	-	-	1.1	2.4	9.0	10.0	2.0	3.0	3.0	(0.3)	-
Sb	-	-	1.7	-	-	0.8	5.0	2.0	5.7	2.0	(5.6)	(1.7)

Continued

Table III.7.1 Chemical composition of ferromanganese concretions and crusts and sapropelic carbonate–manganese mud (only the last column) of the Baltic Sea—*continued*

Element	Gulf of Riga			Gulf of Finland			Eland Basin		Bornholm Basin		Gotland Basin	
	PD-3158 (36 m)	AK-2578 (30 m)	Mean*)	Sh.a. 27-53m)	D.a. (61-102m)	Mean*)	PSh-2525 (61 m)	Mean (118-155m)	PSh-2551*) (73 m)	Mean (79-93m)	PSh-2526 (60 m)	Mud***) (210-248m) (mean)
Cs	-	-	2.0	-	-	2.5	-	-	1.1	-	(6.4)	(7.9)
Ba	-	-	9900	-	-	2800	2200	770	1300	900	1200	(520)
Hf	-	-	2.1	-	-	2.4	-	-	1.4	-	(4.4)	(2.4)
Ta	-	-	-	-	-	0.4	-	-	0.4	-	(0.7)	(1.2)
Pt	-	-	0.003	-	-	-	-	-	0.002	-	(0.005)	-
Pb	-	-	(9)	-	-	60	100	41	30	44	50	-
Th	-	-	4.6	-	-	5.0	-	-	1.0	-	(12.0)	(12.7)
U	-	-	3.8	-	-	7.2	-	-	4.1	-	(7.1)	(3.5)
La	-	-	57	-	-	52	-	-	21	-	(46)	45(48)
Ce	-	-	126	-	-	72	-	-	25	-	(69)	72(78)
Nd	-	-	65	-	-	56	-	-	20	-	(44)	42(51)
Sm	-	-	9.2	-	-	6.4	-	-	2.6	-	(6.1)	5.6(6.4)
Eu	-	-	1.9	-	-	1.3	-	-	0.6	-	(1.3)	1.2(1.3)
Tb	-	-	1.2	-	-	0.8	-	-	0.5	-	(1.4)	0.8(1.1)
Yb	-	-	3.9	-	-	4.7	-	-	2.6	-	(4.9)	3.6(4.2)
Lu	-	-	0.39	-	-	0.32	-	-	0.13	-	(0.48)	0.40

Sh.a. – shallow area (27-53m); D.a. – deep area (61 - 102 m); Fe–P in %, Li–Lu in ppm.
*)After Baturin et al., 1995 (elements Rd–Lu at station PSh-2526 – also after Baturin et al., 1995).
**)In brackets – old data (Emelyanov, 1995_1).
***)Only samples with Mn content of >1.0% in mud (0-5 cm) (7 samples).

The Oxygen Minimum Layer and Development of Cobalt–Manganese Crusts

As shown in Part II.10, the oxygen minimum layer (OML) not only is of crucial importance for the biological and geoecologal situation in the ocean, contributing thus to geological processes there, but also a factor controlling, under specific facial conditions, the process of ore formation in seas and oceans.

Vertical zonation in the sediment sequence, in application to ore formation, has been the subject of considerable interest for a long time, and the problem was studied by many specialists in geochemistry and sedimentology (Piper and Williamson, 1977, 1984; Emelyanov, 1979, 1982; Andreev et al., 1989; Bogdanov et al., 1990).

Specific facial situations developed in the past (and are being developed now) on seamounts, the tops and slopes of which come in contact with the OML (see Fig. II.10.4). In some areas, there is the occasional accumulation of phosphorites both on the continental slopes or in shelf areas. Although the process of accumulation of manganese crusts on the tops and slopes of seamounts, especially below the level of the core of the OML, is a subject of considerable interest now, no detailed sedimentological or geochemical studies of such Mn deposits have been undertaken. It is evident that the rate of accumulation of manganese crusts on the slopes and tops of seamounts within the limits of the OML has diminished (Fig. III.8.1).

As mentioned earlier in Part II.10, waters of the OML are markedly enriched in dissolved Mn (Emelyanov, 1974, 1977; Berner, 1977).

Furthermore it was shown that dissolved forms of Ni, Zn and Cu also accumulate in this layer (Bruland, 1983); moreover, the distribution of these elements is of biogenic type, similarly to Mn.

The Mn/Fe ratio in the ocean water for the OML is maximum (Bruland, 1983; Baturin, 1993, p. 119). In the OML, manganese occurs in the form of Mn^{2+} and $MnCl^{-}$.

A distinguishing feature of Mn crusts on seamounts is that elements such as Co, Ni and probably Ba, Pb, Yb and Ln accumulate in these sediments (Piper, 1974; Cronan, 1980). Several scientists (Piper and Williamson, 1977; Halbach and Putenaus, 1984) pointed out a relationship between high contents of Mn in crusts and nodules of seamounts, at depths of no more than 1000–2000 m, and an increased concentration of dissolved Mn^{2+} in the OML.

Earlier, the author (Emelyanov, 1974, 1977) proposed that the dissolution of a majority of organic detritus that sinks from the photic layer is the reason for increased concentrations of dissolved Mn in the OML. Based on studies of elemental composition in the ocean, Halbach and Putenaus (1984) discovered further

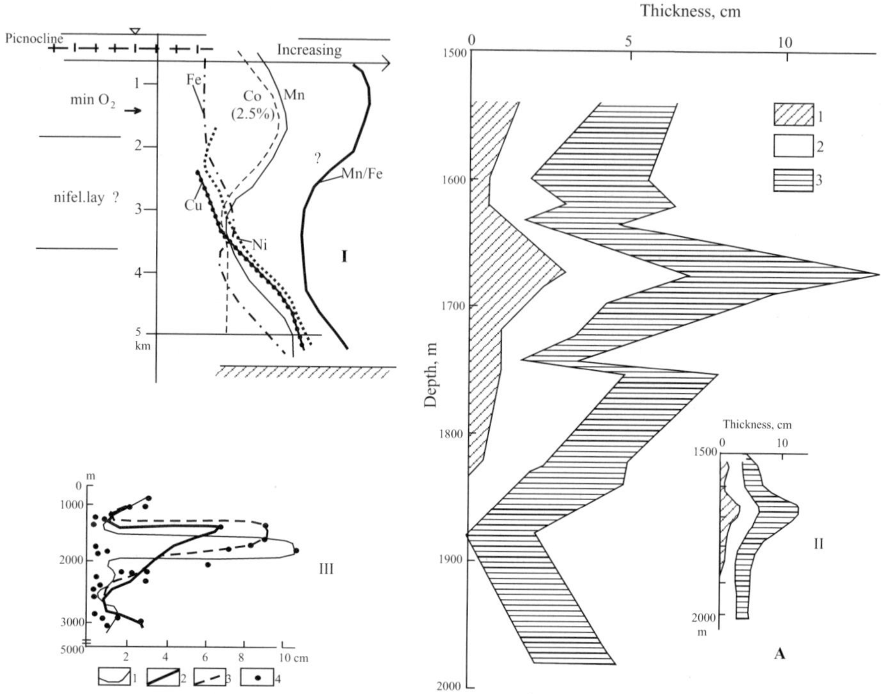

Fig. III.8.1. Relation between thickness of Mn crusts on ocean mountains and position of the OML.
I. Principal scheme of ore element distribution in the ferromanganese formations (in crusts and nodules) on sea mountains. Min. O_2—oxygen minimum layer;
II. Principal scheme of changes in crust thickness of three horizons of cobalt-bearing crusts of sea mountains in the Nekker Ridge region in the Pacific Ocean versus the sea depth. After Bogdanov et al., 1990, p. 116.
1–3—horizons: 1—first; 2—second; and 3—third. A—general scheme.
III. Relation of crust thickness (in cm) to depth in the Pacific Ocean.
After Andreyev et al., 1989. 1–3—substrate (1—basalts; 2—gravellites; 3—limestone).

gradation of elements by their relative abundance (compositional ratios) in marine plankton: Fe > Zn > Cu > Ni > Mn. They believe that skeletal carbonate remains are major contributors of these elements, instead of the organic detritus of plankton. These remains, as well as soft parts of phytoplankton bodies, begin to dissolve in the OML, that is, in the space between the aragonite and calcite lysoclines, where saturation of waters with respect to aragonite does not exceed 80% and concentrations of pCO_2 are markedly increased (see Fig. I.18, II.10.3). This is because the skeletal remains (pteropods, coccoliths, and some types of planktonic foraminifera) are composed of aragonite instead of calcite.

So, the OML is responsible not only for intensive dissolution of organic detritus, but also aragonite. When organic detritus dissolves, many microelements pass into solution, with the result that concentrations of dissolved forms of these elements in water become markedly increased.

Fe hydroxides, which form owing to dissolution of soft and solid (aragonite) parts of plankton, is "a substrate which is favorable for catalitic oxidation of dissolved divalent manganese and its further transition into solid phase" (Bogdanov et al., 1990, p. 143). As is well known, fresh hydroxides of Fe and Mn are good sorbents for Cu, Zn, Ni, Co and other elements. Thus, the OML is predominated by sedimentary conditions (especially in its lower part, where the amount of dissolved oxygen again begins to increase) favorable for intensive precipitation of hydroxides. (Fig. III.8.2). The action of this potential is especially evident on the slopes of volcanic seamounts (including guyots), which are found within the OML or just below this layer. The rate of growth of manganese crusts, as well as the thicknesses of these deposits at the above-mentioned bathymetric levels, is commonly twice as large as that at smaller or larger depths. In the Pacific Ocean, the thickest crusts of the same age are generally found at depths of about 1700 m (Fig. III.8.1, III.8.2).

A characteristic feature of FMN crusts is that the thickness of these crusts depends on the depth of the water layer in various areas of the ocean. So, in the region of Land-Pacific—the Nekker Ridge in the Pacific Ocean—the thickest crusts (8–10 cm) have been found at depths of 1500–2100 m (Andreev et al., 1989), whereas at depths of no more than 1500 m, the thickness of crusts was 2–3.5 cm, and at depths greater than 2000 m, 0.5–3.5 cm (Fig. III.8.1). Recall that the OML in the

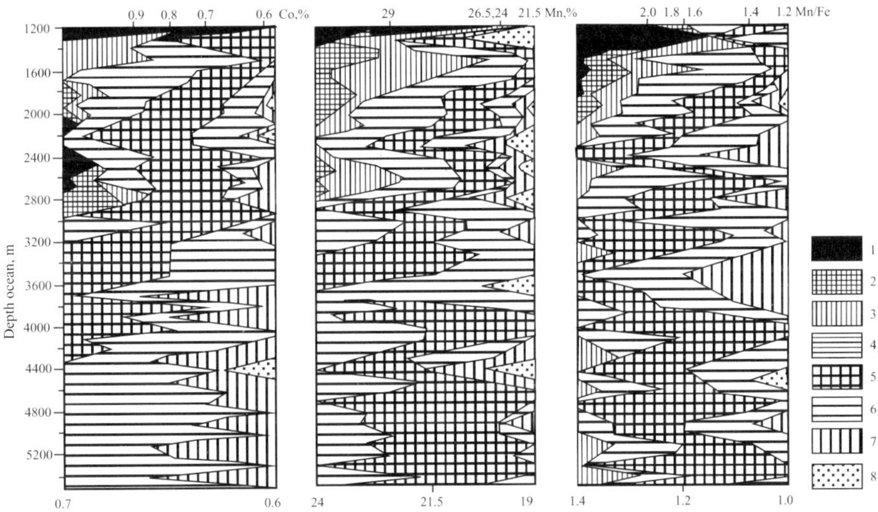

Fig. III.8.2. Correlation of distribution of mean contents (in %) of Co and Mn and mean Mn/Fe ratio in ferromanganese formations of guyots on the Markus–Wake elevation of the Magellan mountains, northwestern part of Pacific Ocean (10–23°N, 153–165°E) with depth. After Melnikov et al., 1995.
1—Co >0.9, Mn >29.0, Mn/Fe >2; 2—Co 0.8–0.9, Mn 26.5–29, Mn/Fe 1.8–2.0; 3—Co 0.7–0.8, Mn 24.0–26.5. Mn/Fe 1.6–1.8; 4—Co 0.6–0.7, Mn 21.5–24.0, Mn/Fe 1.4–1.6; 5—Co 0.5–0.6, Mn 19.0–21.5, Mn/Fe 1.2–1.4; 6—Co 0.4–0.5, Mn 16.5–19.0, Mn/Fe 1.0-1.2; 7—Co 0.3-0.4, Mn 14.0-16.5, Mn/Fe 0.8-1.0; 8—Co <0.3, Mn <14.0, Mn/Fe <0.8.

vicinity of the Nekker Ridge is presently found at depths of 600–800 m. Consequently, the thickest crusts are found below the present-day OML.

Within the OML, as well as just below it (down to 2.5 km), the Mn/Fe ratio in crusts is highest (1–4) compared with the values of this parameter in crusts from larger depths (which are 0.5–1.5) (Baturin, 1993, p. 119) (Fig. III.8.1).

Large fields of FMN crusts have been found on the Sierra Leone Rise in the Atlantic Ocean (down to depths of 2.5 km) (Emelyanov and Kharin, 1993). At the tops of flat underwater mountains, where the sedimentary cap has advanced, Fe/Mn crusts have not been noted on ancient (Eocene and Miocene) limestones, but these crusts are as thick or dense in comparison with crusts on magmatic rocks. The latter have a thickness of 8–10 cm.

We have discovered a direct correlation of the Fe, Ti, Ca, Mg, P, Na, Cr, Ni concentrations in crusts (Table III.8.1) with their contents in rocks/underlayers (Asavin et al., 2002, Table III.8.2). For Mn, Zn and Co there is no such correlation. Sedimentation of Mn, Zn and Co proceeded without a noticeable influence from matrix rock. It is typical that on biogenic limestones, which are usually characterized by a low Co concentration, a crust with very high contents of Co formed. This attests to the fact that crusts are likely a hydrogenic formation of the Atlantic Ridge. In addition, the high concentration of rare elements of the iron group (Table III.8.3) samples are characterized by higher contents of rare elements–Zr and Th, which are typomorphic for alkaline rocks. The high concentration of rare-earth elements is of great interest. Nodules are also enriched in Ag, Au, elements of platinum group, and precious metals (Table III.8.4)

As a whole, our data (Asavin et al., 2002) testify to the outlook of this object as a new source of rare-earth elements.

Table III.8.1 Mean content (Fe–Al in %, Zn–Cr in 10^{-4}%) of elements in crusts and concretions of the Ita-Mitan guyot in the Pacific ocean. After Bogdanov et al., 1990, p. 71.

Element	Content	Concentration*
Fe	13.43	1.14
Mn	18.61	1.04
Ti	0.81	1.01
Si	3.74	0.27
Al	0.9	0.27
Zn	822	0.98
Cu	1014	0.24
Ni	4556	0.77
Co	4008	1.21
Cr	76	7.6

* The degree of relative concentration of elements in ore of guyots (the ratio of ore concentration in the guyots to pelagic concretions of the Pacific Ocean).

Table III.8.2 Elemental composition of ferramanganese crusts of the Sierra Leone Rise, Atlantic Ocean. After Asavin et al., 2002.

Sample	Element, wt %							
	Fe	Mn	Ti	P	Ca	Mg	K	Na
1	21.50	15.40	0.85	0.62	1.82	1.12	0.17	1.92
2	21.50	15.20	0.94	0.59	1.82	1.21	0.64	1.92
3	15.91	7.97	0.74	0.42	-	-	-	-
4	23.80	16.20	0.72	0.62	1.93	1.11	0.34	1.76
5	25.50	24.38	0.67	0.62	1.36	1.01	0.29	1.36
6	9.60	0.54	2.81	1.90	4.47	2.97	2.01	1.28
7	13.97	0.71	3.00	0.59	2.42	2.75	1.69	1.28
8	20.10	13.70	1.38	0.78	3.22	1.56	0.55	1.40
9	33.07	16.70	1.09	0.67	-	-	-	-
10	-	-	-	-	-	-	-	-
11	29.09	18.11	0.77	0.62	-	-	-	-
12	24.79	17.82	0.66	0.94	-	-	-	-
13	17.86	21.29	-	-	3.41	1.38	0.20	1.26
14	17.13	22.22	-	0.20	2.33	1.52	0.27	1.19

NOTES: 1-3 – station 866, coordinates 6°23.12′N, 22°38.26′W, depth 2300 m; ferromanganese crusts on hydrothermal changed trachyte and their tuffs; 4-7 – station 874, coordinates 9°01.09′N, 19°45.25′W, depth 2118-1880 m; ferromanganese crusts on trachybasalts; 4 – top; 5 – medium crusts; 6 and 7 – transitive zone to trachybasalt; 8–9 – station 876, coordinates 9°07.44′N, 21°10.3′W, depth 2500-1900 m; ferromanganese crusts on tuffs of trachybasalt; 10 – station 877, coordinates 9°03.89′N, 21°10.09′W, depth 1880-1300 m; ferromanganese crusts on tuffs of trachybasalt; 11 – station 880, coordinates 9°19.76′N, 21°18.6′W, depth 1500-1020 m; ferromanganese crusts on tuffs of trachy-basalt; 12-14 – station 1225, coordinates 9°01.4′N, 21°02.4′W, depth 969–699 m.

Since the level of the OML in the ocean and also the geographical position of seamounts have varied with time as a result of global sea level fluctuations and geodynamic movements of lithospheric plates, the bathymetric positions of crusts having maximum thickness have also correspondingly changed. Subsequent geochemical studies of seamounts in the Pacific Ocean carried out by Y.A. Bogdanov and coauthors (1990) have confirmed this conclusion. For instance, time intervals responsible for highest rates of crust formation on slopes of Ita-Matan guyot and corresponding bathymetric levels of these deposits (relative to the present-day ocean level) are as follows: O Ma*—1300–1900 m; 5 Ma—1100–1700 m; 10–15 Ma—1400–2000 m; 25 Ma—1000–1500 m; 75 Ma—800–1300 m; 100 Ma—200–600 m.

Contents of Co and Ni in crusts were found to be maximum at depths of 1.0–1.5 km (Halbah and Manheim, 1984), i.e., either in the lowest part of the OML or directly below the OML (Table III.8.4). In the Atlantic Ocean, there are no signs of increasing contents of Co and Mn in manganese crusts directly below the OML. The reason for this contradiction is that calculations of average contents of chemical elements were based on data characteristic of the ocean as a whole. At the same time, the position of the OML in the ocean experiences variations within a considerable depth interval, from 400 m in the equatorial zone to

*Ma-millions ans ago

Table III.8.3. Co and Mn content (%) in manganese crusts of submarine mountains in the Pacific Ocean depending of the position of the oxygen minimum layer (OML). After F. Manheim, 1986 (OML data after V.N. Ivanenkov and A.M. Chernyakova, 1979).

Depth, m	A		B		C		D		A.o	
	Co	Mn	Co	Mn	Co	Mn	Co	Mn	Co	Mn
0–1000	0.01	23.85	1.84	33.38	-	-	1.98	28.09	0.85	22.89
1000–1500	0.94	24.24	0.84	23.87	1.05	27.08	1.25	26.97	0.80	19.87
1500–2000	0.66	22.02	0.74	24.36	0.93	26.01	1.41	24.87	0.53	19.99
2000–2500	0.70	21.68	0.87	26.57	0.59	21.94	-	-	0.55	17.90
Total depth	0.76	22.22	0.77	23.80	0.77	23.82	0.78	20.49	0.50	20.15
	OML depth, m									
-	800–1000		600–800		600–800		600–700		600–1200	

A–D – the regions of the Pacific Ocean: A – Hawaiin Ridge in the area of Musicians Mountain (20°–30° N); B – Marshall Islands–western Mid-Pacific mountains (5°–20° N, 165°W–160°E); C – northern Line Islands–Mid-Pacific mountains; D – southern Line Islands–French Polynesia; A.O. – Atlantic Ocean as a whole.

Table III.8.4 Contents of rare elements in investigated samples ferromanganese crusts of the Sierra Leone Rise, Atlantic Ocean (neutron-activation analysis) (ppm). After Asavin et al., 2002

sample	6744	836-5	836-5A	836-5B	836-5B	839-1	839-2	846-1	866-1	866-2	874-1	874-2	874-3	880-A	844
Cr	202.2	12.2	5.29	1.08	2.52	9.91	537.5	26.8	4.69	1.8	55.2	4.1	42.7	24.7	24.4
Co	111.7	2028.8	12.7	31	55.1	27.4 2	113.3	2732	2667	2458	3342	3269	3423	4188	2349
Ni	320	910	690	1080	3360	70	40	3920	2190	3230	550	1190	2900	920	1320
Zn	340	450	480	630	1100	460	190	390	390	180	180	100	220	50	460
Sr		765	295	215	61	790	175	2155	745	1205	255	985	4275	285	2350
Zr	650	440	160			1100	95	2600	1520	1700	1700	620	430		
Au					0.049		0.017	0.1	0.09	0.04		0.08		0.08	
Ba	410	2095	605	1000	460	1685	385	685	1285	2010	960	1075	2070	2170	2085
As	45.1	54 6. 2	101. 9	4 2. 2	234.6	444	181. 6	379	331	329	418	444	409	429	330. 7
Th	13.8	16.4	0.58	1.15	0.083	35.4	1.82	19.9	36.2	23.1	62.3	55	54	38.1	28.9
U	4.73	4.89	0.56	0.62	0.78	11.4	2.75	4.88	6.23	11.4	8.44	5.71	5.59	9.2	3.04
La	38.9	3 05. 8	5.47	10.8	17.6	224	36.1	245	302	280	297	245	241	227	255. 6
Ce	74	54 8. 5	12.1	19.9	34.7	400	78.5	419	547	499	522	422	420	399	432
Pr	7.96	56.6	1.56	2.24	3.74	44.1	9.27	44	56.6	52.1	55	45.1	449	42.1	44.5
Nd	30.1	209	6.87	8.28	14.5	168	39.2	151	207	191	199	165	165	150	152
Sm	7.11	47.7	2	2.18	3.54	40	10.6	36.1	47.7	44.2	45.3	38.9	38.7	36.3	36.5
Eu	2.38	11.5	0.24	0.1	0.88	9.68	2.91	12.1	10.8	11.7	11.6	11.2	8.38	8.86	10.3
Gd	7.6	55.5	2.63	2.6	4.5	48.2	13.4	43.1	55.7	50.7	51	44	45 5	44	44
Tb	1.07	7.9	0.4	0.38	0.67	7	2.08	6.33	8	7.35	7.17	6.3	6.66	6.46	6.5
Dy	5.62	44.7	2.31	2.26	4	40	12	37	45.3	42.2	40.1	35.8	37 9	37.1	38
Ho	1.18	9	0.5	0.51	0.85	8	2.62	7.55	9.6	8.48	7.92	7.3	7.76	7.59	7.9
Er	2.98	23.2	1.39	1.41	2.27	21.7	7.2	20.7	24.9	22.5	20.3	19	20 5	20.3	21.7
Tm	0.42	3.3	0.21	0.22	0.34	2.97	1.07	2.85	3.5	3.19	2.75	2.69	2.8	2.8	3
Yb	1.81	17.3	1.14	1.18	1.75	15.8	5.7	15.7	17.8	17	14	13.9	15	15.1	16.6
Lu	0.3	2.6	0.18	0.19	0.29	2.35	0.95	2.38	2.8	2.55	2.02	2.08	2.26	2.27	2.53
Sc	28.3	18.3	0.67	2.15	3.32	14	29.3	8.67	9.35	7.68	13.6	11.5	13.8	10.9	7.95

>800 m in northern arid zones to>1400 m in the southern moderate zones (see Fig. II.10.1).

To recognize these regularities, one should consider those areas in the Atlantic Ocean where the CCD is found at the same depth.

A distinctive feature is that the content of ore elements in ferromanganese crusts depends not only on the depth of the ocean but also on distance from shore, where there is a strong source of terrigenous ore material (river). Such a phenomenon was found, for instance, in the profile of the Angola Basin (Fig. III.8.3). The greater the distance from the rift zone of the North Atlantic Ridge and the less the distance from land, the higher the contents of Ni, Mn, and especially Co in crusts.

Some hypotheses explain the occurrence of underdeveloped Mn crusts at depths less than 800 m as being due to the abundance of biogenic sediments (Manheim, 1986, p. 604), whereas according to the author, this situation can be attributed to a large extent to the effect of the OML, which is found within the same depth interval (1000–600 m) in this region.

Under certain environmental conditions, it may be expected that, if a number of hydrotherms exist on seamounts and the exits of these hydrotherms are found within the OML, sulfates (of barite) or sulfides of Fe, Cu, and Zn may provide the components of hydrothermal deposits. If the top of the seamount with the hydrotherm is found below (or above) the level of the OML, then the chance of discovering sulfates and sulfides in hydrothermal deposits becomes much lower.

Whereas the dissolved organic matter of seawater demonstrates the ability to reduce Mn^{4+} (Sunda et al., 1983), suspended matter is able to extract Mn from solution (Hunt, 1983). Organomineral aggregates in suspended matter in pelagic areas of the World Ocean were found to contain the smallest authigenic particles of sulfur,

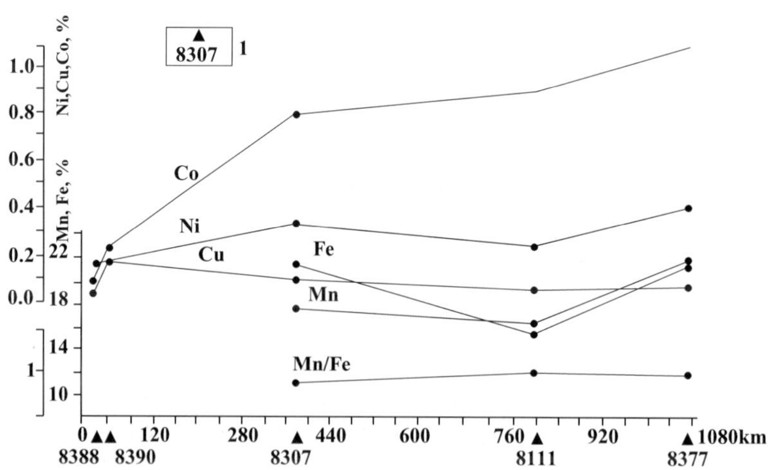

Fig. III.8.3. Changes in the contents of ore elements in ferromanganese crusts with increasing distance from the axis of the South Atlantic Ridge (Angola Basin). After Kornyshkov and Smetannikova, 1986.

pyrite, covellite, and sphalerite (Jedwab, 1979, 1980). It is evident that their origin is a result of biochemical processes and low rates of oxidation (especially within the OML). It is not clear whether aggregates of such type are present in lower water layers of the ocean (no evidence of these deposits have yet been found), and if they do reach the seabed (this seems probable, if this material settles onto the bottom as a component of, or together with, pellets), could these sediments be preserved under the available sedimentary conditions. The preservation of this material may be assumed to occur under conditions when the particle flux of organic matter (mainly as a component of pellets) settling onto the seabed is very strong: in this case, micro-centers dominated by reducing reactions could exist on the bottom for a certain period of time, as well. Due to the rain of particulate organic matter sinking from surface waters to the seabed, the surface layer of FMNs is inhabited by organisms of benthic micro-fauna (Greenslate, 1974; Wendt, 1974; Dugolinsky et al., 1978; Jhiel, 1978; Dudley, 1976), and the development of a viable population of heterotrophic organisms is possible (Sorokin, 1971).

Co is removed from seawater to be deposited onto the surface of crusts in the form of oxides (Heath, 1981). Owing to the fact that Mn crusts accumulate predominantly on seamounts, a certain portion of Co may come from volcanic sources (but not due to the effect of cold halmyrolysis of basalts). At the same time, high contents of Mn and low contents of Co are used as indicators of rapid hydrothermal sedimentation (Manheim, 1986, p. 605).

A distinguishing feature is that the gradation of abundance of chemical elements in crusts of guyots is similar to that of plankton: Co > Fe > Mn > Ti > Zn > Ni > Cu. This is indirect evidence for inferring that plankton is the most important means of transporting ore elements down to the bottom, being thus a major contributor of these elements to Mn crusts. The biomechanical mechanism gives no detailed physical explanation for accumulation of Co in manganese crusts. As for cobalt, the rate at which this element is assimilated and transported by phytoplankton is not as high as that for other transitive metals (Manheim, 1986; Gromov and Starodubtsev, 1973).

In seawater oxides, Co is oxidized to a trivalent state, with the result that Mn is substituted for Co. Because soluble Co passes into insoluble form (due to the oxidizing reaction), crusts and nodules, which are found within oxidizing environments like those occurring around seamounts, become enriched in Co (Manheim, 1986, p.606).

Because FMN crusts lie shallowest, in terms of bathymetry, near the position of the equator (it is in this zone that the position of the OML is highest), maximum contents of Co and Mn were found in crusts from minimal depths. High contents of Co (>2%) were found in three samples collected from slopes of the southern islands in Line–French Polynesia, from depths of not more than 1000 m. The maximum Co content (2.5%) was found in thin manganese crusts from seamount S.P. Lee, Pacific Ocean (8°N), which exists on the fringe of a distinct OML (where it is found at an elevation of 600 m) between the equatorial current and the countercurrent

Such a feature as the maximum enrichment of Co in crusts was recognized as being related to the equatorial zone of high productivity (Alpin and Cronan, 1985, p. 427) (see Fig. II.10.3).

Within the depth interval from 1000 to 1500 m, contents of Co very frequently amount to >1% (Manheim, 1986). The largest concentrations of Co were found in crusts in central parts of the Pacific Ocean, predominantly at depths of 800–1200 m. FMN crusts, which are present in sediments of the continental shelf of the ocean, within the same depth interval, tend to contain smaller amounts of Co, and this feature makes them very different from typical oceanic crusts.

Crusts and nodules of seamounts are composed mostly of vernadite (δ—Mno_2). Vernadite is enriched in Co and Ni. There is a direct correlation between these elements and Mn in crusts and FMNs of seamounts. It is evident that cobalt and nickel are adsorbed in the form of hydrated Co^{2+} and Ni^{2+} ions. Then, cobalt is oxidized and provides the component of vernadite (Burns and Burns, 1977; Murray, 1975). Thus, manganese crusts and FMNs present on seamounts are markedly enriched in Co (to values of 0.6–1.0% in the Atlantic Ocean, and to values of 1.0–2.5% in the Pacific). On distribution maps of Co in crusts and FMNs in the Pacific, there is a distinct maximum in Co contents at depths of 500–2000 m (see Fig. III.8.2). Also, there is a tendency toward a notable decrease in contents of Co at bathymetrically lower levels, where contents of this element do not exceed 0.5% (Skornyakova and Murdmaa, 1989). In contrast to Co, Ni and Mn in crusts and FMNs have two maximums, one is smaller and the other is greater, at depths of 500–2000 m and 4500–5500, respectively.

At present, the tops of guyots fall predominantly within the depth interval of 1000–2500 m (Baturin and Bezrukov, 1979). However, formerly, many of them were present at much higher levels, even probably within the upper active layer. This is evidenced by the fact that breccia of reef limestones and other types of shallow-water deposits of the Cretaceous and Eocene occur on their flat tops (Murdmaa, 1987, p. 206). There are a number of guyots (including their flat tops and steep slopes) where there occur phosphorites or sedimentary, volcanogenic-sedimentary rocks enriched in phosphates (Bezrukov et al., 1969). In our opinion, a major source of phosphatic material for the formation of phosphorites is biogenic (planktonogenic) material that formed under the sedimentary conditions of moderate depths, in the presence of a distinct OML.

In addition to Mn crusts, FMNs also accumulate on the slopes of guyots. Such FMNs have been observed, for example, on the surface of Pleistocene foraminiferal oozes on one guyot, the Mid-Pacific (Murdmaa, 1987, p. 205). There, FMNs are found at depths of 3000–4000 m, which are far above the CCD level. Also, FMNs were found to occur on the tops of seamounts belonging to the Ji-Anomaly Ridge, Atlantic Ocean (Tucholke et al., 1979): here, FMNs occur on the surface of various nano-oozes, the age of which dates back to Late Cretaceous and Eocene times.

In the geological past, the tops of guyots also were within zone of the CCD. A greater portion of phosphorus was delivered to the seabed together with organic detritus. Later, as a result of decomposition of this element, P was liberated from the organic matter matrix and entered interstitial water. At the water–bottom boundary, phosphorus was deposited from water in the form of solid phosphates (leading to the substitution of skeletons and scales of fish, bones of animals, and biogenic carbonates with this element). The formation of phosphorites as such is a secondary process, which appears to be related to events involving rewash of sedimentary material due to the effects of bottom currents, evacuation of non-ore material, and further decrystallization of phosphatic minerals. Moreover, the tops of guyots (and other seamounts) existed in physicochemical situations that are thought to have varied with time, as a result of variations in global sea level, the span of which only in the Pleistocene amounted to 320 m [from −200m to +200 m, (Menard, 1966)]. At different times, the highest seamounts were either above the OML, being within the action of the photic zone, or they sink out of the photic zone to larger depths, to become below the OML.

Events shown in Fig. II.10.3 were very abundant in times when stagnant conditions were prevalent in the ocean. During the Triassic and Jurassic, mineralization of bottom sediments and formation of hardgrounds occurred predominatly at the expense of hydroxides of iron and manganese, whereas Fe hydroxides, phosphates and glauconite are a typical association for mineralization in times of the Cretaceous when the degree of aeration of oceanic waters was much lower. The abundance of phosphate and organic matter is greater in Cretaceous sediments. There is a good correlation, in terms of geological time, between spacious transgressions and periods of stagnant conditions (Arthur and Jenkyns, 1981).

At present, phosphates and glauconite accumulate under moderate reducing conditions (see Parts III.4–5). The initial and end periods of stagnancy are thought to be responsible for global peaks in the deposition of phosphates. However, it is quite possible that local upwellings on shelves and seamounts did not follow this global regularity (Jenkyns, 1986, p. 393).

Geochemical Barrier Zones and Processes of Accumulation of Carbonate–Oxic Manganese Ores

The deeps of the Baltic Sea, the bottom waters of which are periodically supplied with H_2S and in which the density (pycnocline) and redox (Eh) barriers in the water strata are very distinct, are ideal traps for C_{org}, S, Mn, P, Mo, V, partly for Fe, Ba, Cu, Zn, Ni, Co, Cr and some other elements (see Part II.11). In these deeps, the production of carbonate–manganese and sapropel-like oozes takes place. Accumulation of these oozes is thought to have occurred during the last 7.8–8.0 thousand years, i.e., since the time when saline North Sea waters broke into the Baltic (Ancylus) lake.

Before looking at these processes in more detail, we first compare carbonate–manganese sediments of the Baltic Sea to Lower Oligocene manganese ore beds from the Nikopol, Grushevsko-Baksanky (Dnepropetrovsk Province) (Table III.9.1), and Chiatura deposits.

Ores of the Lower Oligocene are rich in Ni, Co, Cu, Ba, Mo, Pb, Zn and other elements (Strakhov et al., 1968). In the Baltic Sea, carbonate–manganese ores are also characterized by increased contents of Ni, Co, Cu, Ba, Mo, Zn, SiO_{2am}, and C_{org}. Ores of the Lower Oligocene (and sediments of the Baltic Sea) show rapid (to the extent of the first few kilometers to a few tens of kilometers) lateral transition into ordinary rocks (sediments), where Mn contents are within the clarke values of this element. The larger the Mn content (in the form of Mn^{2+}), the larger the depth at which maximum accumulations of this element occur within the sediment column.

The process of ore formation taking place in deeps of the Baltic Sea is superimposed on the usual clarke geochemical process, which is prevalent on the remaining portion of the Baltic Sea bottom. The process of ore formation in the Baltic Sea develops not as a continuation of the usual clarke process, but is superposed on the clarke sedimentation due to a favorable combination of exogenic factors including the inflow of saline Atlantic waters, resulting in layering of the water column, rapid development of phytoplankton, the accumulation of their remains on the seabed as organic mud, the appearance of H_2S in near-bottom water layer, an active supply of Mn to the sea, and diffusion of Mn from sediments into seawater.

The rates of accumulation of Mn in deeps of the Baltic Sea varied within the limits from $10–50 \times 10^{-4}$ mg Mn/1cm^2 to $>1000 \times 10^{-4}$ mg Mn/1cm^2 over the period of the marine Holocene (a timespan of 7800 years) (Fig. III.9.1, Table III.9.2). Or, in terms of square meters and tons, these values range from 10 mg to >1000 mg

Table III.9.1 Comparison of carbonate–manganese mud of the Baltic Sea and manganese ore of the South Ukrainian Basin (mainly, Nikopol deposits). After Gryaznov, 1964; Strakhov et al., 1967.

	Carbonate–manganese mud	Mn ores
Duration of ore formation process	Holocene, 8000 years	Lower Oligocene
Tectonic	Platform sea	Platform-sea bays
Geomorphological depth characteristic, m	Part of deeps, 100–250 m	Central parts of bays 50–200 m
Ore formation area	Mud with content >0.2% (up to 13%); Mn, 15000 km^2	-
Stratum thickness	1–3 m (sediments)	1–2 m (ore)
Stratum structure	Microlaminar (0.1–1 mm, sometimes up to 1 cm) clay, sapropelic carbonate–manganese, sapropelic mud, 3–10% C_{org}. (or 5–20% OM), 1–5% SiO_{2am}.	Microlaminar (less than 1 mm to 1 cm or more); composition of primary sediments is unknown
Mn content, %	Up to 13%; after recalculation on cf, 1.5–20%	17–19% (carbonate ore) (in the Mangyshlak deposit, primary sediments contained 1.8–2.6% Mn)
Content of clayey material, %	30–50	-

(or 1 g)/m^2 and from 0.0001 to >0.01 tons/m^2, respectively. As is known, the process of ore formation has been the most active in the Holocene (that is, during the last 7.8–8 Ka) and only in those areas where the Mn content in mud in deeps of the Baltic Sea is not less than 1%. This is why average values of Mn abundance (or absolute mass) in Holocene mud for these areas are about 0.008 tons/m^2. Division of this average value by 7.8 Ka yields 0.001 tons/m^2 Mn per thousand years (Table III.9.3). According to estimates, absolute masses of Mn in the ore bed at the Grushevsky-Baksank field of the Southern Ukrainian basin range from <0.25 to >2.25 tons/m^2 (1.5–2 tons/m^2 on average) (Strakhov et al., 1967, p. 521), and this is up to 1500–2000 times as much as those in the Baltic Sea. If we assume that the rate of accumulation of Mn at the Grushevsky-Baksank field was the same as that in deeps of the Baltic Sea, then a period of time of about 1.5–2 Ma years would be sufficient for the deposition of the Grushevsky-Baksank Mn deposit.

The sedimentation rate of carbonate–manganese muds with >1% Mn content in deeps of the Baltic Sea is estimated at about 10–20 cm per thousand years, on average. This rate provides for 150–400 m of sediments over a time period of 1.5–2 Ma. These muds contain up to 70% moisture. Removal of moisture from the muds would

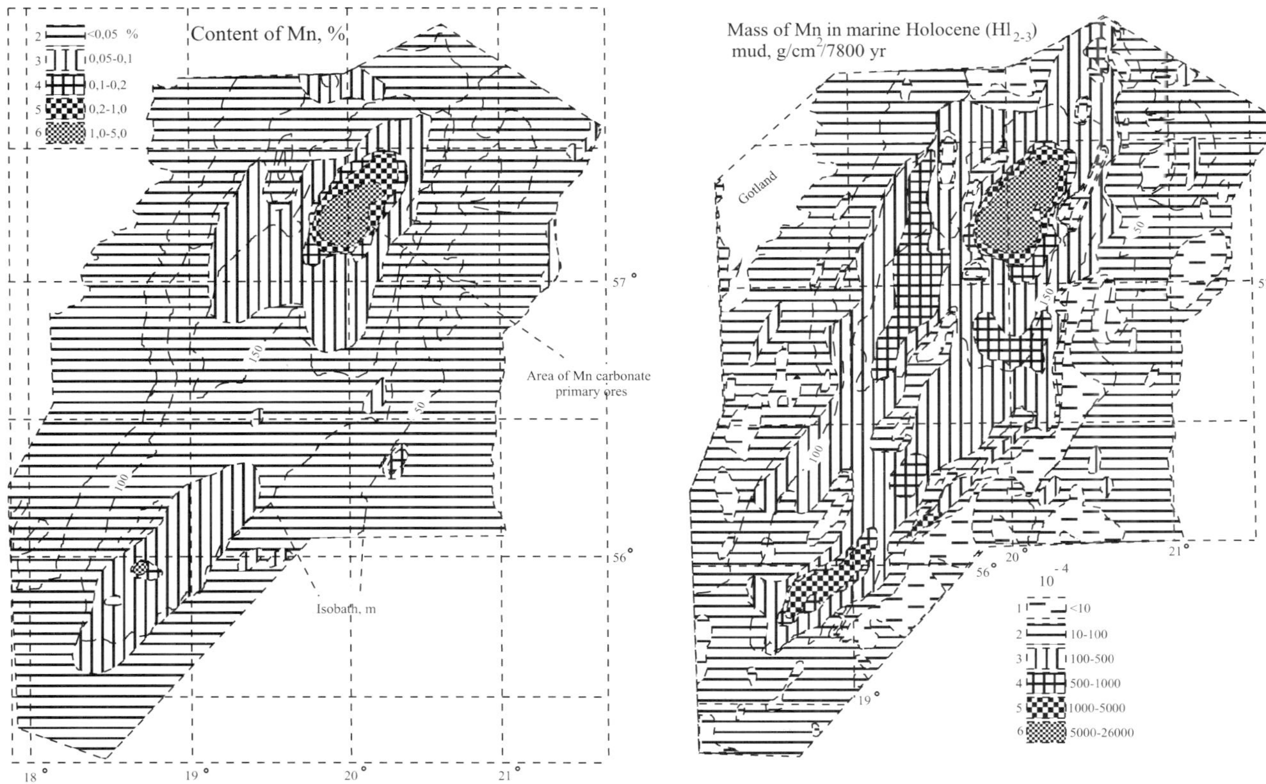

Fig. III.9.1. Baltic Sea deeps (190–248 m)—only region in the world where primary manganese carbonate sapropelic muds form. y contain up to 20% organic matter (5–10 % C_{org}) and up to 10–30% $MnCO_3$ (up to 13.01 % Mn).

Table III.9.2 Average annual authigenic deposits of Mn in the Baltic Sea. After to A.I. Blazhchishin, 1984.

Sea region	Sedimentary basins			Only concretionary zones			Total average annual anthigenic Mn deposits, tons
	Area, km^2	Authigenic Mn per 8000 years, thousands tons	Annual mean, t	Area, km^2	In bulk per 1000 years	Annual mean, tons	
Baltic Sea	37092	9934	1242	7095	2329	233	1533
Gulf of Riga	4824	488	61	2817	2180	218	279
Gulf of Finland	8216	400	50	5694	3787	379	249
The whole sea except for Gulf of Bothnia	50131	109822	1353	15606	8296	830	2241

Table III.9.3 Mean authigenic Mn content in mud of Baltic Sea shallows (basins)* over 7800 years, (in thousands of tons)

Basin	Area, km^2	Mn in carbonate form	Mn in sulfides, hydroxyls, phosphates	Average rate of authigenic Mn accumulation, g/cm^2/1000 years
Bornholm basin	8100	867	573	5.1
Gotland basin	4577	8183	1751	6.6
Lanswort basin	2374	1030	100	8.7
Total :	15051	10080	2424	-

x) Only basins with overclarke (more than 0.2%) contents of Mn in mud are considered.

lead to a decrease in thickness of these deposits by 70%, which in this case would amount to 45–120 m. In addition, present-day mud contain 5% C_{org} (or 10% organic matter) and about 5% SiO_{2am}, on average. At the same time, manganese ores contain less than 1% of these biogenic components (biogenic components are removed from sediments due to diagenetic processes and rewashing). The average thickness of the ore bed at the Grushevsky-Baksank Mn field is 1–3 m; this is 20–25 times less than our estimates. Recalculation of these data with respect to abiogenic material evidences that Baltic Sea muds contain 1.5–17.0% Mn. Some estimates suggest that only a 1.5- to 2-fold enrichment of these muds would be required to transform them into ore deposits with a Mn content generally similar to those found at the Grushevsky-Baksank Mn field (20–35% Mn). Following the model of Strakhov et al. (1967), the enrichment of some elements in muds can be attributed to a large extent to the effects of sediment rewashing and removal of clayey minerals from the ore bed. Clayey minerals in muds of the Baltic Sea make up about 40% of the total sediment. As a result of hypothetical removal of these minerals from muds, the thickness of the sediment layer would become twice as small, i.e., approximately 7–2 m thick. Nevertheless, this value is still twice as high as that for the thickness of the ore bed at the Grushevsky-Baksank field. From this it follows that (1) either Mn contents in primary muds of the Grushevsky-Baksank basin were twice as large as those in muds of the Baltic Sea, or (2) Baltic muds would have been transformed into manganese ores only in those depositional situations when the Mn contents in these muds were twice as large as those used in our calculations (1.5–7% Mn, after recalculation for abiogenic matter). In such cases, according to the model of Strakhov and co-authors (1967), as well as our model (Emelyanov, 1988; Emelyanov and Lisitzin, 1986), only those Baltic muds which contain 5–10% Mn can be transformed into manganese ores. Muds of such type have been found only in two small regions of the Gotland and Farö deeps (Fig. III.9.1). They are found at depths of 180–250 m. The surface area of each of these mud-fill regions is a few tens of square kilometers, comparable to the area of the Nikopol ore deposit.

and transformation of manganese sediments and ores may be distinguished (Fig. III.9.2).

Two stages are characteristic of the Baltic Sea. Stage 1 is the period when layering of the water strata was absent and deeps were well aerated (the stage of the Baltic Ice Lake). At the same stage, sedimentary conditions below the thermocline were characterized by dispersal of Mn over the area, so that features such as ferruginous crusts and flat and drop-like FMNs began to accumulate there (zone of oxidized ferromanganese ore formation).

Stage 2 is the period of Holocene transgression, when communication of the Baltic Sea with the North Sea (i.e., with the World Ocean) was established as a result of eustatic rising of the global sea level. As a result of penetration of saline ocean waters into the Baltic Sea, deep-sea basins were filled with this denser water. Layering of water strata was formed, and a distinct halocline was established. The halocline (pycnocline), which is a density barrier, became an obstacle for the vertical mixing of waters, with the result that bottom waters become deficient in oxygen and hydrogen sulfide appeared there. In times of strong storms and inflows of saline North Sea waters, there was a mixing of water layers, so that stagnant conditions were absent for several years. The periodicity in conditions of stagnancy and aeration provided for accumulation of sapropelic microlaminated manganese–carbonate oozes containing amorphous carbonates of Mn. At stage 2, the first thousands to tens of thousands of years, there was a transformation of amorphous carbonate into rhodochrosite as a result of diagenetic reactions.

Stage 3 is the period when strong diagenetic reactions continued after burial of the carbonate–manganese sediments bed that had developed. At this time, further diagenetic transformation of rhodochrosite and manganocalcite occurs. Although

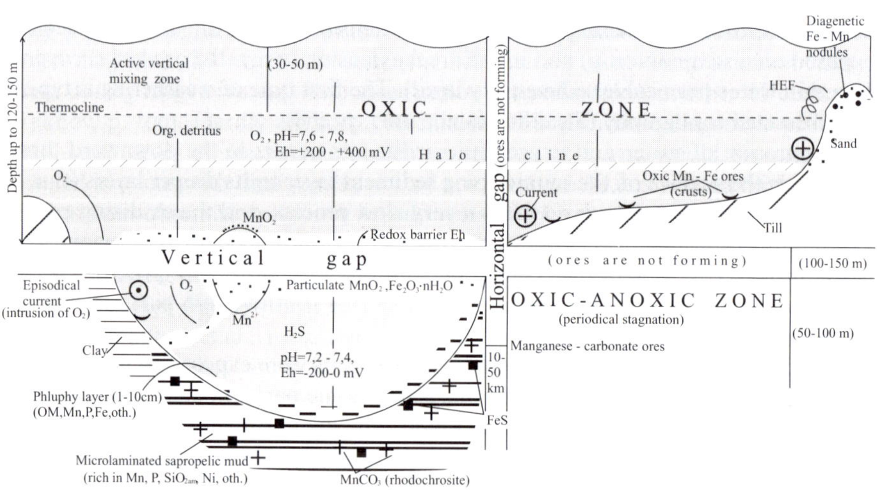

Fig. III.9.2. Model of formation of the oxic-carbonate Mn ores. HEF—high energy facies, OM—organic matter.

this stage of manganese ore formation in nature is a subject of discussion, no evidence of this process has yet been found.

Stage 4 is the period when the diagenetically reworked ore bed again becomes exposed in the frontal wave zone (zone of sediment reworking and strong bottom currents). The ore bed becomes exposed to the effects of strong weathering, its oxidation, dilution and redeposition. Manganese—carbonates enter manganite, and then into pyrolusite. Formation of FMNs occurs. Redeposited ore material becomes enriched in Mn. Clearly, no evidence of this stage has yet been found.

Stage 5 is the period when the ore bed happened to be above sea level, under subaerial conditions. Ores experienced further oxidation and a transition from carbonate into oxic form as a result of penetration of oxygen (together with ground waters) into the ore bed.

There also exist some other ideas on the formation of manganese–carbonate ores. According to Kalinenko (1990), the Laban deposit in the Northern Caucasus is partially localized within the once-existing early Miocene foredelta of an old river.

In the Maikop suite of the early Miocene, there are localities of rhodochrosite ores containing up to 34.63% Mn. There are two reasons why these ores formed: (1) input of managanese, which was delivered by the paleoriver and then accumulated in sands of the outer part of the paeleodelta, and (2) because the main mechanism responsible for growth of manganese concentrations to ore values was the influx of this element via water solution in the lateral extension during early diagenesis. As for the mechanism proposed by Kalinenko (1990), there are no indications (not even one reference) that the Mn content in the sediments of foredeltas of present-day rivers are higher than the clarke of this element. From the above it follows, see first of all Part II.1, that in contrast to Fe, Mn is transported from the mouths of rivers to the open sea to distances of a few hundreds of kilometers, where the deposits of the foredelta are dominated by reducing conditions, which in no way encourage the preservation of mobile forms of Mn^{2+} and Mn^{+4}. The author believes that the accumulation of Mn might have been able to occur within the foredelta of a paleoriver only under conditions when this river initially entered a small bight, being a hydrodynamical trap, the bottom waters of which were periodically exposed to contamination by hydrogen sulfide. Over-clarke values of Mn may have been concentrated in sediments of this bight, as a result of a process consistent with the pattern described for the Baltic Sea (Emelyanov, 1981, 1982, $1988_{1,2}$). The secondary enrichment of manganese in sands of the foredelta may have been able to occur during a regressive sea (stage 3, see Fig. III.9.2). It can be suggested that several regressive events took place in this area. This conclusion is borne out by geological facts. In particular, in the sediment sequence of Maikop deposits there are rapid changes in facial conditions. Also, the zone of wave reworking consists of several manganese sediment beds, and this is evidence that sediments were repeatedly sorted, as confirmed by the oblique layering signature. The secondary chemical (process) reactions that led to the enrichment of Mn in sediments and transformation of oxic compounds of Mn, which were formed under oxidizing conditions at the stage of regression, into manganese–carbonates, come under the general heading of diagenesis and could follow the theory proposed by Kalinenko (1990). The only exception to this scheme is the intensive influx of Mn with interstitial waters in the lateral extension, i.e., from adjacent layers of the same age. Kalinenko has given no explanation for the mechanism that causes such influx of this metal, especially to distances of as great as hundreds of meters or even kilometers.

The model proposed by the author (Emelyanov, 1982, 1986) describes the development process for carbonate–manganese (rhodochrosite) muds in deeps of the Baltic Sea, which are periodically exposed to contamination by hydrogen sulfide. This is a present-day analog of old carbonate–manganese ores (in the given case, Oligocene ores of the Nikopol deposit in the south of the Russian Platform), which is why this model has been applied for explaining depositional processes in many other European ore deposits, including those of the Lower Carbonaceous and Lower Jurassic (Fig. III.9.3, Table III.9.4) (Huckriede and Meischner, 1996). Moreover, the explanations of these authors for the mechanisms that cause the accumulation of rhodochrosite sediments have essentially confirmed the ideas discussed above (Emelyanov, 1977, $1979_{1,2}$, 1981, 1982, 1986, 1988, 1999; Emelyanov and Lisitzin, 1986; Emelyanov et al., 1987). On the other hand, Huckriede and Meischner (1996) studied the composition of manganese–carbonates in microlaminations of Baltic Sea muds and in old microlaminated manganese ores using more advanced and precise methods. Owing to these research efforts, this model has become more persuasive.

In previous works on the subject (Emelyanov, 1979, 1981, pp. 149–150), the author wrote that in the upper layer of the sedimentary sequence development of manganese–carbonate of complex composition occurs, which is subsequently transformed, in a matter of hundreds or even thousands of years, into well-crystallized rhodochrosite. Particles of non-crystallized manganese–carbonates are 1–5 μm in size, which can be seen under a microscope as semi-transparent clayey aggregates with a refraction index of about 1.6. Crystallized rhodochrosite begins to exist in sediment cores only at depths of 0.5–1.0 m from the sediment surface. Some geologists engaged in the study of sediments of Panama basin (Pedersen and Price, 1982) have reached the same conclusion. According to these authors, the formation of rhodochrosite occurs within the sediment column, at depths from several centimeters to a few tens of centimeters below the seafloor. Some other researchers believe that rhodochrosite forms directly in the upper film of mud (Huckriede and Meischner, 1996). However, no reliable proof in favor of this theory has been presented.

Enrichment in rhodochrosite is greatest in microlaminations (Fig. III.9.4) which were formed in 2–5 years, and this is consistent with what is known about the periodicity of inflows of salt waters over the past few decades. In these microlaminations, which are enriched in rhodochrosite, there is an abundance of remains of bottom fauna, first of all, foraminifera with calcareous skeletons of *Elphidium excavatum* (Saidova, 1981). These features are found exclusively in rhodochrosite microlaminations (Huckriede and Meischner, 1996, p. 1407). This is evidence that the formation of rhodochrosite in sediments occurs mostly due to rapid changes in depositional conditions, from a stagnant sedimentation regime into an environment characterized by a well-aerated situation, as has been repeatedly reported by the author (Emelyanov, 1979, 1981, 1986). Accumulation of rhodochrosite in sediments occurs as a result of its precipitation under (reducing) conditions of strong

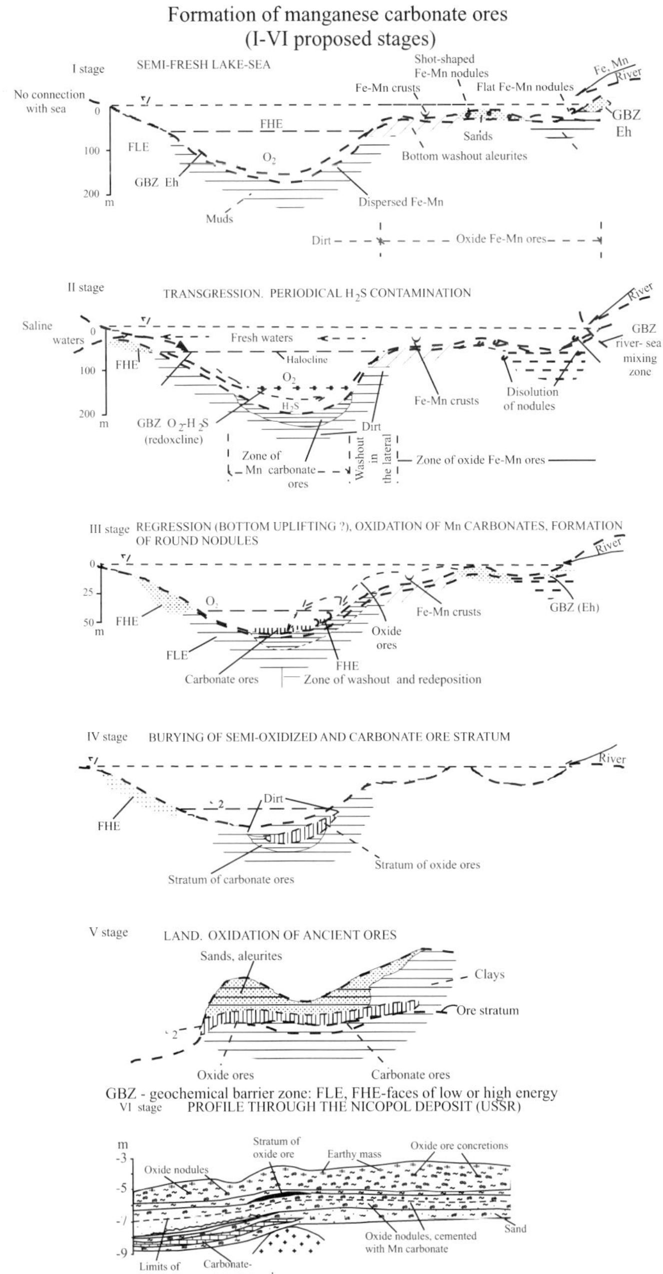

Fig. III.9.3. Formation of manganese carbonate ores (I–VI proposed stages).
Inferred stages of carbonate manganese sediments and manganese ore accumulation (the Baltic Sea and Lower Oligocene ores of the Nikopol deposit used as examples). I–V—various stages of sediment formation and ore accumulation. V—stage is after N.M. Strakhov et al., 1968. VI—the section through the Nikopol deposit is after N.A. Panchenko, 1982. (p. 96)

Table III.9.4 Comparison of recent (Baltic Sea deeps) and ancient manganese deposits within basins of black shales. After Emelyanov, 1981, 1982, 1986, 1998; Huckriede and Meischner, 1996.

	Holocene (Baltic Sea)	Lower Jurassic (Austria and Hungary)	Lower Carboniferous
1. Depths of sedimentation	200-450	?	?
2. Primary manganese concretions	Calcium–rhodochrosite (15-40 mol % of $CaCO_3$)	Calcium–rhodochrosite (20-0 mol % of $CaCO_3$)	Calcium–rhodochrosite (20-0 mol % of $CaCO_3$)
3. $\Delta^{13}C$ relation in the Ca-rhodocrosite	–6 to –14	–5 to –22	–6 to –10.5
4. Sediment type (rock)	Mud. Rich in organic matter and iron sulfides. Interbeds of microlaminated and bioturbated muds	Shales. Rich in organic matter with pyrite. Interbeds of microlaminated and bioturbated horizons	Shales and cherts. Rich in organic matter and pyrite. Microlaminated in most cases, some layers are bioturbated
5. Environment	Centers of sea deeps. Strong water stratification, periodic near-bottom water stagnation (every 5-11 years) due to inflows of saline seawater	Center of basin. Stratified water column. Near-bottom waters were often stagnant	Center of the basin. Stratified water column. Near-bottom waters were often stagnant
6. Short-term aeration of the near-bottom water	Rhodochrositic layers with the shells of benthic foraminifera (*Elphidium excavatum*) remnants of planktonic foraminifera (brought from the North Sea)	Rhodochrositic layers with benthic foraminifera	Rhodochrositic layers with the benthic foraminifera
7. Long-term aeration of the near-bottom water	Enrichment by manganese of the upper horizons of bioturbated interbeds	Bioturbated, massive rhodochrositic layers	Crusts of manganese oxides in the root of rhodochrositic layers
8. Type of the manganese ores in the periphery of the basins	Iron-manganese spheroidal (1-5 cm) concretions and flat crusts	Manganese concretions and crusts	Manganese concretions with the nucleus of macrofauna

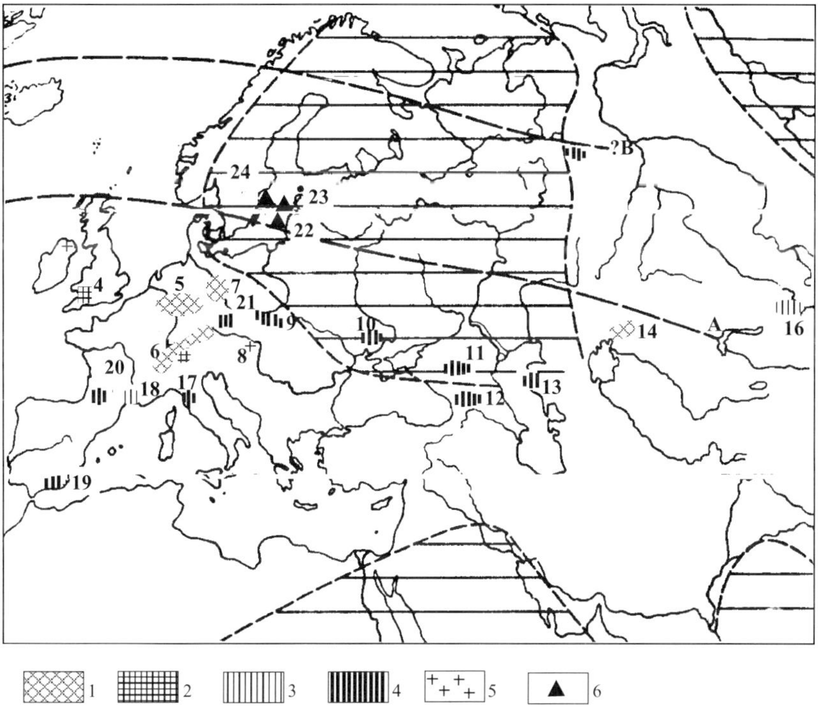

Fig. III.9.4. Location of hypergenic (ferruginous and manganese) ores and bauxites of Paleogene and other ages in Europe. After Strakhov, 1947; Emelyanov, 1981, 1986; Huckriede and Meishner, 1996.

1—continental ferruginous and manganese ores; 2—marine ferruginous ores; 3, 4—manganese ores; 5—bauxites; 6—recent rhodochrositic mud.

Northern border of the regions in the Eocene: A-A—tropical; B-B—subtropical (in the Paleocene and, especially, the Oligocene, both these borders were located to the south).

1—the primary shamositic ores in Texas and Louisiana; 2—Mississippi, Georgia and Alabama; 3—Appalachian zone (1, 2 and 3 are not shown on the map); 4—deposits of Ferness and Meedoni; 5—Rhein shale mountains; 6—pisolitic ore formation (Bohnerzenformation) in Switzerland in the South Germany; 7—Harz; 8—Zhynt, Hungaria; 9—Ukrainian; 10—Nikopol; 11—Labinskoye; 12—Chiatura; 13—Mangyshlak; 14—Near-Ural ferruginous ore region; 15—Northern Ural; 16—foothills of Altai; 17—Pisa (Italy); 18—Southern France; 19-21—after Huckriede and Meishner, 1996; 19—Banos de la Fuensenta; 20—Upper Pyrenees; 21—Arkut; 22-24—after Emelyanov, 1982,1995, the deeps of the Baltic Sea (recent primary manganese carbonate ore): 22—Gotland; 23—Faröe; 24—Landsort.

alkalinity (which is explained by bacterial sulfate-reducing reactions) and due to high concentrations of dissolved Mn^{2+} in interstitial waters, which can be as large as 48.2 mg/l (Table III.9.5). Manganese precipitates as Ca-rich rhodochrosite (Huckriede and Meischner, 1996). This mineral forms a microbed on the surface of a sapropelic microlamination. The low iron content in the rhodochrosite microbed is explained by the chemical reactions that reduce reactionable Fe^{2+} into sulfide.

Table III.9.5 Content of microelements in the sea and interstitial waters of the Baltic Sea. After Shishkina et al., 1981, p. 208.

Component, element	Units	Surface sea water	Mud from the bottom below the redox (Eh) barrier in the water		Stagnant water from the Gotland Deep				
					Near bottom water	Interstitial water			
						2-5 cm		5-15 cm	
			Limit	Average	Average	Limit	Average	Limit	Average
Eh	m	-	–50 to –220	-	-	–200 to –220	-	–190 to –230	-
pH	-	7.8–8.0	7.4–7.6	-	7.1	7.4	-	-	-
AlK	mg/eq/kg	-	6.4–6.7	-	1.8	12.2	-	-	-
Mn	mg/l	0.1–0.4	0.18–3.7	1.2	0.34	7.1–17.3	13	7.0–18.2	13.4
Fe	mg/l	0.7–1.9	0.21–0.42	0.3	0.06	0.04–1.1	0.5	0.17–1.65	0.7
Zn	µg/l	0.7–7.9	45–420	169	22	50–150	98	10–120	81
Cu	"	0.6–7.6	45–90	61	16	9.6–37	22	8–20.2	16
Ni	"	-	<6	-	8	<8–29	-	<5-8	-
Li	"	-	-	102	105	-	117	-	

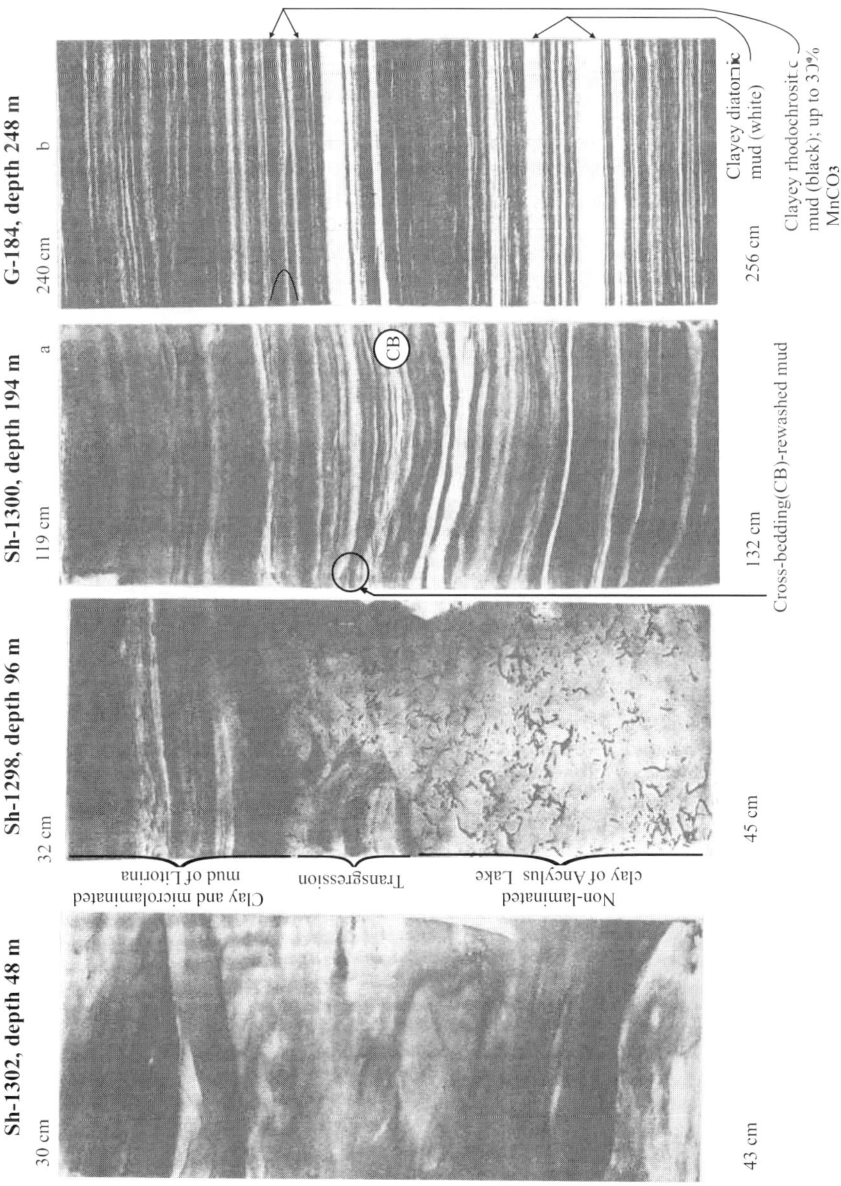

Caption see page 524

Fig. III.9.5. X-ray photos of clay and mud of the Gotland Basin, Baltic Sea. Horizons of the cores are shown in cm. First three photos were taken by F. Kögler and H. Kassens, University of Kiel, Germany.

Station Sh-1302, depth 48 m, located near the coast of Latvia. Glacial clay (BIL), redeposited many times. There are many lenses, brownish gray in color, viscous.

Station Sh-1298, depth 96 m, southeastern part of Gotland Basin. Transgressive layering of microlaminated Litorina (HL_2) mud on rewashed Ancylus lake clay (Hl_1) during Litorina transgression. Thickness of microlayers is ~ 1 mm.

Station Sh-1300, depth 194 m, central part of North Gotland Deep. Marine microlaminated mud. Clayey-diatomic microlayers are white, clayey-manganese microlayers are black. Wavy form of microlayering may point to rewashing of sediments by periodic near-bottom currents.

Station G-184, depth 248 m. Most deep of the North Gotland Deep. After Huckriede and Meishner, 1996. Horizontal layering of microlamination without evidences of rewashing. Black microlayers—clayey-rhodochrositic, white—clayey-diatomic.

Ore Formation at the Redox (Eh) Barrier in Sediments

The position of the redox (Eh) barrier in sediments may be different in various areas of the World Ocean. Only 16 types of depositional situations thought to be of greatest importance in the author's opinion will be described here (Fig. III.10.1).

Situation 1. This type of sedimentary environment is found in the water strata a short distance above the water–bottom interface (evidently, within the first few millimeters above the bottom surface). This situation is typical for seas and peripheral (near-continental) areas of oceans, where the environment is characterized by a normal oxygen regime and high rates of sedimentation. The sedimentary material (including ore components) is delivered both by rivers from land and by products of coastal erosion. The C_{org} content in the sediments is more than 0.5% and generally amounts to 1–3%. Diagenesis is of aerobic–anaerobic type (which is quite typical). In sediments, development of single grains of Fe sulfides occurs (chemical reactions nos. 5, 7 ,8 (and occasionally 9) are characteristic of this sedimentary environment; see Table II.15.1). There is a flux of Mn^{2+} to the near-bottom water from sediments. As a result, sediments become depleted in Mn and contain clarke values of Fe. Ore formation at this stage is absent.

Situation 2. The barrier is found within 1–10 mm below the seafloor. The situation is typical for those areas which are well-supplied with muddy material from land (generally, in the vicinities of the mouths of rivers). The oxygen regime is normal. The rates of sedimentation in these areas are high, and the C_{org} content in sediments is 1–4%. At the same time, the rates of accumulation of sediments (commonly clayey muds) lag behind the rates of oxidizing processes, which is why the upper layer of muds has time to become oxidized. There is an intensive inflow of dissolved forms of Fe^{2+} and Mn^{2+} to the Eh barrier from below (from interstitial water). At the barrier, these substances precipitate in the form of FeOOH (goethite) or Fe_2O_3 (limonite–hematite) and MnO_2. The abundance of these substances tends to become aggregated in the form of coin-like FMNs, "floating" in the uppermost layer of oxidized mud. After burial, these aggregates appear in the zone of reducing reactions, the intensity of which is sufficient to dissolve these aggregates. Thus, situation 2 is responsible only for ephemeral ore formation. The barrier is dominated by reaction nos. 4 and 6, and the environment below this barrier is dominated by reactions nos. 5, 7, and 9 (Table II.15.1). The process of formation of nodules on the seabed fall under the heading of diagenetic reactions.

Situation 2–type reactions are characteristic of bays of the Baltic Sea (the Gulfs of Riga, Finland and Bothnia) and the Black Sea (Kalamits Bay), as well as certain

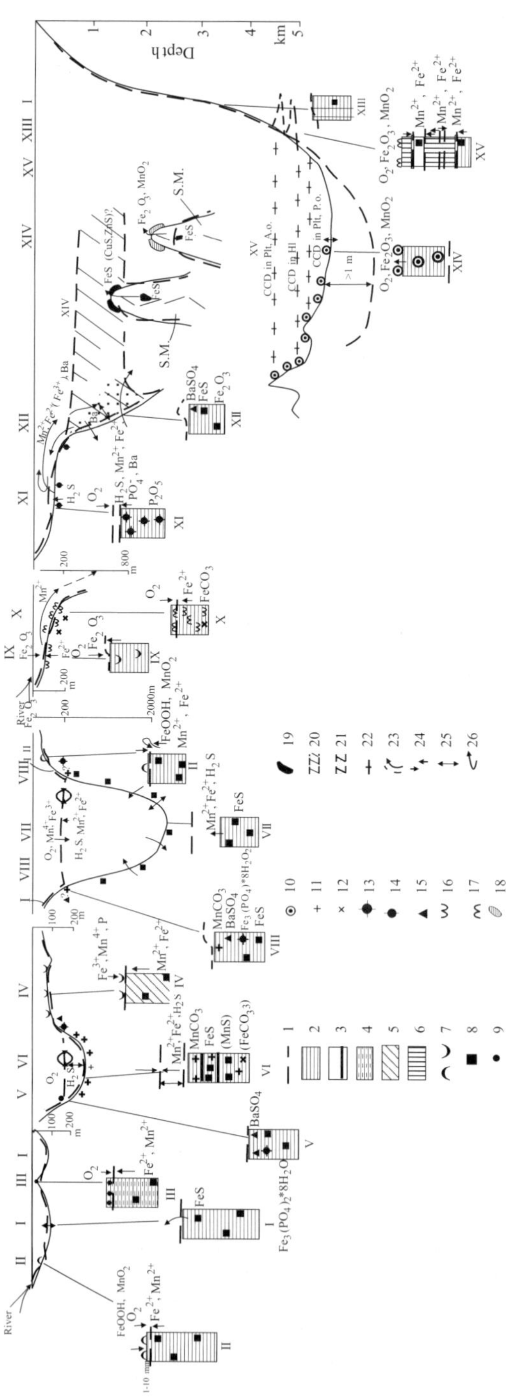

Caption see page 527

other areas of the World Ocean. Nodules are characterized by high Fe contents and decreased Mn contents (see Table III.10.1).

Situation 3. This barrier is found at depths of 1–10 cm below the sediment surface. There is an abundance of shallow-water facies, sands and aleurites. The type of processes occurring in this environment are essentially analogous to what occurs in situation 2, but due to the effect of active wave processes and repeated elutriation of sediments (the seafloor is washed by the upper mixed water layer), the layer of oxidized sediments consists of sandy-aleuritic material, in which development of small, round (drop-like) FMNs occurs. These nodules mostly lie partly buried in the upper 0–10 centimeters of sands and aleurites, rather then completely on top of them. Two possibilities have been suggested to explain why these nodules have a rounded shape. One is that they are periodically rolled by bottom currents (providing also for redeposition and reworking of nodule-bearing sands); and the second is that the nodules are being supplied with ore components (Fe^{2+} and Mn^{2+}), derived from interstitial waters, more or less uniformly from all sides. The process of

Table III.10.1 Average chemical composition of ferromanganese concretions from various basins of the Baltic Sea, Mn–Fe in %, Zn–Cu in 10^{-4}%. After Glasby et al, 1997

Element	Gulf of Bothnia	Gulf of Finland	Gulf of Riga	Gotland Region	Gdansk Bay	Kiel Bay	Kiel Bay
Mn	14.6	13.3	9.7	14.0	8.7	29.3	26.4
Fe	16.6	19.7	22.8	22.5	18.5	10.1	10.0
Zn	200	113	135	80	137	340	527
Co	140	96	64	160	91	77	96
Ni	260	35	47	750	148	97	61
Cu	80	9	17	48	42	21	27
Mn/Fe	0.88	0.68	0.43	0.62	0.47	2.90	2.64

Fig. III.10.1. Ore formation processes in the seas and ocean on the redox-barrier Eh in the water and in the sediments.

I–XV—different facial environments (on the basis of the Atlantic Ocean basin). After Emelyanov, 1982, 2002, with some additions. bottom relief, cores of sediments and position of the barrier Eh are given.

Legend. 1—position of the barrier Eh (redoxcline); 2—reduced sediments (mainly terrigenous mud), >1 % C_{org}; 3—sapropelic microlaminated carbonate manganese mud; 4—weakly reduced sand and aleurite; 5—weakly reduced (or weakly oxidized) ancient clay; 6—oxidized pelagic mud (red clay); 7–10—authigenic-diagenetic formations that occur in the sediment of the mentioned (1–5) facial environments; 7—Fe crusts and flat concretions; 8—iron sulfides; 9—small drop-shaped ferromangnese concretions; 10—round FMNs; 11—carbonate manganese formations (mainly rhodochrosite); 12—siderite; 13—vivianite; 14—authigenic phosphate (ftorapatite, francolite); 15—barite; 16—shamosite; 17—hydrogethite; 18—hydrothermal terrigenous (oxic) sediments; 19—the same with iron sulfides; 20—OML (in the open ocean); 21—the same near the continental slope (active exchange processes in bottom water occur); 22—CCD during the Holocene; Plt A.O.—CCD in the Atlantic Ocean during the Pleistocene: during glacial stages the CCD was much higher than now; 23—hydrotherms (in fracture zones); 24—exchange of elements (the most common elements only); 25—active exchange at the boundaries ; 26—near-shore upwelling. SM—submarine mountain.

formation of drop-like nodules is diagenetic. Reactions continue after burial of these nodules, when they are found within sediments in a reducing environment, leading to their complete dissolution. The process of ore formation is ephemeral.

Situation 3 is characteristic of some banks and island slopes in the Baltic Sea (Gulf of Riga) (Varentsov, Blazhchishin, 1976; Emelyanov, 1982), and evidently in some other shallow-water areas of the World Ocean.

Situation 4. This barrier occurs either at the water–bottom interface (which is represented by hard rocky ground and analogous, for example, to what occurs at the water–basalt or water–granite barriers) or several millimeters (centimeters) below the sea floor. No accumulation of sediments occurs in this situation, but events such as periodic erosion of these sediments and accumulation of residual sandy-aleuritic material are possible. There are strong near-bottom currents. Supply of ore components to the barrier occurs from layers lying both above and below this barrier, that is, both from the overlying seawater and from interstitial waters (if sediments are present on the sea floor). The former is the case for areas with rocky ground, whereas the latter is for a bottom made up of soft ground (old clays, moraines, etc.). The Eh barrier is dominated by reaction nos. 4 and 6 (Table II.15.1). On this barrier, formation of either flat ferromanganese crusts (predominantly on rocky grounds) or flat oval (in the form of small cakes) FMNs (on a clayey substrate) occurs. The crusts consist mostly of Fe_2O_3 and are enriched in Mn and P; the nodules are composed of Fe and Mn and are enriched in P.

Situation 4 is very characteristic of the Baltic Sea (bottom locales occurring between two vertical barriers: the halocline and the O_2–H_2S barrier). Situation 4 has also been found to occur in the Black, Mediterranean, Barents, Kara and some other seas, as well as in some areas of oceans [for example, in the Atlantic Ocean (McGeary and Damuth, 1973)].

Situation 4 does not lead to accumulation of an ore bed because there is a tendency for ore formations (like crusts and nodules) to dissolve as they are buried in sediments.

Situation 5. The position of the Eh barrier is unstable: it is periodically found either at the seawater–bottom interface or several meters (or a few tens of meters) above the sediment surface (i.e., in the water strata). These events have a frequency of about 1 in every 5–10 years. The barrier is always underlain by sediments (which are characterized by reducing conditions); the surface of the seafloor is found periodically either within reducing or oxidizing conditions (when oxygenated waters come in contact with the bottom). The most distinguishing feature of this situation are low concentrations of O_2 in water (in general, 1.0–0.2 ml O_2/l). Sedimentation rates are high: the bottom is covered by gray muds. Chemical reactions active in this environment are nos. 5, 7, 8, 9 (see Table II.15.1). An anaerobic type of diagenesis is prevalent. Mn^{2+} is actively derived from muds. This situation is characteristic of fringe areas of deeps in the Baltic Sea.

The most abundant minerals in the sediments are iron sulfides, pyrite (FeS), and iron phosphates, vivianite [$Fe_3(PO_4).8H_2O$]; rarer are manganese phosphates, $Mn_3(PO_4)_2$. Vivianite forms under weakly reducing conditions, which force Fe^{3+} to convert into Fe^{2+}, but the sulfate-reducing process, which is able to consume virtually all reactionable Fe, remains absent. Vivianite is very abundant in lacustrine deposits, where reactions of sulfate reduction are limited. In addition to the Baltic Sea (situation 5), this situation has also been found in the Ionic Sea (Sevastyanov et al., 1979) and in some other areas (Postma, 1981; Baturin, 1978).

It is evident that the reason for the formation of Mn phosphate in muds of the Landsort deep (Suess, 1979) is the high Mn content in the sediments, that is, in amounts sufficient for the development of all reduced forms of Mn (carbonates, phosphates and sulfides) to occur.

Situation 5 is responsible for the most favorable physiochemical conditions for the formation of barite ($BaSO_4$).

Situation 6. The depth of occurrence of the Eh barrier is not continuous, and there is a depth interval of 10–100 m between its lowest and highest positions. Conditions of stagnancy also appear to be periodical: these events occur about once every 5–10 years. At this stage, stagnation is stronger than that in situation 5. H_2S, CO_2, Mn^{2+}, Fe^{2+}, methane and other components diffuse from sediments (muds) into near-bottom waters. In the water strata with H_2S, accumulation of Mn^{2+}, Fe^{2+} and microelements occurs. Being found within the water layer, the Eh barrier is a zone where reaction nos. 4 and 6 (the top of the barrier), and also of 5 and 6 (in the bottom part of the barrier, as well as throughout the entire depth of the water layer and sediments with H_2S, Table II.15.1). One of the most dramatic events in the Baltic Sea is the intrusion of well-mixed saline oceanic waters into the deeps of the Baltic sea, for it changes the reducing environment in the sea into an oxidizing one, and, in this case, it takes only one to several months for the entire mass of Mn^{2+} dissolved in the near-bottom waters (up to one million tons in the entire Baltic Sea) to become oxidized, before finally settling onto the seafloor in the form of gels. When the basin again becomes stagnant, the action of chemical reaction no. 5 results in the renewal of reducing conditions. Such fluctuations in sedimentary environments have been occurring over the last thousand years. In interstitial waters trapped in the muds of deeps, Mn^{2+} contents are 30 mg/l . In the muds, formation of Mn carbonates of complex composition occurs ($(Mn_{56.8}.Ca_{25.5}.Mg_{9.7}.Fe_8)(CO_3)$) (Hartman, 1964), which are recrystallized into rhodochrosites during early diagenesis. There is rapid enrichment of manganese in these muds (the sediment samples contain up to 12.88% Mn, or as much as 29.94% $MnCO_3$).

Situation 6 is one of the most favorable for the formation Fe sulfides, Mn carbonates and sulfides. As a result of chemical reactions, H_2S is released from muds, and sulfur from this substance becomes fixed in the form of sulfides. Release of free H_2S begins when all Fe^{2+} is combined in the form of sulfides. In deeps of the Baltic Sea, muds have been found to contain up to 45 ml H_2S/l of sediment (Emelyanov et al., 1986). The rate of the sulfate-reducing process is greatly diminished. A certain portion of non-mineralized organic matter buried in the sediments is used for the formation of methane. However, the majority of organic matter is preserved in sediments to be used later in the process of authigenic ore formation during diagenesis and catagenesis.

Among other types of sulfides, pyrite is the most abundant mineral in sediments where situation 6 is prevalent. It forms as a result of reaction between hydrotroillite (FeS) and elementary sulfur (S):

$$FeS + FeS_2 \text{ (cubic structure).}$$

Marcasite (a rhombic modification of FeS_2) is rare or entirely absent. Minerals such as mackinawite (FeS, tetragonale structure), greigite (FeS_4, cubic singony) and Fe trisulfide (Fe_2S_3, only in the Black Sea) can be periodically found in muds of deeps where there is H_2S contamination.

Pyrite forms predominantly at the initial stage of the sulfate-reducing process (mainly, in situation 5), i.e., during aerobic–anaerobic diagenesis (Volkov, 1984):

$$2H_2S + O_2 = 2\ H_2O + 2S°;\ FeS + S° = FeS_2$$

Conversion of Mn^{4+}_{react} and Fe^{3+}_{react} into divalent forms begins under aerobic conditions (above the Eh barrier), due to interaction with organic matter. The appearance of free H_2S results in a considerable increase in the rate of these reactions.

Mn sulfide (βMnS) is formed diagenetically in a sedimentary environment in which there are very high concentrations of Mn^{2+} and S^{2-} in interstitial waters (up to 3–5 mg Mn^{2+}/l). In a sharply reducing environment, which exists, for example, in the Landsort deep of the Baltic Sea (Suess, 1979), this manganese sulfide (MnS, hexagonal structure) is unstable and can easily be oxidized in the presence of oxygen to then be converted into MnO_2.

In individual cases, siderite ($FeCO_3$) forms in sediments where situation 6 exists (Blazhchishin, 1976; Blazhchishin and Emelyanov, 1977).

Situation 6 is the most favorable for diagenetic formation of manganese–carbonates in mud, as well as for accumulation of sapropel. This process is facilitated by the presence of CO_3^{2-} ions, which form as a result of oxidation of organic matter and combine with Mn^{2+} and Fe^{2+}. The ratio of manganese–carbonate to iron–carbonate depends on their ability to dissolve. Solubility products of these species are given by the following formula:

$$FeCO_3 = [Fe^{2+}]\,[CO_3^{2-}] = 3.5 \times 10^{-11}$$

$$MnCO_3 = [Mn^{2+}][\,CO_3^{2-}] = 1.8 \times 10^{-11} \qquad \text{(Naumov et al., 1971)}$$

Mn sulfide /Mn^{2+}/ /S^{2-}/, the solubility product of which is 10^{-11}, is a "competitor" neither of $MnCO_3$ nor $FeCO_3$ (Rozanov, 1988).

In sediments where situation 6 is prevalent, various elements combine with each other to form isomorphic mixtures between Mn, Fe and Ca, Mn, and with each other (Manheim, 1961; Hartmann, 1964), and these evolutions are represented by the following successions:

Rhodochrosite–manganocalcite–calcite; siderite–magnesite; siderite–rhodochrosite (Rosanov, 1988, 1995).

Species which are mostly characteristic of muds in deeps of the Baltic Sea (situation VI) are authigenic minerals: pyrite, Mn carbonates of complex composition, rhodochrosite, and also Mn sulfide and siderite.

Situation 6 is very characteristic of Baltic Sea deeps, where accumulation of microlayered sapropelic manganese–carbonate muds occurs (Emelyanov, 1981, 1995). We believe that these deeps (and oozes) should be regarded as the most adequate model, the application of which will make it possible to appreciate more correctly the depositional processes of Oligocene manganese ores in the southern part of the Russian Platform (for example, with respect to the Nikopol manganese deposit).

Situation 7. The Eh barrier occurs within the water strata, at a large distance above the bottom, and its positioning is constant. Below this barrier, there is strong reducing environment; on the seafloor, there are abundant gray muds enriched in organic matter. Mn^{2+}, H_2S and, in part, Fe^{2+} diffuse from the muds into near-bottom waters. In the water layer below the barrier, the concentrations of these elements are strongly increased. In the water strata, there is an intensive influx of H_2S, Mn^{2+} and Fe^{2+} to the Eh barrier from below, and of O_2, Mn^{4+} and Fe^{3+} from above. Reactions are the same as those in situation 6. As a result, the water suspension within a short distance above the Eh barrier is strongly enriched in Mn hydroxides (the suspension contains up to 45% Mn). In contrast to situation 6, in situation 7 muds become depleted in Mn (relative to the clarke of this element), rather than enriched in it. In the muds, reactionable iron combines in the form of sulfides and the Fe contents are normal (clarke values) for muds. Fe sulfides are markedly enriched in heavy metals (Butuzova, 1969), and some of these metals are isomorphically substituted for Fe. In situation 7, muds are enriched in C_{org}, SiO_{2am}, Mo, Fe, As.

The above-mentioned chemical elements are present in these sediments, but none of them in concentrations that make them potential ore for these metals, so that the process of ore formation is absent. Accumulation of only organic carbon-rich sapropel muds occurs.

In those cases when the redox Eh barrier lies within the water strata and comes in contact with the bottom of the shelf, where formation of nodules on the surface of muds occurs, a deep turbidity layer—the suspended matter of which contains up to 45% Mn and up to 10–15%Fe—may become a source of ore material, providing for growth of FMNs. Such conditions exist in Kalamits Bay of the Black Sea (95–115 m), to the west of the Crimea (Krasovskiy, 1984). Against a background of very low sedimentation rates, rapid growth of nodules occurs (at rates 100–1000 times (!) higher than those in oceans). The elemental composition of these nodules is markedly different from that of oceanic FMNs (Krasovskiy, 1984).

Situation 8 is characteristic of the Black Sea (Strakhov, 1976; Brewer and Spencer, 1974; Emelyanov et al., 1992) and cerain other areas of the World Ocean where contamination of bottom waters by H_2S is continuous (the Carjac Trench in Caribbean Sea, some fjords of Norway). In the geological past (Pliocene–Pleistocene), situation 7 was characteristic of the Eastern Mediterranean (Emelyanov and Shimkus, 1986).

Situation 8. The Eh barrier coincides with the bottom surface. At the same time, in times of strong storm (tide and ebb) flows and river floods, when turbid shelf waters (rich in oxygen) begin to move down the continental slope in the form of suspension (or turbidite) currents, the barrier may depart from the bottom surface, to be found in the water layer a few tens of meters above the seabed. This process, in contrast to situation 5, is not periodic: it depends on both the weather conditions and the season of the year.

Situation 8 is characteristic of the upper parts of the continental slope of the Black Sea. The geochemistry of this situation is not known with any certainty. It can be expected that processes occurring in muds there fall under the general heading of authigenic reactions similar to those found in situation 5. Processes of ore formation are absent in situation 8.

Situation 9. In this case, the Eh barrier may coincide with the bottom surface, but (more rarely) can be a few millimeters (or centimeters) below the bottom surface. A characteristic feature of this situation is calm hydrodynamic conditions (or hydrodynamic or geomorphologiccal traps for muddy material) that favor accumulation of muds. Considerable amounts of sedimentary material (including Fe_2O_3) are supplied by rivers, together with the riverborne load. Deposition of this sedimentary material occurs in so-called hydrodynamic (or geomorphological) traps in the shelf area. Waters and muds are inhabited by abundant mud-eaters and burrowing organisms. The mud is taken by these organisms as food and is repackaged into oval coprolites, which then settle onto the seabed to become very abundant in sediments. These coprolites have diameters of a millimeter to a centimeter. Coprolites are exposed to ferrugination and gradually change into chamosite $(Fe_5^{2+}Al)(Si_3Al)O_{10}(OH)_{18}$ (Emelyanov and Senin, 1969; Emelyanov, 1972).

These evolutions, which are characteristic of situation 9, are represented by the following succession: chlorite–glauconite–pyrite–siderite (Kholodov and Nedumov, 1981).

Situation 10. The Eh barrier is 1–10 cm below the water–bottom interface and the facial situation is represented by the outer part of shelf and its edge.

Distinguishing features of this situation are high-energy hydrodynamics, erosive events and elutriation of sediments that accumulated in situation IX but are presently in situation 10 as a result of a global fall in sea level. Clayey matter is derived from sediments to be evacuated beyond the limits of this facial situation, whereas the sediments become enriched in coprolites. This stage is responsible for the oxidation of chamosite and its conversion into hydrogoethite.

Situation 11. The Eh barrier may coincide with the surface of the seafloor or is within 0.1–10 m above it. The facial conditions are represented by the middle part of shelf in arid climatic zones under conditions of strong coastal upwelling. For a description of processes of sediment- and ore genesis in situation 11, see Part II.4.

Situation 12. This situation is characteristic of slopes in the presence of the OML. Processes occurring in this situation are described in Part II.10.

Situation 13 is characteristic of pelagic areas of the ocean, above the CCD (a major part of its description is given in Part II.13).

This situation is most evident in the eastern part of the Angola deep, where there is a combination (which took place in the Holocene and late Pleistocene) of sedimentary conditions favorable for ferromanganese ore formation, including (1) strong biological productivity in the upper water layer as a result of coastal upwelling; (2) intensive supply of very fine riverborne load with large contents of Mn- and Fe hydroxides (from the mouth of the Congo River) to this region, and accumulation of sandy-aleuritic turbidites, which are a reservoir for interstitial waters with high concentrations of ore components, and (3) fluctuations in the CCD levels due to climatic changes during the Pleistocene–Holocene. All these factors combine to cause the accumulation of deep-sea sedimentary strata composed of muds, which are in total 1.5 m thick and comprised of beds of sandy-aleuritic turbidites. Each of these beds composed of oxidized muds occurring within this sedimentary strata have a thickness ranging from 0.5 to a few tens of centimeters (Fig. III.10.2). The main zone of this sedimentary-diagenetic ore formation, which

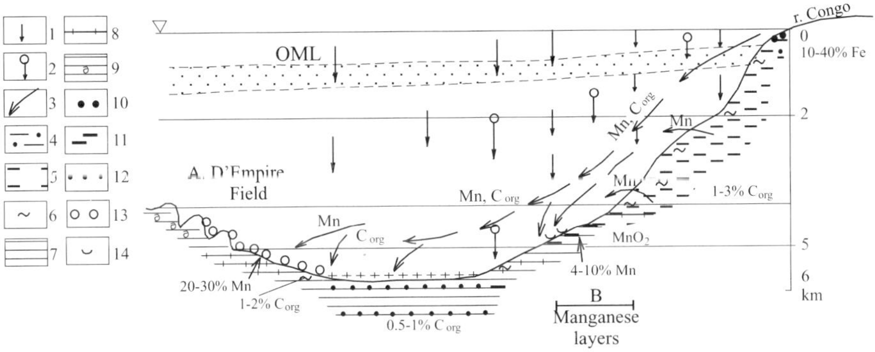

Fig. III.10.2. Scheme of the formation of the ferruginous and manganese ores—nodules (field d'Empire A, situation XV) and manganese sediments (B, situation XIII) in the Angola profile in the South Atlantic (from the Congo mouth to the flank of the Mid-Atlantic Ridge).
1–3—delivery of material: 1—organic detritus; 2—diatomic and terrigenous clayey material, C_{org} and Mn; 3—direction of supply; 4–12—bottom sediments: 4—terrigenous aleurites; 5—terrigenous aleuro-pelitic and pelitic (clayey) mud, gray (1-3 % C_{org}); 6—sapropelic (gray) mud; 7—pelagic (gray and brownish gray) mud (below the CCD level, <10% $CaCO_3$); 8—mixed calcareous (nano-foraminiferal)-terrigenous mud (10–50 % $CaCO_3$); 9—nano-foraminiferal ooze (>50% $CaCO_3$); 10—shamosite–hydrogethite aleurites (10–40% Fe); 11—manganese (mainly oxidized) mud (4–10 % Mn); 12—interlayers of oxidized (brown) pelagic clay with Mn microconcretions; 13—oval FMNs; 14—mainly flat FMNs (crusts).
OML—oxygen minimum layer.

extends along the base of the continental slope from north to south, is a feature which can be detected over a distance of 200 km and at depths from 4200 to 4400 m (Kornyshkov, 1986). This zone of ore formation occurs at the boundary between the zone of pelagic oxidizing conditions and the hemipelagic zone of weakly reducing conditions. Manganese beds are restricted to a strong gradient (barrier) of Eh values (Fig. III.10.2) and contain about 4–10% Mn. These oxidized beds may alternate with beds of terrigenous (aleuro-sandy) turbidites formed by sedimentary material from the Congo River.

On the surface of weakly oxidized clayey muds with turbidite beds of aleurite and sand (10–30% $CaCO_3$, 10–30% SiO_{2am}, see Fig.II.1.7), containing 0.5–1.0C_{org}% (1–4% C_{org} in reduced beds), flat, tabular, mainly concretionary FMNs are abundant. Individual nodules are typically about 1 cm across, and tabular nodules have dimensions of 10 × 10 × 2 cm (Kornyshkov and Smetannikova, 1986). On average, these nodules consist of 14.89% Mn, 19.84% Fe, 0.16% Cu, 0.59% Ni, and 0.27% Co. Such nodules can be referred to as Ni–Co–Fe type.

There are several reasons for the occurrence of FMNs on the continental rise (mainly within the depth interval of the 4200- and 4500-m isobaths) in mixed biogenic–clayey oozes lying at the margins of debris cone of the Congo River (see Fig. III.1.7). These reasons are as follows: (1) supply of fine sediments (enriched both in Fe- and Mn hydroxides and in organic matter) in this region from the Congo River through its canyon, and (2) high biological productivity and supply of ore

Fe^{2+} to the upper layer of red clays from the underlying reduced layer is hindered, if nothing else.

Pelagic red clays are composed of many authigenic minerals. They are represented mostly by oxide and hydroxide compounds of Mn and Fe. Whereas some of them form at the Eh barrier, certain others form at the water–bottom barrier and in the upper active layer of red clays during early diagenesis. The most exotic and characteristic feature related to red clays is FMNs. The nodules contain not only large contents of Fe and Mn, but also of many microelements such as Ni, Co, Cu, P, and As (Table III.10.2). However, Mn and Fe produce their own minerals in the nodules, while microelements do not produce their own authigenic materials.

Mn and Fe oxides and hydroxides accumulated in the FMNs are characterized, as a rule, by poor crystallization; they are unstable and have variable elemental composition, which makes X-ray (diffraction) analysis a problem of considerable difficulty (Cronan, 1982; Baturin, 1986). The same is true with respect to Mn- and Fe oxides and hydroxides, which exist in red clays in a dispersed state (in the form of micronodules, single grains, flakes and films), providing for considerably increased total concentrations of Mn and Fe. The mineral which is thought (Gramm-Osipov and Volkova, 1979; Murray, 1979) most likely to be formed in situation 14 is goethite ($\alpha FeOOH$). Coarse grains of this mineral are more resistant to exogenic (oceanic and marine) conditions, as opposed to hematite (αFe_2O_3). It is evident that hematite is the second mineral by its total abundance in red clays. In addition, these clays were found to contain lepidocrocite ($\gamma FeOOH$), magnetite (Fe_2O_3), and maghemite (FeOOH). All these minerals can be found in FMNs as well. FMNs were also found to contain ferrigidite ($Fe_2O_3 \cdot 9H_2O$) and akaganeite ($\beta FeOOH$) (Rozanov, 1988; Baturin, 1986).

Magnetite (at least ultra-thin grains of it) is thought to form as a result of bacterial activity (Chukhrov et al., 1981).

A variety of minerals accumulate in pelagic red clays exposed to the effects of hydrotherms (Chukhrov et al., 1981, pp. 5–21).

Minerals such as vernadite (β-MnO_2), todorokite (10Å MnOOH), bernessite (7Å MnOOH), manganite, and buzerite (10 Å MnOOH) have been found in FMNs.

Accumulation of vernadite occurs commonly under conditions of rapid oxidation of Mn^{2+}. This substance is supplied from land both in the form of Mn^{2+} and inorganic (Cl^-, HCO_3^-, SO_4^{-2}) or organic complexes (Rosanov, 1988).

Todorokite, which is commonly found in the upper layers of nodules, contains Mn^{2+}. This is why this mineral tends to become oxidized with time and converts into a more stable substance—vernadite. Such a behavior of this mineral is probably facilitated by the abundance of its other constituents, such as Ni^{2+}, Cu^{2+}, Zn^{2+}, and Mg^{2+} (Rozanov, 1988).

Increased amounts of reactionable Fe concentrate within the sediment column either above the Eh barrier (authigenic compounds of Fe^{3+} are predominant) or markedly below the zone of sulfate-reducing reactions (authigenic compounds of Fe^{2+}– sulfides, are prevalent).

Mn is derived from reduced muds. There, where oxidized film is absent, Mn^{2+} diffuses into the near-bottom water layer, where it is oxidized and transported by currents towards the open ocean.

Table III.10.2 Chemical composition of the ferromanganese nodules of the World Ocean According to various authors Summerized by Baturin.*)

Element	Pacific Ocean			Indian Ocean			Atlantic Ocean		
	max.	min.	average	max.	min.	average	max.	min.	average
B	0.06	0.007	0.029	-	-	-	0.05	0.000	0.03
Na	4.7	0.46	2.06	-	-	-	3.5	1.4	2.3
Mg	2.4	0.42	1.76	2.68	0.89	1.54	2.4	1.4	1.7
Al	7.93	0.48	3.27	4.80	1.04	2.41	5.50	1.37	3.20
Si	20.56	0.52	8.27	20.75	1.08	7.46	25.66	0.93	7.36
P	0.48	0.031	0.196	5.80	0.043	1.34	2.09	0.028	0.299
K	2.41	0.20	0.74	-	-	-	0.95	0.30	0.56
Ca	12.16	0.60	1.98	14.96	0.61	4.10	11.10	0.41	2.84
Se	0.003	0.001	-	-	-	-	0.003	0.002	0.002
Ti	2.65	0.02	0.80	1.76	0.14	0.63	0.91	0.053	0.434
V	0.15	0.009	0.056	0.091	0.017	0.049	0.11	0.02	0.07
Cr	0.007	0.001	0.001	0.0110	0.0003	0.0019	0.003	0.001	0.002
Mn	42.3	0.47	17.94	37.00	1.60	14.74	23.9	2.4	14.93
Fe	33.17	0.60	11.72	22.48	1.20	13.05	25.9	1.54	13.08
Co	2.53	-	0.33	0.998	0.014	0.254	0.91	0.09	0.323
Ni	2.48	0.025	0.59	1.42	0.040	0.441	1.3	0.013	0.484
Cu	1.90	0.01	0.39	1.37	0.009	0.173	0.50	0.030	0.155
Zn	0.31	0.019	0.084	0.23	0.017	0.061	0.144	0.035	0.066
Ga	0.003	0.0002	0.001	0.0042	0.0006	0.0014	-	-	-
Sr	0.16	0.024	0.088	-	-	-	0.19	0.058	0.11-
Zr	0.12	0.009	0.050	0.149	0.002	0.035	0.045	0.021	0.035
Mo	0.15	0.0007	0.033	0.091	0.0016	0.0313	0.055	0.014	0.0369
Ag	0.0006	-	0.0003	-	-	-	-	-	-
Ba	0.98	0.02	0.39	0.50	0.030	0.158	0.80	0.14	0.498
Ge	0.00013	0.00005	0.00009	0.00011	0.00008	0.00009	-	-	-
S	0.306	0.071	0.188	0.20	0.07	0.14	-	-	-
As	0.0289	0.0002	0.0114	0.049	0.0039	0.0174	0.021	0.009	0.014
W	0.058	0.0023	0.0163	0.0127	0.0010	0.0067	-	-	-

Continued

Table III.10.2 Chemical composition of the ferromanganese nodules of the World Ocean According to various authors. Summerized by Baturin.*⁾—*continued*

Element	Pacific Ocean			Indian Ocean			Atlantic Ocean		
	max.	min.	average	max.	min.	average	max.	min.	average
Ti	0.039	0.0001	0.0102	0.0290	0.0077	0.0134	-	-	-
Pb	0.36	0.005	0.11	0.25	0.005	0.076	0.230	0.070	0.134
Be	0.0005	0.0002	0.0003	-	-	-	-	-	-
Nb	0.0150	0.0030	0.0085	0.0060	0.0002	0.0031	0.0028	0.0009	0.0021
U	0.0047	0.0003	0.00128	0.00122	0.00016	0.00059	0.00111	0.00088	0.00098
Tb	0.0135	0.0003	0.0044	0.0075	0.0025	0.0044	0.0152	0.0010	0.0064
C_{org}	0.33	0.06	0.15	-	-	-	-	-	-

*⁾ Summerized data of Bezrukov and Andruschenko,1973; Volkov et al., 1976; Glagoleva, 1972; Mero, 1965; Skornyakova, 1976; Skornyakova and Andruschenko, 1976; Strakhov et al, 1968; Cronan and Tooms, 1967, 1969; Glasby, 1981; Glasby and Lawrence, 1974; Summerhayes and Willis, 1975; Calvert and Price, 1977. All citations after Baturin, 1983.

In situations when an oxidized film is present on the sediment surface, Mn accumulates in this film (as dispersed MnO_2) in the form of authigenic compounds, thus producing an increase in the content of this elements, up to 4% (cfb).

Based on carbon-14 dating, amounts of CO_2 produced by sulfate-reducing reactions have been calculated (Ivanov and Lein, 1980). The input of CO_2 (per kilogram of dry mud) that reaches the bottom water from a 90- to 91-cm-deep sediment layer, station DM-669 in the Pacific ocean (see Fig. II.15.2), is 52 g (or 15.5% of the total CO_2 formed in the sediments); the input from the 180- to 181-cm sediment layer (station DM-670) is 57g CO_2 per kilogram of dry mud (or 8.4%). In general, in the studied sediments of the Californian profile, about 60% of the CO_2 produced by microbial genesis escapes into bottom water. The remainder is fixed in sediments as newly formed minerals.

Microbial activity is continuous not only at the water–bottom GB or at the active layer of sediments GBZ, but extends to deeper layers; i.e., the process appears to occur throughout the complete length of the sediment column of gravity cores (4–5 m depth) or deeper. This is why the depth of the lower boundary of the biosphere within the sediment column remains to be determined. In marginal areas of the ocean, it is expected to extend to a depth of 1–6 m below the sediment surface.

Situation 15 is characteristic of marginal areas of pelagic deeps in oceans, where pelagic red clays are substituted by biogenic carbonate sediments. One example is the western fringe of Angola deep–eastern flank of the North Atlantic Ridge, where the D'Empire ore field is located (field 16 in Fig. III.6.1; see also Fig. II.1.7). Here, nodules (generally 3–5 cm across) occur mostly as a narrow nodular field orientated north to south, which is restricted to the boundary of the marginal hilly–mountainous area with the abyssal plain of the Angola deep. Here, their productivity amounts to 15–25 kg/m^2. The nodules are abundant mostly on gently sloping hills 5100–5600 m deep (i.e., at depths somewhat lower than the CCD), where low-calcareous and calcareous pelagic muds are abundant (Kornyshkov and Smetannikova, 1980). There is a correlation between $CaCO_3$ contents in sediments and the relative abundance of FMNs in muds. The frequency distribution of nodules fall within the following range: if the $CaCO_3$ content in sediments is $<10\%$, FMNs are essentially absent; for 10–30% $CaCO_3$, the abundance of nodules greater than 1 kg/m^2 accounts for 25–30% of the entire nodular area covered; for 30–50% $CaCO_3$, the abundance of nodules of more than 1 kg/m^2 accounts for 60–70% of the entire nodular area covered; for 50–70% $CaCO_3$, FMNs are essentially absent.

The FMNs from the D'Empire field are characterized by large Mn contents (18.73–29.58%), low Fe contents (16.0–21.0%), and increased contents of Ni (up to 2.0%) and Cu (up to 0.58%). The Mn/Fe ratio varies from 1.2 to 2.0. These nodules are of Ni–Cu–Mn type. The contents of Co and Ni in the FMNs tend to increase away from ore strip of the D'Empire field, both eastward and westward, whereas Cu contents decrease (see Fig. III.8.3).

There are several reasons why the nodule-producing D'Empire ore field is restricted to pelagic muds, in which the $CaCO_3$ contents fall within the ranges of 30–50% and 10–30%: (1) supply of very fine suspended matter from the Congo River steadily enriched in Mn hydroxides (mostly in the lower nepheloid layer) as it moves through the water column; (2) supply of ore components transported from the photic layer together with calcareous skeletons and organic detritus; (3) occurrence of low-oxidizing conditions in which contents of ore elements are increased due to increased contents of C_{org} in the interstitial waters of muds. The reason for the high Mn/Fe ratio in the nodules (1.2–2.0) is most likely that Fe from the Congo River is deposited predominantly on the shelf (in the form of hydroxides and pellets) and at the base of the continental slope (as flat, mainly concretionary FMNs), whereas smaller gel-like particles of Mn oxides (with Ni and Cu sorbed on their surfaces) are transported together with light particles of organic detritus and clayey minerals to that part of Angola deep furthest from the mouth of the Congo River (Fig. III.10.2). The canyon of the Congo is believed to be responsible for rapid transportation of ore elements of clayey minerals and C_{org} from the mouth of this river to the flank of the Mid-Atlantic Ridge.

Throughout geological history, the depth of the occurrence of the CCD underwent considerable fluctuations: one time it sunk to a larger depths, but then it rose toward the water surface. A reflection of these processes is variations in surface areas of deposited red clays, which become either more spacious (when the CCD was shallower) or smaller (the CCD depth was greater). In such situations, in the zone of facial migration, accumulation of the sequence of layered sediments occurred, comprising beds of pelagic red clays and low-calcareous and calcareous pelagic muds. Such a bedded structure of sediments was found to occur in the Angola, Canary, and Cabo Verde deeps of the Atlantic Ocean (Emelyanov et al., 1975: Emelyanov, 1982; Emelyanov and Kruglikova, 1990). These beds within red clays are composed of 0.10–0.24% C_{org}, 4–6% Fe, and 0.20–2.27 Mn. In the Pacific, the coefficients of enrichment are different for oxidized pelagic sediments and reduced sediments (i.e., sediments found on opposite sides of the Eh barrier), and for the former they range from 0 to 316.5 (Volkov, 1980, p. 164). These values are maximum for Mo (316.5 ppm) and Mn (9.8%). In the oxidized sediments from one of the stations, the Mn content reaches 11.75%.

Elements, including Fe, Mn, P, Co, Cu, Mo, As, Cr, accumulate only on that side of the Eh barrier characterized by large positive values of this parameter, whereas V, and occasionally Ni, are found on both sides of this barrier.

Under open ocean conditions, rhodochrosite forms diagenetically in the zone of transition from hemipelagic sedimentary conditions (characterized by low-oxidizing and low-reducing reactions) to myopelagic (oxidizing) conditions (Volkov et al., 1979; Svalnov, 1991). The processes most likely to produce it are as follows. After burial of oxidized muds enriched in Mn (up to 0.5% Mn, and the in Guatemala deep of the Pacific, up to 4.5% Mn), chemical reactions reduce Mn^{4+}. This process is very slow, and the manganese-reducing conditions are far below the sediment surface. The speed of reducing reactions is greater for Mn^{4+} than for Fe^{3+}. Therefore, it is evident that Mn in the form of $Mn(HCO_3)_2$ begins to migrate earlier and moves to a greater distance than for Fe^{2+}. This upward migration of Mn towards the Eh barrier continues until the depth at which the process of decomposition of $Mn(HCO)_2$ begins to occur, as a result of a decrease in partial pressure of CO_2 in interstitial water; the result of these reactions is the crystallization of the cores of $MnCO_3$ (Volkov et al., 1970, p. 90). To maintain the process of dissolution of Mn compounds in sediments, a considerable reserve of organic matter is required. Such a reserve of organic matter was evidently available in the Rio Grande transform fault in the South Atlantic, because burials of sediments include not only skeletons of diatoms, but also notable amounts of organic matter [probably more than 1% (Emlyanov et al., 1986)].

In cases when FMNs began to accumulate on the surface of red clays that formed during interglacial periods, they became buried below the layer of reduced calcareous-clayey or silica-clayey muds in glacial times. Such burials of FMNs have been observed in Late-Pleistocene pelagic low-calcareous and calcareous muds from the flank of the Mid-Atlantic Ridge in the Canary basin (Emelyanov et al., 1990).

The main reason for the accumulation of bedded red clays and reduced calcareous or mixed muds on the margins of pelagic basins of oceans is climate. During interglacials, including the Holocene, the level of the CCD in the Atlantic was low, in glacial times high, and the depth interval (amplitude) between these extremities reached about 400 m (see Part II.13). In the Pacific, on the contrary, the CCD level was lower for glacials than for the Holocene.

Another reason for the bedded structure of pelagic calcareous muds and red clays are slides and turbidite flows.

Sediments composed of red clays interbedded with pelagic calcareous turbidites have been found, for example, in the Cabo Verde basin (Emelyanov et al., 1990). Such bedded (rhythmic) muds form in depressions of the seafloor between abyssal hills. The tops of these hills break through the CCD level to occur within the depth interval between the CCD and the calcite lysocline or even higher. Calcareous material that accumulates on these peaks is in an unstable situation and likely to slide (flow) down the slope of these hills to become underlain by red clays, where it evidences good preservation in the water column below the CCD.

In the Pacific, bedded muds (red clays and pelagic calcareous muds) occurring on the western flank of the East Pacific Rise (7°30′N) have been studied. In this region of the ocean, the depth of the CCD was about 4 km, and the amplitude of its fluctuations was within 200–300 m (Emelyanov, 1991, 1992).

Situation 16 (not shown in Fig. III.10.1) is characteristic of a facial environment in which there were non-depositional hiatuses within the sediment section or the sedimentation rate was periodically strongly diminished. In such situations, on the surface of seafloor or on the muddy substrate, development of either ferruginous or manganese crusts occurs. The first event was found within the limits of the submarine fan of the Amazon River (McGeary and Kumar, 1975) at depths of 1394–4919 m. Here, the sequence of sediments recovered in 26 sediment cores preserves a reddish ferruginous crust which distinguishes Pleistocene sediments from Holocene sediments. The crust was formed due to the effect of both diagenetic and hydrogenic processes (including migration of Fe^{2+} from interstitial waters and precipitation of hydroxides from seawater) at the water–bottom interface during the early Holocene when, as the oceanic level began to rise, the mouth of the Amazon began to recede towards the continent. All this led to increasingly larger accumulations of riverborne sediments on the shelf and near shore. The submarine fan of the river began to experience a deficit in riverborne sedimentary material; this gave rise to the onset of a non-depositional stage in the sedimentary environment, and this was the reason for the formation of ferruginous crust. The raw material for formation of crusts was reactionable Fe^{2+}, which was delivered from the Amazon.

Table III.10.3. Chemical composition of cemented bottom sediments of the continental slope of the North Atlantic near Cape Hatteras, USA ($Si0_2$-Na_20 in %, Sr-Cd in 10^{-4}%). After Rona et al., 1982.

Component, element	Bulk sample	Pinkish-brown layer	Orange layer
SiO_2	42.57	31.51	40.51
TiO_2	0.36	0.32	0.42
Al_2O_3	5.53	4.55	6.21
Fe_2O_3	20.13	23.05	21.59
MnO	0.16	0.16	0.16
CaO	5.92	7.46	5.48
MgO	2.00	1.77	2.12
Na_2O	1.35	1.30	1.35
Sr	1490	2050	1340
Ba	249	217	228
Cu	14	25	17
Cr	58	50	65
Ni	20	23	25
N	126	141	130
Zr	142	146	249
Y	16	16	17
Zn	48	47	48
As	1090	559	1030
Cd	17	19	14
Σ of oxides, %	78.00	70.10	77.80
Soluble in the HCl fraction, %	51.40	62.80	50.00

In cases when sediments on the oceanic bottom are either strongly oxidized muds (for example, pelagic red clays) or the same, but old and non-solidified muds, no diagenetic processes providing for the movement of reactionable Fe^{2+} from sediments towards the water–bottom interface occur. This is because essentially all Fe is fixed in the form of hydroxides. In such a situation, the predominant hydrogenic process is the accumulation of manganese on the bottom surface, where it forms a black manganese crust.

In some cases, ferruginous lightly cemented sediments may form on the bottom surface due to oxidation of old pyritized beds. Such an event was found on the continental slope off the United States near Cape Hatteras at depths of 2100–2350 m (Rona et al., 1982). Slabs of terrigenous sediments cemented to each other by Fe hydroxides containing up to 23.05% Fe_2O_3 have been found there (Table III.10.3). According to these authors, the most likely process to form these hydroxides is oxidation of pyritized beds, which occur on the sediment surface (due to the downslope creeping of upper layers of the sediment sequence), by the following formulas:

$$3.75O_2 + 2.5H_2O + FeS \rightarrow FeOOH + 2SO_4^{\ 2-} + 4H^+.$$

This is evidently true, because ferruginated slabs are very impoverished in MnO, but they are enriched in those components that accumulate in sediments in parallel to sulfidic Fe, including As, Cd, and V.

Geochemical Barrier Zones and Formation of Siderite

Siderite is a mineral which occurs in sediments of Fanerozoic geological time. It forms mostly at a water–sediment interface or during early diagenesis, as a result of precipitation of this mineral from interstitial water. Siderite occurs in the recent sediments of the Baltic Sea (Blazhchishin and Emelyanov, 1977) and in Pliocene–Pleistocene deposits of the Black Sea (Emelyanov et al., 1978).

Miocene–Pleistocene sediments of the Black Sea contain up to 35.30% Fe (Table III.11.1), or up to 45.01% when recalculated for carbonate-free and silica-free material (cfb) (Emelyanov et al., 1982, p. 114). In some cases, these ferruginous sediments contain 1.84–5.43%Mn. Fe occurs predominantly in the form of siderite and sulfides. Dolomite and calcite are also there. Manganese occurs predominantly in the form of hydroxides and rhodochrosite. It is evident that Mn is an isomorphic component of diagenetic siderite: a clear correlation between Mn and siderite in sediments has been revealed (Emelyanov et al., 1982).

Siderite accumulated in warm climatic conditions, when the water had increased salinity. This time saw an intensive supply of Fe to the sea from the plains of its watershed area, where rocks were exposed to intensive chemical weathering (Hsü and Kelts, 1978). Diatomaceous phytoplankton developed in seawater, and values of Eh in eutrophicated water masses were low. In such a situation, the Ca/Fe ratio was lower than 20 : 1. This is why development of iron carbonate instead of calcium carbonate occurred here.

In accordance with a thermodynamical model (Rajan et al., 1996), calculations were carried out to determine the concentrations of different components in the water, which are required to remove (precipitate) each of carbonate phases from water. It has been shown that evaporation of only one-twelfth of the water volume from the sea in warm climatic conditions (during interglacials) would have provided for precipitation of clean siderite only. Application of this model led us to the inference that siderite-rich sediments could be formed in the Black Sea both as single thick layers and laminated sediments. The laminated structure of accumulated ferruginous sideritic ores is one of the most characteristic structural features of such deposits (Rajan et al., 1996).

Figs. II.6.8, II.6.9, Table II.6.9). In contrast to oceanic sediment, Fe ore deposits of Marsili volcano contain an increased content of Fe, as well as increased contents of Mn (up to 9.12%) in individual layers, whereas the contents of Zn, Cu, Ni, Co and other microelements are within the clarke values of these elements.

There is much evidence that hydrothermal process not only contribute to the accumulation of ore deposits near hydrotherms, but also markedly affect the growth and chemical composition of pelagic FMNs. We will be able only to describe how these hydrothermal processes influence the geochemical situation that occurs in the area of Clarion–Clipperton fracture zone and the Guatemala Basin, located in the equatorial part of the East Pacific Rise (Emelyanov, 1991, 1992; Emelyanov et al., 2002).

As noted in Part II.6, many hydrotherms in a rift zone in the ocean have a temperature of about 300–350°C. Hot waters are expelled from oceanic trenches (or vents) at great speeds. When the diameter of such a vent is 20 cm and the speed of a hot flow is very high, then the amount of heat energy produced is estimated to be

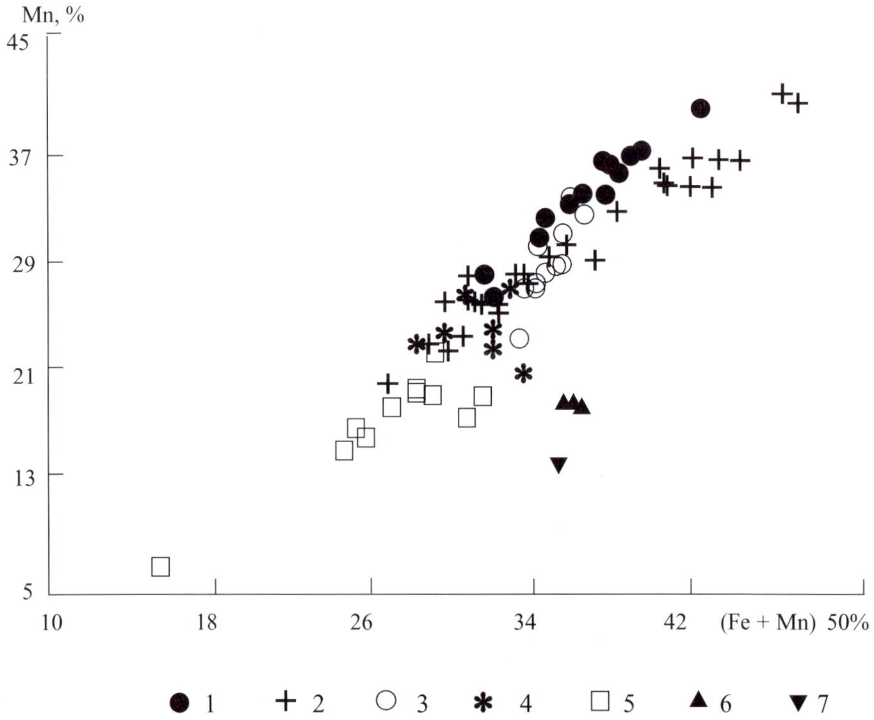

Fig. III.12.3. Dependence between Mn and (Fe+Mn) in the studied nodules. After Emelyanov et al., 2002.

(1) Guatemala basin; (2-7) Clarion-Clipperton province: (2) field 1, stations 3911 and 3913; (3) field 2, stations 3914–3916; (4) field 2, stations 3918–3926; (5) field 3, station 3940; (6) field 4, station 3838; (7) field 4, station 3839. For the location of test areas see Fig. II.6.5; II.6.6.

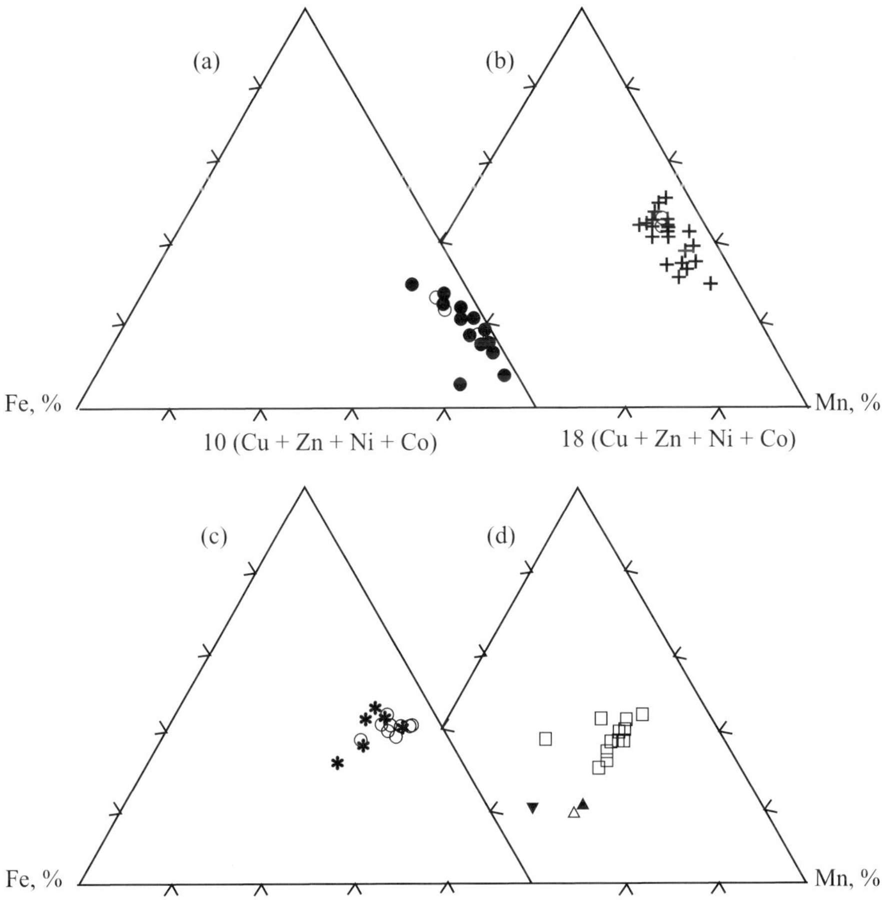

Fig. III.12.4. Compositions of nodules on a ferromanganese-10(Cu + Zn + Ni + Co) diagram. After Emelyanov et al., 2002.

(a) Guatemala basin; (b) field 1 between the Clarion and Clipperton fracture zones; (c) field 2 between the fracture zones; (d) the northern flank of the Clarion fracture zone and the southern flank of the Clipperton fracture zone (fields 3 and 4).

about 10^8 V (Exon et al., 1992). An isotherm of medium size produces about 10^9 V. Armed with information that there are hundreds (or thousands!) of such hydrotherms on the ocean floor, we may calculate that the energy produced on the ocean floor is equivalent to those produced by 1000 large nuclear power plants, or 25% of the total energy produced by the burning of oil products (Exon et al., 1992). It is possible that in the foreseeable future the energy of hydrotherms will be converted into electric power directly on the ocean bottom to be then transported to the surface via cable.

Rocks as such are also strongly affected by hydrothermal processes: they become enriched, depending on certain conditions, n Mn, Fe, or certain rare elements (Table

III.12.0.2). It is characteristic of such deposition situations that earth smectite occurs in the form of breccia with Fe- and Mn oxides and that the As content in basalt with sulfides periodically increases to values of $(220–350) \times 10^{-4}\%$, whereas the clarke value of this element in clays and shales is $10 \times 10^{-4}\%$.

The northeastern equatorial part of the Pacific Ocean is among the territories that are particularly abundant in FMNs. There, in the region of the Clarion–Clipperton fracture zones, the FMN reserves are extremely abundant (up to 12–13 billion tons (Pilipchuk, 2003) and many countries (including Russia) perform licensed exploration to estimate the commercial value these FMNs. This part of the ocean comprises numerous hydrothermal springs and metalliferous deposits of hydrothermal origin. The level of knowledge on FMNs and various metalliferous sediments in this region is extremely high. There are hundreds of publications devoted to the composition, genesis, and reserves of FMNs. One of these problems is the effect of hydrothermal springs on FMN composition.

The formation of most FMNs in the ocean occurs in complex sedimentation in which biogenic matter is involved, as well as in the redistribution of deposited elements during early diagenetic processes. The components and ore elements supplied to oceanic water from various sources are mixed. However, some geochemical signatures are preserved, which indicates how each source influences and contributes to FMN composition. Comparison of FMNs from the Guatemala basin and Clarion–Clipperton province reveals the influence of hydrothermal solutions on FMN composition. The results obtained in our study (Emelyanov et al., 2002) support and amplify the conclusions of previous researchers.

(1) The degree of differentation of Fe and Mn, the major ore components of the nodules, generated nearer to the East Pacific Rise and the Clarion–Clippeton fracture zone, depends on the distance from hydrothermal vents. The nodules richest in Fe represent zones, while the nodules richest in Mn are found apart from these regions, in the center of the Guatemala basin and within a depression between the above fracture zones in the western part of the study region.

(2) At low total concentrations of trace elements typical of FMNs and crusts directly related to hydrothermal springs, some Mn-rich nodules of the Guatemala basin preserve elevated contents of Li, whose high concentrations are found in hydrothermal vents on the East Pacific Rise. The nodules richest in Fe (Clipperton fracture zone) have elevated Co concentrations. These specific geochemical features of the nodules can disappear during later diagenetic transformations.

(3) The decreasing influence of hydrothermal solutions is expressed by variations in trace element concentrations and their correlation with major elements. The concentrations of some trace elements (Ni, Cu, and Zn) increase in the nodules. However, their distribution grows irregular and all geochemical features related to hydrothermal influence become less pronounced with increasing distance from the spreading and fracture zones (Emelyanov et al., 2002).

Calcium Carbonate Compensation Depth and Formation of FMNs

FMNs in the World Ocean have been found to occur within a depth interval from several meters (lakes) to 6–7 km. In oceans as such, nodules are abundant throughout the entire depth interval, starting at 300–400 m (Blake Plateau in the Atlantic Ocean). In pelagic areas of oceans, this depth interval is even narrower: from 3300 (Guatemala basin) to 6160 m (*Pacific Ocean Basins: Ferromanganese Nodules...*, 1976, p. 44).

FMNs in pelagic areas of oceans are commonly restricted to depths close to that of the CCD. Such a phenomenon is facilitated by (1) lower rates of delivery of terrigenous material (diluting of ore components) to this level; (2) very low rates of sedimentation; (3) input of very fine pelitic material, which is commonly enriched in Fe- and Mn hydroxides (for example, delivery of riverborne sediments from the Congo River to the western part of the Angola basin, that is, already to the flank of the Mid-Atlantic Ridge); (4) input of large amounts of hydrothermal material from mid-oceanic ridges. FMNs are found both in mixed pelagic calcareous muds (50–10% $CaCO_3$) and in pelagic red clays (0–10% $CaCO_3$). According to the author (Emelyanov, 1982), environmental conditions most favorable for the growth of FMNs occur in the water column a few hundred meters above the CCD, i.e., within the CCD range (300–500 m). The reason is probably that the sedimentation rates here are somewhat higher (due to incomplete solution of calcium carbonate biogenic remains) than those in the water column below the CCD, but the degree of oxidation of muds is lower compared with red clays. That is why a considerable portion of ore components (Fe, Mn and microelements) have time to be buried in sediments, whence they migrate to interstitial waters, where they will provide a considerable reserve for following ore deposits. The mechanism that causes the formation of FMNs in this zone, lying deepest in the depth interval between the lysocline and the CCD, is most likely a combination of two processes: physicochemical (sedimentation) and diagenetic (i.e., combined sedimentation and diagenetic). Consequently, in the given sedimentary environment, ore matter may reach the active surface of particles, which have catalytic properties, both from near-bottom waters and underground solutions, to be sorbed by the active surfaces of particles or to precipitate in the form of $Fe(OH)_3$, MnO_2 and other hydroxides. When ore material reaches particles from ground solutions, the type of the nodule-formation process remains the same: physicochemical [electrochemical (*Ferromanganese Nodules...*, 1961)]. However, it is evident that supply of ore material from two sources (near-bottom water and ground solutions) is a contributory factor in maintaining a more rapid growth of nodules and their greater abundance at the boundary between two litho-geochemical areas (facies): mixed calcite-clayey sediments and pelagic oxidic-clayey manganese sediments (Emelyanov, 1982) (i.e., notably below the lysocline, but slightly above the CCD). This is why nodules are most frequently found at depths of about 5 km in the Atlantic Ocean and at depths of about 4.5–5 km in the Pacific.

In pelagic areas below the level of the CCD, clays commonly become oxidized throughout the entire vertical profile; the dissolution of calcareous material (as well as destruction and dissolution of all biogenic material delivered to the ocean floor as fecal pellets) occurs on the seabed surface. Because the process is fairly rapid, ore microelements do not have enough time to enter the porous waters of clays. Here, the environmental situation is unfavorable for supply of abundant Fe and Mn to nodules. The only input of these elements comes from near-bottom waters. Deposition of these elements occurs due to physicochemical (sedimentary) processes: settling of Fe, Mn and microelements on the active surfaces of particles (sorption). There, near-bottom water is the source of Fe, Mn and microelements. The finest colloidal aggregates of Fe^{3+} and Mn^{4+} hydroxides migrate from seawater by being sorbed by the active surfaces of particles having catalytic properties (Goldberg and Arrhenius, 1958). Such active surfaces are typical for Fe^{3+} (Mn^{4+} and Mn^{2+}) hydroxides, Fe^{3+} and Mn^{4+} hydroxides (microelements) and various debris.

The above physicochemical mechanism is confirmed by the author's findings of FMNs in the Guyana basin, the Atlantic Ocean (Emelyanov, 1982, 1984). There, the FMNs are restricted to gently sloping hills, which are covered by yellowish gray, low-calcareous (10–30% $CaCO_3$) or calcareous (30–50% $CaCO_3$) muds at depths of 4900–5100 m. The amount of Mn in manganese-bearing muds appears to be low, 0.10–0.15%. At greater depths (>5000 m), as well as in much shallower areas, FMNs are either absent completely or found as a unique species. It is evident that in cases when the heights of hills are considerable, FMNs have been deposited in concentric layers around these hills. Such an abundance of nodules in areas with hilly bottom topography not only in the Atlantic but also in the Pacific (*Ferromanganese Nodules...*, 1976, pp. 52–64) can be explained by increased hydrodynamic activity of near-bottom waters above these hilly areas, resulting in low rates of sediment accumulation.

It is a matter of common knowledge that the level of the CCD has been exposed to cyclic fluctuations with time: periods of drops and rises in the CCD level have periodically alternated in the course of the geological evolution of oceans. The time interval required for the growth of FMNs is on the order of a few millions of years. This is suggestive of the following situation: when FMNs were far below the level of the CCD, their growth rates decreased, and vice versa, when these nodules appeared far above the CCD, their growth rates increased. The amplitude of oscillations in the CCD level in the Atlantic, for example, during the Quaternary, reached 400–500 m (Emelyanov et al., 1989), whereas in the Meso-Cenozoic it was as large as 1–2 km. Therefore, the depth interval within which FMNs are abundant in the present-day ocean is much greater than the narrow depth range where the process of accumulation of FMNs is considered to be the most active [(CCD ± (300–500 m)] and most favorable for their accumulation in the Holocene.

Geochemical Barrier Zones and Formation of Sapropel

Sapropel-like oozes and sapropels were deposited intensively in the geological past during transgressive stages in the history of sedimentation basins, including periods of warm and wet climatic conditions (Emelyanov et al., 1982; Shimkus and Emelyanov, 1991). Such sediments may accumulate in semiclosed marine basins with various salinity, but which are linked with the ocean through a system of straits. The depths of these basins may be different: from a few hundred meters to several kilometers.

In inland seas having an established link with the ocean, rise in sea level was associated with transgressions (climatic optimums); this time is responsible not only for abrupt variations in the hydrodynamic and hydrobiological situations of these seas, but also for changes in the whole process of sediment accumulation there. Reflections of these climatic conditions, particularly in such seas as the Baltic, Black and Mediterranean, were (1) the development of a well-defined halocline and pycnocline, which became an obstacle for effective vertical mixing of the water strata of basins, and this feature, in turn, has been a contributing factor (2) in maintaining the stagnancy of basin waters which became contaminated with hydrogen sulfide, and the establishment of a redox (Eh) barrier in the water strata. In times of stagnation, the water strata was characterized by a low-energy dynamics and the environment was very unfavorable for bottom organisms. Such environmental conditions were responsible for the accumulation of microlaminated sapropelic and sapropel-like oozes with various contents of C_{org}, SiO_{2am}, Mn and other chemical elements on the seabed, depending on such factors as the intensity of the biological production of organic matter in the photic layer and the deposition rates of organic matter, SiO_{2am}, $CaCO_3$ (biogenic) and clayey substance. The total C_{org} content in muds exceeded 3–5% and was periodically as large as 16% or even 22% (40% organic matter, equivalently).

Accumulation of sapropels and sapropel-like oozes, as well as of gray (black) clays enriched in organic matter, also occurred in the geological past. Moreover, such accumulation of these deposits (including black clays of the Aptian and Albian) in the Northern Atlantic occurred at different stages of the geological history of present-day (Baltic, Black and Mediterranean) seas and the Atlantic, as well as in seas and oceans that had vanished long ago [the Bazhenov suite in Western Siberia (*Collectors of Oil ....*, 1983) domanikites and black shale in many places around Europe, etc.]. Based on geochemical studies, domanikites have been identified as being heterogeneous facies from shallow-water and deep basins, where enrichment

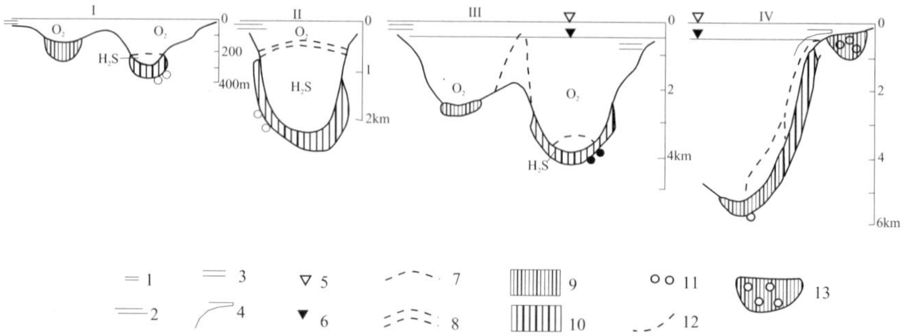

Fig. III.14.1. Proposed models of the deposition of sapropelic mud (black clay and shale) on the recent-sea and ancient-sea bottom in the Atlantic (models are compiled on the basis of the facies of the Atlantic Ocean). After Emelyanov et al., 1989.

I—platform (shelf) sea with periodic stagnation (with H_2S) of near-bottom water (Baltic Sea deeps); II—deep sea with stable stagnation of deep water (Black Sea); III—deep seas with separated basins (Mediterranean Sea); IV—basins in the arid zones of the ocean with well-expressed near-shore upwelling and with a big river nearby as the source of terrigenous and biogenic (organic detritus, clayey particles) matter (Angola Basin).

1—inflow of saline water into the sea with low salinity (I); 2—inflow of saline water into the sea with moderate salinity (II); 3—inflow of freshened water into the sea with high salinity (during regression) (III); 4—near-shore upwelling; 5–6—the level of the ocean: 5—during transgression; 6—during regression; 7—the upper boundary of the near-bottom water with periodic stagnation; 8—the upper boundary of the water with stable stagnation; 9—homogeneous gray mud with the hightened content of C_{org} (3-5 %); 10—microlaminated sapropelic mud (from 5 to 20 % of C_{org}); 11—occurrence of siliceous remains of diatomic algae (up to 10–20 % of SiO_{2am}); 12—delivery from the near-shore zone into pelagic areas of basins of organic matter and diatoms during ocean regression; 13—diatomic sapropelic ooze (up to 10–20% SiO_{2am}).

in organic matter is a distinguishing feature common to all these facies. The time of accumulation of sapropelic sediments in such basins appears to be closely related to the following factors: (a) the character of water exchange between a basin and ocean, including the water balance of a basin as a whole; (b) geomorphological isolation of these basins and their deeps; (c) stability of a certain circulation pattern of surface currents. This led to the conclusion that accumulation of domanikites occurred under sedimentary conditions very similar to those for sapropel in recent sediments of the Black and Baltic seas.

Prolonged accumulation of domanikites rich in organic matter was possible in geomorphologically isolated basins under conditions of stationary stratified water strata. The physical mechanism that formed and maintained these stratified water strata was the balance between the inflow of oceanic waters and river runoff. A stratified water column was continuous only at certain depths in each given basin.

At present, gray mud with 2–3% C_{org} (about 5% organic matter) accumulate not only in seas (see Part II.1) but also at large depths in the open ocean (at depths of 5–6 km in the Angola deep).

Based on previous data (presented in sections of this book that describe the formation and distribution of sediments rich in organic matter), a new model of sapropel formation is presented, in addition to presently existing models (Shimkus, 1981; Shimkus et al., 1975; Emelyanov et al., 1982; Emelyanov and Shimkus, 1986).

Proposed models of sapropel formation (Fig. III.14.1) in the Black and Baltic seas (see Emelyanov, 1995; Emelyanov et al., 1989, Fig. 51, p. 264) may be used not only for explaining the genesis of Jurassic domanikites in Europe, but also for a proper understanding of accumulation processes of black clays (or black shale) during the Cretaceous (Aptian–Albian–Turonian–Senomanian) in deeps of the North and South Atlantic, as well as in deeps of the Paleotethys and Indian Ocean (Murdmaa, 1979).

Sapropel-like and sapropelic muds (domanikites, black clays and shales) accumulated in the course of geological evolution on the Earth at various places around the world. Moreover, deposition of these sediments occurred predominantly either under conditions of hindered water exchange or in the presence of hydrogen-sulfide in near-bottom waters. Domanikites (mainly domanikites from Europe) have been described in detail in geological literature (Strakhov, 1960; Ghurari, 1981). Sapropelic beds of the Pleistocene from the Mediterranean Sea and black clays of the Cretaceous from the Atlantic Ocean have also been studied quite well (Murdmaa, 1979). The model of sapropel formation in the Black Sea is one of considerable complexity, and we will be able to consider only some aspects of it (Shimkus et al., 1975; Emelyanov et al., 1982; Shimkus and Emelyanov, 1991). Not only the Holocene but also Pleistocene and Neogene are responsible for the formation of sapropel-like and sapropelic muds there. The sediments of the Neogene, especially those in sections of the Meotian–Pontian–Kimmerian, uncovered in the southwest part of the Black Sea in the shallow-water area of the Eastern Paratethys, are enriched not only in organic matter but also in biogenic silica: C_{org} and SiO_{2am} contents in these sediments reach values of 13.68% and 18.99%, respectively (Emelyanov et al., 1982).

Oozes rich in organic matter are interbedded with muds depleted in organic matter. However, such interlayering is not always distinctive there. Therefore, a part of the sediment sequence corresponding to the Meotian–Pontian may be referred to the type of underdeveloped domanikites, because the degree of enrichment in organic matter in these deposits is smaller than that in typical domanikites. Average C_{org} contents in sediments fall within the limits of 1.0–3.68%. The thickness of the section is about 150 m.

In the Quaternary section, notable enrichment in organic matter in deep-sea muds is restricted to interglacial suites, where series of interlayers with increased (up to 5–10%) and high (up to 14.75%) C_{org} contents were found to occur (Emelyanov et al., 1978; Emelyanov et al., 1982). However, all these interlayers have small thicknesses, and contents of organic matter, such as those are characteristic of domanikites, are rare in these sediments. Finally, the most essential thing is that there is no continuous enrichment of organic matter in a considerable part of this sediment sequence. Therefore, average values of C_{org} in interglacial suites fall within the range 1.22–2.53%, i.e. very close to those found in Neogene sediments. Enrichment of organic matter in the greater part of the sediment sequence of the Holocene was continuous only in abyssal areas, where C_{org} contents in individual beds exceeded 20% (Shimkus et al., 1975). The degree of enrichment in organic matter was irregular throughout the sediment sequence.

In paleohalystatic areas of the Black Sea basin during interglacials, nearly continuous sedimentary sections formed rich in organic matter, where the content of organic matter approached its maximum contents in domanikites. The thicknesses of these suites were less than that of sediments recovered in sediment cores collected by the *Glomar Challenger* in the Black Sea (Emelyanov et al., 1978; Emelyanov et al., 1982), because the rate of sedimentation in paleohalystatic regions is almost an order of magnitude lower than that in the peripheral zone of the Black Sea basin.

The sediment sequence of the Quaternary in the deep-sea part of the Black Sea deep contains sedimentary members rich in organic matter, covering an area of about 50,000 km^2 and having a thickness of about 10 m. However, these suites are still intermitted with suites of terrigenous deposits, depleted in organic matter, that accumulated in glacial times. Enrichment in organic matter in intreglacial suites was controlled predominantly by planktogenic organic matter, mainly dinoflagellates (Shimkus and Emelyanov, 1991).

Formation of sediments rich in organic matter in the Black Sea occurred under environmental conditions similar or close to present-day conditions, such as those in the Black Sea in the time of the old Black Sea (Middle Holocene).

Based on an analysis of the processes of sediment accumulation in the Baltic Sea, Black Sea, Sea of Azov, and Mediterranean Sea, Shimkus and Emelyanov reached the following conclusions:

(1) Sapropel-like oozes and sapropels were deposited during transgressive stages in the history of sedimentary basins, including periods of warm and wet climatic conditions.
(2) Such sediments may accumulate in semiclosed marine basins with various salinity, but which have an established link with the ocean through a system of straits. The depths of these basins may be different: from a few hundred meters to several kilometers.
(3) Domanikites are heterogeneous facies from shallow-water and deep-sea basins, where the enrichment of organic matter is a distinguishing feature common to all these facies. There is a close correlation between the time within which there occurs accumulation of sapropelic sediments in such basins and factors including: (a) a character of water exchange between a basin and the ocean, including the water balance of a basin as a whole; (b) geomorphological isolation of these basins and their deeps; (c) stability of a certain circulation pattern of surface currents.
(4) Prolonged accumulation of domanikites rich in organic matter is possible in geomorphologically isolated basins under conditions of stationary stratified water strata. The physical mechanism that forms and maintains this stationary water stratification is the balance between the inflow of oceanic waters and river runoff. A stratified water column in such basins may be continuous only at certain depths. In the present-day Baltic Sea, the depths of deeps (basins) and the present balance of oceanic and river waters do not provide for this condition. To provide for such stratification of Baltic Sea waters, the depths of these deeps would have to be as large as a few hundred meters and the inflow of oceanic waters would have to be more intensive.
(5) In inland seas with an established link with the ocean, sea-level rise associated with transgressions (climatic optimums) was the reason not only for abrupt vari-

ations in the hydrodynamic and hydrobiological situations of these seas, but also caused changes in the entire process of sediment accumulation there. As a consequence, in such seas as the Baltic, Black and Mediterranean, this led to stagnancy of near-bottom waters and their contamination by hydrogen sulfide. In times of stagnation, the water layer was a low-energy environment (calm dynamics) extremely inhospitable to bottom organisms. Such environmental conditions were responsible for the accumulation of microlaminated oozes with various contents of C_{org}, SiO_{2am}, Mn and other chemical elements on the seabed, depending on such factors as the intensity of biological production of organic matter in the photic layer and the deposition rates of organic matter, SiO_{2am}, $CaCO_3$ (biogenic) and clayey substance.

(6) Models of sapropel formation in the Black and Baltic seas (see Emelyanov, 1985; Emelyanov et al., 1989, Fig. 51, p. 264) can be useful not only in explaining the genesis of Jurassic domanikites in Europe but also for a proper understanding of the processes of accumulation of black clays (or black shales) during the Cretaceous (Aptian–Albian–Turonian–Senomanian) in deeps of the North and

Table III.14.1. Locations of the rhodochrosite in the World Ocean

Region	Facial environment	Author
Gulf of Riga, Baltic Sea	Rhodochrosite in flat concretions, depth up to 60-70 m	Shterenberg et al, 1968
Baltic Sea	Deeps with periodic stagnation (H_2S). Depth 100-450 m. Dispersed grains in sopropelic laminated mud	Manheim, 1961 Emelyanov, 1981
Southern Atlantic (station PSh-841)	Transform fault at 29°S. The depth m, horizon cm Dispersed grains in the diatomic oozes	Emelyanov et al, 1986
Pacific Ocean	-	Lynn and Bonatti, 1965
Pacific Ocean (station DM-6171b 33°49′4N, 151°22′06E)	The pelagic clay at depth 6000 m, horison 220-225 cm. Concretions of rhodochrosite	Volkov et al, 1979 Logvinenko et al, 1972
Pacific Ccean near coasts of Peru and Chile	-	Zen-E-an, 1959
Pacific Ocean (Guatemala Deep)	Pelagic clay	Svalnov and Kuleshov, 1992

South Atlantic, as well as in deeps of the Paleotethys and Indian Ocean (Murdmaa, 1979).

In addition to sapropels and sulphides, a variety of other minerals, including barite, vivianite, siderite, amorphous manganese carbonate, and rhodochrosite (Table III.14.1), form in reduced sediments of the uppermost active sediment layer.

Predicting and Prospecting for Ores of Marine Genesis

Taking into consideration the theory of GBs and GBZs, it seems more reasonable to predict the stages of accumulation of various hard mineral resources, instead searching for their final deposits.

To predict ore-bearing fields and beds, it is necessary, in the first place, to perform the paleoreconstruction of seas or oceans which existed at the time when deposition of the ore bed occurred. Such factors as paleohydrodynamic and paleohydrochemical situations, and geomorphological and tectonic features of the seabed, as well as the location of prognostic ore province in relation to topographical features such as shore, shelf, continental slope, mid-oceanic ridge or volcanic seamount—all these influence the types of sediments that accumulate in a particular region. Paleoreconstruction of climatic zonation is also required.

As mentioned earlier, in the ocean there are not only lateral but also vertical (depth-related, bathymetric) zonations in the distribution and relative abundance of types of sediments and chemical elements in the water column, and types of sediments, authigenic minerals and ores on the seafloor (Fig. III.15.1). Phosphorites, pseudoolitic ferruginous ores, shells, building sands, and placers form on shelf; foraminiferal limestone and marlstones are deposits from middle and large depths; nano-limestones and marly clays are deposits from large depths; and pelagic (red) clays are from very large (abyssal) deeps.

In the present-day oceans, three bathymetric levels of accumulation of ferromanganese ores exist: shelf (hydrogoethite–chamosite and carbonate-manganese ores), middle (cobalt–manganese crusts) and pelagic (FMNs) (Fig. III.15.2). Within the framework of proposed classification (hierarchy), the position of hydrothermal–sedimentary ores, which are most frequently found in recent sediments, is somewhere within the depth interval between middle and pelagic depths.

It is necessary to emphasize that changes in bathymetric levels of depth-sensitive (vertical) barriers and GBZs play an essential role in the formation of ores in seas and oceans. This factor is of special importance for the formation of ores on the first, shelf level. In times of transgression, only sediments enriched in ore components accumulate on shelves (Emelyanov et al., 1989). On the contrary, a regression is a period when the main portion of sediments formed during transgressive times undergo reworking by bottom currents (and other hydrodynamic effects) (evacuation of non-ore, clayey material beyond the limits of the shelf) and diagenetic enrichment of ore components in a bed occurs. Such features as vertical oscillations of the OML, lysocline, CCD and other barriers are factors

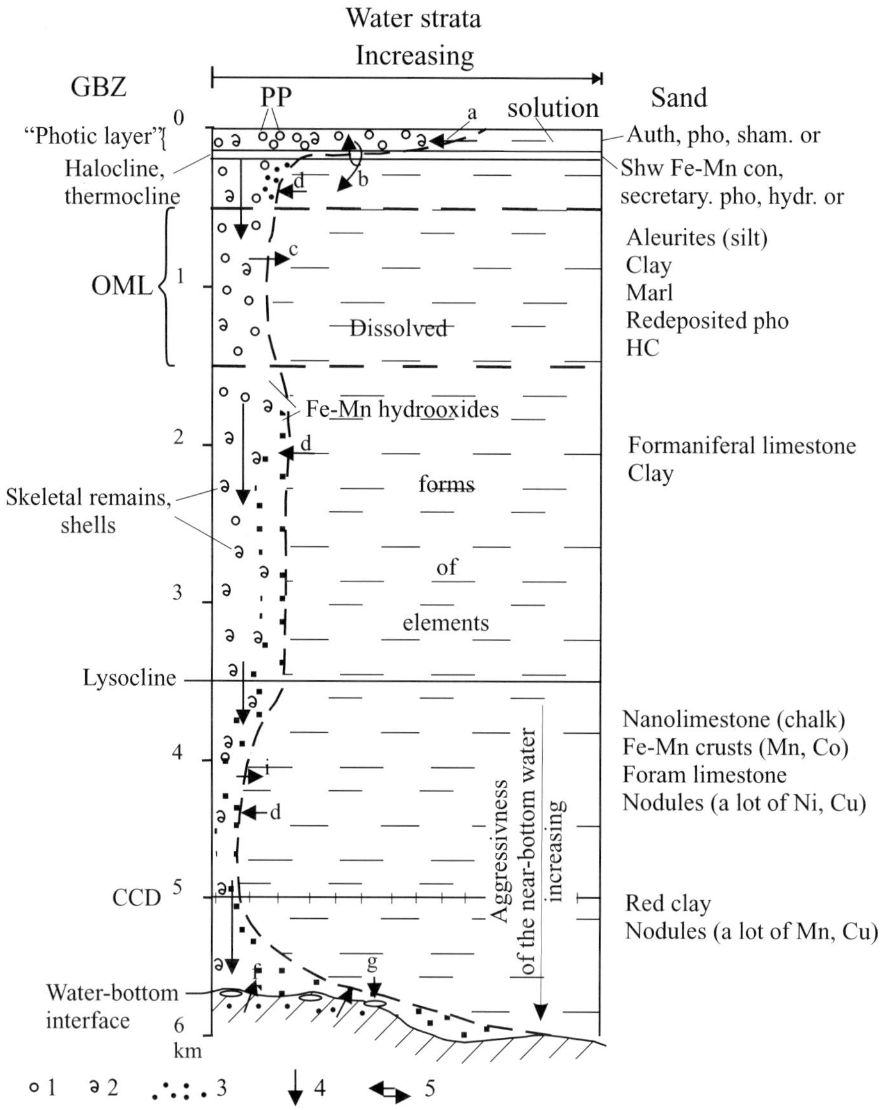

Fig. III.15.1. Principal scheme of changes in the forms of elements in vertical GBZs in the ocean and vertical localization of sediment types and mineral resources.

1-PP—phytoplankton, organic detritus, siliceous skeletal remains; 2—calcareous shells; 3—Fe- and Mn hydroxides; 4—express transport (pellet transport) of sedimentary material in the water strata; 5—preferable direction of flux of sedimentary material; a—from the water to the bodies of phytoplankton (Pb); b—from Pb to seawater and from Pb to zooplankton; c—from organic detritus (Od) to the sea water; d—from water to the suspended matter (Sp) (adsorbtion of microelements by hydroxides); e—from calcareous skeletal remains to water; f—from bottom sediments to the sea–bottom interface and to seawater; g—from seawater to FMNs.

Dotted lines show the principal distribution of particulate and dissolved forms of sedimentary material in the water strata (to the right, increasing; to the left, decreasing).

Abbreviations: m—mineral; aut—authigenic; pho—phosphatic; cham—chamosite; or—ores; shw—shallow water; hydr—hydroghethite; nod—nodules; con—concretion; sec—secondary; foram—foraminiferal; ind—individual; r—red; min—minimal; photo—photosynthesis; HC—hydrocarbon.

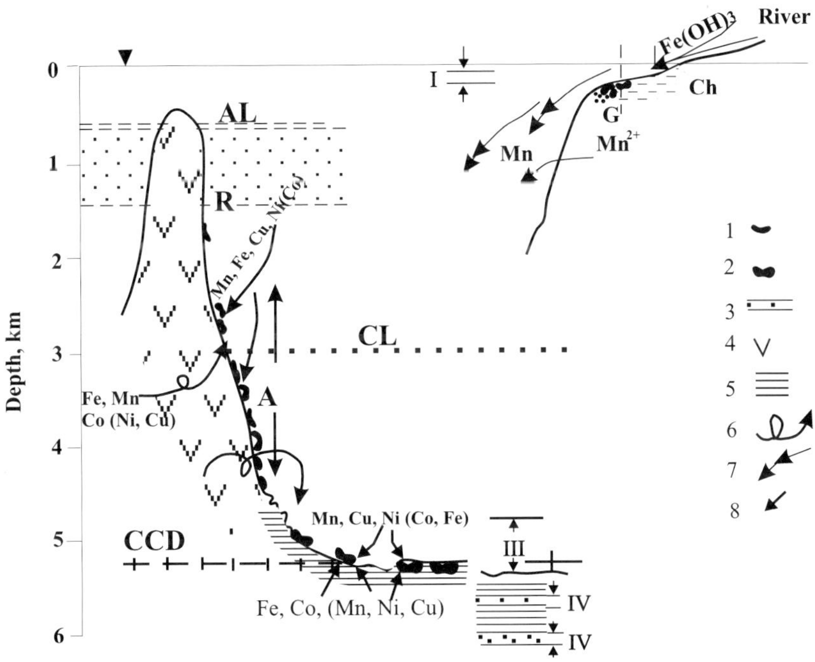

Fig. III.15.2. Bathymetric levels of ferromanganese ore formations in the recent ocean (principal scheme).
I—shelf level (50–200 m): deposition of hydroghetite (H)–chamositic (Ch) (ferruginous) pseudoolithic and primary carbonate manganese ores;
II—the level of medium depths. Deposition of Mn-Fe-Co crusts (R—recent, A—ancient, during the low standing of the ocean);
III—pelagic level. Deposition of FMNs;
IV—pelagic level below the bottom level. Diagenetic deposition of Mn microconcretions (in the ocean), and sideritic and rhodochrositic formations in seas.
Much below the sea bottom—formation of the hydrocarbons.
Arrows on the level I show regression (arrow down) and transgression (arrow up) changes of the sea level in Pleistocene.
AL—aragonite lysocline; CL—calcite lysocline; CCD—carbonate compensation depth.
Legend: 1—crusts and films on the rocks; 2—FMNs; 3—interlayers in the pelagic red clay, enriched in Mn micronodules; 4—basalts; 5—pelagic red clay; 6—strong near bottom currents, rewashing and redeposition of the sediments, delivery to the sea bottom of the ore elements; 7—evacuation of Mn from the shelf (from the rivers mouth); 8—evacuation of Mn or other elements from the sediments.

that contribute to the process of ore formation, having far reaching geological consequences.

Analysis of material presented in Parts II and III made it clear that the search for minerals, for example, pseudo-oolithic (hydrogoethite–chamosite) ferruginous ores, should be conducted in the environmental conditions referred to as type 1, including (a) humid climatic zones, first of all, in the equatorial zone; (b) near the mouths of rivers; (c) shelf facies of sediments (or reworked shelf-edged sediment facies); (d) hydrodynamic traps on shelves of the open ocean or in bays (primary chamosites) or in the zone of initially rewashed ferruginous sediments (hydrogoethite sediments). On the contrary, the type of regions appropriate for prospecting for FMNs are referred to as environmental conditions 2, including, first of all, (a) arid climatic zones; (b) depths close to the CCD (at the time when the nodules were formed); (c) areas far away from continents; (d) if environmental conditions 1 and 2 are absent, the search for nodules should be conducted in areas exposed to the effects of strong bottom currents, where sedimentation rates were very low or equal to zero (at the moment of formation of these nodules), or in areas where prolonged non-depositional gaps (hiatuses) took place in the past. Following the prediction model proposed by the author, the search for primary manganese–carbonate ores must be carried out in shallow-water semiclosed water basins or deeps, where periodical (!) contamination by hydrogen sulfide probably took place. Oxic-carbonate–manganese ores can most probably be found in areas of collision between two opposing situations: purely oxic conditions, when a basin was well supplied with Fe and Mn, and a zone of sharply reducing conditions, which happened periodically. It is evident that such ores might have accumulated only under conditions of transgressive–regressive sedimentogenesis, in moderately humid climatic zones. Phosphorites are commonly restricted to (a) the zone of the middle shelf in areas of upwelling (primary phosphates) or (b) the shelf-edge (secondary, reworked phosphates).

Methods used for searching for mineral resources in oceans must be differentiated. The key elements of the proposed method are (a) facial analysis and (2) lithogeochemical zonation of those old sediments, which are expected to contain the required mineral resources.

In general, a proper understanding of processes occurring at the lithological barriers, where sediments experience rapid variations in mobility of ore material, is becoming increasingly important. Knowledge or understanding of a model that describes the ore-formation environment makes it possible to use a complex of prospecting traces for ores, related to conditions of ore genesis. So, in order to predict the occurrence of hard mineral resources successfully, it is not only essential to know more about ore components as such (including their variability), but part of such modeling includes, to a greater degree, seeking optimal conditions for ore genesis, paying particular attention to geological situations (in the geological past) that caused effective ore formation. When such prediction is focused on specific aspects, certain corrections should be introduced into the method as well: research efforts should be undertaken to reveal not only direct traces of ore material, but also evidence of optimal conditions for ore genesis, such as the moisture of sediments and permeability of rocks of the basaltic layer.

The study of lithogeochemical barriers, all of which lie in a class of natural boundaries, falls naturally within the scope of the already established discipline called limology (the science of geographical and geochemical boundaries). Successive solution of problems of oceanic limology, where the extent to which barrier infrastructure of the ocean can affect ore genesis in the ocean has become a topic of paramount importance, suggests an application of a systematic approach. Such an approach, in addition to other well-known principles—including the establishing of an objective, integrity, an heuristic approach, and logical inconsistency (conflict)—implies physicomathematical modeling of individual elements of the system.

Part IV

SOME GEOLOGICAL ASPECTS OF VARIATIONS IN MARINE ECOLOGICAL SYSTEMS

An ecosystem is a closed, functionally united community of organisms (plants, animals and microbes) inhabiting a common area and able to exist for a long time within a totally closed biochemical cycle (i.e., without any exchange of matter over the boundaries of this system).

Recognition of an ecosystem's boundaries is an important step that assists in understanding the assimilation capacity of the system (Israel and Tsyban, 1987, p. 500). These authors recognized three types of ecosystems: (1) closed (lagoons and bays), (2) semi-closed (river mouths and estuaries), and (3) open, the boundaries of which are determined by the specific ecological and hydrological situation. According to the given classification, and taking into account the definition of an ecosystem, we cannot consider each sea as a single ecosystem (for example, one as complex as the Mediterranean Sea). This is all the more true when it concerns an ocean, or the World Ocean as a whole. However, to simplify the description of Part IV, the author will arbitrarily use the singular term "marine ecosystem."

In the marine ecosystem, as well as in any other ecosystem of a water basin, including the ocean, first- and second-order boundaries, as well as internal boundaries, can be recognized. First-order boundaries are the limits of a water basin as follows: from above, each basin is limited by the atmosphere–water boundary; on each side of such a basin are the shore–sea and river–sea boundaries; and from below, each system is limited by the water–bottom boundary.

Second-order boundaries are not as distinct. Boundaries of this type are thought to exist in the following forms: from above, an ecosystem is limited by the upper boundary of the atmosphere (troposphere), which is responsible for intensive circulation of various aerosols and atmospheric fallout; on each side of the ecosystem, there is a land-based divide, a watershed with river heads; and from below, an ecosystem is limited by subterranean water being discharged at the bottom of the sea (Fig. IV.1).

Internal boundaries are hydrofronts, thermocline–halocline (pycnocline), the OML, the O_2–H_2S transition layer (or redox Eh barrier in water), and certain others.

The sea, as an ecosystem, does not become a closed system due to the effects of first-order boundaries. Matter and energy enter the sea via these boundaries from the geosphere, which is limited by second-order boundaries.

Second-Order Boundaries

The upper second-order boundary is clearly found at an altitude of 10–12 km, i.e., where the troposphere ends. The troposphere, beginning near the surface of water or land, is a layer characterized by all processes that can notably affect a marine ecosystem, including atmospheric dust derived from sediments. Matter such as microorganisms, spores, pollens, soil particles, and pollutants ejected into the atmosphere by industry, are transported from land into the sea.

The dividing part of any drainage basin (a lateral second-order boundary) is found at a distance of 100–1000 km from seacoasts and 100–5000 km from ocean coasts (see Fig. IV.1). This is why some pollutants entering rivers in faraway lands may influence a marine ecosystem.

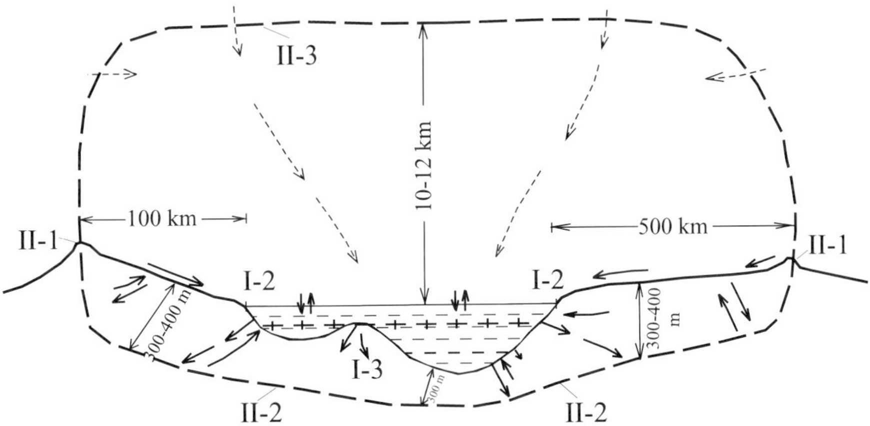

Fig. IV.1. Boundaries of the ecosystem of the Baltic Sea.
A. Outer boundaries. I—boundaries of the first rank:
I-1—atmosphere–hydrosphere (air–water); I-2—shore–sea; I-3—water–bottom;
II—Boundaries of the second rank:
II-1—boundary of the drainage area (watershed) (on the surface of the land); II-2—the lower boundary of the dynamically active layer of groundwater; II-3—upper boundary of the troposphere.
1—supply of aeolian dust through boundary II-3 (from other ecosystems); 2—supply from the drainage area; 3—exchange by water in the ground; 4—exchange at the sea–bottom interface.
B. Inner boundaries. 1–2—boundaries inside the ecosystem: 1—halocline; 2—redox barrier in the water (O_2–H_2S interface).

In a sea, the lower second-order ecosystem boundary appears at various depths below the land surface (within the limits of the watershed divide) or sea bottom. In the Baltic Sea, for example, this boundary occurs at depths of 300–400 m below the surface of the bottom, that is, depths down to which intensive penetration (infiltration) of atmospheric fallout occurs. As revealed by hydrogeological observations, subterranean waters move in different directions from these horizons (also towards the Baltic Sea) at rates from a few tens of meters to several hundreds of meters per year. Below, at depths of 400–1000 m, circulation of subterranean waters also occurs, but these move at very low velocities—only a few tens of meters per year. It is evident that subterranean waters from these depths cannot exert any significant influence on the environmental conditions of shelf seas such as the Baltic. However, in basin-type seas, such as the Black and Mediterranean seas, subterranean waters undoubtedly enter the water basin due to rough bottom relief and the large difference in the depths between watersheds and water basins (which may be as great as 5–8 km). One example of such phenomena is the input of brines into sea basins (for example, into the Red and Mediterranean seas, see Part II.12)—a process which completely changes the ecological situation in these basins, with the result that habitation of any living matter there becomes impossible.

In oceans, exchange between ocean water and subterranean water is as great as that in seas, as evidenced by the numbers presented in Fig. II.14.1. Here, the second-order boundary reaches depths of 3–4 km below the surface of the ocean bottom, that is, down to a depth where seawater penetrates places in the bottom in which its temperature increases to 300–350°C. Then this water again rises in the form of hydrothermal vents to pour out into the ocean bottom water.

Chapter 2

First-Order Boundaries

In practice, first-order boundaries are biological contours because it is at these boundaries where life activity is highest. First-order boundaries are favorable for the accumulation of various pollutants.

In calm weather, the shore–sea boundary corresponds to the position of the water's edge. However, precise fixation of this boundary during severe storms and hurricanes is not always possible: depending on weather conditions, this boundary can move either towards land (as in the case of tide and ebb currents) or towards the sea. It is generally accepted that the shore–sea boundary is a part of land or sea which lies between the maximum climb of a wave and a water depth of 15–20 m, that is, depths at which sediments may experience erosion or redeposition by wave processes.

One dangerous factor that affects environmental conditions unfavorably is land-based extraction of fresh subterranean waters. Extraction of large volumes of subterranean waters leads to penetration of marine waters into coastal land-based aquifers, resulting in their increasing salinity over time—making them unsuitable for drinking. Another dangerous factor is the effect of nuclear power plants in the coastal zone, where their activity may lead to considerable warming of subterranean waters, which in turn may contribute to undesirable warming of seawater. This may lead to rapid development of various microbes and algae harmful to the marine ecosystem.

Change in the coastline due to land-based amber mining is one example of the changing ecosystem of the Baltic Sea as a result of anthropogenic activity. Over the last 100 years, the amount of friable deposits delivered to the Baltic Sea over the shore–sea boundary is estimated at about 60–65 million m^3, as much as this sea has received from all the rivers of its basin over the same period of time. As a result, the shore–sea boundary near the Yantar' factory (Kaliningrad region, Russian Federation) extends 800 m further into the sea. Certain other changes have also occurred in the coastal portion of the Baltic Sea environmental system.

Construction of seaports, coastal protection devices, beaches, artificial islands and new areas for buildings and engineering systems (for example, in Japan) exert a strong influence on the marine ecosystem.

A great deal of anthropogenic wastewater also passes through the shore–sea boundary. Over the last 20 years, the volume of wastewater has increased from 35 km^3/yr to 150 km^3/yr. Wastewater is a very strong polluter: for example, the wastewaters from the city of Los Angeles were found to contain 101 pollutants, 36 of which are hazardous to living organisms (Aibulatov and Artyukhin, 1993, p. 206).

Mass development of bacterial flora, especially in the vicinities of seaports, is a characteristic feature of the coastal zone.

The shore–sea boundary is not a line, but a strip of variable width. This strip corresponds to the zone where river water mixes with seawater. The seaward margin of the river–sea mixing zone in seas and oceans is found in regions where the salinity of seawater is almost beyond the influence of river waters (in oceans, this is mostly the 20–25% isohaline). However, in such seas as the Baltic, the salinity of seawater is different: it is very high near the Dutch Straits and low in coastal areas of the Gulfs of Finland and Bothnia. Therefore, it is difficult to draw a line between the mixing zone and seawater. In general, the first-order boundary of an ecosystem in seas and oceans, in relation to the mouths of rivers, should be considered as corresponding to the 2‰ isohaline.

As mentioned in Part II.1, the World Ocean receives tens of billions of tons of sediment load from rivers, including a supply of sediments into the seas of the Arctic Ocean, which comprises 115×10^6 tons/yr [22×10^6 tons/yr in the Barents and White seas, 33.2×10^6 tons/yr in the Kara Sea, 25.1×10^6 tons/yr in the Laptev Sea, and 33.6×10^6 tons/yr in the East Siberian Sea (Gordeev, 2000)].

As solid sedimentary material moves from land into the sea across the river–sea boundary, it supplies a large amount of substances which are harmful for the marine ecosystem, such as radionuclides, pesticides, surface-active substances (soap, detergents, etc.), petroleum hydrocarbons, mercury, copper, zinc, chrome, phosphorus, etc., that accumulate either in seawater or in sediments, thus having an unfavorable effect on the ecosystem. Rivers and watercourses are arteries whereby almost all pollutants are delivered from land to seas and oceans, including radionuclides from nuclear power plants, petroleum hydrocarbons, polycyclic aromatic hydrocarbons, surface-active matter, chloro-organic products, polychlorinated biphenils (PCBs), heavy metals, and decayed products of mineral fertilizers (such as nitrogen and phosphorus compounds) (Fig. IV.2). The total amount of anthropogenic nitrogen and phosphorus delivered to oceans by rivers annually is 32×10^6 and $3{,}75.10^6$ tons, respectively (GESAMP, 1987). The amounts of chemical elements (metals, which are delivered by rivers into seas and oceans mostly due to mining activities) are as follows (in tons per year): 319,000 for Fe, 1600 for Mn, 4460 for Cu, 358 for Ni, 3930 for Zn, 2330 for Pb, 7 for Mo, 7 for Hg, 166 for Sn, and 40 for Sb (*The Global...*, 1980). According to other estimates, the amount of elements received in the coastal zone of oceans from rivers and watercourses are, on average, as follows (in tons per year): 590,000 for Fe, 10,470 for Mn, 16,750 for Ni, 39,070 for Cr, 33,720 for Cu, 10,370 for Zn, 15,680 for Pb, 2820 for Cd, and 180 for Hg (Anikiev, 1987).

Estimates for the anthropogenic load have been calculated for the marine basin of the Barents Sea (in Ivanov, 2002, p. 13, Table I.3). As can be seen, the majority of various chemical components referred to as polluting substances are of natural origin. The contribution of industrial pollutants, which makes up more than 1% of the natural volume, is characteristic of suspended matter alone, as well as nitrites, sulfates, nickel, copper and fluorine. The main sources of pollutants are wastewaters (Ivanov, 2002).

Products of land-based mineral resource mining, which are washed out by river- and meltwaters, are also toxic. For example, deposits of raw materials from Novaya

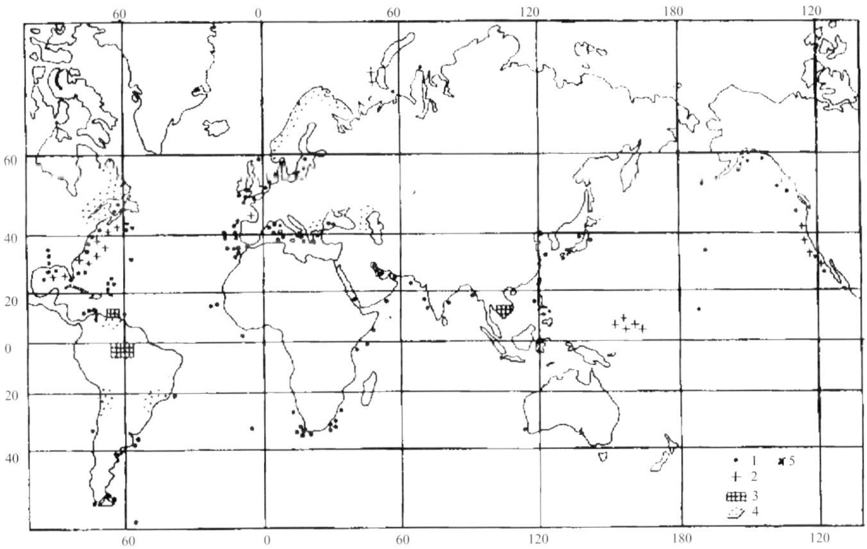

Fig. IV.2. Pollution of the shelf areas of the World Ocean. After Aibulatov and Artiuhin, 1993, with the addition of symbol 5 in the legend:
1—hydrocarbon pollution during wrecks of tankers 1966-1990;
2—places of burial of radioactive waste and areas under conditions of radioactive pollution; 3—regions where defoliants were used;
4—near-shore areas under influence of acid rain;
5—places with buried chemical munitions.

Zemlya are responsible for delivery of such elements as Pb, Zn, Cu, Mn, U, Au, Ag, Se, Cd, Hg, and As into coastal waters (Ivanov, 2002, p. 13). So, the amounts of Zn, Pb, Cu and Cd delivered by the Bezymyannaya River to Bezymyannaya Guba annually are (in tons per year) 29, 8.7, 1.9, and ~0.1, respectively. In general, the total amount of Zn and Pb delivered by the Bezymyannaya River is several times as large as that of the Pechora and Severnaya Dvina rivers.

The increased contents of many of these elements in bottom sediments of the Barents Sea are strictly restricted to ore shows in Novaya Zemlya.

Various radionuclides produced by nuclear power plants (or by nuclear incidents at these plants) enter river waters that then transport them to drainage basins (for example, following the Chernobyl nuclear incident, a large amount of radionuclides were delivered to the Black Sea by the Dnieper). The distribution of certain radionuclides in the area from the mouth of the Columbia River to the Pacific Ocean is shown in Fig. II.1.9.

Radionuclides from nuclear detonations are also transported to seas and oceans by rivers. For example, radioactive clouds that formed as a result of detonations in the atmosphere above Novaya Zemlya were dispersed over the Eurasian continent (Aibulatov, 2002, p. 56), from which radionuclides were delivered to Arctic seas by rivers (the Ob, Enisey, Lena, etc.).

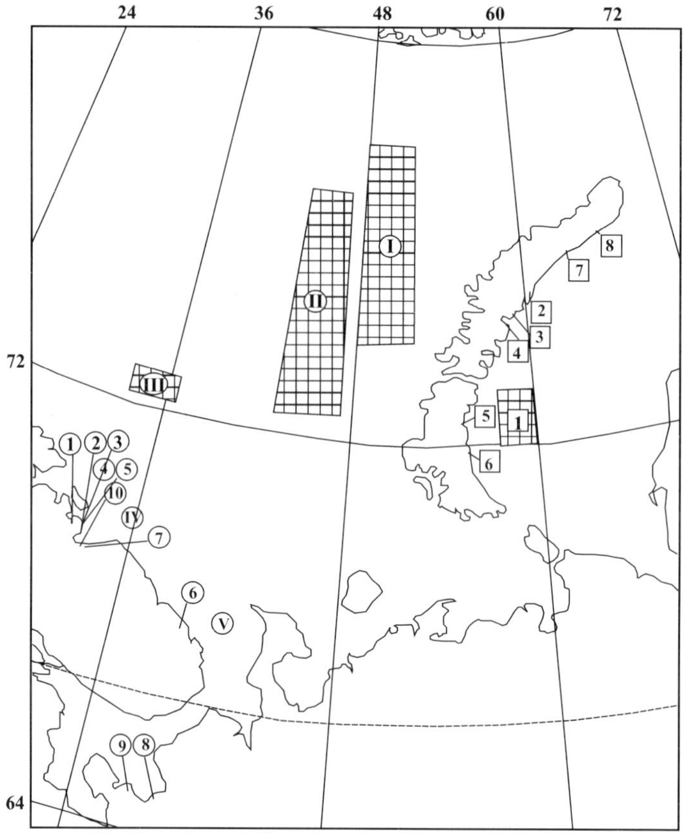

Fig. IV.6. Map illustrating the location of the main sources of radioactive waste and their burial in northern seas (*Facts...*, 1993).
I–V (big blocks): areas of discharge of liquid radioactive waste. 1–8 (in small squares): burials of solid radioactive waste. Naval bases of the Northern Fleet: 1—Zapadnaya Litsa inlet; 2—Olenya inlet and Saida inlet; 3—Ara inlet; 4—Pala inlet; 6—Yokanga. Mooring berths for lay time and utilization of naval ships and ships with nuclear power plants which were decomissioned: 4—Polyarny; 6—Yokanga; 7—Murmansk (repairing and shipping enterprise Atomflot); 8—Severodvinsk. Temporary storage for exhausted fissionable material (fuel) (ENF): 1—Zapadnaya Litsa inlet; 2—Support ship for loading of nuclear reactors of nuclear-powered submarines; 6—Yokanga; 7—universal mother ships Imandra , Lepse , Lotta. Ship repairing factories (SRF) and shipbuilding yards: 4—Polyarny (Naval SRF); 4—Vyuzhniy (SRF Nerpa); 8—Severodvinsk (Sevmashpredpriyatiye industrial corporation, Sever. industr. corp.); 9–10—others; the following are missing: Saint Petersburg (Baltiyskiy Zavod industr. corp., Admiralteyskoye ind.corp.); Nizhniy Novgorod (Krasnoye Sormovo Industr. corp..).

The total average global flux of sedimentary material supplied by aerosols to the bottom of the World Ocean is estimated at 0.24 mg × cm^{-2} (GESAMP, 1989, quotation after Lisitzin, 2001, p. 35). In the Arctic Basin, this flux is much smaller: in different areas of the Beaufort Sea, the supply of aerosol matter is estimated to lie in the range 0.003–0.014 mg × cm^2/y; in the vicinity of Greenland, it is 0.005–0.21 mg × cm^2/y. A considerable portion of aerosols in the Arctic falls onto sea ice, icebergs and glaciers. The chemical composition of sedimentary material in Arctic ice is very similar to that of the upper layer of bottom sediments in the Arctic Ocean (Holeman et al., 1990, see Fig. 1, p. 36 in *Experience...in the Arctic*). Aerosols (and cryosols) are the main sources of such polluting materials as S, In, Zn, As, Sb, I, and Se. According to Shevtchenko et al. (1998), aerosols in the Arctic are markedly enriched (compared with aerosols from other areas) in such chemical elements as Na, K, Cs, Ca, Cr, Co, As, Vr, Zn, Ni, Cu, Au and certain others (ibid., p. 36, Fig. 2).

The water–bottom boundary involves the first few millimeters of the upper layer of sediments and the first few centimeters of the water layer above the seabed. This space interval represents only a few tenths to a few hundredths of a percent of the total thickness of the water strata in seas and oceans. Nevertheless, the role of this barrier zone in geoecological processes is at least similar to that of the SML or the shore–sea boundary. Most pollutants that enter the oceans, especially their pelagic areas, ultimately end up in bottom sediments. All matter able to cross the shore–sea barrier, river–sea barrier, SML and internal boundaries of the ecosystem in a water basin settles onto the seabed. The concentrations of pollutants on the seabed in some seas is 10–1000 times higher than that of the water strata. For example, the concentration of petroleum hydrocarbons at the bottom in the central part of the Black Sea in 1992 was 131.1 mg/l, on average, whereas this concentration in the SML was 0.14 mg/l; in the surface water layer (the uppermost 1-m-thick water layer), 0.02 mg/l; in the bottom water layer, 0.01 mg/l (Medinets, 1994, p. 26).

The concentration of polycyclic aromatic hydrocarbons, heavy metals and many other pollutants occurs in the upper layer of sediments (Medinets, 1994, p. 30). The concentration of heavy metals in sediments of many seas (especially in Europe) began to increase substantially during the rapid post-war economical development, 1945–1950. This situation is especially characteristic of the Baltic Sea, which is the European water basin (Emelyanov, 1992$_3$, 1995).

The bottoms of seas and oceans became a "shelter" for anthropogenic waste material and debris. Typical anthropogenic waste material includes bottles, cans, and polyethylene bags. "The present-day situation in the canyon of the Danube is grievous: the thalweg of this canyon is filled with plastic bags...." (Aibulatov, Artyukhin, 1993. p. 204). There are also such factors as bombs and shells containing poisonous substances (Fig. IV.2), cargo containers, and ships containing spent nuclear reactors and warheads (the Kara and Barents seas).

In the near-bottom water layer, concentrations of inorganic hydrocarbons are also increased. In the Barents Sea, for example, there is a distinct tongue of seawater with concentrations of inorganic hydrocarbons ranging from 0 to 5.0 µg/l, whereas their average concentration in seawater is 0.18 µg (Ivanov, 2002). This tongue is easily traced from the Norwegian Sea and the motion of this water has a direction similar to that of Atlantic currents, penetrating to the Barents Sea through the Bear Island trough (Ivanov, 2003, p. 32, Fig. 3.7). Concentrations of inorganic hydrocarbons are also increased up to 0.5-4.7 µg/l near the outlet of the White Sea and in the Pechora Sea. The reasons for such an increase in concentrations of

At this point several questions arise: Can the products of decaying nuclear warheads (for example, in the Kara and Barents seas) or products of decaying chemical munitions buried in the Baltic (40,000 tons) penetrate into the upper active layer? Or is it possible for them to travel from one deep into another? In many cases, such behavior of these substances is possible, especially in the shallow Baltic and North seas, as well as in Arctic seas. For example, the isotope ^{137}Cs that appeared in the waters of the North Sea, together with waste from nuclear power plants, was transported through the Kattegat Sea into the Baltic, where increased contents of this isotope have been found in the lower water layer of the central Baltic (Emelyanov, 1995_1). The products of hydrolysis of poisonous substances (adamsite, yperite, phosgene, tabun) contained in aircraft bombs and shells are transported to the Stolpe channel by bottom currents during strong inflows of saline North Sea water, whence they enter the Gotland Deep. When saline waters spill over the sides of the channel, which are found at the depth of the halocline or above it, these products may penetrate the upper active layer, where they are involved once again in the biological cycle due to photosynthesis, or even escape into the atmosphere together with air bubbles.

Our studies have revealed that in shallow-water lagoons (3–10 m deep) like the Curonian and Vistula lagoons of the Baltic Sea, the upper 0- to 10-cm layer of muds is disturbed (resuspended) by strong storms, so that suspended matter becomes present in the upper water layer. Because these events have a frequency of at least once a year, muds are also annually elutriated. Due to this, pollutants from muds also become suspended in the upper water layer, whence they are delivered by air bubbles and droplets into the atmosphere and may be carried great distances towards land or sea. The upper layer of muds in such lagoons, which has accumulated throughout the eras of agriculture, steam energy, and the internal combustion engine, is essentially free of anthropogenic pollutants (for example, heavy metals), in contrast to pelagic areas of the Baltic Sea, where muds are not exposed to the effects of wave-induced elutriation.

In oceans, the supply of pollutants from the bottom to the surface is possible only in areas of strong bottom currents or due to upwelling.

In addition, supply of pollutants and bottom sediments to the water column may be caused by anthropogenic processes, including laying of electric cables and oil and gas pipelines onto the seabed, and mining for building materials or hard mineral resources such as FMNs and ferromanganese crusts from bottom. Experimental studies undertaken to estimate geoecological losses due to excavation of FMNs in the Clarion–Clipperton nodule province in the Pacific Ocean have confirmed this conclusion (Pilipchuk, 2003; Fig. IV.9). The most abundant types of sediments in this part of the ocean are radiolarian-clayey oozes (or pelagic clays). The technological problems involved in extracting a thin layer of nodules from the seabed are enormous. Equipment used for this include drags, churners, dredgers, pumps and other devices. Signs of redeposition of sediments were observed on the seabed, at a distance of 1.2–1.3 km from a churner (in the direction of the flow of bottom currents) used to loosen the uppermost 0–5 cm of sediments (or to an even larger depth, Figs. IV.10–12) with FMNs. Experimental studies have shown that the FMNs loosened by the churner are composed of clays, 50–66% of which is represented by aggregates 0.05–0.01 mm in size, and organic detritus (Pilipchuk, 2003, p. 21).

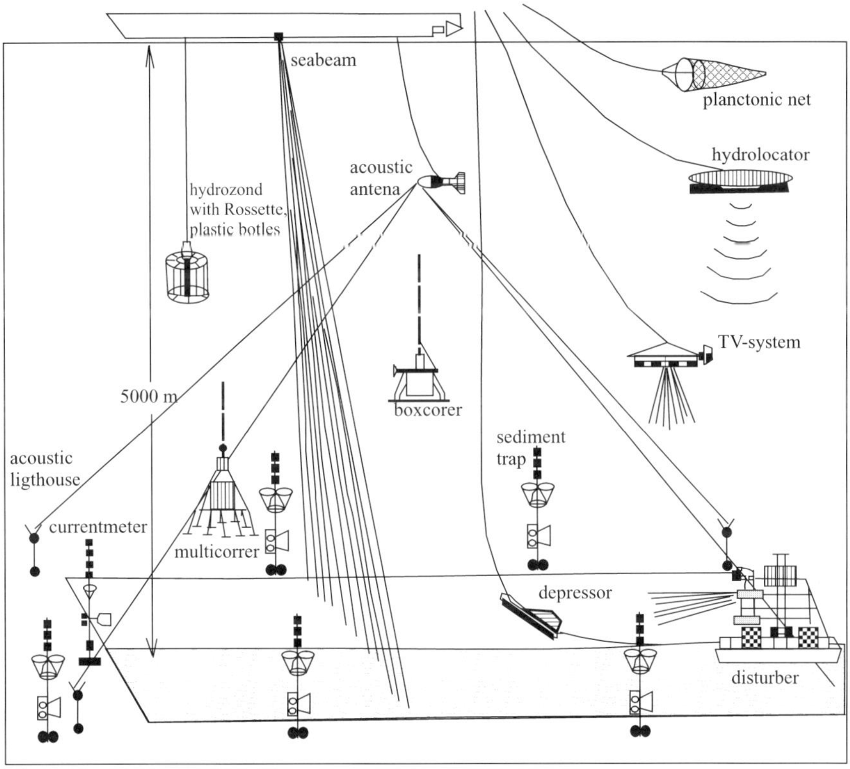

Fig. IV.9. Technical equipment used during BIE experiment in the Pacific Ocean. After Pilipchuk, 2003.

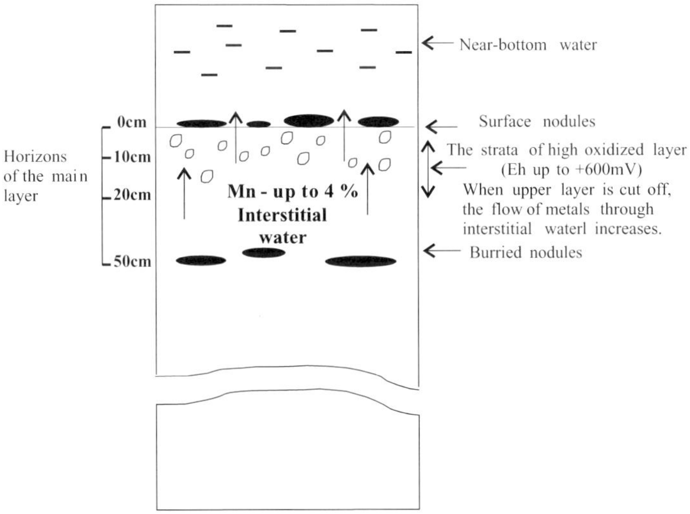

Fig. IV.10. Scheme of flux of metals in interstitial waters after cut of surface layer of the sediments during maintenance operations.

Fig. IV.11. Area of the bottom with the tracks of the runners of a dredger (BIE-II, 1993). After Pilipchuk, 2003.

The upper 10 cm of pelagic clay is a habitat for bottom fauna, including mega- and macrofauna in the top 1-cm sediment layer (these species can be found at depths of no more then 5 cm below the sediment surface), and meyobenthos (in the top 3 cm of clay). Megafauna is represented mostly by sea urchins, holothuria, starfish, ophiurs, sea lilies, polychordic worms, crustaceans, mollusks, fish and squids. Macrofauna (each

Fig. IV.12. Area of the bottom after dumping of mud (BIE-II, 1993). After Pilipchuk, 2002.

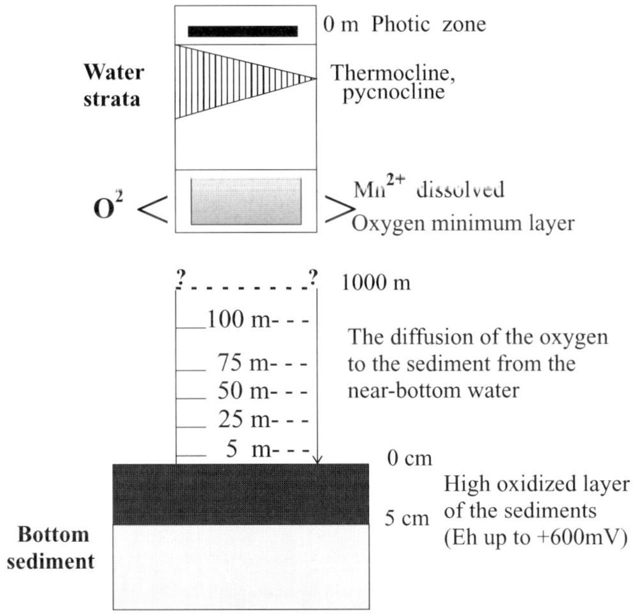

Fig. IV.13. Most sensitive geochemical zones of ocean during maintenance. After Pilipchuk, 2003.

species was 0.3–20 mm in size) is dominated by Polychaeta; Isopoda and Tanaida have also been found there. The most abundant species of meyofauna are multicellular ones (Metazoa), including nematodes, harpacticides and ostracodes; unicellular species are represented by foraminifera. Experiments have shown that a considerable decrease in food is the most dangerous situation for the benthos. Destruction of the upper sediment layer will inevitably lead to rapid changes in the food chain. A lack of food results in either the dying off or suppression of certain types of benthos. Other dangerous situations for benthos are when they become buried in loosened sediments or when there is complete elimination of bottom fauna as a result of excavation of FMNs from the seabed—FMNs represent a natural habitat for benthic fauna.

An increasingly large flux of biogenic elements from interstitial waters of pelagic clays due to extraction of FMNs is also a violation of the oceanic ecosystem. The reason for the abrupt increase in this flux is the absence of a turbidity layer of highly oxidized clay, which is a natural GB. If they reach the surface of the ocean, deep waters enriched in such a way can strongly influence photosynthesis (i.e., growth of primary production), and via phytoplankton they can influence the entire food chain of marine organisms.

The OML is also weak point in terms of geoecology (Fig. IV.13). Major environmental disruption could result not only from dredging of large tracts of the seafloor, but also from disposal of large volumes of "waste" sediments (pulverized wastes of FMNs processed onboard ships). These solid particles would move downwards to become modified within the OML: Mn would be reduced to Mn^{+2}, and when particulate matter is dissolved, Mn and its associated microelements pass into solution, leading to deterioration of oxygen conditions in the OML, up to the appearance of stagnant waters.

Because intensive extraction of FMNs is accompanied by disturbance and resuspension of bottom sediments, one should expect an increase in water turbidity in the photic layer and, consequently, a decrease in primary production.

Internal Boundaries

The internal boundaries of a marine ecosystem play a special role in the ecology of the sea in general and in its geoecology in particular (Figs. IV.14, IV.15). In the author's opinion, these boundaries coincide with such boundaries as the thermocline, halocline (pycnocline), OML, O_2–H_2S transition layer (redox Eh barrier in water), and hydrofronts.

The role of the halocline (pycnocline) was examined in Part II.9. Here, we would only like to stress that the pycnocline divides the marine ecosystem into two zones: upper and lower. The boundary which divides these zones is the depth interval, which includes the layer of the strongest salinity gradient and the lower boundary of mesoplankton accumulations (Fig. IV.16). The upper zone is called aerobiotic, and the lower, chemobiotic (Vinogradov. et al., 1992, p. 86). Such a subdivision is characteristic of such seas as the Black and Baltic. As much as 99% of the total mass of mesoplanktonic organisms is contained in the upper zone, whereas the lower zone contains particular faunal complexes of the simplest organisms, including anaerobic bacteria (Fig. IV.16).

The halocline (pycnocline) is not an impenetrable boundary: communication between the upper and lower zones takes place, although the rate of exchange is small. For example, in the Black Sea, in winter, which is a period in which the upper water layer cools, the rate of vertical exchange notably increases and this leads to the development of water "domes." The main cyclonic gyres are areas where these domes rise toward the surface (Fig. IV.10). As a result, about 3–4 km^3 of waters rich in biogenic components are upwelled to the surface layer (Vinogradov et al., 1992, p. 9),

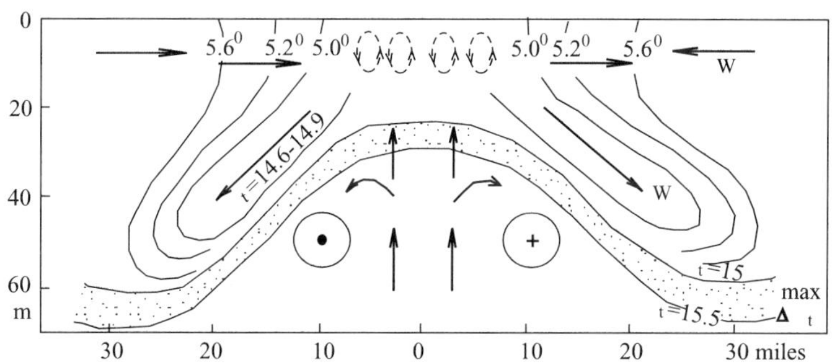

Fig. IV.14. Inner boundaries of the ecosystem in the Black Sea. Circulation scheme of seawater above the cyclonic dome and position of the halocline in the Black Sea. After Latun, 1988.

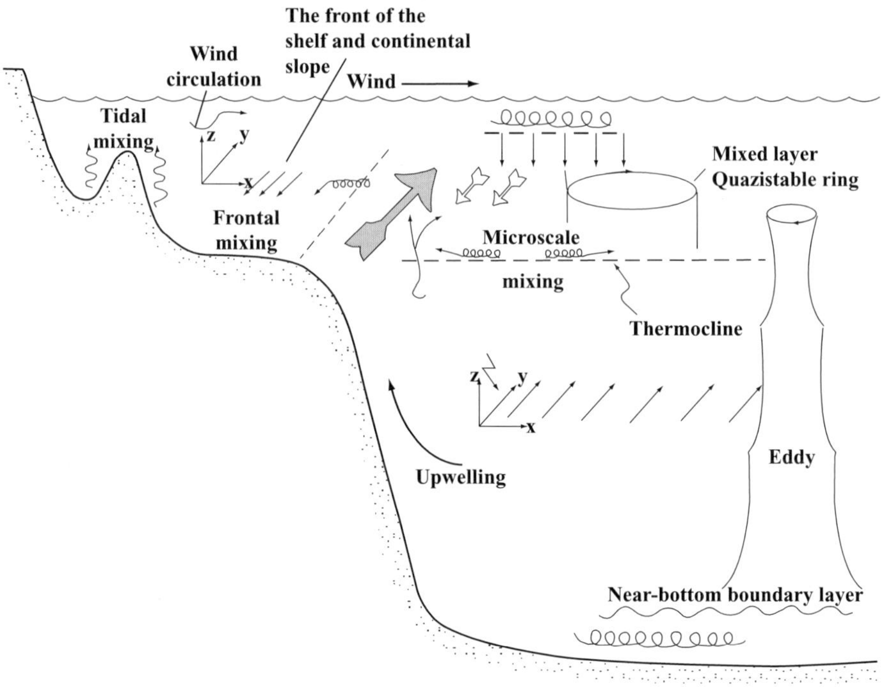

Fig. IV.15. General scheme of transport of pollutants in the ocean. After Izrael and Tsyban, 1989.

and this is an order of magnitude greater than the volume of river runoff. Such upwelling is one of the most important factors causing a notable increase in bioproductivity in the Black Sea.

In frontal zones of coastal upwelling, sharply increased contents of heavy metals, various organic compounds and garbage are frequently observed. So, accumulations of garbage, ammonia, urea, P_{org}, N_{org}, suspended albumen, lipides, hydrocarbons and C_{org}, as well as chloro- and phospho-organic pesticides and polyaromatic hydrocarbons, were found to occur at the hydrochemical barrier in the frontal zone of the Black Sea. Also, there are strongly increased concentrations of Zn, Cu, Cd, Hg, Pb (Sapozhnikov, 1991). Anthropogenic pollutants, which are abundant in frontal coastal waters, are transported by these waters, sinking along a convergence line, to pelagic areas, beyond the limits of the shelf, where their further distribution and geochemical (bio- and hydrochemical) transformation (or deposition) on the seafloor occurs, in accordance with the regularities described in Part II. Because this frontal zone is located between the main Black Sea current and coastal waters and can be observed in marginal areas everywhere in the Black Sea, the biohydrochemical barrier is also found everywhere in peripheral areas of the sea.

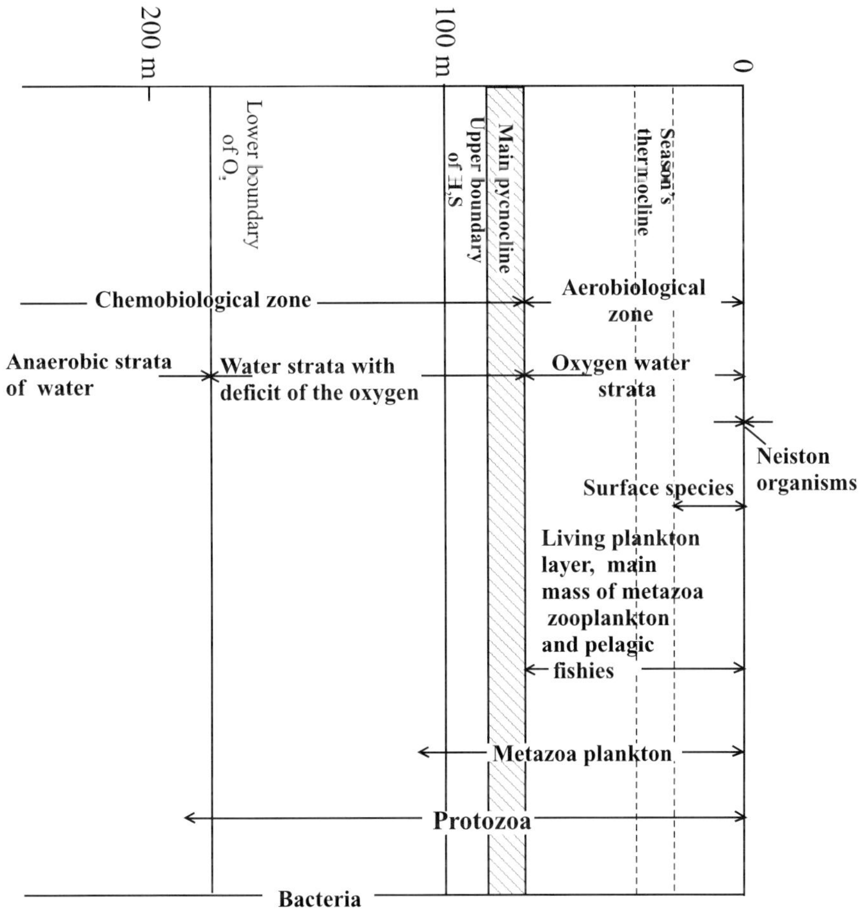

Fig. IV.16. Scheme of vertical zonality of the Black Sea. After Vinogradov et al., 1992.

Conclusion

The reader, together with the author, has traced the migration of a particle and covered thousands of miles—from mountains at the borders of the catchment basin to the pelagic areas of the ocean where the particle settled to the seafloor and became part of the bottom sediment. Other particles began their journey to the oceanic depths from the surface of the ocean. In its descent, the particle encountered many obstacles and barriers along the way. At these barriers, the particle was transformed: it passed through the stomach of a living organism and was partially dissolved, losing certain atoms and molecules while acquiring others. It, however, invariably moved toward its destination—the seafloor, where it joined other particles and became part of what we call bottom sediment or sedimentary rock. This is an extremely complicated route. We understand it in our minds, but could it be represented visually? In the future a block diagram could be created that would visually represent the processes occurring in the ocean. It would show on big screens, in different colors, the barrier zones which have already been discovered and the ones that will probably be discovered. The ocean would be represented in a three-dimensional space with well-defined barrier zones in the form of stripes, lines and planes of different color and shade. The photic layer would be shown in green; the fronts, salinity and physicochemical barriers, and barrier zones would be indicated in blue; stagnation phenomena would be shown in grey and black; and volcanic activity, in red. The clearer the barrier, the more intensive the material inflow and transformation and, consequently, the brighter the color. The three-dimensional model of the ocean would consist of visible stripes, lines, and the invisible water mass between them. The visible part of the ocean would look like a framework representing the "nervous" and "circulatory" systems of the ocean. Bear in mind that the nervous and circulatory systems are alive, pulsating, curving at some places, and expanding or narrowing at others. The water space inside the framework would appear inert and less active. But here, as well as in barrier zones, tiny lights would flash from time to time. These are particles that interact with water, exchange atoms and molecules, and achieve balance with the physicochemical properties of water.

The diagram would also indicate deposits of mineral resources formed on the seafloor by barriers and barrier zones (or near them). Black-and-red stripes would indicate areas where the nervous and circulatory systems are destroyed by detrimental human activity.

Having studied the ocean model, scientists can proceed further. When the features and criteria (indicators) of barriers and barrier zones in the ocean have been defined, scientists will be able to restore the oceans of the past (the framework of barrier zones). This work has already been started, but combined research can be carried out in the future.

References

Ackley, S.F. and Sulivant, C.W., Physical Controls on the Development and Characteristics of Antarctic Sea Ice Biological Communities–a Review and Synthesis, *Deep-Sea Res.*, 1994, vol. 41, no. 10, pp. 1583–1604.

Adelseck, C.D., Dissolution of Deep-Sea Carbonate: Preliminary Calibration of Preservational and Morphologic Aspects, *Deep-Sea Dril. Progr.*, 1977, vol. 24, pp. 1167–1185.

Aghib, F.Sh., Bernoulli, D. and Weissert, H., Hardground Formation in the Bannock Basin (Eastern Mediterranean), *Anoxic Basins of the Eastern Mediterranean*, Cita, M.B., Camerlenghi, A. and Corselli, C., Eds., Milano, Ricerca scient. ed educaz. permanente, 1989, Suppl. no. 72.

Aibulatov, N.A., Suspended Matter Flows of the Continental Margin of the Black Sea, in *Particles Flux in the Ocean*, Hamburg: n.i. 1987, pp. 87–99.

Aibulatov, N.A., *Dinamika tverdogo veshchestva v shelfovoy zone* (Dynamics of Hard Material in Shelf Zone), Leningrad: Gydrometeoizdat, 1990.

Aibulatov, N.A., *Ekologicheskoe ekho kholodnoy voyny v moryakh rossiyskoy Arktiki* (An Ecological Response of the Cold War's Impact in the Russian Arctic Seas), Moscow: GEOS, 2000.

Aibulatov, N.A. and Artyukhin, Yu.V., *Geoekologiya shelfa i beregov Mirovogo okeana* (Geoecology of the Ocean's Shelf and Coasts), St. Petersburg: Hydrometeoizdat, 1993.

Aibulatov, N.A. and Zhindarev, A.L., Processes of Sediment Deposition. Transport of Sedimentary Material, in *Gidrometeorologiya i gidrokhimiya morey SSSR* (Hydrometeorology and Hydrochemistry of Seas in the USSR), vol. 3: Baltic sea, St. Petersburg: Hydrometeoizdat, 1992, pp. 81–83.

Aibulatov, N.A. and Serova, V.V., Eolian Processes on the Coast of the North Africa and their Role in the Sedimentation on the Shelf, *Lithol. and Mineral Resources*, 1983, no. 6, pp. 28–41.

Aibulatov, N.A., Zhindarev, A.L. and Piskareva, M.A., Particularity of Behaviour and Chemical Composition of Suspended Matter in Coastal Area of Sea under Conditions of Deficit of Sedimentary Material, *Water resources*, 1984, no. 2, pp. 62–68.

Ainemer, A.I. and Konshin, G.I., *Rossypi shelfovikh zon Mirovogo okeana* (The Placers of the Shelf Zones of the World Ocean), Leningrad: Nedra, 1982.

Aizatulin, T.A., Lebedev, B.L. and Khailov, K.M., *Okean: fronti, dispersiya, zhizn* (Ocean: Fronts, Dispersion, Life), Leningrad: Hydrometeoizdat, 1984.

Aizatulin, T.A., Lebedev, B.L., Suetova, I.A. and Khailov, K.M., Boundary Surfaces and Geography of the Ocean, *Vestnik MGU. Geography*, 1976, no. 3, pp. 25–35.

Aksionov, A.A., *O rudnom protsesse v verkhney zone shelfa* (About Ore Process at the Upper Zone of the Shelf), Moscow: Nauka, 1972.

Alejnik, D.L. Termohaline Water Structure in the Zone of the Involvement of the Western Boundary Current in the Deep Subtropical Atlantic Circulation, *Okeanologicheskie issledovaniya vo frontalnoy zone golfstrima* (Oceanographic Investigations of the Gulf Stream Frontal Zone), Sagalevitch, A.M., Bogdanov, Yu., Vinogradov, M.E., Eds., Moscow: Nauka, 2002, pp. 25–65.

Alekhin, O.A., Liakhin, Yu.I., Bogoyavlenskiy, A.N. and Ivanenko, V.N., Carbonate System, in *Okeanologiya. Khimiya okeana* (Oceanology. The Chemistry of the Ocean), Moscow: Nauka, 1979, ch. 1, pp. 85–132.

Alekhin, O.A. and Liakhin, Yu.I., *Khimiya okeana* (The Chemistry of the Ocean), Leningrad: Hydrometeoizdat, 1984, pp. 108–113.

Alldredge, A.L., Robinson, B.H., Fleminger, A. et al., Direct Sampling and in situ Observation of a Persistent Copepod Aggregation in the Mesopelagic Zone in the Santa Barbara Basin, *Mar. Biol.*, 1984, vol. 80, no. 1, pp. 75–81.

Allen, F.P. and Kollinson, J.D., Lakes, in *Environments of Sedimentation and Facies,* Reding, Ch., Ed., Moscow: Mir, 1990, vol. 1, pp. 92–96.

Allison, M.A., Nittrouer, C.A. and Faria, L.E.C., Jr., Rates and Mechanisms of Shoreface Progradation and Retreat Downdrift of the Amazon River Mouth, *Mar. Geol.,* 1995, vol. 125, no. 3/4, pp. 373–392.

Almukhamedov, A.I., Bogdanov, Y.A., Kuzmin, M.I. et al., *Riftovaya zona khrebta Reyk'ianes: tektonika, magmatizm i usloviya osadkoobrazovaniya* (Reykjaness Ridge Rift Zone: Tectonics, Magmatism and Conditions of Sedimentology), Kuzmin, M.I., Ed., Moscow: Nauka, 1990.

Aplin, A.C. and Cronan, D.S., Ferromanganese Oxide Deposits from the Central Pacific Ocean, *Geochim. et Cosmochim. Acta,* 1985, vol. 49, no. 2, pp. 437–451.

Anderson, J.D., Nearshore Glacial–Marine Deposition from Modern Sediments of the Weddel Sea, *Nature Physo. Sci.,* 1972, vol. 240, pp. 189–192.

Anderson, J.J., Nutrient Chemistry in the Tropical North Pacific: DOMES Sites A, B, and C in Marine *Geology and Oceanography of the Pacific Manganese Nodule Province,* Brischoff, J.L. and Piper, D.Z., Eds., Plenum Press, 1979, no. 4, pp. 113–162.

Andreev, S.I., Kazmin, Yu.B. and Yegiazarov, B.Y. et al., *Zhelezomargantsevye konkrestii Mirovogo okeana* (Ferromanganese Nodules of the World Ocean), Moscow: Nedra, 1984, pp. 18–61.

Andreev, S.I., Vanshtein, B.G. and Anikeeva, P.I. et al., *Kobaltonosnye zhelezomargantsevye korki okeana* (Cobalt–Bearing Ferromanganese Crusts of the Ocean), Review, Moscow: VIEMS, 1989.

Andrie, C., Temon, J.-F., Messias, M.-J., Memory, L. and Bowles, B., Chlorofluoromethane Distributions in the Deep Equatorial Atlantic during January–March 1993, *Deep–Sea Res.,* 1998, vol. 145, pp. 903–930.

Anikeev, V.V., *Korotkoperiodniye geokhimicheskiye protsessy i zagryazneniye okeana* (Short–Term Geochemical Processes and Ocean Pollution), Moscow: Nauka, 1987.

Anonymous, Water Balance of the Baltic Sea, *Baltic Sea Environment Proc.,* 1986, p. 16.

Aoki, S., Kokhyama, N. and Sudo, T., An Iron–Rich Montmorillonite in a Sediment Core from the Northeastern Pacific, *Deep–Sea Res.,* 1974, vol. 21, pp. 865–875.

Arkhangelsky, A.D. and Strakhov, N.M., *Geologicheskoye stroenie i istoriya razvitiya Chornogo morya* (Geological Structure and History of the Black Sea), Moscow: Izd. Akad. Nauk SSSR, 1938.

Arnold, M. and Shepard, S., East Pacific Rise at Latitude 21°N: Isotopic Composition and Origin of the Hydrothermal Sulphur, Earth. *Pl. Sci. Letters,* 1981, vol. 56, no. 12.

Arrhenius, G., Sediment Cores from the East Pacific, *Cöteborg. Rep. Swed. Deep–Sea Exped. 1947–1948,* 1952, vol. 5, no. 1.

Artemyev, V.E., *Geokhimiya organicheskogo veshchestva v sisteme reka–more* (Geochemistry of Organic Matter in River–Sea System), Moscow: Nauka, 1993.

Arthur, M.A., Secular Variations in Amounts and Environments of Organic Carbon Burial the Phanerozoic, in Mar. Petrol Source Rocks, *Geol. Soc. Lond. Meeting,* 1983.

Arthur, M.A. and Jenkyns, H.C., Phosphorites and Paleoceanography, in Geology of Oceans, *Proc. 26th Int. Geol. Cong.,* Paris, Suppl. Oceanol. Acta, 1981, pp. 83–96.

Asavin, A.M., Emelyanov, E.M., Kolesov, G.M. and Kogarko, L.N., Geochemistry of Trace Elements in Ferromanganese Ores of Sierra Leone Seamounts (the Atlantic ocean), *Int. Conference "Minerals of the Ocean",* St. Petersburg, 2002, pp. 99–100.

Bacon, M.P. and Edmond, J.M., Barium at Geosecs II in the Southwest Pacific, *Earth and Planet. Lett.,* 1972, no. 16, pp. 66–77.

Baker, E.T. and Feely, R.A., Chemical Composition, Size Distribution and Partial Morphology of Suspended Matter at Domes Sites A, B, and C: Relationships with Local Sediment Composition, in *Marine Geology and Oceanography of the Pacific Manganese Nodule Province,* Bischoff, J.L. and Piper, D.Z., Eds., N.-Y., Plenum Press., 1979, pp. 163–201.

Balzer, W., On the Distribution of Iron and Manganese at the Sediment/Water Interface: Thermodynamic Versus Kinetic Control, *Geochim. Cosmochim. Acta,* 1982, vol. 46, pp. 1153–1161.

Balzer W., Organic Matter Degradation and Biogenic Element Cycling in a Nearshore Sediment (Kiel Bight), *Limnol. Oceanogr.,* 1984, no. 9, pp. 1231–1246.

Balzer, W., Erlenkeuser, H., Hartmann, M., Müller, P. J. and Pollehne, F., Diagenesis and Exchange Processes at the Benthic Boundary. Seawater Sediments Interactions in Coastal Waters, Rumohr, J., Ed., *Marine Geology and Oceanography of the Pacific Manganese Nodule Province,* Bischoff, J.L. and Piper, D.Z., Eds., Plenum Press., N.-Y., Berlin: Springer–Ver. coop., 1987, pp. 111–161.

Balzer, W. and Wefer, Dissolution of Carbonate Minerals in a Subtropical Shallow Marine Environment, *Marine Chemistry,* 1981, v. 10, pp. 545–558.

Baranov, E. I., Main Features of Dynamics of Waters in the Gulf Stream System. Conditions of Sedimentation in the Atlantic ocean, in *Okeanologicheskiye issledovaniya* (Oceanological Researches), Moscow: Nauka, 1971, vol. 21, pp. 60–79.

Barikko (Baricco) A., *Oceano mare* (Ocean Sea), 1993, Rizzoli (in Italio).

Baskov, E.A. and Surikov, S.N., *Gidrotermy zemli* (Hydrotherms of the Earth), Leningrad: Nedra, Leningrad Branch, 1989.

Baturin, G.N., Authigenic Phosphorite Concretions in the Sediments of the Shelf of the S-W Africa, *Doklady Akad. Nauk SSSR* (Proc. AS USSR), 1969, vol. 189, no. 6, pp. 1359–1362.

Baturin, G.N., *Fosfority na dne okeanov* (Phosphorites on the Ocean Bottom), Moscow: Nauka, 1978.

Baturin, G.N., *Geohimiya zhelezomargantsevykh konkretsiy okeana* (Geochemistry of Iron–Manganese Nodules of the Ocean), Moscow: Nauka, 1986.

Baturin, G.N., *Rudy okeana* (Oceanic Ores), Moscow: Nauka, 1993.

Baturin, G.N., Emelyanov, E.M. and Kunzendorf, H., Authigenous Deposits in the Sediments, in Geology of the Bornholm Basin, Emelyanov, E.M., Christiansen and Mishelsen, O., Eds., *Aarhus Geoscience,* 1995, vol. 5, pp. 189–194.

Baturin, G.N., Emelyanov, E.M. and Yushina, I., Chemical Composition of Suspended Matter, *Okeanologiya,* 1995, vol. 35, no. 1, pp. 114–120.

Baturin, G.N., Gordeev, V.V. and Kosov, A.T., Metals in the Interstitial Waters, in *Trudy IOAN* (Proc. IOAS), Moscow: Nauka, 1986, vol. 122, pp. 251–269.

Baturin, G.N., Kochenov, A.V. and Petelin, V.P., The Formation of the Phosphorites on West–African Shelf, *Lithol. and Mineral Res.,* 1970, no. 3, pp. 15–26.

Bazilevskaya, E.S., Mobile Forms of Manganese and Associated Elements in Ferromanganese Nodules and Sediments being Derived from Sediments by Using the Acidic Extraction Method, in *Zhelezomargantseviye konkretsii tsentralnoy chasti Tikhogo okeana* (Ferromanganese Nodules in the Central Part of the Pacific Ocean), Moscow: Nauka, 1986, pp. 238–250.

Be, A.W.H., Ecology of Recent Planktonic Foraminifera, Part 1. Areal Distribution in the Western North Atlantic, *Micropaleo,* 1959, no. 5, pp. 77–100.

Belyaeva, N.V. and Burmistrova, I.I., About the Origin of the Critical Level of the Deposition of Carbonates, III *s'yezd sovetskikh okeanografov* (III Conference of the Soviet Oceanographers), Leningrad: Hydrometeoizdat, 1987.

Bender, M.L., Ku, T.L. and Broeder, W.S., Manganese Nodules: Their Evolution, *Science,* 1966, vol. 151, pp. 325–328.

Berger, R.F., Sedimentation of Planktonic Foraminifera, *Marine Geology,* 1971, no. 11, pp. 325–358.

Berger, W.H., Planktonic Foraminifera: Selective Solution and Paleoclimatic Interpretation, *Deep–Sea Res.,* 1968, n.v., pp. 31–43.

Berger, W.H., Planktonic Foraminifera: Selective Solution and the Lysocline, *Marine Geology,* 1970, no. 8, pp. 111–138.

Berger, W.H., Deep–Sea Sedimentation, *The Geology of Continental Margins* Burk, C.A. and Drake, C.L., Eds., N.-Y.: Springer–Verlag, 1974, pp. 213–241.

Berger, W.H., Carbon Dioxide Excursions and the Deep–Sea Record: Apects of the Problem, *The Fate of Fossil Fuel* CO_2 *in the Ocean,* Andersen, R. and Malahoff, A., Eds., N.-Y.: Plenum Publishing Corporation, 1977, pp. 505–542.

Berger, W.H., Paleoceanography: the Deep–Sea Record, *The Sea,* Emiliani, C., Ed., n.p.: John Wiley and Sons, 1981, vol. 7, pp. 1437–1519.

Berger, W.H. and Winterer, E.L., Plate Stratigraphy and the Fluctuating Carbonate Line, *Pelagic Sediments on Land and under the Sea,* Hsü, K.J. and Jenkins, H.C., Eds., Spec. Publ. no. 1. Intern. Assoc. Sedimentologists, Oxford: n.i., 1974, pp. 11–48.

Berner, R.A., *Early Diagenesis: A Theoretical Approach,* Prinston (N.J.): Univ. Press, 1980.

Berner, H. and Wefer, G., Physiographic and Biological Factors Controlling Surface Sediment Distribution in the Fram Strait, *Geological History of the Polar Oceans: Arctic Versus Antarctic,* Bleil, U. and Thiede, J., Eds., Dordrecht–Boston – London: Kluwer Academic Publishers, 1990, pp. 317–335.

Betekhtin, A.G., Chiatura Manganese Deposit and its Commercial Characteristics, *Trudy TSNIGRI* (Preceedings of CSIGRI), Moscow: ONTI, 1936, vypusk 1.

Betekhtin, A.G., About the Origin of the Chiatura Manganese Deposit, *Trudy conferentsii po genezisu rud zheleza, margantsa i alluminiya* (Proceedings about the Origin of the Iron, Manganese and Aluminum Ores), Moscow: Izd. Akad. Nauk SSSR, 1937.

Betekhtin, A.G., Avaliani, G.A. and Gavascheli, A. B. et al., *Chiaturskoye margantsevoye mestorozhdeniye* (Chiatura Manganese Deposit), Moscow: Nedra, 1964.

Betzer, R.R., Bolger, G.W., McGregor, B.A. and Rona, P.A., The Mid–Atlantic Ridge and its Effect on the Composition of Particulate Matter in the Deep Ocean, *EOS. Trans. Amer. Geophys. Union,* 1974, vol. 55.

Bezborodov, A.A. and Eremeev, V.N., About Formation of the Composition of the Atmosphere over the Sea, *Doklady Akad. Nauk SSSR,* 1982, ser. B, no. 8, pp. 3–6.

Bezborodov, A.A. and Eremeev, V.N., *Fiziko–khimicheskiye aspekty vzaimodeystviya okeana i atmosfery* (Physiochemical Aspects of Interaction of the Ocean and Atmosphere), Sobotovich, E.V., Ed., Kiev: Naukova Dumka, 1984.

Bezrukov, P.L., Problems of the Zonality in the Sedimentation, *Preprints of Intern. Oceanogr. Congr.,* Washington, DC, 1959, pp. 448–451.

Bezrukov, P.L., Some Problems of Zonality of the Sedimentation in the World Ocean, *Trudy Okeanograf. Kommissii,* 1962, vol. 10, no. 3, pp. 3–9.

Bezrukov, P.L., Zonality and Irreqularity of the Sedimentation in the Ocean, in *Sovremenniye problemy geografii* (Recent Problems of Geography), Moscow: Nauka, 1964, pp. 245–249.

Bezrukov, P.L., *Zhelezomargantseviye konkretsii Tikhogo okeana* (Ferromanganese Nodules of the Pacific Ocean), Moscow: Nauka, 1976.

Bezrukov, P.L., Andrushchenko, P.F., Murdmaa, I.O. and Skornyakova, N.S., Phosphorites on the Bottom of the Central Part of the Pacific Ocean, *Doklady Akad. Nauk SSSR,* 1969, vol. 185, no. 4, pp. 913–916.

Bezrukov, P.L. and Andrushchenko, P.F., About Geochemistry of the Ferromanganese Nodules of the Indian Ocean, *Izvestiya Akad. Nauk SSSR,* geol. ser., 1973, no. 9, pp. 18–37.

Bezrukov, P.L., Emelyanov, E.M., Lisitzin, A.P. and Romankevich, E.A., Organic Carbon in the Surface Layer of the Sediments of the World Ocean, *Okeanologiya,* 1977, vol. 17, no. 5, pp. 850–855.

Bezrukov, P.L., Skornyakova, N.S. and Murdmaa, I.O. Problems of Ferromanganese Nodule Genesis, in *Zhelezomargantsevye konkretsii Tikhogo okeana* (Ferromanganese Nodules of the Pacific Ocean), Bezrukov, P.L., Ed., Moscow: Nauka, 1976, pp.241–250.

Biogeokhimiya okeana (Biogeochemistry of the Ocean), Monin, A.S. and Lisitzin, A.P., Eds., Moscow: Nauka, 1983.

Biscaye, P.E. and Eittreim, S.L., Suspended Particulate Loads and Transport on the Nepheloid Layer of the Abyssal Atlantic Ocean, *Marine Geology,* 1977, vol. 23, pp. 155–172.

Bishoff, J. L., Red Sea Geothermal Brine Deposits: Their Mineralogy, Chemistry and Genesis in *Hot Brines and Recent Heavy Metal Deposits in the Red Sea,* N.–Y.: Springer, 1969, pp. 368–401.

Bischoff, J.L. and Ku, T.L., Pore Fluids of Recent Marine Sediments, Part 1. Oxidizing Sediments of 20°N, Continental Rise to Mid–Atlantic Ridge, *J. Sed. Petrology,* 1970, vol. 40, pp. 960–972.

Bischoff, J. L. and Piper, D.Z., Marine Geology and Oceanography of the Pacific Manganese Nodule Province, Bischoff, J. L. and Piper, D.Z., Eds., *Marine Science,* 9, Plenum Press, N.–Y. and London, 1979.

Blanchard, D.C., Transport of Marine Organic Matter into the Atmosphere, *EOS Trans.,* 1975, no. 56.

Blazhchishin, A.I., Mineral Composition of Bottom Sediments, *Geology of the Baltic Sea.* Gudelis, V.R. and Emelyanov, E.M., Eds., Vilnius: Mokslas, 1976, pp. 221–254.

Blazhchishin, A.I., Main Chemical Components in Bottom Sediments, *Geology of the Baltic Sea,* Gudelis, V.R. and Emelyanov, E.M., Eds., Vilnius: Mokslas, 1976, pp. 255–287.

Blazhchishin, A.I., The Mass and Balance of Manganese in the Baltic Sea, in *Geologiya i geokhimiya margantsa* (Geology and Geochemistry of Manganese), Moskow: Nauka, 1982, pp. 252–252.

Blazhchishin, A.I., Balance of Fe, Mn, and P in the Baltic Sea, in *Gelogicheskaya istoriya i geohimiya Baltiyskogo morya* (Geological History and Geochemistry of the Baltic Sea), Lisitzin, A.P., Ed., Moskow: Nauka, 1984, pp. 122–130.

Blazhchishin, A.I., *Paleogeografiya i evolutsiya pozdnechetvertichnogo osadkonakopleniya v Baltiyskom more* (Paleogeography and Evolution of Late Quaternary Sedimentation in the Baltic Sea), Gaygolas, A.A., Ed., Kaliningrad: Yantarny skaz, 1998.

Blazhchishin, A.I. and Emelyanov, E.M., Main Traits of the Geochemistry of the Baltic Sea, *Geokhimicheskiye metody poiskov v Pribaltike i Belorussii* (Geochemical Researches and Explorations on the Territory of Byelorussia and Baltic), Minsk: Nauka I Tekhnika, 1977, pp. 60–157.

Blazhchishin, A. I. and Lukashina, N. P., 1994. Sedimentation and Paleogeography of the Cariaco Trench Deep during the Late Plestocene and in Holocene, *Okeanologiya,* vol. 34, no. 6, pp. 905–910.

Blashchishin, A.I. and Pustelnikov, O., Biogenic Components (C_{org} and SiO_{2am}) in Water, Suspension and Bottom Sediments of the Baltic Sea, *Baltica,* 1977, no. 6, pp. 161–172.

Blazhchishin, A.I., Szczepanska, T., Jäbel, W., Dublitz, G., Yuspina, L.F. and Lukashov, V.K., Comprehensive Lithostratigraphic, Geochemical and Petrophysical Investigations on the Indicator Core Sample of Late Quaternary Deposits from the Gdansk Basin, *Kwartalnik Geologiczny* (Sopot), 1990, vol. 34, no. 4, pp. 725–732.

Bleil, U. and Thiede, J., Eds., *Geological History of the Polar Oceans: Arctic Versus Antarctic,* Dordrecht–Boston – London: Kluwer Academic Publ., 1990.

Boden, P. and Backman, J., A Laminated Sediment Sequence from the Northern North Atlantic Ocean and its Climatic Record, *Geology,* 1996, vol. 24, no. 6, p. 507–510.

Bogdanov, Yu.A., Gurwich, E.G. and Butuzova, G.Yu. et al., *Metallonosnye osadki Krasnogo morya* (Metalliferrous sediments of the Red Sea), Moscow: Nauka, 1986.

Bogdanov, Yu.A. and Lisitzin, A.P., Hydrothermal Formations in the Rift Valleys of the Ocean, in *Podvodnye geologicheskiye issledovaniya s obitayemykh apparatov* (Undersea Researches with the Manual Apparatus), Monin, A.S., and Lisitzin, A.P., Eds., Moscow: Nauka, 1985, pp. 191–206.

Bogdanov, Yu.A. and Lisitzin, A.P., Peculiarities of Chemical Elemental Composition of Hydrothermal Formations (Features). The Reasons for Differentiation of Hydrothermal Material, in *Geologicheskoe stroenie i gidrotermalnye obrazovaniya na khrebte Khuan–de–Fuka* (Geological Structure and Hydrothermal Formations of Juan–de–Fuca Ridge), Moscow: Nauka, 1990, pp. 80–86.

Bogdanov, Yu.A., Lisitzin, A.P. and Migdisov, A.A. et al., About the Origin of the Metalliferrous Sediments, Metellonosnye osadki ugo–vostochnoy chasti Tikhogo okeana (Metalliferrous Sediments in the S–E Part of the Pacific Ocean), Moscow: Nauka, 1979, pp. 249–275.

Bogdanov, Yu.A., Sorokhtin, O.G., Zonenshtain, N.A. and Podrazhanskiy, A.M., *Zhelezo–margantsevye korki i konkretsii podvodnykh gor Tikhogo okeana* (Ferromanganese Crusts and Nodules of the Sea Mountains of the Pacific Ocean), Moscow: Nauka, 1990.

Bogorov, V.G., About the Quantity of the Mattes in the Living Organisms of the World Ocean, in *Organicheskoe veshchestvo v sovremennykh i iskopaemykh osadkakh* (Organic Matters in the Recent and Ancient Deposits), Moscow: Nauka, 1971, pp. 12–15.

Bogorov, V.G., *Plankton Mirovogo okeana* (Plankton of the World Ocean), Moscow: Nauka, 1974.

Bogorov, V.G., Vinogradov, M.E., Voronina, N.M., Kanayeva, I.P. and Suetova, I.A., The Distribution of the Biomass of Zooplankton in the Surface Layer of the World Ocean, *Doklady Akad. Nauk SSSR* (Proc. AS USSR), 1968, vol. 182, no. 5, pp. 1205–1207.

Bonatti, E., Honnorez, J., Joensuu., o. and Rydell, H., Submarine Iron Deposits from the Mediterranean Sea, in *The Mediterranean Sea: a Natural Sedimentation Laboratory,* Stanley, D.J., Ed., Stroudsburg (Pa): Dowden, Hutchinson and Ross. Inc., 1972, p. 701–710.

Bonsand, B., Nguyen, B.C., Gaudru, A. and Lambert, G., Sulfate Enrichment in Marine Aerosols Due to Biogenic Gaseous Sulfur Compounds, *J. Geophys Res,* 1980, vol. 85, no. 12, p.7361

Borchert, H., On the Ore–Deposition and Geochemistry of Manganese, *Miner. deposits.,* 1970, no. 5, pp. 300–314.

Bordovsky, O. K., *Nakopleniye i transformatsiya organicheskogo veshchestva v morskikh osadkakh* (Accumulation and Transformation of Organic Matter in the Sea Sediments), Moscow: Nedra, 1964.

Bordovskiy, O.K. and Ivanenkov, V.N., Eds., *Okeanologiya. Khimiya okeana. Khimiya vod okeana* (Oceanology. Chemistry of the Ocean. Chemistry of the Ocean Water), Moscow: Nauka, 1979.

Boström, K., Geochemical Evidence for Ocean Spreading in South Atlantic Ocean, *Nature,* 1970, vol. 227, p. 1041.

Boström K., Submarine Volcanism as a Source for Iron, *Earth and Planet. Sci. Lett.,* 1970, vol. 9, p. 348.

Boström, K., The Origin and Fate of Ferromanganoan Active Ridge Sediments, *Stockholm Contrib. Geol.,* 1973, vol. 27, no. 2, pp. 148–243.

Boström, K. and Peterson, M.N.A., Precipitates from Hydrothermal Exhalations on the East Pacific Rige, *Econ. Geol.,* 1966, vol. 61, pp. 1258–1265.

Boström, K. and Peterson, M.N.A., The Origin of Aluminium–Poor Ferromanganoan Sediments in Areas of High Heat of the East Pacific Rise, *Mar. Jeil.*, 1969, vol. 7, no. 5, pp. 427–448.

Boudreau, B. P. and Guinasso, N.L., The Influence of a Diffusive Sublayer on Accretion, Dissolution, and Diagenesis at the Sea Floor, *The Dynamic Environment of the Ocean Floor,* Fanning, K.A. and Manheim, F.T., Eds., Toronto: Lexingron Books, 1982, pp. 115–142.

Bradshaw, J. S., Ecology of the Living Planktonic Foraminifera in the North and Equatorial Pacific Ocean, *Contribs. Cushamn. Found. Foraminiferal Res.*, 1959, vol. 10 (2), pp. 25–64.

Brewer, P.G. and Spencer, D.W., Distribution of Some Trace Elements in Black Sea and their Flux between Dissolved and Particulate Phases, in *The Black Sea: Geology, Chemistry and Biology,* Tulsa (Okla), 1974, pp. 137–143.

Brinton, E., Parameters Relating to the Distribution of Planktonic Organisms Especially Euphausiids in the Eastern Tropical Pacific, *Progr. Oceanogr.*, 1979, vol. 8, pp. 125–189.

Broadus, J.M., Seabed Materials, *Science,* 1987, vol. 235, no. 4791, pp. 853–860.

Broecker, W.S. and Broecker, S., Carbonate Dissolution on the Western Flank of the East Pacific Rise, in Studies in Paleooceanography, Hay, W.W., Ed., *Spec. Publ. Soc. Econ. Paleont. Miner.* (Tulsa), 1974, vol. 20, pp. 447–457.

Bruland, K.W., Trace Elements in Sea–Water, in *Khimicheskaya okeanografiya* (Chemical Oceanography), Leningrad: Acad. Press, 1983, vol. 8, pp. 157–220.

Bubnova, N.P., 1972. The Feeding of the Detritus–Eating Mollusks Baltica (C) and Portlandica Arctica (Gray) and their Influence on the Bottom Sediments, *Okeanologia,* vol. 12, no. 6, pp. 1024–1090.

Buat–Menard, P. and Arnnold, M., The Heavy Metal Chemistry of Atmospheric Particulate Matter Emitted by Mount Etna Volcano, *Geophys. Res. Lett.*, 1978, vol. 5, no. 4, pp. 245–248.

Burckle, L.H., Diatom Distribution and Paleooceanographic Reconstruction in the Southern Ocean – Present and Last Glacial Maximum, *Mar. Micropaleont.*, 1984, no. 9, pp. 241–261.

Burckle, L.H., Rudolph, S. and Mortlock, R., Evidence for an Early Pliocene Cold Event in the Southern Ocean, *Antarc. J. U. S.*, 1992, vol. 28, pp. 110–111.

Burkov, V.A., *Obshchaya tsirkulyatsiya Mirovogo okeana* (The Circulation of the World Ocean), Leningrad: Hydrometeoizdat, 1980.

Burkov, V.A., Koshlyakov, M.H. and Stepanov, V.N., The Common Data about the World Ocean, in *Fizika okeana. Gidrofizika okeana* (Physics of the Ocean. Hydrophysics of the Ocean), Kamenkovich, V.M. and Monin A.S., Eds., Moscow: Nauka, 1978, pp. 11–62.

Burns, R.G. and Burns, V.M., Mechanism for Nucleation and Growth of Manganese Nodules, *Nature,* 1975, vol. 255, pp. 130–131.

Burns, R.G. and Burns, V.M., Mineralogy of Manganese Nodules, in *Marine Manganese Deposits,* Glasby, G.P., Ed., New–York: Elsevier, 1977, vol. 7, pp. 185–248.

Burst, I.F., Mineral Heterogeneity of «Glauconite»–Pellets, *Amer. Miner.*, 1958, vol. 43, no. 56.

Butuzova, G.Y., Recent Volcanogenic–Sedimentary Iron–Ore Process in the Santorini Volcano Caldera (the Aegean Sea) and its Influence on Sediment Geochemistry in *Trudy Geol. Inst.* (Proc. Geol. Inst.), Moscow: Nauka, 1969, vol. 194.

Butuzova, G.Y., *Gidrotermalno–osadochnoye rudoobrazovaniye v riftovoy zone Krasnogo morya* (Hydrothermal–Sedimentary Ore–Forming Processes in the Red Sea Rift Zone), Moscow: GEOS, 1998.

Butylin,V.P., Zhamoida, V.A. and Kosin, M.B., Distribution of Chemical Elements and Concretion Formation in the Quaternary Deposits of the Gulf of Finland, in *Glyatsialnye shelfy, problemy geologii i metody issledovaniy* (Glacial Shelves, Problems of Geology and Methods of Investigation), Leningrad: VSEGEI, Ministry of Geology, 1985, pp. 43–54.

Bychkov, A.S., Pavlova, G.Yu. and Kropotov, V.A., The Carbonate System, in *Khimiya morskoy vody I autigennoye obrazovaniye mineralov* (Chemistry of the Sea Water and Authigenic Formation of Minerals), Moscow: Nauka, 1989, pp. 49–111.

Callender, E., Transition Metal Geochemistry of Interstitial Fluids Extracted from Manganese Nodule Rich Pelagic Sediments of the North–Eastern Equatorial Pacific Ocean, in *Abstr. Joint. Oceanogr. Assembl.* (Edinburgh), 1976, p. 85.

Callender, E. and Bowser, C.J., Manganese and Copper Geochemistry of Interstitial Fluid from Manganese–Nodule Rich Pelagic Sediments of the North–Eastern Equatorial Pacific Ocean, *Amer. J. Sci.* (New Haven), 1980, no. 10, pp. 1063–1096.

Calvert, S.E. and Price, N.B., Geochemical Variation in Ferromanganese Nodules and Associated Sediment from the Pacific Ocean, *Mar. Chem.*, 1977. vol. 5, no. 1, pp. 43–74.

Campbell, A.C., Palmer, M.R. and Klinkhammer, G.P. et al., Chemistry of hot Springs on the Mid Atlantic Ridge, *Nature,* 1988, vol. 335, pp. 514–519.
Cambray, R.S., Jeffris, D.F. and Topping, G., An Estimate of the Input of Atmospheric Trace Elements into the North Sea and the Clyde Sea, *UK Atom. Energy Auth. Harwell Rep.,* 1975, no. 7733, 30 p.
Campbell, A.C. and Edmond, J.M., Comparative Geochemistries of Plum Particles from 21°, Guyymas Basin and Mid–Atlantic Ridge, *EOS,* 1986, vol. 67, no. 44, p. 1022.
Cappellen, P.Van. and Wang, Y., Cycling of Iron and Manganese in Surface Sediments: a General Theory for the Coupled Transport and Reaction of Carbon, Oxygen, Nitrogen, Sulfur, Iron, and Manganese, *Amer. Journal. Science,* 1996, vol. 296, no. 3, pp. 197–243.
Castellarin, A. and Sartori, R., Quaternary Iron–Manganese Deposits and Associated Pelagic Sediments (Radiolarian Clay and Chert, Gypsiferous Mud) from the Tyrrhenian Sea, *Sedimentology,* 1978, vol. 25, pp. 801–821.
Catalano, G., Bregant, D., Civitarese, G. and Lucchelta, A., Anoxic Hypersaline Mediterranean Basins: Bannock (with its Subbasins Libeccio, Ostro and Scirocco) and Tyro Basins. Chemical Characteristics and Comparison among their Basins, *Anoxic Basins of the Eastern Mediterranean,* Cita, M.B., Camerlenghi, A. and Corselli, C., Eds., Milano, Ricerca scient. ed educaz. permanente, 1989, Suppl. no. 72, pp. 76–79.
Cavonough, C.M., Gardiner, S.U. and Jones, M. T. et al., Prokaryotic Cells in the Hydrothermal Vent Tube–Worm Rifia Pachyptila Jones: Possible Chemoautotrophic Simbiouts, Science, 1981, vol. 213.
Cheng, L., *Marine Pleuston–Animals at the Sea–Air Interface, in Oceanography and Marine Biology,* Barnes, H.L., Ed., n.p.: George Allen and Unwin Ltd., 1975, pp. 181–212.
Cherkashev, G.A., Metalliferous Sediments of the Sulphidic Ore Forming Areas in the Ocean (Northern Part of the East Pacific Rise), *Extended Abstract of Cand. Sci. Dissertation,* Leningrad: VNII Okeangeologiya, 1990.
Cherkashev, G.A., Krasnov, S.G. and Egiazarov, B.M. et al., *Gidrotermalno–osadochnye i gidrotermalnye rudnye obrazovaniya v Mirovom okeane* (Hydrothermal–Sedimentary and Hydrothermal Ore Formations in the World Ocean), Moscow: VIEMS, 1985.
Chester, R., The Marine Mineral Aerosol, in *The Role of Air–Sea Exchange in Geochemical Cycling,* Buat–Menard, P., Ed., n.p., 1984, pp. 43–76.
Chester, R., *Marine Geochemistry,* London: Unwin Hyman, 1990.
Chillingar, J., Y. Bissel, G. and Fairbridge, P., Eds., *Karbonatnye porody. Genezis, rasprostraneniye, klassifikatsiya* (Carbonate Rocks. Genesis, Distribution, Classification), Moscow: Mir, vol.1, 1970.
Chizhevsky, A.A., *Vsya Zhizn* (All the life), Moscow: Sov. Russia, 1974.
Chou, I. and Wollast, R., Biogeochemical Behavior and Mass Balance of Dissolved Aluminum in the Western Mediterranean Sea, *Deep–Sea Research II,* 1997, vol. 44, no. 3–4, pp. 741–768.
Chukhrov, F.V., Gorshkov, A.I. and Ermilova, L.P. et al., 1981. Mineral Forms of Manganese and Iron in the Ocean Sediments, *Izv. Akad. Nauk SSSR,* geol. Ser., no. 4, pp. 5–21.
Cifelli, R., Evolution of Ocean Climate and the Record of Planktonic Foraminifera, *Nature,* 1976, vol. 264, pp. 431–432.
Cifuentas, L.A., Corselli, C. and Halring, T.C., Isotopic and Chemical Evidence on the Origin of the «Gypsum Garden» Gypsum Crystals, Bacino Bannock, Eastern Mediterranean Sea, Report of the F.O.G. Croup, *Anoxic Basins of the Eastern Mediterranean,* Cita, M.B., Camerlenghi, A. and Corselli, C., Eds., Milano, Ricerca scient. ed educaz. permanente, 1989, Suppl. no. 72, pp. 63–66.
Cita, M.B., Camerlenghi, A. and Corselli, C., Eds., *Anoxic Basins of the Eastern Mediterranean,* Milano, Ricerca scient. ed educaz. permanente, 1989, Suppl. no. 72.
Conomos, T.J. and Gross, M.G., Summer Mixing of Columbia River and Ocean Waters, *J. Sanit. Div.* (Amer. Soc. Civil. Eng.) 1968, vol. 946, pp. 979–995.
Corliss, J.B., Dymond, J. and Gordon, L.J. et al., Submarine Thermal Spring on the Galapagos Rift, *Science,* 1979, vol. 203, pp. 1073–1083.
Correns, C.W., Die Sedimente des äquatorialen Atlantischen Ozeans, *Wiss. Erg. D–tsch. Atlant. Exped. "Meteor", Bd. III,* 1937, vol. 3, no. 2b.
Corselli, C. and McCoy, F., Sedimentation of Organic Matter, Bacino Bannock, Anoxic Basins of the *Eastern Mediterranean,* Cita, M.B., Camerlenghi, A. and Corselli, C., Eds., Milano, Ricerca scient. ed educaz. permanente, 1989, Suppl. no. 72, pp. 50–52.
Craig, H., Chung, Y. and Fladeiro, M., A Benthic Front in the South Pacific, *Earth and Planet Sci. Lett.,* 1972, vol. 16, pp. 50–65.

Craig, J.D., Geological Investigation of the Equatorial North Pacific Sea–Floor: a Discussion of Sediment Redistribution, *Marine Geology and Oceanography of the Pacific Manganese Nodule Province,* Bischoff, J.L. and Piper, D.Z., Eds., N–Y and London: Plenum Press, 1979, pp. 529–559.

Crane, K., Hydrothermal Stress Drops and Convective Patterns at there Mid Oceanic Spreading Centres, *Tectonophysics,* 1979, vol. 5, nos. 1–2, pp. 215–238.

Cronan, D.S., Underwater Minerals, Leningrad: Acad.press. Inc., 1980, pp. 72–190, (Russian translation – 1982).

Cronan, D.S. and Toomas, J. S., The Geochemistry of Manganese Nodules and Associated Pelagic Deposits from the Pacific and Indian Ocean, *Deep–Sea Res.,* 1969, vol. 16, pp. 335–359.

Dames, R. and Deionge, J., Chemical Composition of Swiss Aerosols from the Jungtraijoch., *Atmos. Environ.,* 1976, vol. 10, p. 1079.

Damuth, J.E. and Kumar, N., Amazon Cone: Morphology, Sediments, Age and Growth Pattern, *Bull. Geol. Soc. Amer.,* 1975, vol. 86, pp. 863–878.

De Domenico, M. and De Domenico, E., Bannock Basin: First Approach to Mater Microbiology, in *Anoxic Basins of the Eastern Mediterranean,* Cita, M.B., Camerlenghi, A. and Corselli, C., Eds., Milano, Ricerca scient. ed educaz. permanente, 1989, Suppl. no. 72, pp. 100–102.

De Lange, G.J., Ham, M. and Van Loon et al., Some Result of Sulphate–Related Elements in Tyro and Bannock Basins, Eastern Mediterranean *Anoxic Basins of the Eastern Mediterranean,* Cita, M.B., Camerlenghi, A. and Corselli, C., Eds., Milano, Ricerca scient. ed educaz. permanente, 1989, Suppl. no. 72, pp. 83–85.

De Lange, G.J., Middeiburg, J.J. and Van der Weijden, C.H., Major Element Composition of the Brines in Tyro and Bannock Basin, Eastern Mediterranean, *Anoxic Basins of the Eastern Mediterranean,* Cita, M.B., Camerlenghi, A. and Corselli, C., Eds., Milano, Ricerca scient. ed educaz. permanente, 1989, Suppl. no. 72, pp. 80–81.

De Madron, X.D. and Weatherly, G., Circulation, Transport and Bottom Boundary Layers in the Brazil Basin, *J. of Mar. Res.,* 1994, vol. 52, pp. 583–638.

Defant, A., *A Physical Oceanography,* N.–Y.: Pergamon press, 1961, vol.1.

Dekov, V.M., *Gidrotermalnye obrazovaniya v Tikhom okeane* (Hydrothermal Formations in the Pacific Ocean), Moscow: Nauka, 1994.

Delaloye, M.F., and Odin, G.S., Chamosite, the Green Marine Clay from Chamoson: a Study of Swiss Oolitic Ironstones, *Green Marine Clays,* Odin, G.S., Ed., Amsterdam–Oxford–N.-Y.-Tokyo: Elsevier, 1988, pp. 7–28.

Denis–Clocchiatti, M., Sedimentation carbonate et paleoenvironment dans l'ocean au Cenozoique, *Mem. Soc. Geol.* (Fr.), 1982, vol. 143.

Dhakar, S.P. and Burdige, D.J., A Coupled, Non–Linear Steady State Model for Early Diagenetic Processes in Pelagic Sediments, Amer. J. of Sci., 1996, vol. 296, pp. 296–330.

Djafari, D., Mangan–Eisen–Akkumulate in der Kieler Bucht, *Meyniana,* 1977, vol. 29, pp. 1–9.

Dyamalov, R.G., Underground Water Exchange between Land and Sea its Regularities, *Extended Abstract of Cand. Sci. Dissertation,* Leningrad Mountain (Gorny) Institute, Leningrad, 1991.

Dmitrenko, Eicken, H., Hubberten, H.-W., Melles, M., Thiede, J. and Timokhov, I., Eds., *Land–Ocean Systems in the Siberian Arctic. Dynamics and History.* Berlin: Springer–Verlag, 1999, pp. 125–140.

Dolotov, J.S., *Dinamicheskiye obstanovki pribrezhno–morskogo reliefoobrazovaniya i osadkonakopleniya* (Dynamic Relief–Forming and Depositional Environments in the Near–Shore Marine Area), Moscow: Nauka, 1989.

Domack, E.W., Biogenic Facies in the Antarctic Glacial–Marine Environment: Basis for a Polar Glaciomarine Summary, *Palaeogeogr., Palaeclimatol., Palaeoecol.,* 1988, vol. 63, pp. 357–372.

Draft Annual Report on Liquid Discharges from Nuclear Installations, 1995, RAD 97/12/1.

Drever, J.I., Lawrence, J.R. and Antweller, R.C., Gypsum and Halite from the Mid–Atlantic Ridge, DSDP Site 395, *Earth plant. Sci. Lett.,* 1979, vol. 42, pp. 98–102.

Droz, L., Rigaut, F., Cochonat, P. and Tofani, R., Morphology and Recent Evolution of the Zaire Turbidite System (Gulf of Guinea), *GSA Bulletin,* 1996, vol. 108, pp. 253–269.

Druzkov, N.V., Druzkova, E.I., Larionov, V.V. and Turakov, A.B., The Composition and Vertical Distribution of the Ice Algae Communities in the Northern Barents Sea at the Beginning of the Winter 1998 (1), *Opyt sistemnykh okeanologicheskikh issledovaniy v Arktike* (Experience of System Oceanologic Studies in the Arctic), Lisitzin, A.P., Vinogradov, M.E., Romankevich, E.A., Eds., Moscow: Scientific World, 2001, pp. 325–339.

Duce, R.A, Hoffman, G.L. and Ray, B.J. et al., Trace Metals in the Marine Atmosphere: Sources and Fluxes, *Marine Pollutant Transfer,* Windom, H., Duce, R. and Heath, D.C., Eds., n.p., 1976, pp. 77–120.

Duce, R.A., Quinn, J.G. and Olney, C.B. et al., Enrichment of Heavy Metals and Organic Compounds in the Surface Microlayer of Harragansett Bay, Rhode Island, *Science,* 1972, vol. 176, pp. 161–163.

Dudley, W.C., Cementation and Iron Concentration in Foraminifera on Manganese Nodules, *J. Foram. Res.,* 1976, vol. 6, pp. 202–207.

Dugolinsky, B.K., Margolis, S.V. and Dudley, W.C., Biogenic Influence on Growth of Manganese Nodules, *J. Sed. Petrol.,* 1977, vol. 47, no. 1, pp. 428–445.

Dzhonenidze, G.S., *Vliyaniye vulkanizma na obrazovaniye osadkov* (The Influence of Volcanism on the Formation of Sediments), Moscow: Nedra, 1965.

Eaton, A., The Impact of Anoxia on Mn fluxes in the Chesapeake Bay, *Geochim. Cosmochim. Acta.,* 1979, vol. 43, no. 3, pp. 429–432.

Edmond, J. M., Boyle, E.A., Grant, B. and Stallard, R.F., The Chemical Mass Balance in the Amazone Plume, *Deep–Sea Res.,* 1981, vol. 28, no. 11, pp. 1339–1374.

Edmond, J. M., Measures, C.I. and McDuff, R.E., Ridge Crest Hydrothermal Activity and the Balance of the Major and Minor Elements in the Ocean: the Galapagos Data, *Earth and Planet. Sci. Lett.,* 1979, vol. 46, no. 1. pp. 1–18.

Edmond, J.M., Von Damm, K.L., McDuff, R.E. and Measures, C.J., Chemistry of Hot Springs on the East Pacific Rise and their Influent Disperd, *Nature,* 1982, vol. 297, no. 5863, pp. 187–191.

Edmont, J. and Von Damm, C., Hot Springs on the Ocean Floor, *The world of science,* 1983, no. 6, pp. 46–60.

Edmond, J.M. and Von Damm, K.L., Chemistry of Ridge Crest Hot Springs, *Bul. Soc. Wash. Bull.,* 1985, no. 6, pp. 43–47.

Eisma, D., Augustinus, P.G.E.F., and Alexander, C., Recent and Subrecent Changes in the Dispersal of Amazon Mud, *Netherlands J. of Sea Res.,* 1991, vol. 28(3), pp. 181–192.

Eisma, D., Kalf, J. and Van der Goast, S.J., Suspended Matter in the Zaire Estuary and the Adjacent Atlantic Ocean, *Netherlands J. of Sea. Res.,* 1978, vol. 12, no. 3/4, pp. 382–406.

Eisma, D. and Van Bennekom, A.J., The Zaire River and Estuary and the Zaire Outflow in the Atlantic Ocean, *Netherlands J. of Sea. Res.,* 1978, vol. 12, no. 3/4, pp. 255–272.

Elderfield, H., Manganese Fluxes to the Oceans, *Marine Chemistry,* 1976, vol. 4, pp. 103–132.

El–Sayed, S.Z. and Taguchi, S., Phytoplankton Standing Crop and Primary Productivity in the Tropical Pacific, *Marine Geology and Oceanography of the Pacific Manganese Nodule Province,* Bischoff, J.L. and Piper, D.Z., Eds., Marine Science, 9, Plenum Press, New–York and London, 1979, pp. 241–286.

Elverhoi, A., Glaciogenic and Associated Marine Sediments in the Weddell Sea, Fjords of Spitsbergen and the Barents Sea, A review, *Mar. Geol.,* 1984, vol. 57, p. 53–88.

Emelyanov, E.M., New Data on the Mediterranean Sea Sediments, *Doklady Akad. Nauk SSSR* (Proc. AS USSR), 1961, vol. 137, no. 6, pp. 1437–1440.

Emelyanov, E.M., 1965_1. The Grain–Size Distribution of Recent Sediments and some Features of their Formation in the Mediterranean Sea, in *Osnovnye cherty geologicheskogo stroeniya gidrologicheskogo rezhima i biologii Sredizemnogo morya* (Basic Properties of Geological Structure, Hydrological Regime and Biology of the Mediterranean Sea), Moscow: Nauka, pp. 42–67.

Emelyanov, E.M., The Carbonate Content of the Mediterranean Recent Bottom Sediments, in *Osnovnye cherty geologicheskogo stroeniya gidrologicheskogo rezhima i biologii Sredizemnogo morya* (Basic Properties of Geological Structure, Hydrological Regime and Biology of the Mediterranean Sea), Moscow: Nauka, 1965_2, pp. 71–84.

Emelyanov, E.M., Quantitative Distribution of the Marine Suspended Matter near Sambian Peninsula, in *Okeanologicheskiye issledovaniya* (Oceanol. Investigations), Moscow: Nauka, 1968_1, pp. 221–238, vol. 18.

Emelyanov, E.M., Mineralogy of Sandy–Aleuritic Fractions of Recent Sediments of the Mediterranean Sea, *Litologia i Poleznye Iskopaemye,* 1968_2, no. 2, pp. 2–21.

Emelyanov, E.M., The Composition of the Glauconitic and Hydrogoethite–Chamosite–Glauconite Sediments of the West African Shelf, *The Geology of the East Atlantic Continental Margin. ICSU/SCOR W. P. 31 Symp., Cambridge, 1970,* London, 1971, vol. 4, pp. 97–104.

Emelyanov, E.M., Principal Types of Recent Bottom Sediments in the Mediterranean Sea, their Mineralogy and Geochemistry, *Mediterranean Sea: a Natural Sedimentation Laboratory,* Stanley, D.J., Ed., Stroudsburg: Dowden, Hutchinson and Ross Inc., 1972, pp. 355–386.

Emelyanov, E.M., Distribution and Composition of the Mud on Shelf of South–Western Africa, *Trudy Instituta okeanologii Akad. Nauk SSSR* (Proc. of Institute Oceanology AS USSR), Kaliningrad, 1973_1, vol. 95, pp. 211–238.

Emelyanov, E.M., Composition of Low–Phosphatic and Phosphatic Sediments of the West African Shelf in *Formation of Biological Productivity and Bottom Sediments as Related to Ocean Circulation in the South–Eastern Atlantic,* Transactions of the P.P. Shirsov Institute of Oceanology, AS USSR, vol. 95, Kaliningrad Book Publishers, 1973_2, pp. 239–260.

Emelyanov, E.M., The Main Types of Recent Bottom Sediments in the Mediterranean Sea, their Mineralogy and Geochemistry, *Litol. Mineral resources,* 1973_3, no. 1, pp. 29–46.

Emelyanov, E.M., Manganese in Water and Sediments of Atlantic Ocean, *Let. Commis. on Manganese (IAGOD), Acta miner. petrogr.* (Szeged), 1974, vol. 21, no. 2, pp. 310–311.

Emelyanov, E.M., Fe, Mn, and Ti in the Atlantic Ocean Sediments, *Lithol. Mineral Resources,* 1975_1, no. 3, pp. 3–19.

Emelyanov, E.M., Sedimentation on Atlantic Shelf, in *Problemy geologii shelfa* (Problems of the shelf geology), Moskow: Nauka, 1975_2, pp. 183–186.

Emelyanov, E.M., Bottom Sediments of the Mediterranean Sea, in *Osadkonakopleniye v Atlanticheskom okeane* (Sedimentation in the Atlantic Ocean), Kaliningrad: Kaliningradskaya Pravda, 1975_3, pp. 309–426.

Emelyanov, E.M., Amorphous Silica in the Atlantic Ocean Sediments, *Doklady Akad. Nauk SSSR* (Proc. AS USSR), 1975_4, vol. 225, no. 5, pp. 1183–1186.

Emelyanov, E.M., Die Entstehung des chemischen Gehalts von Beimengungen und Ablagerungen in der Ostsee, *Beitrage zur Meereskunde* (Berlin), 1976, issue 38, pp. 61–79.

Emelyanov, E.M. The Manganese in Water and in Particulate Matter of the Atlantic Ocean Basin, in *Problems of Ore–Formation, Proc. 6 th conf. IAGOD,* Varna, 1977_1, vol. 3, pp. 45–63.

Emelyanov, E.M., Lithological–Geochemical Zones of the Sedimentology in the Atlantic Ocean, Abstracts of Papers, *I Conf. sovetskikh okeanologov* (I Conference of the Soviet Oceanologists), Moscow: Nauka, 1977_2, vol. 3, p.147.

Emelyanov, E.M., 1977_3. Some Lithological–Geochemical Indicators of the Supply Products of Denudation into end Basin (on the Basis of Atlantic Ocean), in *Kontinentalny i pribrezhno–morskoy litogenez* (Continental and Near–Shore–Sea Lithogenesis), Novosibirsk, pp. 103–111.

Emelyanov, E.M., On Probability of Application of Coefficient of Stagnation (Concentration of Hydrogen Sulfide) for Paleogeographic Reconstructions, *Oceanologiya,* 1978, vol. 48, no. 3, pp. 508–511.

Emelyanov, E.M., Mineral Composition of the Mediterranean Sea Sediments, in *Okeanologicheskiye issledovaniya* (Oceanol. Investigations), Moscow: Sov. Radio, 1968_1, pp. 61–108, vol. 26.

Emelyanov, E.M., Organic matter and phosphorus, in Emelyanov, E.M., Romankevich, E.A. *Geokhimiya Atlanticheskogo okeana* (Geochemistry of the Atlantic Ocean), Moscow: Nauka, 1979_2, pp. 102–208.

Emelyanov, E.M., The Polyvalent Metals in Waters and in Particulate Matter of the Atlantic Ocean Basin, in *Vzaimodeysyviye mezhdu vodoy i zhivym veshchestvom* (Interaction Between Water and Living Matter), Moscow: Nauka, 1979_3, vol. 1, pp. 59–65.

Emelyanov, E.M., Sedimentation in the Atlantic Ocean Basin and Features of the Zonation, *Extended Abstracts of Doctoral Dissertation,* Moscow, IORAN USSR, 1979_4.

Emelyanov, E.M., The Role of the Geochemical Barrier Zones in the Sedimentology on the Basis of Atlantic Ocean, *Tezisy dokladov 4–oy vsesouznoy shkoly morskoy geologii* (Abstract of the 4th All–Union Conference on Marine Geology), Moscow, 1980, vol. 2, pp. 30–31.

Emelyanov, E.M., Function on the Geochemical Barrier Zones in the Sedimentogenesis, in *Geograficheskiye aspekty izucheniya gidrologii i gidrokhimii Azovskogo morya* (Geographic Aspects of Studying of the Hydrology and Hydrochemistry of the Azov Sea), Leningrad: Geograph. Society of the USSR, 1981_1, pp. 137–151.

Emelyanov, E.M., Alumosilicate Carbonate–Manganese Lithologic Geochemical Region of the Gotland and Landsort Deeps, in *Osadkonakopleniye v Baltiyskom more* (Sedimentation in the Baltic sea), Lisitzin, A.P. and Emelyanov, E.M., Eds., Moscow: Nauka, 1981_2, pp. 130–180.

Emelyanov, E.M., Zonality of the Sedimentogenesis in the Atlantic Ocean, in *Klimaticheskaya zonalnost i sedimentologiya* (Climatic Zonality and Sedimentology), Lisitzin, A.P. and Gershanovich, D.E., Eds., Moscow: Nauka, 1981_3, pp. 113–126.

Emelyanov, E.M., *Sedimentogenez v basseyne Atlanticheskogo okeana* (Sedimentation in the Atlantic Ocean Basin), Moscow: Nauka, 1982_1.

Emelyanov, E.M., About the Concentrations of Manganese in the Atlantic Ocean Basin, in *Geologiya i geokhimiya margantsa* (Geology and Geochemistry of the Manganese), Moscow: Nauka, 1982_2, pp. 236–243.

Emelyanov, E.M., Le Role des Zones de Barriers Geochemiques dans la Sedimentation (Exemple du Basin Nord–Atlantic), (The Role of Geochemical Barrier Zones in Sedimentation), *Bull. de l'Institut de Geologie du Bassin d'Aquitaine* (Bordeaux), 1982_3, nos. 31–32, pp. 361–364.

Emelyanov, E.M., The Most Important Geochemical Barrier Zones in the Ocean (on the Basis of Atlantic Ocean), *Izv. Akad. Nauk SSSR* (Proc. AS USSR), geogr. Ser., 1984_1, no. 3, pp. 39–53.

Emelyanov, E.M., The Role of the Geochemical Barrier Zones during Formation of the Mineral Resources in the Seas and Oceans, in Geology of the Oceans and Seas, *Tezisy 6–oy vsesouznoy shkoly morskoy geologii* (Abstracts of 6th All–Union Marine Geol. School), Moscow, 1984_2, vol. 1, pp. 218–219.

Emelyanov, E.M., Basins of the Baltic Sea – Traps for Elements, *Finnish Marine Research,* 1986_1, vol. 253, pp. 79–96.

Emelyanov, E.M., Geochemical Barriers and Barrier Zones and their Function in the Sedimentogenesis, in *Geokhimiya osadochnogo protsessa v Baltiyskom more* (Geochemistry of the Sedimentary Process in the Baltic Sea), Emelyanov, E.M. and Lukashin, V.N., Eds., Moscow: Nauka, 1986_2, pp. 3–25.

Emelyanov, E.M., Geochemistry of Suspended Matter and Sediments of Gdansk Basin and Sedimentation Processes, in (Geochemistry of the Sedimentary Process in the Baltic Sea), Emelyanov, E.M. and Lukashin, V.N., Eds., Moscow: Nauka, 1986_3, pp. 67–115.

Emelyanov, E.M., The Composition of the Metalliferous Sediment of the Brazil Basin, *Lithol. and Mineral. Res.,* 1986_4, no. 6, pp. 125–130.

Emelyanov, E.M., The Distribution of the Chemical Elements and Components in the Bottom Sediments and Some Features of their Diagenesis, in *Protsessy sedimentatsii Gdanskogo basseyna (Baltiyskoe more)* (Processes of the Sedimentation in the Gdansk Basin (Baltic Sea)), Emelyanov, E.M. and Vypyh, K., Eds., Moscow: Institute of Oceanology AS USSR, 1987_1, pp. 217–242.

Emelyanov, E.M., Conclusion. The Main Features of the Processes of Sedimentation, in *Protsessy sedimentatsii Gdanskogo basseyna (Baltiyskoe more)* (Processes of the Sedimentation in the Gdansk Basin (Baltic Sea)), Emelyanov, E.M. and Vypyh, K., Eds., Moscow: Institute of Oceanology AS USSR, 1987_2, pp. 248–259.

Emelyanov, E.M., The Baltic Sea Deeps – Traps for the Chemical Elements, *Geological Juornal* (Kiev), 1988_1, no. 2, pp. 70–76.

Emelyanov, E.M., Biogenic Sedimentation in the Baltic Sea and its Consequences, in The Baltic Sea, Winterhalter, B., Ed., Espoo: Geologian tutkimuskeskus, 1988_2, pp. 127–136.

Emelyanov, E.M., Hydrothermal Ferruginous Deposits on the Marsilli Mountain (Tyrrhenian Sea), in Geology of the Seas and Oceans, *Tezisy 8–oy vsesouznoy shkoly morskoy geologii* (Abstracts, 8th All–Union Conference of Marine Geology), Moscow, 1988_3, vol. 3, pp. 26–27.

Emelyanov, E.M., The Similarities and Differences in the Properties of the Sediments of the Shelf and Continental Slope of the Atlantic Ocean, *Abstracts of Intern. Symp. "Engineering Geology of the Shelf and Continental Slope of the Seas and Oceans of the World,* Tbilisi–Batumi, 1988_4, pp. 39–41.

Emelyanov, E.M., "The Iron Cap" on the Marsilli Mountain, *Priroda,* The Science news, 1989, no. 1.

Emelyanov, E.M., The Composition of the Metalliferous Sediments and Polymetallic Nodules of the Brazil Basin, in *Geofizicheskiye polya i stroeniye dna okeanskikh kotlovin* (Geophysical Field and Bottom Structure of the Basins), Neprochnov, Yu.P., Ed., Moscow: Nauka, 1990, pp. 156–163.

Emelyanov, E.M., The World Ocean: Traps of Chemical Elements, *Nauka i Chelovechestvo, Annual review,* MOSCOW: Znaniye, 1991_1, pp. 178–187.

Emelyanov, E.M., Biogenic Components in Bottom Sediments of East Equatorial Part of the Pacific Ocean, *Lithol. and mineral. resources,* 1991_2, no. 5, pp. 42–60.

Emelyanov, E.M., Ore Elements in Bottom Sediments of East Equatorial Part of the Pacific Ocean, Okeanologiya, 1991_3, vol.31, no. 4, pp. 638–646.

Emelyanov, E.M., Copper, Nickel and Cobalt in Atlantic Ocean Bottom Deposits, 1992_1, Translated from *Litologiya i Poleznye iskopaemye,* no. 5, Original Article Submitted January–5, 1991, pp. 15–34.

Emelyanov, E.M., The Sediments of the Anoxic Basins of the Mediterranean Sea, *Okeanologiya,* 1992_2, vol. 32, no. 3, pp. 567–575.

Emelyanov, E.M., Middle and Upper Holocene Sediments, in *Gidrometeorologiya i gidrokhimiya morey SSSR* (Hydrometeorology and Hydrochemistry of Seas of the USSR), St. Petersburg, 1992_3, vol. 3, Baltic Sea, no. 1, Hydrometeorological Conditions, pp. 50–72.

Emelyanov, E.M., Recent Bottom Sediments of the Levantine Sea: their Composition and Processes of Formation, in Geological Structure of the North–Eastern Mediterranean, Krasheninnikov, V.A. and Hall, J.K., Eds., Jerusalem, 1994_1, pp. 141–158.

Emelyanov, E.M., Sedimentation in the Norvegian–Greenland Basin and in the Region of Iceland, in Berichte zur Polarforshung, 144, Russian–German Cooperation in the Siberian Seas: Geosystem Laptev Sea, Kassens, H., Hubberten, N.-W., Pryamikov, S.M. and Stein, R., Eds., Bremenhaven, 1994_2, pp. 117–133.

Emelyanov, E.M., Baltic Sea: Geology, Geochemistry, Paleoceanograraphy, Pollution, P.P. Shirshov Institute of Oceanology RAS, Atlantic Branch, Kaliningrad: Yantarny Skaz, 1995_1.

Emelyanov, E.M., Processes of Sedimentation in the Gulf of Finland, *Okeanologiya,* 1995_2, vol.35, no. 3, pp. 700–779.

Emelyanov, E.M., The Processes of Sedimentation in the Gulf of Finland, in Prace Panstwowego Institutu Geologicznego, CXLIX, *Proc. of the Third Marine Geological Conference "The Baltic",* Warshawa, 1995_3, pp. 138–144.

Emelyanov, E.M., Chemical Composition and Elements in the Suspended Matter and Sediments of the Western Baltic, *Baltica,* 1995_4, vol. 9, pp. 5–15.

Emelyanov, E.M., *Bariernye zony v okeane:osadko–i rudoobrazovaniye, geoecologiya* (The Barrier Zones in the Ocean: Sedimentation and Ore Formation, Geoecology), Kaliningrad: Yantarny skaz, 1998_1.

Emelyanov, E.M., Biogenic Components and Elements in Sediments of the Central Baltic and their Redistribution, *Marine Geology,* 2001, vol. 172, no. 12, pp. 23–41.

Emelyanov, E.M., Ed., *Geologiya Gdanskogo basseyna, Baltiyskoe more* (Geology of the Gdansk Basin, Baltic Sea), Kaliningrad: Yantarny Skaz, 2002.

Emelyanov, E.M., Aibulatov, N.A. and Yurevichus, L.Yu., Aluminum, Titan and Manganese in the Suspended Particulate Matter of the Shelf of the North–Eastern Part of the Black Sea, in *Hydrometeologicheskyje Materially,* vol. XCV, Leningrad: Hydrometeoizdat, 1989, pp. 178–190.

Emelyanov, E.M., Aibulatov, N.A. and Zhindarev, A.L. et al., Processes of Sedimentation, in *Gidrometeorologiya i gidrokhimiya morey SSSR* (Hydrometeorology and Hydrochemistry of Seas of the USSR), St. Petersburg, 1992_3, vol. 3, Baltic Sea, no. 1, Hydrometeorological Conditions, pp. 73–83.

Emelyanov, E.M. and Baturin, G.N., Distribution and Composition of Diatom Oozes in Shelf Area of the South–Western Africa, in *Problemy geologii shelfa* (Problems of the Shelf Geology), Moscow: Nauka, 1975, pp. 206–210.

Emelyanov, E.M., Baturin, G.N. and Kunzendorf, H., Distribution of Chemical Elements in Recent and Late Quaternary Deposits of the Bornholm Basin, Geology of the Bornholm Basin, Emelyanov, E.M., Christiansen and Mishelsen, O., Eds., *Aarhus Geoscience,* 1995, vol. 5, pp. 175–189.

Emelyanov, E.M., Blazhis, I.K., Yuryavichus, R.Yu., Paeda, R.N., Valyukyavichus, P.A. and Yankauskas, I.I., Determination of Microquantities of Iron, Cobalt, Titanium in Sea Water and Suspended Matter (on the Baltic Sea and the Atlantic Ocean Basins), *Okeanologiya,* 1971, vol. 11, no. 6, pp. 1116–1125.

Emelyanov, E.M., Boström, K., Trimonis, E.C., Bublitz, G., Vestman, P., Kuptsov, V.M., Leipe, T., Lukashina, N.P. and Slobodyanik, V.M., Stratigraphy and Composition of Strato–Type Core from the Gotland Deep (the Baltic Sea), Okeanologiya, 1995, vol. 35, no. 1, pp. 108–113.

Emelyanov, E.M. and Chumakov, V.D., Some Data on Study of Interstitial Water of the Sea of Marmara and Mediterranean Sea, *Doklady Akad. Nauk SSSR* (Proc. AS USSR), 1962, vol. 143, no. 3, pp. 701–704.

Emelyanov, E.M., Dmitriev, L.V., Elnikov, I.N., Trimonis, E.S., Kharin, G.S., Geological–Geophysical Expedition in the Central and Southern Atlantic, *Okeanologiya,* 1983, vol. 23, no. 5, pp. 900–903.

Emelyanov, E.M., Evsiukov, Yu.D., Ilyin, A.V., Rudenko, M.V. and Slobodyanik, V.M., Bottom Relief, Sediments and Polymetallic Nodules in the Deep Basins of the Atlantic Ocean, in *Geofizicheskiye polya i stroeniye dna okeanskikh kotlovin* (The Geophysical Fields and Structure of Ocean's Basins Bottom), Neprochnov, Yu.P., Ed., Moscow: Nauka, 1990, pp. 136–148.

Emelyanov, E.M., Gordeev, V.V., Djumailo, N.A. and Katargin, N.V., Lithologic–Geochemical Province of the Gulf of Riga, in *Osadkoobrazovaniye v Balyiyskom more* (Sediment Formation in the Baltic Sea), Lisitzin, A.P. and Emelyanov, E.M., Eds., Moscow: Nauka, 1981, pp. 108–127.

Emelyanov, E.M. and Gritsenko, V.A., About the Role of the Near–Bottom Currents in the Formation of the Bottom Sediments in the Gotland Basin, Baltic Sea, *Okeanologiya,* 1999, vol. 39, no. 5, pp. 776–786.

Emelyanov, E.M. and Kharin, G.S., Sedimentation in Guinea and North–American Basins under the Influence of the Solid Load of Amazon and Orinoco Rivers, *Lithol. and mineral. res.*, 1974, no. 2, pp. 151–186.

Emelyanov, E.M. and Kharin, G.S., Geology of the Atlantic in the Iceland Region, *Rezultaty issledovaniy po MGP* (Results of Researches of Internat. Geoph. Projects), Moscow, 1987

Emelyanov, E.M. and Kharin, G.S., Ferromanganese Con the Sea Mountains of the Sierra–Leone Rise (Equatorial Atlantic), *Okeanologiya,* 1993, vol. 33, no. 4, pp. 615–622.

Emelyanov, E.M. and Kharin, G.S., Formation of the Mineral Resources on the Ocean Bottom, *Priroda,* 1993, no. 1 (209), pp. 32–40.

Emelyanov, E.M., Kharin, G.S. and Chernysheva, E.A., The Influence of Hydrothermal Springs on the Composition of Iron–Manganese Nodules in the Eastern Equatorial part of the Pacific Ocean, in *Geochemistry International,* 2002, vol. 40, no. 5, pp. 481–491.

Emelyanov, E.M. and Kool, L.V., Transport of Dust and its Role in the Process of Sedimentation in the Atlantic Ocean, *Lithol. and Mineral Resources,* 1979, no. 2, pp. 3–15.

Emelyanov, E.M. and Kravtsov, V.A., The Role of the Redox–Barrier "O_2–H_2S" on the Boundary of Water Layers in Concentration and Precipitation Trace Metals in the Baltic Sea, in *Trudy mezhdunarodnoy konferentsii "EkoBaltika–91"* (Proc. of International Conference EcoBaltic–91), Kaliningrad, 1991, pp. 27–29.

Emelyanov, E.M., Kravtsov, V.A. and Paka, V.T., Danger to Life of Areas of Dumped Chemical Munitions Skaggerrak Sea and in Bornholm Basin, Baltic Sea, in WACRA Europe, *Abstracts of XVI International Conference "Local Agenda 21"*, München, Mering, 2000, pp. 58–64.

Emelyanov, E.M., Kruglikova, S.B., Bottom Sediments of a Rise in the Central Indian Basin (Polygon V), in *Geofizicheskiye polya i stroeniye dna okeanskikh kotlovin* (Geophysical Fields and Structure of the Bottom of the Ocean Basins), Neprochnov, Yu.P., Ed., Moscow: Nauka, 1990, pp. 163–171.

Emelyanov, E.M., Kruglikova, S.B., Pelagic Clay of the Atlantic Ocean Deep Basins, in *Geofizicheskiye polya i stroeniye dna okeanskikh kotlovin* (Geophysical Fields and Structure of the Bottom of the Ocean Basins), Neprochnov, Yu.P., Ed., Moscow: Nauka, 1990, pp. 148–156.

Emelyanov, E.M., Lisitzin, A.P. and Iliyn, A.V., Types of Bottom Sediments of the Atlantic Ocean, *Kaliningradskaya pravda,* Kaliningrad, 1975.

Emelyanov, E.M. and Lisitzin, A.P., Silica in the Atlantic Ocean, in *Geokhimiya osadkov Atlanticheskogo okeana, Karbonaty i kremnezem* (Geochemistry of the Sediments of the Atlantic Ocean, Carbonates and Silica), Lisitzin, A.P., Emelyanov, E.M. and Eltsina, G.N., Eds., Moscow: Nauka, 1977, pp. 191–234.

Emelyanov, E.M., Lisitzin, A.P., Koshelev, B.A., Distribution and Composition of Carbonates in the Upper Layer of the Sediments of the Atlantic Ocean, *Doklady Akad. Nauk SSSR,* 1971, vol. 196, no. 1, pp. 207–210.

Emelyanov, E.M., Lisitzin, A.P., Shimkus, K.M., Trimonis, E.S., Lukashev, V.K., Lukashin, V.N., Mitropolskiy, A.Yu. and Pilipchuk, M.F., Geochemistry of Late Cenozoic Sediments of the Black Sea, in *Initial Reports of DSDP,* Ross, D. and Neprochnov, Yu., Eds., Washington (us Government Printing office), 1978, vol. 428, pp. 543–605.

Emelyanov, E.M., Lisitzin, A.P., Trimonis, E.S., Shimkus, K.M., Philipchuck, M.F., Lukashev, V.K., Lukashin, V.N., Mitropolsky, A.Yu. and Katargin, N.V., *Geokhimiya pozdnekaynozoyskikh otlozheniy Chernogo morya* (Geochemistry of the Late Cenozoic Sediments of the Black Sea), MOSCOW: Nauka, 1982.

Emelyanov, E.M. and Lisitzin, A.P., Conclusion, in *Geokhimiya osadochnogo protsessa v Baltiyskom more* (Geochemistry of the Sedimentation Process in the Baltic Sea), Emelyanov, E.M. and Lukashin, N.N., Eds., Moscow: Nauka, 1986, pp. 209–220.

Emelyanov, E.M. and Lukashin, V.N., Eds., in *Geokhimiya osadochnogo protsessa v Baltiyskom more* (Geochemistry of the Sedimentation in the Baltic Sea), Moscow: Nauka, 1986.

Emelyanov, E.M. and Lukashina, N.P., Stratigraphy, in Geology of the Bornholm Basin, Emelyanov, E.M., Christiansen and Mishelsen, O., Eds., *Aarhus Geoscience,* 1995, vol. 5, pp. 37–41.

Emelyanov, E.M. and Lyakhin, Y.I., On Factors Contributing to Carbonatization of Bottom Sediments on Shelf of the South–Western Africa, Collected Papers: Formation of Bioproductivity and Bottom

Sediments under Specific Conditions of Water Circulation in the Southeastern Atlantic, *Trudy IORAN SSSR* (Proc. of IORAS USSR), 1973, vol. 95, pp. 268–274.

Emelyanov, E.M. and Marakuyev, V.I., The Researches of the Bottom Surface of the Mediterranean Sea with the Submarine Video Camera, *Trudy IORAN SSSR* (Proc. of IORAS USSR), 1962, vol. 55, pp. 107–116.

Emelyanov, E.M., Mitropolsky, A.Yu., Shimkus, K.M. and Moussa, A.A., *Geokhimiya Sredizemnogo morya* (Geochemistry of the Mediterranean Sea), Kiev: Naukova Dumka, 1979.

Emelyanov, E.M., Moussa, A.A. and Mitropolsky, A.Yu., Mineral and Chemical Composition of River Nile Alluvium, *Lithol. and Mineral Res.*, 1978, no. 4, pp. 134–139.

Emelyanov, E.M., Pilipchuk, M.F., Volostnykh, B.V., Khandros, G.S. and Shaidurov, Yu.O., Fe and Mn Forms in Sediments on the Geochemical Profile of the Baltic Sea, *Baltica* (Vilnius, Mokslas), 1982, vol. 7, pp. 153–171.

Emelyanov, E.M. and Pustelnikov, J.S., Amount of Suspended Forms of Elements (C_{org}, SiO_{2am}, Fe, Al, Ti, Mn, Ni, Co, Cu) in the Waters of the Baltic Sea, *Geokhimiya,* 1975, no. 7, pp. 1049–1063.

Emelyanov, E.M. and Pustelnikov, J.S., Chemical Composition of River and Sea Suspended Matter of the Baltic Sea, *Geokhimiya,* 1975, no. 6, pp. 918–932.

Emelyanov, E.M. and Romankevich, E.A., *Geokhimiya Atlanticheskogo okeana, organicheskoe veshchestvo i fosfor* (Geochemistry of the Atlantic Ocean, Organic Matter and Phosphorus), Moscow: Nauka, 1979.

Emelyanov, E.M. and Senin, Yu.M., Peculiarities of the Composition of the Sediments of the South–Western African Shelf, *Lithol. and Mineral Res.,* 1969, no. 2, pp. 10–26.

Emelyanov, E.M. and Shimkus, K.M., Suspended Matter in the Mediterranean Sea, *in The Mediterranean Sea: a Natural Sedimentation Laboratory,* Stanley, D.J., Ed., Dowden, Stroudsburg: Hutchinson and Ross Inc., 1972, pp. 417–440.

Emelyanov, E.M. and Shimkus, K.M., *Geochemistry and Sedimentology of the Mediterranean Sea,* Dordrecht: D. Reidel Publ. Co., 1986.

Emelyanov, E.M., Shimkus, K.M. and Kuprin, P.N., *Unconsolidated Bottom Sediments of the Mediterranean and Black Seas,* Intergovernmental Oceanographic Commission (UNESCO), IBCM Geol. Geoph. Series, Scale 1:1000000, 10 sheets, St. Petersburg, 1996–1998.

Emelyanov, E.M. and Stryuk, V.L., Water Suspended Particulate Matter, in *Osadkonakopleniye v Baltiyskom more* (Sedimentation in the Baltic sea), Lisitzin, A.P. and Emelyanov, E.M., Eds., Moscow: Nauka, 1981, pp. 79–106.

Emelyanov, E.M. and Trimonis, E.S., On the Problem of Supply of River Sedimentary Material to the Atlantic Ocean, *Okeanologiya,* 1977, vol. 17, no. 1, pp. 94–98.

Emelyanov, E.M. and Trimonis, E.S., Geochemical Investigation of Sediments from the Brazil Basin and the Rio Grande Rise, in *Initial Reports of the Deep Sea Drilling Project,* Barker, P.F., Carlson, R.L. and Johnson, D.A. et al, Washington: U.S. Government Printing Office, 1983, vol. 72, pp. 421–442.

Emelyanov, E.M., Trimonis, E.S. and Kharin, G.S., *Paleogeografiya Atlanticheskogo okeana* (Paleooceanography of the Atlantic Ocean), Leningrad: Nedra, 1989.

Emelyanov, E.M., Trimonis, E.S. and Kharin, G.S., Bottom Structure of the Northern Baltic, *Okeanologiya,* 1996, vol. 36, no. 6, pp. 910–918.

Emelyanov, E.M. and Vlasenko, N.B., Concentration of Dissolved Forms of Fe, Mn and Cu in Marine Pore and Hydrothermal Waters in the Basin of the Atlantic Ocean, *Geochemistry,* 1972, no. 10, pp. 1268–1277.

Emelyanov, E.M., Volkov, I.I., Rosanov, A.G., Khandros, G.S. and Zhabina, N.N., Processes of Reducing Diagenesis in the Sediments of the Deeps, in *Geokhimiya osadochnogo protsessa v Baltiyskom more* (Geochemistry of the Sedimentation Process in the Baltic Sea), Emelyanov, E.M. and Lukashin, N.N., Eds., Moscow: Nauka, 1986, pp. 131–135.

Emery, K.O., Wigley, A.S. and Rubin, M.A., Submerged Peat Deposit off the Atlantic Coast of the United States, *Limnology and Oceanography,* 1965, Supp. 10.

Emery, K.O., Relict Sediments on Continental Shelves of World, *Bull. Amer. Assoc. Petrol. Geol.,* 1968, vol. 52, no. 3, pp. 445–464.

Emery, K.O. and Uchupi, E., *The Geology of the Atlantic Ocean,* D., Springer, 1984.

Erba, E., Stratigraphy of Cruise Tyro–87 Cores from the Bannock Tyro and Kretheus Basins, *Anoxic Basins of the Eastern Mediterranean,* Cita, M.B., Camerlenghi, A. and Corselli, C., Eds., Milano, Ricerca scient. ed educaz. permanente, 1989, Suppl. no. 72, pp. 20–23.

Erickson, D.J., Mervil, J.I. and Duce, R.A., Seasonal Estimates of Global Atmospheric Sea–Salt Distributions, *J. Geophys. Res.*, 1986, vol. 91, pp. 1067–1072.

Erickson, D.J. and Duce, R.A., The Global Flux of Atmospheric Sea Salt, *J. Geophys. Res*, 1988, vol. 93, pp. 14–88.

Etchichury, M.C. and Remiro, J.R., Muestras de fondo de la Platforma Continental, *Ciencias Geologicas* (Buenos Aires), 1960, vol. 4, no. 4.

Exon, N.F., Bogdanov, N.A., Franchetau, Garret, C., Hsü, K.J., Mienert, J., Richen, W., Scott, S.D., Stein, R.H., Thiede, J. and Stackelberg, U. von, Group Report. What is the Resource Potential of the Deep Ocean, in *Use and Missuse of the Ocean Seafloor*, Hsü, K.J., and Thiede, J., Eds., Chichester: John Willey and Sons, John Willey and Sons, pp. 7–28.

Opyt sistemnykh okeanologicheskikh issledovaniy v Arktike (Experience of System Oceanologic Studies in the Arctic), Lisitzin, A.P., Vinogradov, M.E., Romankevich, E.A., Eds., Moscow: Scientific World, 2001.

Facts and Problems Related to Damping of Radioactive Wastes in Seas Bordering upon Russian Federation *Materialy doklada Pravitelstvennoy Komissii po voprosam, svyazannym s zakhoroneniyem v more radioaktivnykh otkhodov, sozdannoy rasporyazheniyem Prezidenta RF* (Report of Governmental Commission on Problems Related to Burials of Radioactive Wastes in Seas, which has been Established by Decision of President of RF from 24 October, 1992, no. 613), Moscow: Administration of President of RF, 1993.

Fanning, K.A. and Manheim, F.T., Eds., The Dynamic Environment of the Ocean Floor, Toronto: Lexington Books, 1982.

Field, M.E. and Pilkey, O.H., Lithification of Deep Sea Sediments by Pyrite, *Nature*, 1970, vol. 226, pp. 836–837.

Figueres, G., Martin J.-M. and Meybeck, Iron Behaviour in the Zaire Estuary, *Netherlands Journal of Sea. Res.*, 1978, vol. 12, pp. 329–338.

Finney, M.L., Heath, G.R. and Lyle, M., Growth Rates of Manganese–Rich Nodules at MANOP Site H (Eastern North Pacific), *Geochim. Cosmochim. Acta*, 1984, vol. 48, pp. 911–919.

Fedorov (Fiodorov), K.N., *Fizicheskaya priroda i struktura okeanicheskikh frontov* (Physical Nature and Structure of the Oceanic Fronts), Leningrad: Hydrometeoizdat, 1983.

Fischer, A.G. and Arthur, M., Secular Variations in the Pelagic Realm, in Deep–Water Carbonate Environments, Coor, H.E. and Enos, P., Eds., *Spec. Publ. Soc. Econ. Paleont. Miner.*, Tulsa, 1977, vol. 25, pp. 19–50.

Flint, M.V., Afanasyev, K.I. and Fetisov, A.N., Population–Genetic Studies of Mass Species of Copepodes in the Pacific Ocean, in Ecosystems of the Eastern Boundary Currents and Central Areas of the Pacific Ocean, Vinogradov, M.E. and Musayeva, E.I., Eds., Moscow: Nauka, 1990, pp. 152–268.

Flint, M.B. and Kolosova, E.G., Mesoplankton of the Peruvian Coastal Waters, in *Izmenchivost ekosistemy Chernogo morya* (The Changeable of the Black Sea Ecosystem), Vinogradov, M.E., Ed., Moscow: Nauka, 1990, pp. 213–229.

Flood, R.D., Sea Beam and Sediment Waves in the Argentine Basin, *Eos*, 1985, vol. 66.

Franklin, H.S., Atmosphere Aerosols, *NASA Contract Rep.*, 1976, no. 2626, 4.

Froelich, P.N., Klinkhammer, G.P. and Bender, M.L. et al., Early Oxidation of Organic Matter in Pelagic Sediments of the Eastern Equatorial Atlantic Diagenesis, *Geochim. Cosmochim. Acta*, 1979, vol. 43, no. 7, pp. 1075–1090.

Froidefond, J.M., Pujos, M. and Andre, X., Migration of Mud Banks and Changing Cin French Guiana, *Mar. Geol.*, 1988, vol. 84, pp. 19–30.

Further Corrections to the Draft Annual Report on Discharges from Nuclear Installations, 1995, *PRAM*, 97/6/4 add 2.

Fütterer, D.K., The Modern Upwelling Record off Northwest Africa, in *Coastal upwelling, Its Sedimentary Record*, Thiede, J. and Suess, E., Eds., Part B, NATO Conference Series, Series 4, Marine Sciences, Plenume Press, N.–Y. and Londom, 1983, pp. 105–123.

Galimov, E.M., Laverov, N.P., Stepanets, O.V. and Kodiana, L.A., Preliminary Results of Ecologic–Geochemical Studies of Arctic Seas of Russia (Based on Information Collected by R/V "Akademik Boris Petrov", cruise 22), *Geochemistry*, 1996, no. 7, pp. 579–597.

Galkus, A. and Joksas, K., *Nuosedine Medziaga Tranzitinije Akvasistemoje* (Sedimentary Material in the Transit Aquasystem), Vilnius, 1997, (in Lithuanian).

Gardner, J.V., Late Pleistocene Carbonate Dissolution Cycles in the Eastern Equatorial Atlantic, in Dissolution of Deep–sea Carbonates, Sliter, W.V., Be, A.H.H. and Berger, W.H., Eds., *Spec. Publ. Geol. Soc. A. M.*, 1975, vol. 83, pp. 143–156.

Garrels, R.M. and Christ, C.L., *Solutions, Minerals and Equilibria,* N.–Y., Evanston, London, Weatherhill, Tokio: Harper & Row, 1965.

Garret, W.D., Impact of Natural and Manmade Surface Forms on the Properties of the Air–Sea Interface, in *The Changing Chemistry of the Ocean,* Dyrssen, D., and Jagner, D., Eds., New–York: Willey, 1972, pp.75–91.

Gathman, S., Climatology, in *The Nordic Seas,* Hurdle, B.G., Ed., New–York, Berlin, Heidelberg, Tokyo: Springer–Verlag, 1986, pp. 1–20.

Gavshin, V.M. and Zakharov, V.A., Geochemistry of the Upper Jurassic Lower Cretaceous Bazhenov Formation, West Siberia, *Economic geology,* 1996, vol. 91, pp. 122–133.

Geokhimiya diageneza donnykh osadkov Tikhogo okeana (Geochemistry of Diagenesis of the Bottom Sediments in the Pacific Ocean), Ostroumov, E.A., Ed., Moscow: Nauka, 1986.

Geodekian, A.A., Trotsyuk, V.Ya., Verhovska, Z.I. and Avilov, V.I., Gases in the Recent Sediments, in *Okeanologiya, Khimiya okeana, Geokhimiya donnykh osadkov* (Oceanology, Chemistry of the Ocean, Geochemistry of the Bottom Sediments), Moscow: Nauka, 1979, pp. 231–312.

Geodekian, A.A. and Zabanbark, A., *Geologiya i razmeshcheniye neftegazovykh resursov v Mirovom okeane* (Geology and Location of the Oil Resources in the Sedimentary Strata of the World Ocean), Moscow: Nauka, 1985.

Geologiya i geomorfologiya Baltiyskogo morya (Geology and Geomorphology of the Baltic Sea), Explanatory Note of the Geological Maps, Scale 1: 500 000, Grigelis, A.A., Ed., Leningrad: Nedra, (Leningrad Branch), 1991.

Geology of the Baltic Sea, Gudelis, V. and Emelyanov, E., Eds.,Vilnius: Mokslas, 1976.

Gershanovich, D.Y., Zonality of the Distribution of the Near–Shore Sediments in the Upwelling Areas of the World Ocean, in *Klimaticheskaya zonalnost i sedimentatsiya* (Climatic Zonality and Sedimentation), Lisitzin, A.P. and Gershanovich, D.Y., Eds., Moscow: Nauka, 1981, pp. 73–84.

GESAMP, Principles for Developing Coastal Water Quality criteria, *Report and Studies,* 1976, no. 5.

GESAMP, Land/Sea Boundary Flux of Contaminants, Contributions from Rivers, *Report and Studies,* 1989, no. 32.

Gershanovich, D.E., Gorshkova, T.I. and Koniuhov, A.I., Organic Matter of the Recent Sediments of the Submarine Continental Margins, in *Organicheskoye veshchestvo sovremennykh osadkov i metody ikh issledovaniya* (Organic Matter of the Recent and Fossil Sediments and the Methods of their Research), Moscow: Nauka, 1974.

Gibbs, R.J., The Geochemistry of the Amazon River System, Part 1, The Factors and the Composition and Concentration of the Suspended Solids, *Bull. Geol. Soc. Amer.*, 1967, vol. 78, pp. 1203–1232.

Gibbs, R.J., The Bottom Sediments of the Amazon Shelf and Tropical Atlantic Ocean, *Mar. Geol.*, 1973, vol. 14, no. 5, pp. 39–45.

Gibbs, R.J., Amazon River Sediment Transport in the Atlantic Ocean, *Geology.*, 1976, vol. 4, no. 1, pp. 45–48.

Gibbs, R.G., Currents on the Shelf of North–Eastern South America, *Estuar. Coast. Shelf Sci.*, 1982, vol. 14, no. 1, pp. 283–300.

Giresse, P., Oolithes Ferrugineuses en vole de Formation au large du Cap Lopez (Gabon), *C.r. Acad. sci.* (Paris), 1965, vol. 260, no. 9, pp. 2550–2552.

Glagoleva, M.A., The Regularities of the Changes of the Ferromanganese Chemical Composition in the North–West Pacifics Sediments, *Lithol. Mineral Resources,* 1972, no. 4, pp. 40–49.

Glasby, G.P., When Manganese Modules Remain at the Sediment Water Inferface, *N.Z. Sci,* 1977, vol. 20, no. 2, pp. 187–190.

Glasby, G.P., Manganese Nodule Studies in the Southwest Pacific, 1975–1980: A review, *S. Pacif. Mar. Geol. Notes.*, 1981, vol. 2, no. 3, pp. 37–46.

Glasby, G.P., Emelyanov, E.M., Zhamoida, V.A., Baturin, G.N., Leipe, T., Banglo, R. and Bonacker, P., Environments of Formation of Ferromanganese Concretions in the Baltic Sea: a Critical Review, in Manganese Mineralisation: Geochemistry and Mineralogy of Terrestrial and Marine Deposits, Nicolson, K., Hein, J.R., Buhn, B. and Dasgupta, S., Eds., *Geological Society Special Publication,* 1997, no. 119, pp. 213–237.

Glasby, G.P. and Lawrence, P., *Manganese Deposits in the South Pacific Ocean: Maps for Ni, Cu, Co and Mn,* N.Z. Oceanogr. Inst., 1974.
Golovin, P., Dmitrenko, I., Kassens, H. and Höleman, J.A., Frazil Ice Formation during Flood and its Role in Transport of Sediments to the Ice Cover, in *Land–Ocean Systems in the Siberian Arctic: Dynamics and History,* Kassens, H., Bauch, H.A., Dmitrenko, I., Eicken, H., Hubberten, H.-W., Melles, M., Thiede, J. and Timokhov, I., Eds., Berlin: Springer–Verlag, 1999, pp. 125–140.
Gordeev, V.V., *Rechnoy stok v okean i cherty ego geokhimii* (River's Load to the Ocean and it Geochemical Features), Moscow: Nauka, 1983.
Gordeev, V.V., River Input of Water, Sediment, Major Ions, Nutrients and Trace Metals from Russian Territory to the Arctic Ocean, in *Fresh Water Budget of the Arctic Ocean,* Lewis, E.L., Ed., Dordrecht: NATO Sci. Ser. Kluwer Acad. Publ., 2000, pp.297–322.
Gordeev, V.V., Emelyanov, E.M., Kuzmina, T.G. and Turanskya, N.V., Geochemical Features of the Sedimentary Suites, in *Geologicheskiye formatsii severo–zapadnoy chasti Atlanticheskogo okeana* (Geological Formations of the North–Western Part of the Atlantic Ocean), Moscow: Nauka, 1979, pp. 96–147.
Gordeev, V.V. and Lisitzin, A.P., Microelements, in *Okeanologiya, Khimiya okeana* (Okeanology, The Chemistry of the ocean), Moscow: Nauka, 1979, vol. 1, pp. 337–375.
Goryainov, I.N. and Goryainova, G.I, To the Problem of "Unsinkability" of Ferromanganese Nodules, *Doklady Akad. Nauk SSSR* (Proc. AS USSR), 1983, vol. 272, no. 2, pp. 432–437.
Gourary, P.G., Ed., *Usloviya formirovaniya nefti i metody poiska neftenosnykh ilistykh otlozheniy Bazhenovskoy formatsii* (Conditions of Oil Formation and Methods of Petroleum Deposits Prospecting in Argillites of the Bazhenow Formation), Moscow: Nedra Press, 1988.
Graham, W.F. and Duce, R.A., *Geochim. et Cosmochim. Acta.,* 1979, vol. 21, no. 5, pp. 1195–1208.
Greenslate, J.L., Microorganism Particle in the Construction of Manganese Nodules, *Nature,* 1974, vol. 249, no. 5453, pp. 181–183.
Greenslate, J.L., Frazer, J.Z. and Arrhenius, G., Origin and Deposition of Elements in the Seabed, in *Origin and Distribution of Manganese Nodules in the Pacific and Prospects for Exploration,* Morgenstein, M., Ed., Honolulu, 1973, pp. 45–60.
Grichuk, D.V., Borisov, M.V. and Melnikova, G.L., Behavior of Heavy Metals in Hydrothermal System of Mid–Ocean Ridge (Results of Thermodynamical Modeling), in Geology of Oceans and Seas, *Tezisy dokladov 6–oy vsesouznoy shkoly morskoy geologii* (Abstract of the 6th All–Union Conference on Marine Geology), 1984, vol. 3, pp. 169–170.
Griffin, J.J., Windom, H. and Goldberg, E.D., The Distribution of Clay Minerals in the World Ocean, *Deep–Sea Res.,* 1968, vol. 15, no. 10, pp. 433–461.
Gromov, V.V. and Starodubtsev, E.G., Assimilation of the Elements of the Iron Group by Phytoplankton, *Okeanologiya,* 1974, vol. 14, no. 6, pp. 1006–1011.
Gromov, V.V. and Spitzin, V.I., *Iskusstvennye radionuklidy v morskoy srede* (Hand–Made (Artificial) Radionuclides in the Sea Environment), Moscow: Atomizdat, 1975.
Gross, M.G., *Oceanography: A View of the Earth,* N.–Y.: Prentice–Hall, Englewood Cliffs, 1977, 2nd ed.
Gryaznov, V.I, Baranova, N.M. and Baas, Y.B. et al., *Nikopolskiy margantsevorudny basseyn* (Nikopol Manganese Ore Basin), Moscow: Nedra, 1964.
Gurvich, E.G., *Metallonosnye osadki v Mirovom okeane* (Metalliferous Sediments of the Word Ocean), Moscow: Scientific World, 1998.
Haese, R.R., The Reactivity of Iron, in *Marine Geochemistry,* Schulz, H.D. and Zabel, M., Eds., n.p.: Springer, 2000, pp. 233–261.
Halbach, P. and Puteanus, D., Cobalt–Rich Ferromanganese Crusts from the Area of Seamounts in the Central Part of the Pacific Ocean–Composition and Formation, in Geology of the World Ocean, Proc. 27th Geol. Congr., Moscow, 1984, vol. 6, part 1, pp. 19–27.
Hall, M.M., McCartney, M.S. and Whitehead, J.A., Antarctic Bottom Waters Flux in the Equatorial Western Atlantic, *J. of Physical Oceanography,* 1997, vol. 27, pp. 1903–1926.
Hart, T.J. and Currie, R.I., The Benguela Current, *Discovery Reports,* 1960, vol. 31, pp. 123–289.
Hartman, M., Sur Geochemie von Mangen und Eisen in der Ostsee, *Meyniana* (Berlin), 1964, vol. 14, no. 53, pp. 3–20.
Hartmann, H., Kögler, F.C., Müller, P. and Sues, E., Preliminary Results of Geochemical and Soilmechanical Investigations on Pacific Ocean Sediments, *in The Origin and Distribution of*

Jorgensen, B., Bacteria and Marine Biogeochemistry, in *Marine Geochemistry,* Shulz, H.D. and Zabel, M., Eds., Springer, 2000, pp. 173–208.

Jouse, A.P., Kozlova, O.G. and Muhina, V.V., Distribution of Diatoms in the Surface Layer of Sediments from the Pacific Ocean, in *The Micropaleontology of the Oceans,* Funnell, B.M. and Riedel, W.R., Eds., London: Cambridge University Press, 1971, pp. 263–269.

Jouse, A.P., Koroliova, G.S. and Nechaev, G.A., Stratigraphic and Paleogeographic Researches in the Indian Sector of the Southern Ocean, *Oceanological researches,* 1963, no. 8, pp. 17–26.

Junge, C.E., Our Knowledge of the Physicochemistry of Aerosols in the Undisturbed Marine Environment, *J. Geophys. Res.,* 1972, vol. 77, no. 27, pp. 5183–5200.

Kagan, B.A., Ryabchenko, V.A., and Sofrai, A.S., *Reaktsiya sistemy okean–atmosfera na vneshniye vozdeystviya* (Behavior of the Ocean–Atmosphere System in Response to External Effects), Leningrad: Hydrometeoizdat, 1990.

Kahru, M., Horstmann, U. and Rud, O., Satellite Detection of Increased Cyanobacteria Blooms in the Baltic Sea: Natural Fluctuation or Ecosystem Change? *Ambio,* 1994, vol. 23, no. 8, pp. 469–471.

Kalhorn, S. and Emerson, S., The Oxidation State of Manganese in Surface Sediments of the Deep Sea, *Geochim. Cosmohim. Acta,* 1984, vol. 48, no. 5, pp. 897–907.

Kalinenko, V.V., *Geokhimiya i rudonosnost morskikh otlozheniy miotsenovogo vozrasta Cevernogo Kavkaza* (Geochemistry and Signs of Ore in Marine Deposits of Early Miocene Age in the North Caucasus), Moscow: Nauka, 1990.

Kantor, M.I., Genesis of Kerch Iron Ore Deposits, in *Trudy conf. Po genezisu rud zheleza, margantsa, alluminiya* (Proceedings of the Conference on Genesis of Iron, Manganese and Aluminum Ores), Moscow: Izd. Akad. Nauk SSSR, 1937, pp. 119–130.

Karavayeva, E.V., Islamova, M.P., Tverdyslov, V.A. and Khunjua, G.G., Non–Equilibrium Manner of Fractionation of Ions in the Surface Layer of Seawater, Okeanologiya, 1990, vol. 30, no. 2, pp.228–233.

Kassens, H., Bauch, H.A., Dmitrenko, I., Eicken, H., Hubberten, H.-W., Melles, M., Thiede, I. and Timokhov, L., *Land–Ocean Systems in the Siberian Arctic: Dynamics and History,* Berlin: Springer–Verlag, 1999, pp. 189–195.

Kautsky, H., Determination of Distribution Processes, Transport Routes and Transport Times in the North Sea and the Northern North Atlantic using Artificial Radionuclides as Transfers Radionuclides, in *A tool for oceanography,* London, New–York, 1988, pp. 271–280.

Kazakov, A.V., Phosphatic Facies, *Transactions of the NIUIF,* 1939, no. 145.

Kellog, T.B. and Kellog, D.E., Antarctic Cryogenic Sediments: Biotics, and Inorganic Facies of Ice Shelf and Marine–Based Ice Sheet Environments, *Palaeoclimatol., Palaeoecol.,* 1988, vol. 67, pp. 51–74.

Kemp Alan, E.A., and Baldauf, J.G., Vast Neogene Laminated Diatom Mat Deposits from the Eastern Equatorial Pacific Ocean, *Nature,* 1993, vol. 362, pp. 141–144.

Kemper, E. and Zimmerle, W., Facies Patters of a Cretaceous Tertiary Subtropical Upwelling System (Great Syrian desert) and an Aptian Albian Boreal Upwelling System (NW Germany), Coastal Upwelling, in *Sediment Record,* Part B., Suess, E. and Thiede, J., Eds., Plenum Press–New–York–London, 1983, pp. 501–534.

Kennet, Y.P., *Morskaya geologiya* (Marine Geology), Moscow: Mir, 1987, vol. 2. (Russian translation from English, Prentic – Hall, Englewood Cliffs N. J, 07632, 1982).

Kharin, G.S. and Emelyanov, E.M., *Geologiya Atlantiki v Islandskom regione* (Geology of the Atlantic in Iceland Region), Moscow: Result of Res., Intern. Geoph. Projects, 1987.

Kharin, G.S. and Soldatov, A.V., The Lithological Features of the Phosphorites of the Atlantic Shelf of Africa, *Lithol. Mineral Resource.,* 1975, no. 2, pp. 14–22.

Kholodov, V.N. and Nedumov, R.I., Lithology and Geochemistry of Middle Miocene East Predcaucasus, *Trudy Geol. Inst. Akad. Nauk SSSR* (Proc. Geol. Inst. AS USSR), 1981, vol. 358.

Khunjua, G.G., Gusev, A.M. and Andreev et al., On the Structure of Cold Surface Film of the Ocean and Heat Exchange between Ocean Atmosphere, *Izv. Akad. Nauk SSSR,* Physics of the Atmosphere and Ocean, 1977, vol. 13, no. 7, pp.753–758.

Kilps, J.R., Logan, B.E. and Alldredge, A.L., Fracral Dimensions of Marine Snow Determined from Image Analysis of in situ Photographs, *Deep–Sea–Res,* 1994, vol. 41, no. 8, pp. 1159–1169.

Kineke, G.C. and Sternberg, R.W., Distribution of Fluid Muds on the Amazon Continental Shelf, Marine *Geology,* 1995, vol. 125, no. 3/4, pp. 193–234.

Kipp, N.G., New Transfer Function for Estimating Past Sea–Surface Conditions From Seabed Distribution of Planktonic Foraminiferal Assemblages in the North Atlantic, in *Geol. Soc. Am.*

Memoir 145, Cline, R.M. and Hays, J.D., Eds., N.-Y.: Geological Society of America, 1976, pp. 3–41.

Klemas, V., Remote Sensing of Coastal Fronts and their Effects on Oil Dispersion, Int. *J. Remote Sensing,* 1980, vol. 1, no. 1, pp.11–28.

Klenova, M.V., *Geologiya morya* (Geology of the Sea), Moscow: Utchpedgiz, 1948.

Klinkhammer, G.P., Heggie, D.T. and Graham, D.W., Metal Diagenesis in Oxic Marine Elements, *Earth and Planet. Sci. Pett.,* 1982, vol. 61, no. 2, pp. 211–219.

Klinkhammer, G., Rona, P., Greaves, M. and Elderfield, H., Hydrothermal Manganese Plumes in the Mid-Atlantic Ridge Rift Valley, *Nature,* 1985, vol. 314, pp. 727–731.

Klinkhammer, G.P. and Bender, M.L., The Distribution of Manganese in the Pacific Ocean, *Earth and Planetary Science Letters,* 1980, vol. 46, pp. 361–384.

Knauer, D.A. and Martin, D.G., Study of Biological Transport of Substance from Surface Waters to Deep Waters of Ocean, in *Chelovek i biosfera* (A Man and Biosphere), Moscow: State University, 1980, no. 5.

Koblentz-Mishke, O.I., Primary Production, in *Okeanologiya, biologiya okeana* (Oceanology, Biology of the Ocean), Moscow: Nauka, 1977, vol. 1, pp. 62–64.

Koblentz-Mishke, O.I., Photosynthetic in *Pervichnaya produktsiya, Biologiya okeana* (Primary Production, Biological Resources of the Ocean), Moscow: Atomizdat, 1985, pp. 86–90.

Koblentz-Mishke, O.I., Konovalov, B.V., Pavlov, V.M. and Radkhakrishna, K., Primary Production Pigment and Hydrooptical Conditions in the South-Eastern Atlantic in April-June 1968, in Formation of Biological Productivity and Bottom Sediments as Related to Ocean Circulation in the South-Eastern Atlantic, *Proc. of the P.P. Shirsov Institute of Oceanology, AS USSR,* 1973, vol. 95, Kaliningrad Book Publishers, pp. 115–137.

Kohfeld, K. and Fairbanks, R.G., Neogloboquadrina Pachyderma (Sinistral Coiling) as Paleoceanographic Traces in Polar Oceans: Evidence from Northeast Water Polynya Plankton Tows, Sediment Traps, and Surface Sediments, *Paleoceanography,* 1996, vol. 11, no. 6, pp. 679–699.

Kollektory Bazhenovskoy svity Zapadnoy Sibiri (The Oil Collectors of Bazhenov Suite of the Western Siberia), Dorofeeva, T.V., Ed., VNIGNI Mingeo, Leningrad: Nedra, 1983.

Komar, P.D., Morse, A.P., Small, L.F. and Fowler, S.W., An Analysis of Sinking Rates of Natural Copepod and Euphausiids Fecal Pellets, *Limnol. and Oceanography,* 1981, vol. 26 (1), pp. 172–180.

Konyuhov, A.I., *Osadochnye formatsii v zonakh perekhoda ot kontinenta k okeanu* (Sedimentary Formations in the Transition Zones Continent-Ocean), Moscow: Nedra, 1987.

Kornyshkov, I.V. and Smetannikova, I.V., Chemical and Mineral Composition of Ferromanganese Ore Signs in the South-Eastern Atlantic, in *Litosfera Angolskoy kotloviny i vostochnogo sklona Uzhno-Atlanticheskogo khrebta* (Lithosphere of the Angola Basin and the Eastern Slope of the South-Atlantic Ridge), Mingeology of the USSR, "Sevmorgeologiya" Geological corporation, Leningrad, pp. 150–163.

Kos'yan, P.D. and Pyhov, N.V., *Gidrogennye peremeshcheniya osadkov v beregovoy sone morya* (Hydrogenic Displacement of the Sediments in the Near-Shore Zone of the Sea), Moscow: Nauka, 1991.

Krasnov, S.G., German, N.E. and Tcherkashev, G.A., The Distribution and Main Factors Controlling the Formation of Composition of Metalliferous Deposits, in *Gidrotermalnye rudy i metallonosnye osadki okeana* (Hydrothermal Sulphidic Ores and Metalliferous Sediments of the Ocean), St. Petersburg: Nedra, 1992, pp. 129–188.

Krasovskiy, K.S., *Zhelezomargantsevye konkretsii* (Ferromanganese Nodules), in Shnyukov, E.F., Ed., Kiev: Naukova Dumka, 1984, pp. 140–146, part 1B.

Krishnaswami, S., Mangini, A. and Thomes, J.H. et al., Be and Th Isotopes in Manganese Nodules and Adjacent Sediments: Nodule Growth Histories and Nuclide Behavior, *Earth and Planet. Sci. Lett.,* 1982, vol. 59, no. 2, pp. 217–234.

Kuenen, Ph.H., Marine Geology, N.-Y.: John Willy and sons, 1950.

Kunzendorf, H., Walter, P., Stoffers, P., and Gwozdz, R., Metal Variations in Divergent Plate-Boundary Sediments from the Pacific, *Chem. Geol.,* 1984/1985, vol. 47, no. 1/2, pp. 113–133.

Kuptsov, V.M., Geochronological Studies of Bottom Sediments, in *Biogeokhimiya pogranichnykh zon Atlanticheskogo okeana* (Biogeochemistry of the Boundary Zones in the Atlantic Ocean), Romankevich, E.A., Ed., Moscow: Nauka, 1994.

Lalou, C., Brichet, E., Perez-Leclaire, H., The Galapagos Hydrothermal Mounds: History from about 600000 Years to Present, *Oceanol. Acta,* 1984, vol. 7, no. 3, pp. 261–270.

La Ferla, R. and Crisafi, E., Direct Microbiotic Count of Water Samples from the Bannock Basin by Epifluorescence Technique, *Anoxic Basins of the Eastern Mediterranean,* Cita, M.B., Camerlenghi, A. and Corselli, C., Eds., Milano, Ricerca scient. ed educaz. permanente, 1989, Suppl. no. 72, pp. 103–104.

Lappo, S.S., Lozovatskii, I.D., Morozov, E.G., Sokov, A.V. and Shapovalov, S.M., Variability of Water Structure in the Equatorial Atlantic, *Doklady Akad. Nauk SSSR* (Proc. AS USSR), 2001, vol. 379, no. 5, pp. 686–690.

Latun, B.S., The Mechanism of Winter Intensification of the Cyclonic Gyre of the Water, in *Protsessy formirovaniya i godovaya izmenchivost gidrofizicheskikh i gidrokhimicheskikh poley* (Processes of Formation and Annual Changeability of the Hydrophysical and Hydrochemical Fields), Sevastopol: Marine Hydrophys Inst. AS USSR, 1988, pp. 5–16.

Lebedev, V.L., *Granichnye poverkhnosti v okeane* (Border Surface in the Ocean), Moscow: MSU Publ., 1986.

Lee, H. and Schwartz, R.S., Biological Processes Affecting the Distribution of Pollutants in Marine Sediments, "2" Biodeposition and Bioturbation in Contaminants and Sediments, *Am. Arbor (Mich.),* 1980, v. 2, pp. 555–605.

Lein, A.Yu., Authigenic Carbonate Formation in the Ocean, *Lithology and Mineral Resources,* 2004, no. 1, pp. 3–35.

Lein, A.Yu., Vanshtein, M.B. and Kashparova, E.M. et al., Biochemistry of the Anaerobic Diagenesis and Isotopic Balance of the Sulfur and Carbon in the Baltic Sea Sediments, in *Geokhimiya osadochnogo protsessa v Baltiyskom more* (Geochemistry of the Sedimentation Process in the Baltic Sea), Emelyanov, E.M. and Lukashin, V.N., Eds., Moscow: Nauka, 1986, pp. 155–176.

Leinen, M. and Stakes, D., Metal Accumulation Rates in the Central Equatorial Pacific during the Cenozoic, *Geol. Soc. Am. Bull,* 1981.

Leontiyev, O.K., The Bottom Abyssal Currents as a Geomorphological Factor, *Geomorphology,* 1987, no.1, pp. 3–16.

Leontiyev, I.O., *Dinamika priboynoy zony* (Dynamic of the Surface Zone), Moscow: Izd. IO Akad. Nauk SSSR, 1989.

Lerman, A. and Lietzke, T.A., Fluxes in a Growing Sediment Layer, *Amer. J. Sci.,* 1977, vol. 277, no. 1, pp. 25–37.

Letnikov, F.A., Logachev, N.A., Emelyanov, E.M., Kharin, G.S., Kiselev, A.N., Gaptimurova, T.P. and Shkarupa, T.A., Fluid Regime of the Rift Zones, in *Problemy riftogeneza* (Main Problems of the Riftogenesis), Irkutsk: Nauka (Sibirian Dept.), 1977, pp. 51–60.

Liakhin, Yu.I., The Saturation of the Pacific Ocean Water by Calcium Carbonate, *Okeanologiya,* 1968, vol. 8, no. 2, pp. 336–340.

Liakhin, Yu.I., The Saturation of the Atlantic Ocean Water by Calcium Carbonate, *Okeanologiya,* 1972, vol. 12, no. 6, pp. 1010–1019.

Lion, L.Q. and Leckie, J.O., Chemical Speciation of Trace Metals at the Air–Sea Interface the Application of an Equilibrium Model, *Environm.Geol.,* 1981, no. 3, pp.293–314.

Lisitzin, A.P., Distribution and Chemical Composition of Suspension in Waters of the Indian Ocean, *Oceanologic researches,* Moscow: Nauka, 1964, no. 10.

Lisitzin, A.P., Sedimentation Rates in the Oceans, *Okeanologiya,* 1971, vol. 11, no. 6, pp. 975–969.

Lisitzin, A.P., *Protsessy terrigennoy sedimentatsii v moryakh i okeanakh* (The Processes of the Terrigeneous Sedimentation in the Seas and Oceans), Moscow: Nauka, 1971.

Lisitzin, A.P., Sedimentation in World Ocean, Tulsa: Bante Press, 1972, vol. 3.

Lisitzin, A.P., *Osadkoobrazovaniye v okeanakh* (Sedimentogenesis in Oceans), Moscow: Nauka, 1974.

Lisitzin, A.P., Biogenic Sedimentation in the Ocean and Zonality, *Lithol. and Mineral Res.,* 1977, no.1, pp. 3–24.

Lisitzin, A.P., The Terrigeneous Sedimentation, Climatic Zonality and Interaction of the Terrigenous and Biogenous Materials in the Oceans, *Lithol. and Mineral Res.,* 1977_2, no. 6, pp. 3–22.

Lisitzin, A.P., *Protsessy okeanskoy sedimentatsii, Litologiya i geokhimiya* (Processes of Ocean Sedimentation, Lithology and Geochemistry), Moscow: Nauka, 1978, pp. 65–110.

Lisitzin, A.P., The Global Zones of the Sedimentogenesis, in *Uspehi sov. okeanologii* (Successes of the Soviet Oceanology), Moscow: Nauka, 1979, pp.118–136.

Lisitzin, A.P., The Main Concept of the Ocean's Biogeochemistry in *Biokhimiya okeana* (Biochemistry of the Ocean), Monin, A.S. and Lisitzin, A.P., Eds., Moscow: Nauka, 1983_1, pp. 9–32.

Lisitzin, A.P., The Flux of Material and Energy in the Ocean and their Biochemical Significance, in *Biokhimiya okeana* (Biochemistry of the Ocean), Monin, A.S. and Lisitzin, A.P., Eds., Moscow: Nauka, 1983_2, pp. 201–273.

Lisitzin, A.P., Biogenic Sedimentation in the Ocean and Zonality, *Lithol. Mineral Res.*, 1986_1, no. 1, pp. 3–24.

Lisitzin, A.P., Biodifferentiation of the Matter in the Ocean and Sedimentation Process, in *Biodifferentsiatsiya osadochnogo materiala v moryakh i okeanakh* (Biodifferentiation of the Sedimentary Matter in the Seas and Oceans), Lisitzin, A.P., Ed., Rostov University, 1986_2, pp. 3–66.

Lisitzin, A.P., *Lavinnaya sedimentatsiya i pereryvy v osadkonakoplenii v moryakh i okeanakh* (The Avalanche Sedimentation and the Hiatuses in the Sedimentation in the Seas and Oceans), Moscow: Nauka, 1988.

Lisitzin, A.P., *Prosessy terrigennoy sedimentatsii v moryakh i okeanakh* (The Processes of the Terrigeneous Sedimentation in the Seas and Oceans), Moscow: Nauka, 1991.

Lisitzin, A.P., Ed., *Gidrotermalnye obrazovaniya Sredinnogo khrebta Atlanticheskogo okeana* (Hydrothermal Formations on the Mid–Oceanic Ridge of the Atlantic Ocean (TAG field)), Moscow: Nauka, 1992.

Lisitzin, A.P., *Ledovaya sedimentatsiya v Mirovom okeane* (Ice sedimentation in the World Ocean), Moscow: Nauka, 1994.

Lisitzin, A.P., Nonresolved Problems of the Arctic Oceanology, in *Opyt sistemnykh okeanologicheskikh issledovaniy v Arktike* (Experience of System Oceanologic Studies in the Arctic), Lisitzin, A.P., Vinogradov, M.E., Romankevich, E.A., Eds., Moscow: Scientific World, 2001, pp. 31–75.

Lisitzin, A.P., Bogdanov, Yu.A., Emelyanov, E.M., Maksimov, A.N., Pustelnikov, O.S. and Serova, V.V., Suspended Matter in the Atlantic Waters, in *Osadkonakopleniye v Atlanticheskom okeane* (Sedimentation in the Atlantic Ocean), Kaliningrad: Kaliningradskaya Pravda, 1975.

Lisitzin, A.P., Bogdanov, Yu.A. and Murdmaa, I.O. et. al., *Metallonosnye osadki i ikh genezis* (Metaliferous Sediments and their Genesis), Smirnov, V.I., Ed., Moscow: Nauka, 1976.

Lisitzin, A.P., Bogdanov, Yu.A., Zonenshain, A.P., Kuzmin, M.I. and Sagalevich, A.M., Hydrothermal Manifestations on the Mid–Atlantic Ridge on 26°N (Hydrothermal Field TAG), *Izv. Akad. Nauk SSSR,* Geol. Ser., 1989, no. 12, pp. 3–20.

Lisitzin, A.P., Bogdanov, Yu.A. and Vorobyov, P.V. et. al., *Gidrotermalnye sistemy i osadochnye formatsii Sredinno–okeanicheskikh khrebtov Atlantiki* (Hydrothermal Systems and Sedimentary Formations of the Atlantic Mid–Oceanic Ridges), Moscow: Nauka, 1993.

Lisitzin, A.P., Bogdanov, Yu.A. and Gurvich, E.G., *Gidrotermalnye obrazovaniya riftovykh zon okeana* (Hydrothermal Formations in the Rift Valleys of the Ocean), Moscow: Nauka, 1990.

Lisitzin, A.P., Gordeev, V.V. and Bogdanov, Yu.A., Geochemistry of the Metalliferous Sediments of the Indian Ocean, in *Metallonosnye osadki v Indiyskom okeane* (Metalliferous Sediments of the Indian Ocean), Moscow: Nauka, 1987, pp. 100–127.

Lisitzin, A.P., Emelyanov, E.M. and Eltsina, G.N., Geochemistry of the Sediments of the Atlantic Ocean, Carbonates and Silica, *Results Res. Intern. Geoph. Projects,* Moscow: Nauka, 1977.

Lisitzin, A.P. and Emelyanov, E.M., Geochemical Point of Oceanic, Marine Geochemical Barriers and Barrier Zones, their Classification and Role in Sedimentogenesis and Ore Formation, in Geology of Oceans and Seas, *Tezisy dokladov 6–oy vsesouznoy shkoly morskoy geologii* (Abstract of the 6th All–Union Conference on Marine Geology), 1984, vol. 1, Moscow: pp. 220–222.

Lisitzin, A.P. and Vinogradov, M.E., Global Regularities of the Distribution of the Life in the Ocean and their Reflection in the Bottom Sediments, *Izv. Akad. Nauk SSSR,* Geol. ser., 1982, no. 4, pp. 5–24.

Lisitzin, A.P. and Vinogradov, M.E., Global Regularities of the Distribution of the Life in the Ocean and Biochemistry of Suspended Matter and Bottom Sediments, in *Biokhimiya okeana* (Biochemistry of the ocean), Monin, A.S. and Lisitzin, A.P., Eds., Moscow: Nauka, 1983, pp. 112–127.

Logvinenko, N.V., Volkov, I.I. and Sokolova, E.G., Rhodochrosite in the Deep Sediments of the Pacific Ocean, *Doklady Akad. Nauk SSSR* (Proc. AS USSR), 1972, vol. 203, no. 1, pp. 204–207.

Logvinenko, N.V., Diagenesis of the Carbonate Sediments, *Okeanologiya, Khimiya okeana* (Oceanology, Chemistry of the Ocean), 1979, vol. 2, Moscow: Nauka, pp. 350–362.

Lohman, H., The Coccolithophridae, *Arch. Protistenk.,* 1902, vol. 2.

Longinov, V.V., *Ocherki litodinamiki okeana* (Essays of the Lithodynamics of the Ocean), Moscow: Nauka, 1973.

Lonsdale, P.F., Clustering of Suspension–Feeding Macrobenthos near Abyssal Hydrothermal Vents at Oceanic Spreading Centers, *Deep–Sea Res.*, 1977, vol. 24, pp. 857–863.

Lonsdale, P.F. and Malfait, B., Abyssal Dunes of Foraminiferal Sand on the Carnegie Ridge, *Bull. Geol. Soc. AM.*, 1974, vol. 85, pp. 1697–1712.

Lonsdale, P. and Southard, J.B., Experimental Erosion of North Pacific Red Clay, *Marin. Geol.*, 1974, vol. 17, pp. M51–M60.

Lukashin, V.N., Trace Elements in the Bottom Sediments of the Baltic Sea, in *Geokhimiya osadochnogo protsessa v Baltiyskom more* (Geochemistry of the Sedimentation in the Baltic Sea), Moscow: Nauka, 1986, pp. 194–201

Lukashin, V.N., Ivanov, G.V., Isaeva, A.B. and Michaylov, S.V., On Geochemistry of Bottom Sediments, in *Biogeokhimiya pogranichnykh zon Atlanticheskogo okeana* (Biogeochemistry of the Boundary Zones in the Atlantic Ocean), Romankevich, E.A., Ed., Moscow: Nauka, 1994.

Lukashin, V.N., Shevchenko, V.P., Romankevich, E.A., Arashkevich, E.A., Borodkin, S.O., Korneeva, G.A., Oskina, N.S. and Pimenov, N.B., The Sedimentary Matter Fluxes, in *Biogeokhimiya pogranichnykh zon Atlanticheskogo okeana* (Biogeochemistry of the Boundary Zones in the Atlantic Ocean), Romankevich, E.A., Ed., Moscow: Nauka, 1994.

Lukashina, N.P., Communities of Benthic Foraminifera and Water Masses of the Northern Atlantic and Norwegian–Greenland Basin, Okeanologiya, 1988, vol. 28, no. 5, pp. 790–796.

Lutz, R.A., Shank, T.M., Fornari, D.J., Haymon, R.M., Lilley, M.D., Von Damm and Desbrueeres, D., Rapid Growth at Deep–Sea Vents, *Nature,* 1994, vol. 371, pp. 663–664.

Lykyanova, T.S., The Estimation of Summary Reserves of Benthic Fauna in the World Ocean, *Cand. Sci. Dissertation,* Moscow: MGU, 1975.

Lyle, M.L., *The Formation and Growing of Ferromanganese Oxides on the Nazca Plate,* Thes., P.D., Corvalis (Oreg), Oregon State Univ., 1978.

Lynn, D.C. and Bonatti, E., Mobility of Manganese in the Diagenesis of Deep–Sea Sediments, *Mar. Geol.*, 1965, vol. 3, no. 6, pp. 457–474.

Mac Intyre, F., *Ph. Sci. Dis.*, M.I.T., 1965.

Mackenzie, F.T., Stoffyn, M. and Wollast, R., Aluminum in Seawater: Control by Biological Activity, *Science,* 1978, vol. 199.

Malfait, B.T. and Van Andel, T.H., A Modern Oceanic Hard Ground on the Carnegie Ridge in the Oceanic Equatorial Pacific, *Sedimentology,* 1980, vol. 27, pp. 467–496.

Mamaeva, M.E., Microzooplankton of the Open Part of the Black Sea, *Ekosistemy pelagiali morya* (Ecosystems of Pelagic Areas of the Black Sea), Vinogradov, M.E., Ed., Moscow: Nauka, 1980, pp. 168–173.

Mamaeva, N.V., Infusoria in the Deep Areas of the Baltic in May–June, 1984, in *Ekosistemy Baltiki v maye–iune* (The Baltic Ecosystems in May–June), Koblentz–Mishke, O.J. and Beliaev, G.A., Eds., Institute of Oceanology RAS, Moscow: 1987, pp. 152–160.

Manheim, F.T., Geochemical Profile in the Baltic Sea, *Geochim. Cosmochim. Acta,* 1961, vol. 25, no. 1, pp. 52–70.

Manheim, F.T., Interstitial Waters of Marine Sediments, in *Chemical Oceanography,* Riley, J.P., Chester, R., Eds., London: Academic Press, 1976, 2nd ed., pp. 115–185.

Manheim, F.T., Marine Cobalt Resources, *Science,* 1986, vol. 232, no. 4750, pp. 606–608.

Manheim, F.T., Pratt, R.M. and McFarlin, Composition and Origin of Phosphorite Deposits of the Blake Plateau, in *Marine Phosphorites,* Bentor, Y.K., Ed., Spec. Publ. Soc. Econ. Paleont. Miner., 29, Tulsa, 1980, pp. 117–137.

Marchese, P.J. and Gordon, A.L., The Eastern Boundary of the Gulf Stream Recirculation, *J. of Marine Research.*, 1996, vol. 54, pp. 521–540.

Marine Geochemistry, Schulz, H.D. and Zabel, M., Eds., Springer, 2000.

Martin, L., Etude des “faecal–pellets” Mineralises des Sediments du Plateau Continental de Cote d’Ivoire, *Cah. ORSTOM. Geol.*, 1972, vol. 4, no. 2, pp. 105–120.

Martin, J.M. and Meybeck, M., *The Content of Major Elements in the Dissolved and Particulate Load of Rivers, Biogeochemistry of Estuarine Sediments,* Goldberg, E.D., Ed., Paris: UNESCO, 1978, pp. 95–110.

Martin, J.M. and Meyback, M., Elemental Mass–Balance of Material Carried by Major World Rivers, *Mar. Chem.*, 1979, vol. 7, no. 2, pp. 173–206.

Matishov, G.G., *Dno okeana v lednikoviy period* (The Ocean Bottom in Glacial Period), Leningrad: Nauka, 1984.

Matishov, G.G., *Mirovoy okean i oledeneniye Zemli* (The World Ocean and Glaciation of the Earth), Moscow: Nauka, 1987.
Matthäus, W., Natural Variability and Human Impacts Reflected in Long–Term Changes in the Baltic Deep Water Conditions, *A Brief Review,* 1995, vol. 4, no. 1.
Matul, A.G., Yushina, I.G. and Emelyanov, E.M., On the Late Quaternary Paleohydrological Parameters of the Labrador Sea Based on Radiolarians, Okeanologiya, 2001, vol. 42, no. 2, pp. 262–266.
McCave, I.N. and Tucholke, B.E., *Deep Current–Controlled Sedimentation in the Western North Atlantic Region,* Wash. (DC): Geol. Soc. of AM, 1986, pp. 451–468.
McGeary, D.F.R. and Damuth, J.E., Postglacial Iron–Rich Crusts in Hemipelagic Deep–Sea Sediments, *Bull. Geol. Soc. Amer.,* 1973, vol. 84, pp. 1201–1272.
McIntyre, F., The Top Millimeter of the ocean, *Science,* 1974, vol. 230, no. 5, pp. 62–77.
McIntyre, F., The Absorption of Fossile Fuel CO_2 by the Ocean, *Oceanol. Acta,* 1980, vol. 3, no. 4, pp. 505–516.
McIntyre, F., Alldredge, A.L. and Gotschalk, C.C., Accumulation of Marine Snow at Density Discontinuities in the Water Column, *Limnol. Oceanogr.,* 1995, vol. 40(3), pp. 449–468.
Meade, R.H., Nordin, C.F. and Curtis, W.F., Sediment in Rio Amazonas and Some of its Principle Tributaries During the High Water Seasons of 1976 and 1977, in *Assos. Brasil. Hydrol. et Recur. Hydrocos lll Simp. Brasil. Hydrol.: Hydrol. Amazon. Anais,* 1979, vol. 2. pp. 472–485.
Medinets, V.I., *Issledovaniya ekosistemy morya* (Researches of the Black Sea Ecosystem), in Collective Papers, Odessa: "Iron–poligraf", 1994, vol. 1.
Melnikov, I.A., *Ekosistema morskogo lda Arktiki* (The Ecosystem of the Arctic Sea Ice), Moscow: IOAN USSR, 1989.
Melnikov, M.E., Zadornov, M.M., Hershberg, L.B. and Mechetin, A.V., Ore Controls, Prospecting Premises and Prognostic Appraisal of Cobalt–Manganese Metallization of Guyots at Markus–Wake Rise and Magellan Mountains, in *Gayoty zapadnoy Patsifiki i ikh rudonosnost* (Guyots in the Western Pacific and their Mineral Reserves), Govorov, I.N. and Baturin, G.N., Eds., Moscow: Nauka, 1995, pp. 337–346.
Menard, H.W., Small Ocean Basins and Continental Growth, *Trans. Amer. Geophys. Union,* 1966, vol. 47, no. 1.
Menzel, D.W., Particulate Organic Carbon in the Deep Sea, *Deep–Sea Res.,* 1967, vol. 14, no. 2.
Menzel, D.W. and Ryther, J.H., Organic Carbon and the Oxygen Minimum in the Souh Atlantic Ocean, *Deep–Sea Res.,* 1968, vol. 15, no. 3.
Mero, J.L, *The Mineral Resources of the Sea,* Amsterdam etc.: Elsevier, 1965.
Mertz, W., The Essential Trace Elements, *Science,* 1981, vol. 213, pp. 1332–1338.
Metallonosnye osadki severo–vostochnoy chasti Tikhogo okeana (Metalliferous Sediments in the S–E Part of the Pacific Ocean), Smirnov, V.I., Ed., Moscow: Nauka, 1981.
Michard, G., L'action de mer les Basaltes, Source Possible de Manganese: Etude Thermodynamique Preliminaire, *C. r. Acad. sci.,* 1975, vol. D 280, no. 103, pp. 1213–1216.
Miklishansky, A.Z., Abiogenic System of the Atmosphere, in *Biogeokhimiya okeana* (Biogeochemistry of the Ocean), Monin, A.S. and Lisitzin, A.P., Eds., Moscow: Nauka, 1983, pp. 72–86.
Miklishansky, A.Z., Biosolid System of the Atmosphere, *Biogeokhimiya okeana* (Biogeochemistry of the Ocean), Monin, A.S. and Lisitzin, A.P., Eds., Moscow: Nauka, 1983, pp. 60–89.
Miliman, J.D., Carbonate Lithification in the Deep Sea, in Carbonate Cements, Bricker, P.O., Ed., *Studies in Geology,* 1971, vol. 19.
Miller, A.R., Densmore, C.D. and Degens, E.T. et al., Hot Brines and Recent Iron Deposits in Deeps in the Red Sea, *Geochem. Cosmochim. Acta,* 1966, vol. 30, pp. 341–359.
Milliman, J.D. and Müller, J., Precipitation and Lithification of Magnesian Calcite in the Deep–Sea Sediments of the Eastern Mediterranean Sea, *Sedimentology,* 1973, vol. 20, no. 1, pp. 29–46.
Mineralnye resursy okeana, Moscow: Progress, 1969. Translated under the title Mero, J.L., *The Mineral Resources of the Sea,* Amssterdam–London–New–York: Elsevier, 1965.
Mitropolskiy, A.Yu., Bezborodov, A.A., Ovsianyi, E.I., *Geokhimiya Chornogo morya* (Geochemistry of the Black Sea), Kiev: Naukova Dumka, 1982.
Moiseev, E.V., Zooflagellates of the Open Part of the Black Sea, in *Ekosistemy pelagiali Chornogo morya* (Ecosystems of the Pelagic Area of the Black Sea), Vinogradov, M.E., Ed., Moscow: Nauka, 1980, pp. 174–178.
Mokievskaya, V.V., Manganese in the Black Sea Water, *Doklady Akad. Nauk SSSR* (Proc. AS USSR), 1961, vol. 137, no. 6, pp. 1445–1447.

Molinari, R.L., Fine, R.A. and Johns, E., The Deep Western Boundary Current in the Tropical North Atlantic Ocean, *Deep-Sea Res.,* 1992, vol. 39, 1967–1984.

Monin, A.S. and Ozmidov, R.V., About the Bboundary Layers above the Seamounts, *Doklady Akad. Nauk SSSR* (Proc. AS USSR), 1986, vol. 287, no. 6, pp. 1470–1473.

Monin, A.S. and Lisitzin, A.P., Eds., *Biogeokhimiya okeana* (Biogeochemistry of the Ocean), Moscow: Science Publisher, 1983.

Monin, A.S., Gordeev, V.V. and Kopelevich, O.V. et al., The Regularities of the Distribution and Transformation of the Amazon Water in Nearly Areas of the Atlantic ocean, *Preprint of IORAN, Moscow,* 1986.

Monin, A.S. and Lisitzin, A.P., Eds., *Okeanologiya, Geologiya okeana, Geologicheskaya istoriya okeana* (Oceanology, Geology of the Ocean, Geological History of the Ocean), Moscow: Nauka, 1980.

Monin, A.S., Romankevich, E.A., Problems of the Biochemistry of the World Ocean, in *Sovremennye zadachi i problemy biogeokhimii* (Recent Tasks and Problems of the Biochemistry), Moscow: Nauka, 1979, pp. 74–83.

Monin, A.S. and Gordeev, V.V., *Amazonia,* Moscow: Nauka, 1988.

Morozov, N.P., Chemical Elements in the Hydrobionts and Food Chains, in *Biokhimiya okeana* (Biochemistry of the ocean), Monin, A.S. and Lisitzin, A.P., Eds., Moscow: Nauka, 1983, pp. 127–167.

Mottl, M.J. and Seyfried, W.E. Jr., Sub-Seafloor Hydrothermal Systems: Rock-vs. Seawater Dominanted, Seafloor Spreading Centers Hydrothermal Systems, *Benchmark Papersin Geology,* 1980, vol. 56. pp. 66–82.

Müller, J. and Fabricius, F., Magnesian Calcite Nodules in the Ionian Deep Sea, An Actualistic Model for the Formation of Some Nodular Limestones, in *Pelagic Sediments: Land and Sea,* Oxford: Blackwell Scientific Publications, 1971, pp. 235–247.

Murdmaa, I.O., The Ocean Facies, in *Okeanologiya, Geologiya okeana, Osadkoobrazovaniye i magmatizm* (Oceanology, Geology of the Ocean, Sedimentation and Magmatizm), Moscow: Nauka, 1979, pp. 269–306.

Murdmaa, I.O., *Fatsii okeana* (The Ocean Facies), Moscow: Nauka, 1987.

Murdmaa, I.O., Oceanic Pelagic Lithogenesis, *Lithology and Miner. Res.,* 1991, no. 5, pp. 3–18.

Murdmaa, I.O., Gordeev, V.V., Bazilevskaya, E.S. and Emelyanov, E.M., Inorganic Geochemistry of the Leg 44 Sediments, *Init. Rep. DSDP,* Washington: U. S. Govern. Print. Off., 1978, vol. 43, pp. 575–583.

Murray, J. and Renard, A.F., Deep-Sea Deposits, *Reports on the Scientific Results of the Voyage of H.M.S."Challenger",* London, 1891.

Murray, J. and Chumley, J., The Deep-Sea Deposits of the Atlantic Ocean, *Trans. Roy. Soc. Edin.,* 1924, vol. 54, part I.

Murray, J.W., The Interaction of Metalions at the Manganese Dioxide-Solution Interface, *Geochim. Cosmochim. Acta,* 1975, vol. 39, pp. 505–519.

Neshyba, S., *Oceanography,* N.-Y.: John Wiley Sons, 1991.

Nevesskiy, E.M., *Protsessy osadkoobrazovaniya v pribrezhnoy zone morya* (Sedimentation Processes in the Near-Shore Sea Zone), Moscow: Nauka, 1967.

Nikolaev, S.D. and Oskina, H.S., Lithology of Bottom Sediments, in *Biogeokhimiya pogranichnykh zon Atlanticheskogo okeana* (Biogeochemistry of the Boundary Zones in the Atlantic Ocean), Romankevich, E.A., Ed., Moscow: Nauka, 1994.

Nikolayeva, I.V., Senin, Y.M. and Golubeva, G.A., Facial Variability of Authigenic Silicates in Connection with Particularities of Sediment Deposition on Shelf of the Eastern Africa, in *Glaukonit v sovremennykh nizhnepaleozoyskikh i dokembriyskikh otlozheniyakh* (Glauconite in the Recent, Low-Paleozoic and pre-Cambrian Deposits), Moscow: Nauka, 1971, pp. 7–51.

Noorany, J., Engineering Properties of Submarine Clays from the Pacific, *Proceed. First Internat. Conf. on Port and Ocean Engunder Arctic Cond.,* Trondheim, 1972.

Ocean Dumping of Chemical Munition: Environmental Effects in Arctic Seas, MEDEA, USA, 1997.

Odin, G.S., Ed., Green Marine Clays, Amsterdam, Oxford, New-York, Tokyo: Elsevier, 1988.

Odin, G.S. and Sen Gupta, B.K., Geological Significance of the Verdine Facies, in *Green Marine Clays,* Odin, S., Ed., Amsterdam-Oxford-New-York-Tokyo: Elsevier, 1988, pp. 205–221.

Odin, G.S., Fullagar, P.D., Geological Significance of the Glaucony Facies, in *Green Marine Clays,* Odin, S., Ed., Amsterdam-Oxford-New-York-Tokyo: Elsevier, 1988, pp. 295–332.

Okeanologiya. Geologiya okeana. Osadkoobrazovaniye i magmatizm okeana (Oceanology. Geology of the Ocean. Sedimentation and Magmatism), Bezrukov, P.L., Ed., Moscow: Nauka, 1979.

Olausson, J., Volcanic Influence on Sea Water at Heimaey, *Nature,* 1975, vol. 255, no. 5504, pp. 138–141.
Ottman, F. and Urien, C.M., *Observaciones Preliminaries Sobre la Distribucion de los Sedimentos en la Zona Exterma del Rio la Plata,* Academia Brasileira de Ciencias, Rio de Janeiro, 1965.
Packer, T., *Survey of Foreign Development Activities for Offshore Nonfuel Mineral Resources,* Part 2, Ottawa: Dept of Energy, Mines and Resources, 1988.
Paffenhöfer, G.A. and Knowles, S.C., Ecological Implications of Fecal Pellet Ooze, Production and Consumption by Copepods, *J. Mar Res.,* 1979, vol. 37, no. 1, pp. 3–19.
Paka, V.T. and Kravtsov, V.A., New Information about the Sunked Chemical Munition, in *Materially mezhdunarodnoy nauchno–prakticheskoy konferentsii "Ekologiya i pravo: zadachi organov gosudarstvennoy vlasti"* (Paper of the International Scientific–Practical Conference "Ecology and the Law: Tasks of Office of the State Authority"), Kaliningrad, KYuI MVD RF, 2003, pp. 95–99.
Paka, V.T. and Spiridonov, M., Research of the Dumped Chemical Weapons Made by R/V "Professor Shtokman" in the Gotland, Bornholm and Skagerrak Dump Sites, in *Chemical Munition Dump Sites in Coastal Environments,* Missiaen, T. and Henriet, J.P., Eds., Renard Centre of Marine Geology, University of Gent, Belgium, 1979, pp. 27–42.
Panchenko, N.A., The Washout within the Ore Stratum of the Nicopol Manganese Deposit, in *Geologiya i geikhimiya margantsa* (Geology and Geochemistry of Manganese), Moscow, 1982, pp. 151–154.
Parker, F.L. and Berger, W.H., Faunal and Solution Patterns of Foraminifera in Surface Sediments of the South Pacific, *Deep–Sea Res.,* 1971, vol. 18, pp. 73–107.
Patin, S.A., *Vliyaniye zagryazneniya na biologicheskiye resursy i produktivnost Mirovogo okeana* (The Influence of the Pollution on the Biological Resources and the Production of the World Ocean), Moscow: Pischevaya promyshlennost, 1979.
Patin, S.A., *Neft i ekologiya kontinentalnogo shelfa* (Oil and Continental Shelf Ecology), Moscow: VNIRO Pabl., 2001.
Patyk–Kara, N.G. and Ivanova, A.M., *Geokhimicheskiye poiski mestorozhdeniy tvyordykh poleznykh iskopayemykh na kontinentalnom shelfe* (Geochemical Prospecting for Solid Mineral Deposits on the Shelf), Moscow: Scientific World, 2003.
Pedersen, T.F. and Price, N.B., The Geochemistry of Manganese Carbonate in Panama Basin Sediments, *Geochim. Cosmochim. Acta.,* 1982, vol. 46, pp. 59–68.
Pellenbarg, R.E. and Church, J.M., The Estuarine Surface Microlayer and Trace Metal Cycling in a Salt Marsh, *Science,* 1979, vol. 203, pp. 1010–1012.
Perelman, A.I., *Geokhimiya landshafta* (Geochemistry of the Landscape), Moscow: Visshaya Shkola, 1966.
Perelman, A.I., *Geokhimiya* (Geochemistry), Moscow: Visshaya Shkola, 1979.
Perelman, A.I., *Geokhimiya* (Geochemistry), Moscow: Visshaya Shkola, 1989.
Petipa, T.S, Pavlova, E.V. and Mironov, G.N., Structure of Food Chains, Transfer and Use of Substances and Energy in Planktonic Communities of the Black sea, in *Biologiya morya* (Biology of Sea), Kiev: Naukova Dumka, 1970, no. 19, p.11–15
Pia, J., *Die Rezenten Kalkasteine,* Leipzig: Akad. Verl., 1970.
Pilipchuk, M.F., Theoretical Principles, Methods and Practice of the Exploring and Exploitation of the Deep–Sea Deposits of the Polymetal Nodules in the World Ocean, *Extended Abstract of Doctoral Dissertation,* JUZH–MORGEOLOGIA, Gelendzhik, 2003.
Piper, D.Z., Rare Earth Elements in Ferromanganese Nodules and other Marine Phases, Geochim. *Cosmochim. Acta.,* 1974, vol. 38, no. 71, pp. 1007–1022.
Piper, D.Z. and Fowler, B., New Constraint on the Maintenance of Mn Nodules at the Sediment Surface, *Nature,* 1974, vol. 286, no. 5776, pp. 880–883.
Piper, D.Z. and Williamson, M.E., Composition of Pacific Ocean Ferromanganese Nodules, *Mar. Geol.,* 1974, vol. 23, no. 4, pp. 285–303.
Piper, D.Z. and Williamson, M.E., Mineralogy and Composition of Concentric Layers within a Manganese Nodule from the North Pacific Ocean, *Mar. Geol.,* 1981, vol. 40, no. 3/4, pp. 255–268.
Pokras, E.M. and Molfino, B., Oceanographic Control of Diatom Abundances and Speciaes Distributions in Surface Sediments of the Tropical and Southeast Atlantic, *Marine Micropaleontology,* 1986, vol. 10, pp. 165–188.
Polynov, B.B., Selected Works, Moscow: USSR Publ., 1986.
Porrenga, D.H., Glauconite and Chamosite as Depth Indicators in the Marine Environment, *Mar. Geol.,* 1967, vol. 5, no. 5/6, pp. 495–503.

Postma, H., Sediment Transport and Sedimentation in the Estuarine Environment, in *Estuaries,* Lauff, G.H., Ed., AAAS, Washington: D.C. Publ., 1967, vol. 83, pp. 158–179.

Prospero, J.M., Eolian Transport to the World Ocean, in *The Sea,* Emeliani, C., Ed., Wiley, 1981, vol. 7, pp. 801–874.

Prospero, J.M., Uematsu M. and Savoie, P.L., Mineral Aerosol Transport to the Pacific Ocean, *Chemical Oceanography,* 1989, vol. 10, pp.188–218.

Prospero, J.M. Mineral and Sea Salt Aerosol Concentration in Various Ocean Regions, *J. Geophys. Res.,* 1979, vol. 84, no. C2, pp. 725–731.

Pushkina, Z.V., Interstitial Waters of the Bottom Sediments on the Transoceanic Profile and its Changes in the Diagenesis, in *Geokhimiya diageneza osadkov Tikhogo okeana (Transokeanskiy profil)* (Geochemistry of the Sediments of the Pacific Ocean (Transoceanic Profile)), Moscow: Nauka, 1980, pp. 70–98.

Pustelnikov, O.S., Delivery of Particulate Material from the Curonian Lagoon to the Baltic Sea, in *Geographia Lituanica,* Vilnius, 1976, pp.123–129.

Pytkowicz, R.M., Calcium Carbonate Saturation in the Ocean, *Limnol. and Oceanogr.,* 1965, vol. 10, no. 2, pp. 220–225.

Pytkowicz, R.M., On the Carbonate Compensation Depth in the Pacific Ocean, *Geochim. Cosmochim. Acta,* 1970, vol. 34, no. 7, pp. 836–839.

Rabiti, S. and Boldrin, A., Vertical Structure of Temperature, Conductivity and Transmittance in Eastern Mediterranean Anoxic Basins, in *Anoxic Basins of the Eastern Mediterranean,* Cita, M.B., Camerlenghi, A. and Corselli, C., Eds., Milano, Ricerca scient. ed educaz. permanente, 1989, Suppl. no. 72, pp. 67–69.

Rad, U. von, Great Meteor and Josephine Sea–Mounts (Eastern North Atlantic): Composition and Origin of Bioclastic Sands, Carbonate and Pyroclastic Rocks, *«Meteor» Forschungsergebnisse,* 1974, vol. 19, pp. 1–61.

Rajan, S., Mackenzie, F.T. and Glenn, C.R., A Thermodynamic Model for Water Column Precipitation of Siderite in the Plio–Pleistocene Black Sea, *American Journal of Science,* 1996, vol. 296, pp. 506–548.

Rancher, J., Etude du cycle, du chlore, du brome et de l'iode dans l'atmosphere an–dessus de l'ocean, *J. Reach.Oceanogr.,* 1979, vol. 4c, no. 3, p. 3.

Rateev, M.A., Emelyanov, E.M. and Kheirov, M.B., Peculiarities of Formation of Clay Minerals in Contemporaneous Sediments of the Mediterranean Sea, *Litol. Mineral. Res.,* 1966, no. 4, pp. 6–23.

Reeburgh, W.S., Methane Consumption in Cariaco Trench Waters and Sediments, *Earth Planet. Sci. Lett.,* 1976, vol. 28, pp. 337–344.

Reeburgh, W.S., A Major Sink and Flux Control for Methane in Marine Sediments: Anaerobic Consumption, in *The Dynamic Environment of the Ocean Floor,* Fanning, K.A. and Manheim, F.T., Eds., Toronto: Lexington Books, 1982, pp. 203–208.

Revelle, R.R., Scientific Results of Cruise VII of the Carnegie during 1928–1929 under Command of Captain J.P. Ault, in *Marine Bottom Samples Collected in the Pacific Ocean by the Carnegie on its Seventh Cruise: Carnegie Inst.,* Washington Pub., 1944, vol. 556, pp. 1–180.

Rhein, M. and Stramma, L., Send, U., The Atlantic Deep Western Boundary Current: Water Masses and Transport near the Equator, *J. Geoph. Res.* 1995, vol. 100, pp. 2441–2457.

Richards, F.A., Anoxic Basins and Fjords, in *Chemical Oceanography,* London: Academic Press, 1965, pp. 611–646.

Richards, F.A., The Cariaco Basin, *Oceanogr. and Marine Vill. Annu.Rev.,* 1975, vol. 18.

Richards, F.A. and Vaccuro, R.F., The Cariaco Trench, an Anaerobic Bin the Caribbean Sea, *Deep–Sea Res.,* 1956, vol. 3, no. 3, p. 214.

Riedle, R.J. and Ott, J.A., Water Movement through Porous Sediments, in *The Dynamic Environment of the Ocean Floor,* Fanning, K.A. and Manheim, F.T., Eds., Toronto: Lexington Books, 1982, pp. 29–57.

Riley, G.A., Organic Aggregates in Seawater and Dynamics of their Formation and Utilization, *Limnol. and Oceanogr,* 1963, vol. 8, pp. 372–381.

Romankevich, B.A., *Geokhimiya organicheskogo veshchestva v okeane* (Geochemistry of Organic Substance in the Ocean), Moscow: Nauka, 1977.

Romankevich, E.A., Ed., *Biogeokhimiya pogranichnykh zon Atlanticheskogo okeana* (Biogeochemistry of the Boundary Zones in the Atlantic Ocean), MOSCOW: Nauka, 1994.

Romankevich, E.N., Emelyanov, E.M. and Bobyleva, N.V., Chemical Composition of Bottom Sediments in Areas of Deposition of Particulate Matter from Fluid Flows in the Baltic Sea, in *Collected Papers,*

Geoacoustic and Gasolithogeochemical Studies in the Baltic sea, Moscow: IOAN USSR, 1990, pp. 141–147.

Romankevich, E.A., Lisitzin, A.P. and Vinogradov, M.E.(eds.) *Pechorskoye more, Sistemnye issledovaniya* (The Pechora Sea: Integrated Research), Moscow: Publ. H. "More", 2003.

Romankevich, E.A., Vetrov, A.A., Peresipkin, V.I., The Organic Carbon in the Bottom Sediments, in *Biogeokhimiya pogranichnykh zon Atlanticheskogo okeana* (Biogeochemistry of the Boundary Zones in the Atlantic Ocean), Romankevich, E.A., Ed., Moscow: Nauka, 1994.

Rona, P.A., Hydrothermal Mineralization at Seafloor Spreading Centers, *Earth Science Review,* 1984, vol. 20, no. 1, pp. 1–104.

Rona, P., Hydrothermal Mineralization in the Ocean Spreading Area, Moscow: Mir, 1986 (Russian translation).

Rona, P.A., Boström, K. and Stanley, D.J., Iron–Cemented Sediment on Lower Continental Slope off Cape Hatteras, *Geo–Marine Letters,* 1982, vol. 2, pp. 89–94.

Rona, P.A., McGregor, B.A. and Betzer, P.R. et al., Anomalous Water Temperatures over the Mid–Atlantic Ridge Crest at 26°N Latitude, *Deep–Sea Res.,* 1975, vol. 22, pp. 611–618.

Ronov, A.B. and Yaroshevsky, Chemical Structure of the Earth Crust, *Geochemistry,* 1967, no. 11, pp. 1285–1309.

Rossi, P.L., Doodu, G., Lucchini, P. and Valera, R., A Manganese Deposit from the Tyrrhenian Region, *Oceanologica Acta,* 1980, no. 3, pp. 107–113.

Rozanov, A.G., Physiochemical Eh Barriers and Their Role in the Redistribution of Sedimentary Material in Seas and Oceans, in Geology of Oceans and Seas, *Tezisy dokladov 6–oy vsesouznoy shkoly morskoy geologii* (Abstract of the 6th All–Union Conference on Marine Geology), 1984, vol. 1, pp. 224–225.

Rozanov, A.G., Oxidation–Reducing Processes in Marine Sediments and Methods of their Studies, in *Khimicheskiy analiz morskikh osadkov* (Chemical Analysis of Marine Sediments), Moscow: Nauka, 1988, pp. 5–44.

Rozanov, A.G., Redox Stratification of Black Sea Waters, *Okeanologiya,* 1995, vol. 35, no. 4, pp. 544–549.

Rozanov, A.G. and Volkov, I.I., Manganese in the Black Sea, in *Mezhdistsiplinarnye issledovaniya severo–vostochnoy chasti Chornogo morya* (Multidisciplinary Investigations of the Northeast Part of the Black Sea), Zatsepin, A.G. and Flint, M.V., Eds., Moscow: Nauka, 2002, pp. 190–200.

Rozanov, A.G., Volkov, I.I. and Sokolov, V.S. et al., Redox Processes in Sediments of the Gulf of California and in Adjacent Area of the Ocean (Compounds of Iron and Manganese), in *Biokhimiya diageneza osadkov okeana* (Biochemistry of Diagenesis of Oceanic Sediments), Volkov, I.I., Ed., Moscow: Nauka, 1976, pp. 136–170.

Ruddiman, W.F., Late Quaternary Deposition of Ice Rafted Sand in the Subpolar North Atlantic (Lat. 40 to 65 N), *Bull. Geol. Soc. Amer.,* 1977, vol. 88, pp. 1813–1827.

Ruddiman, W.F., Heezen, B.C., Differential Solution of Planktonic Foraminifera, *Deep–Sea Res.,* 1967, vol. 14, no. 6, pp. 801–808.

Saidova, H.M., The Recent Biocenosis of the Benthic Foraminifera, Stratigraphy and Paleoceanography of the Baltic Sea Holocene, in *Osadkonakopleniye v Baltiyskom more* (Sedimentation in the Baltic sea), Lisitzin, A.P. and Emelyanov, E.M., Eds., Moscow: Nauka, 1981, pp. 215–231.

Sapozhnikov, V.V., Reserve of Phosphorus in Euphotic Layer of the Pacific Ocean, *Okeanologiya,* 1981, vol. 21, no. 4, pp. 639–644.

Sapozhnikov, V.V., Biohydrochemical Barrier on the Boundary of Shelf Waters in the Black Sea, *Okeanologiya,* 1991, vol. 31, no. 4, pp. 577–584.

Sapozhnikov, V.V., Rudyakov, Y.A. and Agatova, A.I., Regeneration of Biogenic Elements as a Result of Decomposition of Mesoplankton, in *Biologiya, fizika i khimiya frontalnykh zon ugo–vostochnoy chacti Tikhogo okeana* (Biology, Physics and Chemistry of Frontal Zones in the South–Eastern Part of the Pacific), Moscow: Nauka, 1984, pp. 84–91.

Sarnthein, M., Stattegger, K. and Dreger, D. et. al, Fundamental Modes and Abrupt Changes in North Atlantic Circulation and Climate over the last 60 ky – Concepts, Reconstruction and Numerical Modeling, in *The Northern North Atlantic: A Changing Environment,* Schufer, P., Ritzrau, W., Schliiter, M. and Thiede, Y., Eds., Berlin: Springer, 2001, pp. 365–410.

Savenko, V.S., Elemental Chemical Composition of the Ocean Aerosols, *Doklady Akad. Nauk SSSR* (Proc. AS USSR), 1988, vol. 299, no. 2, pp. 465–468.

Savenko, V.S. and Baturin, G.N., The Problems of the Modeling of the Manganese Precipitation from the Sea Water and Ferromanganese Nodules Origin, *Lithol. and Mineral Resources,* 1981, no. 5, pp. 64–70.

Sayles, F.L., The Composition and Diagenesis of Interstitial Solution, Fluxes Across the Sea Water – Sediment Interface in the Atlantic Ocean, *Geochim. Cosmochim. Acta,* 1979, vol. 43, no. 4, pp. 527–54.

Sazhin, A.F., The Rate of Formation of Detritus and its Flow from Surface Waters to Deep Waters, in *Biodifferentsiatsiya osadochnogo materiala v moryakh i okeanakh* (Biodifferentiation of the Sedimentary Matter in the Seas and Oceans), Lisitzin, A.P., Ed., Rostov University, 1986, pp. 103–110.

Schluter, M., Fluid Flow in Continental Margin Sediments, in Wefer, G., Billett, D., Hebbeln, D., Jorgensen, B.B., Schluter, M. and Weering, T. Van, Eds., *Ocean Margin Systems,* Berlin, Heidelbtrg: Springer–Verlag, 2002, pp. 205–217.

Schott, G., *Weltkarte zur Ubersicht der Meeresströmungen, "Ann. Hydrogr. Und Marit. Meteorol.",* 1943, 71.

Scott, R.B., Rona, P.A., McGregor, B.A. and Scott, M.R., The TAG Hydrothermal Field, *Nature,* 1974, vol. 251, pp. 301–302.

Sellwood, B.C., The Shallow Marine Carbonate Environments, in *Obstanovki osadkonakopleniya i fatsii* (Environments of Sedimentation and Facies), Moscow: Mir, 1990, pp. 5–74.

Semina, G.I., Phytoplankton, in *Okeanologiya, Biologiya okeana* (Oceanology, Biology of the Ocean), vol. 1, Moscow: Nauka, 1977, vol. 1, pp. 117–124.

Senin, Yu.M., The Sediment Types and their Distribution on the North–African Shelf, *Trudy AtlantNIRO,* 1970, vol. 27, pp. 216–247.

Shcherbina, V.V., *Osnovy geokhimii* (Principles of Geochemistry), Moscow: Nauka, 1972.

Shepard, F.P., *Submarine Geology,* New–York: Harper and Row, 1948 (Russian translation).

Shepard, F.P., *Submarine Geology,* New York: Harper and Row, 1963, 2nd ed.

Shevchenko, V.P., Ivanov, G.I., Burovkin, A.A., Djinaridze, R.N., Zernova, V.V., Poliak, L.V. and Syanin, S.S., The Flux of Sedimentary Matter in the Snt. Anna Strait and in the Eastern Part of the Barents Sea, *Doklady Akad. Nauk SSSR* (Proc. AS USSR), 1998, vol. 359A, pp. 400–403.

Shevchenko, V.P., Lisitzin, A.P., Vinogradova, A.A., Serova V.V. and Stein, R., Fluxes of Aerosols to the Arctic Ocean Surface and their Role in Sedimentation and in Formation of the Arctic Environment, in *Opyt sistemnykh okeanologicheskikh issledovaniy v Arktike* (Experience of System Oceanologic Studies in the Arctic), Lisitzin, A.P., Vinogradov, M.E., Romankevich, E.A., Eds., Moscow: Scientific World, 2001, pp. 385–393.

Shilo, N.A., *Osnovy ucheniya o rossypyakh* (Foundations of Science of Placers), Moscow: Nauka, 1985.

Shimkus, K.M., *Osadkoobrazovaniye v Sredizemnom more v pozdnechetvertichnoye vremya* (Sedimentogenesis in the Mediterranean Sea during the Late Quaternary), Moscow: Nauka, 1981.

Shimkus, K.M., Emelyanov, E.M. and Trimonis, E.S., Bottom Deposits and some Features of the History of the Black Sea, in *Zemnaya kora i istoriya razvitiya Chernomorskoy vpadiny* (Earth Cand the History of the Development of the Black Sea), Moscow: Nauka, 1975, pp. 138–161.

Shishkina, O.V., Gordeev, V.V., Blashchishin, A.I. and Mitropolsky, A.Yu., Microelements in Interstitial Waters of the Baltic Sea, in *Osadkonakopleniye v Baltiyskom more* (Sedimentation in the Baltic sea), Lisitzin, A.P. and Emelyanov, E.M., Eds., Moscow: Nauka, 1981, pp. 207–215.

Shnyukov, E.F., Beloded, P.M., Tsemko, V.P., *Minetaknye resursy Mirovogo okeana* (Mineral Resources of the World Ocean), Kiev: Naukova Dumka, 1974.

Sholkovitz, E.R., Flocculation of Dissolved Organic and Inorganic Matter during the Mixing of River Water and Seawater, *Geochim. Cosmochim. Acta,* 1976, vol. 40, no. 7, pp. 831–835.

Sholkovitz, E.R., The Flocculation of Dissolved Fe, Mn, Al, Cu, Ni, Co and Cd during Estuarine Mixing, *Earth Planet. Sci. Lett.,* 1978, vol. 41.

Sholkovitz, E.R., Price, N.B., The Major–Element Chemistry of Suspended Matter in the Amazon Estuary, *Geochim. Cosmochim. Acta,* 1980, vol. 44, no. 2. pp. 163–171.

Sholkovitz, E.R., van Grieken and Eisma, D., The Major–Element Composition of Suspended Matter in the Zaire River and Estuary, *Netherlands J. of Sea Research,* 1978, vol. 12, pp. 407–413.

Shterenberg, L.E., Gorshkova, T.I., Naktinas, E.M., Manganese Carbonates in the Ferro–Manganese Concretions of the Riga Gulf, *Lithol. Mineral Res.,* 1968, no. 4, pp. 63–69.

Shyisky, Yu.D., *Problemy issledovaniya balansa nanosov v beregovoy zone morey* (Problems of Drifts Balance Investigation in Coastal Zone), Leningrad: Hydrometeoizdat, 1986.

Simonov, A.I. and Michailov, V.I., Chemical Pollution of thin Surface Layer of the World Ocean, *Trudy GOIN,* 1979, vol. 49, pp. 4–13.

Skopintsev, B.A., *Formirovaniye sovremennogo khimicheskogo sostava vod Chornogo morya* (The Formation of the Recent Chemical Composition of the Black Sea Water), Leningrad: Hydrometeoizdat, 1975.

Skopintzev, B.A. and Popova, T.P., About the Concentration of Manganese in the Basins with Hydrogene Sulphide (on the Example of the Black Sea), *Trudy Geol. Inst. Akad. Nauk SSSR,* 1963, vol. 97, pp. 165–178.
Skornyakova, N.S., Dispersed Iron and Manganese in the Sediments of the Pacific Ocean, *Lithol. Mineral Res.,* 1964, no. 5, pp. 3–20.
Skornyakova, N.S., Baturin, G.N. and Murdmaa, I.O., Ferromanganese Nodules of the Near–Equator Zone of the Radioliarian Oozes of the Pacific Ocean, *Reports Intern. Geol. Congress.,* 27 ses., Moscow, 1983, pp.19–27.
Skornyakova, N.S. and Murdmaa, I.O., Processes of the Formation of Ferromanganese Nodules of Radiolarian Belt, in *Zhelezomargantsevye konkretsii ekvatorialnoy chasti Tikhogo okeana* (Ferromanganese Nodules of the Equatorial Part of the Pacific Ocean), Moscow: Nauka, 1986, pp. 297–320.
Skornyakova, N.S., Murdmaa, I.O., Prokoptsev, N.G. and Marakuev, V.I., Bottom Deposits and Volcanogenic Rocks of One Region of Southern Basin of the Pacific Ocean, *Lithol. Mineral. Res.,* 1973, no. 1, pp. 17–28.
Skornyakova, N.S., Murdmaa, I.O. and Zayikin, V.N., Cobalt in the Ferromanganese Crusts and Nodules of the Pacific Ocean, *Lithol. Mineral. Res.,* 1989, no. 2, pp. 106–121.
Skornyakova, N.S. and Andrushchenko, P.F., *Zhelezomargantsevye konkretsii Tikhogo okeana* (Ferromanganese Nodules of the Pacific Ocean), Moscow: Nauka, 1976, pp. 76–122.
Smith, P.A. and Cronan, D.S., Chemical Composition of Aegean Sea Sediments, *Marine Geology,* 1975, vol. 18, M7–M–11.
Sorokin, Yu.I., Experimental Data about the Oxidation Velocity of the Hydrogen in the Black Sea, *Okeanologiya,* 1971, vol. 11, no. 3, pp. 423–431.
Sorokin, Yu.I., Production of the Microflora, in *Okeanologiya, Biologiya okeana* (Oceanology, Biology of the Ocean), Moscow: Nauka, 1977, vol. 2, pp. 209–231.
Sorokin, Yu.N., Bacterioplankton, in *Okeanologiya, Biologiya okeana* (Oceanology, Biology of the Ocean), Moscow: Nauka, 1977, vol. 1, pp. 124–132.
Sorokin, Y.I., *Chornoye more* (The Black Sea), Moscow: Nauka, 1982.
Sorokin, Yu.I. and Kovalevskaya, R.Z., Biomass and Production of Bacterioplankton of the Oxygen Zone of the Black Sea, *Ekosistemy pelagiali morya* (Ecosystems of Pelagic Areas of the Black Sea), Vinogradov, M.E., Ed., Moscow: Nauka, 1980, pp. 162–167.
Steeman–Nielsen, E. and Jensen, E.A., Primary Oceanic Production, The Authtropic Production of Organic Matter in the Oceans, *Sci. Rept. Danish Deep–Sea Exped., 1952, Galathea Rep.,* 1957, vol. 1, pp. 49–135.
Stepanov, V.N., *Mirovoy okean* (The World Ocean), Moscow: Znanie, 1974.
Stow, D.A.V. and Holbrook, J.A., North Atlantic Contourites: An Overview, in *Fine Grained Sediments: Deep Water Processes and Fabrics,* Piper, D.J. and Stow, D.A.V., Eds., Geol. Soc. Amer. Publ., 1984, vol. 15, pp. 245–256.
Strakhov, N.M., Fe–Ore Facies and their Analoge in Earth's History (an Attempt to Make Historic and Geological Analysis of the Process of Sedimentation), Moscow, *Trudy Akad. Nauk SSSR* (Proc. of USSR Acad. Sc.), 1947, issue 73, geol.series, no. 22, 286 p.
Strakhov, N.M., *Osnovy teorii litogeneza* (Foundations of the Theory of Lithogenesis), Moscow: USSR Acad. Sc. Publ. House, 1960–1962, vols. 1–3.
Strakhov, N.M., *Problemy geokhimii sovremennogo okeanskogo litogeneza* (Geochemical Problems of the Recent Ocean Lithogenesis), Moscow: Nauka, 1976.
Strakhov, N.M., *Izbrannye trudy, Problemy osadochnogo rudoobrazovaniya* (Selected Works, Problems of Sedimentary Ore Formation), Moscow: Nauka, 1986.
Strakhov, N.M., Fe–Ore Facies and their Analogs in the Earth's History, in Izbrannye trudy, *Problemy osadochnogo rudoobrazovaniya* (Selected Works, Problems of Sedimentary Ore Formation), Moscow: Nauka, 1986, pp. 165–199.
Strakhov, N.M., Brodskaya, N.G., Kniazeva, L.M., Razhivina, A.N., Rateev, M.A., Sapozhnikov, D.G. and Shishova, E.E., *Osadkoobrazovaniye v sovremennykh vodoyomakh* (Formation of Sediments in Recent Water Basins), Moscow: USSR Publ. House, 1954.
Strakhov, N.M., Shterenberg, L.E, Kalinenko, V.V. and Tikhomirov, G.S., *Geokhimiya osadochnogo margantsevo–rudnogo protsessa* (The Geochemistry of the Manganese Ore Problems), Moscow: Nauka, 1968.

Strakhov, N.M. and Shterenberg, L.E., To the Problem of Genetic Type of Chiatura Deposit, *Lithol. Mineral Res.*, 1965, no. 1, pp. 18–30.

Strakhov, N.M., Varentsov, I.M., Kalinenko, V.V., Tikhomirova, E.S. and Shterenberg, L.E., On Cognition of Mn–Ore Formation Mechanism on Example of Oligocene Ores of USSR South, in *Margantsevye rudy SSSR* (Manganese Ores of the USSR), Moscow: Nauka, 1967, pp. 34–56.

Strahov, N.M., About the Exhalations on the Mid–Oceanic Ridges as a Source of the Ore Elements in the Ocean Pelagic Sediments, *Lithol. and Mineral. Res.*, 1974, no. 3, pp. 20–37.

Striuk, V.L., Quantitative Distribution of the Suspended Matter in the Baltic Sea, *Cand. Sci. Dissertation,* Kaliningrad, 1994.

Stumm, W. and Baccini, P., Man–Made Chemical Perturbation of Lakes, in *Lakes, Chemistry, Geology, Physics,* Lerman, A., Ed., Heidelberg–Berlin: Springer–Verlag, 1978, pp. 91–126.

Stumm, W. and Morgan, J.J., *Aquatic Chemistry,* New–York: Wiley–Interscience, 1970.

Sturm, M. and Matter, A., Turbidites and Varves in Lake Brienz (Switzerland): Deposition of Clastic Detritus by Densitycurrents, in *Modern and Ancient Lake Sediments,* Matter, A. and Tucker, M.E., Eds., Spec. Publ. Int. Ass. Sediment., 1978, vol. 2, pp. 145–166.

Suess, E., Mineral Phases Formed in Anoxic Sediment by Microbial Decomposition of Organic Matter, *Geochim. Cosmochim. Acta,* 1979, vol. 43, no. 3, pp. 339–352.

Suess, E. and Thiede, J., Eds., *Coastal Upwelling, Its Sediment Record,* Part A: Responses of the Sedimentary Region to Present Coastal Upwelling, N.-Y. and London: Plenum Press, 1983.

Summerhayes, C.P. and Willis J.P., Geochemistry of Manganese Deposits in Relation to Environment on the Sea Floor Around Southern Africa, *Marine Geology,* 1975, vol. 18, pp. 159–173.

Sunda, W.G., Huntsman and Harvey, G.R., Photoreduction of Manganese Oxides in Seawater and its Geochemical and Biological Implications, *Nature,* 1983, vol. 301, no. 5897, pp. 234–236.

Sundby, B., Silverberg, N. and Chesselet, R., *Geochim. et Cosmochim. Acta,* 1981, vol. 45., no. 3, pp. 293–307.

Svalnov, V.N., *Dinamika pelagicheskogo litogeneza* (The Dynamic of Pelagic Lithogenesis), Moscow: Nauka, 1991.

Svalnov, V.N. and Kuleshov, V.N., Calcite Rhodochrosite in the Guatemala Basin Sediments, *Lithol. Mineral Resources,* 1994, no. 3, pp. 20–25.

Svirenko, I.P., Quantitative Distribution of the Water Suspended Matter in the Atlantic Ocean (2nd cruise of R/V "Belogorsk"), Okeanologiya, 1970, vol. 10, no. 3, pp. 101–108.

Syvitski, J.P.M., Fjords, Processes and Products, N.-Y.: Springer–Verlag, 1987.

The Encyclopedia of Oceanography, Fairbridge, W., Ed., N.-Y.: Rinhold Publishing Corporation, 1966 (Russian translation 1974).

The Fifth Marine Geological Conference «The Baltic», 1997, Grigelis, A., Ed., Vilnius: Lithuanian Institute of Geology, pp. 22–24.

The Global 200 Report to the President, Entering the 21st Century (Citation after Israel and Tsyban, 1989), 1980.

Theede, Y., Studies on the Role of Benthic Animals of the Western Baltic in the Flow of Energy and Organic Material, *Kieler Meeresforsch.*, 1981, vol. 5, pp. 434–444.

Thiede, J., Skeletal Plankton and Nekton in Upwelling Water Masses off North–Western South America and Northwest Africa, in *Coastal Upwelling, Its Sediment Record,* Suess, E. and Thiede, J., Eds., N.-Y.-London, pp. 183–208.

Thompson, G., Humphris, S.E. and Schroeder, B. et al., Active Vents and Massive Sulfides at 26° N (TAG) and 23° N (Snakepit) on the Mid–Atlantic Ridge, *Canad. Miner.*, 1988, vol. 26, pp. 697–711.

Thomson, J., Wilson, T.R.S., Culkin, F. and Hydes, D.J., *Earth and Planet. Sci. Letters,* 1984, vol. 71, no. 1, pp. 23–30.

Tikhomirov, V.N., Studies of the State and Sorption Behavior of Metals at the Water–Bottom Boundary by Using Labeled Atoms, in *Zhelezomargantsevye konkretsii ekvatorialnoy chasti Tikhogo okeana* (Ferromanganese Nodules in the Equatorial Part of the Pacific), Moscow: Nauka, 1986, pp. 270–283.

Timonin, F.G. and Flint, M.V., The Features of the Structure of Mezoplankton of Peruvian Upwelling, in *Biologicheskiye osnovy promyslovogo osvoeniya otkrytykh rayonov okeana* (Biological Basis of the Fisheries Assimilation of the Open Region of the Ocean), Moscow: Nauka, 1985, pp. 155–165.

Toth, J.P., Deposition of Submarine Crusts Rich in Manganese and Iron, *Bull. Geol. Soc. Amer.*, 1980, vol. 91, no. 1, pp. 44–54.

Trimonis, E.S., *Terrigennaya sedimentatsiya v Atlanticheskom okeane* (Terrigenous Sedimentation in the Atlantic Ocean), Moscow: Nauka, 1995.

Trimonis, E.S., Emelyanov, E.M. and Kharin, G.S., Serpentinites in the Sediments from the Fracture Zone of the South–Atlantic Ridge, *Okeanologiya,* 1987, vol. 31, no. 2, pp. 280–285.

Tsunogai, Sh. and Kusakabe, M., Migration of Manganese in the Deep–Sea Sediments, in *The Dynamic Environment of the Ocean Floor,* Fanning, K.A. and Manheim, F.T., Eds., Toronto: Lexingron Books, 1982, pp. 257–274.

Twenhofel, W.H., *Treatise on Sedimentation,* N.-Y., 1932, vols. 1–2.

Van Andel, T.H., Thiede, J., Sclater, J.G. and Hay, W.W., Depositional History of the South Atlantic Ocean During the Last 125 Million Years, *J. Geol.,* 1977, vol. 85, pp. 651–698.

Varentsov, I.M. and Blazchishin, A.I., Ferromanganese Deposits, in *Geology of the Baltic Sea,* Gudelis, V.K. and Emelyanov, E.M., Eds., Vilnius: Mokslas, 1976, pp. 307–349.

Varnavas, S.P., Papaioannou, J. and Catani, J., A Hydrothermal Manganese Deposit from the Eratosthenes Seamount, Eastern Mediterranean Sea, *Marine Geology,* 1988, vol. 81, pp. 205–214.

Vasilyev, I.A. et al., The Estimation of Potential Danger of the Dumped Chemical Munition in the Baltic Sea, *Russian Chemical Journal,* 1994, vol. 38, no. 2, pp. 112–116.

Vernadsky, V.I., *Khimicheskoye stroyeniye biosfery Zemli i eye okruzheniya* (The Chemical Structure of the Biosphere of the Earth and Environment), Moscow: Nauka, 1965.

Vernadsky,V. I., *Izbrannye sochineniya* (Selective Works), Moscow: Akad. Nauk SSSR, 1954–1960, vols. 1–5.

Vershinin, A.V. and Rozanov, A.G., *Khimicheskiy obmen na granitse voda–dno v okeanakh I moryakh* (Chemical Exchange Across the Sediment, Water Interface in Oceans and Seas), Moscow: GEOS, 2002.

Vershinin, A.V., Bogdanovskyi, V.V., Borodkin, S.O. and Safarova, E.S., The Diagenesis of the Sediments in the Area of River Congo Influence According to the Chemical Analyses of the Sediments and the Pore Waters, in *Biogeokhimiya pogranichnykh zon Atlanticheskogo okeana* (Biogeochemistry of the Boundary Zones in the Atlantic Ocean), Romankevich, E.A., Ed., Moscow: Nauka, 1994.

Vinogradov, A.P., Chemical Elemental Composition of the Sea Organisms, *Trudy Biokhim. Lab. SSSR* (Proc. Biochem. Lab. USSR), 1935–1944, parts 1–3, Part 1, 1935, pp. 63–278; part 2, 1937, pp. 2–225; part 3, 1944, p. 273.

Vinogradov, A.P., Mean Contents of the Chemical Elements in the Main Types of Effusive Rocks of the Earth Crust, *Geochemistry,* 1962, vol. 7, pp. 555–571.

Vinogradov, A.P., *Vvedeniye v geokhimiyu okeana* (Introduction into Geochemistry of Ocean), Moscow: Nauka, 1967.

Vinogradov, M.E., *Vertikalnoye raspredeleniye zooplanktona v okeane* (Vertical Distribution of the Ocean Zooplankton), Moscow: Nauka, 1968.

Vinogradov, M.E., Zooplankton, in *Okeanologiya, Biologiya okeana* (Oceanology, Biology of the Ocean), Moscow: Nauka, 1977, vol. 1, pp. 65–69.

Vinogradov, M.E., Independence of the Detrital Formation Processes from the Space–Time Changes of the Plankton Associations, in *Biodifferentsiatsiya osadochnogo materiala v moryakh i okeanakh* (Biodifferentiation of the Sedimentary Matter in the Seas and Oceans), Lisitzin, A.P., Ed., Rostov University, 1986, pp. 66–75.

Vinogradov, M.E. and Musaeva, E.I., Eds., Ecosystems of the Eastern Boundary Currents and Central Regions of the Pacific Ocean, Moscow: Nauka, 1990.

Vinogradov, M.E., Sapozhnikov, V.V. and Shushkina, E.A., *Ekosistema Chornogo morya* (The Black Sea Ecosystem), Moscow: Nauka, 1992.

Vinogradov, M.E. and Shushkina, E.A., The Peculiarities of the Zooplankton Distribution, in *Ekosistemy pelagiali morya* (Ecosystems of Pelagic Areas of the Black Sea), Vinogradov, M.E., Ed., Moscow: Nauka, 1980, pp. 15–25.

Vinogradov, M.E. and Shushkina, E.A., Ecosystem of the Arctic Pelagial, in *Opyt sistemnykh okeanologicheskikh issledovaniy v Arktike* (Experience of System Oceanologic Studies in the Arctic), Lisitzin, A.P., Vinogradov, M.E., Romankevich, E.A., Eds., Moscow: Scientific World, 2001, pp. 282–288.

Vogt, P.R., Seafloor Topography, Sediments and Paleoenvironments, in *The Nordic Seas,* Hurdle, B.G., Ed., New-York, Berlin, Heidelberg, Tokyo: Springer–Verlag, 1986, pp. 237–412.

Volkov, I.I., Oxidation–Reduction Diagenetic Processes of the Sediments, in *Okeanologiya, Khimiya okeana* (Oceanology, Chemistry of the Ocean), 1979, vol. 2, Moscow: Nauka, pp. 363–413.

Volkov, I.I., Ferromanganese Nodules, in *Okeanologiya, Khimiya okeana* (Oceanology, Chemistry of the Ocean), 1979, vol. 2, Moscow: Nauka, pp. 414–467.

Volkov, I.I., Redistribution of the Chemical Elements in the Diagenesis of the Sediments, in *Geokhimiya diageneza donnykh osadkov Tikhogo okeana* (Geochemistry of Diagenesis of the Bottom Sediments in the Pacific Ocean), Ostroumov, E.A., Ed., Moscow: Nauka, 1980, pp. 144–168.

Volkov, I.I., The Main Regularities of Metasomatism of Organic Matter During Early Diagenesis of the Recent Sediments, in *Geokhimiya diageneza donnykh osadkov Tikhogo okeana* (Geochemistry of Diagenesis of the Bottom Sediments in the Pacific Ocean), Ostroumov, E.A., Ed., Moscow: Nauka, 1980, pp. 99–116.

Volkov, I.I., Aerobic–Anaerobic Diagenesis of the Sediments at the Water–Bottom Boundary, in Geology of the Ocean and Seas, *Tezisy dokladov 6–oy vsesouznoy shkoly morskoy geologii* (Abstract of the 6th All–Union Conference on Marine Geology), 1984, vol. 1, n.p.

Volkov, I.I., Fomina, L.S. and Yagodinskaya, T.A., Chemical Composition of the Ferro–Manganese Nodules of the Pacific Ocean on the Profile of Atoll Wake–Near–Shore Area of Mexico, in *Biogeokhimiya diageneza osadkov okeanov* (Biogeochemistry of the Sediment Diagenesis of the Ocean), Volkov, J.J., Ed., Moscow: Nauka, 1976, pp. 186–204.

Volkov, I.I., Logvinenko, N.V., Sokolova, E.G. and Leyn, A.Yu., Rodochrozite, in *Litologiya i geokhimiya donnykh osadkov Tikhogo okeana* (Lithology and Geochemistry of the Bottom Sediments of the Pacific Ocean), Holodov, V.I., Ed., Moscow: Nauka, 1979, pp. 85–90.

Volkovinsky, V.V., Zernova, V.V. and Senina, G.I. et al., The Distribution of Phytoplankton in the World Ocean, in *Rybopromyslovaya okeanologiya i podvodnaya tekhnika* (Fishery Oceanology and Underwater Technique), INILTEN of Fishering Ministry of the USSR, Moscow: Nauka, p. 110.

Voronina, N.M., Community of the Moderate and Cold Waters of the Southern Hemisphere, in *Okeanologiya, Biologiya okeana* (Oceanology, Biology of the Ocean), Moscow: Nauka, 1977, vol. 2, pp. 68–90.

Wadhams, P., The Ice Cover, in *The Nordic Seas,* Hurdle, B.G., Ed., New–York, Berlin, Heidelberg, Tokyo: Springer–Verlag, 1986, pp. 21–87.

Wallace, G.T. and Duce, R.A, Transport of Particulate Organic Matter by Bubbles in Marine Waters, *Limnol. and Oceanogr.,* 1978, vol. 23(b), pp. 1155–1167.

Wallast, T.R., Billen, G. and Mackenzie, F.T., Behavior of Mercury in Natural Systems and its Global Cycle, in *Ecolog. Loxicol. Res.,* 1976, pp. 145–160.

Walter, P. and Stoffers, P., Chemical Characteristics of Metalliferous Sediments from Eight Areas of Galapagos Rift and East Pacific Rise between 2° N and 42° S, *Mar. Geol.,* 1985, vol. 65, no. 3/4, pp. 271–287.

Warren, B.A., Deep Water Circulation in the World Ocean, in *Evolution of Physical Oceanography,* Warren, B.A. and Wunsch, C., Eds., Scientific Surveys in Honor of Henry Stommel, The MIT Press, Cambridge, MA, 1981, vol. 26, pp. 6–41.

Wattenberg, H., Kalziumkarbonat und Kohlensauregehalt des Meerwasers, in *Wiss. Ergebh. Dtsch. Atlant. Exp. "Meteor"* 1925–1927, 1933, vol. 8.

Wendt, J., Encrusting Organisms in Deep–Sea Manganese Nodules, Spec. *Publ. Intern. Assoc. Sedimentol.,* 1974, vol. 1, pp. 437–447.

Weppering, R., Schlosser, P., Khatiwala, S. and Fairbanks, R.G., Isotope Data from Ice Station Weddell: Implications for Deep Water Formation in the Weddell Sea, *J. of Geoph. Res.,* 1996, vol. 101, nos. C10, 25, pp. 723–739.

Willer, A., *Studien uber das Frische Haff. I. Die allgemeinen hydrographischen und biologischen Verhaltnisse das Frischen Haffes,* 23, Berlin, 1925, pp. 317–349.

Willer, A., *Studien uber das Frische Haff. II. Die Haffkrankheit. Zeitschrift für Fischerei,* 23, Berlin, 1925, pp. 349–379.

Wimbush, M. and Munk, W., The Benthic Boundary Layer, in *The Sea,* Maxwell, A.E., Ed., Wiley, N.-Y., 1970, vol. 4(1).

Yang, H., Coombs, N. and Sokolov, I., Free–Standing and Oriented Mesoporous Silica Films Grown at the Air–Water Interface, *Nature,* 1996, vol. 381, pp. 589–592.

Yeats et al., 1995, March, 49, pp.283–293.

Yoder, J.F., Ackleson, S.G., Barber, R.T., Flament, P. and Balch, W.V., A Line in the Sea, *Nature,* 1994, vol. 371, pp. 689–692.

Zachowicz, J., Laban, S., Uscinowicz, S., Ebbing, J. and Emelyanov, E.M., Recent Sedimentation in the Gulf of Gdansk and its Geochemical Expression, in *Geology of the Gdansk Basin,* Baltic Sea, Emelyanov, E.M., Ed., Kaliningrad: Yantarny skaz, 2002, pp. 380–388.

Zaitsev, Ju.P., Neuston–a Biological Factor of Influence on the Water Properties in the Hydrosphere–Atmosphere Zone, Interaction between Water and Living Matter, *Proc. of International Symposium,* Odessa, 6–10 October, 1975, Moscow: Nauka, 1979, vol. 1, pp. 21–25.

Zangenberg, N. and Siedler, G., Path of the North Atlantic Deep Water in the Brazil Basin, *Journal of Geographical Research,* 1998, vol. 103, no. C3, pp. 5419–5428.

Zelenov, K.K., *Vulkany kak istochnik rudoobrazuyushchikh komponentov osadochnykh tolshch* (Volcanoes as Resources of the Ore Components of the Sedimentary Strata), Moscow: Nauka, 1972.

Zhelezo–margantsevye konkretsii Tikhogo okeana (Ferromanganese Nodules of the Pacific Ocean), Bezrukov, P.L., Ed., Moscow: Nauka, 1976.

Zhelezo–margantsevye korki i konkretsii podvodnykh gor Tikhogo okeana (Ferromanganese Crusts and Nodules of the Submarine Mountains of the Pacific Ocean), Lisitzin, A.P., Ed., Moscow: Nauka, 1990.

Zen, E.-An., Mineralogy and Petrography of Marine Bottom Sediment Samples of the West Coast Peru and Chile, *J. Sediment. Petrol,* 1959, vol. 29, no. 4, pp. 513–539.

Zenkovitch, L.A., *Special Quantitative Characteristic of the Abyssal Life in the Ocean, Oceanography,* Moscow: Progress, 1960, pp. 254–262.

Zenkevich, L.A., Filatova, Z.A., Beliaev, G.M., Lukiyanova, T.C. and Suetova, J.A., The Quantitative Distribution of the Zoobenthos in the World Ocean, *Bull. MOIP, Biol. Sect.,* 1971, vol. 76, no. 3.

Zenkevich, V.P., *Osnovy ucheniya o razvitii morskikh beregov* (The Foundations of Development of Sea Shores), Moscow: Acad. Sc. of the USSR, 1962.

Zernova, V.V., The Distribution of the Biomass of Plankton, in *Biologicheskiye resursy okeana* (Biological Resources of the Ocean), Moscow: Agropromizdat, 1985, pp. 85–90.

Zhamoida, V., Glasby, G., Grigoriev, A., Manuilov, S., Moskalenko, P. and Spiridonov, M., Distribution, Morphology, Composition and Economic Potential of Ferromanganese Concretions from the Eastern Gulf of Finland, in *The Seventh Marine Geological Conference "Baltic–7",* Kaliningrad, 2002, p. 145.

Zhindarev, L.A., Morpholithodynamic of the Dissected Tide Less Shores of the Seas, *Extended Abstracts of Doctoral Dissertation,* Moscow, MSU.

Index